LOGARITHMS

$y = \log_a x$ means $a^y = x$

$\log_a a^x = x$ $\qquad\qquad$ $a^{\log_a x} = x$

$\log_a 1 = 0$ $\qquad\qquad$ $\log_a a = 1$

$\log x = \log_{10} x$ $\qquad\qquad$ $\ln x = \log_e x$

$\log_a xy = \log_a x + \log_a y$

$\log_a \left(\dfrac{x}{y}\right) = \log_a x - \log_a y$

$\log_a x^b = b \log_a x$ $\qquad\qquad$ $\log_b x = \dfrac{\log_a x}{\log_a b}$

SEQUENCES AND SERIES

Arithmetic:

$$a,\ a + d,\ a + 2d,\ a + 3d,\ a + 4d,\ \dots$$

$$a_n = a + (n - 1)d$$

$$S_n = \sum_{k=1}^{n} a_k = \frac{n}{2}[2a + (n - 1)d]$$

$$= n\left(\frac{a + a_n}{2}\right)$$

Geometric:

$$a,\ ar,\ ar^2,\ ar^3,\ ar^4,\ \dots$$

$$a_n = ar^{n-1}$$

$$S_n = \sum_{k=1}^{n} a_k = a\,\frac{1 - r^n}{1 - r}$$

If $|r| < 1$, then the sum of an infinite geometric series is

$$S = \frac{a}{1 - r}$$

THE BINOMIAL THEOREM

$$(a + b)^n = \binom{n}{0}a^n + \binom{n}{1}a^{n-1}b + \binom{n}{2}a^{n-2}b^2 + \cdots$$

$$+ \binom{n}{n-1}ab^{n-1} + \binom{n}{n}b^n$$

GEOMETRIC FORMULAS

Formulas for area A, circumference C, and volume V:

Rectangle

$A = lw$

Box

$V = lwh$

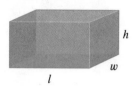

Triangle

$A = \frac{1}{2}bh$

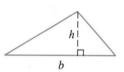

Pyramid

$V = \frac{1}{3}ha^2$

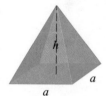

Circle

$A = \pi r^2$

$C = 2\pi r$

Sphere

$V = \frac{4}{3}\pi r^3$

$A = 4\pi r^2$

Cylinder

$V = \pi r^2 h$

Cone

$V = \frac{1}{3}\pi r^2 h$

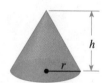

COLLEGE ALGEBRA

ABOUT THE COVER

The College Algebra course is an opportunity to learn about the beauty and practical power of mathematics. The course may also be a stepping stone to the study of calculus.

The violin, with its sound hole in the shape of an integral sign, has become a symbol of lead author James Stewart's calculus textbook series, which includes *Calculus, Third Edition*, and *Calculus: Early Transcendentals, Third Edition*.

The cover of this edition, which shows violin-making as a "work in progress," reflects both the Stewart authorship and the foundational importance of the College Algebra course.

ABOUT THE AUTHORS

James Stewart was educated at the University of Toronto and Stanford University, did research at the University of London, and now teaches at McMaster University. His research field is harmonic analysis.

He is the author of a best-selling calculus textbook series published by Brooks/Cole, including *Calculus, 3rd Ed.*, and *Calculus: Early Transcendentals, 3rd Ed.*, as well as a series of high school mathematics textbooks.

A talented violinist, Stewart was concertmaster of the McMaster Symphony Orchestra for eight years and played professionally in the Hamilton Philharmonic Orchestra. One of his greatest pleasures is playing string quartets.

Lothar Redlin grew up on Vancouver Island, received a Bachelor of Science degree from the University of Victoria, and a Ph.D. from McMaster University in 1978. After completing his education, he did research and taught at the University of Washington, the University of Waterloo, and California State University, Long Beach.

He is currently Associate Professor of Mathematics at The Pennsylvania State University, Abington-Ogontz Campus. His research field is topology.

Saleem Watson received his Bachelor of Science degree from Andrews University in Michigan. He did his graduate studies at Dalhousie University and McMaster University, where he received his Ph.D. in 1978. After completing his education, he did research at the Mathematics Institute of the University of Warsaw in Poland. He subsequently taught and did research at McMaster University and The Pennsylvania State University.

He is currently Professor of Mathematics at California State University, Long Beach. His research field is functional analysis.

The authors have also published *Mathematics for Calculus, Second Edition* (Brooks/Cole, 1993).

COLLEGE ALGEBRA

Second Edition

James Stewart
McMaster University

Lothar Redlin
The Pennsylvania State University

Saleem Watson
California State University, Long Beach

BROOKS/COLE PUBLISHING COMPANY

I(T)P™ An International Thomson Publishing Company

Pacific Grove ■ Albany ■ Bonn ■ Boston ■ Cincinnati ■ Detroit ■ London ■ Madrid ■ Melbourne
Mexico City ■ New York ■ Paris ■ San Francisco ■ Singapore ■ Tokyo ■ Toronto ■ Washington

 A GARY W. OSTEDT BOOK

Publisher: *Gary W. Ostedt*
Assistant Editor: *Elizabeth Barelli Rammel*
Marketing Team: *Patrick Farrant,*
 Margaret Parks, Michele LeBoeuf
Editorial Associate: *Carol Ann Benedict*
Production Editor: *Jamie Sue Brooks*
Production Service: *TECHarts*
Manuscript Editor: *Kathi Townes, TECHarts*

Cover Design: *Vernon T. Boes*
Cover Photo: *David Ash, Tony Stone*
 Worldwide
Illustrations: *TECHarts*
Typesetting: *Beacon Graphics*
Cover Printing: *Phoenix Color Corp.*
Printing and Binding: *R. R. Donnelley &*
 Sons, Crawfordsville

For more information, contact:

BROOKS/COLE PUBLISHING COMPANY
511 Forest Lodge Road
Pacific Grove, CA 93950
USA

International Thomson Publishing Europe
Berkshire House 168-173
High Holborn
London WC1V 7AA
England

Thomas Nelson Australia
102 Dodds Street
South Melbourne, 3205
Victoria, Australia

Nelson Canada
1120 Birchmount Road
Scarborough, Ontario
Canada M1K 5G4

International Thomson Editores
Campos Eliseos 385, Piso 7
Col. Polanco
11560 México D. F. México

International Thomson Publishing GmbH
Königswinterer Strasse 418
53227 Bonn
Germany

International Thomson Publishing Asia
221 Henderson Road
#05-10 Henderson Building
Singapore 0315

International Thomson Publishing Japan
Hirakawacho Kyowa Building, 3F
2-2-1 Hirakawacho
Chiyoda-ku, Tokyo 102
Japan

Printed in the United States of America

10 9 8 7 6 5 4 3 2

LIBRARY OF CONGRESS CATALOGING-IN-PUBLICATION DATA
Stewart, James
  College algebra/James Stewart, Lothar Redlin, Saleem Watson. 2nd ed.
    p.   cm.
  Includes index.
  ISBN 0-534-33983-2
  1. Algebra   I. Redlin, L.   II. Watson, Saleem.   III. Title.
QA152.2.2.S74   1996   95-32697
512.9–dc20   CIP

THIS BOOK IS PRINTED ON ACID-FREE RECYCLED PAPER

TO PHYLLIS

PREFACE

The art of teaching is the art of assisting discovery.
MARK VAN DOREN

For many students a College Algebra course represents the first opportunity to learn about the beauty and practical power of mathematics. Thus, teachers and textbook authors are faced with the challenge of teaching the techniques of the subject while at the same time imparting a concept of the true nature of mathematics. This text represents our view of how the subject can best be taught.

Our main goal in writing this edition was to sharpen the clarity of the exposition, while retaining the main features which have contributed to the success of this book. To assist students in understanding the examples we have included step-by-step comments on solutions, given in the margins so as not to interrupt the flow of the solutions. We have added many graphs and figures to remind students of the geometric meaning behind a calculation and to promote visual insight into formulas and theorems. We continue to present College Algebra as a *problem-solving* activity because we think this is what mathematics is all about. We have included many examples and exercises that make substantial use of college algebra to solve real-life problems. Our historical "vignettes" add interest to the subject and also serve to show the universality of mathematics. We feel that all these features make for a user-friendly book.

Graphing calculators are powerful tools for developing the students' understanding of equations and functions. We have devoted entire sections to the use of this tool so as to make substantial use of it where it is appropriate. Those who choose to teach these sections can follow up by using the subsections and exercises in the rest of the book which are clearly identified with a special icon: ⌁.

Many of the changes in the structure of the present book are a result of our own experience in teaching College Algebra. We have experimented with different ways of incorporating graphing calculators into the course and with different orders of presenting the material. We have benefited greatly from the insightful comments of our colleagues who have used the text and critiqued it section by section as they taught it. We have also benefited from the sharp insight of our reviewers.

SPECIAL FEATURES

Focus on Problem Solving

We are committed to helping students develop mathematical thinking rather than "memorizing all the rules," an endless and pointless task that many students are not taught to avoid. We believe that mathematical thinking is best taught by careful exposition and examples and well-graded exercises, but also by giving guidelines for problem solving. Thus, an emphasis on problem solving is integrated throughout the text. In addition, we have concluded each chapter with a section entitled *Focus on Problem Solving,* which highlights a particular problem-solving technique. The first of these sections, on pages 61–67, gives a general introduction to the principles of problem solving. Each *Focus* section includes problems of a varied nature that encourage students to apply their problem-solving skills. In selecting these problems we have kept in mind the following advice from David Hilbert: "A mathematical problem should be difficult in order to entice us, yet not inaccessible lest it mock our efforts."

Graphing Calculators and Computers

Calculator and computer technology provides completely new ways of visualizing mathematics and is affecting not only how a topic is taught but also what is emphasized. We have integrated the technology carefully in four sections devoted to the fundamentals of graphing devices—Sections 3.3, 4.2, 4.6, and 4.9. In other sections we have included subsections, examples, and exercises to show how the graphing calculator can be used (see, for instance, pages 223 and 351). One example of how technology enhances understanding is that we can use it to draw accurate graphs of *families* of functions, so that students can see how varying a parameter can affect the graph of a function (see pages 191–92 and 271). These graphing calculator sections, subsections, examples, and exercises, all marked with the special logo ▨, are optional and may be skipped without loss of continuity, but we encourage you to try them. You may in fact wish to integrate graphing devices more fully into your course by using them for the "regular" sections of the text as well. It has been our experience that students enjoy working with the graphing calculator, and they learn the material better and with more enthusiasm when we require or recommend its use in our classes.

The availability of graphing calculators makes it not less important but far more important to clearly understand the concepts that underlie the image the calculator produces on its screen. Accordingly, all our calculator-oriented sections and subsections are preceded by sections in which students must sketch graphs by hand and analyze them, so they can understand precisely what the graphing device is doing when they later use it to simplify the routine, mechanical part of their work. We must never lose sight of the fact that we are teaching mathematical ideas, not the use of any particular tool. Thus, we treat the calculator as an aid to understanding, an extension of pencil and paper, not as the central feature of the course.

"Real-World" Applications

We have included substantial applied problems that we believe will capture the attention of students. These are integrated throughout the text in both examples and exercises. Applications from engineering, physics, chemistry, business, biology, environmental studies, and other fields show how mathematics is used to model real-life situations. We have carefully chosen applications that show the relevance of algebra to our daily lives and its remarkable power as a problem-solving tool. (See, for instance, the examples and exercises on energy expended in bird flight, page 110; determining the optimal shape for a can, pages 225–26; terminal velocity of a skydiver, page 350, Exercise 28; establishing time of death, page 387, Exercise 30.)

Mathematical "Vignettes"

Throughout this book we make use of the margins to provide short biographies of interesting mathematicians as well as applications of algebra to the "real world." The biographies often include a key insight that was discovered by the mathematician and is relevant to College Algebra. (See, for instance, the vignettes on Viète, page 91; Salt Lake City, page 140; and radiocarbon dating, page 371.) They serve to enliven the material and show that mathematics is an important, vital activity, and that even at this elementary level, it is fundamental in everyday life.

Review Sections and Chapter Tests

Each chapter ends with an extensive review section, including a Chapter Test designed to help the students gauge their progress. Brief answers to the odd-numbered exercises in each section (including the review exercises), and to all questions in the Chapter Tests, are given at the back of the book.

MAJOR CHANGES FOR THE SECOND EDITION

- Chapter 1, which is a review chapter, has been restructured so that the material on integer and rational exponents appears together in one section, and the material on algebraic expressions and factoring also appears in one section.

- Chapter 2 introduces the *Check Your Answer* feature, which is used everywhere in the book that equations occur, to emphasize the importance of looking back to check whether an answer is reasonable. Section 2.7 now contains the method of sign diagrams for solving nonlinear inequalities.

- Chapter 3 now combines the material on the coordinate plane and the material on functions. This permits earlier introduction of the graphing calculator. The elementary discussion on conics has been deleted—a full treatment of this topic now appears in Chapter 7.

(Major changes continue on page xiii)

CHECK YOUR ANSWER

$x = 4$:

$$\text{LHS} = 2 + \frac{5}{4-4} = 2 + \frac{5}{0}$$

$$\text{RHS} = \frac{4+1}{4-4} = \frac{5}{0}$$

Impossible—can't divide by 0. LHS and RHS are undefined, so $x = 4$ is not a solution. ✗

$$\begin{aligned} 2x - 8 + 5 &= x + 1 && \text{Distributive Property} \\ 2x - 3 &= x + 1 && \text{Simplify} \\ 2x &= x + 4 && \text{Add 3} \\ x &= 4 && \text{Subtract } x \end{aligned}$$

But now if we try to substitute $x = 4$ back into the original equation, we would be dividing by 0, which is impossible. So, this equation has no solution. ■

The first step in the preceding solution, multiplying by $x - 4$, had the effect of multiplying by 0. (Do you see why?) Multiplying each side of an equation by an expression that contains the variable may introduce extraneous solutions. That is why it is important to check every answer.

EXAMPLE 5 ■ Solving for one Variable in Terms of Others

The surface area A of the closed rectangular box shown in Figure 1 can be calculated from the length l, the width w, and the height h according to the formula

$$A = 2lw + 2wh + 2lh$$

Solve for w in terms of the other variables in this equation.

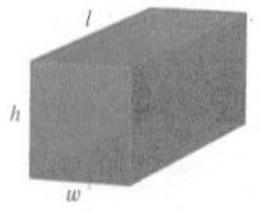

FIGURE 1
A closed rectangular box

SOLUTION

Although this equation involves more than one variable, we solve it as usual by isolating w on one side, treating the other variables as we would numbers.

$$\begin{aligned} A &= (2lw + 2wh) + 2lh && \text{Collect terms involving } w \\ A - 2lh &= 2lw + 2wh && \text{Subtract } 2lh \\ A - 2lh &= (2l + 2h)w && \text{Factor } w \text{ from RHS} \\ \frac{A - 2lh}{2l + 2h} &= w && \text{Divide by } 2l + 2h \end{aligned}$$

The solution is $w = \dfrac{A - 2lh}{2l + 2h}$. ■

It is, of course, irrelevant which letters we use for the variables in an algebra problem—the letters have no effect on the solution. Nevertheless, it is helpful in applied problems to use letters that are somehow related to the quantities they represent. In other problems, it is customary to use letters from the end of the alphabet $(\ldots, x, y, z)$ for unknown quantities that we are solving for, and to use

◀ Author notes provide step-by-step comments on solutions.

◀ *Check Your Answer* emphasizes the importance of looking back at your work to check whether an anwer is reasonable.

This *Warning* symbol points out situations where many students make the same mistake. ▶

SECTION 1.5 FRACTIONAL EXPRESSIONS 53

It is sometimes useful to use the formula

$$\frac{A}{C} + \frac{B}{C} = \frac{A+B}{C}$$

backward to write a quotient as a sum of fractions; that is, we write

$$\frac{A+B}{C} = \frac{A}{C} + \frac{B}{C}$$

For example, in certain calculus problems it is advantageous to write

$$\frac{x+3}{x} = \frac{x}{x} + \frac{3}{x} = 1 + \frac{3}{x}$$

But remember to avoid the following common error:

$$\frac{A}{B+C} \neq \frac{A}{B} + \frac{A}{C}$$

Do not make the mistake of applying properties of multiplication to the operation of addition. Many of the common errors in algebra involve doing just that. The following table states several properties of multiplication and illustrates the error in applying them to addition.

Correct multiplication property	Common error with addition
$(a \cdot b)^2 = a^2 \cdot b^2$	$(a + b)^2 \neq a^2 + b^2$
$\sqrt{a \cdot b} = \sqrt{a}\sqrt{b} \quad (a, b \geq 0)$	$\sqrt{a + b} \neq \sqrt{a} + \sqrt{b}$
$\sqrt{a^2 \cdot b^2} = a \cdot b \quad (a, b \geq 0)$	$\sqrt{a^2 + b^2} \neq a + b$
$\frac{1}{a} \cdot \frac{1}{b} = \frac{1}{a \cdot b}$	$\frac{1}{a} + \frac{1}{b} \neq \frac{1}{a + b}$
$\frac{ab}{a} = b$	$\frac{a + b}{a} \neq b$
$a^{-1} \cdot b^{-1} = \frac{1}{a \cdot b}$	$a^{-1} + b^{-1} \neq \frac{1}{a + b}$

To verify that the formulas in the right-hand column are wrong, simply substitute numbers for a and b and calculate each side. For example, if we take $a = 2$

The coordinates of a point in the xy-plane uniquely determine its location. We can think of the coordinates as the "address" of the point. In Salt Lake City, Utah, the addresses of most buildings are in fact expressed as coordinates. The city is divided into quadrants with Main Street as the vertical (North-South) axis and S. Temple Street as the horizontal (East-West) axis. An address such as

               1760 W 2100 S

indicates a location 17.6 blocks west of Main Street and 21 blocks south of S. Temple Street. (This is the address of the main post office in Salt Lake City.) With this logical system it is possible for someone unfamiliar with the city to locate any address immediately, as easily as one can locate a point in the coordinate plane.

gles APM and MQB are congruent because $d(A, M) = d(M, B)$ and corresponding angles are equal. It follows that $d(A, P) = d(M, Q)$ and so

$$x - x_1 = x_2 - x$$

Solving this equation for x, we get

$$2x = x_1 + x_2$$

$$x = \frac{x_1 + x_2}{2}$$

Similarly,

$$y = \frac{y_1 + y_2}{2}$$

MIDPOINT FORMULA

The midpoint of the line segment from $A(x_1, y_1)$ to $B(x_2, y_2)$ is

$$\left(\frac{x_1 + x_2}{2}, \frac{y_1 + y_2}{2} \right)$$

EXAMPLE 4 ■ Finding the Midpoint

The midpoint of the line segment that joins the points $(-2, 5)$ and $(4, 9)$ is

$$\left(\frac{-2 + 4}{2}, \frac{5 + 9}{2} \right) = (1, 7)$$

See Figure 8.

FIGURE 8

EXAMPLE 5 ■ Applying the Midpoint Formula

Show that the quadrilateral with vertices $P(1, 2)$, $Q(4, 4)$, $R(5, 9)$, and $S(2, 7)$ is a parallelogram by proving that its two diagonals bisect each other.

◀ Mathematical vignettes provide applications of algebra to the "real world" or give short biographies of mathematicians, contemporary as well as historical.

◀ Summary boxes organize and clarify topics and highlight key ideas.

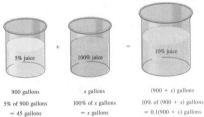

900 gallons	x gallons	$(900 + x)$ gallons
5% of 900 gallons	100% of x gallons	10% of $(900 + x)$ gallons
= 45 gallons	= x gallons	= $0.1(900 + x)$ gallons

Let x be the amount (in gallons) of pure juice to be added. Then $(900 + x)$ gallons of 10% orange juice mixture will result. The key idea to turn the picture into an equation here is to notice that the total amount of juice on both sides of the equal sign is the same. The orange juice in the first vat is 5% of 900 gal, or 45 gal. The second vat contains x gallons of juice, and the third contains $0.1(900 + x)$ gallons.

Equating the total amounts of pure juice before and after mixing, we get the equation

$$45 + x = 0.1(900 + x) \qquad \text{From Figure 2}$$

$$45 + x = 90 + 0.1x \qquad \text{Multiply}$$

$$0.9x = 45 \qquad \text{Subtract } 0.1x \text{ and } 45$$

$$x = \frac{45}{0.9} = 50 \qquad \text{Divide by } 0.9$$

The manufacturer should add 50 gallons of pure orange juice to the soda.

CHECK YOUR ANSWER

amount of juice before mixing = 5% of 900 gal + 50 pure gal

                                        = 45 gal + 50 gal = 95 gal

amount of juice after mixing = 10% of 950 gal = 95 gal

Amounts are equal. ✓

EXAMPLE 7 ■ Time Needed to Do a Job

Because of an anticipated heavy rainstorm, the water level in a reservoir must be lowered by 1 ft. Opening spillway A lowers the level by this amount in 4 h, whereas opening the smaller spillway B does the job in 6 h. How long will it take to lower the water level by 1 ft if both spillways are opened?

Problem-solving notes in margins ▶ apply problem-solving steps to examples in the text.

Check Your Answer ▶

FIGURE 3

Example titles clarify the ▶ purpose of examples.

350   CHAPTER 5 EXPONENTIAL AND LOGARITHMIC FUNCTIONS

25. Under ideal conditions, a certain type of bacteria has a relative growth rate of 220% per hour. A number of these bacteria are introduced accidentally into a food product. Two hours after contamination, a bacteria count shows that there are about 40,000 bacteria in the food.
(a) Find the initial number of bacteria introduced into the food.
(b) Find the estimated number of bacteria in the food three hours after contamination.

26. A radioactive substance decays in such a way that the amount of mass remaining after t days is given by the function

$$m(t) = 13e^{-0.015t}$$

where $m(t)$ is measured in kilograms.
(a) Find the mass at time $t = 0$.
(b) How much of the mass remains after 45 days?

27. Radioactive iodine is used by doctors as a tracer in diagnosing certain thyroid gland disorders. This type of iodine decays in such a way that the mass remaining after t days is given by the function

$$m(t) = 6e^{-0.087t}$$

where $m(t)$ is measured in grams.
(a) Find the mass at time $t = 0$.
(b) How much of the mass remains after 20 days?

28. A sky diver is dropped from a reasonable height above the ground. The air resistance she experiences is proportional to her velocity, and the constant of proportionality is 0.2. It can be shown that the velocity of the sky diver at time t is given by

$$v(t) = 80(e^{-0.2t} - 1)$$

where t is measured in seconds and $v(t)$ is measured in feet per second (ft/s).
(a) Find the initial velocity of the sky diver.
(b) Find the velocity after 5 s and after 10 s.
(c) Draw a graph of the velocity function $v(t)$.
(d) The maximum velocity of a falling object with wind resistance is called its *terminal velocity*. From the graph in part (c), find the terminal velocity of this sky diver.

$v(t) = 80(e^{-0.2t} - 1)$

29. A 50-gallon barrel is filled completely with pure water. Salt water with a concentration of 0.3 lb/gal is then pumped into the barrel, and the resulting mixture overflows at the same rate. The amount of salt in the barrel at time t is given by

$$Q(t) = 15(1 - e^{-0.04t})$$

where t is measured in minutes and $Q(t)$ is measured in pounds.
(a) How much salt is in the barrel after five minutes?
(b) How much salt is in the barrel after ten minutes?
(c) Draw a graph of the function $Q(t)$.
(d) From the graph in part (c), what value does the amount of salt in the barrel approach as t becomes large? Is this what you would expect?

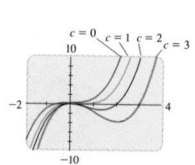

$Q(t) = 15(1 - e^{-0.04t})$

◀ Graphing calculator exercises are indicated with the graphing calculator logo.

◀ Real-world applications show the relevance of algebra to everyday life and indicate its remarkable problem-solving power.

SECTION 4.2 USING GRAPHING DEVICES TO GRAPH POLYNOMIALS   271

SOLVING INEQUALITIES

In the next example, we solve an inequality using the graph of a polynomial.

EXAMPLE 6 ■ Solving an Inequality Using a Graphing Calculator

Solve the inequality

$$2x^4 - 6x^3 - 5x^2 - 3x - 3 \le 0$$

SOLUTION

First we graph the polynomial $P(x) = 2x^4 - 6x^3 - 5x^2 - 3x - 3$, as in Figure 8. To solve the inequality $P(x) \le 0$ we must find all values of x for which the graph of the function lies on or below the x-axis. From the graph we see that the solution is the interval that lies between the two x-intercepts. By zooming in we find that the x-intercepts are -0.79 and 3.79 (correct to two decimal places), so the approximate solution of the inequality is the interval $[-0.79, 3.79]$. ■

-100

FIGURE 8
$P(x) = 2x^4 - 6x^3 - 5x^2 - 3x - 3$

FAMILIES OF POLYNOMIALS

A graphing calculator enables us to quickly draw the graphs of many functions at once, on the same viewing screen. This enables us to see how changing a value in the definition of the functions affects the shape of its graph. In the next example we apply this principle to a family of third-degree polynomials.

EXAMPLE 7 ■ A Family of Polynomials

Sketch the family of polynomials $P(x) = x^3 - cx^2$ for $c = 0, 1, 2,$ and 3. How does changing the value of c affect the graph?

SOLUTION

The polynomials

$$P_0(x) = x^3 \qquad\qquad P_1(x) = x^3 - x^2$$
$$P_2(x) = x^3 - 2x^2 \qquad\qquad P_3(x) = x^3 - 3x^2$$

are graphed in Figure 9. We see that increasing the value of c causes the graph to develop an increasingly deep "valley" to the right of the y-axis, creating a local maximum at the origin and a local minimum at a point in quadrant IV. This local minimum moves lower and further to the right as c increases. ■

Families of functions can be studied ▶ using the graphing calculator, so that students can see how varying a parameter affects the graph of a function.

$c = 0$ $c = 1$ $c = 2$ $c = 3$

FIGURE 9
$P(x) = x^3 - cx^2$

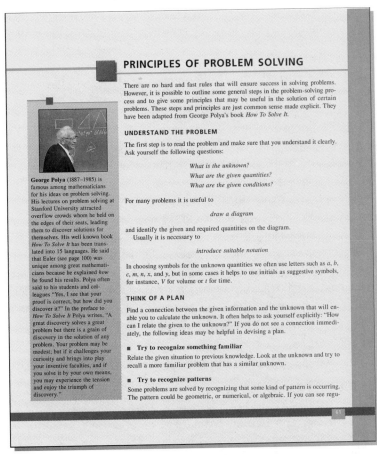

PRINCIPLES OF PROBLEM SOLVING

There are no hard and fast rules that will ensure success in solving problems. However, it is possible to outline some general steps in the problem-solving process and to give some principles that may be useful in the solution of certain problems. These steps and principles are just common sense made explicit. They have been adapted from George Polya's book *How To Solve It*.

UNDERSTAND THE PROBLEM

The first step is to read the problem and make sure that you understand it clearly. Ask yourself the following questions:

> *What is the unknown?*
> *What are the given quantities?*
> *What are the given conditions?*

For many problems it is useful to

> *draw a diagram*

and identify the given and required quantities on the diagram.
Usually it is necessary to

> *introduce suitable notation*

In choosing symbols for the unknown quantities we often use letters such as a, b, c, m, n, x, and y, but in some cases it helps to use initials as suggestive symbols, for instance, V for volume or t for time.

THINK OF A PLAN

Find a connection between the given information and the unknown that will enable you to calculate the unknown. It often helps to ask yourself explicitly: "How can I relate the given to the unknown?" If you do not see a connection immediately, the following ideas may be helpful in devising a plan.

■ **Try to recognize something familiar**

Relate the given situation to previous knowledge. Look at the unknown and try to recall a more familiar problem that has a similar unknown.

■ **Try to recognize patterns**

Some problems are solved by recognizing that some kind of pattern is occurring. The pattern could be geometric, or numerical, or algebraic. If you can see regu-

George Polya (1887–1985) is famous among mathematicians for his ideas on problem solving. His lectures on problem solving at Stanford University attracted overflow crowds whom he held on the edges of their seats, leading them to discover solutions for themselves. His well known book *How To Solve It* has been translated into 15 languages. He said that Euler (see page 100) was unique among great mathematicians because he explained *how* he found his results. Polya often said to his students and colleagues "Yes, I see that your proof is correct, but how did you discover it?" In the preface to *How To Solve It* Polya writes, "A great discovery solves a great problem but there is a grain of discovery in the solution of any problem. Your problem may be modest; but if it challenges your curiosity and brings into play your inventive faculties, and if you solve it by your own means, you may experience the tension and enjoy the triumph of discovery."

61

◄ *Focus on Problem Solving* sections include many new problems that are related to the problem-solving principle discussed in the section or to the material of the chapter itself.

◄ Mathematical vignettes provide short biographies of interesting mathematicians, contemporary as well as historical, or give applications of algebra to the real world.

(Major changes continued from page ix)

■ Chapter 4 on polynomial and rational functions has been restructured to begin with the idea of the graph of a polynomial function; the material on the theory of equations now comes later in the chapter. Many of the exercise sets in this chapter have been rewritten to include problems that involve more straightforward factorizations.

■ Chapter 5 has been completely rewritten to emphasize the importance of the natural exponential function and its role in applications. Exponential and logarithmic equations are now treated in a separate section, Section 5.5.

■ In Chapter 6, the concept of echelon form has been introduced to clarify the Gaussian elimination process, and many of the Gaussian elimination exercises have been simplified from a computational point of view.

■ Chapter 7 is a new chapter—it contains a complete treatment of conic sections, material that in the first edition was divided between Chapter 3 and an appendix. The summary boxes for each conic have been restructured and clarified, and we now summarize guidelines for sketching hyperbolas.

■ Chapters 8 and 9 retain the structure of the first edition, with all the new features and enhancements.

- The emphasis on problem solving has been enhanced throughout the text. Numerous margin comments referring to the principles of problem solving have been added, and "guideline" boxes that apply these principles to the topic at hand have been expanded and improved.

- Where possible, "prose-style" explanations in examples have been replaced by author notes to improve readability.

- We have made greater use of summary boxes to clarify the structure of some topics.

- Numerous graphs and figures have been added to enhance understanding through visualization. In particular, figures have been added to summary boxes wherever appropriate.

- Each example has been given a title to clarify its purpose.

- Several new "vignettes" have been added and existing ones have been updated.

- The *Focus on Problem Solving* sections include many new problems that are related to the problem-solving principle discussed in the section or to the material of the chapter itself.

- This edition employs additional colors, which are used to clarify the relationships between different parts of figures.

ACKNOWLEDGMENTS

We thank the following people for their thoughtful and constructive comments.

Reviewers of the First Edition

Barry W. Brunson, *Western Kentucky University*; Gay Ellis, *Southwest Missouri State University*; Martha Ann Larkin, *Southern Utah University*; Franklin A. Michello, *Middle Tennessee State University*; Kathryn Wetzel, *Amarillo College*.

Reviewers of the Second Edition

David Watson, *Rutgers University*; Floyd Downs, *Arizona State University at Tempe*; Muserref Wiggins, *University of Akron*; Marjorie Kreienbrink, *University of Wisconsin*; Richard Dodge, *Jackson Community College*; Christine Panoff, *University of Michigan at Flint*; Arnold Volbach, *University of Houston, University Park*; Keith Oberlander, *Pasadena City College*; Tom Walsh, *City College of San Francisco*; and George Wang, *Whittier College*.

We thank Kathi Townes and the staff of TECHarts, our production service, for their excellent work and their unflagging energy in attending to all the details of design and production. We also thank Diana Gerardi, for checking the accuracy of examples and exercises, and Phyllis Panman, for working all the exercises. At Brooks/Cole, our thanks go to production editor Jamie Sue Brooks, assistant editor Elizabeth Rammel, editorial associate Carol Ann Benedict, marketing manager Patrick Farrant, marketing communications director Margaret Parks, and art director Vernon Boes. They have all done an outstanding job.

We are especially grateful to mathematics publisher Gary W. Ostedt. His extensive experience and keen editorial insight were invaluable resources in the writing of this second edition.

ANCILLARIES FOR COLLEGE ALGEBRA, SECOND EDITION

Instructor's Ancillaries

PRINTED

Instructor's Solutions Manual
by John Banks (San Jose City College)
- Solutions to all even-numbered text exercises

Test Items
by Andrew Bulman-Fleming
- Over 3000 multiple-choice and short-answer test items
- Computerized versions also available

SOFTWARE

Computerized Test Items
- Over 3000 multiple-choice and short-answer test items
- Available for DOS, DOS/Windows, and Mac platforms
- Print version also available

Technical support
- Toll-free technical support available for any Brooks/Cole software product:
  (800) 214-2661 or E-mail: support@brookscole.com

Students' Ancillaries

PRINTED

Student Solutions Manual
by John Banks (San Jose City College)
- Solutions to all odd-numbered text exercises

Study Guide
by John Banks (San Jose City College)
- Detailed explanations
- Worked practice problems

SOFTWARE

Brooks/Cole Exerciser (BCX) 2.0
by Laurel Technical Services
- Text-specific tutorial software available for DOS, DOS/Windows, and Mac platforms
- Examples/problems similar to those in the text
- Some problems algorithmically generated
- Monitors student progress; prints reports

Technical support
- Toll-free technical support available for any Brooks/Cole software product:
  (800) 214-2661 or E-mail: support@brookscole.com

Maple V, Release 3, Student Edition
- Macintosh or DOS/Windows software
- Provides science, math, and engineering students all the power they need to graph and compute algebraic and numeric problems they are likely to encounter in practically all of their undergraduate courses.

Exploring Calculus with Math T/L3
by Doug Child and Don Small
- Macintosh software offering explorations of precalculus and calculus concepts and topics in a friendly, easy-to-use program
- Based on popular Calculus T/L II program

TEMATH 1.5
by Robert Kowalczyk and Adam Hausknecht
- Macintosh software for graphing functions, finding roots and asymptotes, graphing tangent lines, studying properties of definite integrals, curve fitting, and more

GraphPlay 1.0
by Laurence Harris
- Macintosh software that allows rapid creation of over 400 graphs, showing relationship between functions in algebraic form and their graphs
- Math is displayed on a screen in natural form, rather than via a special set of key commands

TO THE STUDENT

This textbook was written for you to use as a guide to mastering algebra. Here are some suggestions to help you get the most out of your course.

First of all, you should read the appropriate section of text *before* you attempt your homework problems. Reading a mathematics text is quite different from reading a novel, a newspaper, or even another textbook. You may find that you have to reread a passage several times before you understand it. Pay special attention to the examples, and work them out yourself with pencil and paper as you read. With this kind of preparation you will be able to do your homework much more quickly and with more understanding.

Don't make the mistake of trying to memorize every single rule or fact you may come across. Mathematics does not consist simply of memorization. Mathematics is a *problem-solving art,* not just a collection of facts. To master the subject you must solve problems—lots of problems. Do as many of the exercises as you can. Be sure to write your solutions in a logical, step-by-step fashion. Don't give up on a problem if you can't solve it right away. Try to understand the problem more clearly—reread it thoughtfully and relate it to what you have learned from your teacher and from the examples in the text. Struggle with it until you solve it. Once you have done this a few times you will begin to understand what mathematics is really all about.

Answers to the odd-numbered exercises, as well as all the answers to each chapter test, appear at the back of the book. If your answer differs from the one given, don't immediately assume that you are wrong. There may be a calculation that connects the two answers and makes both correct. For example, if you get $1/(\sqrt{2} - 1)$ but the answer given is $1 + \sqrt{2}$, your answer *is* correct, because you can multiply both numerator and denominator of your answer by $\sqrt{2} + 1$ to change it to the given answer.

The symbol ⊘ is used to warn against committing an error. We have placed this symbol in the margin to point out situations where we have found that many of our students make the same mistake.

CALCULATORS AND CALCULATIONS

Calculators are essential in most mathematics and science subjects. They free us from performing routine tasks, so we can focus more clearly on the concepts we are studying. Calculators are powerful tools but their results need to be interpreted with care. In what follows, we describe the features a calculator suitable for a College Algebra course should have, and we give guidelines for interpreting the results of its calculations.

SCIENTIFIC AND GRAPHING CALCULATORS

For this course you will need a *scientific* calculator—one that has, as a minimum, the usual arithmetic operations ($+$, $-$, $\times$, $\div$) and exponential and logarithmic functions (e^x, 10^x, $\ln x$, log). In addition, a memory and at least some degree of programmability will be useful. Many scientific calculators can perform operations on matrices and determinants, which are studied in Chapter 6.

Your instructor may recommend or require that you purchase a graphing calculator. This book has optional sections and exercises that require the use of a graphing calculator or a computer with graphing software. These special sections and exercises are indicated by the symbol ⌐⌐.

It is important to realize that, because of limited resolution, a graphing calculator gives only an *approximation* to the graph of a function. It can plot only a finite number of points and then connect them to form a *representation* of the graph. In Chapters 3, 4, and 5 we point out that we sometimes have to be careful when interpreting graphs produced by calculators.

CALCULATIONS AND SIGNIFICANT FIGURES

Most of the applied examples and exercises in this book involve approximate values. For example, one exercise states that the moon has a radius of 1074 miles. This does not mean that the moon's radius is exactly 1074 miles but simply that this is the radius rounded to the nearest mile.

One simple method for specifying the accuracy of a number is to state how many **significant digits** it has. The significant digits in a number are the ones from the first nonzero digit to the last nonzero digit (reading from left to right). Thus, 1074 has four significant digits, 1070 has three, 1100 has two, and 1000 has one significant digit. This rule may sometimes lead to ambiguities. For example, if a distance is 200 km to the nearest kilometer, then the number 200 really has three significant digits, not just one. This ambiguity is avoided if we use scientific notation—that is, if we express the number as a multiple of a power of 10:

$$2.00 \times 10^2$$

When working with approximate values, students often make the mistake of giving a final answer with *more* significant digits than the original data. This is incorrect because you cannot "create" precision by using a calculator. The final result can be no more accurate than the measurements given in the problem. For example, suppose we are told that the two shorter sides of a right triangle are measured to be 1.25 and 2.33 inches long. By the Pythagorean Theorem, we find, using a calculator, that the hypotenuse has length

$$\sqrt{1.25^2 + 2.33^2} \approx 2.644125564 \text{ in.}$$

But since the given lengths were expressed to three significant digits, the answer cannot be any more accurate. We can therefore say only that the hypotenuse is 2.64 in. long, rounding to the nearest hundredth.

In general, the final answer should be expressed with the same accuracy as the *least*-accurate measurement given in the statement of the problem. The following rules make this principle more precise.

RULES FOR WORKING WITH APPROXIMATE DATA

1. When multiplying or dividing, round off the final result so that it has as many *significant digits* as the given value with the fewest number of significant digits.

2. When adding or subtracting, round off the final result so that it has its last significant digit in the *decimal place* in which the least-accurate given value has its last significant digit.

3. When taking powers or roots, round off the final result so that it has the same number of *significant digits* as the given value.

As an example, suppose that a rectangular table top is measured to be 122.64 in. by 37.3 in. We express its area and perimeter as follows:

Area = length × width = 122.64 × 37.3 ≈ 4570 in²   Three significant digits

Perimeter = 2(length + width) = 2(122.64 + 37.3) ≈ 319.9 in   Tenths digit

Note that in the formula for the perimeter, the value 2 is an exact value, not an approximate measurement. It therefore does not affect the accuracy of the final result. In general, if a problem involves only exact values, we may express the final answer with as many significant digits as we wish.

Note also that to make the final result as accurate as possible, you should wait until the last step to round off your answer. If necessary, use the memory feature of your calculator to retain the results of intermediate calculations.

ABBREVIATIONS

cm	centimeter		**MHz**	megahertz
dB	decibel		**mi**	mile
F	farad		**min**	minute
Ft	foot		**mL**	milliliter
g	gram		**mm**	millimeter
gal	gallon		**N**	Newton
h	hour		**qt**	quart
H	henry		**oz**	ounce
Hz	Hertz		**s**	second
in.	inch		**Ω**	ohm
J	Joule		**V**	volt
kcal	kilocalorie		**W**	watt
kg	kilogram		**yd**	yard
km	kilometer		**yr**	year
kPa	kilopascal		**°C**	degree Celsius
L	liter		**°F**	degree Fahrenheit
lb	pound		**K**	Kelvin
M	mole of solute per liter of solution		$\Rightarrow$	implies
m	meter		$\Leftrightarrow$	is equivalent to

CONTENTS

COLLEGE
ALGEBRA

BASIC ALGEBRA

For the ancient Egyptian priests, calculating the volume of a pyramid was a complicated process because the language of algebra had not yet been invented. With algebraic notation, however, we can easily find the volume of a pyramid using the formula $V = \frac{1}{3}b^2h$.

One learns by doing the thing; for though you think you know it, you have no certainty until you try.

SOPHOCLES

The power of algebra stems from its ability to say many things at once.

Let us begin with a simple example. Consider the following list of facts about addition:

$$3 + 4 = 4 + 3$$

$$5 + 7 = 7 + 5$$

$$11 + 2 = 2 + 11$$

$$51032 + 873 = 873 + 51032$$

The order in which we add two numbers appears to make no difference—we get the same answer in either case. A little thought will convince you that this is *always* true. One way to express this fact is to complete the list to include all possible pairs of numbers; however, this would take forever because there are infinitely many pairs of numbers. A better way to express this fact is to say that when adding two numbers we have

(first number) + (second number) = (second number) + (first number)

This sentence covers all possible cases at once. If we use letters such as a, b, c, ... to stand for numbers, we can write this even more briefly as

$$a + b = b + a$$

This is the shortest way of saying what we mean.

This example illustrates the basic idea in all of algebra. Instead of working with specific numbers, as we do in arithmetic, we work with letters that can stand for any number. The crucial advantage in writing things this way is that we can see patterns more easily. This, in turn, helps us to write formulas and solve problems. We illustrate these key ideas with some examples.

> The word *algebra* comes from the ninth century Arabic book *Hisâb al-Jabr w'al-Muqabala*, written by al-Khowarizmi. The title refers to transposing and combining terms, two processes used in solving equations. In Latin translations the title was shortened to *Aljabr*, from which we get the word *algebra*. The author's name itself made its way into the English language in the form of our word *algorithm*.

WRITING FORMULAS

We begin with examples that illustrate how to write formulas in algebra.

EXAMPLE 1 ■ Finding a Formula for Average

Find a formula for the average of three numbers.

SOLUTION

To find the average of three numbers, we add them and divide by 3. For example, if your scores on three tests in a college algebra course are 78 and

81 and 93, then your average test score is

$$\frac{78 + 81 + 93}{3} = 84$$

Intuitively, the average is where the numbers "balance."

We see that the formula for the average A of any three numbers a, b, and c is

$$A = \frac{a + b + c}{3}$$

∎

EXAMPLE 2 ■ Finding a Formula for Weekly Pay

Find a formula for a worker's weekly pay in terms of the hourly wage and the number of hours worked.

SOLUTION

If you are paid $6 an hour and if you work 20 hours in one week, then your pay (which we will call P) for that week is

$$P = 6 \times 20$$

If you work only 18 hours the following week, then your pay for that week is $P = 6 \times 18$. The general rule is that your pay is $6 times the number of hours worked. Letting H stand for the number of hours worked, we have

$$P = 6H$$

The formula would be different for a worker who makes $10 or $14 an hour. But if we let W stand for the hourly wage, we can write

$$P = WH$$

∎

The simple formula in Example 2 gives the pattern for calculating the weekly pay of any worker, with any hourly wage, for any number of hours worked. For example, the weekly pay of a worker who is paid $8 per hour and has worked 32 hours is

$$P = WH = 8 \times 32 = \$256$$

The ability to discover and use formulas as we have done here is fundamental to understanding what algebra is all about. You will be asked in exercises throughout this book to do precisely that.

SOLVING PROBLEMS

The methods of algebra enable us to solve many types of problems. Here are examples that use the formulas we have just discovered.

EXAMPLE 3 ■ Solving an Algebra Problem involving Averages

If your scores on the first two college algebra tests are 76 and 87, what score must you earn on the third test in order for your average for the three tests to be 85?

SOLUTION

The formula we discovered in Example 1 tells us that the average of three numbers is

$$A = \frac{a + b + c}{3}$$

Since we want $A = 85$ and we know that $a = 76$ and $b = 87$, we substitute all these values in our formula. Thus

$$85 = \frac{76 + 87 + c}{3}$$

$$85 = \frac{163 + c}{3}$$

To find the necessary test score on the third test, we solve for c:

$$85 \times 3 = \frac{163 + c}{3} \times 3 \qquad \text{Multiply each side by 3}$$

$$255 = 163 + c$$

$$92 = c \qquad \text{Subtract 163 from each side}$$

You must earn a test score of 92 on the third test in order to have an average test score of 85.

■

EXAMPLE 4 ■ Solving an Algebra Problem involving Pay

James makes $5.50 an hour working at a car wash. He wants to earn enough money to buy a portable CD player that costs $77 (including tax). How many hours does he need to work to earn this amount?

SOLUTION

Let us first write down what we know. James's hourly wage is $W = \$5.50$ and the CD player costs $77. So, the amount of pay he needs is $P = \$77$. From the formula in Example 2 we know that

$$P = WH$$

Diophantus lived in Alexandria about 250 A.D. He wrote a book called *Arithmetica,* which is considered the first book on algebra. It contains methods for finding integer solutions of algebraic equations. The *Arithmetica* continued to be read for more than a thousand years. Fermat (see page 532) made some of his most important discoveries while studying this book. Diophantus' major contribution is the use of symbols to stand for the unknowns in a problem. Although his symbolism is not as simple as that used today, it was a major advance over writing everything in words. In Diophantus' notation the equation

$$x^5 - 7x^2 + 8x - 5 = 24$$

is written

$$\Delta K^{\gamma} \alpha \zeta \theta \phi \Delta^{\gamma} \eta \dot{M} \varepsilon \iota \kappa \delta$$

Our modern algebraic notation did not come into common use until the 17th century.

We now substitute the values we know for P and W and solve for H:

$$77 = 5.50H$$

$$H = \frac{77}{5.50} \qquad \text{Divide each side by 5.50}$$

$$H = 14$$

So, James needs to work 14 hours in order to earn $77. ■

You are probably familiar with the formulas for the area and volume of geometric shapes; these are listed on the inside of the front cover of this book. For example, the formula for the area A of a rectangle is

$$A = lw$$

where l is the length and w is the width. In the next example we use this formula to derive another formula.

EXAMPLE 5 ■ An Algebra Problem involving Area

(a) Find a formula for the area of a rectangle whose length is twice its width.
(b) A manufacturing company requires a rectangular sheet of metal whose length is twice its width and whose total area is 50 ft². What are the dimensions of this sheet?

SOLUTION

(a) If x is the width of the rectangle, then its length is $2x$ (see Figure 1). So, from the formula for the area of a rectangle we get the formula

$$A = lw = (2x)x = 2x^2$$

(b) We know that $A = 50$, and we want to find x. Using the formula in part (a), we get

$$50 = 2x^2$$

$$25 = x^2 \qquad \text{Divide each side by 2}$$

$$5 = x \qquad \text{Take the square root of each side}$$

Thus, the width of the sheet is 5 ft and its length is 10 ft. ■

The problems we have solved in the preceding examples are simple, but the methods are far-reaching. In fact, algebra applies to a wide variety of problems and is an indispensable tool in the study of any exact science. In algebra we use letters to stand for numbers in order to talk concisely about numbers, to see patterns (or formulas), and to manipulate these patterns easily in order to solve problems.

w

l

Area $= lw$

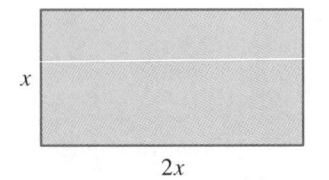

x

$2x$

FIGURE 1

| 1.1 | **EXERCISES** |

1–8 ■ The list of facts illustrates a general principle about numbers. State the principle in algebraic notation.

1. $3 \cdot 5 = 5 \cdot 3, \quad 6 \cdot 2 = 2 \cdot 6, \quad 4 \cdot 7 = 7 \cdot 4$

2. $6(2 + 3) = 6 \cdot 2 + 6 \cdot 3, \quad 2(3 + 5) = 2 \cdot 3 + 2 \cdot 5$

3. $(5 - 2)(5 + 2) = 5^2 - 2^2, \quad (6 - 4)(6 + 4) = 6^2 - 4^2$

4. $(3 + 5)^2 = 3^2 + 2(3 \cdot 5) + 5^2$
$(4 + 7)^2 = 4^2 + 2(4 \cdot 7) + 7^2$

5. $(2 + 3) + 4 = 2 + (3 + 4)$
$(4 + 6) + 2 = 4 + (6 + 2)$

6. $(2 \cdot 3) \cdot 4 = 2 \cdot (3 \cdot 4)$
$(4 \cdot 6) \cdot 2 = 4 \cdot (6 \cdot 2)$

7. $(2 \cdot 3)^2 = 2^2 \cdot 3^2, \quad (5 \cdot 7)^2 = 5^2 \cdot 7^2$

8. $(5 - 3) = -(3 - 5), \quad (9 - 4) = -(4 - 9)$

9–20 ■ Write an algebraic formula for the given quantity.

9. The average of two numbers

10. The average of four numbers

11. The sum of a number and twice its square

12. The area of a rectangle whose length is 4 ft more than its width

13. The number of days in w weeks

14. The number of cents in n quarters

15. The product of two consecutive integers

16. The sum of three consecutive integers

17. The sum of the squares of two numbers

18. The distance in miles that a car travels in t hours at a speed of r miles per hour

19. The time it takes an airplane to travel d miles if its speed is r miles per hour

20. The speed of a boat that travels d miles in t hours

21–32 ■ Write an algebraic formula for the given quantity. You may need to consult the formulas for area and volume listed on the inside of the front cover of this book.

21. The area of a square of side x

22. The volume of a cube of side x

23. The volume of a box with a square base of side x and height $2x$

24. The surface area of a cube of side x

25. The surface area of a box of dimensions l, w, and h

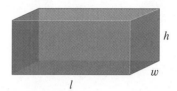

26. The surface area of a box with open top whose base has dimensions x and $2x$ and whose height is h

27. The area of a triangle whose base is twice its height

28. The area of what remains when a circle is cut out of a square, as shown in the figure

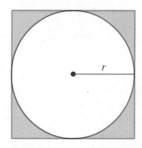

29. The length of the race track shown in the figure

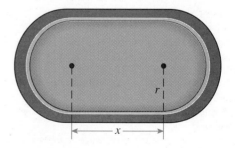

30. The area enclosed by the race track shown in the preceding figure

31. The volume of what remains of a solid ball of radius R after a smaller solid ball of radius r is removed from inside the larger ball, as shown in the figure

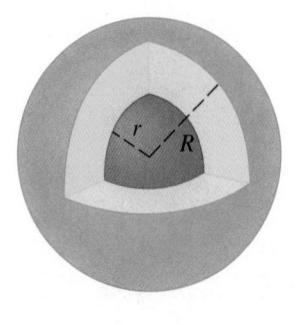

32. The volume of a cylindrical can whose height is twice its radius, as shown in the figure

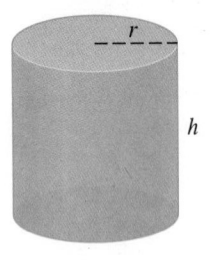

33. Suppose that your scores on two college algebra tests are 79 and 83, and the only test left to take is the final exam. This exam, however, counts double (this means that your test score on the final exam counts twice in the formula for average).
  (a) If you earn a score of 88 on the final exam, what is your average test score for the course?
  (b) Write a formula for a student's average test score in terms of the two test scores a and b and the test score f on the final exam.
  (c) What test score do you need to earn on the final exam in order for your average test score for the course to be 85?

34. Suppose that you walk at a constant rate of 4 feet per second.
  (a) If you walk for 20 seconds, how far will you have walked?
  (b) Find a formula for the distance d you walk in t seconds.

  (c) How long do you have to walk to cover a distance of one mile?

35. Maria can mow a lawn at a rate of 150 square feet per minute.
  (a) What area can she mow in half an hour?
  (b) Find a formula for the area she can mow in T minutes.
  (c) How long would it take Maria to mow a lawn that is 80 ft wide and 120 ft long?
  (d) If she wants to mow the lawn in part (c) in one hour, what would her rate of mowing have to be?

36. A breakfast cereal company needs to manufacture boxes for their product. For esthetic purposes, the box must have the following properties: its width is four times its depth and its height is six times its depth.
  (a) Find the volume of the box if its depth is 2 inches.
  (b) Find a formula for the volume V of such a box in terms of its depth x.
  (c) If the company needs to produce a box whose volume is 648 in^3, what dimensions must the box have?

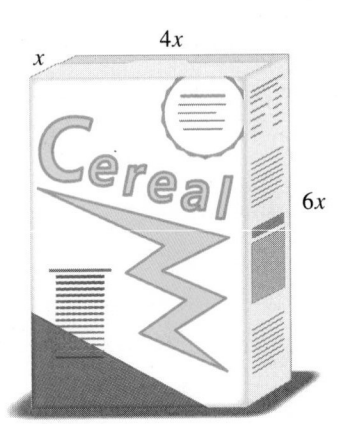

37. In many universities a student is given grade points for each credit unit according to the following scale:

A	4 points
B	3 points
C	2 points
D	1 point
F	0 point

For example, a grade of A in a 3-unit course earns $4 \times 3 = 12$ grade points and a grade of B in a 5-unit course earns $3 \times 5 = 15$ grade points. A student's grade point average (GPA) for these two courses is the total number of grade points earned divided by the number of units; in this case the GPA is $(12 + 15)/8 = 3.375$.

(a) Find a formula for the GPA of a student who earns a grade of A in a units of course work, B in b units, C in c units, D in d units, and F in f units.

(b) Find the GPA of a student who has earned a grade of A in two 3-unit courses, B in one 4-unit course, and C in three 3-unit courses.

1.2 REAL NUMBERS

Let us recall the types of numbers that make up the real number system. We start with the **natural numbers**:

$$1, 2, 3, 4, \ldots$$

The **integers** consist of the natural numbers together with their negatives and 0:

$$\ldots, -3, -2, -1, 0, 1, 2, 3, 4, \ldots$$

We construct the **rational numbers** by taking ratios of integers. Thus, any rational number r can be expressed as

$$r = \frac{m}{n} \qquad \text{where } m \text{ and } n \text{ are integers and } n \neq 0$$

Examples are

$$\frac{1}{2} \qquad -\frac{3}{7} \qquad 46 = \frac{46}{1} \qquad 0.17 = \frac{17}{100}$$

(Recall that division by 0 is always ruled out, so expressions like $\frac{3}{0}$ and $\frac{0}{0}$ are undefined.) There are also real numbers, such as $\sqrt{2}$, that cannot be expressed as a ratio of integers and are therefore called **irrational numbers.** It can be shown, with varying degrees of difficulty, that each of the following numbers is also an irrational number:

$$\sqrt{3} \qquad \sqrt{5} \qquad \sqrt[3]{2} \qquad \pi \qquad \frac{3}{\pi^2}$$

The set of all real numbers is usually denoted by the symbol $\mathbb{R}$. When we use the word *number* without qualification, we will mean "real number." Figure 1 is a schematic diagram of the types of real numbers that we work with in this book.

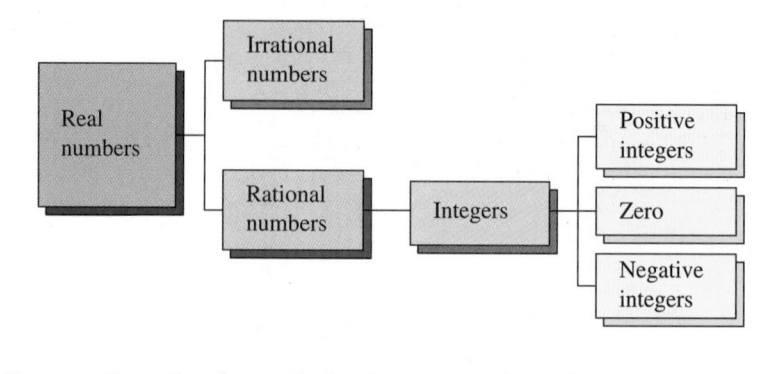

FIGURE 1

The real number system

Every real number has a decimal representation. If the number is rational, then its corresponding decimal is repeating. For example,

$$\tfrac{1}{2} = 0.5000\ldots = 0.5\overline{0} \qquad\qquad \tfrac{2}{3} = 0.66666\ldots = 0.\overline{6}$$

$$\tfrac{157}{495} = 0.3171717\ldots = 0.3\overline{17} \qquad \tfrac{9}{7} = 1.285714285714\ldots = 1.\overline{285714}$$

A repeating decimal such as

$$x = 3.5474747\ldots$$

is a rational number. To convert it to a ratio of two integers we write:

$$1000x = 3547.47474747\ldots$$
$$\underline{\quad 10x = \quad\; 35.47474747\ldots}$$
$$\;\,990x = 3512.0$$

Thus $x = \tfrac{3512}{990}$. (The idea is to multiply x by appropriate powers of 10, and then subtract to eliminate the repeating part.)

(The bar indicates that the sequence of digits repeats forever.) If the number is irrational, the decimal representation is nonrepeating:

$$\sqrt{2} = 1.414213562373095\ldots \qquad \pi = 3.141592653589793\ldots$$

If we stop the decimal expansion of any number at a certain place, we get an approximation to the number. For instance, we can write

$$\pi \approx 3.14159265$$

where the symbol $\approx$ is read "is approximately equal to." The more decimal places we retain, the better the approximation we get.

The real numbers can be represented by points on a line, as shown in Figure 2. The positive direction (toward the right) is indicated by an arrow. We choose an arbitrary reference point O, called the **origin,** which corresponds to the real number 0. Given any convenient unit of measurement, each positive number x is represented by the point on the line a distance of x units to the right of the origin, and each negative number $-x$ is represented by the point x units to the left of the origin. Thus, every real number is represented by a point on the line, and every point P on the line corresponds to exactly one real number. The number associated with the point P is called the coordinate of P and the line is then called a **coordinate line,** or a **real number line,** or simply a **real line.** Often we identify the point with its coordinate and think of a number as being a point on the real line.

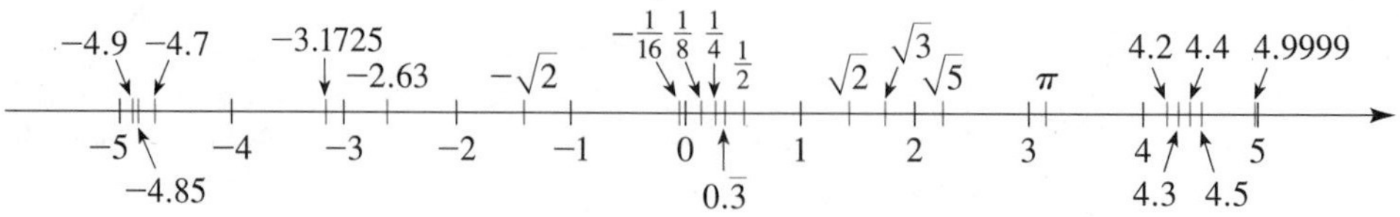

FIGURE 2

The real line

The real numbers are *ordered*. We say that *a* **is less than** *b* and write $a < b$ if $b - a$ is a positive number. Geometrically, this means that *a* lies to the left of *b* on the number line. (Equivalently, we can say that *b* is greater than *a* and write $b > a$.) The symbol $a \leq b$ (or $b \geq a$) means that either $a < b$ or $a = b$ and is read "*a* is less than or equal to *b*." For instance, the following are true inequalities (see Figure 3):

$$7 < 7.4 < 7.5 \qquad -\pi < -3 \qquad \sqrt{2} < 2 \qquad 2 \leq 2$$

FIGURE 3

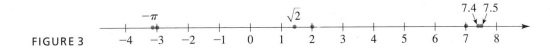

PROPERTIES OF REAL NUMBERS

In combining real numbers using the familiar operations of addition and multiplication, we use the following properties of real numbers.

PROPERTIES OF REAL NUMBERS

Property	Example	Name and Description
$a + b = b + a$	$7 + 3 = 3 + 7$	**Commutative Property for addition** When we add two numbers, order doesn't matter.
$ab = ba$	$3 \cdot 5 = 5 \cdot 3$	**Commutative Property for multiplication** When we multiply two numbers, order doesn't matter.
$(a + b) + c = a + (b + c)$	$(2 + 4) + 7 = 2 + (4 + 7)$	**Associative Property for addition** When we add three numbers, it doesn't matter which two we add first.
$(ab)c = a(bc)$	$(3 \cdot 7) \cdot 5 = 3 \cdot (7 \cdot 5)$	**Associative Property for multiplication** When we multiply three numbers, it doesn't matter which two we multiply first.
$a(b + c) = ab + ac$ $(b + c)a = ab + ac$	$2 \cdot (3 + 5) = 2 \cdot 3 + 2 \cdot 5$ $(3 + 5) \cdot 2 = 2 \cdot 3 + 2 \cdot 5$	**Distributive Property** When we multiply a number by a sum of two numbers, we get the same result as multiplying the number by each of the terms and then adding the results.

From our experience with numbers we know intuitively that these properties are valid. To reacquaint yourself with these properties, calculate the expressions on both sides of the equal sign in each of the examples in the table. The operations in the parentheses are to be performed first. So, for example,

$$(2 + 4) + 7 = 6 + 7 = 13 \quad \text{and} \quad 2 + (4 + 7) = 2 + 11 = 13$$

2(3 + 5)

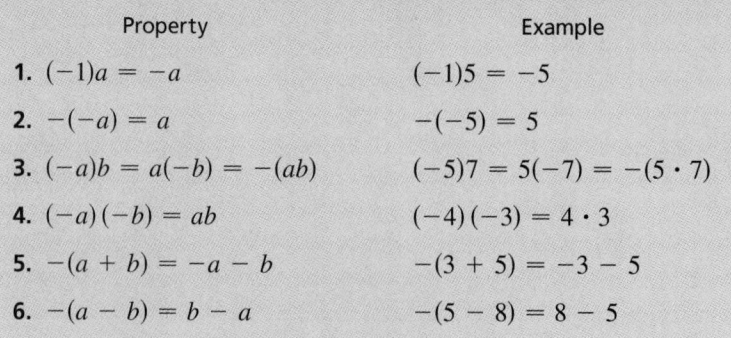

2 · 3   2 · 5

FIGURE 4
The Distributive Property

The Distributive Property is crucial in algebra because it describes the way addition and multiplication interact with each other. It applies whenever we multiply a number by a sum. Figure 4 explains why this property works for the case in which all the numbers are positive integers, but the property is true for any real numbers a, b, and c.

EXAMPLE 1 ■ Using the Properties of the Real Numbers

Let x, y, z, and w be real numbers.

(a) $(x + y)(2zw) = (2zw)(x + y)$ Commutative Property for multiplication

(b) $(x + y)(z + w) = (x + y)z + (x + y)w$ Distributive Property (with $a = x + y$)

$\qquad\qquad\quad = (zx + zy) + (wx + wy)$ Distributive Property

$\qquad\qquad\quad = zx + zy + wx + wy$ Associative Property of addition

In the last step we removed the parentheses because, according to the Associative Property, the order of addition doesn't matter.   ■

Don't make the mistake of assuming that $-a$ is a negative number. Whether $-a$ is negative or positive depends on the value of a. For example, if $a = 5$, then $-a = -5$, a negative number; but if $a = -5$, then $-a = -(-5) = 5$ (Property 2), a positive number.

The number 0 is special for addition; it is called the **additive identity** because $a + 0 = a$ for any real number a. Every real number a has a **negative**, $-a$, that satisfies $a + (-a) = 0$. **Subtraction** is the operation that undoes addition; to subtract a number from another, we simply add the negative of that number. By definition,

$$a - b = a + (-b)$$

To combine real numbers involving negatives we use the following properties.

PROPERTIES OF NEGATIVES

Property	Example
1. $(-1)a = -a$	$(-1)5 = -5$
2. $-(-a) = a$	$-(-5) = 5$
3. $(-a)b = a(-b) = -(ab)$	$(-5)7 = 5(-7) = -(5 \cdot 7)$
4. $(-a)(-b) = ab$	$(-4)(-3) = 4 \cdot 3$
5. $-(a + b) = -a - b$	$-(3 + 5) = -3 - 5$
6. $-(a - b) = b - a$	$-(5 - 8) = 8 - 5$

Property 6 states the intuitive fact that $a - b$ and $b - a$ are negatives of each other. Property 5 is often used with more than two terms:

$$-(a + b + c) = -a - b - c$$

EXAMPLE 2 ■ **Using Properties of Negatives**

Let x, y, and z be real numbers.

(a) $-(x + 2) = -x - 2$ Property 5

(b) $-(x + y - z) = -x - y - (-z)$ Property 5

$\qquad\qquad\quad = -x - y + z$ Property 2 ■

The number 1 is special for multiplication; it is called the **multiplicative identity** because $a \cdot 1 = a$ for any real number a. Every nonzero real number a has an **inverse**, $1/a$, that satisfies $a \cdot (1/a) = 1$. **Division** is the operation that undoes multiplication; to divide by a number, we multiply by the inverse of that number. If $b \neq 0$, then, by definition,

$$a \div b = a \cdot \frac{1}{b}$$

We write $a \cdot (1/b)$ as simply a/b. We refer to a/b as the **quotient** of a and b or as the **fraction** a over b; a is the **numerator** and b is the **denominator** (or **divisor**). To combine real numbers using the operation of division we use the following properties.

PROPERTIES OF FRACTIONS

Property	Example	Description
1. $\dfrac{a}{b} \cdot \dfrac{c}{d} = \dfrac{ac}{bd}$	$\dfrac{2}{3} \cdot \dfrac{5}{7} = \dfrac{2 \cdot 5}{3 \cdot 7} = \dfrac{10}{21}$	To multiply fractions, multiply numerators and denominators.
2. $\dfrac{a}{b} \div \dfrac{c}{d} = \dfrac{a}{b} \cdot \dfrac{d}{c}$	$\dfrac{2}{3} \div \dfrac{5}{7} = \dfrac{2}{3} \cdot \dfrac{7}{5} = \dfrac{14}{15}$	To divide fractions, invert the divisor and multiply.
3. $\dfrac{a}{c} + \dfrac{b}{c} = \dfrac{a + b}{c}$	$\dfrac{2}{5} + \dfrac{7}{5} = \dfrac{2 + 7}{5} = \dfrac{9}{5}$	To add fractions having the same denominator, add the numerators.
4. $\dfrac{a}{b} + \dfrac{c}{d} = \dfrac{ad + bc}{bd}$	$\dfrac{2}{5} + \dfrac{3}{7} = \dfrac{2 \cdot 7 + 3 \cdot 5}{35} = \dfrac{29}{35}$	To add fractions having different denominators, find a common denominator. Then add the numerators.
5. $\dfrac{ac}{bc} = \dfrac{a}{b}$	$\dfrac{2 \cdot 5}{3 \cdot 5} = \dfrac{2}{3}$	Cancel numbers that are common factors in numerator and denominator.
6. If $\dfrac{a}{b} = \dfrac{c}{d}$, then $ad = bc$.	$\dfrac{2}{3} = \dfrac{6}{9}$, so $2 \cdot 9 = 3 \cdot 6$	Cross multiply.

When adding fractions with different denominators, we don't usually use Property 4. Instead we rewrite the fractions so that they have a common denominator (often smaller than the product of the denominators), and then we use Property 3. This denominator is the **Least Common Denominator (LCD)** described in the next example.

EXAMPLE 3 ■ **Using the LCD to Add Fractions**

Evaluate: $\dfrac{5}{36} + \dfrac{7}{120}$

SOLUTION

Factoring each denominator into prime factors gives

$$36 = 2^2 \cdot 3^2 \quad \text{and} \quad 120 = 2^3 \cdot 3 \cdot 5$$

We find the least common denominator (LCD) by forming the product of all the factors that occur in these factorizations, using the highest power of each factor. Thus the LCD is $2^3 \cdot 3^2 \cdot 5 = 360$. So

$$\frac{5}{36} + \frac{7}{120} = \frac{5 \cdot 10}{36 \cdot 10} + \frac{7 \cdot 3}{120 \cdot 3} \qquad \text{Property 5}$$

$$= \frac{50}{360} + \frac{21}{360}$$

$$= \frac{71}{360} \qquad \text{Property 3} \qquad ■$$

SETS AND INTERVALS

In the discussion which follows we need to use set notation. A **set** is a collection of objects, and these objects are called the **elements** of the set. If S is a set, the notation $a \in S$ means that a is an element of S, and $b \notin S$ means that b is not an element of S. For example, if Z represents the set of integers, then $-3 \in Z$ but $\pi \notin Z$.

Some sets can be described by listing their elements within braces. For instance, the set A that consists of all positive integers less than 7 can be written as

$$A = \{1, 2, 3, 4, 5, 6\}$$

We could also write A in **set-builder notation** as

$$A = \{x \mid x \text{ is an integer and } 0 < x < 7\}$$

which is read "A is the set of all x such that x is an integer and $0 < x < 7$."

If S and T are sets, then their **union** $S \cup T$ is the set that consists of all elements that are in S *or* T (or in both). The **intersection** of S and T is the set $S \cap T$ consisting of all elements that are in both S *and* T. In other words, $S \cap T$ is the common part of S and T. The **empty set,** denoted by $\varnothing$, is the set that contains no element.

EXAMPLE 4 ■ Union and Intersection of Sets

If $S = \{1, 2, 3, 4, 5\}$, $T = \{4, 5, 6, 7\}$, and $V = \{6, 7, 8\}$, find the sets $S \cup T$, $S \cap T$, and $S \cap V$.

SOLUTION

$$S \cup T = \{1, 2, 3, 4, 5, 6, 7\} \qquad \text{All elements in } S \text{ or } T$$

$$S \cap T = \{4, 5\} \qquad \text{Elements common to both } S \text{ and } T$$

$$S \cap V = \varnothing \qquad S \text{ and } V \text{ have no element in common} \qquad ■$$

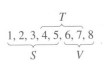

Certain sets of real numbers, called **intervals,** occur frequently in calculus and correspond geometrically to line segments. For example, if $a < b$, then the **open interval** from a to b consists of all numbers between a and b and is denoted by the symbol (a, b). Using set-builder notation, we can write

$$(a, b) = \{x \mid a < x < b\}$$

a   *b*

FIGURE 5

The open interval (a, b)

Note that the endpoints, a and b, are excluded from this interval. This fact is indicated by the round brackets ( ) in the interval notation and the open circles on the graph of the interval in Figure 5.

The **closed interval** from a to b is the set

$$[a, b] = \{x \mid a \le x \le b\}$$

a   *b*

FIGURE 6

The closed interval $[a, b]$

Here the endpoints of the interval are included. This is indicated by the square brackets [ ] in the interval notation and the solid circles on the graph of the interval in Figure 6. It is also possible to include only one endpoint in an interval, as shown in the table of intervals on page 16.

We also need to consider infinite intervals, such as

$$(a, \infty) = \{x \mid x > a\}$$

This does not mean that ∞ ("infinity") is a number. The notation (a, ∞) stands for the set of all numbers that are greater than a, so the symbol ∞ simply indicates that the interval extends indefinitely far in the positive direction.

The following table lists the nine possible types of intervals. When these intervals are discussed, we will always assume that $a < b$.

No Smallest or Largest Number In An Open Interval

Any interval contains infinitely many numbers—every point on the graph of an interval corresponds to a real number. In the closed interval $[0, 1]$ the smallest number is 0 and the largest is 1. But the open interval $(0, 1)$ contains no smallest or largest number. To see this, note that 0.01 is close to zero, but 0.001 is closer, 0.0001 closer yet, and so on. So we can always find a number in the interval $(0, 1)$ closer to zero than any given number. Since 0 itself is not in the interval, the interval contains no smallest number. Similarly, 0.99 is close to 1, but 0.999 is closer, 0.9999 closer yet, and so on. Since 1 itself is not in the interval, the interval has no largest number.

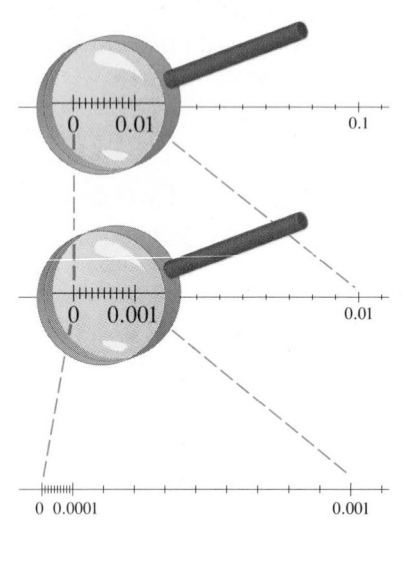

Notation	Set description	Graph
(a, b)	$\{x \mid a < x < b\}$	
$[a, b]$	$\{x \mid a \le x \le b\}$	
$[a, b)$	$\{x \mid a \le x < b\}$	
$(a, b]$	$\{x \mid a < x \le b\}$	
(a, ∞)	$\{x \mid a < x\}$	
$[a, \infty)$	$\{x \mid a \le x\}$	
$(-\infty, b)$	$\{x \mid x < b\}$	
$(-\infty, b]$	$\{x \mid x \le b\}$	
$(-\infty, \infty)$	$\mathbb{R}$ (set of all real numbers)	

EXAMPLE 5 ■ Graphing Intervals

Express each interval in terms of inequalities, and then graph the interval.

(a) $[-1, 2)$ (b) $[1.5, 4]$ (c) $(-3, \infty)$

SOLUTION

(a) $[-1, 2) = \{x \mid -1 \le x < 2\}$

(b) $[1.5, 4] = \{x \mid 1.5 \le x \le 4\}$

(c) $(-3, \infty) = \{x \mid -3 < x\}$

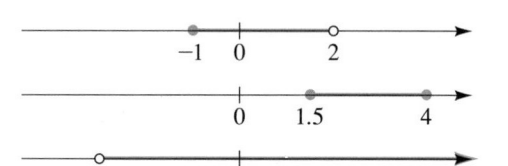

EXAMPLE 6 ■ Finding Unions and Intersections of Intervals

Graph each set.

(a) $(1, 3) \cap [2, 7]$ (b) $(-2, -1) \cup (1, 2)$

SOLUTION

(a) The intersection of two intervals consists of the numbers that are in both intervals. Therefore

$$(1, 3) \cap [2, 7] = \{x \mid 1 < x < 3 \quad \text{and} \quad 2 \le x \le 7\}$$

$$= \{x \mid 2 \le x < 3\}$$

$$= [2, 3)$$

This set is illustrated in Figure 7.

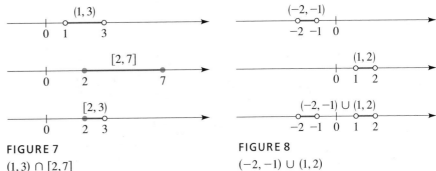

FIGURE 7
$(1, 3) \cap [2, 7]$

FIGURE 8
$(-2, -1) \cup (1, 2)$

(b) The union of the intervals $(-2, -1)$ and $(1, 2)$ consists of the numbers that are in either $(-2, -1)$ or $(1, 2)$, so

$$(-2, -1) \cup (1, 2) = \{x \mid -2 < x < -1 \quad \text{or} \quad 1 < x < 2\}$$

This set is illustrated in Figure 8. ■

ABSOLUTE VALUE AND DISTANCE

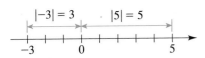

FIGURE 9

The **absolute value** of a number a, denoted by $|a|$, is the distance from a to 0 on the real number line (see Figure 9). Distance is always positive or zero, so we have $|a| \geq 0$ for every number a. Remembering that $-a$ is positive when a is negative, we have the following definition.

DEFINITION OF ABSOLUTE VALUE
If a is a real number, then the **absolute value** of a is $$|a| = \begin{cases} a & \text{if } a \geq 0 \\ -a & \text{if } a < 0 \end{cases}$$

EXAMPLE 7 ■ **Evaluating Absolute Values of Numbers**

(a) $|3| = 3$

(b) $|-3| = -(-3) = 3$

(c) $|0| = 0$

(d) $|\sqrt{2} - 1| = \sqrt{2} - 1$ (since $\sqrt{2} > 1 \Rightarrow \sqrt{2} - 1 > 0$)

(e) $|3 - \pi| = -(3 - \pi) = \pi - 3$ (since $\pi > 3 \Rightarrow 3 - \pi < 0$) ■

FIGURE 10

What is the distance on the real line between the numbers -2 and 11? From Figure 10 we see that the distance is 13. We arrive at this by finding either

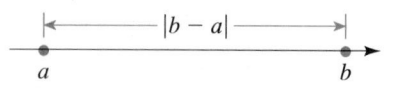

FIGURE 11

Length of a line segment $= |b - a|$

$|11 - (-2)| = 13$ or $|(-2) - 11| = 13$. From this observation we make the following definition (see Figure 11).

> ### DEFINITION OF DISTANCE BETWEEN POINTS ON THE REAL LINE
>
> If a and b are real numbers, then the **distance** between the points a and b on the real line is
>
> $$d(a, b) = |b - a|$$

From Property 6 of negatives it follows that

$$|b - a| = |a - b|$$

This confirms that, as we would expect, the distance from a to b is the same as the distance from b to a.

EXAMPLE 8 ■ Distance between Points on the Real Line

The distance between the numbers -8 and 2 is

$$d(a, b) = |-8 - 2| = |-10| = 10$$

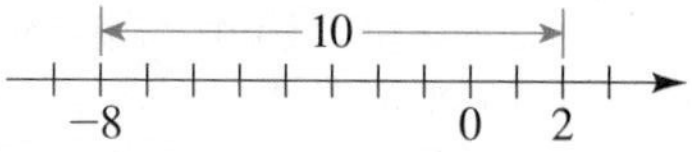

FIGURE 12

We can check this calculation geometrically, as shown in Figure 12.   ■

1.2 EXERCISES

1–8 ■ State the property of real numbers being used.

1. $2x + y = y + 2x$

2. $c(a + b) = (a + b)c$

3. $(x + y) + 5z = x + (y + 5z)$

4. $2(w + x) = 2w + 2x$

5. $3(5x + 1) = 15x + 3$

6. $(xy)S = x(yS)$

7. $(x + a)(x + b) = (x + a)x + (x + a)b$

8. $a(x + y + z) = ax + ay + az$

9–16 ■ Use properties of real numbers to write the expression without parentheses.

9. $3(x + y)$

10. $8(a - b)$

11. $4(2m)$

12. $\frac{1}{2}(10z)$

13. $\frac{4}{3}(-6y)$

14. $-2(r + s)$

15. $-\frac{5}{2}(2x - 4y)$

16. $(3a)(b + c - 2d)$

17–20 ■ Perform the indicated operations.

17. (a) $\dfrac{4}{13} + \dfrac{3}{13}$ (b) $\dfrac{3}{10} + \dfrac{7}{15}$

18. (a) $\dfrac{7}{45} + \dfrac{2}{25}$ (b) $\dfrac{5}{14} - \dfrac{1}{21} + 1$

19. (a) $\dfrac{2}{5} \div \dfrac{9}{10}$ (b) $\left(4 \div \dfrac{1}{2}\right) - \dfrac{1}{2}$

20. (a) $\left(\dfrac{1}{8} - \dfrac{1}{9}\right) \div \dfrac{1}{72}$ (b) $\left(2 \div \dfrac{2}{3}\right) - \left(\dfrac{2}{3} \div 2\right)$

21–26 ■ State whether each inequality is true or false.

21. (a) $-6 < -10$ (b) $0.66 < \frac{2}{3}$

22. (a) $-3 < -1$ (b) $\sqrt{2} > 1.41$

23. (a) $\frac{10}{11} < \frac{12}{13}$ (b) $8 \leq 8$

24. (a) $-\pi > -3$ (b) $8 \leq 9$

25. (a) $\pi \geq 3$ (b) $-2 < -\sqrt{2}$

26. (a) $1.1 > 1.\overline{1}$ (b) $-\frac{1}{2} < -1$

27–28 ■ Write each statement in terms of inequalities.

27. (a) x is positive
(b) t is less than 4
(c) a is greater than or equal to π
(d) x is less than $\frac{1}{3}$ and is greater than -5
(e) The distance from p to 3 is at most 5

28. (a) y is negative
(b) z is greater than 1
(c) b is at most 8
(d) w is positive and is less than or equal to 17
(e) y is at least 2 units from π

29–32 ■ Find the indicated set if

$$A = \{1, 2, 3, 4, 5, 6\} \qquad B = \{2, 4, 6, 8\}$$

and

$$C = \{7, 8, 9, 10\}$$

29. (a) $A \cup B$ (b) $A \cap B$

30. (a) $B \cup C$ (b) $B \cap C$

31. (a) $A \cup C$ (b) $A \cap C$

32. (a) $A \cup B \cup C$ (b) $A \cap B \cap C$

33–34 ■ Find the indicated set if

$$A = \{x \mid x \geq -2\} \qquad B = \{x \mid x < 4\}$$

and

$$C = \{x \mid -1 < x \leq 5\}$$

33. (a) $B \cup C$ (b) $B \cap C$

34. (a) $A \cap C$ (b) $A \cap B$

35–44 ■ Express the interval in terms of inequalities, and then graph the interval.

35. $(-3, 0)$ **36.** $(2, 8]$

37. $[2, 8)$ **38.** $\left[-6, -\frac{1}{2}\right]$

39. $[-1, 1]$ **40.** $(-4, \infty)$

41. $[2, \infty)$ **42.** $(-\infty, 1)$

43. $(-\infty, -2]$ **44.** $(0, \pi)$

45–52 ■ Express the inequality in interval notation, and then graph the corresponding interval.

45. $x \leq 1$ **46.** $0 < x < 8$

47. $1 \leq x \leq 2$ **48.** $x < 3$

49. $-2 < x \leq 1$ **50.** $x \geq -5$

51. $x > -1$ **52.** $-5 < x < 2$

53–60 ■ Graph the set.

53. $(-2, 0) \cup (-1, 1)$ **54.** $(-2, 0) \cap (-1, 1)$

55. $[-4, 6] \cap [0, 8]$ **56.** $[-4, 6) \cup [0, 8)$

57. $(-\infty, -4) \cup (4, \infty)$ **58.** $(-\infty, 6] \cap (2, 10)$

59. $(0, 7) \cap [2, \infty)$ **60.** $(-1, 0] \cup [1, 2)$

61–64 ■ Evaluate each expression.

61. (a) $|100|$
(b) $|-73|$
(c) $|2 - 6|$

62. (a) $|-8 - (-23)|$
(b) $|-\pi|$
(c) $|\pi - 10|$

63. (a) $\left|\sqrt{3} - 3\right|$
(b) $\left||-6| - |-4|\right|$
(c) $\dfrac{-1}{|-1|}$

64. (a) $|-26 - 14|$
(b) $\left|2 - |-12|\right|$
(c) $-1 - \left|1 - |-1|\right|$

65–66 ■ Find the distance between the given numbers.

65. (a) 2 and 17
(b) -3 and 21
(c) $\frac{11}{8}$ and $-\frac{3}{10}$

66. (a) $\frac{7}{15}$ and $-\frac{1}{21}$
(b) -38 and -57
(c) -2.6 and -1.8

67–68 ■ Express each repeating decimal as a fraction.

67. (a) $0.\overline{7}$ (b) $0.2\overline{8}$ (c) $0.\overline{57}$

68. (a) $5.\overline{23}$ (b) $1.3\overline{7}$ (c) $2.1\overline{35}$

69. Use the given figure to locate each point on the real line.

(a) $\sqrt{2}$ (b) $\sqrt{2}-1$ (c) $\sqrt{8}$ (d) $\dfrac{1}{\sqrt{2}}$

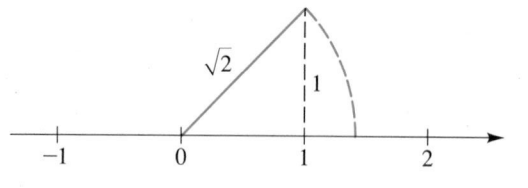

70. (a) Which of the numbers 0.000231 or 0.0000231 is closer to 0?

(b) Give a convincing argument which proves that there is no positive real number closest to zero.

71. (a) Complete the following tables.

(b) Describe what happens to the fraction $\dfrac{1}{x}$ as x gets large.

(c) Describe what happens to the fraction $\dfrac{1}{x}$ as x gets close to zero.

x	$\dfrac{1}{x}$
1	
2	
10	
100	
1000	

x	$\dfrac{1}{x}$
1.0	
0.5	
0.1	
0.01	
0.001	

72. Show that the sum, difference, and product of two rational numbers is a rational number.

73. (a) Is the sum of two irrational numbers always an irrational number?

(b) Is the product of two irrational numbers always an irrational number?

1.3 EXPONENTS AND RADICALS

In this section we give meaning to expressions such as $a^{m/n}$ in which the exponent m/n is a rational number. In order to do this we need to recall some facts about radicals and nth roots.

INTEGER EXPONENTS

A product of identical numbers is usually written in exponential notation. For example, $5 \cdot 5 \cdot 5$ is written as 5^3. In general, we have the following definition.

EXPONENTIAL NOTATION

If a is any real number and n is a positive integer, then the **nth power of** a is

$$a^n = \underbrace{a \cdot a \cdot \cdots \cdot a}_{n \text{ factors}}$$

The number a is called the **base** and n is called the **exponent.**

EXAMPLE 1 ■ **Exponential Notation**

(a) $\left(\dfrac{1}{2}\right)^5 = \left(\dfrac{1}{2}\right)\left(\dfrac{1}{2}\right)\left(\dfrac{1}{2}\right)\left(\dfrac{1}{2}\right)\left(\dfrac{1}{2}\right) = \dfrac{1}{32}$

(b) $(-3)^4 = (-3) \cdot (-3) \cdot (-3) \cdot (-3) = 81$

(c) $-3^4 = -(3 \cdot 3 \cdot 3 \cdot 3) = -81$ ■

In Example 1, note the distinction between $(-3)^4$ and -3^4. In part (b), $(-3)^4$ means that the exponent 4 applies to the negative sign as well as to the 3. In part (c), -3^4 means that the exponent applies only to the 3. Similarly, in $(3x)^n = 3^n x^n$, the n applies to both 3 and x; but in $3x^n = 3(x^n)$, the n applies only to x.

We can state several useful rules for working with exponential notation. To discover the rule for multiplication, we multiply 5^4 by 5^2:

$$5^4 \cdot 5^2 = \underbrace{(5 \cdot 5 \cdot 5 \cdot 5)}_{4 \text{ factors}} \underbrace{(5 \cdot 5)}_{2 \text{ factors}} = \underbrace{5 \cdot 5 \cdot 5 \cdot 5 \cdot 5 \cdot 5}_{6 \text{ factors}} = 5^6 = 5^{4+2}$$

It appears that *to multiply two powers of the same base, we add their exponents.* In general, we can confirm this by considering any real number a and any positive integers m and n:

$$a^m a^n = \underbrace{(a \cdot a \cdot \cdots \cdot a)}_{m \text{ factors}} \underbrace{(a \cdot a \cdot \cdots \cdot a)}_{n \text{ factors}} = \underbrace{a \cdot a \cdot a \cdot \cdots \cdot a}_{m + n \text{ factors}} = a^{m+n}$$

Thus, we have shown that

$$\boxed{a^m a^n = a^{m+n}}$$

whenever m and n are positive integers.

We would like this rule to be true even when m and n are 0 or negative integers. Thus, for instance, we must have

$$2^0 \cdot 2^3 = 2^{0+3} = 2^3$$

But this can happen only if $2^0 = 1$. Likewise, we want to have

$$5^4 \cdot 5^{-4} = 5^{4+(-4)} = 5^{4-4} = 5^0 = 1$$

and this will be true if $5^{-4} = 1/5^4$. These observations lead to the following definition.

ZERO AND NEGATIVE EXPONENTS

If $a \neq 0$ is any real number and n is a positive integer, then

$$\text{(a) } a^0 = 1 \qquad \text{(b) } a^{-n} = \frac{1}{a^n}$$

EXAMPLE 2 ■ Zero and Negative Exponents

(a) $\left(\frac{4}{7}\right)^0 = 1$

(b) $x^{-1} = \frac{1}{x^1} = \frac{1}{x}$

(c) $(-2)^{-3} = \frac{1}{(-2)^3} = \frac{1}{-8} = -\frac{1}{8}$ ■

Familiarity with the following rules is essential for our for work with exponents and bases. In the table, the bases a and b are real numbers and the exponents m and n are integers.

LAWS OF EXPONENTS

Law	Description
1. $a^m a^n = a^{m+n}$	To multiply two powers of the same number, add the exponents.
2. $\dfrac{a^m}{a^n} = a^{m-n}$	To divide two powers of the same number, subtract the exponents.
3. $(a^m)^n = a^{mn}$	To raise a power to a new power, multiply the exponents.
4. $(ab)^n = a^n b^n$	To raise a product to a power, raise each factor to the power.
5. $\left(\dfrac{a}{b}\right)^n = \dfrac{a^n}{b^n}$	To raise a quotient to a power, raise both numerator and denominator to the power.

We have already seen how to prove Law 1 whenever m and n are positive integers, but it can be proved for any integer exponents by using the definition of negative exponents. We now give the proofs of Laws 3 and 4. The proofs of Laws 2 and 5 are requested in Exercise 88.

■ **Proof of Law 3** If m and n are positive integers, we have

$$(a^m)^n = (\underbrace{a \cdot a \cdot \cdots \cdot a}_{m \text{ factors}})^n$$

$$= \underbrace{(\underbrace{a \cdot a \cdot \cdots \cdot a}_{m \text{ factors}})(\underbrace{a \cdot a \cdot \cdots \cdot a}_{m \text{ factors}}) \cdots (\underbrace{a \cdot a \cdot \cdots \cdot a}_{m \text{ factors}})}_{n \text{ groups of factors}}$$

$$= \underbrace{a \cdot a \cdot \cdots \cdot a}_{mn \text{ factors}} = a^{mn}$$

The cases for which $m \le 0$ or $n \le 0$ can be proved using the definition of negative exponents. □

■ **Proof of Law 4** For the case in which m and n are positive integers, we have

$$(ab)^n = \underbrace{(ab)(ab) \cdots (ab)}_{n \text{ factors}} = (\underbrace{a \cdot a \cdot \cdots \cdot a}_{n \text{ factors}}) \cdot (\underbrace{b \cdot b \cdot \cdots \cdot b}_{n \text{ factors}}) = a^n b^n$$

Here we have used the Commutative and Associative Properties repeatedly. If $m \le 0$ or $n \le 0$, Law 4 can be proved using the definition of negative exponents. □

EXAMPLE 3 ■ Using Laws of Exponents

(a) $x^4 x^7 = x^{4+7} = x^{11}$ — Law 1

(b) $y^4 y^{-7} = y^{4-7} = y^{-3} = \dfrac{1}{y^3}$ — Law 1

(c) $\dfrac{c^9}{c^5} = c^{9-5} = c^4$ — Law 2

(d) $\dfrac{d^2}{d^{10}} = d^{2-10} = d^{-8} = \dfrac{1}{d^8}$ — Law 2

(e) $(b^4)^5 = b^{4 \cdot 5} = b^{20}$ — Law 3

(f) $(2^m)^3 = 2^{3m}$ — Law 3

(g) $(xy)^3 = x^3 y^3$ — Law 4

(h) $\left(\dfrac{x}{2}\right)^5 = \dfrac{x^5}{2^5} = \dfrac{x^5}{32}$ — Law 5

■

EXAMPLE 4 ■ Simplifying Expressions with Exponents

Simplify: (a) $(2a^3b^2)(3ab^4)^3$ (b) $\left(\dfrac{x}{y}\right)^3\left(\dfrac{y^2x}{z}\right)^4$

SOLUTION

(a)
$$
\begin{aligned}
(2a^3b^2)(3ab^4)^3 &= (2a^3b^2)[3^3a^3(b^4)^3] && \text{Law 4}\\
&= (2a^3b^2)(27a^3b^{12}) && \text{Law 3}\\
&= (2)(27)a^3a^3b^2b^{12} && \text{Group together factors with the same base}\\
&= 54a^6b^{14} && \text{Law 1}
\end{aligned}
$$

(b)
$$
\begin{aligned}
\left(\frac{x}{y}\right)^3\left(\frac{y^2x}{z}\right)^4 &= \frac{x^3}{y^3}\,\frac{(y^2)^4x^4}{z^4} && \text{Laws 5 and 4}\\
&= \frac{x^3}{y^3}\,\frac{y^8x^4}{z^4} && \text{Law 3}\\
&= (x^3x^4)\left(\frac{y^8}{y^3}\right)\frac{1}{z^4} && \text{Group together factors with the same base}\\
&= \frac{x^7y^5}{z^4} && \text{Laws 1 and 2}
\end{aligned}
$$

■

When simplifying an expression, you will find that many different methods will lead to the same result; you should feel free to use any of the rules of exponents to arrive at your own method. We now give two additional laws that are useful in simplifying expressions with negative exponents.

LAWS OF EXPONENTS

6. $\left(\dfrac{a}{b}\right)^{-n} = \left(\dfrac{b}{a}\right)^{n}$ To raise a fraction to a negative power, invert the fraction and change the sign of the exponent.

7. $\dfrac{a^{-n}}{b^{-m}} = \dfrac{b^m}{a^n}$ To move a number raised to a power from numerator to denominator or from denominator to numerator, change the sign of the exponent.

■ **Proof of Law 7** Using first the definition of negative exponents and then Property 2 of fractions, we have

$$
\frac{a^{-n}}{b^{-m}} = \frac{1/a^n}{1/b^m} = \frac{1}{a^n}\cdot\frac{b^m}{1} = \frac{b^m}{a^n}
$$

□

The proof of Law 6 is left as an exercise. In the next example we show how these laws are used.

EXAMPLE 5 ■ Simplifying Expressions with Negative Exponents

Eliminate negative exponents and simplify each expression.

(a) $\dfrac{6st^{-4}}{2s^{-2}t^2}$ (b) $\left(\dfrac{y}{3z^2}\right)^{-2}$

SOLUTION

(a) We use Law 7, which allows us to move a number raised to a power from the numerator to the denominator (or vice versa) by changing the sign of the exponent.

$$\frac{6st^{-4}}{2s^{-2}t^2} = \frac{6ss^2}{2t^4t^2} \qquad \text{Law 7}$$

$$= \frac{3s^3}{t^6} \qquad \text{Law 1}$$

(b) We use Law 6, which allows us to change the sign of the exponent of a fraction by inverting the fraction.

$$\left(\frac{y}{3z^2}\right)^{-2} = \left(\frac{3z^2}{y}\right)^2 \qquad \text{Law 6}$$

$$= \frac{9z^4}{y^2} \qquad \text{Laws 5 and 4}$$ ■

SCIENTIFIC NOTATION

Exponential notation is used by scientists as a compact way of writing very large numbers and very small numbers that occur in science. For example, the nearest star beyond the sun, Proxima Centauri, is approximately 40,000,000,000,000 km away. The mass of a hydrogen atom is about 0.00000000000000000000000166 g. Scientists usually write such numbers in a more convenient way, called *scientific notation*. For example they would write

$$40,000,000,000,000 = 4 \times 10^{13}$$

and $\qquad 0.00000000000000000000000166 = 1.66 \times 10^{-24}$

In general, a positive number x is said to be written in **scientific notation** if it is expressed as follows:

$$x = a \times 10^n \qquad \text{where } 1 \le a < 10 \text{ and } n \text{ is an integer}$$

For instance, when we state that the distance to the star Proxima Centauri is 4×10^{13} km, the positive exponent 13 indicates that the decimal point should be moved 13 places to the *right*:

$$4 \times 10^{13} = \underbrace{40{,}000{,}000{,}000{,}000}_{13 \text{ places}}$$

When we state that the mass of a hydrogen atom is 1.66×10^{-24} g, the exponent -24 indicates that the decimal point should be moved 24 places to the *left*:

$$1.66 \times 10^{-24} = \underbrace{0.0000000000000000000000166}_{24 \text{ places}}$$

EXAMPLE 6 ■ Writing Numbers in Scientific Notation

(a) $56920 = 5.692 \times 10^4$ (b) $0.000093 = 9.3 \times 10^{-5}$ ■

To use a calculator for scientific notation, use a key labeled $\boxed{\text{EE}}$ or $\boxed{\text{EXP}}$ or $\boxed{\text{EEX}}$ to enter the exponent. For example, to enter the number 3.629×10^{15} on an HP15 calculator, we enter

$$3.629 \quad \boxed{\text{EEX}} \quad 15$$

and the display reads

$$\boxed{3.629 \quad 15}$$

Scientific notation is often used on a calculator to display a very large or very small number. For instance, if we use a calculator to square the number 1,111,111, the display panel may show (depending on the calculator model) the approximation

$$\boxed{1.234568 \quad 12} \quad \text{ or } \quad \boxed{1.234568 \quad \text{E12}}$$

Here the final digits indicate the power of 10, and we interpret the result as

$$1.234568 \times 10^{12}$$

EXAMPLE 7 ■ Using Scientific Notation

If $a \approx 0.00046$, $b \approx 1.697 \times 10^{22}$, and $c \approx 2.91 \times 10^{-18}$, use a calculator to approximate the quotient ab/c.

SOLUTION

We could enter the data using scientific notation, or we could use laws of exponents as follows:

$$\frac{ab}{c} \approx \frac{(4.6 \times 10^{-4})(1.697 \times 10^{22})}{2.91 \times 10^{-18}}$$

$$= \frac{(4.6)(1.697)}{2.91} \times 10^{-4+22+18}$$

$$\approx 2.7 \times 10^{36}$$

We state the answer correct to two significant figures because the least accurate of the given numbers is stated to two significant figures.   ■

RADICALS

We know what 2^n means whenever n is an integer. We would like to give meaning to a power, like $2^{4/5}$, whose exponent, $\frac{4}{5}$, is a rational number. To do this, we first need to discuss radicals.

The symbol $\sqrt{\ \ }$ means "the positive square root of." Thus

$$\sqrt{a} = b \qquad \text{means} \qquad b^2 = a \ \text{ and } \ b \geq 0$$

Since $a = b^2 \geq 0$, the symbol $\sqrt{a}$ makes sense only when $a \geq 0$. For instance,

$$\sqrt{9} = 3 \qquad \text{because} \qquad 3^2 = 9 \ \text{ and } \ 3 \geq 0$$

It is true that the number 9 has two square roots, 3 and -3, but the notation $\sqrt{9}$ is reserved for the *positive* square root of 9 (sometimes called the *principal square root* of 9). If we want the negative root, we *must* write $-\sqrt{9}$, which is -3.

Square roots are special cases of nth roots. The nth root of x is the number which, when raised to the nth power, gives x.

DEFINITION OF nth ROOT

If n is any positive integer, then the **principal nth root** of a is defined as follows:

$$\sqrt[n]{a} = b \qquad \text{means} \qquad b^n = a$$

If n is even, we must have $a \geq 0$ and $b \geq 0$.

Thus

$$\sqrt[4]{81} = 3 \qquad \text{because} \qquad 3^4 = 81 \ \text{ and } \ 3 \geq 0$$

$$\sqrt[3]{-8} = -2 \qquad \text{because} \qquad (-2)^3 = -8$$

But $\sqrt{-8}$, $\sqrt[4]{-8}$, and $\sqrt[6]{-8}$ are not defined. (For instance, $\sqrt{-8}$ is not defined because the square of every real number is nonnegative.) Odd roots are unique, but even roots are not:

The equation $x^5 = 31$ has only one real solution: $x = \sqrt[5]{31}$.

The equation $x^4 = 31$ has two real solutions: $x = \pm\sqrt[4]{31}$.

Notice that

$$\sqrt{4^2} = \sqrt{16} = 4 \qquad \text{but} \qquad \sqrt{(-4)^2} = \sqrt{16} = 4 = |-4|$$

Thus the equation $\sqrt{a^2} = a$ is not always true; it is true only when $a \geq 0$. However, we can always write $\sqrt{a^2} = |a|$. This last equation is true not only for square roots, but for any even root. This and other rules used in working with nth roots are listed in the following box. In each property we assume that all the given roots exist.

PROPERTIES OF nth ROOTS

Property	Examples
1. $\sqrt[n]{ab} = \sqrt[n]{a}\,\sqrt[n]{b}$	$\sqrt[3]{-8 \cdot 27} = \sqrt[3]{-8}\,\sqrt[3]{27} = (-2)(3) = 6$
	$\sqrt{250} = \sqrt{25 \cdot 10} = \sqrt{25}\,\sqrt{10} = 5\sqrt{10}$
2. $\sqrt[n]{\dfrac{a}{b}} = \dfrac{\sqrt[n]{a}}{\sqrt[n]{b}}$	$\sqrt[4]{\dfrac{16}{81}} = \dfrac{\sqrt[4]{16}}{\sqrt[4]{81}} = \dfrac{2}{3}$
	$\sqrt{\dfrac{100}{25}} = \dfrac{\sqrt{100}}{\sqrt{25}} = \dfrac{10}{5} = 2$
3. $\sqrt[m]{\sqrt[n]{a}} = \sqrt[mn]{a}$	$\sqrt{\sqrt[3]{729}} = \sqrt[6]{729} = 3$
4. $\sqrt[n]{a^n} = a$ if n is odd	$\sqrt[3]{(-5)^3} = -5, \quad \sqrt[5]{2^5} = 2$
5. $\sqrt[n]{a^n} = |a|$ if n is even	$\sqrt[4]{(-3)^4} = |-3| = 3$

EXAMPLE 8 ■ Using Properties of nth Roots

(a) $\sqrt{20} \cdot \sqrt{5} = \sqrt{20 \cdot 5} = \sqrt{100} = 10$ Property 1

(b) $\dfrac{\sqrt[5]{64}}{\sqrt[5]{2}} = \sqrt[5]{\dfrac{64}{2}} = \sqrt[5]{32} = 2$ Property 2

(c) $\sqrt[6]{64} = \sqrt[3]{\sqrt{64}} = \sqrt[3]{8} = 2$ Property 3   ■

EXAMPLE 9 ■ Simplifying Expressions involving nth Roots

(a)
$$\sqrt[3]{x^4} = \sqrt[3]{x^3 x}$$ Factor out the largest cube

$$= \sqrt[3]{x^3}\,\sqrt[3]{x}$$ Property 1

$$= x\sqrt[3]{x}$$ Property 4

(b)
$$\sqrt[4]{81x^8y^4} = \sqrt[4]{81}\,\sqrt[4]{x^8}\,\sqrt[4]{y^4}$$ Property 1

$$= 3\sqrt[4]{(x^2)^4}\,|y|$$ Property 5

$$= 3x^2|y|$$ Property 5

(c) If x is any real number and y is a positive number, then

$$\sqrt{x^2y} = \sqrt{x^2}\,\sqrt{y}$$ Property 1

$$= |x|\sqrt{y}$$ Property 5   ■

It is frequently useful to combine like radicals in an expression such as $2\sqrt{3} + 5\sqrt{3}$. This can be done by using the Distributive Property. Thus

$$2\sqrt{3} + 5\sqrt{3} = (2 + 5)\sqrt{3} = 7\sqrt{3}$$

The next example further illustrates this process.

Avoid making the following common error:

$$\sqrt{a + b} \neq \sqrt{a} + \sqrt{b}$$

For instance, if we let $a = 9$ and $b = 16$, then we see the error:

$$\sqrt{9 + 16} \stackrel{?}{=} \sqrt{9} + \sqrt{16}$$
$$\sqrt{25} \stackrel{?}{=} 3 + 4$$
$$5 \stackrel{?}{=} 7 \quad \text{Wrong!}$$

EXAMPLE 10 ■ **Combining Radicals**

(a) $10\sqrt[3]{4} + 7\sqrt[3]{4} - 2\sqrt[3]{4} = (10 + 7 - 2)\sqrt[3]{4}$ Distributive Property

$$= 15\sqrt[3]{4}$$

(b) $\sqrt{32} + \sqrt{200} = \sqrt{16 \cdot 2} + \sqrt{100 \cdot 2}$ Factor out the largest squares

$$= \sqrt{16}\sqrt{2} + \sqrt{100}\sqrt{2}$$ Property 1

$$= 4\sqrt{2} + 10\sqrt{2} = 14\sqrt{2}$$ Distributive Property

(c) If $b > 0$, then

$$b\sqrt{25b} - \sqrt{b^3} = b\sqrt{25}\sqrt{b} - \sqrt{b^2}\sqrt{b}$$ Property 1

$$= 5b\sqrt{b} - b\sqrt{b}$$ Property 5

$$= (5b - b)\sqrt{b}$$ Distributive Property

$$= 4b\sqrt{b}$$

(d)

$$\sqrt[3]{8a^3b^8} + \sqrt[3]{27a^3b^5} = \sqrt[3]{(8a^3b^6)b^2} + \sqrt[3]{(27a^3b^3)b^2}$$ Factor out the largest cubes

$$= \sqrt[3]{8a^3b^6}\sqrt[3]{b^2} + \sqrt[3]{27a^3b^3}\sqrt[3]{b^2}$$ Property 1

$$= 2ab^2\sqrt[3]{b^2} + 3ab\sqrt[3]{b^2}$$ Find cube roots

$$= ab\sqrt[3]{b^2}(2b + 3)$$ Distributive Property ■

RATIONAL EXPONENTS

To define what is meant by a *rational exponent* or, equivalently, a *fractional exponent* such as $a^{1/3}$, we need to use radicals. In order to give meaning to the symbol $a^{1/n}$ in a way that is consistent with the Laws of Exponents, we would have to have

$$(a^{1/n})^n = a^{(1/n)n} = a^1 = a$$

So, by the definition of nth root,

$$a^{1/n} = \sqrt[n]{a}$$

If n is even, then we require that $a \geqslant 0$. In general, we define rational exponents as follows.

DEFINITION OF RATIONAL EXPONENTS

For any rational exponent m/n in lowest terms, where m and n are integers and $n > 0$, we define

$$a^{m/n} = \left(\sqrt[n]{a} \right)^m$$

or, equivalently, $a^{m/n} = \sqrt[n]{a^m}$

If n is even, then we require that $a \geqslant 0$.

With this definition it can be proved that *the Laws of Exponents also hold for rational exponents.*

EXAMPLE 11 ■ Using the Definition of Rational Exponents

(a) $64^{1/3} = \sqrt[3]{64} = 4$

(b) $4^{1/2} = \sqrt{4} = 2$

(c) $4^{3/2} = \left(\sqrt{4} \right)^3 = 2^3 = 8$ Alternate solution: $4^{3/2} = \sqrt{4^3} = \sqrt{64} = 8$

(d) $(125)^{-1/3} = \dfrac{1}{(125)^{1/3}} = \dfrac{1}{\sqrt[3]{125}} = \dfrac{1}{5}$

(e) $\dfrac{1}{\sqrt[3]{x^4}} = \dfrac{1}{x^{4/3}} = x^{-4/3}$ ■

EXAMPLE 12 ■ Using the Laws of Exponents with Rational Exponents

(a) $a^{1/3} a^{7/3} = a^{8/3}$ Law 1

(b) $\dfrac{a^{2/5} a^{7/5}}{a^{3/5}} = a^{2/5 + 7/5 - 3/5} = a^{6/5}$ Laws 1 and 2

(c) $(2a^3 b^4)^{3/2} = 2^{3/2} (a^3)^{3/2} (b^4)^{3/2}$ Law 4

$\qquad\qquad = \left(\sqrt{2} \right)^3 a^{3(3/2)} b^{4(3/2)}$ Law 3

$\qquad\qquad = 2\sqrt{2} \, a^{9/2} b^6$

(d) $\left(\dfrac{2x^{3/4}}{y^{1/3}} \right)^3 \left(\dfrac{y^4}{x^{-1/2}} \right) = \dfrac{2^3 (x^{3/4})^3}{(y^{1/3})^3} \cdot (y^4 x^{1/2})$ Laws 5, 4, and 7

$\qquad\qquad\qquad = \dfrac{8x^{9/4}}{y} \cdot y^4 x^{1/2}$ Law 3

$\qquad\qquad\qquad = 8x^{11/4} y^3$ Laws 1 and 2   ■

EXAMPLE 13 ■ **Simplifying by Writing Radicals as Rational Exponents**

(a) $\left(2\sqrt{x}\right)\left(3\sqrt[3]{x}\right) = \left(2x^{1/2}\right)\left(3x^{1/3}\right)$ Definition of rational exponents

$\qquad\qquad\qquad = 6x^{1/2+1/3} = 6x^{5/6}$ Law 1

(b) $\sqrt{x\sqrt{x\sqrt{x}}} = \left[x(xx^{1/2})^{1/2}\right]^{1/2}$ Definition of rational exponents

$\qquad\qquad = \left[x(x^{3/2})^{1/2}\right]^{1/2}$ Law 1

$\qquad\qquad = (x \cdot x^{3/4})^{1/2}$ Law 3

$\qquad\qquad = (x^{7/4})^{1/2}$ Law 1

$\qquad\qquad = x^{7/8}$ Law 3   ■

It is often useful for us to eliminate the radical in a denominator by multiplying both numerator and denominator by an appropriate expression. This procedure is called **rationalizing the denominator.** If the denominator is of the form $\sqrt{a}$, we multiply numerator and denominator by $\sqrt{a}$. In doing so we multiply the given quantity by 1, so we do not change its value. For instance,

$$\frac{1}{\sqrt{a}} = \frac{1}{\sqrt{a}} \cdot 1 = \frac{1}{\sqrt{a}} \cdot \frac{\sqrt{a}}{\sqrt{a}} = \frac{\sqrt{a}}{a}$$

Note that the denominator in the last fraction contains no radical. In general, if the denominator is of the form $\sqrt[n]{a^m}$ with $m < n$, then multiplying numerator and denominator by $\sqrt[n]{a^{n-m}}$ will rationalize the denominator, because

$$\sqrt[n]{a^m}\,\sqrt[n]{a^{n-m}} = \sqrt[n]{a^{m+n-m}} = \sqrt[n]{a^n} = a$$

EXAMPLE 14 ■ **Rationalizing Denominators**

Rationalize the denominator in each expression.

(a) $\dfrac{2}{\sqrt{3}}$ (b) $\dfrac{1}{\sqrt[3]{x^2}}$ (c) $\sqrt[7]{\dfrac{1}{a^2}}$ (d) $\sqrt[5]{\dfrac{a}{b^3}}$

SOLUTION

(a) $\dfrac{2}{\sqrt{3}} = \dfrac{2}{\sqrt{3}} \cdot \dfrac{\sqrt{3}}{\sqrt{3}} = \dfrac{2\sqrt{3}}{3}$

(b) $\dfrac{1}{\sqrt[3]{x^2}} = \dfrac{1}{\sqrt[3]{x^2}}\,\dfrac{\sqrt[3]{x}}{\sqrt[3]{x}} = \dfrac{\sqrt[3]{x}}{\sqrt[3]{x^3}} = \dfrac{\sqrt[3]{x}}{x}$

(c) $\sqrt[7]{\dfrac{1}{a^2}} = \dfrac{1}{\sqrt[7]{a^2}} = \dfrac{1}{\sqrt[7]{a^2}}\,\dfrac{\sqrt[7]{a^5}}{\sqrt[7]{a^5}} = \dfrac{\sqrt[7]{a^5}}{\sqrt[7]{a^7}} = \dfrac{\sqrt[7]{a^5}}{a}$

(d) $\sqrt[5]{\dfrac{a}{b^3}} = \dfrac{\sqrt[5]{a}}{\sqrt[5]{b^3}} = \dfrac{\sqrt[5]{a}}{\sqrt[5]{b^3}}\,\dfrac{\sqrt[5]{b^2}}{\sqrt[5]{b^2}} = \dfrac{\sqrt[5]{ab^2}}{\sqrt[5]{b^5}} = \dfrac{\sqrt[5]{ab^2}}{b}$ ■

| 1.3 | EXERCISES |

1–10 ■ Evaluate each of the given numbers.

1. (a) $(-2)^4$ (b) -2^4 (c) π^0

2. (a) $\left(\frac{1}{3}\right)^4 4^{-3}$ (b) $\left(\frac{1}{4}\right)^{-2}$ (c) $\left(\frac{4}{9}\right)^0 \cdot 2^{-3}$

3. (a) $2^{-3} 5^4$ (b) $\dfrac{10^9}{10^4}$ (c) $(2^4 \cdot 2^2)^2$

4. (a) $\left(\frac{2}{3}\right)^{-1}$ (b) $\sqrt[3]{-64}$ (c) $\sqrt[5]{-32}$

5. (a) $\sqrt{\frac{4}{9}}$ (b) $\sqrt[4]{256}$ (c) $\sqrt[6]{\frac{1}{64}}$

6. (a) $\sqrt{7}\,\sqrt{28}$ (b) $\sqrt[3]{3}\,\sqrt[3]{9}$ (c) $\sqrt[4]{24}\,\sqrt[4]{54}$

7. (a) $\dfrac{\sqrt{72}}{\sqrt{2}}$ (b) $\dfrac{\sqrt{48}}{\sqrt{3}}$ (c) $\sqrt{\dfrac{9}{25}}$

8. (a) $9^{7/2}$ (b) $(-32)^{2/5}$ (c) $(-125)^{-1/3}$

9. (a) $\left(\frac{4}{9}\right)^{-1/2}$ (b) $\left(-\frac{27}{8}\right)^{2/3}$ (c) $\left(\frac{25}{64}\right)^{3/2}$

10. (a) $1024^{-0.1}$ (b) $3^{2/7} 3^{5/7}$ (c) $3^{1/2} 9^{1/4}$

11–14 ■ Simplify the expression.

11. $\sqrt[3]{108} - \sqrt[3]{32}$

12. $\sqrt{8} + \sqrt{50}$

13. $\sqrt{245} - \sqrt{125}$

14. $\sqrt[3]{54} - \sqrt[3]{16}$

15–18 ■ Write each number as a power of 2.

15. (a) 128   (b) 128^2 (c) $(2^9)^4$

16. (a) $\dfrac{1}{4}$ (b) $\dfrac{2^{12}}{2^{18}}$ (c) 1

17. (a) $2^8 \cdot 4^{-7}$ (b) $4\sqrt{2}$ (c) $1/\sqrt{2}$

18. (a) $\left(\sqrt{2}\right)^4$ (b) $\sqrt{2^4}$ (c) $\sqrt{2^{-4}}$

19–36 ■ Simplify the expression and eliminate any negative exponent(s).

19. $t^7 t^{-2}$

20. $(4x^2)(6x^7)$

21. $(12x^2y^4)\left(\frac{1}{2}x^5y\right)$

22. $(6y)^3$

23. $\dfrac{x^9(2x)^4}{x^3}$

24. $\dfrac{a^{-3}b^4}{a^{-5}b^5}$

25. $b^4\left(\frac{1}{3}b^2\right)(12b^{-8})$

26. $(2s^3t^{-1})\left(\frac{1}{4}s^6\right)(16t^4)$

27. $(rs)^3(2s)^{-2}(4r)^4$

28. $(2u^2v^3)^3(3u^3v)^{-2}$

29. $\dfrac{(6y^3)^4}{2y^5}$

30. $\dfrac{(2x^3)^2(3x^4)}{(x^3)^4}$

31. $\dfrac{(x^2y^3)^4(xy^4)^{-3}}{x^2y}$

32. $\left(\dfrac{c^4d^3}{cd^2}\right)\left(\dfrac{d^2}{c^3}\right)^3$

33. $\dfrac{(xy^2z^3)^4}{(x^3y^2z)^3}$

34. $\left(\dfrac{xy^{-2}z^{-3}}{x^2y^3z^{-4}}\right)^{-3}$

35. $\left(\dfrac{q^{-1}rs^{-2}}{r^{-5}sq^{-8}}\right)^{-1}$

36. $(3ab^2c)\left(\dfrac{2a^2b}{c^3}\right)^{-2}$

37–52 ■ Simplify the expression and eliminate any negative exponent(s). Assume that all letters denote positive numbers.

37. $x^{2/3}x^{1/5}$

38. $(-2a^{3/4})(5a^{3/2})$

39. $(4b)^{1/2}(8b^{2/5})$

40. $(8x^6)^{-2/3}$

41. $(c^2d^3)^{-1/3}$

42. $(4x^6y^8)^{3/2}$

43. $(y^{3/4})^{2/3}$

44. $(a^{2/5})^{-3/4}$

45. $(2x^4y^{-4/5})^3(8y^2)^{2/3}$

46. $(x^{-5}y^3z^{10})^{-3/5}$

47. $\left(\dfrac{x^6y}{y^4}\right)^{5/2}$

48. $\left(\dfrac{-2x^{1/3}}{y^{1/2}z^{1/6}}\right)^4$

49. $\left(\dfrac{3a^{-2}}{4b^{-1/3}}\right)^{-1}$

50. $\dfrac{(y^{10}z^{-5})^{1/5}}{(y^{-2}z^3)^{1/3}}$

51. $\dfrac{(9st)^{3/2}}{(27s^3t^{-4})^{2/3}}$

52. $\left(\dfrac{a^2b^{-3}}{x^{-1}y^2}\right)^3\left(\dfrac{x^{-2}b^{-1}}{a^{3/2}y^{1/3}}\right)$

53–62 ■ Simplify the expression. Assume the letters denote any real numbers.

53. $\sqrt[4]{x^4}$

54. $\sqrt[3]{x^3y^6}$

55. $\sqrt[3]{x^3y}$

56. $\sqrt{x^4y^4}$

57. $\sqrt[5]{a^6b^7}$

58. $\sqrt[3]{a^2b}\,\sqrt[3]{a^4b}$

59. $\sqrt{x^2y^6}$

60. $\sqrt[4]{x^4y^2z^2}$

61. $\sqrt[3]{\sqrt{64x^6}}$

62. $\sqrt{x^2y^2z^2}$

63–66 ■ Rationalize the denominator.

63. (a) $\dfrac{1}{\sqrt{6}}$ (b) $\sqrt{\dfrac{x}{3y}}$ (c) $\sqrt{\dfrac{3}{20}}$

64. (a) $\sqrt{\dfrac{x^5}{2}}$ (b) $\sqrt{\dfrac{2}{3}}$ (c) $\sqrt{\dfrac{1}{2x^3y^5}}$

65. (a) $\dfrac{1}{\sqrt[3]{x}}$ (b) $\dfrac{1}{\sqrt[5]{x^2}}$ (c) $\dfrac{1}{\sqrt[7]{x^3}}$

66. (a) $\dfrac{1}{\sqrt[3]{x^2}}$ (b) $\dfrac{1}{\sqrt[4]{x^3}}$ (c) $\dfrac{1}{\sqrt[3]{x^4}}$

67–68 ▪ Write each number in scientific notation.

67. (a) 69,300,000
    (b) 0.000028536
    (c) 129,540,000

68. (a) 7,259,000,000
    (b) 0.0000000014
    (c) 0.0007029

69–70 ▪ Write each number in ordinary decimal notation.

69. (a) 3.19×10^5
    (b) 2.670×10^{-8}
    (c) 7.1×10^{14}

70. (a) 8.55×10^{-3}
    (b) 6×10^{12}
    (c) 6.257×10^{-10}

71–72 ▪ Write the number indicated in each statement in scientific notation.

71. (a) A light year, the distance that light travels in one year, is about 5,900,000,000,000 mi.
    (b) The diameter of an electron is about 0.0000000000004 cm.
    (c) A drop of water contains more than 33 billion billion molecules.

72. (a) The distance from the earth to the sun is about 93 million miles.
    (b) The mass of an oxygen molecule is about 0.000000000000000000000053 g.
    (c) The mass of the earth is about 5,970,000,000,000,000,000,000,000 kg.

73–78 ▪ Use scientific notation, Laws of Exponents, and a calculator to perform the indicated operations. State your answer correct to the number of significant digits indicated by the given data.

73. $(7.2 \times 10^{-9})(1.806 \times 10^{-12})$

74. $(1.062 \times 10^{24})(8.61 \times 10^{19})$

75. $\dfrac{1.295643 \times 10^9}{(3.610 \times 10^{-17})(2.511 \times 10^6)}$

76. $\dfrac{(73.1)(1.6341 \times 10^{28})}{0.0000000019}$

77. $\dfrac{(0.0000162)(0.01582)}{(594621000)(0.0058)}$

78. $\dfrac{(3.542 \times 10^{-6})^9}{(5.05 \times 10^4)^{12}}$

79. The speed of light is about 186,000 mi/s. Use the information in Exercise 72(a) to find how long it takes for a light ray from the sun to reach the earth.

80. Without using a calculator, determine which number is larger in each pair of numbers:
    (a) $7^{1/4}$ and $4^{1/3}$ (b) $\sqrt[3]{5}$ and $\sqrt{3}$

81. Which of the following pairs of numbers is closer together?

$$10^{10} \quad \text{and} \quad 10^{53} \qquad 10^{100} \quad \text{and} \quad 10^{101}$$

82. Complete the following tables, and then use your results to answer the following questions.

n	$2^{1/n}$
1	
2	
5	
10	
100	

n	$\left(\frac{1}{2}\right)^{1/n}$
1	
2	
5	
10	
100	

    (a) Describe what happens to the nth root of 2 as n gets large.
    (b) Describe what happens to the nth root of $\frac{1}{2}$ as n gets large.

83. The theory of relativity states that as an object travels with velocity v, its rest mass m_0 changes to a mass m given by the formula

$$m = \frac{m_0}{\sqrt{1 - \dfrac{v^2}{c^2}}}$$

where $c \approx 3.0 \times 10^5$ km/s is the speed of light.
    (a) By what factor is the rest mass of a spaceship multiplied if the ship travels at half the speed of light?
    (b) How does the mass of the spaceship change as it travels at a speed very close to the speed of light?

84. Due to the curvature of the earth, the maximum distance D that you can see from the top of a tall building of height h is estimated by the formula $D = \sqrt{2rh + h^2}$, where $r = 3960$ mi is the radius of the earth and D and h are also measured in miles. How far can you see from the observation deck of the Toronto CN Tower, 1135 ft above the ground?

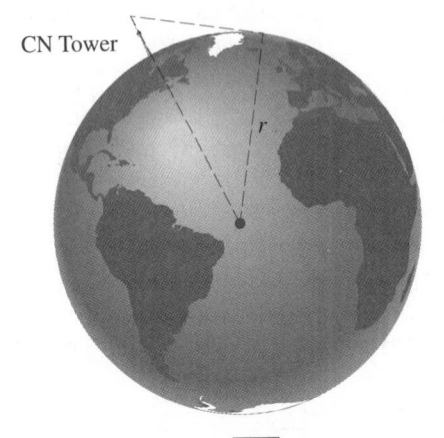

CN Tower

85. Police use the formula $s = \sqrt{30fd}$ to estimate the speed s (in mi/h) at which a car is traveling if it skids d feet after the brakes are applied suddenly. The number f is the coefficient of friction of the road, which is a measure of the "slipperiness" of the road. The table gives some typical estimates for f.

	Tar	Concrete	Gravel
Dry	1.0	0.8	0.2
Wet	0.5	0.4	0.1

(a) If a car skids 65 ft on wet concrete, how fast was it moving when the brakes were applied?

(b) If a car is traveling at 50 mi/h, how far will it skid on wet tar?

86. The astronomer Johannes Kepler (1571–1630) discovered a remarkable relationship between the mean distances d of the planets from the sun (taking the unit of measurement to be the distance from the earth to the sun) and their periods T in years (time required for one revolution). Find the relationship between d and T by completing the following table. (The table includes only the planets known at the time of Kepler; however, the relationship holds for all the planets.)

Planet	d	$\sqrt{d}$	T	$\sqrt[3]{T}$
Mercury	0.387		0.241	
Venus	0.723		0.615	
Earth	1.000		1.000	
Mars	1.523		1.881	
Jupiter	5.203		11.861	
Saturn	9.541		29.457	

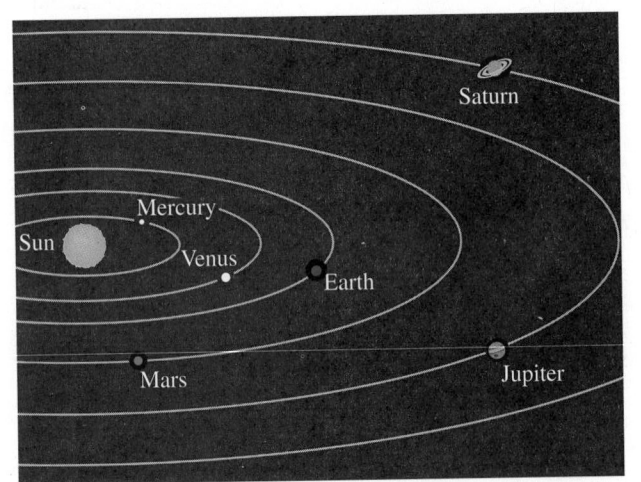

87. It follows from Kepler's Third Law of planetary motion that the average distance from a planet to the sun (in meters) is

$$d = \left(\frac{GM}{4\pi^2}\right)^{1/3} T^{2/3}$$

where $M = 1.99 \times 10^{30}$ kg is the mass of the sun, $G = 6.67 \times 10^{-11}$ N·m²/kg² is the gravitational constant, and T is the period of the planet's orbit (in seconds). Use the fact that the period of the earth's orbit is about 365.25 days to find the distance from the earth to the sun.

88. Prove each of the given Laws of Exponents for the case in which m and n are positive integers and $m > n$.
(a) Law 2   (b) Law 5   (c) Law 6

ALGEBRAIC EXPRESSIONS

Algebraic expressions such as

$$2x^2 - 3x + 4 \qquad ax + b$$

$$\frac{y - 1}{y^2 + 2} \qquad \frac{cx^2y + dy^2z}{\sqrt{x^2 + y^2 + z^2}}$$

are obtained by starting with variables such as x, y, and z and constants such as 2, -3, a, b, c, and d, and combining them using addition, subtraction, multiplication, division, and roots. A **variable** is a letter that can represent any number in a given set of numbers, whereas a **constant** represents a fixed (or specific) number. The **domain** of a variable is the set of numbers that the variable is permitted to have. For instance, in the expression $\sqrt{x}$ the domain of x is $\{x \mid x \geq 0\}$, whereas in the expression $2/(x - 3)$ the domain of x is $\{x \mid x \neq 3\}$.

The simplest types of algebraic expressions use only addition, subtraction, and multiplication. Such expressions are called **polynomials.** The general form of a polynomial of degree n in the variable x is

$$a_n x^n + a_{n-1} x^{n-1} + \cdots + a_1 x + a_0$$

where $a_0, a_1, \ldots, a_n$ are constants and $a_n \neq 0$. The **degree** of a polynomial is the highest power of the variable. Any polynomial is a sum of **terms** of the form ax^k, called **monomials,** where a is a constant and k is a nonnegative integer. A **binomial** is a sum of two monomials, a **trinomial** is the sum of three monomials, and so on. Thus $2x^2 - 3x + 4$, $ax + b$, and $x^4 + 2x^3$ are polynomials of degree 2, 1, and 4, respectively; the first is a trinomial, the other two are binomials.

We **add** and **subtract** polynomials using the properties of real numbers that were discussed in Section 1.2. The idea is to combine **like terms** (that is, terms with the same variables raised to the same powers) using the Distributive Property. For instance,

$ac + bc = (a + b)c$
$$5x^7 + 3x^7 = (5 + 3)x^7 = 8x^7$$

In subtracting polynomials we have to remember that if a minus sign precedes an expression in parentheses, then the sign of every term within the parentheses is changed when we remove the parentheses:

$$-(b + c) = -b - c$$

[This is just a case of the Distributive Property, $a(b + c) = ab + ac$, with $a = -1$.]

EXAMPLE 1 ■ Adding and Subtracting Polynomials

(a) Find the sum $(x^3 - 6x^2 + 2x + 4) + (3x^3 + 5x^2 - 4x)$.
(b) Find the difference $(x^3 - 6x^2 + 2x + 4) - (3x^3 + 5x^2 - 4x)$.

SOLUTION

(a) $(x^3 - 6x^2 + 2x + 4) + (3x^3 + 5x^2 - 4x)$

$\qquad = (x^3 + 3x^3) + (-6x^2 + 5x^2) + (2x - 4x) + 4 \qquad$ Group like terms

$\qquad = 4x^3 - x^2 - 2x + 4 \qquad$ Combine like terms

(b) $(x^3 - 6x^2 + 2x + 4) - (3x^3 + 5x^2 - 4x)$

$\qquad = x^3 - 6x^2 + 2x + 4 - 3x^3 - 5x^2 + 4x \qquad$ Distributive Property

$\qquad = (x^3 - 3x^3) + (-6x^2 - 5x^2) + (2x + 4x) + 4 \qquad$ Group like terms

$\qquad = -2x^3 - 11x^2 + 6x + 4 \qquad$ Combine like terms

■

To find the **product** of polynomials or other algebraic expressions, we need to use the Distributive Property repeatedly. In particular, using it three times on the product of two binomials, we get

$$(a + b)(c + d) = (a + b)c + (a + b)d = ac + bc + ad + bd$$

This says that we multiply the two factors by multiplying each term in one factor by each term in the other factor and adding these products. Schematically, we have

$$(a + b)(c + d)$$

EXAMPLE 2 ■ Multiplying Binomials

(a) $(2x + 1)(3x - 5) = 6x^2 + 3x - 10x - 5 \qquad$ Distributive Property

$\qquad\qquad\qquad\quad = 6x^2 - 7x - 5 \qquad$ Combine like terms

(b) $3(x - 1)(4x + 3) = 3(4x^2 - 4x + 3x - 3) \qquad$ Distributive Property

$\qquad\qquad\qquad\quad = 3(4x^2 - x - 3) \qquad$ Combine like terms

$\qquad\qquad\qquad\quad = 12x^2 - 3x - 9 \qquad$ Distributive Property ■

In general, we can multiply any two polynomials by using the Distributive Property and the Laws of Exponents.

EXAMPLE 3 ■ Multiplying Polynomials

Find the product $(x^2 - 3)(x^3 + 2x + 1)$.

SOLUTION

We start by regarding the first factor as a single number and using the Distributive Property:

$$(x^2 - 3)(x^3 + 2x + 1) = (x^2 - 3)x^3 + (x^2 - 3)2x + (x^2 - 3)1 \quad \text{Distributive Property}$$

$$= x^5 - 3x^3 + 2x^3 - 6x + x^2 - 3 \quad \text{Distributive Property}$$

$$= x^5 - x^3 + x^2 - 6x - 3 \quad \text{Combine like terms}$$

■

The next example shows that the methods we have used for multiplying polynomials also apply to other algebraic expressions.

EXAMPLE 4 ■ Multiplying Algebraic Expressions

(a) $\sqrt{x}\,(x^2 + 2x + \sqrt{x}\,) = x^2\sqrt{x} + 2x\sqrt{x} + \sqrt{x}\,\sqrt{x}$ Distributive Property

$$= x^{5/2} + 2x^{3/2} + x \quad \text{Laws of Exponents}$$

(b) $(1 + \sqrt{x}\,)(2 - 3\sqrt{x}\,) = 2 - 3\sqrt{x} + 2\sqrt{x} - 3(\sqrt{x}\,)^2$ Distributive Property

$$= 2 - \sqrt{x} - 3x \quad \text{Combine like terms}$$

■

We can also consider polynomials in two or more variables. For instance,

$$4x^2y^3 - 2xy + 6$$

is a polynomial in the variables x and y. The techniques for combining such polynomials are similar to those for polynomials in one variable.

EXAMPLE 5 ■ Multiplying Polynomials with More Than One Variable

Find the product $(x^2 - xy + y^2)(x - y)$.

SOLUTION

$$(x^2 - xy + y^2)(x - y) = (x^2 - xy + y^2)x - (x^2 - xy + y^2)y \quad \text{Distributive Property}$$

$$= x^3 - x^2y + xy^2 - x^2y + xy^2 - y^3 \quad \text{Distributive Property}$$

$$= x^3 - 2x^2y + 2xy^2 - y^3 \quad \text{Combine like terms}$$

■

Certain types of products occur so frequently that you should memorize them. You can verify each of the following formulas by performing the multiplications.

The figures give a geometric interpretation of the formula
$$(A + B)^2 = A^2 + 2AB + B^2$$

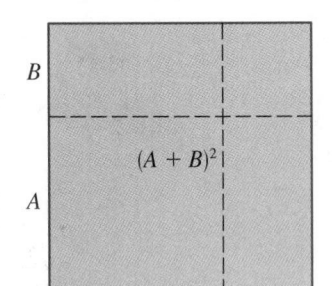

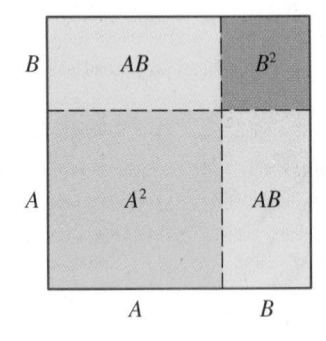

SPECIAL PRODUCT FORMULAS

1. $(A - B)(A + B) = A^2 - B^2$

2. $(A + B)^2 = A^2 + 2AB + B^2$

3. $(A - B)^2 = A^2 - 2AB + B^2$

4. $(A + B)^3 = A^3 + 3A^2B + 3AB^2 + B^3$

5. $(A - B)^3 = A^3 - 3A^2B + 3AB^2 - B^3$

The key idea in using these formulas (or any other formula in algebra) is the *Principle of Substitution*: We may substitute any algebraic expression for any letter in a formula. For example, to find $(x^2 + y^3)^2$ we use Formula 2 and substitute x^2 for A and y^3 for B. Thus, using the formula

$$(A + B)^2 = A^2 + 2AB + B^2$$

we can substitute these values for A and B to determine the product:

$$(x^2 + y^3)^2 = (x^2)^2 + 2(x^2)(y^3) + (y^3)^2$$

This type of substitution is valid because every algebraic expression (in this case, x^2 or y^3) represents a number.

EXAMPLE 6 ■ Using the Special Product Formulas

Use the special product formulas to find each product.

(a) $(2x + 5)^2$

(b) $\left(2\sqrt{y} - \dfrac{1}{\sqrt{x}}\right)\left(2\sqrt{y} + \dfrac{1}{\sqrt{x}}\right)$

(c) $(x^2 - 2)^3$

SOLUTION

(a) Product Formula 2, with $A = 2x$ and $B = 5$, gives

$$(2x + 5)^2 = (2x)^2 + 2(2x)(5) + 5^2 = 4x^2 + 20x + 25$$

(b) Using Product Formula 1 with $A = 2\sqrt{y}$ and $B = 1/\sqrt{x}$, we have

$$\left(2\sqrt{y} - \frac{1}{\sqrt{x}}\right)\left(2\sqrt{y} + \frac{1}{\sqrt{x}}\right) = (2\sqrt{y})^2 - \left(\frac{1}{\sqrt{x}}\right)^2 = 4y - \frac{1}{x}$$

(c) Putting $A = x^2$ and $B = 2$ in Product Formula 5, we get

$$(x^2 - 2)^3 = (x^2)^3 - 3(x^2)^2(2) + 3(x^2)(2)^2 - (2)^3$$
$$= x^6 - 6x^4 + 12x^2 - 8$$

■

FACTORING

We used the Distributive Property to expand algebraic expressions. We sometimes need to reverse this process (again using the Distributive Property) by **factoring** an expression as a product of simpler ones. For example, we can write

$$\xleftarrow{\qquad} \text{EXPANDING} \xrightarrow{\qquad}$$

$$x^2 - 4 = (x - 2)(x + 2)$$

$$\xrightarrow{\qquad} \text{FACTORING} \xrightarrow{\qquad}$$

and we say that $x - 2$ and $x + 2$ are **factors** of $x^2 - 4$. The easiest type of factoring occurs when the terms have a common factor.

EXAMPLE 7 ■ **Factoring Out Common Factors**

Factor each expression.

(a) $3x^2 - 6x$ (b) $8x^4y^2 + 6x^3y^3 - 2xy^4$

SOLUTION

(a) The greatest common factor of the terms $3x^2$ and $-6x$ is $3x$, so we have

$$3x^2 - 6x = 3x(x - 2)$$

(b) We note that

8, 6, and -2 have the greatest common factor 2

x^4, x^3, and x have the greatest common factor x

y^2, y^3, and y^4 have the greatest common factor y^2

So the greatest common factor of the three terms in the polynomial is $2xy^2$ and we have

$$8x^4y^2 + 6x^3y^3 - 2xy^4 = (2xy^2)(4x^3) + (2xy^2)(3x^2y) + (2xy^2)(-y^2)$$

$$= 2xy^2(4x^3 + 3x^2y - y^2) \qquad ■$$

To factor a second-degree polynomial, or **quadratic,** of the form $x^2 + bx + c$, we note that

$$(x + r)(x + s) = x^2 + (r + s)x + rs$$

so we need to choose numbers r and s so that $r + s = b$ and $rs = c$.

EXAMPLE 8 ■ Factoring $x^2 + bx + c$ by Trial and Error

Factor each quadratic expression.

(a) $x^2 + 7x + 12$ (b) $x^2 + 5x - 24$

SOLUTION

(a) In this case $rs = 12$, so r and s must be factors of 12 and their sum must be 7. We enumerate the trial factors of 12 as shown in the margin.

r	1	2	3
s	12	6	4
Sum	13	8	7

Thus $r = 3$ and $s = 4$ are the factors of 12 whose sum is 7. The factorization is

$$x^2 + 7x + 12 = (x + 3)(x + 4)$$

CHECK YOUR ANSWER

Multiplying gives

$(x + 3)(x + 4) = x^2 + 7x + 12$ ✓

(b) The trial factors of -24 are listed in the following table.

r	1	-1	2	-2	3	-3	4	-4
s	-24	24	-12	12	-8	8	-6	6
Sum	-23	23	-10	10	-5	5	-2	2

We see that the two integers with product -24 and sum 5 are -3 and 8. Therefore,

CHECK YOUR ANSWER

Multiplying gives

$(x - 3)(x + 8) = x^2 + 5x - 24$ ✓

$$x^2 + 5x - 24 = (x - 3)(x + 8)$$ ■

To factor a quadratic of the form $ax^2 + bx + c$ with $a \neq 1$, we look for factors of the form $px + r$ and $qx + s$:

$$ax^2 + bx + c = (px + r)(qx + s) = pqx^2 + (ps + qr)x + rs$$

factors of a
↓   ↓
$ax^2 + bx + c = (px + r)(qx + s)$
↑   ↑
factors of c

Therefore, we try to find numbers p, q, r, and s such that

$$pq = a \qquad rs = c \qquad ps + qr = b$$

If these numbers are all integers, then we will have a limited number of possibilities to try for p, q, r, and s.

EXAMPLE 9 ■ Factoring $ax^2 + bx + c$ by Trial and Error

Factor each quadratic expression.

(a) $2x^2 - 7x - 4$ (b) $6x^2 + 7x - 5$

CHECK YOUR ANSWER

Multiplying gives

$(2x + 1)(x - 4) = 2x^2 - 7x - 4$ ✓

SOLUTION

(a) Since the coefficient of x^2 is 2, we look for factors of the form $2x + r$ and $x + s$, where $rs = -4$. By trial and error we find that

$$2x^2 - 7x - 4 = (2x + 1)(x - 4)$$

Multiplying gives
$$(3x + 5)(2x - 1) = 6x^2 + 7x - 5$$

(b) We can factor 6 as $6 \cdot 1$ or $3 \cdot 2$. We can factor -5 as $-5 \cdot 1$ or $5 \cdot (-1)$. By trying these possibilities, we arrive at the factorization

$$6x^2 + 7x - 5 = (3x + 5)(2x - 1)$$ ■

Some special algebraic expressions can be factored using the following formulas. The first three are just the special product formulas, but written backward.

FACTORING FORMULAS

Formula	Name
1. $A^2 - B^2 = (A - B)(A + B)$	Difference of squares
2. $A^2 + 2AB + B^2 = (A + B)^2$	Perfect square
3. $A^2 - 2AB + B^2 = (A - B)^2$	Perfect square
4. $A^3 - B^3 = (A - B)(A^2 + AB + B^2)$	Difference of cubes
5. $A^3 + B^3 = (A + B)(A^2 - AB + B^2)$	Sum of cubes

EXAMPLE 10 ■ Factoring Differences of Squares

Factor each polynomial.

(a) $4x^2 - 25$ (b) $9x^4 - y^6$ (c) $(x + y)^2 - x^2$

SOLUTION

(a) Using the formula for a difference of squares with $A = 2x$ and $B = 5$, we have

$$4x^2 - 25 = (2x)^2 - 5^2$$
$$= (2x - 5)(2x + 5)$$

(b) We recognize that $9x^4 = (3x^2)^2$ and $y^6 = (y^3)^2$ and so the expression is a difference of squares. We may use the formula for the difference of squares with $A = 3x^2$ and $B = y^3$. We have

$$9x^4 - y^6 = (3x^2)^2 - (y^3)^2$$
$$= (3x^2 - y^3)(3x^2 + y^3)$$

(c) We use the formula for the difference of squares with $A = x + y$ and $B = x$.

$$(x + y)^2 - x^2 = [(x + y) - x][(x + y) + x]$$
$$= y(2x + y)$$ ■

EXAMPLE 11 ■ Factoring Perfect Squares

Factor each trinomial: (a) $x^2 + 6x + 9$ (b) $x^2 - xy + \frac{1}{4}y^2$

SOLUTION

(a) Using Formula 2 with $a = x$ and $b = 3$, we get

$$x^2 + 6x + 9 = x^2 + 2(3x) + 3^2 = (x + 3)^2$$

(b) Here we take $a = x$ and $b = \frac{1}{2}y$ in Formula 3:

$$x^2 - xy + \frac{1}{4}y^2 = x^2 - 2x(\tfrac{1}{2}y) + (\tfrac{1}{2}y)^2$$
$$= (x - \tfrac{1}{2}y)^2 \qquad ■$$

EXAMPLE 12 ■ Factoring Differences and Sums of Cubes

Factor each polynomial: (a) $27x^3 - 1$ (b) $x^6 + 8$

SOLUTION

(a) Using the formula for a difference of cubes with $A = 3x$ and $B = 1$, we get

$$27x^3 - 1 = (3x)^3 - 1^3 = (3x - 1)[(3x)^2 + (3x)(1) + 1^2]$$
$$= (3x - 1)(9x^2 + 3x + 1)$$

(b) Using the formula for a sum of cubes with $A = x^2$ and $B = 2$, we have

$$x^6 + 8 = (x^2)^3 + 2^3 = (x^2 + 2)(x^4 - 2x^2 + 4) \qquad ■$$

When we factor an expression the result can sometimes be factored further. In general, we first factor out common factors, then inspect the result to see if it can be factored by any of the other methods of this section. We repeat this process until we have factored the expression completely.

EXAMPLE 13 ■ Factoring an Expression Completely

Factor completely each expression.

(a) $2x^4 - 8x^2$ (b) $x^5y^2 - xy^6$

SOLUTION

(a) We first factor out the power of x with the smallest exponent.

$$2x^4 - 8x^2 = 2x^2(x^2 - 4) \qquad \text{Common factor is } 2x^2$$
$$= 2x^2(x - 2)(x + 2) \qquad \text{Factor } x^2 - 4 \text{ as a difference of squares}$$

FACTORING NUMBERS

To factor a whole number completely means to write it as a product of smaller whole numbers that cannot themselves be factored, that is, as a product of primes. For example, $420 = 2 \cdot 2 \cdot 3 \cdot 5 \cdot 7$. To factor a very large number can be a difficult task. High-speed computers employing the fastest known methods would take about a day to factor an arbitrary 30-digit number and about a million years to factor a 40-digit number. Ted Rivest, Adi Shamir, and Leonard Adleman used this fact in the 1970s to devise the RSA code for sending secret messages. This code uses an extremely large number to encode a message, but requires knowledge of the factors to decode it. Since multiplying numbers is easy but factoring the result is hard, this code is very difficult to break. It was at first thought that a carefully selected 80-digit number would provide an unbreakable code, but recent advances in the study of factoring have made numbers with many more digits necessary to assure complete security.

(b) We first factor out the powers of x and y with the smallest exponents.

$$x^5 y^2 - xy^6 = xy^2(x^4 - y^4) \qquad \text{Common factor is } xy^2$$

$$= xy^2(x^2 + y^2)(x^2 - y^2) \qquad \text{Factor } x^4 - y^4 \text{ as a difference of squares}$$

$$= xy^2(x^2 + y^2)(x + y)(x - y) \qquad \text{Factor } x^2 - y^2 \text{ as a difference of squares} \quad \blacksquare$$

In the next example we factor out variables with fractional exponents. This type of factoring occurs in calculus.

EXAMPLE 14 ■ **Factoring Variables with Fractional Exponents**

Factor each expression.

(a) $3x^{3/2} - 9x^{1/2} + 6x^{-1/2}$ (b) $(1 + x)^{-2/3}x + (1 + x)^{1/3}$

SOLUTION

(a) Factor out the power of x with the *smallest exponent*, that is, $x^{-1/2}$.

$$3x^{3/2} - 9x^{1/2} + 6x^{-1/2} = 3x^{-1/2}(x^2 - 3x + 2) \qquad \text{Factor out } 3x^{-1/2}$$

$$= 3x^{-1/2}(x - 1)(x - 2) \qquad \text{Factor the quadratic } x^2 - 3x + 2$$

(b) Factor out the power of $1 + x$ with the smallest exponent, that is, $(1 + x)^{-2/3}$.

$$(1 + x)^{-2/3}x + (1 + x)^{1/3} = (1 + x)^{-2/3}[x + (1 + x)] \qquad \text{Factor out } (1 + x)^{-2/3}$$

$$= (1 + x)^{-2/3}(1 + 2x) \qquad\qquad \blacksquare$$

Polynomials with at least four terms can sometimes be factored by grouping terms. The following example illustrates the idea.

EXAMPLE 15 ■ **Factoring by Grouping**

Factor each polynomial.

(a) $x^3 + x^2 + 4x + 4$ (b) $x^3 - 2x^2 - 3x + 6$

SOLUTION

(a) $x^3 + x^2 + 4x + 4 = (x^3 + x^2) + (4x + 4) \qquad \text{Group terms}$

$$= x^2(x + 1) + 4(x + 1) \qquad \text{Factor out common factors}$$

$$= (x^2 + 4)(x + 1) \qquad \text{Factor out } x + 1 \text{ from each term}$$

(b) $x^3 - 2x^2 - 3x + 6 = (x^3 - 2x^2) - (3x - 6) \qquad \text{Group terms}$

$$= x^2(x - 2) - 3(x - 2) \qquad \text{Factor out common factors}$$

$$= (x^2 - 3)(x - 2) \qquad \text{Factor out } x - 2 \text{ from each term} \quad \blacksquare$$

1.4 EXERCISES

1-40 ■ Perform the indicated operations and simplify.

1. $2(x - 1) + 4(x + 2)$

2. $5(2x + 3) - 7(2x - 3)$

3. $(2x^2 + x + 1) + (x^2 - 3x + 5)$

4. $(2x^2 + x + 1) - (x^2 - 3x + 5)$

5. $(x^3 + 6x^2 - 4x + 7) - (3x^2 + 2x - 4)$

6. $4(x^2 - x + 2) - 5(x^2 - 2x + 1)$

7. $2(2 - 5t) + t^2(t - 1) - (t^4 - 1)$

8. $5(3t - 4) - (t^2 + 2) - 2t(t - 3)$

9. $\sqrt{x}\,(x - \sqrt{x})$

10. $x^{3/2}(\sqrt{x} - 1/\sqrt{x})$

11. $\sqrt[3]{y}\,(y^2 - 1)$

12. $(4x - 1)(3x + 7)$

13. $(3t - 2)(7t - 5)$

14. $(t + 6)(t + 5) - 3(t + 4)$

15. $(x + 2y)(3x - y)$

16. $(4x - 3y)(2x + 5y)$

17. $(1 - 2y)^2$

18. $(3x + 4)^2$

19. $(2x - 5)(x^2 - x + 1)$

20. $(x^2 + 3)(5x - 6)$

21. $x(x - 1)(x + 2)$

22. $(1 + 2x)(x^2 - 3x + 1)$

23. $y^4(6 - y)(5 + y)$

24. $(t - 5)^2 - 2(t + 3)(8t - 1)$

25. $(2x^2 + 3y^2)^2$

26. $(x^{1/2} + y^{1/2})(x^{1/2} - y^{1/2})$

27. $(x^2 - a^2)(x^2 + a^2)$

28. $(\sqrt{h^2 + 1} + 1)(\sqrt{h^2 + 1} - 1)$

29. $(1 + a^3)^3$

30. $(x - 1)(x^2 + x + 1)$

31. $\left(\sqrt{a} - \dfrac{1}{\sqrt{b}}\right)\left(\sqrt{a} + \dfrac{1}{\sqrt{b}}\right)$

32. $\left(c + \dfrac{1}{c}\right)^2$

33. $(x^2 + x - 2)(x^3 - x + 1)$

34. $(1 + x + x^2)(1 - x + x^2)$

35. $(1 + x^{4/3})(1 - x^{2/3})$

36. $(x^{3/2} - x + 1)(x^2 + x^{1/2} - 2)$

37. $(1 - b)^2(1 + b)^2$

38. $(1 + x - x^2)^2$

39. $(3x^2y + 7xy^2)(x^2y^3 - 2y^2)$

40. $(x^4y - y^5)(x^2 + xy + y^2)$

41-90 ■ Factor the expression completely.

41. $2x + 12x^3$

42. $8x^5 + 4x^3$

43. $6y^4 - 15y^3$

44. $5ab - 8abc$

45. $x^2 + 7x + 6$

46. $x^2 - x - 6$

47. $x^2 - 2x - 8$

48. $x^2 - 14x + 48$

49. $y^2 - 8y + 15$

50. $z^2 + 6z - 16$

51. $2x^2 + 5x + 3$

52. $2x^2 + 7x - 4$

53. $9x^2 - 36$

54. $8x^2 + 10x + 3$

55. $6x^2 - 5x - 6$

56. $6 + 5t - 6t^2$

57. $(x - 1)(x + 2)^2 - (x - 1)^2(x + 2)$

58. $(x + 1)^3x^2 - 2(x + 1)x^2 + x^3(x + 1)$

59. $y^4(y + 2)^3 + y^5(y + 2)^4$

60. $n(x - y) + (n - 1)(y - x)$

61. $(a^2 - 1)b^2 - 4(a^2 - 1)$

62. $(a + b)^2 - (a - b)^2$

63. $t^3 + 1$

64. $4t^2 - 9s^2$

65. $4t^2 - 12t + 9$

66. $x^3 - 27$

67. $x^3 + 2x^2 + x$

68. $3x^3 - 27x$

69. $4x^2 + 4xy + y^2$

70. $4r^2 - 12rs + 9s^2$

71. $x^4 + 2x^3 - 3x^2$

72. $x^6 + 64$

73. $8x^3 - 125$

74. $x^4 + 2x^2 + 1$

75. $x^4 + x^2 - 2$

76. $x^3 + 3x^2 - x - 3$

77. $y^3 - 3y^2 - 4y + 12$

78. $y^3 - y^2 + y - 1$

79. $2x^3 + 4x^2 + x + 2$

80. $3x^3 + 5x^2 - 6x - 10$

81. $x^6 - y^6$

82. $x^8 - 1$

83. $x^{5/2} - x^{1/2}$

84. $3x^{-1/2} + 4x^{1/2} + x^{3/2}$

85. $x^{-3/2} + 2x^{-1/2} + x^{1/2}$

86. $(x - 1)^{7/2} - (x - 1)^{3/2}$

87. $(x^2 + 1)^{1/2} + 2(x^2 + 1)^{-1/2}$

88. $x^{-1/2}(x + 1)^{1/2} + x^{1/2}(x + 1)^{-1/2}$

89. $(a^2 + 1)^2 - 7(a^2 + 1) + 10$

90. $(a^2 + 2a)^2 - 2(a^2 + 2a) - 3$

91. Factor $x^4 + 3x^2 + 4$. [*Hint:* Write the expression as $(x^4 + 4x^2 + 4) - x^2$ and note that this is a difference of squares.]

92. Show that $ab = \frac{1}{2}[(a + b)^2 - (a^2 + b^2)]$.

93. Show that
$$(a^2 + b^2)(c^2 + d^2) = (ac + bd)^2 + (ad - bc)^2$$

94. Show that $(a^2 + b^2)^2 - (a^2 - b^2)^2$ is a perfect square.

95. Factor the expression $4a^2c^2 - (c^2 - b^2 + a^2)^2$ completely.

96. Verify each of the following formulas algebraically.
(a) Special Product Formulas 1 and 2
(b) Special Product Formulas 3 and 4
(c) The formula for a difference of cubes
(d) The formula for a sum of cubes

97–99 ■ The figure gives a geometrical interpretation of the given product or factoring formula. Explain how the figure verifies the formula. In each case, $a > b > 0$.

97. $a^2 - b^2 = (a + b)(a - b)$

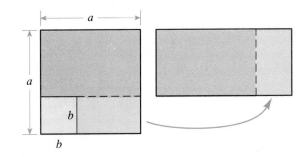

98. $(a - b)^2 = a^2 - 2ab + b^2$

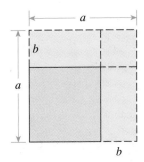

99. $(a + b)^3 = a^3 + 3a^2b + 3ab^2 + b^3$

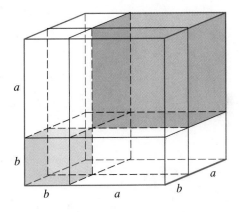

1.5 FRACTIONAL EXPRESSIONS

A quotient of two algebraic expressions is called a **fractional expression.** We assume that all fractions are defined; that is, *we deal only with values of the variables such that the denominators are not zero.*

A common type of fractional expression occurs when both the numerator and denominator are polynomials. This is called a **rational expression.** For instance,

$$\frac{4x^3 + 2x + 5}{x + 3}$$

is a rational expression whose denominator is 0 when $x = -3$. So, in dealing with this expression we implicitly assume that $x \neq -3$.

In simplifying rational expressions, we factor both numerator and denominator and **cancel** any common factors. In doing this we are using the following property of fractions:

$$\frac{AC}{BC} = \frac{A}{B}$$

This says that we can cancel common factors from numerator and denominator.

EXAMPLE 1 ■ Simplifying Fractional Expressions by Cancellation

Simplify: (a) $\dfrac{x^2 - 1}{x^2 + x - 2}$ (b) $\dfrac{2x^3 + 5x^2 - 3x}{6 - x - x^2}$

SOLUTION

We cannot cancel the x^2's in
$\dfrac{x^2 - 1}{x^2 + x - 2}$ because x^2 is not a factor.

(a) $\dfrac{x^2 - 1}{x^2 + x - 2} = \dfrac{(x - 1)(x + 1)}{(x - 1)(x + 2)}$ Factor

$\qquad\qquad = \dfrac{x + 1}{x + 2}$ Cancel common factors

(b) $\dfrac{2x^3 + 5x^2 - 3x}{6 - x - x^2} = \dfrac{x(2x^2 + 5x - 3)}{-(x^2 + x - 6)} = \dfrac{x(2x - 1)(x + 3)}{-(x - 2)(x + 3)}$ Factor

$\qquad\qquad = -\dfrac{x(2x - 1)}{x - 2}$ Cancel common factors

■

When multiplying fractional expressions we use the following property of fractions:

$$\frac{A}{B} \cdot \frac{C}{D} = \frac{AC}{BD}$$

This says that to multiply two fractions we multiply their numerators and multiply their denominators.

EXAMPLE 2 ■ **Multiplying Fractional Expressions**

Perform the indicated multiplication and simplify: $\dfrac{x^2 + 2x - 3}{x^2 + 8x + 16} \cdot \dfrac{3x + 12}{x - 1}$

SOLUTION

We first factor:

$$\frac{x^2 + 2x - 3}{x^2 + 8x + 16} \cdot \frac{3x + 12}{x - 1} = \frac{(x - 1)(x + 3)}{(x + 4)^2} \cdot \frac{3(x + 4)}{x - 1} \qquad \text{Factor}$$

$$= \frac{3(x - 1)(x + 3)(x + 4)}{(x - 1)(x + 4)^2} \qquad \text{Property of fractions}$$

$$= \frac{3(x + 3)}{x + 4} \qquad \text{Cancel common factors}$$

■

When dividing fractional expressions we use the following property of fractions:

$$\frac{A}{B} \div \frac{C}{D} = \frac{A}{B} \cdot \frac{D}{C}$$

This says that to divide a fraction by another fraction we invert the divisor and multiply.

EXAMPLE 3 ■ **Dividing Fractional Expressions**

Perform the indicated division and simplify: $\dfrac{x - 4}{x^2 - 4} \div \dfrac{x^2 - 3x - 4}{x^2 + 5x + 6}$

SOLUTION

$$\frac{x - 4}{x^2 - 4} \div \frac{x^2 - 3x - 4}{x^2 + 5x + 6} = \frac{x - 4}{x^2 - 4} \cdot \frac{x^2 + 5x + 6}{x^2 - 3x - 4} \qquad \text{Invert and multiply}$$

$$= \frac{(x - 4)(x + 2)(x + 3)}{(x - 2)(x + 2)(x - 4)(x + 1)} \qquad \text{Factor}$$

$$= \frac{x + 3}{(x - 2)(x + 1)} \qquad \text{Cancel common factors}$$

■

In adding and subtracting rational expressions, we first find a common denominator and then use the following property of fractions:

$$\frac{A}{C} + \frac{B}{C} = \frac{A + B}{C}$$

Although any common denominator will work, it is best to use the **least common denominator** (LCD) as explained in Section 1.2. The LCD is found by factoring each denominator and taking the product of the distinct factors, using the highest power that appears in any of the factors.

EXAMPLE 4 ■ Adding and Subtracting Fractional Expressions

Perform the indicated operations and simplify.

(a) $\dfrac{3}{x - 1} + \dfrac{x}{x + 2}$ (b) $\dfrac{1}{x^2 - 1} - \dfrac{2}{(x + 1)^2}$

(c) $\dfrac{1}{x^2 + 4x - 5} - \dfrac{1}{2x} + \dfrac{x + 1}{x^2 - x}$

SOLUTION

(a) Here the LCD is simply the product $(x - 1)(x + 2)$, so we have

$$\frac{3}{x - 1} + \frac{x}{x + 2} = \frac{3(x + 2)}{(x - 1)(x + 2)} + \frac{x(x - 1)}{(x - 1)(x + 2)} \qquad \text{Write fractions using LCD}$$

$$= \frac{3x + 6 + x^2 - x}{(x - 1)(x + 2)} \qquad \text{Add fractions}$$

$$= \frac{x^2 + 2x + 6}{(x - 1)(x + 2)} \qquad \text{Combine terms in numerator}$$

(b) The LCD of $x^2 - 1 = (x - 1)(x + 1)$ and $(x + 1)^2$ is $(x - 1)(x + 1)^2$, so we have

$$\frac{1}{x^2 - 1} - \frac{2}{(x + 1)^2} = \frac{1}{(x - 1)(x + 1)} - \frac{2}{(x + 1)^2} \qquad \text{Factor}$$

$$= \frac{(x + 1) - 2(x - 1)}{(x - 1)(x + 1)^2} \qquad \text{Combine fractions using LCD}$$

$$= \frac{x + 1 - 2x + 2}{(x - 1)(x + 1)^2} \qquad \text{Distributive Property}$$

$$= \frac{3 - x}{(x - 1)(x + 1)^2} \qquad \text{Combine terms in numerator}$$

(c) We start by factoring the denominators:

$$\frac{1}{x^2 + 4x - 5} - \frac{1}{2x} + \frac{x + 1}{x^2 - x} = \frac{1}{(x - 1)(x + 5)} - \frac{1}{2x} + \frac{x + 1}{x(x - 1)}$$

The LCD of the denominators is $2x(x - 1)(x + 5)$, so we can combine the fractions:

$$\frac{1}{(x - 1)(x + 5)} - \frac{1}{2x} + \frac{x + 1}{x(x - 1)}$$

$$= \frac{2x - (x - 1)(x + 5) + (x + 1)(2)(x + 5)}{2x(x - 1)(x + 5)}$$

$$= \frac{2x - (x^2 + 4x - 5) + (2x^2 + 12x + 10)}{2x(x - 1)(x + 5)}$$

$$= \frac{x^2 + 10x + 15}{2x(x - 1)(x + 5)}$$ ■

In the next example we simplify a **compound fraction,** which contains a fraction in both the numerator and the denominator.

EXAMPLE 5 ■ **Simplifying a Compound Fraction**

Simplify: $\dfrac{\dfrac{x}{y} + 1}{1 - \dfrac{y}{x}}$

SOLUTION 1

We combine the terms in the numerator into a single fraction. We do the same in the denominator. Then we invert and multiply.

$$\frac{\dfrac{x}{y} + 1}{1 - \dfrac{y}{x}} = \frac{\dfrac{x + y}{y}}{\dfrac{x - y}{x}} = \frac{x + y}{y} \cdot \frac{x}{x - y}$$

$$= \frac{x(x + y)}{y(x - y)}$$

SOLUTION 2

We find the least common denominator (LCD) of all the fractions in the expression, then multiply numerator and denominator by it. In this example the

LCD of all the fractions is xy. Thus

$$\frac{\dfrac{x}{y} + 1}{1 - \dfrac{y}{x}} = \frac{\dfrac{x}{y} + 1}{1 - \dfrac{y}{x}} \cdot \frac{xy}{xy}$$

$$= \frac{x^2 + xy}{xy - y^2}$$

$$= \frac{x(x + y)}{y(x - y)}$$ ∎

The remaining examples show situations in calculus in which facility with fractional expressions is required.

EXAMPLE 6 ■ Simplifying a Compound Fraction

Simplify: $\dfrac{\dfrac{1}{(a + h)^2} - \dfrac{1}{a^2}}{h}$

SOLUTION

As in the first solution to Example 5, we begin by combining the fractions in the numerator using a common denominator:

$$\frac{\dfrac{1}{(a + h)^2} - \dfrac{1}{a^2}}{h} = \frac{\dfrac{a^2 - (a + h)^2}{(a + h)^2 a^2}}{h} \qquad \text{Combine fractions in the numerator}$$

$$= \frac{a^2 - (a + h)^2}{(a + h)^2 a^2} \cdot \frac{1}{h} \qquad \text{Property 2 of fractions (invert divisor and multiply)}$$

$$= \frac{a^2 - (a^2 + 2ah + h^2)}{(a + h)^2 a^2} \cdot \frac{1}{h} \qquad \text{Special product formula 2}$$

$$= \frac{-2ah - h^2}{h(a + h)^2 a^2} \qquad \text{Subtract}$$

$$= \frac{h(-2a - h)}{h(a + h)^2 a^2} \qquad \text{Factor out } h$$

$$= -\frac{2a + h}{(a + h)^2 a^2} \qquad \text{Property 5 of fractions (cancel common factors)}$$ ∎

EXAMPLE 7 ■ Simplifying a Compound Fraction

Simplify: $\dfrac{(1 + x^2)^{1/2} - x^2(1 + x^2)^{-1/2}}{1 + x^2}$

SOLUTION 1

Factor $(1 + x^2)^{-1/2}$ from the numerator.

$$\frac{(1 + x^2)^{1/2} - x^2(1 + x^2)^{-1/2}}{1 + x^2} = \frac{(1 + x^2)^{-1/2}[(1 + x^2) - x^2]}{1 + x^2}$$

$$= \frac{(1 + x^2)^{-1/2}}{1 + x^2}$$

$$= \frac{1}{(1 + x^2)^{3/2}}$$

SOLUTION 2

Since $(1 + x^2)^{-1/2} = 1/(1 + x^2)^{1/2}$ is a fraction, we may clear all fractions by multiplying numerator and denominator by $(1 + x^2)^{1/2}$.

$$\frac{(1 + x^2)^{1/2} - x^2(1 + x^2)^{-1/2}}{1 + x^2} = \frac{(1 + x^2)^{1/2} - x^2(1 + x^2)^{-1/2}}{1 + x^2} \cdot \frac{(1 + x^2)^{1/2}}{(1 + x^2)^{1/2}}$$

$$= \frac{(1 + x^2) - x^2}{(1 + x^2)^{3/2}}$$

$$= \frac{1}{(1 + x^2)^{3/2}}$$ ■

If a fraction has a denominator of the form $A + B\sqrt{C}$, we may rationalize the denominator by multiplying numerator and denominator by the **conjugate radical** $A - B\sqrt{C}$. This is effective because, by Product Formula 1 in Section 1.4, the product of the denominator and its conjugate radical does not contain a radical:

$$(A + B\sqrt{C})(A - B\sqrt{C}) = A^2 - B^2C$$

EXAMPLE 8 ■ Rationalizing the Denominator

Rationalize the denominator in each expression.

(a) $\dfrac{1}{1 + \sqrt{2}}$ (b) $\dfrac{1}{\sqrt{x} - \sqrt{3}}$

SOLUTION

(a) We multiply both the numerator and the denominator by the conjugate radical of $1 + \sqrt{2}$, which is $1 - \sqrt{2}$:

$$\frac{1}{1 + \sqrt{2}} = \frac{1}{1 + \sqrt{2}} \cdot \frac{1 - \sqrt{2}}{1 - \sqrt{2}}$$ Multiply numerator and denominator by the conjugate radical

Product Formula 1
$(a + b)(a - b) = a^2 - b^2$

$$= \frac{1 - \sqrt{2}}{1^2 - (\sqrt{2})^2}$$ Product Formula 1

$$= \frac{1 - \sqrt{2}}{1 - 2} = \frac{1 - \sqrt{2}}{-1} = \sqrt{2} - 1$$

(b) The conjugate radical of $\sqrt{x} - \sqrt{3}$ is $\sqrt{x} + \sqrt{3}$, so we have

$$\frac{1}{\sqrt{x} - \sqrt{3}} = \frac{1}{\sqrt{x} - \sqrt{3}} \cdot \frac{\sqrt{x} + \sqrt{3}}{\sqrt{x} + \sqrt{3}}$$ Multiply numerator and denominator by the conjugate radical

$$= \frac{\sqrt{x} + \sqrt{3}}{(\sqrt{x})^2 - (\sqrt{3})^2}$$ Product Formula 1

$$= \frac{\sqrt{x} + \sqrt{3}}{x - 3}$$ ∎

EXAMPLE 9 ■ Rationalizing a Numerator

Rationalize the numerator: $\dfrac{\sqrt{4 + h} - 2}{h}$

SOLUTION

We multiply numerator and denominator by the conjugate radical $\sqrt{4 + h} + 2$. The advantage of doing this is that we can use Product Formula 1 and the radicals then disappear from the numerator.

$$\frac{\sqrt{4 + h} - 2}{h} = \frac{\sqrt{4 + h} - 2}{h} \cdot \frac{\sqrt{4 + h} + 2}{\sqrt{4 + h} + 2}$$ Multiply numerator and denominator by the conjugate radical

$$= \frac{(\sqrt{4 + h})^2 - 2^2}{h(\sqrt{4 + h} + 2)}$$ Product Formula 1

$$= \frac{4 + h - 4}{h(\sqrt{4 + h} + 2)}$$

$$= \frac{h}{h(\sqrt{4 + h} + 2)} = \frac{1}{\sqrt{4 + h} + 2}$$ Property 5 of fractions (cancel common factors)

∎

It is sometimes useful to use the formula

$$\frac{A}{C} + \frac{B}{C} = \frac{A + B}{C}$$

backward to write a quotient as a sum of fractions; that is, we write

$$\frac{A + B}{C} = \frac{A}{C} + \frac{B}{C}$$

For example, in certain calculus problems it is advantageous to write

$$\frac{x + 3}{x} = \frac{x}{x} + \frac{3}{x} = 1 + \frac{3}{x}$$

But remember to avoid the following common error:

$$\frac{A}{B + C} \neq \frac{A}{B} + \frac{A}{C}$$

⊘　　Do not make the mistake of applying properties of multiplication to the operation of addition. Many of the common errors in algebra involve doing just that. The following table states several properties of multiplication and illustrates the error in applying them to addition.

Correct multiplication property	Common error with addition
$(a \cdot b)^2 = a^2 \cdot b^2$	$(a + b)^2 \neq a^2 + b^2$
$\sqrt{a \cdot b} = \sqrt{a}\,\sqrt{b} \quad (a, b \geqslant 0)$	$\sqrt{a + b} \neq \sqrt{a} + \sqrt{b}$
$\sqrt{a^2 \cdot b^2} = a \cdot b \quad (a, b \geqslant 0)$	$\sqrt{a^2 + b^2} \neq a + b$
$\dfrac{1}{a} \cdot \dfrac{1}{b} = \dfrac{1}{a \cdot b}$	$\dfrac{1}{a} + \dfrac{1}{b} \neq \dfrac{1}{a + b}$
$\dfrac{ab}{a} = b$	$\dfrac{a + b}{a} \neq b$
$a^{-1} \cdot b^{-1} = \dfrac{1}{a \cdot b}$	$a^{-1} + b^{-1} \neq \dfrac{1}{a + b}$

To verify that the formulas in the right-hand column are wrong, simply substitute numbers for a and b and calculate each side. For example, if we take $a = 2$

and $b = 2$ in the fourth property, we find that the left-hand side is

$$\frac{1}{a} + \frac{1}{b} = \frac{1}{2} + \frac{1}{2} = 1$$

whereas the right-hand side is

$$\frac{1}{a + b} = \frac{1}{2 + 2} = \frac{1}{4}$$

Since $1 \neq \frac{1}{4}$, the stated formula is wrong. You should similarly convince yourself of the error in each of the other formulas.

1.5   EXERCISES

1–48 ■ Simplify the expression.

1. $\dfrac{x^2 + 3x + 2}{x^2 + 5x + 6}$

2. $\dfrac{x^2 + x - 6}{x^2 - 4}$

3. $\dfrac{y - y^2}{y^2 - 1}$

4. $\dfrac{2y^2 - 9y - 18}{4y^2 + 16y + 15}$

5. $\dfrac{2x^3 - x^2 - 6x}{2x^2 - 7x + 6}$

6. $\dfrac{1 - x^2}{x^3 - 1}$

7. $\dfrac{t - 3}{t^2 + 9} \cdot \dfrac{t + 3}{t^2 - 9}$

8. $\dfrac{x^2 - x - 6}{x^2 + 2x} \cdot \dfrac{x^3 + x^2}{x^2 - 2x - 3}$

9. $\dfrac{x^2 + 7x + 12}{x^2 + 3x + 2} \cdot \dfrac{x^2 + 5x + 6}{x^2 + 6x + 9}$

10. $\dfrac{x^2 + 2xy + y^2}{x^2 - y^2} \cdot \dfrac{2x^2 - xy - y^2}{x^2 - xy - 2y^2}$

11. $\dfrac{2x^2 + 3x + 1}{x^2 + 2x - 15} \div \dfrac{x^2 + 6x + 5}{2x^2 - 7x + 3}$

12. $\dfrac{4y^2 - 9}{2y^2 + 9y - 18} \div \dfrac{2y^2 + y - 3}{y^2 + 5y - 6}$

13. $\dfrac{\dfrac{x^3}{x + 1}}{\dfrac{x}{x^2 + 2x + 1}}$

14. $\dfrac{\dfrac{2x^2 - 3x - 2}{x^2 - 1}}{\dfrac{2x^2 + 5x + 2}{x^2 + x - 2}}$

15. $\dfrac{x/y}{z}$

16. $\dfrac{x}{y/z}$

17. $\dfrac{1}{x + 5} + \dfrac{2}{x - 3}$

18. $\dfrac{1}{x + 1} + \dfrac{1}{x - 1}$

19. $\dfrac{1}{x + 1} - \dfrac{1}{x + 2}$

20. $\dfrac{x}{x - 4} - \dfrac{3}{x + 6}$

21. $\dfrac{x}{(x + 1)^2} + \dfrac{2}{x + 1}$

22. $\dfrac{5}{2x - 3} - \dfrac{3}{(2x - 3)^2}$

23. $u + 1 + \dfrac{u}{u + 1}$

24. $\dfrac{2}{a^2} - \dfrac{3}{ab} + \dfrac{4}{b^2}$

25. $\dfrac{1}{x^2} + \dfrac{1}{x^2 + x}$

26. $\dfrac{1}{x} + \dfrac{1}{x^2} + \dfrac{1}{x^3}$

27. $\dfrac{2}{x + 3} - \dfrac{1}{x^2 + 7x + 12}$

28. $\dfrac{x}{x^2 - 4} + \dfrac{1}{x - 2}$

29. $\dfrac{1}{x + 3} + \dfrac{1}{x^2 - 9}$

30. $\dfrac{x}{x^2 + x - 2} - \dfrac{2}{x^2 - 5x + 4}$

31. $\dfrac{2}{x} + \dfrac{3}{x - 1} - \dfrac{4}{x^2 - x}$

32. $\dfrac{x}{x^2 - x - 6} - \dfrac{1}{x + 2} - \dfrac{2}{x - 3}$

33. $\dfrac{1}{x^2 + 3x + 2} - \dfrac{1}{x^2 - 2x - 3}$

34. $\dfrac{1}{x + 1} - \dfrac{2}{(x + 1)^2} + \dfrac{3}{x^2 - 1}$

35. $\dfrac{\dfrac{x}{y} - \dfrac{y}{x}}{\dfrac{1}{x^2} - \dfrac{1}{y^2}}$

36. $x - \dfrac{y}{\dfrac{x}{y} + \dfrac{y}{x}}$

37. $\dfrac{1 + \dfrac{1}{c - 1}}{1 - \dfrac{1}{c - 1}}$

38. $1 + \dfrac{1}{1 + \dfrac{1}{1 + x}}$

39. $\dfrac{\dfrac{5}{x - 1} - \dfrac{2}{x + 1}}{\dfrac{x}{x - 1} + \dfrac{1}{x + 1}}$

40. $\dfrac{\dfrac{a - b}{a} - \dfrac{a + b}{b}}{\dfrac{a - b}{b} + \dfrac{a + b}{a}}$

41. $\dfrac{x^{-2} - y^{-2}}{x^{-1} + y^{-1}}$

42. $\dfrac{x^{-1} + y^{-1}}{(x + y)^{-1}}$

43. $\dfrac{\dfrac{1}{a + h} - \dfrac{1}{a}}{h}$

44. $\dfrac{(x + h)^{-3} - x^{-3}}{h}$

45. $\dfrac{\dfrac{1 - (x + h)}{2 + (x + h)} - \dfrac{1 - x}{2 + x}}{h}$

46. $\dfrac{(x + h)^3 - 7(x + h) - (x^3 - 7x)}{h}$

47. $\sqrt{1 + \left(\dfrac{x}{\sqrt{1 - x^2}}\right)^2}$

48. $\sqrt{1 + \left(x^3 - \dfrac{1}{4x^3}\right)^2}$

49–52 ■ Simplify the expression.

49. $\dfrac{2(1 + x)^{1/2} - x(1 + x)^{-1/2}}{x + 1}$

50. $\dfrac{(1 - x^2)^{1/2} + x^2(1 - x^2)^{-1/2}}{1 - x^2}$

51. $\dfrac{3(1 + x)^{1/3} - x(1 + x)^{-2/3}}{(1 + x)^{2/3}}$

52. $\dfrac{(7 - 3x)^{1/2} + \frac{3}{2}x(7 - 3x)^{-1/2}}{7 - 3x}$

53–56 ■ Rationalize the denominator.

53. $\dfrac{2}{3 + \sqrt{5}}$

54. $\dfrac{1}{\sqrt{x} + 1}$

55. $\dfrac{2}{\sqrt{2} + \sqrt{7}}$

56. $\dfrac{y}{\sqrt{3} + \sqrt{y}}$

57–64 ■ Rationalize the numerator.

57. $\dfrac{1 - \sqrt{5}}{3}$

58. $\dfrac{\sqrt{3} + \sqrt{5}}{2}$

59. $\dfrac{\sqrt{r} + \sqrt{2}}{5}$

60. $\dfrac{\sqrt{3(x + h) + 5} - \sqrt{3x + 5}}{h}$

61. $\dfrac{\sqrt{x} - \sqrt{x + h}}{h\sqrt{x}\sqrt{x + h}}$

62. $\sqrt{x^2 + 1} - x$

63. $\sqrt{x^2 + x + 1} + x$

64. $\sqrt{x + 1} - \sqrt{x}$

65–74 ■ State whether the given equation is true for all values of the variables. (Disregard any value that makes a denominator 0).

65. $\dfrac{16 + a}{16} = 1 + \dfrac{a}{16}$

66. $\dfrac{b}{b - c} = 1 - \dfrac{b}{c}$

67. $\dfrac{2}{4 + x} = \dfrac{1}{2} + \dfrac{2}{x}$

68. $\dfrac{x + 1}{y + 1} = \dfrac{x}{y}$

69. $\dfrac{x}{x + y} = \dfrac{1}{1 + y}$

70. $2\left(\dfrac{a}{b}\right) = \dfrac{2a}{2b}$

71. $\dfrac{-a}{b} = -\dfrac{a}{b}$

72. $\dfrac{1 + x + x^2}{x} = \dfrac{1}{x} + 1 + x$

73. $\dfrac{x^2 + 1}{x^2 + x - 1} = \dfrac{1}{x - 1}$

74. $\dfrac{x^2 - 1}{x - 1} = x + 1$

75. If two electrical resistors with resistances R_1 and R_2 are connected in parallel (see the figure) then the total resistance R is given by

$$R = \dfrac{1}{\dfrac{1}{R_1} + \dfrac{1}{R_2}}$$

(a) Simplify the expression for R.
(b) If $R_1 = 10$ ohms and $R_2 = 20$ ohms, what is the total resistance R?

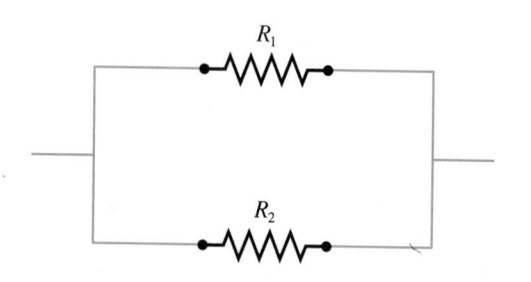

76. The rational expression

$$\dfrac{x^2 - 9}{x - 3}$$

is not defined for $x = 3$. In this exercise we investigate this expression for values of x very close to 3.
(a) Complete the table and determine what value the expression approaches as x gets closer and closer to 3.

x	$\dfrac{x^2 - 9}{x - 3}$		x	$\dfrac{x^2 - 9}{x - 3}$
2.80			3.20	
2.90			3.10	
2.95			3.05	
2.99			3.01	
2.999			3.001	

(b) Simplify the given expression by factoring the numerator.
(c) How does the simplified form in part (b) explain your answer to part (a)?

---

1   REVIEW

KEY TOPICS ■ Define, state, or discuss each of the following.

1. Integers
2. Rational and irrational numbers
3. Real number line
4. Commutative Property
5. Associative Property
6. Distributive Property
7. Properties of negatives
8. Properties of fractions
9. Union of sets
10. Intersection of sets
11. Open interval, closed interval

12. Absolute value of a number
13. Distance between points on the real line
14. Base and exponent
15. Laws of exponents
16. Scientific notation
17. Principal nth root
18. Properties of nth roots
19. Rationalizing a denominator
20. Rationalizing a numerator
21. Rational exponents
22. Variable and constant

23. Special product formulas for $(a + b)^2$, $(a - b)^2$, $(a + b)^3$, $(a - b)^3$

24. Factoring formulas

25. Difference of squares formula

26. Difference of cubes formula

27. Sum of cubes formula

EXERCISES

1–4 ■ State the property of real numbers being used.

1. $x + 5 = 5 + x$

2. $(a + b)(a - b) = (a - b)(a + b)$

3. $A(x + y) = Ax + Ay$

4. $(A + 1)(x + y) = (A + 1)x + (A + 1)y$

5–6 ■ Express the interval in terms of inequalities, and then graph the interval.

5. $(-1, 3]$

6. $(-\infty, 4]$

7–8 ■ Express the inequality in interval notation, and then graph the corresponding interval.

7. $x > 2$

8. $1 \leq x \leq 6$

9–18 ■ Evaluate the expression.

9. $\big|3 - |-9|\big|$

10. $1 - \big|1 - |-1|\big|$

11. $2^{-3} - 3^{-2}$

12. $\sqrt[3]{-125}$

13. $216^{-1/3}$

14. $64^{2/3}$

15. $\dfrac{\sqrt{242}}{\sqrt{2}}$

16. $\sqrt[4]{4}\,\sqrt[4]{324}$

17. $2^{1/2}8^{1/2}$

18. $\sqrt{2}\,\sqrt{50}$

19–26 ■ Write the expression as a power of x.

19. $\dfrac{1}{x^2}$

20. $x\sqrt{x}$

21. $x^2 x^m (x^3)^m$

22. $((x^m)^2)^n$

23. $x^a x^b x^c$

24. $((x^a)^b)^c$

25. $x^{c+1}(x^{2c-1})^2$

26. $\dfrac{(x^2)^n x^5}{x^n}$

27–36 ■ Simplify the expression.

27. $(2x^3 y)^2(3x^{-1}y^2)$

28. $(a^2)^{-3}(a^3 b)^2(b^3)^4$

29. $\dfrac{x^4(3x)^2}{x^3}$

30. $\left(\dfrac{r^2 s^{4/3}}{r^{1/3} s}\right)^6$

31. $\sqrt[3]{(x^3 y)^2 y^4}$

32. $\sqrt{x^2 y^4}$

33. $\dfrac{x}{2 + \sqrt{x}}$

34. $\dfrac{\sqrt{x} + 1}{\sqrt{x} - 1}$

35. $\dfrac{8r^{1/2}s^{-3}}{2r^{-2}s^4}$

36. $\left(\dfrac{ab^2 c^{-3}}{2a^3 b^{-4}}\right)^{-2}$

37. Write the number 78,250,000,000 in scientific notation.

38. Write the number 2.08×10^{-8} in ordinary decimal notation.

39. If $a \approx 0.00000293$, $b \approx 1.582 \times 10^{-14}$, and $c \approx 2.8064 \times 10^{12}$, use a calculator to approximate the number ab/c.

40. If your heart beats 80 times per minute and you live to be 90 years old, estimate the number of times your heart beats during your lifetime. State your answer in scientific notation.

41–60 ■ Factor the expression.

41. $12x^2 y^4 - 3xy^5 + 9x^3 y^2$

42. $x^2 - 9x + 18$

43. $x^2 + 3x - 10$

44. $6x^2 + x - 12$

45. $4t^2 - 13t - 12$

46. $x^4 - 2x^2 + 1$

47. $25 - 16t^2$

48. $2y^6 - 32y^2$

49. $x^6 - 1$

50. $y^3 - 2y^2 - y + 2$

51. $x^{-1/2} - 2x^{1/2} + x^{3/2}$

52. $a^4 b^2 + ab^5$

53. $4x^3 - 8x^2 + 3x - 6$

54. $8x^3 + y^6$

55. $(x^2 + 2)^{5/2} + 2x(x^2 + 2)^{3/2} + x^2\sqrt{x^2 + 2}$

56. $3x^3 - 2x^2 + 18x - 12$

57. $a^2 y - b^2 y$

58. $ax^2 + bx^2 - a - b$

59. $(x + 1)^2 - 2(x + 1) + 1$

60. $(a + b)^2 + 2(a + b) - 15$

61–84 ■ Perform the indicated operations.

61. $(2x + 1)(3x - 2) - 5(4x - 1)$

62. $(2y - 7)(2y + 7)$

63. $(2a^2 - b)^2$

64. $(1 + x)(2 - x) - (3 - x)(3 + x)$

65. $(x - 1)(x - 2)(x - 3)$

66. $(2x + 1)^3$

67. $\sqrt{x}\left(\sqrt{x} + 1\right)\left(2\sqrt{x} - 1\right)$

68. $x^3(x - 6)^2 + x^4(x - 6)$

69. $x^2(x - 2) + x(x - 2)^2$

70. $\dfrac{x^3 + 2x^2 + 3x}{x}$

71. $\dfrac{x^2 - 2x - 3}{2x^2 + 5x + 3}$

72. $\dfrac{t^3 - 1}{t^2 - 1}$

73. $\dfrac{x^2 + 2x - 3}{x^2 + 8x + 16} \cdot \dfrac{3x + 12}{x - 1}$

74. $\dfrac{x^3/(x - 1)}{x^2/(x^3 - 1)}$

75. $\dfrac{x^2 - 2x - 15}{x^2 - 6x + 5} \div \dfrac{x^2 - x - 12}{x^2 - 1}$

76. $x - \dfrac{1}{x + 1}$

77. $\dfrac{1}{x - 1} - \dfrac{x}{x^2 + 1}$

78. $\dfrac{2}{x} + \dfrac{1}{x - 2} + \dfrac{3}{(x - 2)^2}$

79. $\dfrac{1}{x - 1} - \dfrac{2}{x^2 - 1}$

80. $\dfrac{1}{x + 2} + \dfrac{1}{x^2 - 4} - \dfrac{2}{x^2 - x - 2}$

81. $\dfrac{\dfrac{1}{x} - \dfrac{1}{2}}{x - 2}$

82. $\dfrac{\dfrac{1}{x} - \dfrac{1}{x + 1}}{\dfrac{1}{x} + \dfrac{1}{x + 1}}$

83. $\dfrac{3(x + h)^2 - 5(x + h) - (3x^2 - 5x)}{h}$

84. $\dfrac{\sqrt{x + h} - \sqrt{x}}{h}$ (rationalize the numerator)

85. Suppose an automobile's fuel consumption is 28 mi/gal in city driving and 34 mi/gal in highway driving. If x denotes the number of city miles and y the number of highway miles, then the total miles this car can travel on a 15-gallon tank of fuel must satisfy the inequality

$$\tfrac{1}{28}x + \tfrac{1}{34}y \le 15$$

Use this inequality to answer the following questions. (Assume the car has a full tank of fuel.)
(a) Can the car travel 165 city miles and 230 highway miles before running out of gas?
(b) If the car has been driven 280 miles in the city, how many highway miles can it be driven before running out of fuel?

86. The speed that a sailboat is capable of sailing is determined by three factors: its total length L, the surface area A of its sails, and its displacement V (the volume of water it displaces), as shown in the sketch.

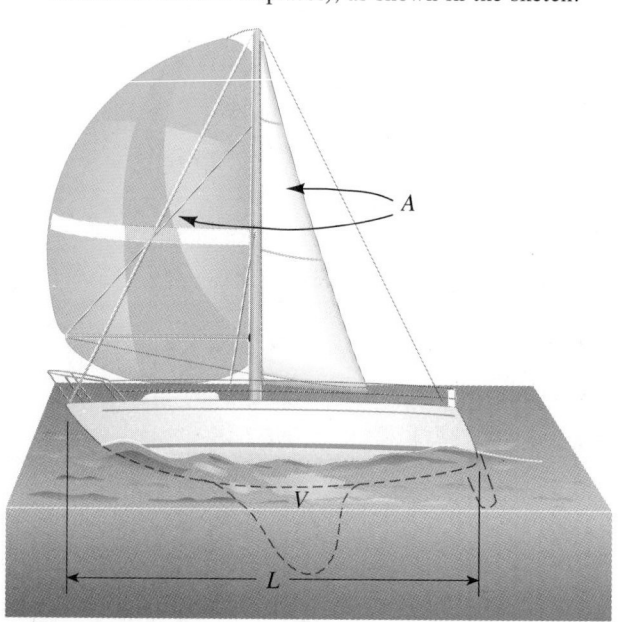

In general, a sailboat is capable of greater speed if it is longer, has a larger sail area, or displaces less water. In order to make sailing races fair, only boats in the same "class" can qualify to race together. For a certain race a boat is considered to qualify if

$$0.30L + 0.38A^{1/2} - 3V^{1/3} \leq 16$$

where L is measured in feet, A in square feet, and V in cubic feet. Use this inequality to answer the following questions.

(a) A sailboat has length 60 ft, sail area 3400 ft^2, and displacement 650 ft^3. Does this boat qualify for the race?

(b) Suppose a sailboat has length 65 ft and displaces 600 ft^3. What is the largest possible sail area that could be used and still allow the boat to qualify for this race?

87–93 ■ State whether the given equation is true for all values of the variables. (Disregard any value that makes a denominator 0.)

87. $(x + y)^3 = x^3 + y^3$

88. $\dfrac{1 + \sqrt{a}}{1 - a} = \dfrac{1}{1 - \sqrt{a}}$

89. $\dfrac{12 + y}{y} = \dfrac{12}{y} + 1$

90. $\sqrt[3]{a + b} = \sqrt[3]{a} + \sqrt[3]{b}$

91. $\sqrt{a^2} = a$

92. $\dfrac{1}{x + 4} = \dfrac{1}{x} + \dfrac{1}{4}$

93. $x^3 + y^3 = (x + y)(x^2 + xy + y^2)$

94. If $m > n > 0$ and $a = 2mn$, $b = m^2 - n^2$, $c = m^2 + n^2$, show that $a^2 + b^2 = c^2$.

95. If $t = \dfrac{1}{2}\left(x^3 - \dfrac{1}{x^3}\right)$ and $x > 0$, show that

$$\sqrt{1 + t^2} = \frac{1}{2}\left(x^3 + \frac{1}{x^3}\right)$$

96. Assume $a = b$. What is wrong with the following argument?

$$a = b$$
$$a^2 = ab$$
$$a^2 - b^2 = ab - b^2$$
$$(a + b)(a - b) = b(a - b)$$
$$a + b = b$$

Now put $a = b = 1$: $\qquad 2 = 1$

1. (a) Graph the intervals $[-3, 2]$ and $(4, \infty)$ on a real number line.
   (b) Express the inequalities $x < 5$ and $-2 \leq x \leq 1$ in interval notation.
   (c) Find the distance between -22 and 31 on the number line.

2. Evaluate each expression.

   (a) $(-3)^4$ (b) 2^{-4} (c) $\dfrac{5^{18}}{5^{12}}$

3. Evaluate each expression.

   (a) $\left(\frac{2}{3}\right)^{-1}$ (b) $\dfrac{\sqrt{32}}{\sqrt{8}}$ (c) $16^{-3/4}$

4. Express $\dfrac{(x^2)^a(\sqrt{x})^b}{x^{a+b}x^{a-b}}$ as a power of x.

5. Simplify each expression.

   (a) $\sqrt{200} - \sqrt{8}$ (b) $(2a^3b^2)(3ab^4)^3$
   (c) $\left(\dfrac{x^2y^{-3}}{y^5}\right)^{-4}$ (d) $\left(\dfrac{2x^{1/4}}{y^{1/3}x^{1/6}}\right)^3$

6. Simplify each expression.

   (a) $\dfrac{x^2 + 3x + 2}{x^2 - x - 2}$ (b) $\dfrac{x^2}{x^2 - 4} - \dfrac{x + 1}{x + 2}$ (c) $\dfrac{\frac{y}{x} - \frac{x}{y}}{\frac{1}{y} - \frac{1}{x}}$

7. Write each number in scientific notation.
   (a) 325,000,000,000   (b) 0.000008931

8. Perform the indicated operations and simplify.
   (a) $4(3 - x) - 3(x + 5)$ (b) $(x - 5)(2x + 3)$ (c) $(\sqrt{x} + \sqrt{y})(\sqrt{x} - \sqrt{y})$
   (d) $(3t + 4)^2$ (e) $(2 - x^2)^3$

9. Factor completely each expression.
   (a) $9x^2 - 25$ (b) $6x^2 + 7x - 5$ (c) $x^3 - 4x^2 - 3x + 12$
   (d) $x^4 + 27x$ (e) $3x^{3/2} - 9x^{1/2} + 6x^{-1/2}$

10. Rationalize the denominator: $\dfrac{x}{\sqrt{x} - 2}$

PRINCIPLES OF PROBLEM SOLVING

There are no hard and fast rules that will ensure success in solving problems. However, it is possible to outline some general steps in the problem-solving process and to give some principles that may be useful in the solution of certain problems. These steps and principles are just common sense made explicit. They have been adapted from George Polya's book *How To Solve It*.

UNDERSTAND THE PROBLEM

The first step is to read the problem and make sure that you understand it clearly. Ask yourself the following questions:

What is the unknown?

What are the given quantities?

What are the given conditions?

For many problems it is useful to

draw a diagram

and identify the given and required quantities on the diagram.
   Usually it is necessary to

introduce suitable notation

In choosing symbols for the unknown quantities we often use letters such as a, b, c, m, n, x, and y, but in some cases it helps to use initials as suggestive symbols, for instance, V for volume or t for time.

THINK OF A PLAN

Find a connection between the given information and the unknown that will enable you to calculate the unknown. It often helps to ask yourself explicitly: "How can I relate the given to the unknown?" If you do not see a connection immediately, the following ideas may be helpful in devising a plan.

■ Try to recognize something familiar

Relate the given situation to previous knowledge. Look at the unknown and try to recall a more familiar problem that has a similar unknown.

■ Try to recognize patterns

Some problems are solved by recognizing that some kind of pattern is occurring. The pattern could be geometric, or numerical, or algebraic. If you can see regu-

George Polya (1887–1985) is famous among mathematicians for his ideas on problem solving. His lectures on problem solving at Stanford University attracted overflow crowds whom he held on the edges of their seats, leading them to discover solutions for themselves. He was able to do this because of his deep insight into the psychology of problem solving. His well known book *How To Solve It* has been translated into 15 languages. He said that Euler (see page 100) was unique among great mathematicians because he explained *how* he found his results. Polya often said to his students and colleagues "Yes, I see that your proof is correct, but how did you discover it?" In the preface to *How To Solve It* Polya writes, "A great discovery solves a great problem but there is a grain of discovery in the solution of any problem. Your problem may be modest; but if it challenges your curiosity and brings into play your inventive faculties, and if you solve it by your own means, you may experience the tension and enjoy the triumph of discovery."

larity or repetition in a problem, then you might be able to guess what the continuing pattern is and then prove it.

■ Use analogy

Try to think of an analogous problem, that is, a similar problem, a related problem, but one that is easier than the original problem. If you can solve the similar, simpler problem, then it might give you the clues you need to solve the original, more difficult problem. For instance, if a problem involves very large numbers, you could first try a similar problem with smaller numbers. Or if the problem is in three-dimensional geometry, you could look for a similar problem in two-dimensional geometry. Or if the problem you start with is a general one, you could first try a special case.

■ Introduce something extra

You may sometimes need to introduce something new—an auxiliary aid—to help make the connection between the given and the unknown. For instance, in a problem for which a diagram is useful, the auxiliary aid could be a new line drawn in the diagram. In a more algebraic problem the aid could be a new unknown that is related to the original unknown.

■ Take cases

You may sometimes have to split a problem into several cases and give a different argument for each of the cases. For instance, we often have to use this strategy in dealing with absolute value.

■ Work backward

Sometimes it is useful to imagine that your problem is solved and work backward, step by step, until you arrive at the given data. Then you may be able to reverse your steps and thereby construct a solution to the original problem. This procedure is commonly used in solving equations. For instance, in solving the equation $3x - 5 = 7$, we suppose that x is a number that satisfies $3x - 5 = 7$ and work backward. We add 5 to each side of the equation and then divide each side by 3 to get $x = 4$. Since each of these steps can be reversed, we have solved the problem.

■ Establish subgoals

In a complex problem it is often useful to set subgoals (in which the desired situation is only partially fulfilled). If you can first reach these subgoals, then you may be able to build on them to reach your final goal.

■ Indirect reasoning

Sometimes it is appropriate to attack a problem indirectly. In using **proof by contradiction** to prove that P implies Q, we assume that P is true and Q is false and try to see why this cannot happen. Somehow we have to use this information and arrive at a contradiction to what we absolutely know is true.

■ Mathematical induction

In proving statements that involve a positive integer n, it is frequently helpful to use the Principle of Mathematical Induction, which is discussed in Section 9.7.

CARRY OUT THE PLAN

In Step 2 a plan was devised. In carrying out that plan, you must check each stage of the plan and write the details that prove that each stage is correct.

LOOK BACK

Having completed your solution, it is wise to look back over it, partly to see if errors have been made in the solution and partly to see if you can discover an easier way to solve the problem. Another reason for looking back is that it will familiarize you with the method of solution and this may be useful for solving a future problem. Descartes said, "Every problem that I solved became a rule which served afterwards to solve other problems."

We illustrate some of these principles of problem solving in an example. Further illustrations of these principles will be presented at the end of every chapter.

PROBLEM ■ Average Speed

A driver sets out on a journey. For the first half of the distance she drives at the leisurely pace of 30 mi/h; during the second half she drives 60 mi/h. What is her average speed on this trip?

PRELIMINARY THOUGHTS   It is tempting to take the average of the speeds and say that the average speed for the entire trip is

$$\frac{30 + 60}{2} = 45 \text{ mi/h}$$

But is this simple-minded approach really correct?

Let us look at an easily calculated special case. Suppose that the total distance traveled is 120 mi. Since the first 60 mi is traveled at 30 mi/h, it takes 2 h. The second 60 mi is traveled at 60 mi/h, so it takes 1 h. Thus, the total time is $2 + 1 = 3$ h and the average speed is

$$\frac{120}{3} = 40 \text{ mi/h}$$

So our guess of 45 mi/h was wrong.

SOLUTION

Understand the problem

We need to look more carefully at the meaning of average speed. It is defined as

$$\text{average speed} = \frac{\text{distance traveled}}{\text{time elapsed}}$$

Introduce notation

Let d be the distance traveled on each half of the trip. Let t_1 and t_2 be the times taken for the first and second halves of the trip. Now, we can write down

State what is given

the information that we have been given. For the first half of the trip, we have

(1) $$30 = \frac{d}{t_1}$$

and, for the second half, we have

(2) $$60 = \frac{d}{t_2}$$

Identify the unknown

Now we identify the quantity we are asked to find:

$$\text{average speed for entire trip} = \frac{\text{total distance}}{\text{total time}} = \frac{2d}{t_1 + t_2}$$

Connect the given with the unknown

To calculate this quantity we need to know t_1 and t_2, so we solve Equations 1 and 2 for these times:

$$t_1 = \frac{d}{30} \qquad t_2 = \frac{d}{60}$$

Now we have the ingredients needed to calculate the desired quantity:

$$\text{average speed} = \frac{2d}{t_1 + t_2} = \frac{2d}{\dfrac{d}{30} + \dfrac{d}{60}}$$

$$= \frac{60(2d)}{60\left(\dfrac{d}{30} + \dfrac{d}{60}\right)}$$

Multiply numerator and denominator by 60

$$= \frac{120d}{2d + d} = \frac{120d}{3d} = 40$$

So, the average speed for the entire trip is 40 mi/h. ∎

PROBLEMS

1. A man drives from home to work at a speed of 50 mi/h. The return trip from work to home is traveled at the more leisurely pace of 30 mi/h. What is the man's average speed for the round-trip?

2. An old car has to travel a 2-mile route, uphill and down. Because it is so old, the car can climb the first mile—the ascent—no faster than an average speed of 15 mi/h. How fast does the car have to travel the second mile—on the descent it can go faster, of course—in order to achieve an average speed of 30 mi/h for the trip?

3. A car and a van are parked 120 mi apart on a straight road. The drivers start driving toward each other at noon, each at a speed of 40 mi/h. A fly starts from the front bumper of the van at noon and flies to the bumper of the car, then immediately back to the bumper of the van, back to the car, and so on, until the car and the van meet. If the fly flies at a speed of 100 mi/h, what is the total distance it travels?

4. Which price is better for the buyer, a 40% discount or two successive discounts of 20%?

5. Use a calculator to find the value of the expression
$$\sqrt{3 + 2\sqrt{2}} - \sqrt{3 - 2\sqrt{2}}$$
The number looks very simple. Show that the calculated value is correct.

6. Use a calculator to evaluate
$$\frac{\sqrt{2} + \sqrt{6}}{\sqrt{2 + \sqrt{3}}}$$
Show that the calculated value is correct.

7. An amoeba propagates by simple division; each split takes three minutes to complete. When I put such an amoeba into a glass container with a nutrient fluid, the rate of multiplication rises, of course—in one hour the vessel is full of amoebas. How long would it take for the vessel to be filled if I start with not one amoeba, but two?

8. Two runners start running laps at the same time, from the same starting position. George runs a lap in 50 s; Sue runs a lap in 30 s. When will the runners next be side by side?

9. Player A has a higher batting average than Player B for the first half of the baseball season. Player A also has a higher batting average than Player B for the second half of the season. Is it true that Player A has a higher batting average than Player B for the entire season?

10. A person starts at a point P on the earth's surface and walks 1 mi south, then 1 mi east, then 1 mi north, and finds herself back at P, the starting point. Describe all points P for which this is possible (there are infinitely many).

11. A spoonful of cream is taken from a cup of cream and put into a cup of coffee. The coffee is then stirred. Then a spoonful of this mixture is put into the cup of cream. Is there now more cream in the coffee cup or more coffee in the cup of cream?

12. An ice cube is floating in a cup of water, full to the brim, as shown in the sketch. As the ice melts, what happens? Does the cup overflow, or does the water level drop, or does it remain the same? (You need to know Archimedes' principle: A floating object displaces a volume of water whose weight equals the weight of the object.)

13. An extended family consists of the members

> {Baby, Son, Daughter, Mother, Father, Uncle, Aunt, Grandma, Grandpa}

listed in increasing order of age. If x and y are members of this family, define $x + y$ to be the older of x and y, and define $x \cdot y$ to be the younger of x and y.
(a) Find (Baby + Uncle) + Mother.
(b) Find Father $\cdot$ (Grandpa + Aunt).

(c) Show that the operations $+$ and $\cdot$ that we have defined satisfy the same Commutative, Associative, and Distributive Properties as do the real numbers with regular addition and multiplication.

(d) Show that the operations also satisfy the property

$$x + (y \cdot z) = (x + y) \cdot (x + z)$$

Do the operations of regular addition and multiplication with real numbers also satisfy this property?

14. The ancient Egyptians, as a result of their pyramid-building, knew that the volume of a pyramid with height h and square base of side length a is $V = \frac{1}{3}ha^2$. They were able to use this fact to prove that the volume of a truncated pyramid is $V = \frac{1}{3}h(a^2 + ab + b^2)$, where h is the height and a and b are the lengths of the sides of the square top and bottom, as shown in the figure. Prove the truncated pyramid volume formula.

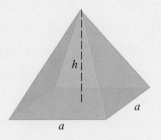

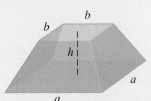

15. The ancient Babylonians developed the following process for finding the square root of a number N.

First they made a guess at the square root—let's call this first guess r_1. Noting that $r_1 \cdot \left(\dfrac{N}{r_1} \right) = N$, they concluded that the actual square root must be somewhere between r_1 and N/r_1, so their next guess for the square root, r_2, was the average of these two numbers: $r_2 = \dfrac{1}{2}\left(r_1 + \dfrac{N}{r_1} \right)$. Continuing in this way, their next approximation was given by $r_3 = \dfrac{1}{2}\left(r_2 + \dfrac{N}{r_2} \right)$, and so on. In general, once we have the nth approximation to the square root of N, we find the $(n + 1)$st using

$$r_{n+1} = \frac{1}{2}\left(r_n + \frac{N}{r_n} \right)$$

Use this procedure to find $\sqrt{72}$, correct to 2 decimal places.

16. This is a problem adapted from a tenth-century Arab manuscript. In his will, a nobleman leaves half of his horses to his oldest son, a third to his second son, and a ninth to his youngest. At his death he has 17 horses, so none of these bequests results in a whole number of horses for any of the sons. The executor of the will solves this dilemma by adding one of his own horses to the estate, making a total of 18. Now, according to the provisions of the will, he gives 9 to the oldest son, 6 to the second, and 2 to the third. Thus each son inherits a little more than he was originally entitled to, and one horse is left over, which the executor takes back again. Everybody is happy, but something seems wrong. What, in fact, *is* wrong?

2

EQUATIONS AND INEQUALITIES

When studying motion, we must often solve equations to determine the speed of an object, or the distance and time it has traveled.

I keep the subject constantly before me and wait till the first dawnings open little by little into the full light.

ISAAC NEWTON

An equation is a statement that two mathematical expressions are equal. For example,

$$3 + 5 = 8$$

is an equation. But it is not a very interesting one—it just states a simple arithmetic fact. Most equations that we study in algebra contain **variables,** which are symbols (usually letters) that stand for numbers. In the equations

$$(w - 4)(w + 4) = w^2 - 16 \quad \text{and} \quad 4x + 7 = 19$$

the letters w and x are variables. In the first of these equations, the equation is true no matter what value the variable w stands for. This equation is the "difference of squares" formula from Chapter 1; it is true for all w, so we say it is an **identity.** The second equation is *not* true for all values of the variable x. The values of x that make the equation true are called the **solutions** or **roots** of the equation, and the process of finding the solutions is called **solving the equation.**

Two equations with exactly the same solutions are called **equivalent equations.** To solve an equation, we try to find a simpler, equivalent equation in which the variable stands alone on one side of the "equal" sign. Here are the rules that we use to solve an equation. (In these rules, A, B, and c stand for any algebraic expressions and the symbol $\Longleftrightarrow$ means "is equivalent to.")

PROPERTIES OF EQUALITY

Property	Description
1. $A = B \Longleftrightarrow A + c = B + c$	Adding the same quantity to both sides of an equation gives an equivalent equation.
2. $A = B \Longleftrightarrow cA = cB \quad (c \neq 0)$	Multiplying both sides of an equation by the same nonzero quantity gives an equivalent equation.

These rules require that you *perform the same operation to both sides of an equation* when solving it. Thus, if we say "*add* -7" when solving an equation, this is just a short way of saying "*add* -7 to each side of the equation."

This is how we use the properties of equality to solve the equation $4x + 7 = 19$:

$$4x + 7 + (-7) = 19 + (-7) \quad \text{Add } -7$$

$$4x = 12 \quad \text{Simplify}$$

$$\tfrac{1}{4} \cdot 4x = \tfrac{1}{4} \cdot 12 \quad \text{Multiply by } \tfrac{1}{4}$$

$$x = 3 \quad \text{Simplify}$$

So the solution of this equation is $x = 3$. To verify this, we check our answer by substituting $x = 3$ to make sure that this value of x does indeed make the equation true:

$$\boxed{x = 3}$$
$$\downarrow$$
$$4(3) + 7 \overset{?}{=} 19$$

$$19 = 19 \qquad \text{Correct!}$$

In Section 2.6 we learn how to solve **inequalities,** which look like equations except that the equality symbol $=$ is replaced by one of the inequality symbols $<$, $>$, $\leq$, or $\geq$. We will see that the solutions of inequalities are, in general, *intervals* of numbers on the real line instead of just a single number or a small set of numbers. The following illustration indicates this difference for one inequality:

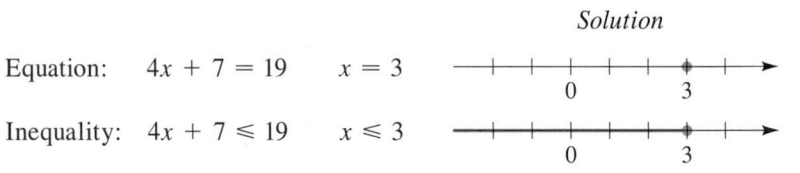

2.1 LINEAR EQUATIONS

Linear Equations

$4x - 5 = 3$

$2x = \tfrac{1}{2}x - 5$

Nonlinear Equations

$x^2 + 2x = 8$

$\sqrt{x} - \dfrac{3}{x} = 6x - 1$

The simplest type of equation is a **linear equation,** or first-degree equation, which is an equation in which each term is either a constant or a nonzero multiple of the variable. This means that it is equivalent to an equation of the form $ax + b = 0$. Here a and b represent real numbers with $a \neq 0$, and x is the unknown variable that we are solving for. The equation in the following example is linear.

EXAMPLE 1 ■ Solving a Linear Equation

Solve the equation $7x - 4 = 3x + 8$.

SOLUTION

We solve this by changing it to an equivalent equation with all terms that have the variable x on one side and all constant terms on the other.

$$7x - 4 = 3x + 8$$
$$(7x - 4) + 4 = (3x + 8) + 4 \qquad \text{Add 4}$$
$$7x = 3x + 12 \qquad \text{Simplify}$$
$$7x - 3x = (3x + 12) - 3x \qquad \text{Subtract } 3x$$
$$4x = 12 \qquad \text{Simplify}$$
$$\tfrac{1}{4} \cdot 4x = \tfrac{1}{4} \cdot 12 \qquad \text{Multiply by } \tfrac{1}{4}$$
$$x = 3 \qquad \text{Simplify}$$

Since all of these equations are equivalent, the solution is 3. To verify this answer, we *look back* (one of the principles of problem solving introduced on pages 61–63) by substituting $x = 3$ into the original equation.

$$7(3) - 4 \stackrel{?}{=} 3(3) + 8$$

$$17 = 17$$

The last statement is true, so $x = 3$ is the solution. ∎

Because checking the answer is so important, we do this frequently in the remaining examples. In these checks, LHS stands for "left-hand side" of the original equation, and RHS stands for "right-hand side."

When a linear equation involves fractions, solving the equation is usually easier if we first multiply each side by the lowest common denominator (LCD) of the fractions, as we see in the next example.

EXAMPLE 2 ■ **Solving an Equation that involves Fractions**

Solve the equation $\dfrac{x}{6} + \dfrac{2}{3} = \dfrac{3}{4}x$.

SOLUTION

The LCD of the denominators 6, 3, and 4 is 12, so we first multiply each side of the equation by 12 to clear the denominators.

$$12 \cdot \left(\frac{x}{6} + \frac{2}{3} \right) = 12 \cdot \frac{3}{4}x \qquad \text{Multiply by LCD}$$

$$2x + 8 = 9x \qquad \text{Distributive Property}$$

$$(2x + 8) - 2x = 9x - 2x \qquad \text{Subtract } 2x$$

$$8 = 7x \qquad \text{Simplify}$$

$$\frac{8}{7} = \frac{7x}{7} \qquad \text{Divide by 7}$$

$$\frac{8}{7} = x \qquad \text{Simplify}$$

The solution is $x = \frac{8}{7}$.

CHECK YOUR ANSWER

$x = \frac{8}{7}$: $\quad$ LHS $= \dfrac{\frac{8}{7}}{6} + \dfrac{2}{3} = \dfrac{4}{21} + \dfrac{2}{3}$ $\qquad$ RHS $= \dfrac{3}{4}\left(\dfrac{8}{7}\right) = \dfrac{24}{28} = \dfrac{6}{7}$

$\qquad\qquad\qquad = \dfrac{4 + 14}{21} = \dfrac{18}{21} = \dfrac{6}{7}$

LHS = RHS ✓

∎

Euclid (circa 300 B.C.) taught in Alexandria. His *Elements* is the most widely influential scientific book in history. For 2000 years it was the standard introduction to geometry in the schools and for many generations was considered the best way to develop logical reasoning. Abraham Lincoln, for instance, studied the *Elements* as a way to sharpen his mind. The story is told that King Ptolemy once asked Euclid if there was a faster way to learn geometry than through the *Elements*. Euclid replied that there is "no royal road to geometry"—meaning by this that mathematics does not respect wealth or social status. Euclid was revered in his own time and was referred to by the title "The Geometer" or "The Writer of the *Elements*." The greatness of the *Elements* stems from its precise, logical, and systematic treatment of geometry. For dealing with equality, Euclid lists the following rules, which he calls "common notions."

1. Things that are equal to the same thing are equal to each other.

2. If equals are added to equals, the sums are equal.

3. If equals are subtracted from equals, the remainders are equal.

4. Things which coincide with one another are equal.

5. The whole is greater than the part.

In the next example we solve an equation that doesn't look like a linear equation, but it does simplify to an equivalent linear equation.

EXAMPLE 3 ■ An Equation that Reduces to a Linear Equation

Solve the equation $\dfrac{x}{x + 1} = \dfrac{2x + 1}{2x - 3}$.

SOLUTION

If $x \neq -1$ and $x \neq \frac{3}{2}$, then the denominators of the fractions in this equation are not 0, so we can multiply each side of the equation by the LCD, which is $(x + 1)(2x - 3)$.

$$(x + 1)(2x - 3)\left(\frac{x}{x + 1}\right) = (x + 1)(2x - 3)\left(\frac{2x + 1}{2x - 3}\right) \qquad \text{Multiply by LCD}$$

$$(2x - 3)x = (x + 1)(2x + 1) \qquad \text{Simplify}$$

$$2x^2 - 3x = 2x^2 + 3x + 1 \qquad \text{Expand}$$

$$-3x = 3x + 1 \qquad \text{Subtract } 2x^2$$

$$-6x = 1 \qquad \text{Subtract } 3x$$

$$x = -\frac{1}{6} \qquad \text{Divide by } -6$$

The solution is $-\frac{1}{6}$. ■

CHECK YOUR ANSWER

$x = -\frac{1}{6}$:

$$\text{LHS} = \frac{-\frac{1}{6}}{\left(-\frac{1}{6}\right) + 1} = \frac{-\frac{1}{6}}{\frac{5}{6}} = -\frac{1}{5}$$

$$\text{RHS} = \frac{2\left(-\frac{1}{6}\right) + 1}{2\left(-\frac{1}{6}\right) - 3} = \frac{\frac{4}{6}}{-\frac{20}{6}}$$

$$= -\frac{1}{5}$$

$$\text{LHS} = \text{RHS} \qquad \checkmark$$

It is always important to check your final answer (even if you never make a mistake in your calculations!), because sometimes extraneous (false) solutions can be introduced when equations are simplified. The next example shows how this can happen.

EXAMPLE 4 ■ An Equation with No Solution

Solve $2 + \dfrac{5}{x - 4} = \dfrac{x + 1}{x - 4}$.

SOLUTION

First, we multiply each side by the common denominator, which is $x - 4$.

$$(x - 4)\left(2 + \frac{5}{x - 4}\right) = (x - 4)\left(\frac{x + 1}{x - 4}\right) \qquad \text{Multiply by } x - 4$$

$$2(x - 4) + 5 = x + 1 \qquad \text{Expand}$$

CHECK YOUR ANSWER

$x = 4$:

$$\text{LHS} = 2 + \frac{5}{4 - 4} = 2 + \frac{5}{0}$$

$$\text{RHS} = \frac{4 + 1}{4 - 4} = \frac{5}{0}$$

Impossible—can't divide by 0.
LHS and RHS are undefined, so
$x = 4$ is not a solution.   ✕

$2x - 8 + 5 = x + 1$	Distributive Property
$2x - 3 = x + 1$	Simplify
$2x = x + 4$	Add 3
$x = 4$	Subtract x

But now if we try to substitute $x = 4$ back into the original equation, we would be dividing by 0, which is impossible. So, this equation has no solution.   ■

The first step in the preceding solution, multiplying by $x - 4$, had the effect of multiplying by 0. (Do you see why?) Multiplying each side of an equation by an expression that contains the variable may introduce extraneous solutions. That is why it is important to check every answer.

EXAMPLE 5 ■ **Solving for one Variable in Terms of Others**

The surface area A of the closed rectangular box shown in Figure 1 can be calculated from the length l, the width w, and the height h according to the formula

$$A = 2lw + 2wh + 2lh$$

Solve for w in terms of the other variables in this equation.

SOLUTION

Although this equation involves more than one variable, we solve it as usual by isolating w on one side, treating the other variables as we would numbers.

$A = (2lw + 2wh) + 2lh$	Collect terms involving w
$A - 2lh = 2lw + 2wh$	Subtract $2lh$
$A - 2lh = (2l + 2h)w$	Factor w from RHS
$\dfrac{A - 2lh}{2l + 2h} = w$	Divide by $2l + 2h$

The solution is $w = \dfrac{A - 2lh}{2l + 2h}$.   ■

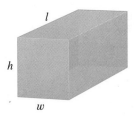

FIGURE 1

A closed rectangular box

It is, of course, irrelevant which letters we use for the variables in an algebra problem—the letters have no effect on the solution. Nevertheless, it is helpful in applied problems to use letters that are somehow related to the quantities they represent. In other problems, it is customary to use letters from the end of the alphabet $(\ldots, x, y, z)$ for unknown quantities that we are solving for, and to use other letters $(a, b, c, \ldots)$ to represent quantities whose values we know or can easily determine.

2.1   **EXERCISES**

1–6 ■ Determine whether each of the given values of the variable is a solution of the equation.

1. $2x - 3 = x + 1$
  (a) $x = 4$ (b) $x = \frac{3}{2}$

2. $4(x - 1) - (2 - x) = 5(x - 2) + 4$
  (a) $x = -3$ (b) $x = 0$

3. $\dfrac{1}{x} - \dfrac{1}{x+3} = \dfrac{1}{6}$
  (a) $x = -3$ (b) $x = 3$

4. $1 - [2 - (3 - x)] = 4x - (6 + x)$
  (a) $x = 2$ (b) $x = 22$

5. $Ax + By + C = 0$ $(A \neq 0, B \neq 0, x \neq 0, y \neq 0)$
  (a) $x = \dfrac{By - C}{A}$ (b) $y = -\dfrac{Ax + C}{B}$

6. $\dfrac{ax - b}{bx - a} = \dfrac{a}{b}$ $(a^2 \neq b^2, a \neq 0, b \neq 0)$
  (a) $x = 1$ (b) $x = b/a$

7–12 ■ Determine whether the equation is an identity.

7. $x^2 + 16 = 32$

8. $4x - 9 = x + 5$

9. $\dfrac{x^2 - 9}{x + 3} = x - 3$ $(x \neq -3)$

10. $\dfrac{1}{x} + \dfrac{1}{2} = \dfrac{1}{x + 2}$

11. $\dfrac{y + 16}{2} - \dfrac{3}{2}y = 8 - y$

12. $\sqrt{4 - x^2} = 2 - x$ $(-2 \leqslant x \leqslant 2)$

13–18 ■ State whether the equation is equivalent to a linear equation. If so, solve the equation.

13. $5x - 3 = 2x + 12$

14. $\dfrac{x}{4} + \dfrac{2}{3} = \dfrac{3x}{2} - \dfrac{1}{6}$

15. $y^2 - 7y + 4 = 2y^2 + 11$

16. $\dfrac{z^2 - 25}{z + 25} = z + 1$

17. $\dfrac{x}{x - 4} = 1 + \dfrac{3}{x + 2}$

18. $2x - 1 = \sqrt{x^2 - 3}$

19–56 ■ Solve the equation.

19. $3x - 5 = 7$ **20.** $4x + 12 = 28$

21. $x - 3 = 2x + 6$ **22.** $4x + 7 = 9x - 13$

23. $-7w = 15 - 2w$ **24.** $5t - 13 = 12 - 5t$

25. $\frac{1}{2}y - 2 = \frac{1}{3}y$ **26.** $\dfrac{z}{5} = \dfrac{3}{10}z + 7$

27. $2(1 - x) = 3(1 + 2x) + 5$

28. $5(x + 3) + 9 = -2(x - 2) - 1$

29. $4\left(y - \frac{1}{2}\right) - y = 6(5 - y)$

30. $\dfrac{2}{3}y + \dfrac{1}{2}(y - 3) = \dfrac{y + 1}{4}$

31. $\dfrac{1}{x} = \dfrac{4}{3x} + 1$ **32.** $\dfrac{2x - 1}{x + 2} = \dfrac{4}{5}$

33. $\dfrac{2}{t + 6} = \dfrac{3}{t - 1}$ **34.** $\dfrac{1}{t - 1} + \dfrac{t}{3t - 2} = \dfrac{1}{3}$

35. $r - 2[1 - 3(2r + 4)] = 61$

36. $(t - 4)^2 = (t + 4)^2 + 32$

37. $\sqrt{3}\,x + \sqrt{12} = \dfrac{x + 5}{\sqrt{3}}$ **38.** $\frac{2}{3}x - \frac{1}{4} = \frac{1}{6}x - \frac{1}{9}$

39. $\dfrac{2}{x} - 5 = \dfrac{6}{x} + 4$ **40.** $\dfrac{6}{x - 3} = \dfrac{5}{x + 4}$

41. $\dfrac{3}{x + 1} - \dfrac{1}{2} = \dfrac{1}{3x + 3}$

42. $\dfrac{4}{x - 1} + \dfrac{2}{x + 1} = \dfrac{35}{x^2 - 1}$

43. $\dfrac{2x - 7}{2x + 4} = \dfrac{2}{3}$

44. $\dfrac{12x - 5}{6x + 3} = 2 - \dfrac{5}{x}$

45. $x - \frac{1}{3}x - \frac{1}{2}x - 5 = 0$

46. $2x - \dfrac{x}{2} + \dfrac{x + 1}{4} = 6x$

47. $\dfrac{1}{z} - \dfrac{1}{2z} - \dfrac{1}{5z} = \dfrac{10}{z + 1}$

48. $\dfrac{1}{1 - \dfrac{3}{2 + w}} = 60$

49. $\dfrac{u}{u - \dfrac{u + 1}{2}} = 4$

50. $\dfrac{1}{3 - t} + \dfrac{4}{3 + t} + \dfrac{16}{9 - t^2} = 0$

51. $\sqrt{x - 4} = \sqrt{2x}$

52. $\sqrt{2x + 8} = \sqrt{6x}$

53. $\dfrac{x}{2x - 4} - 2 = \dfrac{1}{x - 2}$

54. $\dfrac{1}{x + 3} + \dfrac{5}{x^2 - 9} = \dfrac{2}{x - 3}$

55. $\dfrac{3}{x + 4} = \dfrac{1}{x} + \dfrac{6x + 12}{x^2 + 4x}$

56. $\dfrac{1}{x} - \dfrac{2}{2x + 1} = \dfrac{1}{2x^2 + x}$

57–60 ■ Find the solution of the equation correct to two decimals.

57. $2.15x - 4.63 = x + 1.19$

58. $3.95 - x = 2.32x + 2.00$

59. $3.16(x + 4.63) = 4.19(x - 7.24)$

60. $\dfrac{0.26x - 1.94}{3.03 - 2.44x} = 1.76$

61–70 ■ Solve the equation for the indicated variable.

61. $PV = nRT$; for R

62. $F = G\dfrac{mM}{r^2}$; for m

63. $\dfrac{1}{R} = \dfrac{1}{R_1} + \dfrac{1}{R_2}$; for R_1

64. $P = 2l + 2w$; for w

65. $\dfrac{ax + b}{cx + d} = 2$; for x

66. $a - 2[b - 3(c - x)] = 6$; for x

67. $a^2x + (a - 1) = (a + 1)x$; for x

68. $\dfrac{a + 1}{b} = \dfrac{a - 1}{b} + \dfrac{b + 1}{a}$; for a

69. $(4m + a)(6m - a) = (3m + 2a)(8m - 2a)$; for m

70. $\dfrac{x + a}{x + 2a} = \dfrac{x + b}{x + 2b}$; for x (Assume $a \neq b$.)

71–72 ■ Find a value for k that will make the given value of x a solution of the equation.

71. $3x + k - 5 = kx - k + 1$; $x = 2$

72. $\dfrac{kx + 1}{kx - 1} = 2x$; $x = -1$

73. Find the mistake in the following solution, and then solve the equation correctly:

$$x^2 + 6x + 5 = x^2 - 1$$
$$(x + 5)(x + 1) = (x - 1)(x + 1) \qquad \text{Factor}$$
$$x + 5 = x - 1 \qquad \text{Divide by } x + 1$$
$$5 \overset{?}{=} -1 \qquad \text{Subtract } x$$

74. The formula for the volume of a right circular cone is $V = \frac{1}{3}\pi r^2 h$, where r is the radius of the base and h is the height. Find the height (correct to two decimals) of a cone with volume 30.00 in³ and radius 4.00 in.

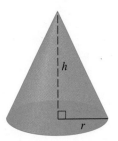

2.2 PROBLEM SOLVING WITH LINEAR EQUATIONS

Many problems in the sciences, economics, finance, medicine, and numerous other fields can be translated into algebra problems: This is one reason that algebra is so useful. In this section we consider "word" problems that lead to linear equations.

EXAMPLE 1 ■ Interest on an Investment

Mary inherits $100,000 and invests it in two certificates of deposit. One certificate pays 6% and the other pays $4\frac{1}{2}$% simple interest annually. If Mary's total interest is $5025 per year, how much money is invested at each rate?

SOLUTION

The key to solving any word problem is to translate the information given into the language of mathematics, that is, into an equation. First, we need to identify the variables. This can usually be done by a careful reading of the question asked in the problem. Here we are asked to find out how much money is invested at each rate. So we let

$$x = \text{amount invested at } 6\%$$

The rest of Mary's money is invested at $4\frac{1}{2}$%, so

$$100{,}000 - x = \text{amount invested at } 4\tfrac{1}{2}\%$$

Now we translate the fact that her total annual interest is $5025 into an equation:

$$(\text{interest at } 6\%) + (\text{interest at } 4\tfrac{1}{2}\%) = 5025$$

Since x dollars are invested at 6%, the annual interest paid by this certificate is 6% of x, or $0.06x$. Similarly, the interest received from the $4\frac{1}{2}$% certificate will be $0.045(100{,}000 - x)$, and so the interest equation is expressed in terms of x as

$$0.06x + 0.045(100{,}000 - x) = 5025$$

Now we solve for x:

$$0.06x + 4500 - 0.045x = 5025 \qquad \text{Multiply}$$
$$0.015x + 4500 = 5025 \qquad \text{Combine the } x \text{ terms}$$
$$0.015x = 525 \qquad \text{Subtract 4500}$$
$$x = \frac{525}{0.015} = 35{,}000$$

So Mary has invested $35,000 at 6% and the remaining $65,000 at $4\frac{1}{2}$%. ■

Simple Interest

Interest = Principal × rate × time

$I = Prt$

CHECK YOUR ANSWER

total interest = 6% of $35,000

\+

$4\frac{1}{2}$% of $65,000

= $2100 + $2925

= $5025   ✓

When students first begin solving word problems they often have difficulty translating a word problem into an equation. Translating from English to mathematics is a skill that can be developed only through practice. After considering a few more examples we will list some general principles to guide your thinking in this translation process.

EXAMPLE 2 ■ Average Test Score

A student has scores of 79, 81, and 72 on his first three tests. He needs an average of at least 80 to earn a grade of B. What score does he need to make on his fourth test to raise his average test score to 80?

SOLUTION

We are asked to find an unknown test score, so we let

$$x = \text{score needed on fourth test to raise average score to 80}$$

Using the fact that

$$\text{average} = \frac{\text{sum of scores}}{\text{number of scores}}$$

we get the equation

$$80 = \frac{79 + 81 + 72 + x}{4}$$

$$320 = 232 + x \qquad \text{Multiply by 4 and simplify}$$

$$88 = x \qquad \text{Subtract 232}$$

The student must earn a score of 88 on his fourth test to raise his average test score to 80. ■

EXAMPLE 3 ■ Dimensions of a Poster

A poster has on it a rectangular printed area 100 cm by 140 cm, with a blank strip of uniform width around the four edges. The perimeter of the poster is $1\frac{1}{2}$ times the perimeter of the printed area. What is the width of the blank strip, and what are the dimensions of the poster?

SOLUTION

In a problem such as this that involves geometry, it is essential to draw a diagram like the one shown in Figure 1. Let

$$x = \text{width of the blank strip}$$

From the figure we see that the poster is $(100 + 2x)$ cm by $(140 + 2x)$ cm, so its perimeter is $2(100 + 2x) + 2(140 + 2x)$. The perimeter of the printed area

---

CHECK YOUR ANSWER

average test score

$$= \frac{79 + 81 + 72 + 88}{4}$$

$$= \frac{320}{4} = 80 \qquad \checkmark$$

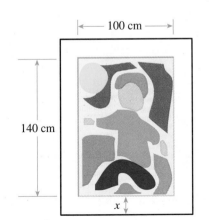

FIGURE 1

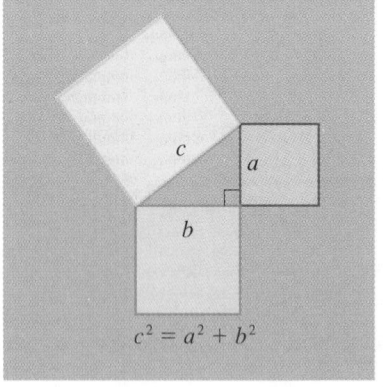

$$c^2 = a^2 + b^2$$

is $2(100) + 2(140) = 480$ cm. We are told that

$$\text{(perimeter of poster)} = \tfrac{3}{2} \times \text{(perimeter of printed part)}$$

so

$$2(100 + 2x) + 2(140 + 2x) = \tfrac{3}{2} \cdot 480$$

$$480 + 8x = 720 \qquad \text{Expand and combine like terms on LHS}$$

$$8x = 240 \qquad \text{Subtract 480}$$

$$x = 30 \qquad \text{Divide by 8}$$

The blank strip is 30 cm wide, so the poster is

$$100 + 30 + 30 = 160 \text{ cm wide}$$

by

$$140 + 30 + 30 = 200 \text{ cm high} \qquad \blacksquare$$

We now adapt the problem-solving principles given on pages 61–63 to the process of translating a word problem from English into mathematics. To solve any problem, the first step is to read it carefully to make sure you understand it. In particular, you must understand exactly what you are being asked to find. The following guidelines should help you to set up the equation that expresses the English statement of a problem in the language of algebra.

GUIDELINES FOR SOLVING WORD PROBLEMS

1. IDENTIFY THE VARIABLE. Identify the quantity that the problem asks you to find. This quantity can usually be determined by a careful reading of the question posed at the end of the problem. Give this quantity a name (x or some other variable). Make sure to write down precisely what the variable represents.

2. EXPRESS ALL UNKNOWN QUANTITIES IN TERMS OF THE VARIABLE. Read each sentence in the problem again, and express all the quantities mentioned in the problem in terms of the variable you defined in Step 1. For instance, in Example 1 we expressed the principal invested at each interest rate and the annual interest paid by each certificate in terms of the variable x. Sometimes it is helpful to organize this information into a chart or diagram, as we did in Example 3.

3. RELATE THE QUANTITIES. Find the crucial fact in the problem that relates two or more of the expressions you listed in Step 2. For instance, in Example 1 we were told that the total interest from the two certificates was $5025. This fact produced the equation for that problem. A statement that one quantity "is", "equals", or "is the same as" another often signals the type of relationship we are looking for in this step.

4. SET UP AN EQUATION. Set up an equation that expresses the crucial fact you found in Step 3 in algebraic form. You will often need to use a formula to translate from English to algebra. For instance, in Example 3 we needed to know that the perimeter of a rectangle is twice the length plus twice the width. This can be expressed as $P = 2L + 2W$.

5. SOLVE THE PROBLEM AND CHECK YOUR ANSWER. Solve the equation, and check to make sure your answer satisfies the original problem posed.

The next example deals with distance, rate (speed), and time. The formula to keep in mind here is

$$\text{distance} = \text{rate} \times \text{time}$$

where the rate is either the constant speed or average speed of a moving object. For example, driving at 60 mi/h for 4 h takes you a distance of $60 \cdot 4 = 240$ mi.

EXAMPLE 4 ■ Distance, Rate, and Time

Bill left his house at 2:00 P.M. and rode his bicycle down Main Street at a speed of 12 mi/h. When his friend Mary arrived at his house at 2:10 P.M., Bill's mother told her the direction in which Bill went, and Mary cycled after him at a speed of 16 mi/h. At what time did Mary catch up with Bill?

SOLUTION

Identify the variable

Let t be the time (in hours) that it took Mary to catch up with Bill. Because Bill had a 10-minute, or $\frac{1}{6}$-hour head start, he cycled for $(t + \frac{1}{6})$ hours.

In problems involving motion it is often helpful to organize the information in a table, using the formula

$$\text{distance} = \text{rate} \times \text{time}$$

First we fill in the "Rate" column in the table, since we are told the speeds at which Mary and Bill cycled. Then we fill in the "Time" column, since we know what their travel times were in terms of t. Finally, we multiply these columns to calculate the entries in the "Distance" column.

Express all unknown quantities in terms of the variable

	Distance (mi)	Rate (mi/h)	Time (h)
Mary	$16t$	16	t
Bill	$12(t + \frac{1}{6})$	12	$t + \frac{1}{6}$

To find t we use the fact that at the instant Mary overtook Bill they had cycled the same distance:

distance traveled by Mary = distance traveled by Bill

$16t = 12\left(t + \frac{1}{6}\right)$	From table
$16t = 12t + 2$	Distributive Property
$4t = 2$	Subtract $12t$
$t = \frac{1}{2}$	Divide by 4

Relate the quantities

Set up an equation

Solve

Mary caught up with Bill after cycling for half an hour, that is, at 2:40 P.M.

CHECK YOUR ANSWER ■ Since Bill traveled for $\frac{1}{2} + \frac{1}{6} = \frac{2}{3}$ h,

distance Bill traveled = 12 mi/h $\times \frac{2}{3}$ h = 8 mi

distance Mary traveled = 16 mi/h $\times \frac{1}{2}$ h = 8 mi

Distances are equal. ✓

EXAMPLE 5 ■ The Sum of Three Integers

The sum of three consecutive even integers is 288. What are the integers?

SOLUTION

Identify the variable

Express all unknown quantities in terms of the variable

Set up an equation

If we call the first integer x, then the others must be $x + 2$ and $x + 4$, since consecutive even integers are two units apart. This means that we must solve the equation

$x + (x + 2) + (x + 4) = 288$	Sum is 288
$3x + 6 = 288$	Combine like terms
$3x = 282$	Subtract 6
$x = 94$	Divide by 3

Solve

The integers are 94, 96, and 98.

EXAMPLE 6 ■ Mixtures and Concentration

A manufacturer of soft drinks makes a type of orange soda that is advertised as "naturally flavored," although it contains only 5% orange juice. A new federal regulation stipulates that to be called "natural," a drink must contain at least 10% fruit juice. How much pure orange juice must this manufacturer add to 900 gal of orange soda to conform to the new regulation?

SOLUTION

In any problem of this type—in which two different substances are to be mixed—a diagram helps us set up the required equation (see Figure 2).

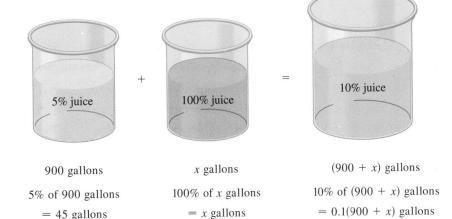

	5% juice	+	100% juice	=	10% juice

Volume	900 gallons	x gallons	$(900 + x)$ gallons
Amount of orange juice	5% of 900 gallons = 45 gallons	100% of x gallons = x gallons	10% of $(900 + x)$ gallons = $0.1(900 + x)$ gallons

FIGURE 2

Identify the variable

Express all unknown quantities in terms of the variable

Set up an equation

Solve

Let x be the amount (in gallons) of pure juice to be added. Then $(900 + x)$ gallons of 10% orange juice mixture will result. The key idea to turn the picture into an equation here is to notice that the total amount of juice on both sides of the equal sign is the same. The orange juice in the first vat is 5% of 900 gal, or 45 gal. The second vat contains x gallons of juice, and the third contains $0.1(900 + x)$ gallons.

Equating the total amounts of pure juice before and after mixing, we get the equation

$$45 + x = 0.1(900 + x) \qquad \text{From Figure 2}$$
$$45 + x = 90 + 0.1x \qquad \text{Multiply}$$
$$0.9x = 45 \qquad \text{Subtract } 0.1x \text{ and } 45$$
$$x = \frac{45}{0.9} = 50 \qquad \text{Divide by } 0.9$$

The manufacturer should add 50 gallons of pure orange juice to the soda.

CHECK YOUR ANSWER

amount of juice before mixing = 5% of 900 gal + 50 pure gal
= 45 gal + 50 gal = 95 gal

amount of juice after mixing = 10% of 950 gal = 95 gal

Amounts are equal. ✓

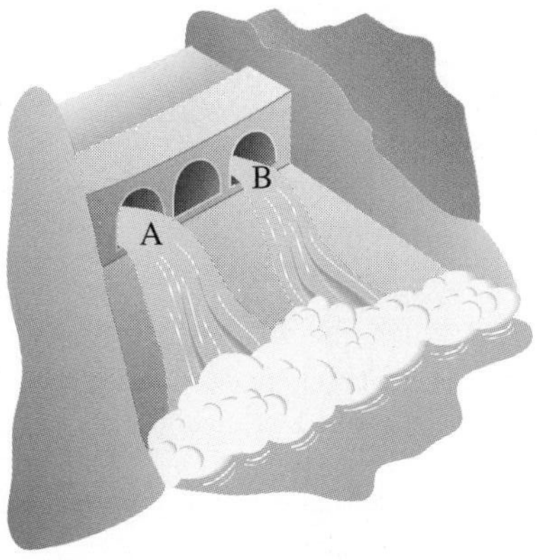

EXAMPLE 7 ■ Time Needed to Do a Job

Because of an anticipated heavy rainstorm, the water level in a reservoir must be lowered by 1 ft. Opening spillway A lowers the level by this amount in 4 h, whereas opening the smaller spillway B does the job in 6 h. How long will it take to lower the water level by 1 ft if both spillways are opened?

FIGURE 3

SOLUTION

As usual, we represent the number we are seeking by x:

Identify the variable

$$x = \text{number of hours it takes to lower the water level}$$
$$\text{by 1 ft if both spillways are open}$$

Finding an equation relating x to the other quantities in this problem is not easy. Certainly x is not simply $4 + 6$, since that would mean that both spillways together require longer to lower the water level than either spillway alone. Instead, *we look at the fraction of the job that can be done in one hour by each spillway.*

Relate the quantities

$$\text{Spillway A lowers the water by } \tfrac{1}{4} \text{ ft in 1 h}$$

$$\text{Spillway B lowers the water by } \tfrac{1}{6} \text{ ft in 1 h}$$

$$\text{Both spillways lower the water by } \frac{1}{x} \text{ ft in 1 h}$$

Therefore, we get the equation

$$(\text{fraction done by A}) + (\text{fraction done by B}) = (\text{fraction done by both})$$

Set up an equation

$$\frac{1}{4} + \frac{1}{6} = \frac{1}{x}$$

$$3x + 2x = 12 \qquad \text{Multiply by the LCD, } 12x$$

$$5x = 12 \qquad \text{Add}$$

Solve

$$x = \frac{12}{5} \qquad \text{Divide by 5}$$

It will take $2\tfrac{2}{5}$ h, or 2 h 24 min to lower the water level by 1 ft if both spillways are open. ∎

2.2 EXERCISES

1–10 ■ Express the given quantity in terms of the indicated variable.

1. The interest obtained after two years on an investment at 7% simple interest per year; x = the number of dollars invested

2. The average of three test scores if the first two scores are 78 and 82; s = the third test score

3. The sum of three consecutive odd integers; n = the first integer of the three

4. The perimeter (in cm) of a rectangle that is 5 cm longer than it is wide; w = width of the rectangle (in cm)

5. The area (in in^2) of a rectangle that is 50 in. long; w = width of the rectangle (in inches)

6. The time (in hours) it takes to travel a given distance at 55 mi/h; d = the given distance (in mi)

7. The distance (in miles) traveled when driving at a certain speed for 2 h, then driving 15 mi/h faster for another hour; s = initial speed (in mi/h)

8. The concentration (in oz/gal) of salt in a mixture of 3 gal of brine containing 25 oz of salt, to which has been added some pure water; x = volume of pure water added (in gal)

9. The average age of three sisters if the second was born 3 years after the first and the third was born 2 years after the second; a = age of firstborn (in years)

10. The value (in cents) of the change in a purse that contains twice as many nickels as pennies, four more dimes than nickels, and as many quarters as dimes and nickels combined; p = number of pennies

11–55 ■ Use the problem-solving principles described in this section to answer the question posed.

11. Phyllis invested $12,000, a portion earning a simple interest rate of $4\frac{1}{2}$% per year and the rest earning a rate of 4% per year. After one year the total interest earned on these investments was $525. How much money did she invest at each rate?

12. If Ben invests $4000 at 4% interest per year, how much additional money must he invest at $5\frac{1}{2}$% annual interest to ensure that the interest he receives each year is $4\frac{1}{2}$% of the total amount invested?

13. A rectangular garden is 25 ft wide. If its area is 1125 ft², what is the length of the garden?

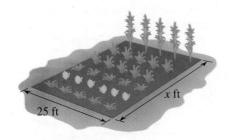

14. During his major league career, Hank Aaron hit 31 more home runs than Babe Ruth hit during his career.

Together they hit 1459 home runs. How many home runs did Babe Ruth hit?

15. A plumber and his assistant work together to replace the pipes in an old house. The plumber charges $45 per hour for his own labor and $25 per hour for his assistant's labor. The plumber works twice as long as his assistant on this job, and the labor charge on the final bill is $4025. How long did the plumber and his assistant work on this job?

16. Helen earns $7.50 per hour at her job, but if she works more than 35 hours in a week she is paid $1\frac{1}{2}$ times her regular salary for the overtime hours worked. One week her gross pay was $352.50. How many overtime hours did she work that week?

17. A change purse contains an equal number of pennies, nickels, and dimes. The total value of the coins is $1.44. How many coins of each type does the purse contain?

18. Mary has $3.00 made up of nickels, dimes, and quarters. If she has twice as many dimes as quarters and five more nickels than dimes, how many coins of each type does she have?

19. The oldest child in a family of four children is twice as old as the youngest. The two middle children are 10 and 11 years old. If the average age of the children is $10\frac{1}{2}$ years, how old is the youngest child?

20. A movie star, unwilling to give his age, posed the following riddle to a gossip columnist. "Seven years ago, I was eleven times as old as my daughter. Now I am four times as old as she is." How old is the star?

21. A merchant blends tea that sells for $3.00 a pound with tea that sells for $2.75 a pound to produce 80 lb of a mixture that sells for $2.90 a pound. How many pounds of each type of tea does the merchant use in the blend?

22. Find three consecutive integers whose sum is 336.

23. Find four consecutive odd integers whose sum is 272.

24. A room is $1\frac{1}{2}$ times as long as it is wide. Its perimeter is 80 feet. How wide is the room?

25. Al paints with watercolors on a sheet of paper 20 in. wide by 15 in. high. He then places this sheet on a mat so that a uniformly wide strip of the mat shows all around the picture. The perimeter of the mat is 102 in. How wide is the strip of the mat showing around the picture?

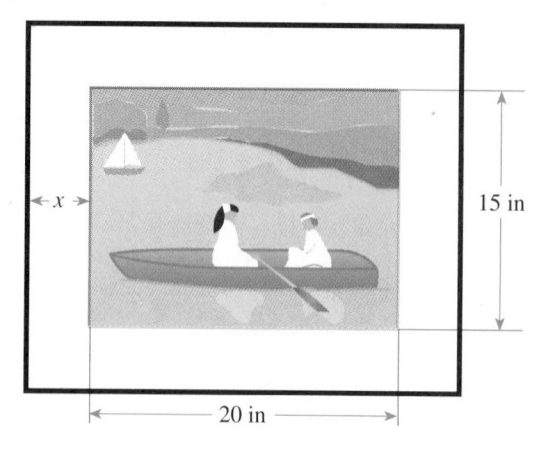

26. Dr. Plath has been raising fruit flies for use in a biology experiment. She has noticed that the number of new baby flies in her colony each day is always half the previous day's total population, but that 200 adult flies die each day. If after three days she has 3100 fruit flies, how many did she start with?

27. A restaurant is open five days per week, from Tuesday through Saturday. During a certain week, gross receipts averaged $1200 per day. If the restaurant sold $650 in meals on Tuesday, $550 on Wednesday, and $300 on Thursday, and if Saturday's gross receipts were twice those of Friday, how much gross income did the restaurant earn on Saturday?

28. Rochelle's psychology professor gives a grade of A for an average of 90 or better and a grade of B for an average of 80 or better in her course. Rochelle has obtained scores of 82, 75, and 71 on her midterm examinations. If the final exam counts twice as much as a midterm, what score must she make on her final exam to earn a grade of B? to earn an A? (Assume that the maximum possible score on each test is 100.)

29. Light travels at about 3.0×10^8 m/s. The distance from the sun to the earth is about 1.5×10^{11} m. How long does it take for the sun's light to reach the earth?

30. The fuel consumption for William's car is 30 mi/gal on the highway and 25 mi/gal in the city. On a vacation trip of 400 mi he used 14 gal of gasoline. How many highway miles did he drive on this trip?

31. A commercial jet took off from Kansas City for San Francisco, a distance of 2550 km, at a speed of 800 km/h. At the same time a private jet, traveling at 900 km/h, left San Francisco for Kansas City. How long after takeoff will the jets pass each other?

32. After robbing a bank in Dodge City, the robber gallops off at 14 mi/h. Ten minutes later the marshal leaves to pursue him at 16 mi/h. How long does it take the marshal to catch up with the bank robber?

33. Wilma drove at an average speed of 50 mi/h from her home in Boston to visit her sister in Buffalo. She stayed in Buffalo 10 h, and on the trip back averaged 45 mi/h. She returned home 29 h after leaving. How many miles is Buffalo from Boston?

34. A coast guard boat that patrols a river separating two countries cruises at 20 knots (nautical miles per hour) in still water. The river flows at 4 knots. For each patrol the captain of the boat is ordered to travel upstream and then return. His trip takes exactly 6 h. How far upstream does the boat travel?

35. Wendy took a trip from Davenport to Omaha, a distance of 300 mi. She traveled part of the way by bus, which arrived at the train station just in time for Wendy to complete her journey by train. The bus averaged 40 mi/h and the train 60 mi/h. The entire trip took $5\frac{1}{2}$ h. How long did Wendy spend on the train?

36. Two cyclists, 90 mi apart, start riding toward each other at the same time. One cycles twice as fast as the other. If they meet 2 h later, at what average speed is each cyclist traveling?

37. A pilot flew a jet from Montreal to Los Angeles, a distance of 2500 mi. On the return trip the average speed was 20% faster than the outbound speed. The round trip took 9 h 10 min. What was the speed from Montreal to Los Angeles?

38. A woman driving a car 14 ft long is passing a truck 30 ft long. The truck is traveling at 50 mi/h. How fast must the woman drive her car so that she can pass the truck completely in 6 s, from the position shown in

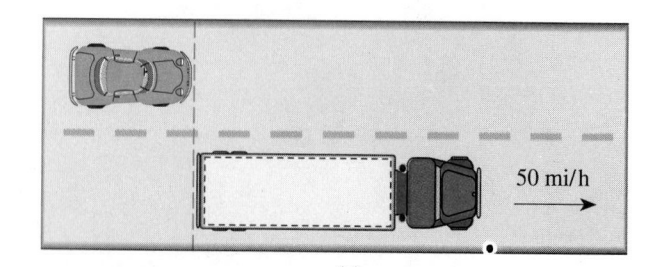

(a)

figure (a) to the position shown in figure (b)?   [*Hint:* Use feet and seconds instead of miles and hours.]

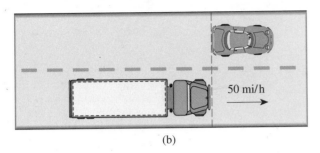

(b)

39. What quantity of a 60% acid solution must be mixed with a 30% solution to produce 300 mL of a 50% solution?

40. A pot contains 6 L of brine at a concentration of 120 g/L. How much of the water should be boiled off to increase the concentration to 200 g/L?

41. A jeweler has five rings, each weighing 18 g, made of an alloy of 10% silver and 90% gold. He wishes to melt down the rings and add enough silver to reduce the gold content to 75%. How much silver should he add?

42. A health clinic uses a solution of bleach to sterilize petri dishes in which cultures are grown. The sterilization tank contains 100 gal of a solution of 2% ordinary household bleach mixed with pure distilled water. New research indicates that the concentration of bleach should be 5% for complete sterilization. How much of the solution should be drained and replaced with bleach to increase the bleach content to the recommended level?

43. The radiator in a car is filled with a solution of 60% antifreeze and 40% water. The manufacturer of the antifreeze suggests that, for summer driving, optimal cooling of the engine is obtained with only 50% antifreeze. If the capacity of the radiator is 3.6 L, how much coolant should be drained and replaced with water to reduce the antifreeze concentration to the recommended level?

44. A bottle contains 750 mL of fruit punch with a concentration of 50% pure fruit juice. Jill drinks 100 mL of the punch and then refills the bottle with an equal amount of a cheaper brand of punch. If the concentration of juice in the bottle is now reduced to 48%, what was the concentration in the punch that Jill added?

45. Candy and Tim share a paper route. It takes Candy 70 min to deliver all the papers, whereas Tim takes 80 min. How long would it take them if they work together?

46. Stan and Hilda can mow the lawn in 40 min if they work together. If Hilda works twice as fast as Stan, how long would it take Stan to mow the lawn alone?

47. Betty and Karen have been hired to paint the houses in a new development. Working together the women can paint a house in two-thirds the time that it takes Karen working alone. Betty takes 6 h to paint a house alone. How long does it take Karen to paint a house working alone?

48. A man is walking away from a lamppost with a light source 6 m above the ground. The man is 2 m tall. How far from the lamppost is the man when his shadow is 5 m long?   [*Hint:* Use similar triangles.]

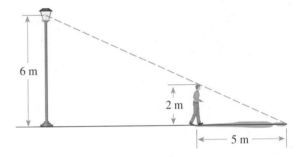

49. A woodcutter determines the height of a tall tree by first measuring a smaller one 125 ft away, then moving so that his eyes are in the line of sight along the tops of the trees, and measuring how far he is standing from the small tree (see the figure). Suppose the small tree is 20 ft tall, the man is 25 ft from the small tree, and his eye level is 5 ft above the ground. How tall is the taller tree?

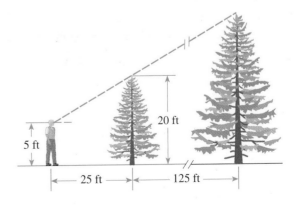

50. The hypotenuse of a right triangle is 3 cm longer than one of the legs, and the other leg is 10 cm long. How long is each of the sides of the triangle? (Express your answer correct to two decimals.)

51. A rectangular parcel of land is 50 ft wide. The length of a diagonal between opposite corners is 10 ft more than the length of the parcel. What is the length of the parcel?

52. A running track has the shape shown in the figure, with straight sides and semicircular ends. If the length of the track is 440 yd and the two straight parts are each 110 yd long, what is the radius of the semicircular parts (to the nearest yard)?

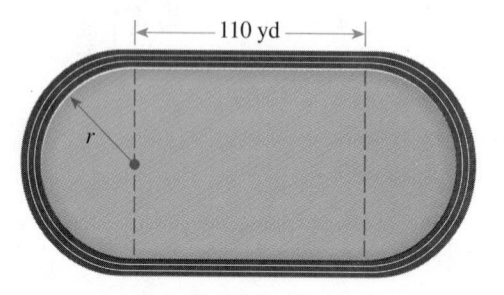

53. A storage bin for corn consists of a cylindrical section made of wire mesh, surmounted by a conical tin roof,

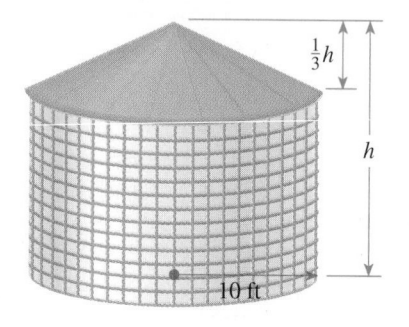

as shown in the figure. The height of the roof is one-third the entire height of the structure. If the total volume of the structure is 1400π ft^3 and its radius is 10 ft, what is its height? [*Hint:* Use the volume formulas on the inside of the front cover of this book.]

54. Next-door neighbors Bob and Jim use hoses from both houses to fill Bob's swimming pool. They know it takes 18 h using both hoses. They also know that Bob's hose, used alone, takes 20% less time than Jim's hose alone. How much time is required to fill the pool by each hose alone?

55. This problem is taken from a Chinese mathematics textbook called *Chui-chang suan-shu*, or *Nine Chapters on the Mathematical Art*, which was written about 250 B.C.

> A 10-ft-long stem of bamboo is broken in such a way that its tip touches the ground 3 ft from the base of the stem, as shown in the figure. What is the height of the break?

[*Hint:* Use the Pythagorean Theorem.]

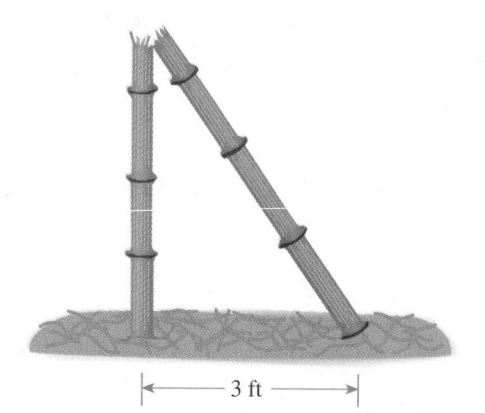

2.3   QUADRATIC EQUATIONS

In the last section we considered linear equations, which are first-degree equations of the form $ax + b = 0$. But applied problems often lead to equations that involve higher powers of the variable. For example, we will see in Example 7 that the motion of a projectile thrown or fired upward (such as a football or bullet) is determined by a second-degree equation. Also, in many cases, distance-rate-time problems—like the ones we considered in the previous section—actually lead to second-degree equations rather than linear equations (see Example 6). In this section we consider quadratic equations, which are second-degree equations; that is, they contain a term involving the square of the variable.

QUADRATIC EQUATIONS

A **quadratic equation** is an equation equivalent to one of the form

$$ax^2 + bx + c = 0$$

where a, b, and c are real numbers with $a \neq 0$.

SOLVING QUADRATIC EQUATIONS BY FACTORING

Some quadratic equations can be solved by factoring and using the following basic property of real numbers.

ZERO-PRODUCT PROPERTY

$$AB = 0 \quad \text{if and only if} \quad A = 0 \quad \text{or} \quad B = 0$$

This means that if we can factor the left-hand side of a quadratic (or other) equation, then we can solve it by setting each factor equal to 0 in turn. This method works only when the right-hand side of the equation is 0.

EXAMPLE 1 ■ Solving a Quadratic Equation by Factoring

Solve the equation $x^2 + 5x = 24$.

SOLUTION

We must first rewrite the equation so that the right-hand side is 0.

$$x^2 + 5x = 24$$

$$x^2 + 5x - 24 = 0 \qquad \text{Subtract 24}$$

$$(x - 3)(x + 8) = 0 \qquad \text{Factor}$$

$$x - 3 = 0 \quad \text{or} \quad x + 8 = 0 \qquad \text{Set each factor equal to 0}$$

$$x = 3 \qquad\qquad x = -8 \qquad \text{Solve}$$

The solutions are $x = 3$ and $x = -8$.

CHECK YOUR ANSWERS

$x = 3$:

$$(3)^2 + 5(3) = 9 + 15 = 24 \quad \checkmark$$

$x = -8$:

$$(-8)^2 + 5(-8) = 64 - 40 = 24 \quad \checkmark$$

Do you see why one side of the equation must be 0 in Example 1? Factoring the equation as $x(x + 5) = 24$ does not help us find the solutions, since 24 can be factored in infinitely many ways, such as $6 \cdot 4$, $\frac{1}{2} \cdot 48$, $(-\frac{2}{5}) \cdot (-60)$, and so on.

If a quadratic equation is of the form $x^2 - c = 0$, where c is a positive constant, then it factors as $(x - \sqrt{c})(x + \sqrt{c}) = 0$, and so the solutions are $x = \sqrt{c}$ and $x = -\sqrt{c}$. We often abbreviate this as $x = \pm\sqrt{c}$. [Remember that for $c > 0$ the symbol $\sqrt{c}$ represents the positive (or *principal*) square root of c.] Thus, such simple quadratic equations can be solved just by taking square roots.

SOLVING A SIMPLE QUADRATIC EQUATION

The solutions of the equation $x^2 = c$ are $x = \sqrt{c}$ and $x = -\sqrt{c}$.

EXAMPLE 2 ■ Solving Two Simple Quadratics

Solve each equation.

(a) $x^2 = 5$ (b) $(x - 4)^2 = 5$

SOLUTION

(a) From the principle in the preceding box, we can solve $x^2 = 5$ as

$$x = \pm\sqrt{5}$$

(b) We can take the square root of each side of this equation as well.

$$(x - 4)^2 = 5$$

$$x - 4 = \pm\sqrt{5} \qquad \text{Take square root}$$

$$x = 4 \pm \sqrt{5} \qquad \text{Add 4}$$

The solutions are $x = 4 + \sqrt{5}$ and $x = 4 - \sqrt{5}$.   ■

SOLVING QUADRATIC EQUATIONS BY COMPLETING THE SQUARE

As we saw in Example 2, if a quadratic equation is of the form $(x + a)^2 = c$, then we can solve it by taking the square root of each side. In an equation of this form, the left-hand side is a *perfect square*: the square of a linear expression in x. So, if a quadratic equation does not factor readily, then we can solve it using the technique of **completing the square.** This means that we add a constant to an expression to make it a perfect square. For example, to make $x^2 - 6x$ a perfect square we must add 9, since $x^2 - 6x + 9 = (x - 3)^2$. In general, from the identity

$$x^2 + bx + \left(\frac{b}{2}\right)^2 = \left(x + \frac{b}{2}\right)^2$$

it follows that to make $x^2 + bx$ a perfect square, we must add the square of half the coefficient of x. (Note that this works whether b is positive or negative.)

COMPLETING THE SQUARE

To make $x^2 + bx$ a perfect square, add $\left(\frac{b}{2}\right)^2$.

$$x^2 + bx + \left(\frac{b}{2}\right)^2 = \left(x + \frac{b}{2}\right)^2$$

Completing the Square

Area of blue region is

$$x^2 + 2\left(\frac{b}{2}\right)x = x^2 + bx$$

Add a small square of area $\left(\frac{b}{2}\right)^2$ to "complete" the square.

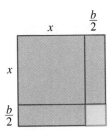

EXAMPLE 3 ■ Solving Quadratic Equations by Completing the Square

Solve each equation.

(a) $x^2 - 8x + 13 = 0$ (b) $3x^2 - 12x + 6 = 0$

SOLUTION

(a) $x^2 - 8x + 13 = 0$

$$x^2 - 8x \qquad = -13 \qquad \text{Subtract 13}$$

$$x^2 - 8x + 16 = -13 + 16 \qquad \text{Complete the square: add } \left(\frac{-8}{2}\right)^2 = 16$$

$$(x - 4)^2 = 3 \qquad \text{Perfect square}$$

$$x - 4 = \pm\sqrt{3} \qquad \text{Take square root}$$

$$x = 4 \pm \sqrt{3} \qquad \text{Add 4}$$

(b) After subtracting 6 from each side of the equation, we must factor the coefficient of x^2 (the 3) from the left side to put the equation in the correct form for completing the square.

$$3x^2 - 12x + 6 = 0$$

$$3x^2 - 12x \qquad = -6 \qquad \text{Subtract 6}$$

$$3(x^2 - 4x \qquad) = -6 \qquad \text{Factor 3 from LHS}$$

Now we complete the square by adding $(-2)^2 = 4$ inside the parentheses. Since everything inside the parentheses is multiplied by 3, this means that we are actually adding $3 \cdot 4 = 12$ to the left side of the equation. Thus, we must add 12 to the right side as well.

$$3(x^2 - 4x + 4) = -6 + 3 \cdot 4 \qquad \text{Complete the square}$$

$$3(x - 2)^2 = 6 \qquad \text{Perfect square}$$

$$(x - 2)^2 = 2 \qquad \text{Divide by 3}$$

$$x - 2 = \pm\sqrt{2} \qquad \text{Take square root}$$

$$x = 2 \pm \sqrt{2} \qquad \text{Add 2} \qquad \blacksquare$$

THE QUADRATIC FORMULA

We can use the technique of completing the square to derive a formula for the roots of the general quadratic equation $ax^2 + bx + c = 0$. First, we divide each side of the equation by a and move the constant to the right side, giving

$$x^2 + \frac{b}{a}x = -\frac{c}{a}$$

We now complete the square by adding $[b/(2a)]^2$ to each side of the equation:

$$x^2 + \frac{b}{a}x + \left(\frac{b}{2a}\right)^2 = -\frac{c}{a} + \left(\frac{b}{2a}\right)^2$$

$$\left(x + \frac{b}{2a}\right)^2 = \frac{-4ac + b^2}{4a^2} \qquad \text{Perfect square}$$

$$x + \frac{b}{2a} = \pm\sqrt{\frac{-4ac + b^2}{4a^2}} = \pm\frac{\sqrt{b^2 - 4ac}}{2a} \qquad \text{Take square root}$$

$$x = \frac{-b \pm \sqrt{b^2 - 4ac}}{2a} \qquad \text{Subtract } \frac{b}{2a}$$

This is the quadratic formula.

THE QUADRATIC FORMULA

The roots of the quadratic equation $ax^2 + bx + c = 0$, where $a \neq 0$, are

$$x = \frac{-b \pm \sqrt{b^2 - 4ac}}{2a}$$

François Viète (1540–1603) was a French mathematician, sometimes known by the Latin form of his name, Vieta. He introduced a new level of abstraction in algebra by using letters to stand for *known* quantities in an equation. Before Viète's time, each equation had to be solved on its own. For instance, the quadratic equations $3x^2 + 2x + 8 = 0$ and $5x^2 - 6x + 4 = 0$ had to be solved separately for the unknown x. Viète's idea was to consider all quadratic equations at once by writing

$$ax^2 + bx + c = 0$$

where a, b, and c are known quantities. Thus, it is possible to write a *formula* (in this case, the quadratic formula) involving a, b, and c that can be used to solve all such equations in one fell swoop.

The quadratic formula could be used to solve the equations in Examples 1, 2, and 3. You should carry out the details of these calculations.

EXAMPLE 4 ■ **Using the Quadratic Formula**

Find all solutions of each equation.

(a) $3x^2 - 5x - 1 = 0$　　(b) $4x^2 + 12x + 9 = 0$　　(c) $x^2 + 2x + 2 = 0$

SOLUTION

(a) Using the quadratic formula with $a = 3$, $b = -5$, and $c = -1$, we get

$$x = \frac{-(-5) \pm \sqrt{(-5)^2 - 4(3)(-1)}}{2(3)} = \frac{5 \pm \sqrt{37}}{6}$$

If approximations are desired, we use a calculator and obtain

$$x = \frac{5 + \sqrt{37}}{6} \approx 1.8471 \quad \text{and} \quad x = \frac{5 - \sqrt{37}}{6} \approx -0.1805$$

(b) Using the quadratic formula with $a = 4$, $b = 12$, and $c = 9$ gives

$$x = \frac{-12 \pm \sqrt{(12)^2 - 4 \cdot 4 \cdot 9}}{2 \cdot 4} = \frac{-12 \pm 0}{8} = -\frac{3}{2}$$

This equation has only one solution, $x = -\frac{3}{2}$.

(c) Using the quadratic formula with $a = 1$, $b = 2$, and $c = 2$ gives

$$x = \frac{-2 \pm \sqrt{2^2 - 4 \cdot 2}}{2} = \frac{-2 \pm \sqrt{-4}}{2} = \frac{-2 \pm 2\sqrt{-1}}{2} = -1 \pm \sqrt{-1}$$

Since the square of any real number is nonnegative, $\sqrt{-1}$ is undefined in the real number system. The equation has no real solution.　■

In the next section we will study the complex number system, in which the square roots of negative numbers do exist. The equation in Example 4(c) does have solutions in the complex number system.

The quantity $b^2 - 4ac$ that appears under the square root sign in the quadratic formula is called the **discriminant** of the equation $ax^2 + bx + c = 0$ and is given the symbol D. If $D < 0$, then $\sqrt{b^2 - 4ac}$ is undefined, so the quadratic equation has no real solution, as in Example 4(c). If $D = 0$, then the equation has only one real solution, as in Example 4(b). Finally, if $D > 0$, then the equation has two distinct real solutions, as in Example 4(a).

The following box summarizes what we have observed about the discriminant.

THE DISCRIMINANT

The discriminant of the general quadratic $ax^2 + bx + c = 0$ ($a \neq 0$) is $D = b^2 - 4ac$.

1. If $D > 0$, then the equation has two distinct real solutions.

2. If $D = 0$, then the equation has exactly one real solution.

3. If $D < 0$, then the equation has no real solution.

EXAMPLE 5 ■ Using the Discriminant

Use the discriminant to determine how many real solutions each equation has.

(a) $x^2 + 2x + 8 = 0$ (b) $3x^2 - 5x + \frac{3}{2} = 0$

SOLUTION

(a) The discriminant is

$$D = 2^2 - 4 \cdot 1 \cdot 8 = -28 < 0$$

Thus, the equation has no real solution.

(b) The discriminant is $D = (-5)^2 - 4 \cdot 3 \cdot \frac{3}{2} = 25 - 18 = 7 > 0$. Thus, the equation has two distinct real solutions. ■

PROBLEM SOLVING WITH QUADRATIC EQUATIONS

Applied problems often lead to quadratic equations. The principles discussed in Section 2.2 for setting up equations are useful here as well.

EXAMPLE 6 ■ A Distance-Rate-Time Problem

A jet flew from New York to Los Angeles, a distance of 4200 km. The speed for the return trip was 100 km/h faster than the outbound speed. If the total trip took 13 h, what was the speed from New York to Los Angeles?

SOLUTION

Identify the variable

Let s = speed from New York to Los Angeles

Then $s + 100$ = speed from Los Angeles to New York

We can organize the given data into the following table.

	Distance (km)	Speed (km/h)	Time (h)
N.Y. to L.A.	4200	s	$\dfrac{4200}{s}$
L.A. to N.Y.	4200	$s + 100$	$\dfrac{4200}{s + 100}$

Relate the quantities

The total trip took 13 h, so we have the equation

$$\frac{4200}{s} + \frac{4200}{s + 100} = 13$$

Set up an equation

Multiplying by the common denominator, $s(s + 100)$, we get

$$4200(s + 100) + 4200s = 13s(s + 100)$$

$$8400s + 420{,}000 = 13s^2 + 1300s$$

$$0 = 13s^2 - 7100s - 420{,}000$$

Although this equation does factor, with numbers this large it is probably quickest to use the quadratic formula and a calculator.

Solve

$$s = \frac{7100 \pm \sqrt{(7100)^2 - 4(13)(-420{,}000)}}{2(13)}$$

$$= \frac{7100 \pm 8500}{26}$$

$$s = 600 \qquad \text{or} \qquad s = \frac{-700}{13} \approx -53.8$$

Since s represents speed, we reject the negative answer and conclude that the jet's speed from New York to Los Angeles was 600 km/h. ■

EXAMPLE 7 ■ The Path of a Projectile

An object thrown or fired straight upward at an initial speed of v_0 ft/s will reach a height of h feet after t seconds, where h and t are related by the formula

$$h = -16t^2 + v_0 t$$

(This formula is derived in elementary physics courses and depends on the fact that the acceleration due to gravity is constant near the surface of the earth. Here we are neglecting the effect of air resistance.)

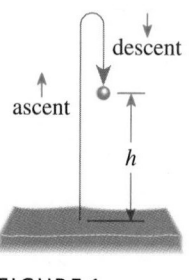

descent

ascent

h

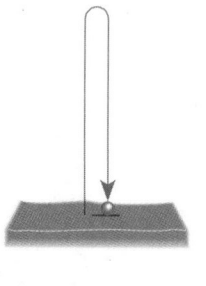

FIGURE 1

Suppose that a bullet is shot straight upward with an initial speed of 800 ft/s. Its path is shown in Figure 1.
(a) When does the bullet fall back to ground level?
(b) When does it reach a height of 6400 ft?
(c) When does it reach a height of 2 mi?
(d) How high is the highest point the bullet reaches?

SOLUTION

Since the initial speed in this case is $v_0 = 800$ ft/s, the formula is $h = -16t^2 + 800t$.

(a) Ground level corresponds to $h = 0$, so we must solve the equation

$$0 = -16t^2 + 800t$$

$$0 = -16t(t - 50)$$

Thus, $t = 0$ or $t = 50$. This means the bullet starts ($t = 0$) at ground level and returns to ground level after 50 s.

(b) Setting $h = 6400$ gives the equation

$$6400 = -16t^2 + 800t$$

$$16t^2 - 800t + 6400 = 0$$

$$t^2 - 50t + 400 = 0 \qquad \text{Divide by 16}$$

$$(t - 10)(t - 40) = 0 \qquad \text{Factor}$$

$$t = 10 \quad \text{or} \quad t = 40$$

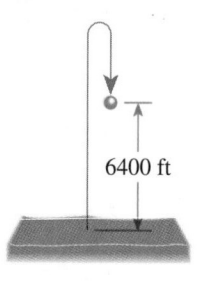

6400 ft

The bullet reaches 6400 ft after 10 s (on its ascent) and again after 40 s (on its descent to earth).

(c) Two miles is $2 \times 5280 = 10{,}560$ ft.

$$10{,}560 = -16t^2 + 800t$$

$$16t^2 - 800t + 10{,}560 = 0$$

$$t^2 - 50t + 660 = 0 \qquad \text{Divide by 16}$$

The discriminant of this equation is $D = (-50)^2 - 4(660) = -140$, which is negative. Thus the equation has no real solution. The bullet never reaches a height of 2 mi.

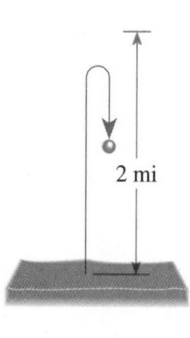

2 mi

(d) Each height the bullet reaches is attained twice, once on its ascent and once on its descent. The only exception is the highest point of its path, which is reached only once. This means that for the highest value of h, the equation

$$h = -16t^2 + 800t$$

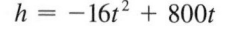

10,000 ft

or $\qquad 16t^2 - 800t + h = 0$

has only one solution for t. This in turn means that the discriminant of the equation is 0, and so

$$D = (-800)^2 - 4(16)h = 0$$

$$640,000 - 64h = 0$$

$$h = 10,000$$

The maximum height reached is 10,000 ft. ■

2.3 EXERCISES

1–6 ■ Find all solutions of the equation.

1. $x^2 = 49$

2. $x^2 = 18$

3. $x^2 - 24 = 0$

4. $x^2 - 7 = 0$

5. $8x^2 - 64 = 0$

6. $5x^2 - 125 = 0$

7–14 ■ Solve the equation by factoring.

7. $x^2 - x - 6 = 0$

8. $x^2 + 2x = 8$

9. $x^2 + 4 = 4x$

10. $x^2 + 6x + 8 = 0$

11. $2y^2 + 7y + 3 = 0$

12. $4w^2 = 4w + 3$

13. $6x^2 + 5x = 4$

14. $a^2x^2 + 2ax + 1 = 0$ $(a > 0)$

15–22 ■ Solve the equation by completing the square.

15. $x^2 + 2x - 2 = 0$

16. $x^2 - 4x + 2 = 0$

17. $x^2 - 6x - 9 = 0$

18. $x^2 + x - \frac{3}{4} = 0$

19. $2x^2 + 8x + 1 = 0$

20. $3x^2 - 6x - 1 = 0$

21. $4x^2 - x = 0$

22. $x^2 = \frac{3}{4}x - \frac{1}{8}$

23–44 ■ Find all real solutions of the equation.

23. $x^2 - 2x - 8 = 0$

24. $2x^2 + x - 3 = 0$

25. $x^2 + 12x - 27 = 0$

26. $8x^2 - 6x - 9 = 0$

27. $3x^2 + 6x - 5 = 0$

28. $x^2 - 6x + 1 = 0$

29. $2y^2 - y - \frac{1}{2} = 0$

30. $\theta^2 - \frac{3}{2}\theta + \frac{9}{16} = 0$

31. $4x^2 + 16x - 9 = 0$

32. $0 = x^2 - 4x + 1$

33. $3 + 5z + z^2 = 0$

34. $w^2 = 3(w - 1)$

35. $x^2 - \sqrt{5}\,x + 1 = 0$

36. $\sqrt{6}\,x^2 + 2x - \sqrt{\frac{3}{2}} = 0$

37. $\dfrac{x^2}{x + 100} = 50$

38. $1 + \dfrac{2x}{(x + 3)(x + 4)} = \dfrac{2}{x + 3} + \dfrac{4}{x + 4}$

39. $\dfrac{x + 5}{x - 2} = \dfrac{5}{x + 2} + \dfrac{28}{x^2 - 4}$

40. $\dfrac{x}{2x + 7} - \dfrac{x + 1}{x + 3} = 1$

41. $\dfrac{1}{x - 1} - \dfrac{2}{x^2} = 0$

42. $\dfrac{x + \dfrac{2}{x}}{3 + \dfrac{4}{x}} = 5x$

43. $ax^2 - (2a + 1)x + (a + 1) = 0$ $(a \neq 0)$

44. $bx^2 + 2x + \dfrac{1}{b} = 0$ $(b \neq 0)$

45–48 ■ Use the quadratic formula and a calculator to find all real solutions, correct to three decimals.

45. $x^2 - 0.011x - 0.064 = 0$

46. $2.232x^2 - 4.112x = 6.219$

47. $x^2 - 2.450x + 1.500 = 0$

48. $x^2 - 2.450x + 1.501 = 0$

49–54 ■ Solve the equation for the indicated variable.

49. $V = \frac{1}{3}\pi r^2 h$;   for r **50.** $F = G\dfrac{mM}{r^2}$;   for r

51. $a^2 + b^2 = c^2$;   for b **52.** $S = \dfrac{n(n+1)}{2}$;   for n

53. $A = P\left(1 + \dfrac{i}{100}\right)^2$;   for i

54. $\dfrac{1}{s+a} + \dfrac{1}{s+b} = \dfrac{1}{c}$;   for s

55–56 ■ Find the value(s) of k that will ensure that the indicated solutions are the solutions of the given quadratic equation.

55. $x^2 - 2x - k = 0$;   solutions $x = -3, 5$

56. $kx^2 + x - 4 = 0$;   solutions $x = -\frac{4}{3}, 1$

57–62 ■ Use the discriminant to determine the number of real solutions of the equation. Do not solve the equation.

57. $x^2 - 6x + 1 = 0$

58. $x^2 = 6x - 9$

59. $x^2 + 2.20x + 1.21 = 0$

60. $x^2 + 2.21x + 1.21 = 0$

61. $x^2 + rx - s = 0$ $(s > 0)$

62. $x^2 - rx + s = 0$ $(s > 0, r > 2\sqrt{s}\,)$

63–64 ■ Find all values for k that ensure that the given equation has exactly one solution.

63. $4x^2 + kx + 25 = 0$

64. $kx^2 + 36x + k = 0$

65. Find two numbers whose sum is 55 and whose product is 684.

66. The sum of the squares of two consecutive even integers is 1252. Find the integers.

67. A rectangular garden is 10 ft longer than it is wide. Its area is 875 ft². What are its dimensions?

68. A small-appliance manufacturer finds that the profit P (in dollars) generated by producing x microwave ovens per week is given by the formula $P = \frac{1}{10}x(300 - x)$ provided that $0 \le x \le 200$. How many ovens must

be manufactured in a given week to generate a profit of $1250?

69. A box with a square base and no top is to be made from a square piece of cardboard by cutting 4-inch squares from each corner and folding up the sides, as shown in the figure. The box is to hold 100 in³. How big a piece of cardboard is needed?

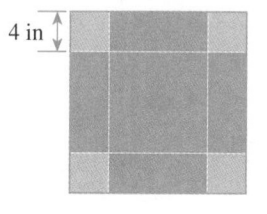
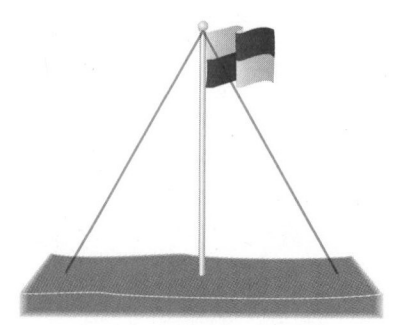

70. A cylindrical can has a volume of 40π cm³ and is 10 cm tall. What is its diameter? [*Hint:* Use the volume formula listed on the inside of the front cover of this book.]

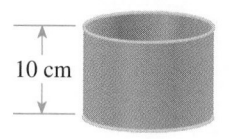

71. A parcel of land is 6 ft longer than it is wide. Each diagonal from one corner to the opposite corner is 174 ft long. What are the dimensions of the parcel?

72. A flagpole is secured on opposite sides by two guy wires, each of which is 5 ft longer than the pole. The distance between the points where the wires are fixed to the ground is equal to the length of one guy wire. How tall is the flagpole (to the nearest inch)?

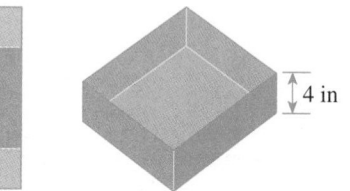

73–74 ■ Use the formula $h = -16t^2 + v_0t$ discussed in Example 7.

73. A ball is thrown straight upward at an initial speed of 40 ft/s.
(a) When does the ball reach a height of 24 ft?
(b) When does it reach a height of 48 ft?
(c) What is the greatest height reached by the ball?
(d) When does the ball reach the highest point of its path?
(e) When does the ball hit the ground?

74. How fast would a ball have to be thrown upward to reach a maximum height of 100 ft? [*Hint:* Use the discriminant of the equation $16t^2 - v_0t + h = 0$.]

75. The fish population in a certain lake rises and falls according to the formula

$$F = 1000(30 + 17t - t^2)$$

Here F is the number of fish at time t, where t is measured in years since January 1, 1992, when the fish population was first estimated.
(a) On what date will the fish population again be the same as on January 1, 1992?
(b) By what date will all the fish in the lake have died?

76. A wire 360 in. long is cut into two pieces. One piece is formed into a square and the other into a circle. If the two figures have the same area, what are the lengths of the two pieces of wire (to the nearest tenth of an inch)?

77. A salesman drives from Ajax to Barrington, a distance of 120 mi, at a steady speed. He then increases his speed by 10 mi/h to drive the 150 mi from Barrington to Collins. If the second leg of his trip took 6 min more time than the first leg, how fast was he driving between Ajax and Barrington?

78. Kiran drove from Tortula to Cactus, a distance of 250 mi. She increased her speed by 10 mi/h for the 360-mi trip from Cactus to Dry Junction. If the total trip took 11 h, what was her speed from Tortula to Cactus?

79. It took a crew 2 h 40 min to row 6 km upstream and back again. If the rate of flow of the stream was 3 km/h, what was the rowing rate of the crew in still water?

80. A factory is to be built on a lot measuring 180 ft by 240 ft. A local building code specifies that a lawn of uniform width and equal in area to the factory must surround the factory. What must the width of this lawn be, and what are the dimensions of the factory?

81. Henry and Irene working together can wash all the windows of their house in 1 h 48 min. Working alone, it takes Henry $1\frac{1}{2}$ h more than Irene to do the job. How long does it take each person working alone to wash all the windows?

82. Jack, Kay, and Lynn deliver advertising flyers in a small town. If each person works alone, it takes Jack 4 h to deliver all the flyers, and it takes Lynn 1 h longer than it takes Kay. Working together, they can deliver all the flyers in 40% of the time it takes Kay working alone. How long does it take Kay to deliver all the flyers alone?

83. If an imaginary line segment is drawn between the centers of the earth and the moon, then the net gravitational force F acting on an object situated on this line segment is

$$F = \frac{-K}{x^2} + \frac{0.012K}{(239 - x)^2}$$

where $K > 0$ is a constant and x is the distance of the object from the center of the earth, measured in thousands of miles. How far from the center of the earth is the "dead spot" where no net gravitational force acts upon the object? (Express your answer to the nearest thousand miles.)

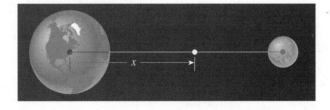

84. A man wishes to determine the water level in a deep well. He drops a stone into the well and hears the stone hit the water 3 s later. If the stone drops $16t^2$ feet after t seconds, and the speed of sound is 1090 ft/s, how far below the opening of the well is the surface of the water (to the nearest foot)?

2.4 COMPLEX NUMBERS

In Section 2.3, we saw that if the discriminant of a quadratic equation is negative, the equation has no real solution. For example, the equation

$$x^2 + 4 = 0$$

has no real solution. If we try to solve this equation, we get $x^2 = -4$, so

$$x = \pm\sqrt{-4}$$

But this is impossible, since the square of any real number is positive. [For example, $(-2)^2 = 4$, a positive number.] Thus, negative numbers don't have real square roots.

To make it possible to solve *all* quadratic equations, mathematicians invented an expanded number system, called the *complex number system*. First they defined the new number

$$i = \sqrt{-1}$$

This means $i^2 = -1$. A complex number is then a number of the form $a + bi$, where a and b are real numbers.

DEFINITION OF COMPLEX NUMBERS

A **complex number** is an expression of the form

$$a + bi$$

where a and b are real numbers and $i^2 = -1$. The **real part** of this complex number is a and the **imaginary part** is b. Two complex numbers are **equal** if and only if their real parts are equal and their imaginary parts are equal.

Note that both the real and imaginary parts of a complex number are real numbers. The following numbers are examples of complex numbers:

$3 + 4i$ Real part 3, imaginary part 4

$\frac{1}{2} - \frac{2}{3}i$ Real part $\frac{1}{2}$, imaginary part $-\frac{2}{3}$

$6i$ Real part 0, imaginary part 6

-7 Real part -7, imaginary part 0

A number such as $6i$, which has real part 0, is called a **pure imaginary number.** A real number like -7 can be thought of as a complex number with imaginary part 0.

In the complex number system every quadratic equation has solutions. The numbers $2i$ and $-2i$ are solutions of $x^2 = -4$ because

$$(2i)^2 = 2^2 i^2 = 4(-1) = -4 \qquad \text{and} \qquad (-2i)^2 = (-2)^2 i^2 = 4(-1) = -4$$

Although we use the term *imaginary* in this context, imaginary numbers should not be thought of as any less "real" (in the ordinary rather than the mathematical sense of that word) than negative numbers or irrational numbers. All numbers (except possibly the positive integers) are creations of the human mind—the numbers -1 and $\sqrt{2}$ as well as the number i. We study complex numbers because they complete, in a useful and elegant fashion, the study of the solutions of polynomial equations. (See Cardano, page 277.) In fact, imaginary numbers are useful not only in algebra and mathematics, but in the other sciences as well. To give just one example, in electrical theory the *reactance* of a circuit is a quantity whose measure is an imaginary number.

ARITHMETIC OPERATIONS ON COMPLEX NUMBERS

Complex numbers are added, subtracted, multiplied, and divided just as we would any number of the form $a + b\sqrt{c}$. The only difference we must keep in mind is that $i^2 = -1$. Thus, in particular, the following calculations are valid.

$$
\begin{aligned}
(a + bi)(c + di) &= ac + (ad + bc)i + bdi^2 && \text{Multiply and collect like terms} \\
&= ac + (ad + bc)i + bd(-1) && i^2 = -1 \\
&= (ac - bd) + (ad + bc)i && \text{Combine real and imaginary parts}
\end{aligned}
$$

We therefore define the sum, difference, and product of complex numbers as follows.

ADDING, SUBTRACTING, AND MULTIPLYING COMPLEX NUMBERS

Definition	Description
Addition	
$(a + bi) + (c + di) = (a + c) + (b + d)i$	To add complex numbers, add the real parts and the imaginary parts.
Subtraction	
$(a + bi) - (c + di) = (a - c) + (b - d)i$	To subtract complex numbers, subtract the real parts and the imaginary parts.
Multiplication	
$(a + bi) \cdot (c + di) = (ac - bd) + (ad + bc)i$	Multiply complex numbers like binomials, using $i^2 = -1$.

Leonhard Euler (1707–1783) was born in Basel, Switzerland, the son of a pastor. At age 13 his father sent him to the University at Basel to study theology, but Euler soon decided to devote himself to the sciences. Besides theology he studied mathematics, medicine, astronomy, physics, and oriental languages. It is said that Euler could calculate as effortlessly as "men breathe or as eagles fly." One hundred years before Euler, Fermat (see page 532) had conjectured that $2^{2^n} + 1$ is a prime number for all n. The first five of these numbers are 5, 17, 257, 65537, and 4,294,967,297. It is easy to show that the first four are prime. The fifth was also thought to be prime until Euler, with his phenomenal calculating ability, showed that it is the product $641 \times 6,700,417$ and so is not prime. Euler published more than any other mathematician in history. His collected works comprise 75 large volumes. Although he was blind for the last 17 years of his life, he continued to work and publish. In his writings he popularized the use of the symbols π, e, and i, which you will find in this textbook. One of Euler's most lasting contributions is the development of complex numbers.

EXAMPLE 1 ■ Adding, Subtracting, and Multiplying Complex Numbers

Express each of the following in the form $a + bi$.

(a) $(3 + 5i) + (4 - 2i)$ (b) $(3 + 5i) - (4 - 2i)$

(c) $(3 + 5i)(4 - 2i)$ (d) i^{23}

SOLUTION

(a) According to the definition, we add the real parts and we add the imaginary parts:

$$(3 + 5i) + (4 - 2i) = (3 + 4) + (5 - 2)i = 7 + 3i$$

(b) $(3 + 5i) - (4 - 2i) = (3 - 4) + [5 - (-2)]i = -1 + 7i$

(c) $(3 + 5i)(4 - 2i) = [3 \cdot 4 - 5(-2)] + [3(-2) + 5 \cdot 4]i = 22 + 14i$

(d) $i^{23} = i^{20+3} = (i^2)^{10}i^3 = (-1)^{10}i^2i = (1)(-1)i = -i$ ■

Division of complex numbers is much like rationalizing the denominator of a radical expression, which we considered in Section 1.5. For the complex number $z = a + bi$ we define its **complex conjugate** to be $\bar{z} = a - bi$. Note that $z \cdot \bar{z} = (a + bi)(a - bi) = a^2 + b^2$. So the product of a complex number and its conjugate is always a nonnegative real number. We use this property to divide complex numbers.

DIVIDING COMPLEX NUMBERS

To simplify the quotient $\dfrac{a + bi}{c + di}$, multiply the numerator and the denominator by the complex conjugate of the denominator:

$$\frac{a + bi}{c + di} = \left(\frac{a + bi}{c + di}\right)\left(\frac{c - di}{c - di}\right) = \frac{(ac + bd) + (bc - ad)i}{c^2 + d^2}$$

Rather than memorize this entire formula, it is best just to remember the first step and then multiply out the numerator and the denominator as usual.

EXAMPLE 2 ■ Dividing Complex Numbers

Express each of the following in the form $a + bi$.

(a) $\dfrac{3 + 5i}{1 - 2i}$ (b) $\dfrac{7 + 3i}{4i}$

SOLUTION

We multiply both numerator and denominator by the complex conjugate of the denominator to make the new denominator a real number.

(a) The complex conjugate of $1 - 2i$ is $\overline{1 - 2i} = 1 + 2i$.

$$\frac{3 + 5i}{1 - 2i} = \left(\frac{3 + 5i}{1 - 2i}\right)\left(\frac{1 + 2i}{1 + 2i}\right) = \frac{-7 + 11i}{5} = -\frac{7}{5} + \frac{11}{5}i$$

(b) The complex conjugate of $4i$ is $-4i$. Therefore,

$$\frac{7 + 3i}{4i} = \left(\frac{7 + 3i}{4i}\right)\left(\frac{-4i}{-4i}\right) = \frac{12 - 28i}{16} = \frac{3}{4} - \frac{7}{4}i \qquad \blacksquare$$

Just as every positive real number r has two square roots ($\sqrt{r}$ and $-\sqrt{r}$), every negative number has two square roots as well. Both square roots are pure imaginary numbers, for if $r > 0$ is real, then

$$(i\sqrt{r})^2 = i^2 r = -r$$

and

$$(-i\sqrt{r})^2 = (-1)^2 i^2 r = -r$$

We call $i\sqrt{r}$ the **principal square root** of $-r$, and we will use the symbol $\sqrt{-r}$ to denote this principal square root. The other square root will then be $-\sqrt{-r} = -i\sqrt{r}$. Note that the two square roots of a negative real number are complex conjugates of each other.

SQUARE ROOTS OF NEGATIVE NUMBERS

If $-r < 0$, then the square roots of $-r$ are

$$i\sqrt{r} \qquad \text{and} \qquad -i\sqrt{r}$$

The **principal** square root of $-r$ is $i\sqrt{r}$.

We usually write $i\sqrt{b}$ instead of $\sqrt{b}\,i$ to avoid confusion with $\sqrt{bi}$.

EXAMPLE 3 ■ Square Roots of Negative Numbers

Evaluate: (a) $\sqrt{-1}$ (b) $\sqrt{-16}$ (c) $\sqrt{-3}$

SOLUTION

(a) $\sqrt{-1} = i\sqrt{1} = i$ (b) $\sqrt{-16} = i\sqrt{16} = 4i$ (c) $\sqrt{-3} = i\sqrt{3}$ $\blacksquare$

Special care must be taken when performing calculations involving square roots of negative numbers. Although $\sqrt{a} \cdot \sqrt{b} = \sqrt{ab}$ when a and b are positive, this is *not* true when both are negative. For example,

$$\sqrt{-2} \cdot \sqrt{-3} = i\sqrt{2} \cdot i\sqrt{3} = i^2\sqrt{6} = -\sqrt{6}$$

but

$$\sqrt{(-2)(-3)} = \sqrt{6}$$

so

$$\sqrt{-2} \cdot \sqrt{-3} \neq \sqrt{(-2)(-3)}$$

When multiplying radicals of negative numbers, express them first in the form $i\sqrt{r}$ (where $r > 0$) to avoid possible errors of this type.

EXAMPLE 4 ■ **Using Square Roots of Negative Numbers**

Evaluate $\left(\sqrt{12} - \sqrt{-3}\right)\left(3 + \sqrt{-4}\right)$ and express in the form $a + bi$.

SOLUTION

$$\left(\sqrt{12} - \sqrt{-3}\right)\left(3 + \sqrt{-4}\right) = \left(\sqrt{12} - i\sqrt{3}\right)\left(3 + i\sqrt{4}\right)$$
$$= \left(2\sqrt{3} - i\sqrt{3}\right)\left(3 + 2i\right)$$
$$= \left(6\sqrt{3} + 2\sqrt{3}\right) + i\left(2 \cdot 2\sqrt{3} - 3\sqrt{3}\right)$$
$$= 8\sqrt{3} + i\sqrt{3}$$

■

IMAGINARY ROOTS OF QUADRATIC EQUATIONS

We have already seen that, if $a \neq 0$, the solutions of the quadratic equation $ax^2 + bx + c = 0$ are

$$x = \frac{-b \pm \sqrt{b^2 - 4ac}}{2a}$$

If $b^2 - 4ac < 0$, then the equation has no real solution. But in the complex number system, this equation will always have solutions, because negative numbers have square roots in this expanded setting.

EXAMPLE 5 ■ **Quadratic Equations with Imaginary Solutions**

Solve each equation: (a) $x^2 + 9 = 0$ (b) $x^2 + 4x + 5 = 0$

SOLUTION

(a) The equation $x^2 + 9 = 0$ means $x^2 = -9$, so

$$x = \pm\sqrt{-9} = \pm i\sqrt{9} = \pm 3i$$

The solutions are therefore $3i$ and $-3i$.

(b) By the quadratic formula, we have

$$x = \frac{-4 \pm \sqrt{4^2 - 4 \cdot 5}}{2}$$

$$= \frac{-4 \pm \sqrt{-4}}{2}$$

$$= \frac{-4 \pm 2i}{2} = \frac{2(-2 \pm i)}{2} = -2 \pm i$$

So, the solutions are $-2 + i$ and $-2 - i$. ■

The solutions of both equations in Example 5 occur in complex conjugate pairs. (The complex conjugate of $3i$ is $-3i$, and the conjugate of $-2 + i$ is $-2 - i$.) This relationship holds for the solutions of any quadratic equation that has real coefficients a, b, and c. (Think about the $\pm$ sign in the quadratic formula to understand why this is true.)

EXAMPLE 6 ■ **Complex Conjugates as Solutions of a Quadratic**

Show that the solutions of the equation

$$4x^2 - 24x + 37 = 0$$

are complex conjugates of each other.

SOLUTION

We use the quadratic formula to get

$$x = \frac{24 \pm \sqrt{(24)^2 - 4(4)(37)}}{2(4)}$$

$$= \frac{24 \pm \sqrt{-16}}{8} = \frac{24 \pm 4i}{8} = 3 \pm \frac{1}{2}i$$

So, the solutions are $3 + \frac{1}{2}i$ and $3 - \frac{1}{2}i$, and these are complex conjugates. ■

EXAMPLE 7 ■ **Equations with Imaginary Coefficients**

Find all solutions of each equation.

(a) $3x + 10 = ix + 20i$ (b) $x^2 - 3ix + 4 = 0$

SOLUTION

(a) This is a linear equation with complex coefficients.

$$3x - ix = -10 + 20i \qquad \text{Move } x \text{ terms to LHS, constants to RHS}$$

$$(3 - i)x = -10 + 20i \qquad \text{Factor } x \text{ from LHS}$$

$$x = \frac{-10 + 20i}{3 - i} \qquad \text{Divide by } 3 - i$$

$$= \left(\frac{-10 + 20i}{3 - i}\right)\left(\frac{3 + i}{3 + i}\right) \qquad \begin{array}{l}\text{Multiply numerator and denominator by} \\ \text{complex conjugate of denominator}\end{array}$$

$$= \frac{-50 + 50i}{10} = -5 + 5i$$

(b) We use the quadratic formula to solve $x^2 - 3ix + 4 = 0$:

$$x = \frac{3i \pm \sqrt{(-3i)^2 - 4(1)(4)}}{2}$$

$$= \frac{3i \pm \sqrt{-25}}{2} = \frac{3i \pm 5i}{2}$$

So, the solutions are $x = 4i$ and $x = -i$.   ■

In Example 7(b) the two solutions of the quadratic equation are not complex conjugates because the equation does not have real coefficients only.

2.4 **EXERCISES**

1–6 ■ Find the real and imaginary part of the complex number.

1. $3 - 5i$

2. $\dfrac{2 + 4i}{3}$

3. $6i$

4. $\frac{1}{2}$

5. $\sqrt{2} + \sqrt{-3}$

6. $\dfrac{2 + 4i}{\sqrt{-16}}$

7–40 ■ Evaluate the expression and write the result in the form $a + bi$.

7. $(4 + 3i) + (5 - 2i)$

8. $(7 - 6i) + (-3 + 7i)$

9. $\left(7 - \frac{1}{2}i\right) + \left(5 + \frac{3}{2}i\right)$

10. $(-4 + i) - (2 - 5i)$

11. $(-12 + 8i) - (7 + 4i)$

12. $6i - (4 - i)$

13. $4(-1 + 2i)$

14. $2i\left(\frac{1}{2} - i\right)$

15. $(7 - i)(4 + 2i)$

16. $(5 - 3i)(1 + i)$

17. $(3 - 4i)(5 - 12i)$

18. $\left(\frac{2}{3} + 12i\right)\left(\frac{1}{6} + 24i\right)$

19. $\dfrac{1}{i}$

20. $\dfrac{1}{1 + i}$

21. $\dfrac{2 - 3i}{1 - 2i}$

22. $\dfrac{5 - i}{3 + 4i}$

23. $\dfrac{26 + 39i}{2 - 3i}$

24. $\dfrac{25}{4 - 3i}$

25. $\dfrac{10i}{1 - 2i}$

26. $(2 - 3i)^{-1}$

27. i^3

28. $(2i)^4$

29. i^{100}

30. i^{1002}

31. $\sqrt{-25}$

32. $\sqrt{\dfrac{-9}{4}}$

33. $\sqrt{-3}\,\sqrt{-12}$

34. $\sqrt{\tfrac{1}{3}}\,\sqrt{-27}$

35. $\left(3 - \sqrt{-5}\right)\left(1 + \sqrt{-1}\right)$

36. $\dfrac{1 - \sqrt{-1}}{1 + \sqrt{-1}}$

37. $\dfrac{2 + \sqrt{-8}}{1 + \sqrt{-2}}$

38. $\left(\sqrt{3} - \sqrt{-4}\right)\left(\sqrt{6} - \sqrt{-8}\right)$

39. $\dfrac{\sqrt{-36}}{\sqrt{-2}\,\sqrt{-9}}$

40. $\dfrac{\sqrt{-7}\,\sqrt{-49}}{\sqrt{28}}$

41–58 ■ Find all solutions of the equation and express them in the form $a + bi$.

41. $x^2 + 9 = 0$

42. $9x^2 + 4 = 0$

43. $x^2 - 4x + 5 = 0$

44. $x^2 + 2x + 2 = 0$

45. $x^2 + x + 1 = 0$

46. $x^2 - 3x + 3 = 0$

47. $2x^2 - 2x + 1 = 0$

48. $2x^2 + 3 = 2x$

49. $t + 3 + \dfrac{3}{t} = 0$

50. $z^2 - iz = 0$

51. $x + 3 - i = 2 + 4i$

52. $3x - 7i = 1 + i$

53. $2x + 5 + 3i = 7i$

54. $4x + i = 7x - 1 + 2i$

55. $ix^2 - 4x + i = 0$

56. $2ix^2 + (4 + i)x + 1 = 0$

57. $\sqrt{2}\,x^2 + \sqrt{3}\,x + \sqrt{8} = 0$

58. $\dfrac{x^2}{i} - x + i = 0$

59–66 ■ Recall that the symbol $\bar{z}$ represents the complex conjugate of z. If $z = a + bi$ and $w = c + di$, prove each statement.

59. $\bar{z} + \bar{w} = \overline{z + w}$

60. $\overline{zw} = \bar{z} \cdot \bar{w}$

61. $(\bar{z})^2 = \overline{z^2}$

62. $\bar{\bar{z}} = z$

63. $z + \bar{z}$ is a real number

64. $z = \bar{z}$ if and only if z is real

65. $\dfrac{z}{\bar{z}} + \dfrac{\bar{z}}{z} = \dfrac{2(a^2 - b^2)}{a^2 + b^2}$

66. $\dfrac{z}{\bar{z}} - \dfrac{\bar{z}}{z} = \dfrac{4abi}{a^2 + b^2}$

67. Suppose that the equation

$$ax^2 + bx + c = 0$$

has two imaginary solutions (where a, b, and c are real numbers). Prove that the solutions are a complex conjugate pair.

2.5 OTHER EQUATIONS

So far we have learned how to solve and apply linear and quadratic equations. In this section we study other types of equations, including those that involve higher powers or radicals.

EXAMPLE 1 ■ Solving Equations Using nth Roots

Find all real solutions of each of the following equations.

(a) $x^3 = -8$ (b) $16x^4 = 81$

SOLUTION

(a) Since every real number has exactly one real cube root, we can solve this equation by taking the cube root of each side.

$$(x^3)^{1/3} = (-8)^{1/3}$$
$$x = -2$$

(b) Here we must remember that if n is even, then every positive real number has *two* real nth roots, a positive one and a negative one.

$$x^4 = \frac{81}{16} \qquad \text{Divide by 16}$$

$$(x^4)^{1/4} = \pm\left(\frac{81}{16}\right)^{1/4} \qquad \text{Take the fourth root}$$

$$x = \pm\tfrac{3}{2} \qquad\qquad\qquad\qquad\qquad\qquad \blacksquare$$

Some equations that at first glance may not appear to be quadratic can be changed into quadratic equations by performing simple algebraic operations on them, such as multiplying each side by a common denominator or squaring each side. Special care must be taken in solving such equations to avoid keeping extraneous solutions that may be introduced when we manipulate the equations.

EXAMPLE 2 ■ Taking Care of Extraneous Solutions

Solve each equation.

(a) $x + 3 = \dfrac{2x^2 - 7x + 3}{3 - x}$ 　　　(b) $x = 1 - \sqrt{2 - \dfrac{x}{2}}$

SOLUTION

(a) We can multiply each side of the equation by $x - 3$ to clear the denominator as long as $x \neq 3$:

$$(x + 3)(x - 3) = \left(\frac{2x^2 - 7x + 3}{3 - x}\right)(x - 3)$$

$$x^2 - 9 = \frac{2x^2 - 7x + 3}{-1} \qquad \text{Expand and cancel}$$

$$3x^2 - 7x - 6 = 0 \qquad \text{Move all terms to LHS and simplify}$$

$$(3x + 2)(x - 3) = 0 \qquad \text{Factor}$$

$$x = -\tfrac{2}{3} \quad \text{or} \quad x = 3 \qquad \text{Set each factor equal to 0}$$

Thus, $x = -\tfrac{2}{3}$ or $x = 3$. But $x = 3$ does not satisfy the original equation (since division by 0 is impossible), so the only solution is $x = -\tfrac{2}{3}$.

CHECK YOUR ANSWERS

$x = -\tfrac{2}{3}$:

$\quad$ LHS $= \left(-\tfrac{2}{3}\right) + 3 = \tfrac{7}{3}$

$\quad$ RHS $= \dfrac{2\left(-\tfrac{2}{3}\right)^2 - 7\left(-\tfrac{2}{3}\right) + 3}{3 - \left(-\tfrac{2}{3}\right)}$

$\qquad = \dfrac{\tfrac{8}{9} + \tfrac{14}{3} + 3}{\tfrac{11}{3}} = \dfrac{\tfrac{77}{9}}{\tfrac{11}{3}} = \tfrac{7}{3}$

$\quad$ LHS $=$ RHS $\qquad\qquad$ ✓

$x = 3$:

$\quad$ LHS $= 3 + 3 = 6$

$\quad$ RHS $= \dfrac{2(3)^2 - 7(3) + 3}{3 - 3} = \dfrac{0}{0}$

RHS is undefined, so $x = 3$ is not a solution. $\qquad\qquad$ ✗

(b) To eliminate the square root, we first isolate it on one side of the equal sign:

$$x - 1 = -\sqrt{2 - \frac{x}{2}}$$

$$(x - 1)^2 = 2 - \frac{x}{2} \qquad \text{Square each side to elimate square root}$$

$$x^2 - 2x + 1 = 2 - \frac{x}{2} \qquad \text{Expand LHS}$$

$$2x^2 - 4x + 2 = 4 - x \qquad \text{Multiply by 2 to clear denominator}$$

$$2x^2 - 3x - 2 = 0 \qquad \text{Move all terms to LHS}$$

$$(x - 2)(2x + 1) = 0 \qquad \text{Factor}$$

CHECK YOUR ANSWERS

$x = -\frac{1}{2}$:

  LHS $= -\frac{1}{2}$

  RHS $= 1 - \sqrt{2 - \dfrac{\left(-\frac{1}{2}\right)}{2}}$

    $= 1 - \sqrt{\frac{9}{4}} = 1 - \frac{3}{2} = -\frac{1}{2}$

  LHS $=$ RHS   ✓

$x = 2$:

  LHS $= 2$

  RHS $= 1 - \sqrt{2 - \frac{2}{2}}$

    $= 1 - 1 = 0$

LHS $\neq$ RHS, so $x = 2$ is not a solution.   ✗

Setting each factor equal to 0 gives us $x = 2$ and $x = -\frac{1}{2}$ as potential solutions. If we substitute these into the original equation (see *Check Your Answers*), we see that $x = -\frac{1}{2}$ is a solution but $x = 2$ is not. The only solution is

$$x = -\frac{1}{2} \qquad\qquad\blacksquare$$

The reason that extraneous solutions are often introduced when we square each side of an equation is that the operation of squaring can turn a false equation into a true one. For example, $-1 \neq 1$, but $(-1)^2 = 1^2$. Thus, the squared equation may be true for more values of the variable than the original equation, so we must always check our answers to make sure that eash satisfies the original equation.

In some cases fourth-degree (or higher-degree) polynomial equations can be changed to quadratic equations by performing algebraic substitutions, as shown in the following example.

EXAMPLE 3 ■ An Equation of Quadratic Type

Find the real solutions of $x^4 - 2x^2 - 2 = 0$.

SOLUTION

If we set $w = x^2$, then we get a quadratic equation in the new variable w:

$$x^4 - 2x^2 - 2 = (x^2)^2 - 2x^2 - 2$$

$$= w^2 - 2w - 2 = 0 \qquad \text{Set } w = x^2$$

From the quadratic formula, we have

$$w = \frac{2 \pm \sqrt{(-2)^2 + 4 \cdot 2}}{2} = 1 \pm \sqrt{3}$$

Because $x^2 = w$, we have $x = \pm\sqrt{w}$. So the potential solutions are $\pm\sqrt{1 \pm \sqrt{3}}$. Since we can't take the square root of a negative number, such as $1 - \sqrt{3}$, in the real number system, we have

$$x = \pm\sqrt{1 + \sqrt{3}}$$

as the only real solutions of the original equation.   ∎

An important method for solving equations is *factoring*, which we first used to solve quadratic equations in Section 2.3. The principle we use here is that if a product is zero, then at least one of the factors must be zero.

EXAMPLE 4 ■ Solving a 6th-Degree Equation by Factoring

Find all roots of the equation $x^6 - 1 = 0$.

SOLUTION

$$x^6 - 1 = (x^3)^2 - 1$$

$$= (x^3 - 1)(x^3 + 1) \qquad \text{Difference of squares}$$

$$= (x - 1)(x^2 + x + 1)(x + 1)(x^2 - x + 1) = 0 \qquad \text{Difference and sum of cubes}$$

Now we set each factor equal to zero.

$$x - 1 = 0 \quad \Rightarrow \quad x = 1$$

$$x^2 + x + 1 = 0 \quad \Rightarrow \quad x = \frac{-1 \pm \sqrt{1 - 4}}{2} = \frac{-1 \pm \sqrt{-3}}{2} \qquad \text{Quadratic formula}$$

$$x + 1 = 0 \quad \Rightarrow \quad x = -1$$

$$x^2 - x + 1 = 0 \quad \Rightarrow \quad x = \frac{1 \pm \sqrt{1 - 4}}{2} = \frac{1 \pm \sqrt{-3}}{2} \qquad \text{Quadratic formula}$$

So, the roots of the equation are

$$1, \quad -1, \quad -\frac{1}{2} + i\frac{\sqrt{3}}{2}, \quad -\frac{1}{2} - i\frac{\sqrt{3}}{2}, \quad \frac{1}{2} + i\frac{\sqrt{3}}{2}, \quad \frac{1}{2} - i\frac{\sqrt{3}}{2} \qquad ∎$$

The technique used in Example 4 can be used to find all solutions of the equations in Example 1. You are asked to carry out the details of these calculations in Exercises 33 and 34.

EXAMPLE 5 ■ Factoring by Grouping

Find all solutions of the equation $12x^3 - 4x^2 + 3x - 1 = 0$.

SOLUTION

The left-hand side of the equation can be factored by grouping the terms in pairs.

$$(12x^3 - 4x^2) + (3x - 1) = 0 \qquad \text{Group terms}$$

$$4x^2(3x - 1) + (3x - 1) = 0 \qquad \text{Factor } 3x - 1 \text{ from each pair}$$

$$(4x^2 + 1)(3x - 1) = 0 \qquad \text{Factor } 3x - 1 \text{ from LHS}$$

So
$$4x^2 + 1 = 0 \qquad \text{or} \qquad 3x - 1 = 0$$

$$x^2 = -\tfrac{1}{4} \qquad\qquad 3x = 1$$

$$x = \pm\tfrac{1}{2}i \qquad\qquad x = \tfrac{1}{3}$$

The solutions are $\tfrac{1}{2}i$, $-\tfrac{1}{2}i$, and $\tfrac{1}{3}$. ■

EXAMPLE 6 ■ Solving an Equation involving Fractional Exponents

Find all solutions of the equation $x^{5/6} + x^{2/3} = 2x^{1/2}$.

SOLUTION

| $x^{1/6} = w$ |
| $x^{1/3} = w^2$ |
| $x^{1/2} = w^3$ |

$$x^{5/6} + x^{2/3} - 2x^{1/2} = 0 \qquad \text{Move all terms to LHS}$$

$$x^{1/2}(x^{1/3} + x^{1/6} - 2) = 0 \qquad \text{Factor } x^{1/2} \text{ (lowest power of } x)$$

$$w^3(w^2 + w - 2) = 0 \qquad \text{Substitute } w = x^{1/6}$$

$$w^3(w - 1)(w + 2) = 0 \qquad \text{Factor the quadratic}$$

So
$$w = 0 \qquad \text{or} \qquad w = 1 \qquad \text{or} \qquad w = -2$$

$$x^{1/6} = 0 \qquad\qquad x^{1/6} = 1 \qquad\qquad x^{1/6} = -2$$

$$x = 0 \qquad\qquad x = 1^6 = 1 \qquad\qquad x = (-2)^6 = 64$$

CHECK YOUR ANSWERS

$x = 0$:

$\quad$ LHS $= 0^{5/6} + 0^{2/3} = 0$

$\quad$ RHS $= 2 \cdot 0^{1/2} = 0 \qquad ✓$

$x = 1$:

$\quad$ LHS $= 1^{5/6} + 1^{2/3} = 2$

$\quad$ RHS $= 2 \cdot 1^{1/2} = 2 \qquad ✓$

$x = 64$:

$\quad$ LHS $= 64^{5/6} + 64^{2/3}$

$\qquad = 32 + 16 = 48$

$\quad$ RHS $= 2 \cdot 64^{1/2} = 2 \cdot 8 = 16$

$\quad$ LHS $\neq$ RHS, so $x = 64$ is not a solution. $\qquad ✗$

Checking these answers, we see that $x = 0$ and $x = 1$ are solutions, but $x = 64$ is not (see *Check Your Answers*). The only solutions are 0 and 1. ■

To divide each side of the original equation in Example 6 by the common factor $x^{1/2}$ would have been wrong, because we would have lost the solution $x = 0$ by doing so. Never divide both sides of an equation by an expression containing the variable (unless you know that expression can never equal 0).

EXAMPLE 7 ■ Energy Expended in Bird Flight

Ornithologists have determined that some species of birds tend to avoid flights over large bodies of water during daylight hours, because air generally rises over land and falls over water in the daytime, so flying over water requires more energy. A bird is released from point A on an island, 5 mi from the nearest point B on a straight shoreline. The bird flies to a point C on the shoreline and then flies along the shoreline to its nesting area D, as shown in Figure 1. Suppose the bird has 170 kcal of energy reserves. It uses 10 kcal/mi flying over land and 14 kcal/mi flying over water.

(a) Where should the point C be located so that the bird uses exactly 170 kcal of energy during its flight?

(b) Does the bird have enough energy reserves to fly directly from A to D?

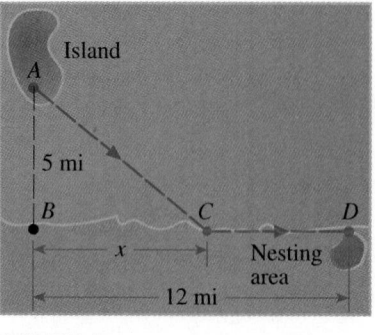

FIGURE 1

SOLUTION

(a) Let x be the distance in miles from B to C. Then

| Identify the variable |

| Relate the quantities |

$$\text{distance flown over water} = \sqrt{x^2 + 25} \qquad \text{Pythagorean Theorem}$$

$$\text{distance flown over land} = 12 - x$$

Since

$$\text{energy used} = (\text{energy per mile}) \times (\text{miles flown})$$

and since

$$\text{total energy} = (\text{energy over water}) + (\text{energy over land})$$

we set up the following equation for the energy that the bird uses on its trip:

| Set up an equation |

$$170 = 14\sqrt{x^2 + 25} + 10(12 - x)$$

To solve this equation, we eliminate the square root by first bringing all other terms to the left of the equal sign and then squaring each side.

| Solve |

$$170 - 10(12 - x) = 14\sqrt{x^2 + 25} \qquad \text{Isolate square root term on RHS}$$

$$50 + 10x = 14\sqrt{x^2 + 25} \qquad \text{Simplify LHS}$$

$$(50 + 10x)^2 = (14)^2(x^2 + 25) \qquad \text{Square each side}$$

$$2500 + 1000x + 100x^2 = 196x^2 + 4900 \qquad \text{Expand}$$

$$0 = 96x^2 - 1000x + 2400 \qquad \text{Move all terms to RHS}$$

This equation could be factored, but because the numbers are so large it is easier to use the quadratic formula:

$$x = \frac{1000 \pm \sqrt{(-1000)^2 - 4(96)(2400)}}{2(96)}$$

$$= \frac{1000 \pm 280}{192} = 6\tfrac{2}{3} \quad \text{or} \quad 3\tfrac{3}{4}$$

Point C should be either $6\tfrac{2}{3}$ mi or $3\tfrac{3}{4}$ mi from B so that the bird uses exactly 170 kcal of energy during its flight.

(b) By the Pythagorean Theorem, the length of the route directly from A to D is $\sqrt{5^2 + 12^2} = 13$ mi, so the energy the bird requires for that route is $14 \times 13 = 182$ kcal. This is more energy than the bird has available, so that route can't be taken by this bird. ∎

2.5 EXERCISES

1–32 ■ Find all real solutions of the equation.

1. $x^3 = 27$

2. $x^5 + 32 = 0$

3. $x^4 - 16 = 0$

4. $64x^6 = 27$

5. $x^4 - x^3 - 2x^2 = 0$

6. $(x - 2)^5 - 9(x - 2)^3 = 0$

7. $x^3 - 5x^2 - 2x + 10 = 0$

8. $2x^3 + x^2 - 18x - 9 = 0$

9. $x^3 - x^2 + x - 1 = x^2 + 1$

10. $7x^3 - x + 1 = x^3 + 3x^2 + x$

11. $\sqrt{2x + 1} + 1 = x$

12. $x - \sqrt{9 - 3x} = 0$

13. $\sqrt{5 - x} + 1 = x - 2$

14. $2x + \sqrt{x + 1} = 8$

15. $\sqrt{\sqrt{x - 5} + x} = 5$

16. $x^4 - 5x^2 + 4 = 0$

17. $2x^4 + 4x^2 + 1 = 0$

18. $x^6 - 2x^3 - 3 = 0$

19. $(x + 5)^2 - 3(x + 5) - 10 = 0$

20. $\left(\dfrac{x + 1}{x}\right)^2 + 4\left(\dfrac{x + 1}{x}\right) + 3 = 0$

21. $4(x + 1)^{1/2} - 5(x + 1)^{3/2} + (x + 1)^{5/2} = 0$

22. $x^{1/2} + 3x^{-1/2} = 10x^{-3/2}$

23. $x^{4/3} - 5x^{2/3} + 6 = 0$

24. $\sqrt{x} - 3\sqrt[4]{x} - 4 = 0$

25. $x^{1/2} - 3x^{1/3} = 3x^{1/6} - 9$

26. $x - 5\sqrt{x} + 6 = 0$

27. $\dfrac{1}{x^3} + \dfrac{4}{x^2} + \dfrac{4}{x} = 0$

28. $4x^{-4} - 16x^{-2} + 4 = 0$

29. $x^2\sqrt{x + 3} = (x + 3)^{3/2}$

30. $\sqrt{11 - x^2} - \dfrac{2}{\sqrt{11 - x^2}} = 1$

31. $\sqrt{x + \sqrt{x + 2}} = 2$

32. $\sqrt{1 + \sqrt{x + \sqrt{2x + 1}}} = \sqrt{5 + \sqrt{x}}$

33. Find all the solutions (including imaginary solutions) of the equation in Example l(a).

34. Find all the solutions of the equation in Example l(b).

35–44 ■ Find all complex number solutions of the equation.

35. $x^3 = 1$

36. $x^4 - 16 = 0$

37. $x^3 + x^2 + x = 0$

38. $x^4 + x^3 + x^2 + x = 0$

39. $x^4 - 6x^2 + 8 = 0$

40. $x^3 + 3x^2 + 9x + 27 = 0$

41. $x^6 - 9x^3 + 8 = 0$

42. $x^6 + 9x^4 - 4x^2 - 36 = 0$

43. $\sqrt{x^2 + 1} + \dfrac{8}{\sqrt{x^2 + 1}} = \sqrt{x^2 + 9}$

44. $\sqrt{1 - \sqrt{x^2 + 7}} = \sqrt{6 - x^2}$

45. A large plywood box has a volume of 180 ft³. Its length is 9 ft greater than its height, and its width is 4 ft less than its height. What are the dimensions of the box?

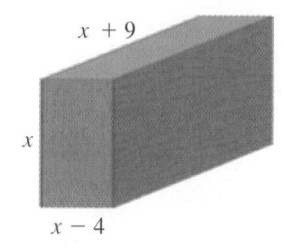

$x + 9$

x

$x - 4$

46. A jeweler has three small solid spheres made of gold, of radius 2 mm, 3 mm, and 4 mm. He wishes to melt these down and make just one sphere out of them. What will the radius of this larger sphere be?

47. The town of Foxton lies 10 mi north of an abandoned east-west road that runs through Grimley, as shown in the figure. The point on the abandoned road closest to Foxton is 40 mi from Grimley. County officials wish to build a new road connecting the two towns. They have determined that restoring the old road would cost $100,000 per mile, while building a new road would cost $200,000 per mile. How much of the abandoned road should be used (as indicated in the figure) if the officials intend to spend exactly $6.8 million? Would it cost less than this amount to build a new road connecting the towns directly?

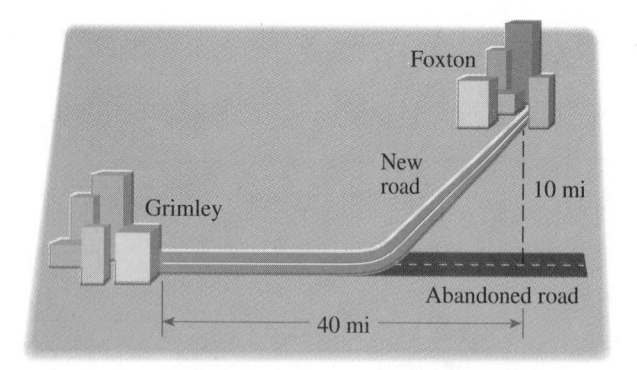

48. A boardwalk is parallel to and 210 ft inland from a straight shoreline. A sandy beach lies between the boardwalk and the shoreline. A man is standing on the boardwalk, exactly 750 ft across the sand from his beach umbrella, which is right at the shoreline. The man walks 4 ft/s on the boardwalk and 2 ft/s on the sand. How far should he walk on the boardwalk before veering off onto the sand if he wishes to reach his umbrella in exactly 4 min 45 s?

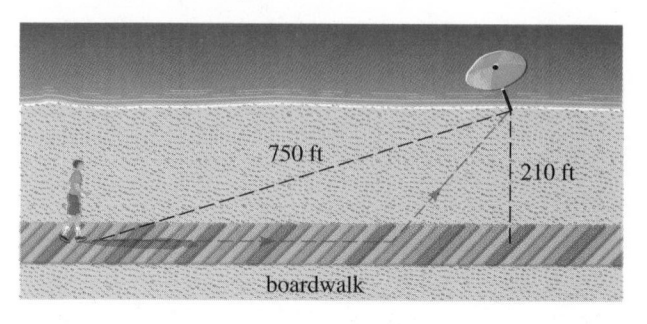

750 ft

210 ft

boardwalk

49. Grain is falling from a chute onto the ground, forming a conical pile whose diameter is always three times its height. How high is the pile (to the nearest hundredth of a foot) when it contains 1000 ft³ of grain?

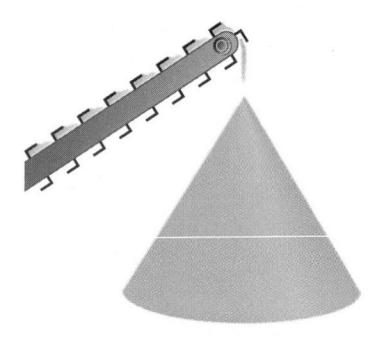

50. A spherical tank has a capacity of 750 gallons. Using the fact that one gallon is about 0.1337 ft³, find the radius of the tank (to the nearest hundredth of a foot).

51. A city lot has the shape of a right triangle whose hypotenuse is 7 ft longer than one of the other sides. The perimeter of the lot is 392 ft. How long is each of the three sides of the lot?

52. Two television monitors sitting beside each other on a shelf in an appliance store have the same screen height. One has a conventional screen, which is 5 in. wider than it is high. The other has a wider, high-definition screen, which is 2.4 times as wide as it is high. The diagonal measure of the wider screen is 14 in. more than the diagonal measure of the smaller. What are the heights of the screens? (Assume that the height is greater than 10 in.)

53–56 ■ Solve the equation for the variable x. The constants a and b represent positive real numbers.

53. $x^4 + 5ax^2 + 4a^2 = 0$

54. $a^3x^3 + b^3 = 0$

55. $\sqrt{x + a} + \sqrt{x - a} = \sqrt{2}\sqrt{x + 6}$

56. $\sqrt{x} + a\sqrt[3]{x} + b\sqrt[6]{x} + ab = 0$

| 2.6 | **LINEAR INEQUALITIES**

Some problems in algebra lead to **inequalities** instead of equations. An inequality looks just like an equation, except that in the place of the equal sign is one of the symbols $<$, $>$, $\leq$, or $\geq$. Here is an example of a linear inequality:

$$3x - 1 \leq 5$$

To **solve** an inequality that contains a variable means to find all values of the variable that make the inequality true. Unlike an equation, an inequality generally has infinitely many solutions, which form an interval or a union of intervals on the real line. (See Section 1.2 for a table that describes intervals.) In Example 1 we solve the inequality $3x - 1 \leq 5$, and we see that the methods we use for solving inequalities are similar to the methods we have used for solving equations.

EXAMPLE 1 ■ Solving a Linear Inequality

Solve the inequality $3x - 1 \leq 5$.

SOLUTION

As when solving an equation, we wish to simplify this inequality to an equivalent one in which the variable x is isolated on one side of the $\leq$ sign.

$$3x - 1 + 1 \leq 5 + 1 \quad \text{Add 1 to each side}$$

$$3x \leq 6 \quad \text{Simplify}$$

$$\tfrac{1}{3}(3x) \leq \tfrac{1}{3} \cdot 6 \quad \text{Multiply by } \tfrac{1}{3}$$

$$x \leq 2 \quad \text{Simplify}$$

FIGURE 1

The inequality is true for all values of x less than or equal to 2. In interval notation, this is the set $(-\infty, 2]$. The solution set is sketched in Figure 1. ■

To solve inequalities we use the following rules to isolate the variable on one side of the inequality sign. These rules tell us when two inequalities are equivalent (the symbol $\Longleftrightarrow$ means "is equivalent to"). In these rules, the symbols A, B, and C stand for real numbers or algebraic expressions. Here we state the rules for inequalities involving the symbol $\leq$, but they apply to each of the four inequality symbols.

RULES FOR INEQUALITIES

Rule	Description
1. $A \leq B \iff A + C \leq B + C$	**Adding** the same quantity to each side of an inequality gives an equivalent inequality.
2. $A \leq B \iff A - C \leq B - C$	**Subtracting** the same quantity from each side of an inequality gives an equivalent inequality.
3. If $C > 0$, then $\qquad A \leq B \iff CA \leq CB$	**Multiplying** each side of an inequality by the same *positive* quantity gives an equivalent inequality.
4. If $C < 0$, then $\qquad A \leq B \iff CA \geq CB$	**Multiplying** each side of an inequality by the same *negative* quantity *reverses the direction* of the inequality.
5. $0 < A \leq B \iff \dfrac{1}{A} \geq \dfrac{1}{B} > 0$	**Taking reciprocals** of each side of an inequality involving *positive* quantities *reverses the direction* of the inequality.
6. If $A \leq B$ and $C \leq D$, then $\qquad A + C \leq B + D$	Inequalities can be added.

Pay special attention to Rules 3 and 4. Rule 3 says that we can multiply (or divide) each side of an inequality by a *positive* number, but Rule 4 says that if we multiply each side of an inequality by a *negative* number, then we reverse the direction of the inequality. For example, if we start with the inequality

$$3 < 5$$

and multiply by 2, we get

$$6 < 10$$

but if we multiply by -2, we get

$$-6 > -10$$

EXAMPLE 2 ■ Solving a Linear Inequality

Solve the inequality $3x < 9x + 4$ and sketch the solution set.

SOLUTION

$$3x < 9x + 4$$

$$3x - 9x < 9x + 4 - 9x \qquad \text{Subtract } 9x$$

$$-6x < 4 \qquad \text{Simplify}$$

$$\left(-\tfrac{1}{6}\right)(-6x) > -\tfrac{1}{6}(4) \qquad \text{Multiply by } -\tfrac{1}{6} \text{ (or divide by } -6)$$

$$x > -\tfrac{2}{3} \qquad \text{Simplify}$$

The solution set consists of all numbers greater than $-\tfrac{2}{3}$. In other words, the solution of the inequality is the interval $\left(-\tfrac{2}{3}, \infty\right)$. It is graphed in Figure 2. ■

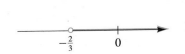

FIGURE 2

EXAMPLE 3 ■ Solving a Pair of Simultaneous Inequalities

Solve the inequalities $4 \leq 3x - 2 < 13$.

SOLUTION

The solution set consists of all values of x that satisfy both inequalities. Using Rules 1 and 3, we see that the following inequalities are equivalent:

$$4 \leq 3x - 2 < 13$$

$$6 \leq 3x < 15 \qquad \text{Add 2}$$

$$2 \leq x < 5 \qquad \text{Divide by 3}$$

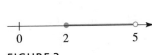

FIGURE 3

Therefore, the solution set is $[2, 5)$. It is shown in Figure 3. ■

EXAMPLE 4 ■ Simultaneous Inequalities that Must Be Solved Separately

Solve the inequalities $2x + 1 \leq 4x - 3 \leq x + 7$.

SOLUTION

Since all three expressions in these inequalities involve the variable x, it is impossible to deal with both inequalities simultaneously, as we did in Example 3. So, here we first solve the inequalities separately:

$$2x + 1 \leq 4x - 3 \qquad \text{and} \qquad 4x - 3 \leq x + 7$$

$$4 \leq 2x \qquad\qquad\qquad 3x \leq 10$$

$$2 \leq x \qquad\qquad\qquad x \leq \tfrac{10}{3}$$

Since x must satisfy both inequalities, we have

$$2 \leq x \leq \tfrac{10}{3}$$

FIGURE 4

Thus, the solution set is the closed interval $\left[2, \tfrac{10}{3}\right]$, as shown in Figure 4. ■

EXAMPLE 5 ■ Relationship between Fahrenheit and Celsius Scales

The instructions on a box of film indicate that the box should be stored at a temperature between 5 °C and 30 °C. What range of temperatures does this correspond to on the Fahrenheit scale?

SOLUTION

The relationship between degrees Celsius (C) and degrees Fahrenheit (F) is given by the equation $C = \frac{5}{9}(F - 32)$. Expressing the statement on the box in terms of inequalities, we have

$$5 < C < 30$$

so the corresponding Fahrenheit temperatures satisfy the inequalities

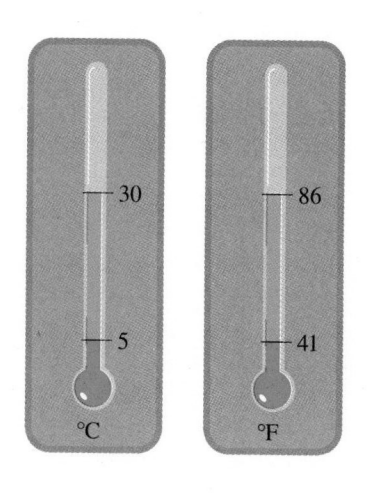

$$5 < \tfrac{5}{9}(F - 32) < 30$$

$$\tfrac{9}{5} \cdot 5 < F - 32 < \tfrac{9}{5} \cdot 30 \qquad \text{Multiply by } \tfrac{9}{5}$$

$$9 < F - 32 < 54 \qquad \text{Simplify}$$

$$9 + 32 < F < 54 + 32 \qquad \text{Add 32}$$

$$41 < F < 86 \qquad \text{Simplify}$$

The film should be stored at a temperature between 41 °F and 86 °F.   ■

Each of the inequalities that we have considered in this section is **linear**; that is, it is equivalent to an inequality of the form

$$ax + b > c$$

where a, b, and c are real numbers, and the symbol $>$ may be replaced by any of the other three inequality signs $<$, $\geq$, or $\leq$. This is the simplest kind of inequality. In the next section we learn to solve quadratic inequalities and other types of more complicated inequalities.

2.6 EXERCISES

1–8 ■ Let $S = \left\{-1, 0, \tfrac{1}{2}, \sqrt{2}, 2\right\}$. Use substitution to determine which of the elements of S satisfy the given inequality.

1. $x + 1 \geq 0$

2. $x - 2 < 0$

3. $2x + 10 > 8$

4. $4x + 1 \leq 2x$

5. $\dfrac{1}{x} \leq \dfrac{1}{2}$

6. $10 \geq \dfrac{5}{2x}$

7. $x^2 + 2 < 4$

8. $\dfrac{1}{2 - x^2} > 0$

9–40 ■ Solve the inequality. Express the solution in interval form and illustrate the solution set on the real number line.

9. $3x \leq 12$

10. $-4x > 16$

11. $20 < -4x$

12. $-1 \geq 7x$

13. $2x - 5 > 3$

14. $3x + 11 < 5$

15. $7 - x \geq 5$

16. $5 - 3x \leq -16$

17. $2x + 1 < 0$

18. $0 < 5 - 2x$

19. $3x + 11 \leq 6x + 8$

20. $6 - x \geq 2x + 9$

21. $1 - x \leq 2$

22. $4 - 3x \geq 6$

23. $4 - 3x \leq -(1 + 8x)$

24. $2(7x - 3) \leq 12x + 16$

25. $-1 < 2x - 5 < 7$

26. $1 < 3x + 4 \leq 16$

27. $0 \leq 1 - x < 1$

28. $-5 \leq 3 - 2x \leq 9$

29. $4x < 2x + 1 \leq 3x + 2$

30. $2x + 3 < x + 4 < 3x - 2$

31. $1 - x \geq 3 - 2x \geq x - 6$

32. $x > 1 - x \geq 3 + 2x$

33. $\dfrac{2}{3} \geq \dfrac{2x - 3}{12} > \dfrac{1}{6}$

34. $-\dfrac{1}{2} < \dfrac{4 - 3x}{5} \leq \dfrac{1}{4}$

35. $\dfrac{x - 1}{2} \geq \dfrac{2}{3} \geq 1 - \dfrac{x}{6}$

36. $4x < \dfrac{1 - x}{12} < \dfrac{2x + 1}{3}$

37. $\dfrac{1}{x} < 4$

38. $-3 < \dfrac{1}{x} \leq 1$

39. $\dfrac{2}{3} \leq \dfrac{1}{x - 2} < 1$

40. $\dfrac{1}{x} > 2 \geq \dfrac{1}{x + 2}$

41. Use the relationship between C and F given in Example 5 to find the interval on the Fahrenheit scale corresponding to the temperature range $20 \leq C \leq 30$.

42. What interval on the Celsius scale corresponds to the temperature range $50 \leq F \leq 95$?

43. A charter airline finds that on its Saturday flights from Philadelphia to London, all 120 seats will be sold if the ticket price is $200. However, for each $3 increase in ticket price, the number of seats sold decreases by one.
  (a) Find a formula for the number of seats sold if the ticket price is P dollars.
  (b) Over a certain period, the number of seats sold for this flight ranged between 90 and 115. What was the corresponding range of ticket prices?

44. As dry air moves upward, it expands and in so doing cools at a rate of about $1\,°C$ for each 100 m rise, up to about 12 km.
  (a) If the ground temperature is $20\,°C$, write a formula for the temperature at height h.
  (b) What range of temperatures can be expected if a plane takes off and reaches a maximum height of 5 km?

45. It is estimated that the annual cost of driving a new Cruiser is given by the formula

$$C = 0.35m + 2200$$

where m represents the number of miles driven per year and C is the cost in dollars. Jane has purchased a Cruiser, and decides to budget between $6400 and $7100 for the next year's driving costs. What is the corresponding range of miles that she can drive her new Cruiser?

46. A coffee merchant sells a customer 3 lb of Hawaiian Kona at $6.50 per pound. His scale is accurate to within $\pm\, 0.03$ lb. By how much could the customer have been overcharged or undercharged because of possible inaccuracy in the scale?

47. Prove Rule 1. [*Hint:* Remember that $a < b$ means that $b - a > 0$.]

48. Prove Rule 3. [See the hint for Exercise 47 and remember that the product of two positive numbers is positive.]

49. Prove Rule 4. [*Hint:* Use the fact that the product of a positive number and a negative number is negative.]

50. Prove Rule 5.

51–52 ■ Solve the inequality for x, assuming that a, b, and c are positive constants.

51. $a(bx - c) \geq bc$

52. $a \leq bx + c < 2a$

53–54 ■ Solve the inequality for x, assuming that a, b, and c are negative constants.

53. $ax + b < c$

54. $\dfrac{ax + b}{c} \leq b$

55. Show that if $a < b$, then $a < \dfrac{a + b}{2} < b$.

56. Show that if $0 < a < b$, then $a^2 < b^2$.

57. Suppose that a, b, c, and d are positive numbers such that

$$\frac{a}{b} < \frac{c}{d}$$

Show that

$$\frac{a}{b} < \frac{a + c}{b + d} < \frac{c}{d}$$

NONLINEAR INEQUALITIES

In the preceding section, we learned how to solve linear inequalities. To solve inequalities involving squares and other powers of the variable we use factoring, together with the fact that the product of two expressions is positive if either (i) both are positive or (ii) both are negative, whereas the product is negative if one factor is positive and the other is negative. This can be generalized to the following principle for any number of factors.

THE SIGN OF A PRODUCT OR QUOTIENT

If a product or a quotient has an *even* number of *negative* factors, then its value is *positive*.

If a product or quotient has an *odd* number of *negative* factors, then its value is *negative*.

EXAMPLE 1 ■ A Quadratic Inequality

Solve the inequality $x^2 - 5x + 6 \leq 0$.

SOLUTION

First we factor the left side:

$$(x - 2)(x - 3) \leq 0$$

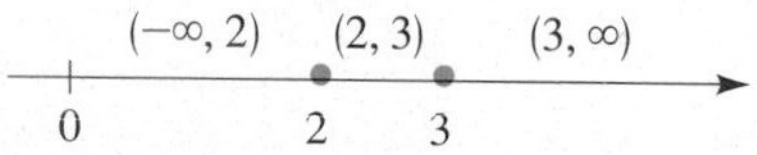

FIGURE 1

We know that the corresponding equation $(x - 2)(x - 3) = 0$ has the solutions 2 and 3. These are the only points at which the expression $(x - 2)(x - 3)$ can change sign. As shown in Figure 1, the numbers 2 and 3 divide the real line into three intervals: $(-\infty, 2)$, $(2, 3)$, and $(3, \infty)$. On each of these intervals we determine the signs of the factors. To do this we use **test values.** We choose a number inside each interval, and check the sign of the factors $x - 2$ and $x - 3$ at the value selected. For instance, if we use the test value $x = 1$ for the interval $(-\infty, 2)$, then substitution in the factors $x - 2$ and $x - 3$ gives

$$x - 2 = 1 - 2 = -1 < 0$$

and

$$x - 3 = 1 - 3 = -2 < 0$$

so both factors are negative on this interval. The factors $x - 2$ and $x - 3$ change sign only at 2 and 3, respectively, so they maintain their signs over the length of each interval. That is why using just a single test value on each interval is sufficient.

Using the test values $x = 2\frac{1}{2}$ and $x = 4$ for the intervals $(2, 3)$ and $(3, \infty)$, respectively, we construct the following sign table. The final row of the table is obtained from the fact that the expression in the last row is the product of the two factors.

Interval	$(-\infty, 2)$	$(2, 3)$	$(3, \infty)$
Sign of $x - 2$	−	+	+
Sign of $x - 3$	−	−	+
Sign of $(x - 2)(x - 3)$	+	−	+

If you prefer, you can represent this information on a real number line, as in the following sign diagram. The vertical lines indicate the points at which the real line is divided into intervals.

		2		3	
Sign of $x - 2$	−		+		+
Sign of $x - 3$	−		−		+
Sign of $(x - 2)(x - 3)$	+		−		+

We read from the table or the diagram that $(x - 2)(x - 3)$ is negative on the interval $(2, 3)$. Thus, the solution of the inequality $(x - 2)(x - 3) \leq 0$ is

$$\{x \mid 2 \leq x \leq 3\} = [2, 3]$$

We have included the endpoints 2 and 3 because we seek values of x such that the product is either less than *or equal to* zero. The solution is illustrated in Figure 2.

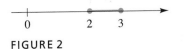

FIGURE 2

We use the following steps to solve an inequality that can be factored.

SOLVING NONLINEAR INEQUALITIES

1. If necessary, rewrite the inequality so that all nonzero terms appear on one side of the inequality sign.

2. If the nonzero side of the inequality involves quotients, bring them to a common denominator.

3. Factor the nonzero side of the inequality.

4. List the intervals determined by the factorization.

5. Make a table or diagram of the signs of each factor on each interval. In the last row of the table we determine the sign of the product (or quotient) of these factors.

6. Determine the solution set from the last row of the table. Be sure to check whether the inequality is satisfied by some or all of the endpoints of the intervals (this may happen if the inequality involves $\geq$ or $\leq$).

⊘   The factoring technique described in these steps works only if all nonzero terms appear on one side of the inequality symbol. If the inequality is not written in this form, first rewrite it, as indicated in Step 1. This technique is illustrated in the examples that follow.

EXAMPLE 2 ■ Solving an Inequality by Factoring

Solve $x^2 + 3x > 4$.

SOLUTION

First we move all nonzero terms to one side of the inequality sign and factor the resulting expression.

$$x^2 + 3x - 4 > 0 \qquad \text{Subtract 4}$$

$$(x - 1)(x + 4) > 0 \qquad \text{Factor}$$

The expression $(x - 1)(x + 4)$ is 0 when $x = 1$ and $x = -4$, so we obtain the three intervals $(-\infty, -4)$, $(-4, 1)$, and $(1, \infty)$. On each interval the product keeps a constant sign, as shown in the following table. The signs in the table could be obtained by using, for example, the test values -5, 0, and 2, respectively, for the three intervals.

Interval	$(-\infty, -4)$	$(-4, 1)$	$(1, \infty)$
Sign of $x - 1$	−	−	+
Sign of $x + 4$	−	+	+
Sign of $(x - 1)(x + 4)$	+	−	+

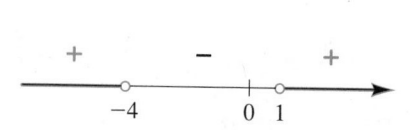

FIGURE 3

We read from the table that the solution set is

$$\{x \mid x < -4 \text{ or } x > 1\} = (-\infty, -4) \cup (1, \infty)$$

The solution is illustrated in Figure 3.   ■

In the next example, we solve an inequality involving a quotient instead of a product.

EXAMPLE 3 ■ An Inequality involving a Quotient

Solve $\dfrac{1 + x}{1 - x} \geq 1$.

SOLUTION 1

First, we move all nonzero terms to the left side, and then we simplify using a common denominator:

$$\frac{1 + x}{1 - x} \geqslant 1$$

$$\frac{1 + x}{1 - x} - 1 \geqslant 0 \qquad \text{Subtract 1}$$

$$\frac{1 + x}{1 - x} - \frac{1 - x}{1 - x} \geqslant 0 \qquad \text{Common denominator } 1 - x$$

$$\frac{1 + x - 1 + x}{1 - x} \geqslant 0 \qquad \text{Combine the fractions}$$

$$\frac{2x}{1 - x} \geqslant 0 \qquad \text{Simplify}$$

The numerator is zero when $x = 0$ and the denominator is zero when $x = 1$, so we construct the following sign table using these values to define intervals on the real line.

Interval	$(-\infty, 0)$	$(0, 1)$	$(1, \infty)$
Sign of $2x$	−	+	+
Sign of $1 - x$	+	+	−
Sign of $\dfrac{2x}{1 - x}$	−	+	−

0   1

FIGURE 4

From the table we see that the solution set is $\{x \mid 0 \leqslant x < 1\} = [0, 1)$. We include the endpoint 0 because the original inequality requires the quotient to be greater than *or equal to* 1. However we do not include the other endpoint 1, since the quotient in the inequality is not defined at 1. Always check the endpoints of solution intervals to determine whether they satisfy the original inequality.

The solution set $[0, 1)$ is illustrated in Figure 4.

SOLUTION 2

Another way to solve the inequality is to simplify it by multiplying each side by $1 - x$. But this must be done with extreme care, because we do not know whether $1 - x$ is positive or negative. If $1 - x$ turns out to be negative for some values of x in the solution set, then multiplying by $1 - x$ will reverse the direction of the inequality. We must take both cases into account.

■ **Case 1** If $1 - x > 0$, then we have

$$\left(\frac{1 + x}{1 - x}\right)(1 - x) \geqslant 1(1 - x) \qquad \text{Multiply by } 1 - x$$

$$1 + x \geqslant 1 - x \qquad \text{Cancel}$$

$$2x \geqslant 0 \qquad \text{Subtract } 1 - x$$

$$x \geqslant 0 \qquad \text{Divide by } 2$$

Thus, since we are assuming $1 - x > 0$, that is, $1 > x$, we get $0 \leqslant x < 1$.

■ **Case 2** If $1 - x < 0$, then we must reverse the direction of the inequality when multiplying by $1 - x$:

$$\left(\frac{1 + x}{1 - x}\right)(1 - x) \leqslant 1(1 - x) \qquad \text{Multiply by } 1 - x$$

$$1 + x \leqslant 1 - x \qquad \text{Cancel}$$

$$2x \leqslant 0 \qquad \text{Subtract } 1 - x$$

$$x \leqslant 0 \qquad \text{Divide by } 2$$

This time we are assuming $1 - x < 0$, that is, $1 < x$. Since it is impossible for x to simultaneously satisfy $x \leqslant 0$ and $1 < x$, there is no solution in this case.

Combining Cases 1 and 2, we see that the solution set is the interval $[0, 1)$. ■

⊘   In Example 3, the second solution shows why we must be extremely careful when multiplying each side of an inequality by an expression involving the variable. We must always consider that the sign of such an expression may be negative or positive. If we had not taken the sign of $1 - x$ into account, we would have ended up with the solution $x \geqslant 0$, which is incorrect. To avoid this problem, we will use the method of Solution 1 in the following examples.

EXAMPLE 4 ■ Solving an Inequality with Three Factors

Solve the inequality $x < \dfrac{2}{x - 1}$.

SOLUTION

After moving all nonzero terms to one side of the inequality, we use a common denominator to combine the terms:

$$x - \frac{2}{x - 1} < 0 \qquad \text{Subtract } \frac{2}{x - 1}$$

$$\frac{x(x - 1)}{x - 1} - \frac{2}{x - 1} < 0 \qquad \text{Common denominator } x - 1$$

$$\frac{x^2 - x - 2}{x - 1} < 0 \qquad \text{Combine the fractions}$$

$$\frac{(x + 1)(x - 2)}{x - 1} < 0 \qquad \text{Factor the numerator}$$

The factors in this quotient change sign at -1, 1, and 2, so we must examine the intervals $(-\infty, -1)$, $(-1, 1)$, $(1, 2)$, and $(2, \infty)$. Using test values, we get the following table.

Interval	$(-\infty, -1)$	$(-1, 1)$	$(1, 2)$	$(2, \infty)$
Sign of $x + 1$	$-$	$+$	$+$	$+$
Sign of $x - 2$	$-$	$-$	$-$	$+$
Sign of $x - 1$	$-$	$-$	$+$	$+$
Sign of $\dfrac{(x + 1)(x - 2)}{x - 1}$	$-$	$+$	$-$	$+$

Since the quotient must be negative, the solution is

$$(-\infty, -1) \cup (1, 2)$$

as illustrated in Figure 5.

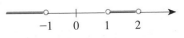

FIGURE 5

EXAMPLE 5 ■ The Path of a Projectile

A stone thrown straight up into the air at 96 ft/s reaches a height of h feet after t seconds, where h and t are related by the formula

$$h = 96t - 32t^2$$

During what time interval will the stone be at least 64 feet above the ground?

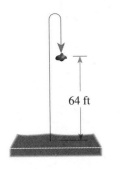

64 ft

SOLUTION

We wish to determine for what times t we have that $h \geq 64$. Thus, we must solve the inequality $96t - 32t^2 \geq 64$. After first dividing by the common factor 32, we get

$$3t - t^2 \geq 2$$

$$0 \geq t^2 - 3t + 2 \qquad \text{Move all terms to RHS}$$

$$0 \geq (t - 1)(t - 2) \qquad \text{Factor}$$

The intervals to check are $(-\infty, 1)$, $(1, 2)$, and $(2, \infty)$. This time we use a sign diagram instead of a sign table.

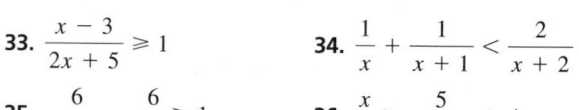

	1		2	
Sign of $t - 1$	$-$		$+$	$+$
Sign of $t - 2$	$-$		$-$	$+$
Sign of $(t - 1)(t - 2)$	$+$		$-$	$+$

Thus, the solution of the inequality is the interval $[1, 2]$. The stone is at least 64 ft high between 1 and 2 seconds (inclusive) after it is thrown. ■

2.7 | **EXERCISES**

1–40 ■ Solve the inequality. Express the solution in interval form and illustrate the solution set on the real number line.

1. $(x - 2)(x - 5) > 0$

2. $(3x + 1)(x - 1) \geq 0$

3. $x^2 - 3x - 18 \leq 0$

4. $x^2 + 5x + 6 > 0$

5. $2x^2 + x \geq 1$

6. $x^2 < x + 2$

7. $3x^2 - 3x < 2x^2 + 4$

8. $5x^2 + 3x \geq 3x^2 + 2$

9. $x^2 > 3(x + 6)$

10. $x^2 + 2x > 3$

11. $x^2 < 4$

12. $x^2 \geq 9$

13. $-2x^2 \leq 4$

14. $(x + 2)(x - 1)(x - 3) \leq 0$

15. $x(x^2 - 4) \geq 0$

16. $x^3 > x$

17. $\dfrac{x - 3}{x + 1} \geq 0$

18. $\dfrac{2x + 6}{x - 2} < 0$

19. $\dfrac{4x}{2x + 3} > 2$

20. $-2 < \dfrac{x + 1}{x - 3}$

21. $\dfrac{2x + 1}{x - 5} \leq 3$

22. $\dfrac{3 + x}{3 - x} \geq 1$

23. $\dfrac{4}{x} < x$

24. $\dfrac{x}{x + 1} > 3x$

25. $\dfrac{x^2 - 4}{x^2 + 4} \geq 0$

26. $\dfrac{x}{x + 2} \leq \dfrac{1}{x}$

27. $1 + \dfrac{2}{x + 1} \leq \dfrac{2}{x}$

28. $\dfrac{3}{x - 1} - \dfrac{4}{x} \geq 1$

29. $\dfrac{1}{1 - x} \leq \dfrac{3}{x}$

30. $\dfrac{(x - 1)^2}{(x + 1)(x + 2)} > 0$

31. $\dfrac{x^2 + 2x - 3}{x^2 - 7x + 6} > 0$

32. $\dfrac{x^2 - 16}{x^4 - 16} < 0$

33. $\dfrac{x - 3}{2x + 5} \geq 1$

34. $\dfrac{1}{x} + \dfrac{1}{x + 1} < \dfrac{2}{x + 2}$

35. $\dfrac{6}{x - 1} - \dfrac{6}{x} \geq 1$

36. $\dfrac{x}{2} \geq \dfrac{5}{x + 1} + 4$

37. $\dfrac{x + 2}{x + 3} < \dfrac{x - 1}{x - 2}$

38. $\dfrac{1}{x + 1} + \dfrac{1}{x + 2} \leq 0$

39. $x^4 > x^2$

40. $x^5 > x^2$

41. Using calculus it can be shown that if a ball is thrown upward with an initial velocity of 16 ft/s from the top of a building 128 ft high, then its height h above the ground t seconds later will be

$$h = 128 + 16t - 16t^2$$

During what time interval will the ball be at least 32 ft above the ground?

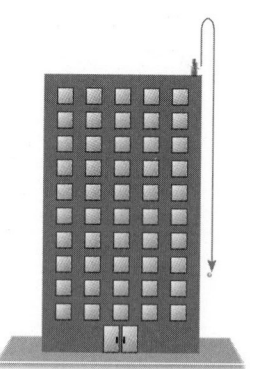

42. The gravitational force F exerted by the earth on an object having a mass of 100 kg is given by the equation

$$F = \dfrac{4,000,000}{d^2}$$

where d is the distance (in km) of the object from the center of the earth, and the force F is measured in newtons (N). For what distances will the gravitational force exerted by the earth on this object be between 0.0004 N and 0.01 N?

43. In the vicinity of a bonfire, the temperature T in °C at a distance of x meters from the center of the fire was given by $T = \dfrac{600{,}000}{x^2 + 300}$. At what range of distances from the fire center was the temperature less than 500 °C?

44. The gas mileage g (measured in mi/gal) for a particular vehicle, driven at v mi/h, is given by the formula

$g = 10 + 0.9v - 0.01v^2$, as long as v is between 10 mi/h and 75 mi/h. For what range of speeds is the the vehicle's mileage 30 mi/gal or better?

45–48 ■ Determine the values of the variable for which the expression is defined as a real number.

45. $\sqrt{16 - 9x^2}$

46. $\sqrt{3x^2 - 5x + 2}$

47. $\left(\dfrac{1}{x^2 - 5x - 14}\right)^{1/2}$

48. $\sqrt[4]{\dfrac{1 - x}{2 + x}}$

49. Solve $\dfrac{x^2 + (a - b)x - ab}{x + c} \leq 0$, where $0 < a < b < c$.

2.8 ABSOLUTE VALUE

Recall from Section 1.2 that the absolute value of a number a is given by

$$|a| = \begin{cases} a & \text{if } a \geq 0 \\ -a & \text{if } a < 0 \end{cases}$$

FIGURE 1

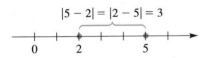

FIGURE 2

and that it represents the distance from a to the origin on the real number line (see Figure 1). Thus, putting the absolute value symbol around an expression leaves its value unchanged if the expression is already positive, or turns a negative expression into a positive one.

More generally, it follows from the definition of absolute value that $|x - a|$ is the distance between x and a on the real number line, as we learned in Section 1.2 (see Figure 2).

EXAMPLE 1 ■ Calculating Absolute Values

Write each of the following expressions without using an absolute value symbol.

(a) $|7 - 10|$ (b) $|a - 6|$ if $a < 6$ (c) $|2x - 1|$

SOLUTION

(a) $|7 - 10| = |-3| = 3$

(b) If $a < 6$, then $a - 6 < 0$. So $|a - 6| = -(a - 6) = 6 - a$.

(c)
$$|2x - 1| = \begin{cases} 2x - 1 & \text{if } 2x - 1 \geq 0 \\ -(2x - 1) & \text{if } 2x - 1 < 0 \end{cases}$$

$$= \begin{cases} 2x - 1 & \text{if } x \geq \frac{1}{2} \\ 1 - 2x & \text{if } x < \frac{1}{2} \end{cases}$$

■

The following properties can be proved using the definition of absolute value. They are useful in solving equations and inequalities and in simplifying expressions that involve absolute values.

PROPERTIES OF ABSOLUTE VALUE

Let a, b, and c be real numbers, with $c > 0$, and let n be an integer.

1. $|ab| = |a||b|$

2. $\left|\dfrac{a}{b}\right| = \dfrac{|a|}{|b|}$ $(b \neq 0)$

3. $|a^n| = |a|^n$

4. $|x| = c$ is equivalent to $x = \pm c$

5. $|x| < c$ is equivalent to $-c < x < c$

6. $|x| > c$ is equivalent to $x > c$ or $x < -c$

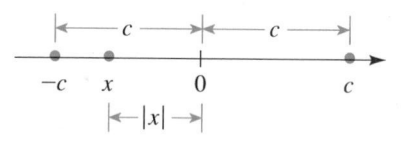

FIGURE 3

To prove Property 5, for example, we see that the inequality $|x| < c$ says that the distance from x to 0 is less than c, and from Figure 3 you can see that this is true if and only if x is between c and $-c$.

EXAMPLE 2 ■ Simplifying an Absolute Value Expression

Simplify the expression $|2x - 6|$.

SOLUTION

Using Property 1, we have

$$|2x - 6| = |2(x - 3)| = |2||x - 3| = 2|x - 3|$$ ∎

EXAMPLE 3 ■ Solving an Equation involving Absolute Value

Solve the equation $|2x - 5| = 3$.

SOLUTION

By Property 4, the equation $|2x - 5| = 3$ is equivalent to

$$2x - 5 = 3 \quad \text{or} \quad 2x - 5 = -3$$
$$2x = 8 \qquad\qquad 2x = 2$$
$$x = 4 \qquad\qquad x = 1$$

The solutions are 1 and 4.   ∎

EXAMPLE 4 ■ Solving an Inequality involving Absolute Value

Solve the inequality $|x - 5| < 2$.

SOLUTION 1

By Property 5, the inequality $|x - 5| < 2$ is equivalent to

$$-2 < x - 5 < 2$$

Therefore, adding 5 to each side, we have

$$3 < x < 7$$

and the solution set is the open interval $(3, 7)$.

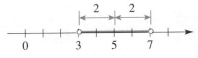

FIGURE 4

SOLUTION 2

Geometrically, the solution set consists of all numbers x whose distance from 5 is less than 2. From Figure 4 we see that this is the interval $(3, 7)$. ■

EXAMPLE 5 ■ Solving an Inequality involving Absolute Value

Solve the inequality $|3x + 2| \geq 4$.

SOLUTION

By Properties 4 and 6, the inequality $|3x + 2| \geq 4$ is equivalent to

$$3x + 2 \geq 4 \qquad \text{or} \qquad 3x + 2 \leq -4$$

In the first case $3x \geq 2$, which gives $x \geq \frac{2}{3}$. In the second case $3x \leq -6$, which gives $x \leq -2$. So, the solution set is

$$\left\{ x \mid x \leq -2 \quad \text{or} \quad x \geq \tfrac{2}{3} \right\} = (-\infty, -2] \cup \left[\tfrac{2}{3}, \infty \right)$$

The set is graphed in Figure 5. ■

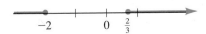

FIGURE 5

EXAMPLE 6 ■ Solving an Inequality involving Absolute Value and a Quotient

Solve the inequality $\dfrac{3}{|2x - 5|} \geq 1$.

SOLUTION

Since the absolute value of any number is always nonnegative, if we multiply each side of the inequality by $|2x - 5|$, we will not reverse the direction of the inequality. So, as long as $2x - 5 \neq 0$, or $x \neq \frac{5}{2}$, each of the following inequalities is equivalent to the one we are solving:

$$3 \geq |2x - 5| \qquad \text{Multiply by } |2x - 5|$$

$$3 \geq 2x - 5 \geq -3 \qquad \text{Property 5}$$

$$8 \geq 2x \geq 2 \qquad \text{Add 5}$$

$$4 \geq x \geq 1 \qquad \text{Divide by 2}$$

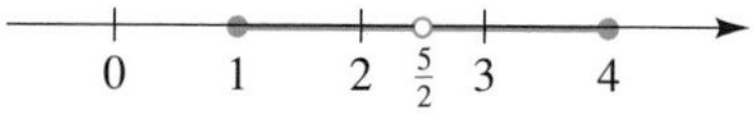

FIGURE 6

Thus, the solution consists of all numbers in the interval $[1, 4]$ except the number $\frac{5}{2}$. We write this set as $\left[1, \frac{5}{2}\right) \cup \left(\frac{5}{2}, 4\right]$. The set is graphed in Figure 6. ∎

Another important property of absolute value—the Triangle Inequality—is used frequently, not only in calculus but throughout mathematics in general.

THE TRIANGLE INEQUALITY

If a and b are any real numbers, then

$$|a + b| \leq |a| + |b|$$

Observe that if the numbers a and b are both positive or negative, then the two sides in the Triangle Inequality are actually equal. However, if a and b have opposite signs, the left side involves a subtraction but the right side involves only addition. This makes the Triangle Inequality seem reasonable, but we can prove it as follows.

■ **Proof of the Triangle Inequality**   Notice that

$$-|a| \leq a \leq |a|$$

is always true because a equals either $|a|$ or $-|a|$. If we write the corresponding statement for b, we have

$$-|b| \leq b \leq |b|$$

and, adding these inequalities, we get

$$-(|a| + |b|) \leq a + b \leq |a| + |b|$$

If we now apply Properties 4 and 5 (with x replaced by $a + b$ and c by $|a| + |b|$), we obtain

$$|a + b| \leq |a| + |b|$$

which is what we wanted to show.   □

EXAMPLE 7 ■ Using the Triangle Inequality

If $|x - 4| < 0.1$ and $|y - 7| < 0.2$, use the Triangle Inequality to estimate $|(x + y) - 11|$.

SOLUTION

In order to use the given information, we use the Triangle Inequality with $a = x - 4$ and $b = y - 7$:

$$|(x + y) - 11| = |(x - 4) + (y - 7)|$$
$$\leq |x - 4| + |y - 7|$$
$$< 0.1 + 0.2 = 0.3$$

Thus $\qquad |(x + y) - 11| < 0.3$ ■

2.8 EXERCISES

1–8 ■ Use the definition of absolute value to write the expression without an absolute value symbol, as in Example 1.

1. $|7 - 3|$

2. $\left|3\frac{1}{2} - 5\right|$

3. $|50 - 50.1|$

4. $|a - 4|$ if $a \geq 4$

5. $|a - 4|$ if $a < 4$

6. $|2x - 8|$ if $x \leq 4$

7. $|x - 3|$

8. $|3x - 12|$

9–14 ■ Simplify the expression, as in Example 2.

9. $|3x + 9|$

10. $|4x - 16|$

11. $\left|\frac{1}{2}x - \frac{5}{2}\right|$

12. $|-2x - 10|$

13. $|-x^2 - 9|$

14. $\left|\frac{x - 1}{1 - x}\right|$

15–22 ■ Solve the equation.

15. $|4x| = 12$

16. $|2x - 1| = 3$

17. $|x - 2| = 0.05$

18. $|x + 5| = -2$

19. $\left|\frac{2x - 1}{x + 1}\right| = 3$

20. $6 + 3|x + 5| = 5$

21. $|x - 1| = |3x + 2|$

22. $|x + 3| = |2x + 1|$

23–44 ■ Solve the inequality. Express the answer using interval notation.

23. $|x| < 2$

24. $|x| \geq 4$

25. $|x - 5| \leq 3$

26. $|x - 9| > 9$

27. $|x + 1| \geq 1$

28. $|x + 4| \leq 0$

29. $|x + 5| \geq 2$

30. $|x + 1| \geq 3$

31. $|2x - 3| \leq 0.4$

32. $|5x - 2| < 6$

33. $\left|\frac{x - 2}{3}\right| < 2$

34. $\left|\frac{x + 1}{2}\right| \geq 4$

35. $|x + 6| < 0.001$

36. $|x - a| < d$

37. $1 \leq |x| \leq 4$

38. $0 < |x - 5| \leq \frac{1}{2}$

39. $\frac{1}{|x + 7|} > 2$

40. $\left|1 - \frac{1}{x}\right| \leq 2$

41. $|x| > |x - 1|$

42. $|2x - 5| \leq |x + 4|$

43. $\left|\frac{x}{2 + x}\right| < 1$

44. $\left|\frac{2 - 3x}{1 + 2x}\right| \leq 4$

45. Prove that $|ab| = |a||b|$.

46. Prove that $\left|\frac{a}{b}\right| = \frac{|a|}{|b|}$.

47. Suppose that $|x - 2| < 0.01$ and $|y - 3| < 0.04$. Show that $|(x + y) - 5| < 0.05$.

48. If $|a - 1| < 2$ and $|b - 1| < 3$, show that $|a + b - 2| < 5$.

49. Show that $|x - y| \leq |x| + |y|$ for all real numbers x and y.

50. Show that $|x - y| \geq |x| - |y|$ for all real numbers x and y. [*Hint:* Use the Triangle Inequality with $a = x - y$ and $b = y$.]

51. Prove that $\qquad |a| = \sqrt{a^2}$ for any real number a.

---

2 REVIEW

KEY TOPICS ■ Define, state, or discuss each of the following.

1. Equation

2. Variable

3. Solution, root

4. Identity

5. Linear equation

6. Quadratic equation

7. Zero-product property

8. Completing the square

9. Quadratic formula

10. Discriminant of a quadratic equation

11. Complex number

12. Real part, imaginary part

13. Imaginary number, pure imaginary number

14. Adding and subtracting complex numbers

15. Multiplying complex numbers

16. Dividing complex numbers

17. Complex conjugate

18. Inequality

19. Linear inequality

20. Solving inequalities using test values

21. Solving inequalities that involve absolute value

22. Triangle Inequality

EXERCISES

1–26 ■ Find all real solutions of the equation.

1. $3x + 12 = 24$

2. $5x - 7 = 42$

3. $7x - 6 = 4x + 9$

4. $8 - 2x = 14 + x$

5. $\frac{1}{3}x - \frac{1}{2} = 2$

6. $\frac{2}{3}x + \frac{3}{5} = \frac{1}{5} - 2x$

7. $2(x + 3) - 4(x - 5) = 8 - 5x$

8. $\dfrac{x - 5}{2} - \dfrac{2x + 5}{3} = \dfrac{5}{6}$

9. $\dfrac{x + 1}{x - 1} = \dfrac{2x - 1}{2x + 1}$

10. $\dfrac{x}{x + 2} - 3 = \dfrac{1}{x + 2}$

11. $x^2 - 9x + 14 = 0$

12. $x^2 + 24x + 144 = 0$

13. $2x^2 + x = 1$

14. $3x^2 + 5x - 2 = 0$

15. $4x^3 - 25x = 0$

16. $x^3 - 2x^2 - 5x + 10 = 0$

17. $3x^2 + 4x - 1 = 0$

18. $x^2 - 3x + 9 = 0$

19. $\dfrac{1}{x} + \dfrac{2}{x - 1} = 3$

20. $\dfrac{x}{x - 2} + \dfrac{1}{x + 2} = \dfrac{8}{x^2 - 4}$

21. $x^4 - 8x^2 - 9 = 0$

22. $x - 4\sqrt{x} = 32$

23. $x^{-1/2} - 2x^{1/2} + x^{2/3} = 0$

24. $|3x| = 18$

25. $|x - 7| = 4$

26. $|2x - 5| = 9$

27. The owner of a store sells raisins for \$3.20 per pound and nuts for \$2.40 per pound. He decides to mix the raisins and nuts and sell 50 lb of the mixture for \$2.72 per pound. What quantities of raisins and nuts should he use?

28. Anthony leaves Kingstown at 2:00 P.M. and drives to Queensville, 160 mi distant, at 45 mi/h. At 2:15 P.M. Helen leaves Queensville and drives to Kingstown at 40 mi/h. At what time do they pass each other on the road?

29. A woman cycles 8 mi/h faster than she runs. Every morning she cycles 4 mi and runs $2\frac{1}{2}$ mi, for a total of one hour of exercise. How fast does she run?

30. The approximate distance d (in feet) that a driver travels after noticing that he or she must come to a sudden stop is given by the following formula, where x is the speed of the car (in miles per hour):

$$d = x + \frac{x^2}{20}$$

If a car travels 75 ft before stopping, what must its speed have been before the brakes were applied?

31. The hypotenuse of a right triangle has length 20 cm. The sum of the lengths of the other two sides is 28 cm. Find the lengths of the other two sides of the triangle.

32. Abbie paints twice as fast as Beth and three times as fast as Cathie. If it takes them 60 min to paint a living room with all three working together, how long would it take Abbie if she works alone?

33. A rectangular swimming pool is 8 ft deep everywhere and twice as long as it is wide. If the pool holds 8464 ft^3 of water, what are its dimensions?

34. A homeowner wishes to fence in three adjoining garden plots, one for each of her children, as shown in the figure. If each plot is to be 80 ft^2 in area, and she has 88 ft of fencing material at hand, what dimensions should each plot have?

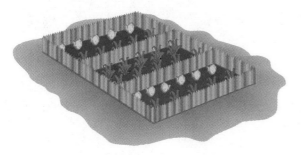

35–44 ■ Evaluate the expression and write the result in the form $a + bi$.

35. $(3 - 5i) - (6 + 4i)$

36. $(-2 + 3i) + \left(\frac{1}{2} - i\right)$

37. $(2 + 7i)(6 - i)$

38. $3(5 - 2i)\dfrac{i}{5}$

39. $\dfrac{2 - 3i}{2 + 3i}$

40. $\dfrac{2 + i}{4 - 3i}$

41. i^{45}

42. $(3 - i)^3$

43. $\left(1 - \sqrt{-3}\,\right)\left(2 + \sqrt{-4}\,\right)$

44. $\sqrt{-5} \cdot \sqrt{-20}$

45–54 ■ Find all real and imaginary solutions of the equation.

45. $x^2 + 16 = 0$

46. $x^2 = -12$

47. $x^2 + 6x + 10 = 0$

48. $2x^2 - 3x + 2 = 0$

49. $(1 + 2i)x - 5 = 2x$

50. $ix^2 - 3x - 2i = 0$

51. $x^4 - 256 = 0$

52. $x^3 - 2x^2 + 4x - 8 = 0$

53. $x^2 + 4x = (2x + 1)^2$

54. $x^3 = 125$

55–68 ■ Solve the inequality. Express the solution using interval notation and graph the solution set on the real number line.

55. $3x - 2 > -11$

56. $12 - x \geqslant 7x$

57. $-1 < 2x + 5 \leqslant 3$

58. $3 - x \leqslant 2x - 7$

59. $x^2 + 4x - 12 > 0$

60. $x^2 \leqslant 1$

61. $\dfrac{2x + 5}{x + 1} \leqslant 1$

62. $2x^2 \geqslant x + 3$

63. $\dfrac{x - 4}{x^2 - 4} \leqslant 0$

64. $\dfrac{5}{x^3 - x^2 - 4x + 4} < 0$

65. $|x - 5| \leqslant 3$

66. $|x - 4| < 0.02$

67. $|2x + 1| \geqslant 1$

68. $|x - 1| < |x - 3|$
[*Hint:* Interpret the quantities as distances.]

69. For what values of x is each of the following algebraic expressions defined as a real number?

(a) $\sqrt{24 - x - 3x^2}$

(b) $\dfrac{1}{\sqrt[4]{x - x^4}}$

70. The volume of a sphere is given by $V = \frac{4}{3}\pi r^3$, where r is the radius. Find the interval of values of the radius so that the volume is between 8 ft^3 and 12 ft^3, inclusive.

71. If $|x - 3| < \dfrac{k}{2}$ and $|y + 3| < \dfrac{k}{2}$, prove that $|x + y| < k$.

1. Solve for x:
   (a) $2x + 7 = 12 + \frac{5}{2}x$
   (b) $\dfrac{2x}{x+1} = \dfrac{2x-1}{x}$

2. Bill drove from Ajax to Bixby at an average speed of 50 mi/h. On the way back, he drove at 60 mi/h. The total trip took $4\frac{2}{5}$ hours of driving time. Find the distance between these two cities.

3. Calculate and write the result in the form $a + bi$:
   (a) $(6 - 2i) - (7 - \frac{1}{2}i)$
   (b) $(1 + i)(3 - 2i)$
   (c) $\dfrac{5 + 10i}{3 - 4i}$
   (d) i^{50}
   (e) $(2 - \sqrt{-2})(\sqrt{8} + \sqrt{-4})$

4. Find all solutions, real and imaginary, of each of the following equations.
   (a) $x^2 - x - 12 = 0$
   (b) $2x^2 + 4x + 3 = 0$
   (c) $\sqrt{3 - \sqrt{x + 5}} = 2$
   (d) $x^{1/2} - 3x^{3/2} + 2x^{5/2} = 0$
   (e) $x^4 + 27x = 0$

5. A rectangular parcel of land is 70 ft longer than it is wide. Each diagonal between opposite corners is 130 ft. What are the dimensions of the parcel?

6. Solve each inequality. Sketch the solution on a real number line, and write the answer using interval notation.
   (a) $-1 \leqslant 5 - 2x < 10$
   (b) $x(x - 1)(x - 2) > 0$
   (c) $|x - 3| < 2$
   (d) $\dfrac{2x + 5}{x + 1} \leqslant 1$

7. A bottle of medicine is to be stored at a temperature between $5\,°C$ and $10\,°C$. What range does this correspond to on a Fahrenheit scale? [*Note:* The Fahrenheit (F) and Celsius (C) scales satisfy the relation $C = \frac{5}{9}(F - 32)$.]

8. For what values of x is the expression $\sqrt{4x - x^2}$ defined as a real number?

FOCUS ON PROBLEM SOLVING

It is sometimes useful to attack a problem by first **working backward.** This means that we assume the conclusion and work backward step by step until we arrive at something that is given or known. Then we may be able to reverse the steps in the argument and proceed forward from the given to the conclusion.

We have already used this procedure in solving equations. For example, in solving the equation $2x + 7 = 23$ we assume that x is a number that satisfies $2x + 7 = 23$ and work backward. We subtract 7 from each side of the equation and then divide each side by 2 to get $x = 8$. Each of these steps can be reversed, so we have solved the problem.

PROBLEM 1 ■ Measuring Water with Two Pails

How is it possible to get from a river exactly 6 gallons of water when you have only two containers, a 9-gallon pail and a 4-gallon pail?

SOLUTION

If you work this problem in the forward direction you might be lucky and discover, out of many possibilities for proceeding, a correct solution. But it is more systematic to work backward.

Imagine that we have 6 gal of water in the 9-gal pail, together with the full 4-gal pail. We could fill the larger pail from the smaller one, leaving just 1 gal in the smaller pail. Then we could empty the large pail and pour the 1 gal of water into it. Finally we could exactly fill the larger pail by adding 8 gal of water using the smaller pail twice.

Now, by reversing the procedure, we have the solution to the problem. Start with a full 9-gal pail. Use it to fill the 4-gal pail, then empty the 4-gal pail. Again fill the 4-gal pail from the larger one and empty the smaller one. This leaves 1 gal in the larger pail. Transfer it to the smaller pail and fill the larger one. Use the larger pail to fill the smaller one. This leaves 6 gal in the larger pail. ■

Another useful problem-solving principle is that of **taking cases,** that is, accounting for all possibilities. We split the problem into several cases and give a different argument for each case. In the next example we combine the technique of working backward with the method of taking cases.

PROBLEM 2 ■ An Absolute Value Equation

Solve the equation $\left| 2x - |3x + 1| \right| = 1$.

SOLUTION

We assume that x is a number such that $\left| 2x - |3x + 1| \right| = 1$. It follows from

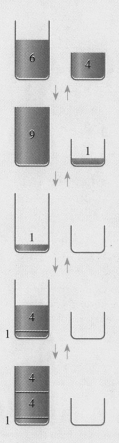

Work backward

Property 4 of absolute values that we must solve two cases:

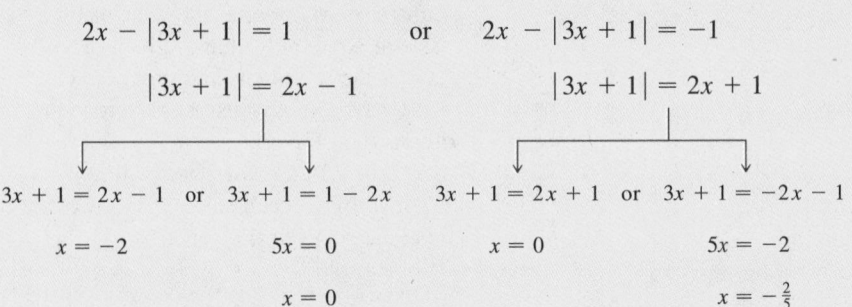

Thus, working backward gives three potential solutions: $x = 0$, $x = -2$, and $x = -\frac{2}{5}$. But is it possible to reverse the steps in each case? Let us check by trying to verify that these are indeed solutions:

Try $x = 0$: $\big|2(0) - |3(0) + 1|\big| = |0 - 1| = 1$ 0 is a solution

Try $x = -2$: $\big|2(-2) - |3(-2) + 1|\big| = |-4 - 5| = 9$ -2 is not a solution

Try $x = -\frac{2}{5}$: $\big|2(-\frac{2}{5}) - |3(-\frac{2}{5}) + 1|\big| = \big|-\frac{4}{5} - \frac{1}{5}\big| = 1$ $-\frac{2}{5}$ is a solution

Therefore, the only solutions are 0 and $-\frac{2}{5}$.   ■

PROBLEMS

1. Three containers hold 19 gal, 13 gal, and 7 gal, respectively. The 19-gallon container is empty. The other two are full. How can you measure out 10 gal using no other container?

2. A farmer has a fox, a goose, and a sack of grain. He wishes to cross a river in a rowboat that can hold only himself and one of these three. The farmer's problem is that foxes eat geese and geese eat grain. How can he safely ferry all three of these possessions to the other side?

3–8 ■ Solve the equation or inequality.

3. $\big|5 - |x - 1|\big| = 3$

4. $\big|4x - |x + 1|\big| = 3$

5. $|2x - 1| - |x + 5| = 3$

6. $|x - 1| + |x - 2| + |x - 3| = 1$

7. $|x + 1| + |x + 4| \leqslant 5$

8. $|x - 1| - |x - 3| \geqslant 5$

9. Each letter in the following multiplication represents a different digit. Find the value of each letter.

$$\begin{array}{r} \text{ABCDE} \\ \underline{\times \quad 4} \\ \text{EDCBA} \end{array}$$

Take cases

Take cases again

THE FOUR-COLOR PROBLEM

A famous example of a problem that was solved with the aid of the method of taking cases is the *four-color problem*: Prove that every map can be colored using no more than four different colors. The requirement is that any two regions that share a common border should have different colors. (If two regions meet only at a single point, as do Utah and New Mexico, for example, then they may have the same color.) The four-color problem was first stated as a conjecture in 1852, but it wasn't until 1976 that Kenneth Appel and Wolfgang Haken were able to prove it. In their proof they considered thousands of cases corresponding to different types of maps, and most of those cases were so complicated that they had to use a high-speed computer for the verification.

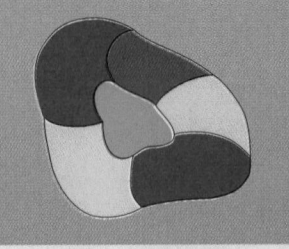

10. Solve the following problem, which was first posed in the 19th century: Every day at noon a ship leaves New York for Le Havre and another ship leaves Le Havre for New York. The trip takes seven full days. How many New York-to-Le Havre ships will the ship leaving Le Havre today meet during its trip to New York?

11. Of nine eggs, eight have exactly the same weight. The ninth egg weighs less than the other eight. How can you determine which is the lighter egg with exactly two weighings on a balance?

12. Of 12 similar coins, one is a counterfeit. It is not known whether the counterfeit coin is lighter or heavier than a genuine coin. Using a balance three times, how can the counterfeit be identified and in the process determined to be lighter or heavier than a genuine coin?

13. A red ribbon is tied tightly around the earth at the equator. How much more ribbon would you need if you raised the ribbon one foot above the earth everywhere? (You don't need to know the radius of the earth to solve this problem.)

14. Find the area of the region between the two concentric circles shown in the figure.

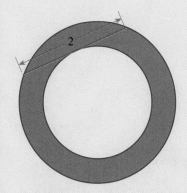

15. Suppose that the equation

$$x^2 + px + 8 = 0$$

has two distinct real roots, r_1 and r_2. Show that

$$|r_1 + r_2| > 4\sqrt{2}$$

16. If x and y are positive real numbers, prove that

$$\sqrt{xy} \le \frac{x + y}{2}$$

3

FUNCTIONS AND GRAPHS

A function is a rule that describes how one quantity depends on another; a graph gives a visual representation of the function. For example, the price of a commodity is a function of time.

That flower of modern mathematical thought—the notion of a function.

THOMAS J. McCORMACK

One of the most basic and important ideas in all of mathematics is that of *function*. Functions are used to describe mathematical relationships between varying quantities. *Graphs* of functions give us visual representations of these relationships. Graphs are drawn in a *coordinate plane*, so we begin this chapter with a discussion of coordinate geometry.

3.1 THE COORDINATE PLANE

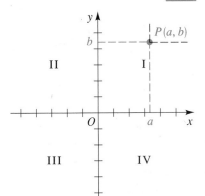

FIGURE 1

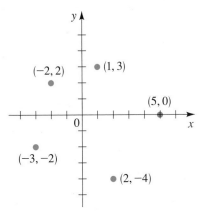

FIGURE 2

Points on a line can be identified with real numbers by assigning them coordinates, as described in Section 1.2. In a similar way, points in a plane can be identified with ordered pairs of real numbers. We start by drawing two perpendicular coordinate lines that intersect at the **origin** O on each line. Usually one line is horizontal with positive direction to the right and is called the **x-axis**; the other line is vertical with positive direction upward and is called the **y-axis.**

Any point P in the plane can be located by a unique ordered pair of numbers as follows. Draw lines through P perpendicular to the x- and y-axes. These lines will intersect the axes in points with coordinates a and b as shown in Figure 1. Then the point P is assigned the ordered pair (a, b). The first number a is called the **x-coordinate** (or **abscissa**) of P; the second number is called the **y-coordinate** (or **ordinate**) of P. We say that P is the point with coordinates (a, b) and we denote the point by the symbol $P(a, b)$. We can think of the coordinates of a point as its "address" in the plane, since the coordinates specify the location of the point. Several points are labeled with their coordinates in Figure 2.

By reversing the preceding process, starting with an ordered pair (a, b), we can arrive at the corresponding point P. Often we identify the point P with the ordered pair (a, b) and refer to "the point (a, b)." [Although the notation used for an open interval (a, b) is the same as the notation used for a point (a, b), you will be able to tell from the context which meaning is intended.]

This coordinate system is called the **rectangular coordinate system** or the **Cartesian coordinate system** in honor of the French mathematician René Descartes (1596–1650), although another Frenchman, Pierre Fermat (1601–1665), also invented the principles of analytic geometry at about the same time as Descartes. The plane used with this coordinate system is called the **coordinate plane** or the **Cartesian plane** and is denoted by $\mathbb{R}^2$.

The x- and y-axes, or the **coordinate axes,** divide the Cartesian plane into four quadrants, which are labeled I, II, III, and IV in Figure 1. Notice that the first quadrant consists of those points whose x- and y-coordinates are both positive. (The points *on* the coordinate axes are not assigned to any quadrant.)

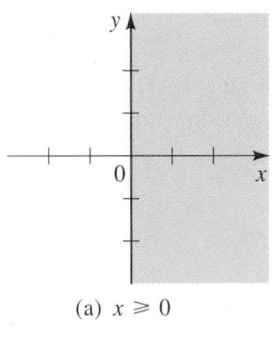

(a) $x \geqslant 0$

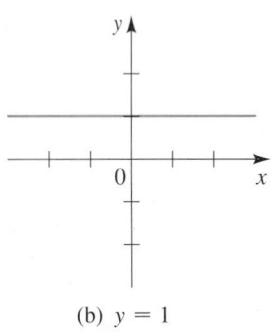

(b) $y = 1$

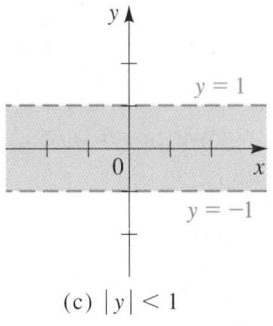

(c) $|y| < 1$

FIGURE 3

EXAMPLE 1 ■ Graphing Regions in the Coordinate Plane

Describe and sketch the regions given by each of the following sets.

(a) $\{(x, y) \mid x \geqslant 0\}$ (b) $\{(x, y) \mid y = 1\}$ (c) $\{(x, y) \mid |y| < 1\}$

SOLUTION

(a) The points whose x-coordinates are 0 or positive lie on the y-axis or to the right of it, as shown in Figure 3(a).

(b) The set of all points with y-coordinate 1 is a horizontal line one unit above the x-axis, as in Figure 3(b).

(c) Recall from Section 2.8 that

$$|y| < 1 \qquad \text{if and only if} \qquad -1 < y < 1$$

The given region consists of those points in the plane whose y-coordinates lie between -1 and 1. Thus, the region consists of all points that lie between (but not on) the horizontal lines $y = 1$ and $y = -1$. These lines are shown as broken lines in Figure 3(c) to indicate that the points on these lines do not lie in the set.   ■

We now find a formula for the distance $d(A, B)$ between two points $A(x_1, y_1)$ and $B(x_2, y_2)$ in the plane. Recall from Section 1.2 that the distance between points a and b on a number line is $d(a, b) = |b - a|$. So, from Figure 4 we see that the distance between the points $A(x_1, y_1)$ and $C(x_2, y_1)$ on a horizontal line must be $|x_2 - x_1|$ and the distance between $B(x_2, y_2)$ and $C(x_2, y_1)$ on a vertical line must be $|y_2 - y_1|$. Since triangle ABC is a right triangle, the Pythagorean Theorem gives

$$d(A, B) = \sqrt{|x_2 - x_1|^2 + |y_2 - y_1|^2}$$

$$= \sqrt{(x_2 - x_1)^2 + (y_2 - y_1)^2}$$

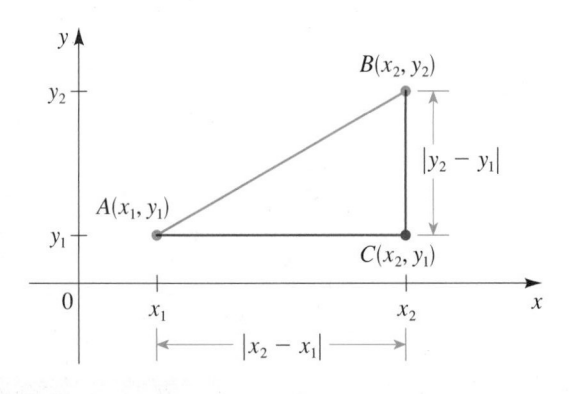

FIGURE 4

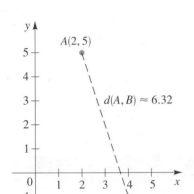

FIGURE 5

DISTANCE FORMULA

The distance between the points $A(x_1, y_1)$ and $B(x_2, y_2)$ in the plane is

$$d(A, B) = \sqrt{(x_2 - x_1)^2 + (y_2 - y_1)^2}$$

EXAMPLE 2 ■ **Finding the Distance between Two Points**

Find the distance between the points $A(2, 5)$ and $B(4, -1)$.

SOLUTION

Using the distance formula, we have

$$d(A, B) = \sqrt{(4 - 2)^2 + (-1 - 5)^2}$$
$$= \sqrt{2^2 + (-6)^2}$$
$$= \sqrt{4 + 36} = \sqrt{40} \approx 6.32$$

See Figure 5.   ■

EXAMPLE 3 ■ **Applying the Distance Formula**

Which of the points $P(1, -2)$ or $Q(8, 9)$ is closer to the point $A(5, 3)$?

SOLUTION

By the distance formula, we have

$$d(P, A) = \sqrt{(5 - 1)^2 + [3 - (-2)]^2} = \sqrt{4^2 + 5^2} = \sqrt{41}$$
$$d(Q, A) = \sqrt{(5 - 8)^2 + (3 - 9)^2} = \sqrt{(-3)^2 + (-6)^2} = \sqrt{45}$$

This shows that $d(P, A) < d(Q, A)$, so P is closer to A (see Figure 6).   ■

Let us now find the coordinates (x, y) of the midpoint M of the line segment that joins the point $A(x_1, y_1)$ to the point $B(x_2, y_2)$. In Figure 7 notice that trian-

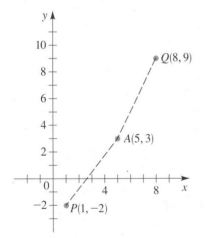

FIGURE 6

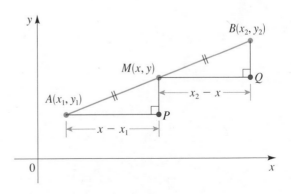

FIGURE 7

The coordinates of a point in the xy-plane uniquely determine its location. We can think of the coordinates as the "address" of the point. In Salt Lake City, Utah, the addresses of most buildings are in fact expressed as coordinates. The city is divided into quadrants with Main Street as the vertical (North-South) axis and S. Temple Street as the horizontal (East-West) axis. An address such as

        1760 W   2100 S

indicates a location 17.6 blocks west of Main Street and 21 blocks south of S. Temple Street. (This is the address of the main post office in Salt Lake City.) With this logical system it is possible for someone unfamiliar with the city to locate any address immediately, as easily as one can locate a point in the coordinate plane.

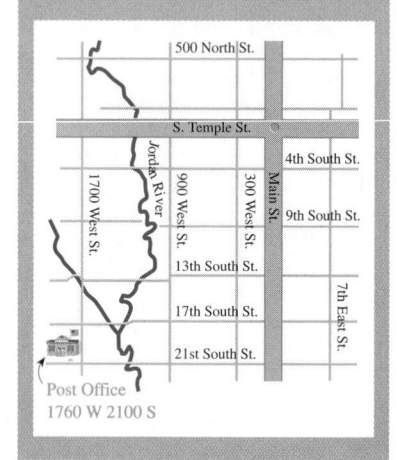

gles APM and MQB are congruent because $d(A, M) = d(M, B)$ and corresponding angles are equal. It follows that $d(A, P) = d(M, Q)$ and so

$$x - x_1 = x_2 - x$$

Solving this equation for x, we get

$$2x = x_1 + x_2$$

$$x = \frac{x_1 + x_2}{2}$$

Similarly, $$y = \frac{y_1 + y_2}{2}$$

MIDPOINT FORMULA

The midpoint of the line segment from $A(x_1, y_1)$ to $B(x_2, y_2)$ is

$$\left(\frac{x_1 + x_2}{2}, \frac{y_1 + y_2}{2} \right)$$

EXAMPLE 4 ■ Finding the Midpoint

The midpoint of the line segment that joins the points $(-2, 5)$ and $(4, 9)$ is

$$\left(\frac{-2 + 4}{2}, \frac{5 + 9}{2} \right) = (1, 7)$$

See Figure 8.

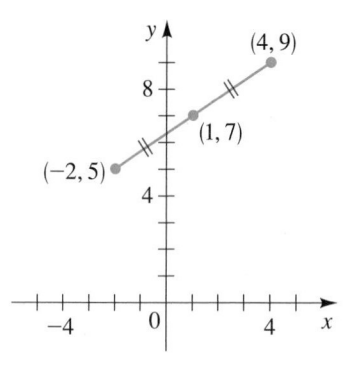

FIGURE 8

EXAMPLE 5 ■ Applying the Midpoint Formula

Show that the quadrilateral with vertices $P(1, 2)$, $Q(4, 4)$, $R(5, 9)$, and $S(2, 7)$ is a parallelogram by proving that its two diagonals bisect each other.

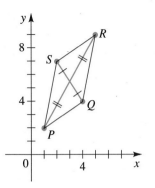

FIGURE 9

SOLUTION

If the two diagonals have the same midpoint, then they must bisect each other. The midpoint of the diagonal *PR* is

$$\left(\frac{1+5}{2}, \frac{2+9}{2}\right) = \left(3, \tfrac{11}{2}\right)$$

and the midpoint of the diagonal *QS* is

$$\left(\frac{4+2}{2}, \frac{4+7}{2}\right) = \left(3, \tfrac{11}{2}\right)$$

so each diagonal bisects the other, as shown in Figure 9. (A theorem from elementary geometry states that the quadrilateral is therefore a parallelogram.)

■

3.1   **EXERCISES**

1–2 ■ Plot the given points in a coordinate plane.

1. $(2,3)$, $(-2,3)$, $(4,5)$, $(4,-5)$, $(-4,5)$, $(-4,-5)$

2. $(1,9)$, $(1,-9)$, $(0,5)$, $(5,0)$, $(-2,2)$, $(-2,-2)$

3–4 ■ Find the coordinates of the points shown in the figure.

3.

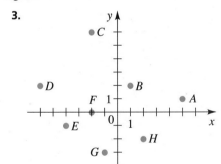

4.

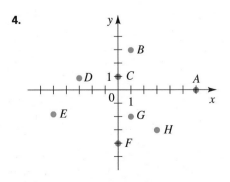

5. Draw the rectangle with vertices $A(1,3)$, $B(5,3)$, $C(1,-3)$, and $D(5,-3)$ on a coordinate plane. Find the area of the rectangle.

6. Draw the parallelogram with vertices $A(1,2)$, $B(5,2)$, $C(3,6)$, and $D(7,6)$ on a coordinate plane. Find the area of the parallelogram.

7–14 ■ A pair of points is given.
(a) Plot the points in a coordinate plane.
(b) Find the distance between them.
(c) Find the midpoint of the segment that joins them.

7. $(2,3)$, $(5,2)$ **8.** $(2,-1)$, $(4,3)$

9. $(6,-2)$, $(-1,3)$ **10.** $(1,-6)$, $(-1,-3)$

11. $(2,5)$, $(4,-7)$ **12.** $\left(-3,\tfrac{1}{2}\right)$, $\left(\tfrac{5}{2},-1\right)$

13. $(3,4)$, $(-3,-4)$ **14.** $(5,0)$, $(0,6)$

15. Plot the points $A(1,0)$, $B(5,0)$, $C(4,3)$, and $D(2,3)$ on a coordinate plane. Draw the segments *AB*, *BC*, *CD*, and *DA*. What kind of quadrilateral is *ABCD*, and what is its area?

16. Plot the points $P(5,1)$, $Q(0,6)$, and $R(-5,1)$ on a coordinate plane. Where must the point *S* be located so that the quadrilateral *PQRS* is a square? Find the area of this square.

17–30 ■ Sketch the region given by the set.

17. $\{(x, y) \mid x \leq 0\}$ **18.** $\{(x, y) \mid y \geq 0\}$

19. $\{(x, y) \mid x = 3\}$ **20.** $\{(x, y) \mid y = -2\}$

21. $\{(x, y) \mid 1 < x < 2\}$

22. $\{(x, y) \mid 0 \le y \le 4\}$

23. $\{(x, y) \mid xy < 0\}$

24. $\{(x, y) \mid xy > 0\}$

25. $\{(x, y) \mid x \ge 1 \text{ and } y < 3\}$

26. $\{(x, y) \mid |y| > 1\}$

27. $\{(x, y) \mid |x| \le 2\}$

28. $\{(x, y) \mid |x| < 3 \text{ and } |y| > 2\}$

29. $\{(x, y) \mid |y - 3| \le 3\}$

30. $\{(x, y) \mid |x - 1| < 2 \text{ and } |y + 1| \le 2\}$

31. Which of the points $A(6, 7)$ or $B(-5, 8)$ is closer to the origin?

32. Which of the points $C(-6, 3)$ or $D(3, 0)$ is closer to the point $E(-2, 1)$?

33. Which of the points $P(3, 1)$ or $Q(-1, 3)$ is closer to the point $R(-1, -1)$?

34. (a) Show that the points $(7, 3)$ and $(3, 7)$ are the same distance from the origin.
    (b) Show that the points (a, b) and (b, a) are the same distance from the origin.

35. Show that the triangle with vertices $A(0, 2)$, $B(-3, -1)$, and $C(-4, 3)$ is isosceles.

36. Find the area of the triangle shown in the figure.

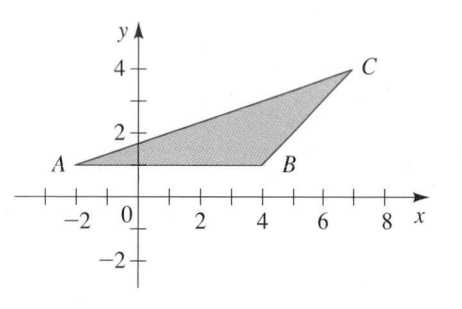

37. Refer to triangle ABC in the figure.
    (a) Show that triangle ABC is a right triangle by using the converse of the Pythagorean Theorem.

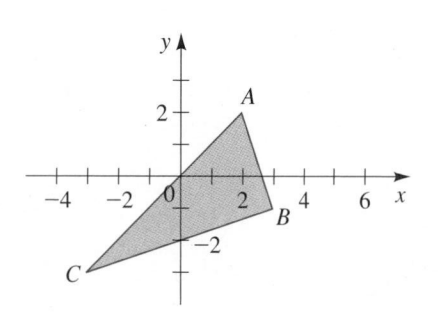

(b) Find the area of triangle ABC.

38. Show that the triangle with vertices $A(6, -7)$, $B(11, -3)$, and $C(2, -2)$ is a right triangle by using the converse of the Pythagorean Theorem. Find the area of the triangle.

39. Show that the points $A(-2, 9)$, $B(4, 6)$, $C(1, 0)$, and $D(-5, 3)$ are the vertices of a square.

40. Show that the points $A(-1, 3)$, $B(3, 11)$, and $C(5, 15)$ are collinear by showing that $d(A, B) + d(B, C) = d(A, C)$.

41. Find a point on the y-axis that is equidistant from the points $(5, -5)$ and $(1, 1)$.

42. Find the lengths of the medians of the triangle with vertices $A(1, 0)$, $B(3, 6)$, and $C(8, 2)$. (A *median* is a line segment from a vertex to the midpoint of the opposite side.)

43. Find the point that is one-fourth of the distance from the point $P(-1, 3)$ to the point $Q(7, 5)$ along the segment PQ.

44. Plot the points $P(-2, 1)$ and $Q(12, -1)$. Which (if either) of the points $A(5, -7)$ and $B(6, 7)$ lies on the perpendicular bisector of the segment PQ?

45. Plot the points $P(-1, -4)$, $Q(1, 1)$, and $R(4, 2)$ on a coordinate plane. Where should the point S be located so that the figure $PQRS$ is a parallelogram?

46. If $M(6, 8)$ is the midpoint of the line segment AB, and if A has coordinates $(2, 3)$, find the coordinates of B.

47. (a) Sketch the parallelogram with vertices $A(-2, -1)$, $B(4, 2)$, $C(7, 7)$, and $D(1, 4)$.
    (b) Find the midpoints of the diagonals of this parallelogram.
    (c) Conclude from part (b) that the diagonals bisect each other.

48. The point M in the figure is the midpoint of the line segment AB. Show that M is equidistant from the vertices of triangle ABC.

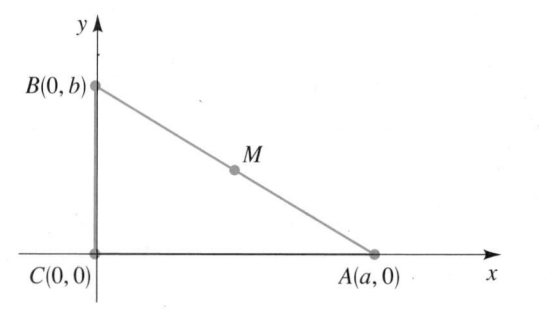

49. Suppose that each point in the coordinate plane is shifted 3 units to the right and 2 units upward.
    (a) The point $(5, 3)$ is shifted to what new point?
    (b) The point (a, b) is shifted to what new point?
    (c) Triangle ABC in the figure has been shifted to triangle $A'B'C'$. Find the coordinates of the points A', B', and C'.

50. Suppose that the y-axis acts as a mirror that reflects each point to the right of it into a point to the left of it.
    (a) The point $(3, 7)$ is reflected to what point?
    (b) The point (a, b) is reflected to what point?
    (c) Triangle ABC in the figure is reflected to triangle $A'B'C'$. Find the coordinates of the points A', B', and C'.

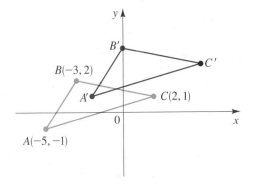

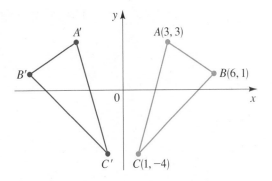

3.2 GRAPHS OF EQUATIONS

Suppose we have an equation involving the variables x and y, such as

$$x^2 + y^2 = 25 \quad \text{or} \quad x = y^2 \quad \text{or} \quad y = \frac{2}{x}$$

Fundamental Principle of Analytic Geometry

A point (x, y) lies on the graph of an equation if and only if its coordinates satisfy the equation.

A point (x, y) **satisfies** the equation if the equation is true when the coordinates of the point are substituted into the equation. For example, the point $(3, 4)$ satisfies the first equation, since $3^2 + 4^2 = 25$, but the point $(2, -3)$ does not, since $2^2 + (-3)^2 = 13 \neq 25$.

THE GRAPH OF AN EQUATION

The **graph** of an equation in x and y is the set of all points (x, y) in the coordinate plane that satisfy the equation.

The graphs of most of the equations that we will encounter are curves. For example, we will see later in this section that the first equation, $x^2 + y^2 = 25$, represents a circle; we will determine the shapes represented by the other equations later in this chapter. Graphs are important in the study of equations because they give visual representations of equations.

EXAMPLE 1 ■ **Sketching a Graph by Plotting Points**

Sketch the graph of the equation $2x - y = 3$.

SOLUTION

We first solve the given equation for y to get

$$y = 2x - 3$$

This helps us calculate the y-coordinates in the following table.

x	$y = 2x - 3$	(x, y)
-1	-5	$(-1, -5)$
0	-3	$(0, -3)$
1	-1	$(1, -1)$
2	1	$(2, 1)$
3	3	$(3, 3)$
4	5	$(4, 5)$

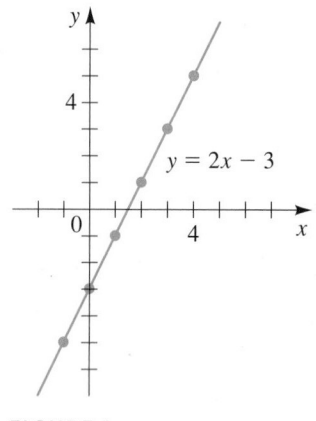

FIGURE 1

Of course, there are infinitely many points on the graph, and it is impossible to plot all of them. But the more points we plot, the better we can imagine what the graph represented by the equation looks like. We plot the points we found in the table in Figure 1; they appear to lie on a line. So, we complete the graph by joining the points by a line. (In Section 3.4 we verify that the graph of this equation is indeed a line.) ■

EXAMPLE 2 ■ **Sketching a Graph by Plotting Points**

Sketch the graph of the equation $y = x^2 - 2$.

SOLUTION

We find some of the points that satisfy the equation in the following table. In Figure 2 we plot these points and then connect them by a smooth curve. A curve with this shape is called a *parabola*.

A detailed discussion of parabolas and their geometric properties is presented in Chapter 7.

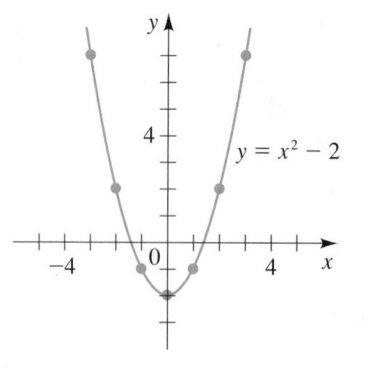

FIGURE 2

x	$y = x^2 - 2$	(x, y)
-3	7	$(-3, 7)$
-2	2	$(-2, 2)$
-1	-1	$(-1, -1)$
0	-2	$(0, -2)$
1	-1	$(1, -1)$
2	2	$(2, 2)$
3	7	$(3, 7)$

■

EXAMPLE 3 ■ Sketching the Graph of an Equation involving Absolute Value

Sketch the graph of the equation $y = |x|$.

SOLUTION

We make a table of values:

x	$y = \|x\|$	(x, y)
-3	3	$(-3, 3)$
-2	2	$(-2, 2)$
-1	1	$(-1, 1)$
0	0	$(0, 0)$
1	1	$(1, 1)$
2	2	$(2, 2)$
3	3	$(3, 3)$

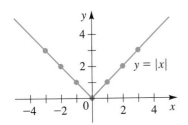

FIGURE 3

In Figure 3 we plot the points with the coordinates given by the table and use these to sketch the graph of the equation. ■

The x-coordinates of the points where a graph intersects the x-axis are called the **x-intercepts** of the graph and are obtained by setting $y = 0$ in the equation of the graph. The y-coordinates of the points where a graph intersects the y-axis are called the **y-intercepts** of the graph and are obtained by setting $x = 0$ in the equation of the graph.

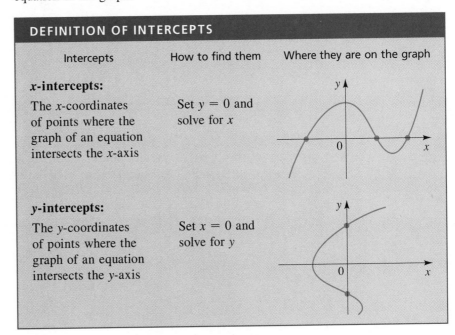

DEFINITION OF INTERCEPTS

Intercepts	How to find them	Where they are on the graph
x-intercepts: The x-coordinates of points where the graph of an equation intersects the x-axis	Set $y = 0$ and solve for x	
y-intercepts: The y-coordinates of points where the graph of an equation intersects the y-axis	Set $x = 0$ and solve for y	

EXAMPLE 4 ■ Finding Intercepts

Find the x- and y-intercepts of the graph of the equation $y = x^2 - 2$.

SOLUTION

To find the x-intercepts we set $y = 0$ and solve for x. Thus

$$0 = x^2 - 2 \qquad \text{Set } y = 0$$

$$x^2 = 2 \qquad \text{Add 2 to each side}$$

$$x = \pm\sqrt{2} \qquad \text{Take the square root}$$

Thus, the x-intercepts are $\sqrt{2}$ and $-\sqrt{2}$.

To find the y-intercepts we set $x = 0$ and solve for y. Thus

$$y = 0^2 - 2 \qquad \text{Set } x = 0$$

$$y = -2$$

Thus, the y-intercept is -2.

The graph of this equation was sketched in Example 2. We sketch it again in Figure 4 and label the x- and y-intercepts.   ■

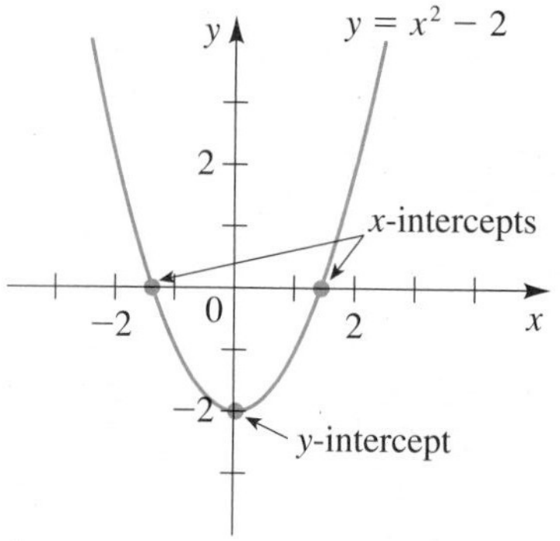

FIGURE 4

CIRCLES

So far we have discussed how to find the graph of an equation in x and y. The converse problem is to find an equation of a graph, that is, an equation that represents a given curve in the xy-plane. Such an equation is satisfied by the coordinates of the points on the curve and by no other point. This is the other half of the basic principle of analytic geometry as formulated by Descartes and Fermat. The idea is that if a geometric curve can be represented by an algebraic equation, then the rules of algebra can be used to analyze the curve.

As an example of this type of problem, let us find the equation of a circle with radius r and center (h, k). By definition, the circle is the set of all points $P(x, y)$ whose distance from the center $C(h, k)$ is r (see Figure 5). Thus, P is on the circle if and only if $d(P, C) = r$. From the distance formula we have

$$\sqrt{(x - h)^2 + (y - k)^2} = r$$

$$(x - h)^2 + (y - k)^2 = r^2 \qquad \text{Square each side}$$

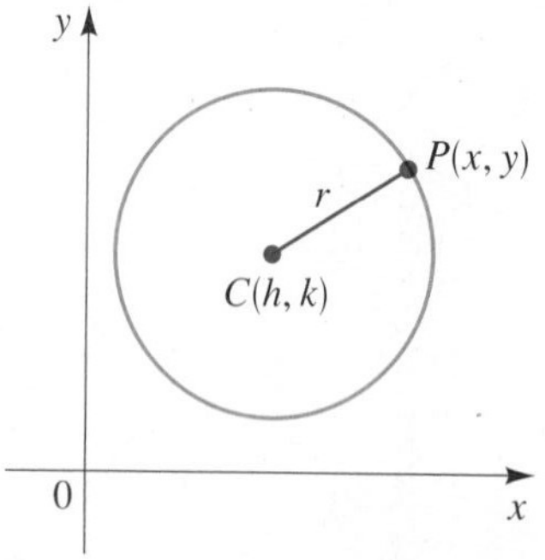

FIGURE 5

This is the desired equation.

EQUATION OF A CIRCLE

An equation of the circle with center (h, k) and radius r is

$$(x - h)^2 + (y - k)^2 = r^2$$

In particular, if the center is the origin $(0, 0)$, then the equation is

$$x^2 + y^2 = r^2$$

EXAMPLE 5 ■ Graphing a Circle

Graph each equation.

(a) $x^2 + y^2 = 25$ (b) $(x - 2)^2 + (y + 1)^2 = 25$

SOLUTION

(a) Rewriting the equation as $x^2 + y^2 = 5^2$, we see that this is an equation of the circle of radius 5 centered at the origin. Its graph is sketched in Figure 6.

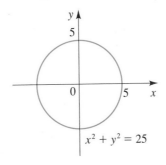

$x^2 + y^2 = 25$

FIGURE 6

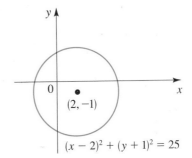

$(x - 2)^2 + (y + 1)^2 = 25$

FIGURE 7

(b) Rewriting the equation as $(x - 2)^2 + (y + 1)^2 = 5^2$, we see that this is an equation of the circle of radius 5 centered at $(2, -1)$. Its graph is sketched in Figure 7. ■

EXAMPLE 6 ■ Finding the Equation of a Circle

(a) Find an equation of the circle with radius 3 and center $(2, -5)$.
(b) Find an equation of the circle that has the points $P(1, 8)$ and $Q(5, -6)$ as the endpoints of a diameter.

SOLUTION

(a) Using the equation of a circle with $r = 3$, $h = 2$, and $k = -5$, we obtain

$$(x - 2)^2 + (y + 5)^2 = 9$$

Its graph is shown in Figure 8.

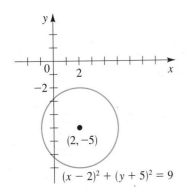

$(x - 2)^2 + (y + 5)^2 = 9$

FIGURE 8

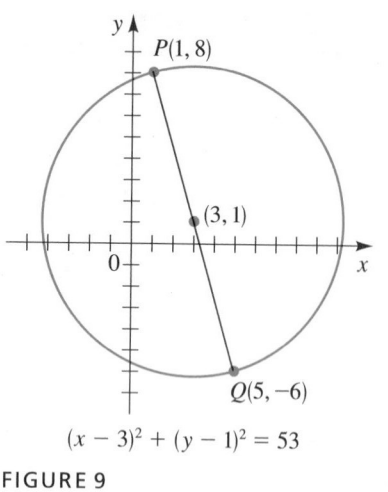

$(x - 3)^2 + (y - 1)^2 = 53$

FIGURE 9

(b) We first observe that the center is the midpoint of the diameter PQ, so by the Midpoint Formula the center is

$$\left(\frac{1 + 5}{2}, \frac{8 - 6}{2} \right) = (3, 1)$$

The radius r is the distance from P to the center, so

$$r^2 = (3 - 1)^2 + (1 - 8)^2 = 2^2 + (-7)^2 = 53$$

Therefore, the equation of the circle is

$$(x - 3)^2 + (y - 1)^2 = 53$$

Its graph is shown in Figure 9.   ∎

EXAMPLE 7 ■ Identifying an Equation of a Circle

Sketch the graph of the equation $x^2 + y^2 + 2x - 6y + 7 = 0$ by first showing that it represents a circle and then finding its center and radius.

SOLUTION

We first group the x-terms and y-terms. Then we complete the square within each grouping (by adding the square of half the coefficient of x and the square of half the coefficient of y to each side of the equation):

$(x^2 + 2x \qquad) + (y^2 - 6y \qquad) = -7$ Group terms

$(x^2 + 2x + 1) + (y^2 - 6y + 9) = -7 + 1 + 9$ Complete the square by adding 1 and 9 to each side

$\qquad (x + 1)^2 + (y - 3)^2 = 3$

Comparing this equation with the standard equation of a circle, we see that $h = -1$, $k = 3$, and $r = \sqrt{3}$, so the given equation represents a circle with center $(-1, 3)$ and radius $\sqrt{3}$. The circle is sketched in Figure 10.   ∎

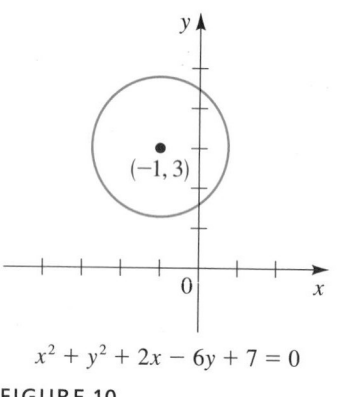

$x^2 + y^2 + 2x - 6y + 7 = 0$

FIGURE 10

SYMMETRY

Figure 11 shows the graph of $y = x^2$. Notice that the part of the graph to the left of the y-axis is the mirror image of the part to the right of the y-axis. The reason is that if the point (x, y) is on the graph, then so is $(-x, y)$, and these points are reflections of each other about the y-axis. In this situation we say the graph is **symmetric with respect to the y-axis.** Similarly, we say a graph is **symmetric with respect to the x-axis** if whenever the point (x, y) is on the graph, then so is $(x, -y)$. A graph is **symmetric with respect to the origin** if whenever (x, y) is on the graph, so is $(-x, -y)$. The following table gives definitions of the three types of symmetry and explains how to determine whether the graph of an equation exhibits each of these symmetries.

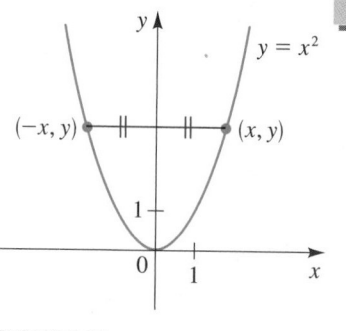

FIGURE 11

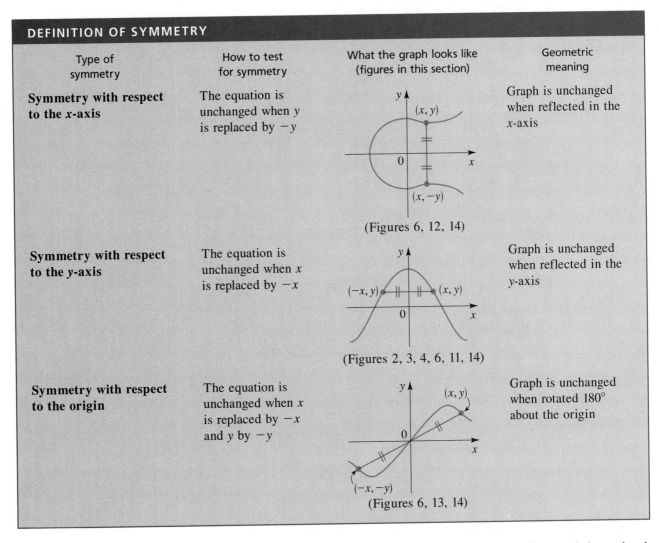

DEFINITION OF SYMMETRY

Type of symmetry	How to test for symmetry	What the graph looks like (figures in this section)	Geometric meaning
Symmetry with respect to the *x*-axis	The equation is unchanged when *y* is replaced by −*y*	*(Figures 6, 12, 14)*	Graph is unchanged when reflected in the *x*-axis
Symmetry with respect to the *y*-axis	The equation is unchanged when *x* is replaced by −*x*	*(Figures 2, 3, 4, 6, 11, 14)*	Graph is unchanged when reflected in the *y*-axis
Symmetry with respect to the origin	The equation is unchanged when *x* is replaced by −*x* and *y* by −*y*	*(Figures 6, 13, 14)*	Graph is unchanged when rotated 180° about the origin

One reason that symmetry is important is that we can use it to help us sketch the graphs of equations. The remaining examples in this section illustrate the use of symmetry in graphing.

EXAMPLE 8 ■ Using Symmetry to Sketch a Graph

Test the equation $x = y^2$ for symmetry and sketch the graph.

SOLUTION

If y is replaced by $-y$ in the equation $x = y^2$, we get

$$x = (-y)^2 = y^2$$

and so the equation is unchanged. Therefore, the graph is symmetric about the x-axis. But changing x to $-x$ gives the equation $-x = y^2$, which is not the same as the original equation, so the graph is not symmetric about the y-axis.

We use the symmetry about the x-axis to sketch the graph by first plotting points just for $y > 0$ and then reflecting the graph in the x-axis, as shown in Figure 12.

René Descartes (1596–1650) was born in the town of La Haye in southern France. From an early age Descartes liked mathematics because of "the certainty of its results and the clarity of its reasoning." He believed that in order to arrive at truth, one must begin by doubting everything, including one's own existence; this led him to formulate perhaps the most well known sentence in all of philosophy: "I think, therefore I am." In his book *Discourse on Method* he described what is now called the Cartesian plane. This idea of combining algebra and geometry allowed mathematicians for the first time to "see" the equations they were studying. The philosopher John Stuart Mill called this invention "the greatest single step ever made in the progress of the exact sciences." Descartes liked to get up late and spend the morning in bed thinking and writing. He invented the coordinate plane while lying in bed watching a fly crawl on the ceiling, and reasoning that he could describe the exact location of the fly by knowing its distance

(continued)

y	$x = y^2$	(x, y)
0	0	$(0, 0)$
1	1	$(1, 1)$
2	4	$(4, 2)$
3	9	$(9, 3)$

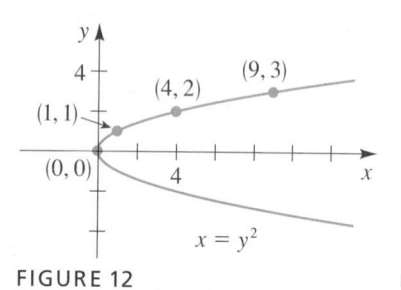

FIGURE 12

EXAMPLE 9 ■ Using Symmetry and Intercepts to Sketch a Graph

Find the x- and y-intercepts of the graph of the equation $y = x^3 - 9x$. Test the equation for symmetry and sketch its graph.

SOLUTION

Setting $x = 0$ in the equation gives $y = 0$, so the y-intercept is 0. Setting $y = 0$ gives $x^3 - 9x = 0$, or $x(x^2 - 9) = x(x - 3)(x + 3) = 0$. Thus, the x-intercepts are 0, 3, and -3.

If we replace x by $-x$ and y by $-y$ in the equation, we get

$$-y = (-x)^3 - 9(-x)$$
$$-y = -x^3 + 9x$$
$$y = x^3 - 9x$$

and so the equation is unchanged. This means that the graph is symmetric with respect to the origin. We sketch it by first plotting points for $x > 0$ and then using symmetry about the origin (see Figure 13).

x	$y = x^3 - 9x$	(x, y)
0	0	$(0, 0)$
1	-8	$(1, -8)$
1.5	-10.125	$(1.5, -10.125)$
2	-10	$(2, -10)$
2.5	-6.875	$(2.5, -6.875)$
3	0	$(3, 0)$
4	28	$(4, 28)$

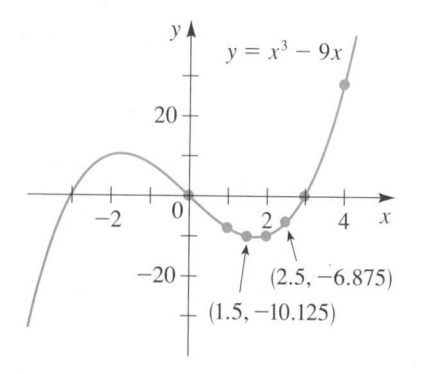

FIGURE 13

from two perpendicular walls. In 1649 Descartes became the tutor of Queen Christina of Sweden. She liked her lessons at 5 o'clock in the morning when, she said, her mind was sharpest. However, the change from his usual habits and the ice-cold library where they studied proved too much for him. In February 1650, after just two months of this, he caught pneumonia and died.

EXAMPLE 10 ■ A Circle that Has All Three Types of Symmetry

Test the equation of the circle $x^2 + y^2 = 4$ for symmetry.

SOLUTION

The equation $x^2 + y^2 = 4$ remains unchanged when x is replaced by $-x$ and y is replaced by $-y$, since $(-x)^2 = x^2$ and $(-y)^2 = y^2$, so the circle exhibits all three types of symmetry. It is symmetric with respect to the x-axis, the y-axis, and the origin. See Figure 14.

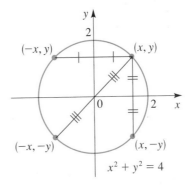

FIGURE 14

3.2 EXERCISES

1–6 ■ Determine whether the given points are on the graph of the equation.

1. $y = 2x + 3$; $(0, 0)$, $\left(\frac{1}{2}, 4\right)$, $(1, 4)$

2. $y = \sqrt{x + 1}$; $(1, 0)$, $(0, 1)$, $(3, 2)$

3. $2y - x + 1 = 0$; $(0, 0)$, $(1, 0)$, $(-1, -1)$

4. $y(x^2 + 1) = 1$; $(1, 1)$, $\left(1, \frac{1}{2}\right)$, $\left(-1, \frac{1}{2}\right)$

5. $x^2 + xy + y^2 = 4$; $(0, -2)$, $(1, -2)$, $(2, -2)$

6. $x^2 + y^2 - 1 = 0$; $(0, 1)$, $\left(\dfrac{1}{\sqrt{2}}, \dfrac{1}{\sqrt{2}}\right)$, $\left(\dfrac{\sqrt{3}}{2}, \dfrac{1}{2}\right)$

7–14 ■ Find the x- and y-intercepts of the graph of the equation.

7. $y = x - 3$

8. $y = x^2 - 5x + 6$

9. $y = x^2 - 9$

10. $y - 2xy + 2x = 1$

11. $x^2 + y^2 = 4$

12. $y = \sqrt{x + 1}$

13. $xy = 5$

14. $x^2 - xy + y = 1$

15–40 ■ Make a table of values and sketch the graph of the equation. Find x- and y-intercepts and test for symmetry.

15. $y = x$

16. $y = -x$

17. $y = x - 1$

18. $y = 2x + 5$

19. $3x - y = 5$

20. $x + y = 3$

21. $y = 1 - x^2$

22. $y = x^2 + 2x$

23. $4y = x^2$

24. $8y = x^3$

25. $y = x^2 - 9$

26. $y = 9 - x^2$

27. $xy = 2$

28. $x + y^2 = 4$

29. $y = \sqrt{x}$

30. $x^2 + y^2 = 9$

31. $y = \sqrt{4 - x^2}$

32. $y = -\sqrt{4 - x^2}$

33. $y = |x|$

34. $x = |y|$

35. $y = 4 - |x|$

36. $y = |4 - x|$

37. $x = y^3$

38. $y = x^3 - 4x$

39. $y = x^4$

40. $y = 16 - x^4$

41–46 ■ Test the equation for symmetry.

41. $y = x^4 + x^2$ **42.** $x = y^4 - y^2$

43. $x^2y^2 + xy = 1$ **44.** $x^4y^4 + x^2y^2 = 1$

45. $y = x^3 + 10x$ **46.** $y = x^2 + |x|$

47–50 ■ Complete the graph using the given symmetry property.

47. Symmetric with respect to the y-axis

48. Symmetric with respect to the x-axis

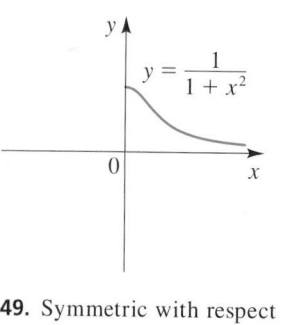

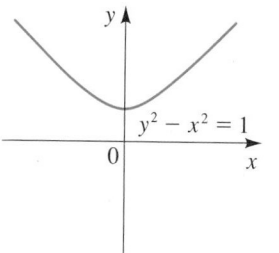

49. Symmetric with respect to the origin

50. Symmetric with respect to the origin

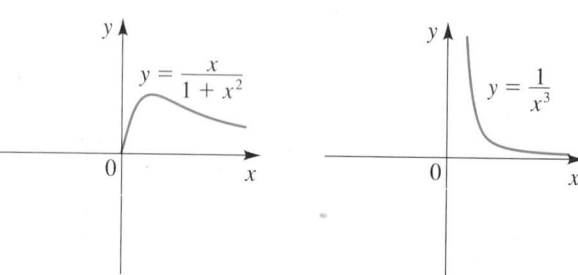

51–58 ■ Find an equation of the circle that satisfies the given conditions.

51. Center $(2, -1)$; radius 3

52. Center $(-1, -4)$; radius 8

53. Center at the origin; passes through $(4, 7)$

54. Center $(-1, 5)$; passes through $(-4, -6)$

55. Endpoints of a diameter are $P(-1, 1)$ and $Q(5, 5)$

56. Endpoints of a diameter are $P(-1, 3)$ and $Q(7, -5)$

57. Center $(7, -3)$; tangent to the x-axis

58. Circle lies in the first quadrant, tangent to both x- and y-axes; radius 5

59–60 ■ Find the equation of the circle shown in the figure.

59.

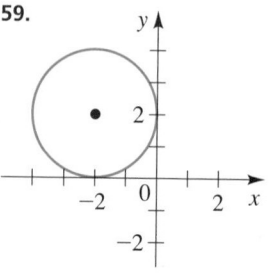

60.

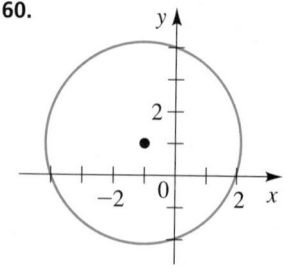

61–68 ■ Show that the equation represents a circle, and find the center and radius of the circle.

61. $x^2 + y^2 - 2x + 4y + 1 = 0$

62. $x^2 + y^2 - 2x - 2y = 2$

63. $x^2 + y^2 - 4x + 10y + 13 = 0$

64. $x^2 + y^2 + 6y + 2 = 0$

65. $x^2 + y^2 + x = 0$

66. $x^2 + y^2 + 2x + y + 1 = 0$

67. $2x^2 + 2y^2 - x + y = 1$

68. $16x^2 + 16y^2 + 8x + 32y + 1 = 0$

69–72 ■ Sketch the graph of the equation.

69. $x^2 + y^2 + 4x - 10y = 21$

70. $4x^2 + 4y^2 + 2x = 0$

71. $x^2 + y^2 + 6x - 12y + 45 = 0$

72. $x^2 + y^2 - 16x + 12y + 200 = 0$

73–76 ■ Sketch the region given by the set.

73. $\{(x, y) \mid x^2 + y^2 \leqslant 1\}$ **74.** $\{(x, y) \mid x^2 + y^2 > 4\}$

75. $\{(x, y) \mid 1 \leqslant x^2 + y^2 < 9\}$

76. $\{(x, y) \mid 2x < x^2 + y^2 \leqslant 4\}$

77. Find the area of the region that lies outside the circle $x^2 + y^2 = 4$ but inside the circle
$$x^2 + y^2 - 4y - 12 = 0$$

78. Sketch the region in the coordinate plane that satisfies both the inequalities $x^2 + y^2 \leqslant 9$ and $y \geqslant |x|$. What is the area of this region?

79. Under what conditions on the coefficients a, b, and c does the equation $x^2 + y^2 + ax + by + c = 0$ represent a circle? When that condition is satisfied, find the center and radius of the circle.

GRAPHING CALCULATORS AND COMPUTERS

In this section we make use of graphing calculators or computers to graph more complicated equations and to solve more complex problems than we would otherwise be able to solve. We begin by describing how these devices work and how to recognize and avoid some common pitfalls of using such devices.

 HOW TO USE GRAPHING DEVICES

A graphing calculator or computer displays a rectangular portion of the graph of an equation in a **display window** or **viewing screen,** which we call a **viewing rectangle.** The default screen often gives an incomplete or misleading picture, so it is important to choose the viewing rectangle with care. If we choose the x-values to range from a minimum value of Xmin $= a$ to a maximum of Xmax $= b$ and the y-values to range from a minimum of Ymin $= c$ to a maximum of Ymax $= d$, then the portion of the graph displayed lies in the rectangle

$$[a, b] \times [c, d] = \{(x, y) \mid a \leqslant x \leqslant b, c \leqslant y \leqslant d\}$$

shown in Figure 1. We refer to this as the $[a, b]$ *by* $[c, d]$ *viewing rectangle.*

The graphing device draws the graph of an equation much as you would. It plots points of the form (x, y) for a certain number of values of x, equally spaced between a and b. If the equation is not defined for an x-value, or if the corresponding y-value lies outside the viewing rectangle, the device ignores this value and moves on to the next x-value. The machine connects each point to the preceding plotted point to form a representation of the graph of the equation.

FIGURE 1
The viewing rectangle $[a, b]$ by $[c, d]$

EXAMPLE 1 ■ Drawing Graphs in Different Viewing Rectangles

Draw the graph of the equation $y = x^2 + 3$ in each viewing rectangle:

(a) $[-2, 2]$ by $[-2, 2]$ (b) $[-4, 4]$ by $[-4, 4]$
(c) $[-10, 10]$ by $[-5, 30]$ (d) $[-50, 50]$ by $[-100, 1000]$

SOLUTION

For part (a) select the range by setting Xmin $= -2$, Xmax $= 2$, Ymin $= -2$, and Ymax $= 2$. The resulting graph is shown in Figure 2(a). The display

FIGURE 2   Graphs of $y = x^2 + 3$

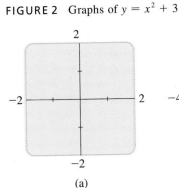

(a)

(b)

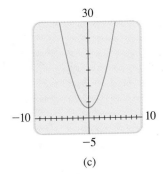

(c)

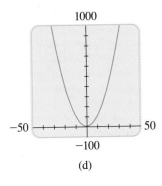

(d)

window is blank! A moment's thought provides the explanation. Notice that $x^2 \geq 0$ for all x, so $y = x^2 + 3 \geq 3$ for all x. This means that the graph of the equation lies totally outside the viewing rectangle $[-2, 2]$ by $[-2, 2]$.

The graphs for the viewing rectangles in parts (b), (c), and (d) are shown in Figure 2, parts (b), (c), and (d). Observe that we get more complete pictures in parts (c) and (d), but the graph in part (d) does not show clearly that the y-intercept is 3. ∎

EXAMPLE 2 ■ Graphing a Cubic Equation

Graph the equation $y = x^3 - 49x$.

SOLUTION

Let us experiment with different viewing rectangles. If we start with the viewing rectangle $[-5, 5]$ by $[-5, 5]$, we get the graph in Figure 3. On most calculators the screen appears to be blank, but it is not quite blank because the point $(0, 0)$ has been plotted. It turns out that for all the other x-values that the calculator chooses between -5 and 5, the value of y is greater than 5 or less than -5, so the corresponding point on the graph lies outside the viewing rectangle.

If we use the zoom-out feature of a graphing calculator to change the viewing rectangle to $[-10, 10]$ by $[-10, 10]$, then we get the picture shown in Figure 4(a). The graph appears to consist of vertical lines, but we know that can't be correct. If we look carefully while the graph is being drawn, we see that the graph leaves the screen and reappears during the graphing process. This indicates that we need to see more of the graph in the vertical direction, so we change the viewing rectangle to $[-10, 10]$ by $[-100, 100]$. The resulting graph is shown in Figure 4(b). It still doesn't reveal all the main features of the equation, so we try $[-10, 10]$ by $[-200, 200]$ in Figure 4(c). Now we are more confident that we have arrived at an appropriate viewing rectangle. In Chapter 4, when we discuss third-degree polynomials, we will see that the graph shown in Figure 4(c) does indeed reveal all the main features of the equation.

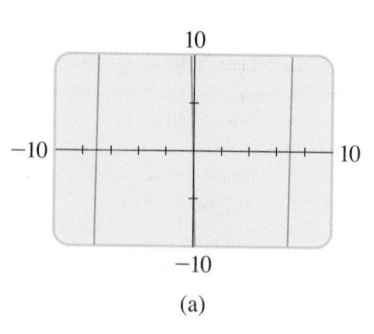

FIGURE 3

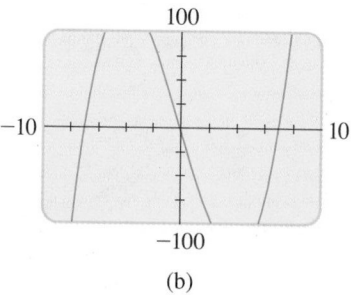

(a)

(b)

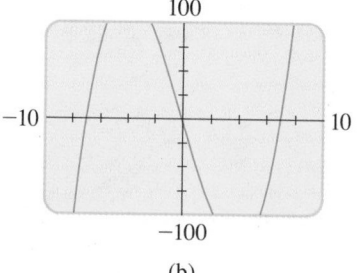

(c)

FIGURE 4

Graphs of $y = x^3 - 49x$

∎

We see from Examples 1 and 2 that the choice of a viewing rectangle can make a big difference in the appearance of a graph. Sometimes it is necessary to change to a larger viewing rectangle to obtain a more complete picture—a more global view—of the graph. In the next example we see how the equation itself can provide us with enough information to select a good viewing rectangle.

EXAMPLE 3 ■ Choosing an Appropriate Viewing Rectangle

Determine an appropriate viewing rectangle for the equation $y = \sqrt{8 - 2x^2}$ and graph the equation.

SOLUTION

The expression for y is defined only when the expression under the square root sign is not negative. So we must have

$$0 \le 8 - 2x^2 \qquad \text{Expression under the root sign must be nonnegative}$$

$$2x^2 \le 8 \qquad \text{Add } 2x^2$$

$$x^2 \le 4 \qquad \text{Divide by 2}$$

$$|x| \le 2 \qquad \text{Take square root}$$

$$-2 \le x \le 2 \qquad \text{Meaning of } |x| \le 2 \text{ (see Section 2.8)}$$

Therefore, the values of x for which y is defined lie in the interval $[-2, 2]$. Also

$$0 \le \sqrt{8 - 2x^2} \le \sqrt{8} = 2\sqrt{2} \approx 2.83$$

so the corresponding values of y lie in the interval $[0, 2.83]$.

We choose the viewing rectangle so that the x-interval and y-interval are somewhat larger than those we found. Taking the viewing rectangle to be $[-3, 3]$ by $[-1, 4]$, we get the graph shown in Figure 5.

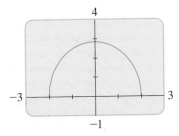

FIGURE 5
$y = \sqrt{8 - 2x^2}$

Graphing calculators are able to easily graph equations in which y can be isolated on one side of the equal sign. The next example shows how to graph equations that do not have this property.

Alan Turing (1912–1954) was at the center of two pivotal events of the 20th century—World War II and the invention of computers. At the age of 23 Turing made his mark on mathematics by solving Hilbert's third problem (see page 490). In this research he invented a theoretical machine, now called a Turing machine, which was the inspiration for modern digital computers. During World War II Turing was in charge of the British effort to decipher secret German codes. His complete success in this endeavor played a decisive role in the Allies' victory. To carry out the numerous logical steps required to break a coded message, Turing developed decision procedures similar to modern computer programs. After the war he helped develop the first electronic computers in Britain. He also did pioneering work on artificial intelligence and computer models of biological processes. At the age of 42 Turing died of cyanide poisoning under mysterious circumstances.

EXAMPLE 4 ■ Graphing a Circle

Graph the circle $x^2 + y^2 = 1$.

SOLUTION

We know from Section 3.2 how to graph this circle by hand; it is a circle with center the origin and radius 1. But to graph the circle with a graphing calculator is not a straightforward task, because we cannot isolate y on one side of the equal sign. If we try to solve the equation of the circle for y, we have

$$y^2 = 1 - x^2$$
$$y = \pm\sqrt{1 - x^2}$$

Therefore, we can regard the circle as being described by the graphs of *two* equations:

$$y = \sqrt{1 - x^2} \qquad \text{and} \qquad y = -\sqrt{1 - x^2}$$

The first equation represents the top half of the circle (because $y \geq 0$), and the second represents the bottom half of the circle (because $y \leq 0$). If we graph the first equation in the viewing rectangle $[-2, 2]$ by $[-2, 2]$, we get the semicircle shown in Figure 6(a). The graph of the second equation is the semicircle in Figure 6(b). Graphing these semicircles together on the same viewing screen, we get the full circle in Figure 6(c).

The graph in Figure 6(c) looks somewhat flattened. Most graphing calculators allow you to set the scales on the axes so that circles really look like circles. On the TI-82, for example, from the Zoom menu, choose "ZSquare" to set the scales appropriately. (On the TI-85 the command is "Zsq.")

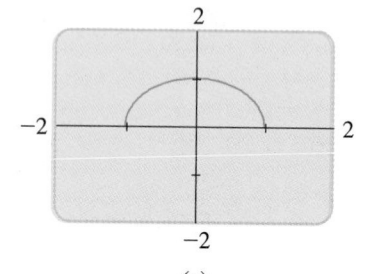

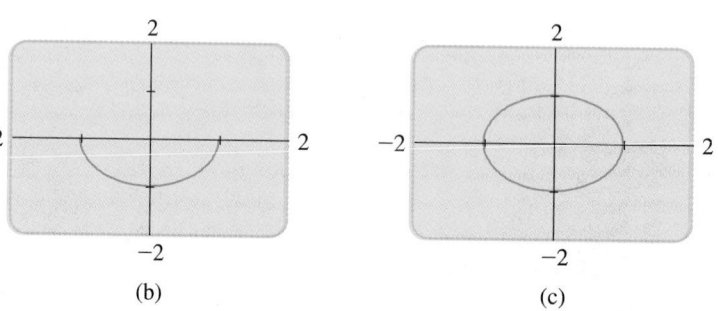

(a)   (b)   (c)

FIGURE 6   Graph of $x^2 + y^2 = 1$

We have already discovered one of the pitfalls of using graphing calculators and computers: Examples 1 and 2 showed that the use of an inappropriate viewing rectangle can give a misleading representation of the graph of an equation. We also saw how to remedy the situation: We included the crucial parts of the graph by changing to a larger viewing rectangle. Another pitfall is illustrated in the next example.

EXAMPLE 5 ■ Avoiding Extraneous Lines in Graphs

Draw the graph of the equation $y = \dfrac{1}{1 - x}$.

SOLUTION

Figure 7(a) shows the graph produced by a graphing calculator with viewing rectangle $[-9, 9]$ by $[-9, 9]$. In connecting successive points on the graph, the calculator produced a steep line segment from the top to the bottom of the screen. That line segment should not be part of the graph. Notice that the right side of the equation $y = 1/(1 - x)$ is not defined for $x = 1$. Sometimes we can get rid of the extraneous near-vertical line by experimenting with a change of scale. Here, for example, when we change to the smaller viewing rectangle $[-5, 5]$ by $[-5, 5]$, we obtain the much better graph in Figure 7(b).

Another way to avoid the extraneous line is to change the graphing mode on the calculator so that the dots are not connected.

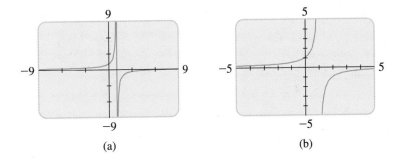

FIGURE 7

Graphs of $y = \dfrac{1}{1 - x}$

(a)　　　　　(b)

SOLVING EQUATIONS AND INEQUALITIES GRAPHICALLY

In Chapter 2 we solved equations and inequalities algebraically. The ability to sketch graphs of equations quickly and accurately gives us another method of finding approximate solutions to equations and inequalities. For example, consider the equation

$$x^2 - 3x - 10 = 0$$

To solve this equation we first graph

$$y = x^2 - 3x - 10$$

as shown in Figure 8(a). We are interested in those values of x for which $y = 0$. But these are precisely the x-intercepts of the graph. By moving the cursor near the x-intercepts, we see that they appear to be at $x = -2$ and $x = 5$. To confirm this we may use the zoom-in feature on the graphing device to get a closer look

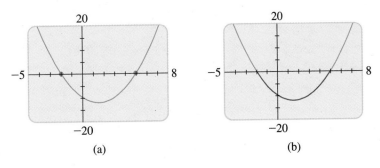

FIGURE 8

Graphs of $y = x^2 - 3x - 10$

(a)　　　　　(b)

at each of the intercepts. In this particular case we can prove that these are the roots by substituting them directly into the equation.

To solve the inequality

$$x^2 - 3x - 10 > 0$$

we again sketch the graph of $y = x^2 - 3x - 10$ and look for the x-coordinates of those points on the graph with y-coordinate greater than 0. These are simply the x-values for which the graph lies above the x-axis. So, from Figure 8(b) we see that the solution of the inequality is $(-\infty, -2) \cup (5, \infty)$. Similarly, the solution of the inequality

$$x^2 - 3x - 10 < 0$$

In Example 6 we can use the quadratic formula to get

$$x = \frac{-3 \pm \sqrt{3^2 - 4 \cdot 2 \cdot (-7)}}{2 \cdot 2}$$
$$= \frac{-3 \pm \sqrt{65}}{4}$$

You can check that, correct to two decimals, these are the solutions we obtained using the graphing calculator.

consists of the x-coordinates of those points on the graph with y-coordinate less than 0, that is, points where the graph lies below the x-axis. So, the solution is the interval $(-2, 5)$.

EXAMPLE 6 ■ Solving an Equation Graphically

Solve the equation $2x^2 + 3x - 7 = 0$.

SOLUTION

We graph the equation

$$y = 2x^2 + 3x - 7$$

in the viewing rectangle $[-5, 5]$ by $[-15, 15]$ (see Figure 9). The solutions occur at the x-intercepts. Moving the cursor to the x-intercepts, we find that $x \approx -2.8$ and $x \approx 1.3$. Using the zoom-in feature to get a closer look at the x-intercepts, we read the more accurate approximations

$$x \approx -2.77 \qquad \text{and} \qquad x \approx 1.27 \qquad \blacksquare$$

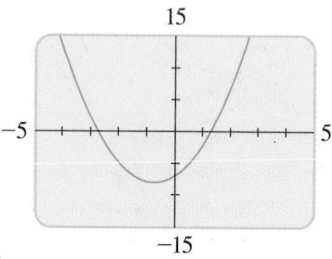

FIGURE 9
$y = 2x^2 + 3x - 7$

EXAMPLE 7 ■ Solving an Inequality Graphically

Solve the inequality $x^3 - 5x^2 \geq -8$.

SOLUTION

We write the inequality as

$$x^3 - 5x^2 + 8 \geq 0$$

and then graph the equation

$$y = x^3 - 5x^2 + 8$$

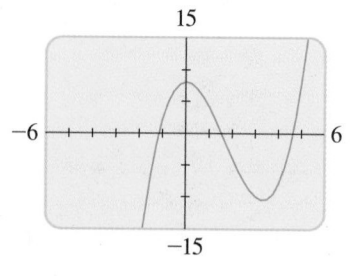

FIGURE 10
$y = x^3 - 5x^2 + 8$

in the viewing rectangle $[-6, 6]$ by $[-15, 15]$, as shown in Figure 10. The solution of the inequality consists of those intervals on which the graph lies on or above the x-axis. By moving the cursor to the x-intercepts we find that, correct to one decimal place, the solution is $[-1.1, 1.5] \cup [4.6, \infty)$. $\blacksquare$

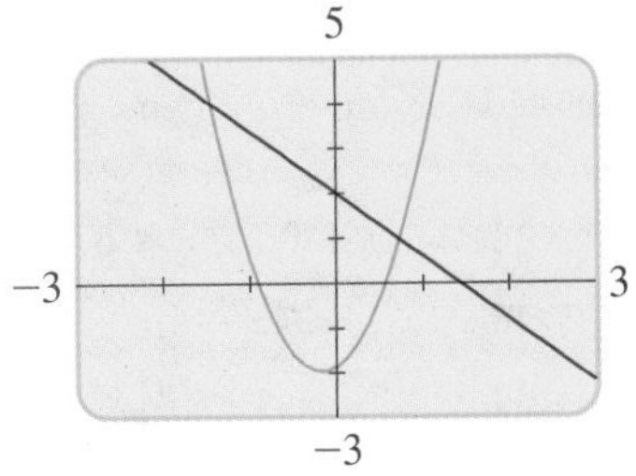

EXAMPLE 8 ■ Solving an Inequality Graphically

Solve the inequality $3.7x^2 + 1.3x - 1.9 \leqslant 2.0 - 1.4x$.

SOLUTION

We graph the equations

$$y_1 = 3.7x^2 + 1.3x - 1.9$$

$$y_2 = 2.0 - 1.4x$$

in the same viewing rectangle in Figure 11. We are interested in those values of x for which $y_1 \leqslant y_2$; these are points for which the graph of y_2 lies on or above the graph of y_1. To determine the appropriate interval, we look for the x-coordinates of points where the graphs intersect. We conclude that the solution is (approximately) the interval $[-1.45, 0.72]$.   ■

FIGURE 11
$y_1 = 3.7x^2 + 1.3x - 1.9$
$y_2 = 2.0 - 1.4x$

   **3.3   EXERCISES**

1–6 ■ Use a graphing calculator or computer to decide which one of the viewing rectangles (a)–(d) produces the most appropriate graph of the equation.

1. $y = x^4 + 2$
   (a) $[-2, 2]$ by $[-2, 2]$ (b) $[0, 4]$ by $[0, 4]$
   (c) $[-8, 8]$ by $[-4, 40]$ (d) $[-40, 40]$ by $[-80, 800]$

2. $y = x^2 + 7x + 6$
   (a) $[-5, 5]$ by $[-5, 5]$ (b) $[0, 10]$ by $[-20, 100]$
   (c) $[-15, 8]$ by $[-20, 100]$ (d) $[-10, 3]$ by $[-100, 20]$

3. $y = 100 - x^2$
   (a) $[-4, 4]$ by $[-4, 4]$ (b) $[-10, 10]$ by $[-10, 10]$
   (c) $[-15, 15]$ by $[-30, 110]$ (d) $[-4, 4]$ by $[-30, 110]$

4. $y = 2x^2 - 1000$
   (a) $[-10, 10]$ by $[-10, 10]$
   (b) $[-10, 10]$ by $[-100, 100]$
   (c) $[-10, 10]$ by $[-1000, 1000]$
   (d) $[-25, 25]$ by $[-1200, 200]$

5. $y = 10 + 25x - x^3$
   (a) $[-4, 4]$ by $[-4, 4]$
   (b) $[-10, 10]$ by $[-10, 10]$
   (c) $[-20, 20]$ by $[-100, 100]$
   (d) $[-100, 100]$ by $[-200, 200]$

6. $y = \sqrt{8x - x^2}$
   (a) $[-4, 4]$ by $[-4, 4]$ (b) $[-5, 5]$ by $[0, 100]$
   (c) $[-10, 10]$ by $[-10, 40]$ (d) $[-2, 10]$ by $[-2, 6]$

7–20 ■ Determine an appropriate viewing rectangle for the equation and use it to draw the graph.

7. $y = 100x^2$ **8.** $y = -100x^2$

9. $y = 4 + 6x - x^2$ **10.** $y = 0.3x^2 + 1.7x - 3$

11. $y = \sqrt[4]{256 - x^2}$ **12.** $y = \sqrt{12x - 17}$

13. $y = 0.01x^3 - x^2 + 5$ **14.** $y = x(x + 6)(x - 9)$

15. $y = \dfrac{1}{x^2 + 25}$ **16.** $y = \dfrac{x}{x^2 + 25}$

17. $y = x^4 - 4x^3$ **18.** $y = x^3 + \dfrac{1}{x}$

19. $y = 1 + |x - 1|$ **20.** $y = 2x - |x^2 - 5|$

21. Graph the circle $x^2 + y^2 = 9$ by solving for y and graphing two equations as in Example 4.

22. Graph the circle $(y - 1)^2 + x^2 = 1$ by solving for y and graphing two equations as in Example 4.

23. Graph the equation $4x^2 + 2y^2 = 1$ by solving for y and graphing two equations corresponding to the negative and positive square roots. (The graph is called an *ellipse*.)

24. Graph the equation $y^2 - 9x^2 = 1$ by solving for y and graphing the two equations corresponding to the positive and negative square roots. (The graph is called a *hyperbola*.)

25–32 ■ Find the solutions of the equation in the given interval by drawing a graph in an appropriate viewing rectangle. State each answer correct to two decimals.

25. $x^2 - 7x + 12 = 0$; $[0, 6]$

26. $x^2 - 0.75x + 0.125 = 0$; $[-2, 2]$

27. $x^3 - 6x^2 + 11x - 6 = 0$; $[-1, 4]$

28. $16x^3 + 16x^2 = x + 1$; $[-2, 2]$

29. $x - \sqrt{x + 1} = 0$; $[-1, 5]$

30. $1 + \sqrt{x} = \sqrt{1 + x^2}$; $[-1, 5]$

31. $x^{1/3} - x = 0$; $[-3, 3]$

32. $x^{1/2} + x^{1/3} - x = 0$; $[-1, 5]$

33–40 ■ Find the solutions of the inequality by drawing appropriate graphs. State each answer correct to two decimals.

33. $x^2 - 3x - 10 \le 0$

34. $0.5x^2 + 0.875x \le 0.25$

35. $x^3 + 11x \le 6x^2 + 6$

36. $16x^3 + 24x^2 > -9x - 1$

37. $x^{1/3} < x$

38. $\sqrt{0.5x^2 + 1} \le 2|x|$

39. $(x + 1)^2 < (x - 1)^2$

40. $(x + 1)^2 \le x^3$

3.4 LINES

In this section we find equations for straight lines lying in a coordinate plane. We first need a way of measuring the "steepness" of a line. From Figure 1 we see that the steepness is determined by how quickly the line rises (or falls) as we move from left to right. Thus we define the slope of a line as the ratio of the change in the y-coordinate (that is, the distance the line rises or falls) to the change in the x-coordinate (that is, the distance we move to the right). The change in the y-coordinate between two points on a line is called the *rise* and the change in the x-coordinate is called the *run*. So, to find the slope of a line we choose two points on the line and find

$$\text{slope} = \frac{\text{rise}}{\text{run}}$$

More precisely, we have the following definition.

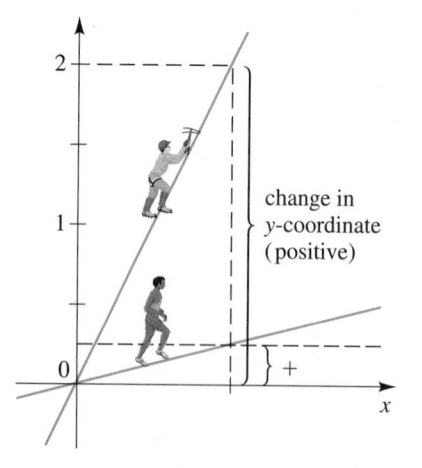
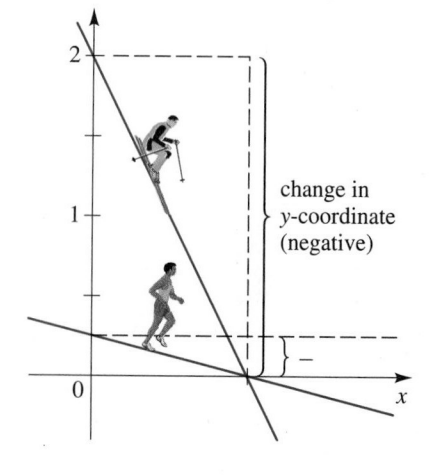

FIGURE 1

SLOPE OF A LINE

The **slope** m of a nonvertical line that passes through the points $A(x_1, y_1)$ and $B(x_2, y_2)$ is

$$m = \frac{y_2 - y_1}{x_2 - x_1}$$

The slope of a vertical line is not defined.

The slope is independent of which two points are chosen on the line. We can see that this is true from the similar triangles in Figure 2:

$$\frac{y_2 - y_1}{x_2 - x_1} = \frac{y_2' - y_1'}{x_2' - x_1'}$$

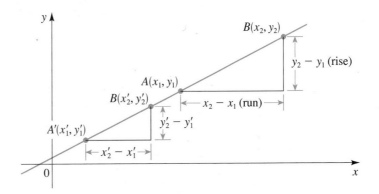

FIGURE 2

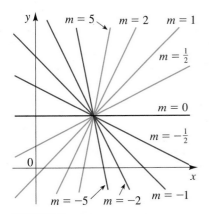

FIGURE 3

Figure 3 shows several lines labeled with their slopes. Notice that lines with positive slope slant upward to the right, whereas lines with negative slope slant downward to the right. The steepest lines are those for which the absolute value of the slope is the largest; a horizontal line has slope zero.

EXAMPLE 1 ■ Finding the Slope of a Line through Two Points

Find the slope of the line that passes through the points $P(2, 1)$ and $Q(8, 5)$.

SOLUTION

Since any two different points determine a line, there is only one line that passes through these two points. From the definition, the slope is

$$m = \frac{y_2 - y_1}{x_2 - x_1} = \frac{5 - 1}{8 - 2} = \frac{4}{6} = \frac{2}{3}$$

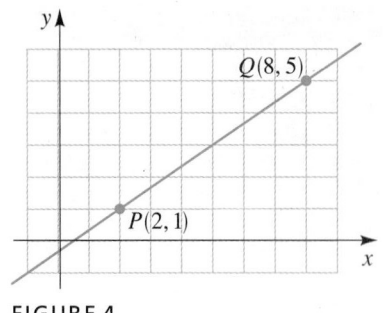

FIGURE 4

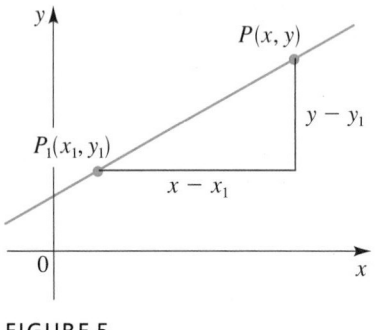

FIGURE 5

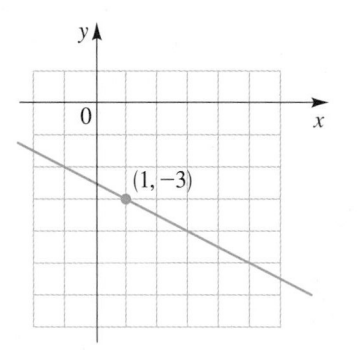

FIGURE 6

This says that for every 3 units we move to the right, the line rises 2 units. The line is drawn in Figure 4. ■

Now let us find the equation of the line that passes through a given point $P(x_1, y_1)$ and has slope m. A point $P(x, y)$ with $x \neq x_1$ lies on this line if and only if the slope of the line through P_1 and P is equal to m (see Figure 5), that is,

$$\frac{y - y_1}{x - x_1} = m$$

This equation can be rewritten in the form

$$y - y_1 = m(x - x_1)$$

and we observe that this equation is also satisfied when $x = x_1$ and $y = y_1$. Therefore, it is an equation of the given line.

POINT-SLOPE FORM OF THE EQUATION OF A LINE

An equation of the line that passes through the point (x_1, y_1) and has slope m is

$$y - y_1 = m(x - x_1)$$

EXAMPLE 2 ■ Finding the Equation of a Line with Given Point and Slope

(a) Find an equation of the line through $(1, -3)$ with slope $-\frac{1}{2}$.
(b) Sketch the line.

SOLUTION

(a) Using the point-slope form with $m = -\frac{1}{2}$, $x_1 = 1$, and $y_1 = -3$, we obtain an equation of the line as

$$y + 3 = -\tfrac{1}{2}(x - 1) \qquad \text{From point-slope equation}$$

$$2y + 6 = -x + 1 \qquad \text{Multiply by 2}$$

$$x + 2y + 5 = 0 \qquad \text{Rearrange}$$

(b) The fact that the slope is $-\frac{1}{2}$ tells us that when we move 2 units to the right, the line drops 1 unit. This enables us to sketch the line in Figure 6. ■

EXAMPLE 3 ■ Finding the Equation of a Line through Two Given Points

Find an equation of the line through the points $(-1, 2)$ and $(3, -4)$.

SOLUTION

The slope of the line is

$$m = \frac{-4 - 2}{3 - (-1)} = -\frac{3}{2}$$

Using the point-slope form with $x_1 = -1$ and $y_1 = 2$, we obtain

$$y - 2 = -\tfrac{3}{2}(x + 1) \qquad \text{From point-slope equation}$$

$$2y - 4 = -3x - 3 \qquad \text{Multiply by 2}$$

$$3x + 2y - 1 = 0 \qquad \text{Rearrange} \qquad \blacksquare$$

Suppose a nonvertical line has slope m and y-intercept b (see Figure 7). This means the line intersects the y-axis at the point $(0, b)$, so the point-slope form of the equation of the line, with $x = 0$ and $y = b$, becomes

$$y - b = m(x - 0)$$

This simplifies to $y = mx + b$, which is called the **slope-intercept form** of the equation of a line.

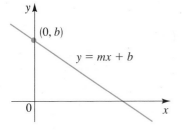

FIGURE 7

SLOPE-INTERCEPT FORM OF THE EQUATION OF A LINE

An equation of the line that has slope m and y-intercept b is

$$y = mx + b$$

EXAMPLE 4 ■ Finding the Slope of a Line from Its Equation

Find the slope and y-intercept of each line.

(a) $y = 3x - 2$ (b) $3y - 2x = 1$

SOLUTION

(a) The equation $y = 3x - 2$ is in the form $y = mx + b$, where $m = 3$ and $b = -2$. So, from the slope-intercept form of the equation, we see that the slope is 3 and the y-intercept is -2.

(b) We first write the equation in the form $y = mx + b$:

$$3y - 2x = 1$$

$$3y = 2x + 1 \qquad \text{Add } 2x$$

$$y = \tfrac{2}{3}x + \tfrac{1}{3} \qquad \text{Divide by 3}$$

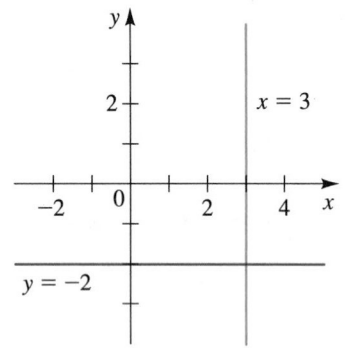

FIGURE 8

From the slope-intercept form of the equation of a line, we see that the slope is $m = \frac{2}{3}$ and the y-intercept is $b = \frac{1}{3}$. ■

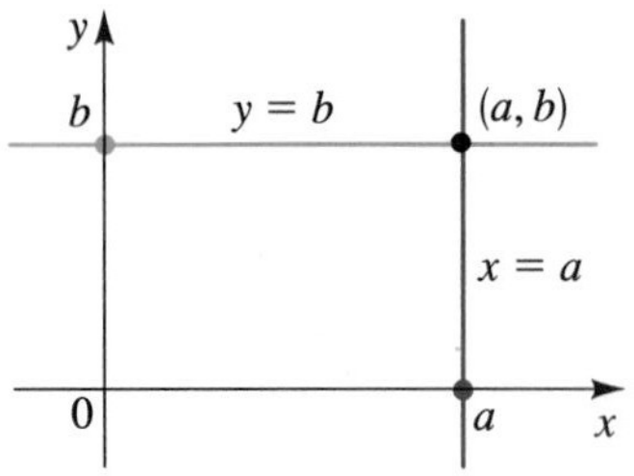

If a line is horizontal, its slope is $m = 0$, so its equation is $y = b$, where b is the y-intercept (see Figure 8). A vertical line does not have a slope, but we can write its equation as $x = a$, where a is the x-intercept, because the x-coordinate of every point on the line is a.

VERTICAL AND HORIZONTAL LINES

The equation of a vertical line through (a, b) is $x = a$.

The equation of a horizontal line through (a, b) is $y = b$.

EXAMPLE 5 ■ **Vertical and Horizontal Lines**

(a) The graph of the equation $x = 3$ is a vertical line with x-intercept 3.
(b) The graph of the equation $y = -2$ is a horizontal line with y-intercept -2.

The lines are graphed in Figure 9. ■

FIGURE 9

A **linear equation** is an equation of the form

$$Ax + By + C = 0$$

where A, B, and C are constants and A and B are not both 0. The equation of a line is a linear equation:

A nonvertical line has the equation $y = mx + b$ or $-mx + y - b = 0$, which is a linear equation with $A = -m$, $B = 1$, and $C = -b$.

A vertical line has the equation $x = a$ or $x - a = 0$, which is a linear equation with $A = 1$, $B = 0$, and $C = -a$.

Conversely, the graph of a linear equation is a line:

If $B \neq 0$, the equation becomes

$$y = -\frac{A}{B} x - \frac{C}{B}$$

and this is the slope-intercept form of the equation of a line (with $m = -A/B$ and $b = -C/B$).

If $B = 0$, the equation becomes $Ax + C = 0$, or $x = -C/A$, which represents a vertical line.

We have proved the following.

GENERAL EQUATION OF A LINE

The graph of every **linear equation**

$$Ax + By + C = 0 \qquad (A, B \text{ not both zero})$$

is a line. Conversely, every line is the graph of a linear equation.

EXAMPLE 6 ■ Graphing a Linear Equation

Sketch the graph of the equation $3x - 5y = 15$.

SOLUTION

Since the equation $3x - 5y = 15$ is linear, its graph is a line. To draw the graph, it is enough to find just two points on the line. The intercepts are the easiest points to find. Substituting $y = 0$ in the given equation, we get $3x = 15$, so $x = 5$ is the x-intercept. Substituting $x = 0$ in the equation, we find that the y-intercept is -3. With these points we can sketch the graph shown in Figure 10.

Another method is to write the equation in slope-intercept form:

$$3x - 5y = 15$$
$$5y = 3x - 15$$
$$y = \tfrac{3}{5}x - 3$$

This equation is in the form $y = mx + b$, so the slope is $m = \tfrac{3}{5}$ and the y-intercept is $b = -3$. This information can be used to sketch the line. ■

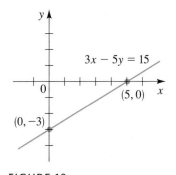

FIGURE 10

PARALLEL AND PERPENDICULAR LINES

Since slope measures the steepness of a line, it seems reasonable that parallel lines should have the same slope. In fact, we can prove this.

PARALLEL LINES

Two nonvertical lines are parallel if and only if they have the same slope.

■ **Proof** Let the lines l_1 and l_2 in Figure 11 have slopes m_1 and m_2. If the lines are parallel, then the right triangles ABC and DEF are similar, so

$$m_1 = \frac{d(B, C)}{d(A, C)} = \frac{d(E, F)}{d(D, F)} = m_2$$

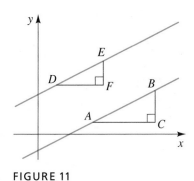

FIGURE 11

Conversely, if the slopes are equal, then the triangles will be similar, so $\angle BAC = \angle EDF$ and the lines are parallel. $\square$

EXAMPLE 7 ■ Finding the Equation of a Line Parallel to a Given Line

Find an equation of the line through the point $(5, 2)$ that is parallel to the line $4x + 6y + 5 = 0$.

SOLUTION

First we write the equation of the given line in slope-intercept form:

$$4x + 6y + 5 = 0$$

$$6y = -4x - 5 \qquad \text{Subtract } 4x - 5$$

$$y = -\tfrac{2}{3}x - \tfrac{5}{6} \qquad \text{Divide by 6}$$

So the line has slope $m = -\tfrac{2}{3}$ and since the required line is parallel to it, it also has slope $m = -\tfrac{2}{3}$. From the point-slope form of the equation of a line, we get

$$y - 2 = -\tfrac{2}{3}(x - 5) \qquad \text{Slope } m = -\tfrac{2}{3}, \text{ point } (5, 2)$$

$$3y - 6 = -2x + 10 \qquad \text{Multiply by 3}$$

$$2x + 3y - 16 = 0 \qquad\qquad \text{Rearrange}$$

So, the equation of the required line is $2x + 3y - 16 = 0$.   ■

The condition for perpendicular lines is not as obvious as that for parallel lines.

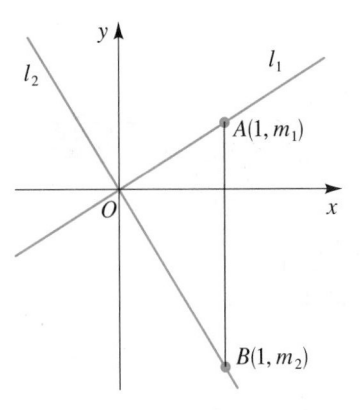

FIGURE 12

PERPENDICULAR LINES

Two lines with slopes m_1 and m_2 are perpendicular if and only if $m_1 m_2 = -1$, that is, their slopes are negative reciprocals:

$$m_2 = -\frac{1}{m_1}$$

Also, a horizontal line (slope 0) is perpendicular to a vertical line (no slope).

■ **Proof** In Figure 12 we show two lines intersecting at the origin. (If the lines intersect at some other point, we consider lines parallel to these that intersect at the origin. These lines have the same slopes as the original lines.)

If the lines l_1 and l_2 have slopes m_1 and m_2, then their equations are $y = m_1 x$ and $y = m_2 x$. Notice that $A(1, m_1)$ lies on l_1 and $B(1, m_2)$ lies on l_2. By the Pythagorean Theorem and its converse, $OA \perp OB$ if and only if

$$[d(O, A)]^2 + [d(O, B)]^2 = [d(A, B)]^2$$

By the Distance Formula, this becomes

$$(1^2 + m_1^2) + (1^2 + m_2^2) = (1 - 1)^2 + (m_2 - m_1)^2$$

$$2 + m_1^2 + m_2^2 = m_2^2 - 2m_1 m_2 + m_1^2$$

$$2 = -2m_1 m_2$$

$$m_1 m_2 = -1 \qquad \square$$

EXAMPLE 8 ■ Perpendicular Lines

Show that the points $P(3, 3)$, $Q(8, 17)$, and $R(11, 5)$ are the vertices of a right triangle.

SOLUTION

The slopes of the lines containing PR and QR are, respectively,

$$m_1 = \frac{5 - 3}{11 - 3} = \frac{1}{4}$$

and

$$m_2 = \frac{5 - 17}{11 - 8} = -4$$

Since $m_1 m_2 = -1$, these lines are perpendicular and so PQR is a right triangle. It is sketched in Figure 13. ■

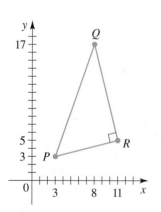

FIGURE 13

EXAMPLE 9 ■ Finding the Equation of a Line Perpendicular to a Given Line

Find the equation of a line that is perpendicular to the line $4x + 6y + 5 = 0$ and passes through the origin.

SOLUTION

In Example 7 we found that the slope of the line $4x + 6y + 5 = 0$ is $-\frac{2}{3}$. Thus, the slope of a perpendicular line is the negative reciprocal, that is, $\frac{3}{2}$. Since the required line passes through $(0, 0)$, the point-slope form gives

$$y - 0 = \tfrac{3}{2}(x - 0)$$

$$y = \tfrac{3}{2}x$$

■

APPLICATIONS OF LINEAR EQUATIONS

The simplest equations relating two variables are linear equations. We now consider how such equations are used in applications to model the relationship between two quantities.

EXAMPLE 10 ■ Linear Relationship between Fahrenheit and Celsius Scales

The relationship between the Fahrenheit (F) and Celsius (C) temperature scales is given by the equation $F = \frac{9}{5}C + 32$.
(a) Sketch a graph of this equation, letting the horizontal axis be the C-axis and the vertical axis be the F-axis.
(b) What is the slope of this graph and what does it represent? What is the F-intercept and what does it represent?

SOLUTION

(a) This is a linear equation, so its graph is a line. (We are calling the coordinates C and F instead of x and y, but this makes no difference in the shape of the graph.) Since two points determine a line, we first find two points that satisfy the equation, and plot those points. Then we draw a straight line through them. When $C = 0$, then $F = \frac{9}{5}(0) + 32 = 32$, and when $C = 5$, then $F = \frac{9}{5}(5) + 32 = 41$. Thus, the points $(0, 32)$ and $(5, 41)$ lie on the line. These points and the line containing them are shown in Figure 14.

(b) Since the equation is given in slope-intercept form, we see that the slope is $\frac{9}{5}$ and the F-intercept is 32. The slope represents the change in °F for every 1 °C change; thus, a 9 °F increase in temperature corresponds to a 5 °C increase. The F-intercept is the F-coordinate of the point on the graph whose C-coordinate is 0. Thus, 32 °F is the same as 0 °C (the freezing point of water). ■

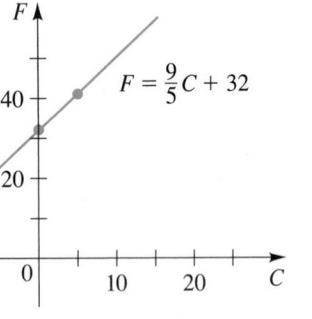

FIGURE 14

EXAMPLE 11 ■ Linear Relationship between Temperature and Elevation

(a) As dry air moves upward, it expands and cools. If the ground temperature is 20 °C and the temperature at a height of 1 km is 10 °C, express the temperature T (in °C) in terms of the height h (in kilometers). (Assume that the expression is linear.)
(b) Draw the graph of the linear equation. What does its slope represent?
(c) What is the temperature at a height of 2.5 km?

SOLUTION

(a) Because we are assuming a linear relationship between h and T, the equation must be of the form

$$T = mh + b$$

where m and b are constants. When $h = 0$, we are given that $T = 20$, so

$$20 = m(0) + b$$

$$b = 20$$

Thus, we have

$$T = mh + 20$$

When $h = 1$, we have $T = 10$ and so

$$10 = m(1) + 20$$

$$m = 10 - 20 = -10$$

The required expression is

$$T = -10h + 20$$

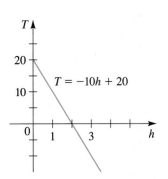

FIGURE 15

(b) The graph is sketched in Figure 15. The slope is $m = -10\ °/\text{km}$, and this represents the rate of change of temperature with respect to distance above the ground.

(c) At a height of $h = 2.5$ km, the temperature is

$$T = -10(2.5) + 20 = -25 + 20 = -5\ °\text{C} \qquad \blacksquare$$

Economists model supply and demand for a commodity by linear equations. For example, for a certain commodity we might have

Supply equation $y = 8p - 10$

Demand equation $y = -3p + 15$

In the supply and demand equations, most economists express p, the price, in terms of y, the amount produced or sold. For clarity, we have chosen to express y in terms of p.

where p is the price of the commodity. In the supply equation y, the *amount produced*, increases as the price increases, because if the price is high more people will produce the commodity. The demand equation indicates that y, the *amount sold,* decreases as the price increases. The **equilibrium point** is the point of intersection of the graphs of the supply and demand equations; at that point the amount produced equals the amount sold.

 EXAMPLE 12 ■ Supply and Demand for Wheat

An economist models the market for wheat by the following equations:

Supply equation $y = 8.33p - 14.58$

Demand equation $y = -1.39p + 23.35$

Here p is the price per bushel (in dollars) and y is the number of bushels produced and sold (in millions).

(a) At what point is the price so low that no wheat is produced?

(b) At what point is the price so high that no wheat is sold?

(c) Draw the graph of the supply and demand lines in the same viewing rectangle and find the equilibrium point. Estimate the equilibrium price and the quantities produced and sold at equilibrium.

SOLUTION

(a) If no wheat is produced, then $y = 0$ in the supply equation.

$$0 = 8.33p - 14.58 \qquad \text{Set } y = 0 \text{ in the supply equation}$$

$$p = 1.75 \qquad\qquad \text{Solve for } p$$

So, at the low price of $1.75 per bushel, the production of wheat halts completely.

(b) If no wheat is sold, then $y = 0$ in the demand equation.

$$0 = -1.39p + 23.35 \qquad \text{Set } y = 0 \text{ in the demand equation}$$

$$p = 16.80 \qquad\qquad \text{Solve for } p$$

So, at the high price of $16.80 per bushel, no one buys wheat.

(c) In Figure 16 we draw the graphs of both the supply and demand equations in the same viewing rectangle. By moving the cursor to their point of intersection we find that the lines intersect (approximately) at the point $(3.9, 17.9)$. Thus, the equilibrium point occurs when the price is $3.90 per bushel and 17.9 million bushels are produced and sold. ∎

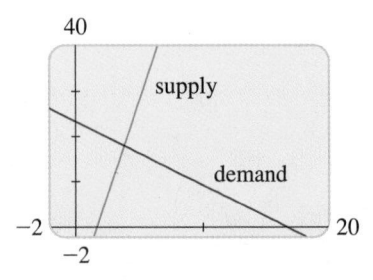

FIGURE 16

3.4 EXERCISES

1–6 ■ Find the slope of the line through P and Q.

1. $P(2, 4)$, $Q(4, 12)$ **2.** $P(-2, 5)$, $Q(4, -3)$

3. $P(-1, 3)$, $Q(1, -6)$ **4.** $P(-1, -4)$, $Q(6, 0)$

5. $P(2, 2)$, $Q(0, 0)$ **6.** $P(-2, 3)$, $Q(5, 5)$

7. Find the slopes of the lines l_1, l_2, l_3, and l_4 in the figure.

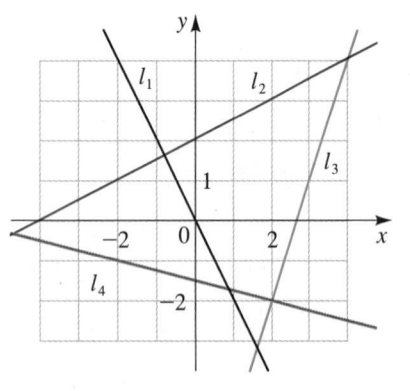

8. (a) Sketch lines through $(0, 0)$ with slopes $1, 0, \frac{1}{2}, 2$, and -1.

(b) Sketch lines through $(0, 0)$ with slopes $\frac{1}{3}, \frac{1}{2}, -\frac{1}{3}$, and 3.

9–12 ■ Find an equation for the line whose graph is sketched.

9.

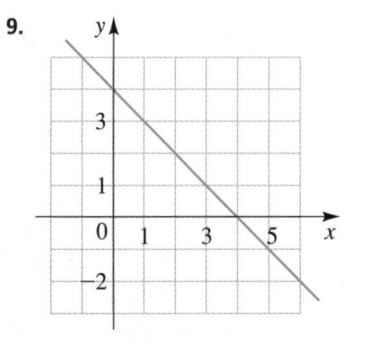

10.

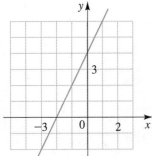

11.

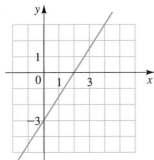

12.

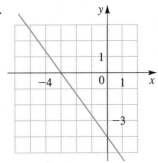

13–32 ■ Find an equation of the line that satisfies the given conditions.

13. Through $(2, 3)$; slope 1

14. Through $(-2, 4)$; slope -1

15. Through $(1, 7)$; slope $\frac{2}{3}$

16. Through $(-3, -5)$; slope $-\frac{7}{2}$

17. Through $(2, 1)$ and $(1, 6)$

18. Through $(-1, -2)$ and $(4, 3)$

19. Slope 3; y-intercept -2

20. Slope $\frac{2}{3}$; y-intercept 4

21. x-intercept 1; y-intercept -3

22. x-intercept -8; y-intercept 6

23. Through $(4, 5)$; parallel to the x-axis

24. Through $(4, 5)$; parallel to the y-axis

25. Through $(1, -6)$; parallel to the line $x + 2y = 6$

26. y-intercept 6; parallel to the line $2x + 3y + 4 = 0$

27. Through $(-1, 2)$; parallel to the line $x = 5$

28. Through $(2, 6)$; perpendicular to the line $y = 1$

29. Through $(-1, -2)$; perpendicular to the line $2x + 5y + 8 = 0$

30. Through $\left(\frac{1}{2}, -\frac{2}{3}\right)$; perpendicular to the line $4x - 8y = 1$

31. Through $(1, 7)$; parallel to the line passing through $(2, 5)$ and $(-2, 1)$

32. Through $(-2, -11)$; perpendicular to the line passing through $(1, 1)$ and $(5, -1)$

33. (a) Sketch the line with slope $\frac{3}{2}$ that passes through the point $(-2, 1)$.
    (b) Find an equation for this line.

34. (a) Sketch the line with slope -2 that passes through the point $(4, -1)$.
    (b) Find an equation for this line.

35–36 ■ Use a graphing calculator or computer to graph the given family of lines on a common screen. What do the lines have in common?

35. $y = 2 + m(x - 3)$ for $m = 0, \pm 0.5, \pm 1, \pm 2, \pm 10$

36. $y = 1.3x + b$ for $b = 0, \pm 1, \pm 2.8$

37–48 ■ Find the slope and y-intercept of the line and draw its graph.

37. $x + y = 3$ **38.** $3x - 2y = 12$

39. $x + 3y = 0$ **40.** $2x - 5y = 0$

41. $\frac{1}{2}x - \frac{1}{3}y + 1 = 0$ **42.** $-3x - 5y + 30 = 0$

43. $y = 4$ **44.** $4y + 8 = 0$

45. $3x - 4y = 12$ **46.** $x = -5$

47. $3x + 4y - 1 = 0$ **48.** $4x + 5y = 10$

49. Show that $A(1, 1)$, $B(7, 4)$, $C(5, 10)$, and $D(-1, 7)$ are vertices of a parallelogram.

50. Show that $A(-3, -1)$, $B(3, 3)$, and $C(-9, 8)$ are vertices of a right triangle.

51. Show that $A(1, 1)$, $B(11, 3)$, $C(10, 8)$, and $D(0, 6)$ are vertices of a rectangle.

52. Use slopes to determine whether the given points are collinear (lie on a line).
(a) $(1, 1)$, $(3, 9)$, $(6, 21)$
(b) $(-1, 3)$, $(1, 7)$, $(4, 15)$

53. Find the equation of the perpendicular bisector of the line segment joining the points $A(1, 4)$ and $B(7, -2)$.

54. Find the area of the triangle formed by the coordinate axes and the line $2y + 3x - 6 = 0$.

55. (a) Show that if the x- and y-intercepts of a line are nonzero numbers a and b, then the equation of the line can be written in the form

$$\frac{x}{a} + \frac{y}{b} = 1$$

This is called the **two-intercept form** of the equation of a line.
(b) Use part (a) to find an equation of the line whose x-intercept is 6 and whose y-intercept is -8.

56. (a) Find an equation for the line tangent to the circle $x^2 + y^2 = 25$ at the point $(3, -4)$. (See the figure.)
(b) At what other point on the circle will a tangent line be parallel to the tangent line in part (a)?

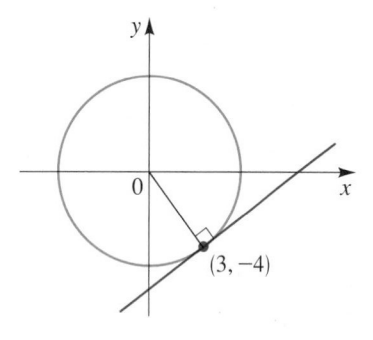

57. West of Albuquerque, New Mexico, Route 40 eastbound is straight and makes a steep descent towards the city. The highway has a 6% grade, which means that its slope is $-\frac{6}{100}$. Driving on this road you notice from elevation signs that you have descended a distance of 1000 ft. What is the change in your horizontal distance?

58. A small business buys a computer for $4000. After four years the value of the computer is expected to be $200. For accounting purposes, the business uses *linear depreciation* to assess the value of the computer at a given time. This means that if V is the value of the computer at time t, then a linear equation is used to relate V and t.
(a) Find a linear equation that relates V and t.
(b) Find the depreciated value of the computer 3 years from the date of purchase.

59. A small-appliance manufacturer finds that if he produces x toaster ovens in a month his production cost is given by the equation $y = 6x + 3000$ (where y is measured in dollars).
(a) Sketch a graph of this linear equation.
(b) What do the slope and y-intercept of the graph represent?

60. The manager of a weekend flea market knows from past experience that if he charges x dollars for a rental space at the flea market, then the number y of spaces he can rent is given by the equation $y = 200 - 4x$.
(a) Sketch a graph of this linear equation. (Remember that the rental charge per space and the number of spaces rented must both be nonnegative quantities.)
(b) What do the slope, the y-intercept, and the x-intercept of the graph represent?

61. Biologists have observed that the chirping rate of crickets of a certain species is related to temperature, and the relationship appears to be very nearly linear. A cricket produces 120 chirps per minute at 70 °F and 168 chirps per minute at 80 °F.
(a) Find the linear equation that relates the temperature t and the number of chirps per minute n.
(b) If the crickets are chirping at 150 chirps per minute, estimate the temperature.

62. The relationship between the Fahrenheit (F) and Celsius (C) temperature scales is given by the equation $F = \frac{9}{5}C + 32$ (see Example 10).

(a) Complete the table to compare the two scales at the given values.

C	F
$-30°$	
$-20°$	
$-10°$	
$0°$	
	$50°$
	$68°$
	$86°$

(b) Find the temperature at which the scales agree. [*Hint:* Suppose that a is the temperature at which the scales agree. Set $F = a$ and $C = a$. Then solve for a.]

63. At the surface of the ocean, the water pressure is the same as the air pressure above the water, 15 pounds per square inch. Below the surface, the water pressure

water pressure increases with depth

increases by 4.34 lb/in² for every 10 feet of descent.

(a) Find an equation for the relationship between pressure and depth below the ocean surface.

(b) At what depth is the pressure 100 lb/in²?

64. Jason and Debbie leave Detroit at 2:00 P.M. and drive at a constant speed, traveling west on I-90. They pass Ann Arbor, 40 mi from Detroit, at 2:50 P.M.

(a) Express the distance traveled in terms of the time elapsed.

(b) Draw the graph of the equation in part (a).

(c) What is the slope of this line? What does it represent?

65. The monthly cost of driving a car depends on the number of miles driven. Lynn found that in May her driving cost was $380 for 480 mi and in June her cost was $460 for 800 mi.

(a) Express the monthly cost C in terms of the distance driven d, assuming that a linear relationship gives a suitable model.

(b) Use part (a) to predict the cost of driving 1500 mi per month.

(c) Draw the graph of the linear equation. What does the slope of the line represent?

(d) What does the y-intercept of the graph represent?

(e) Why is a linear relationship a suitable model for this situation?

66. The manager of a furniture factory finds that it costs $2200 to manufacture 100 chairs in one day and $4800 to produce 300 chairs in one day.

(a) Assuming that the relationship between cost and the number of chairs produced is linear, find an equation that expresses this relationship. Then graph this equation.

(b) What is the slope of the line of part (a), and what does it represent?

(c) What is the y-intercept of this line, and what does it represent?

67–68 ■ Equations for supply and demand are given.

(a) Draw the graphs of the two equations in an appropriate viewing rectangle.

(b) Estimate the equilibrium point from the graph.

(c) Estimate the price and the amount of the commodity produced and sold at equilibrium.

67. Supply: $y = 0.45p + 4$
Demand: $y = -0.65p + 28$

68. Supply: $y = 8.5p + 45$
Demand: $y = -0.6p + 300$

3.5 | WHAT IS A FUNCTION?

In nearly every physical phenomenon, we observe that one quantity depends on another. For example, your height depends on your age, the temperature depends on the date, the cost of mailing a package depends on its weight. These are all examples of functions. We say that your height is a function of your age, the temperature is a function of the date, and the cost of mailing a package is a function of its weight. (See Figure 1.) Although there is no simple rule relating height to age or temperature to date, there is a definite rule (one that the post office actually uses) that relates the cost of mailing a package to its weight.

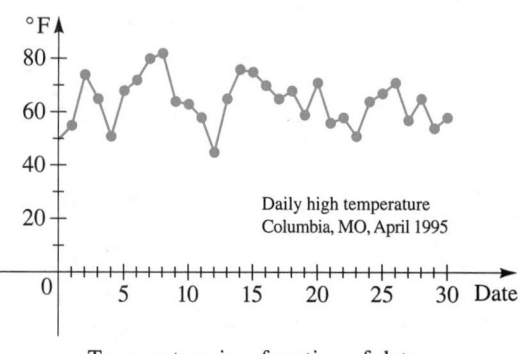

Height is a function of age

Temperature is a function of date

FIGURE 1

Can you think of other examples of functions? Here are two more examples. The area A of a circle depends on its radius r. The rule that connects r and A is given by the equation $A = \pi r^2$. To each positive number r there is associated one value of A, so we say that A is a function of r. If a bacteria culture starts with 5000 bacteria and the population doubles every hour, then the number N of bacteria depends on the time t. The rule that connects t and N is given by the equation $N = (5000)2^t$. For each value of t there is a corresponding value of N, so we say that N is a function of t.

From these examples you can see why functions are important in science. For instance, a biologist observes that the number of bacteria in a culture increases with time. The biologist then tries to find the rule or function that relates the number of bacteria to the time. A physicist observes that the weight of an astronaut depends on her elevation. The physicist then tries to discover the rule or function that relates her weight to her elevation. An economist observes that the price for a certain commodity depends on the demand for that commodity. The economist then tries to discover the rule or function that relates these quantities.

In the following list the numbers on the right are related to the numbers on the left.

$$
\begin{aligned}
1 &\rightarrow 1 \\
2 &\rightarrow 4 \\
3 &\rightarrow 9 \\
4 &\rightarrow 16 \\
&\vdots
\end{aligned}
$$

Can you discover the rule that relates the two? The numbers on the right are the squares of those on the left; so the rule is "square the number." In order to talk about this rule we need to give it a name; we will call it f. Thus, in this example,

$$f \quad \text{means} \quad \text{"square the number"}$$

We have previously used letters to represent numbers. Here we are doing something quite different. We are using the letter f to represent a rule.

Here is how the rule f is used. When we write $f(2)$ we mean "apply the rule to 2"; applying the rule to 2 gives $2^2 = 4$. So we write $f(2) = 4$. Similarly, applying the rule to 3 gives $3^2 = 9$, so we write $f(3) = 9$. In general, applying the rule "square the number" to x gives x^2. So we write

$$f(x) = x^2$$

This expression explains what the rule does to any number x.

The following definition of function concisely captures the ideas explained above.

DEFINITION OF FUNCTION

A **function** f is a rule that assigns to each element x in a set A exactly one element, called $f(x)$, in a set B.

We usually consider functions for which the sets A and B are sets of real numbers. The set A is called the **domain** of the function. The symbol $f(x)$ is read "f of x" or "f at x" and is called the **value of f at x,** or the **image of x under f.** The **range** of f is the set of all possible values of $f(x)$ as x varies throughout the domain, that is,

$$\{f(x) \mid x \in A\}$$

The symbol that represents an arbitrary number in the domain of a function f is called an **independent variable.** The symbol that represents a number in the range of f is called a **dependent variable.** For instance, in the bacteria example, t is the independent variable and N is the dependent variable.

It is helpful to think of a function as a **machine** (see Figure 2). If x is in the domain of the function f, then when x enters the machine, it is accepted as an input and the machine produces an output $f(x)$ according to the rule of the function. Thus, we can think of the domain as the set of all possible inputs and the range as the set of all possible outputs.

The preprogrammed functions in a calculator are good examples of a function as a machine. For example, the $\sqrt{x}$ key on your calculator is such a function. First you input x into the display. Then you press the key labeled $\sqrt{x}$. If $x < 0$, then x is not in the domain of this function; that is, x is not an acceptable input and the calculator will indicate an error. If $x > 0$, then an approximation to $\sqrt{x}$ will appear in the display, correct to a certain number of decimal places. [Thus, the $\sqrt{x}$ key on your calculator is not quite the same as the exact mathematical function f defined by $f(x) = \sqrt{x}$.]

FIGURE 2

Machine diagram of f

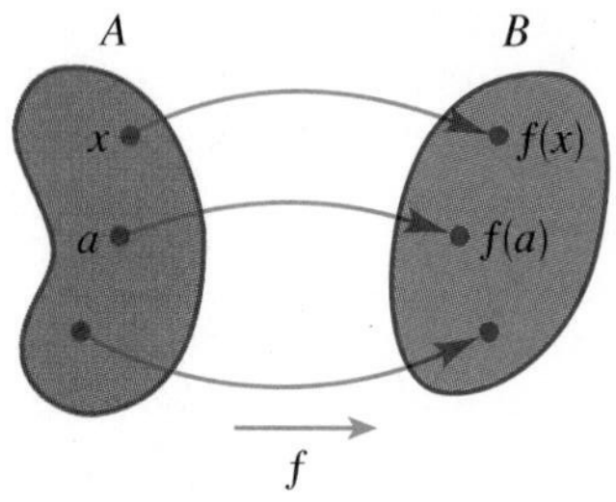

A   *B*

f

FIGURE 3

Arrow diagram of *f*

Another way to picture a function is by an **arrow diagram** as in Figure 3. Each arrow connects an element of *A* to an element of *B*. The arrow indicates that $f(x)$ is associated with x, $f(a)$ is associated with a, and so on.

EXAMPLE 1 ■ The Squaring Function

The squaring function assigns to each real number x its square x^2. It is defined by

$$f(x) = x^2$$

(a) Evaluate $f(3)$, $f(-2)$, and $f(\sqrt{5})$.
(b) Find the domain and range of f.
(c) Draw a machine diagram and an arrow diagram for f.

SOLUTION

(a) The values of f are found by substituting for x in $f(x) = x^2$:

$$f(3) = 3^2 = 9 \qquad f(-2) = (-2)^2 = 4 \qquad f(\sqrt{5}) = (\sqrt{5})^2 = 5$$

(b) The domain of f is the set $\mathbb{R}$ of all real numbers. The range of f consists of all values of $f(x)$, that is, all numbers of the form x^2. But $x^2 \geq 0$ for all numbers x, and any nonnegative number c is a square, since $c = (\sqrt{c})^2 = f(\sqrt{c})$. Therefore, the range of f is $\{y \mid y \geq 0\} = [0, \infty)$.

(c) Machine and arrow diagrams for this function are shown in Figure 4.

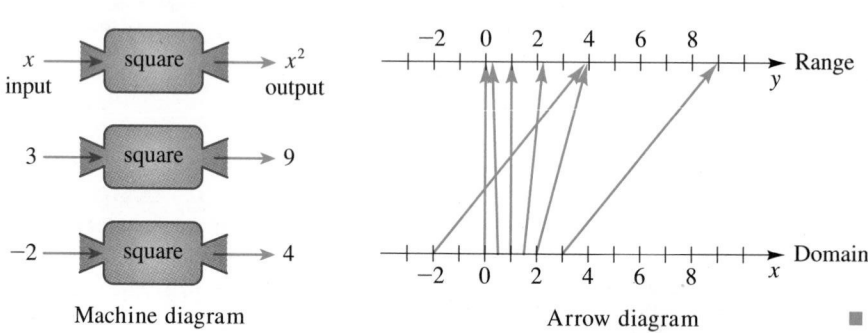

FIGURE 4   Machine diagram   Arrow diagram

EXAMPLE 2 ■ The Domain and Range of a Function

If we define a function g by

$$g(x) = x^2, \qquad 0 \leq x \leq 3$$

then the domain of g is given as the closed interval $[0, 3]$. The function g is different from the function f in Example 1, because in considering g we are restricting our attention to those values of x between 0 and 3. The range of g is

$$\{x^2 \mid 0 \leq x \leq 3\} = \{y \mid 0 \leq y \leq 9\} = [0, 9]$$

Consider the function f defined by the rule

$$f(x) = x^3 + \sqrt{x^2 + 1}$$

To evaluate f at a number we simply substitute that number for x in the definition of f. Thus, we have

$$f(\boxed{5}) = \boxed{5}^3 + \sqrt{\boxed{5}^2 + 1}$$

$$f(\boxed{a}) = \boxed{a}^3 + \sqrt{\boxed{a}^2 + 1}$$

$$f(\boxed{a+h}) = \boxed{(a+h)}^3 + \sqrt{\boxed{(a+h)}^2 + 1}$$

To find $f(a + h)$ we substituted the expression $a + h$ for each occurrence of x because the rule f applies to the number $a + h$.

EXAMPLE 3 ■ Evaluating a Function

If $f(x) = 2x^2 + 3x - 1$, evaluate each function value.

(a) $f(a)$ (b) $f(-a)$

Expressions like the one in part (d) of Example 3 occur frequently in calculus; they are called *difference quotients*.

(c) $f(a + h)$ (d) $\dfrac{f(a + h) - f(a)}{h}$, $h \neq 0$

SOLUTION

(a) $f(a) = 2a^2 + 3a - 1$

(b) $f(-a) = 2(-a)^2 + 3(-a) - 1 = 2a^2 - 3a - 1$

(c) $f(a + h) = 2(a + h)^2 + 3(a + h) - 1$
$$= 2(a^2 + 2ah + h^2) + 3(a + h) - 1$$
$$= 2a^2 + 4ah + 2h^2 + 3a + 3h - 1$$

The *difference quotient* in part (d) represents the average change in the value of f between $x = a$ and $x = a + h$.

(d) Using the results from parts (c) and (a), we have

$$\frac{f(a + h) - f(a)}{h} = \frac{(2a^2 + 4ah + 2h^2 + 3a + 3h - 1) - (2a^2 + 3a - 1)}{h}$$

$$= \frac{4ah + 2h^2 + 3h}{h} = 4a + 2h + 3 \qquad ■$$

We should distinguish between a function f (which is a rule) and the number $f(x)$, which is the value of f at x. Nonetheless, it is common to abbreviate an expression such as

the function f defined by $f(x) = x^2 + x$

to

the function $f(x) = x^2 + x$

In Examples 1 and 2 the domain of the function was given explicitly. But if a function is given by a formula and the domain is not stated explicitly, *the convention is that the domain is the set of all real numbers for which the formula makes sense and defines a real number.* In the next example we find the domain from the formula of a function.

EXAMPLE 4 ■ Finding Domains of Functions

Find the domain of each function.

(a) $f(x) = \dfrac{1}{x^2 - x}$ (b) $g(t) = \dfrac{t}{\sqrt{t + 1}}$ (c) $h(x) = \sqrt{2 - x - x^2}$

SOLUTION

(a) The function is not defined when the denominator is 0. Since

$$f(x) = \frac{1}{x^2 - x} = \frac{1}{x(x - 1)}$$

we see that $f(x)$ is not defined when $x = 0$ or $x = 1$. Thus, the domain of f is

$$\{x \mid x \neq 0, x \neq 1\}$$

The domain may also be written in interval notation as $(-\infty, 0) \cup (0, 1) \cup (1, \infty)$.

(b) The square root of a negative number is not defined (as a real number), so we require that $t + 1 \geqslant 0$. Also, the denominator cannot be 0, that is, $\sqrt{t + 1} \neq 0$, so we also require that $t + 1 \neq 0$. Thus, $g(t)$ exists when $t + 1 > 0$, that is, $t > -1$. So, the domain of g is

$$\{t \mid t > -1\} = (-1, \infty)$$

(c) Since the square root of a negative number is not defined (as a real number), the domain of h consists of all values of x such that

$$2 - x - x^2 \geqslant 0$$

We solve this inequality using the methods of Section 2.7. Since $2 - x - x^2 = (2 + x)(1 - x)$, the product will change sign when $x = -2$ or 1, as indicated in the following chart.

Interval	$x < -2$	$-2 < x < 1$	$1 < x$
Sign of $2 + x$	$-$	$+$	$+$
Sign of $1 - x$	$+$	$+$	$-$
Sign of $(2 + x)(1 - x)$	$-$	$+$	$-$

Therefore, the domain of h is

$$\{x \mid -2 \leqslant x \leqslant 1\} = [-2,1]$$ ■

3.5 **EXERCISES**

1–4 ■ Express the rule in function notation. [For example, the rule "square, then subtract 5" is expressed as the function $f(x) = x^2 - 5$.]

1. Multiply by 3, then add 1

2. Subtract 5, then divide by 7

3. Add 2, then square

4. Square, add 1, then take the square root

5–8 ■ Express the function (or rule) in words.

5. $f(x) = \dfrac{x}{3} - 5$

6. $g(x) = \dfrac{x - 5}{3}$

7. $h(x) = 2x^2 - 3$

8. $k(x) = \sqrt{2x + 1}$

9–10 ■ Draw a machine diagram and an arrow diagram for the function.

9. $f(x) = \sqrt{x}, \quad 0 \leqslant x \leqslant 4$ **10.** $f(x) = \dfrac{2}{x}, \quad 1 \leqslant x \leqslant 4$

11. If $f(x) = 2x + 1$, find $f(1)$, $f(-2)$, $f(\frac{1}{2})$, $f(\sqrt{5})$, $f(a)$, $f(-a)$, and $f(a + b)$.

12. If $f(x) = x^3 + 2x^2 - 3$, find $f(0)$, $f(3)$, $f(-3)$, $f(-x)$, and $f(1/a)$.

13. If $g(x) = \dfrac{1 - x}{1 + x}$, find $g(2)$, $g(-2)$, $g(\pi)$, $g(a)$, $g(a - 1)$, and $g(-a)$.

14. If $h(t) = t + \dfrac{1}{t}$, find $h(1)$, $h(\pi)$, $h(t + 1)$, $h(t) + h(1)$, and $h(x)$.

15. If $f(x) = 2x^2 + 3x - 4$, find $f(0)$, $f(2)$, $f(\sqrt{2})$, $f(1 + \sqrt{2})$, $f(-x)$, $f(x + 1)$, $2f(x)$, and $f(2x)$.

16. If $f(x) = 2 - 3x$, find $f(1)$, $f(-1)$, $f(\frac{1}{3})$, $f(x/3)$, $f(3x)$, $f(x^2)$, and $[f(x)]^2$.

17–22 ■ Find $f(a)$, $f(a) + f(h)$, $f(a + h)$, and $\dfrac{f(a + h) - f(a)}{h}$, where $h \neq 0$.

17. $f(x) = 3x + 2$

18. $f(x) = x^2 + 1$

19. $f(x) = 5$

20. $f(x) = \dfrac{1}{x + 1}$

21. $f(x) = 3 - 5x + 4x^2$

22. $f(x) = x^3 + x + 1$

23–26 ■ Find $f(2 + h)$, $f(x + h)$, and $\dfrac{f(x + h) - f(x)}{h}$, where $h \neq 0$.

23. $f(x) = 8x - 1$

24. $f(x) = x - x^2$

25. $f(x) = \dfrac{1}{x}$

26. $f(x) = \dfrac{x}{x + 1}$

27–38 ■ Find the domain and range of the function.

27. $f(x) = 2x$

28. $f(x) = x^2 + 1$

29. $f(x) = 2x, \quad -1 \leqslant x \leqslant 5$

30. $f(x) = x^2 + 1, \quad 0 \leqslant x \leqslant 5$

31. $f(x) = 6 - 4x, \quad -2 \leqslant x \leqslant 3$

32. $f(x) = 2x + 7, \quad -1 \leqslant x \leqslant 6$

33. $f(x) = 2 - x^2$

34. $g(x) = \sqrt{7 - 3x}$

35. $h(x) = \sqrt{2x - 5}$

36. $h(x) = 1 - \sqrt{x}$

37. $F(x) = 3 + \sqrt{1 - x^2}$

38. $G(x) = \sqrt{x^2 - 9}$

39–60 ■ Find the domain of the function.

39. $f(x) = \dfrac{1}{x - 3}$

40. $f(x) = \dfrac{1}{3x - 6}$

41. $f(x) = \dfrac{x + 2}{x^2 - 1}$

42. $f(x) = \dfrac{x^4}{x^2 + x - 6}$

43. $f(x) = \dfrac{x + 2}{x^2 + 1}$

44. $f(x) = \dfrac{x^3 + 1}{x^3 - x}$

45. $f(x) = \sqrt{x - 5}$

46. $f(x) = \sqrt[4]{x + 9}$

47. $f(t) = \sqrt[3]{t - 1}$

48. $f(t) = \sqrt{t^2 + 1}$

49. $g(x) = \dfrac{\sqrt{2 + x}}{3 - x}$

50. $g(x) = \dfrac{\sqrt{x}}{2x^2 + x - 1}$

51. $F(x) = \dfrac{x}{\sqrt{x - 10}}$

52. $F(x) = x^2 - \dfrac{x}{\sqrt{9 - 2x}}$

53. $G(x) = \sqrt{x} + \sqrt{1 - x}$

54. $G(x) = \sqrt{x + 2} - 2\sqrt{x - 3}$

55. $f(x) = \sqrt{4x^2 - 1}$

56. $f(x) = \sqrt{2 - 3x^2}$

57. $g(x) = \sqrt[4]{x^2 - 6x}$

58. $g(x) = \sqrt{x^2 - 2x - 8}$

59. $\phi(x) = \sqrt{\dfrac{x}{\pi - x}}$

60. $\phi(x) = \sqrt{\dfrac{x^2 - 2x}{x - 1}}$

3.6 GRAPHS OF FUNCTIONS

In the preceding section we saw how to picture functions using machine diagrams and arrow diagrams. A third method for visualizing a function is its graph.

THE GRAPH OF A FUNCTION

If f is a function with domain A, then the **graph** of f is the set of ordered pairs

$$\{(x, f(x)) \mid x \in A\}$$

In other words, the graph of f is the set of all points (x, y) such that $y = f(x)$; that is, the graph of f is the graph of the equation $y = f(x)$.

The graph of a function f gives us a useful picture of the behavior or "life history" of a function. Since the y-coordinate of any point (x, y) on the graph is $y = f(x)$, we can read the value of $f(x)$ from the graph as being the height of the graph above the point x (see Figure 1). The graph of f also allows us to picture the domain and range of f on the x-axis and y-axis as in Figure 2.

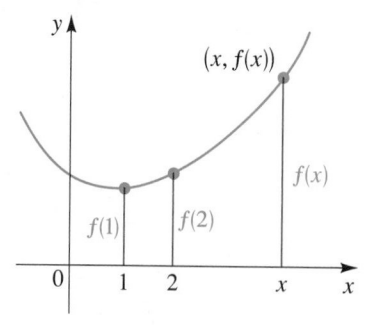

FIGURE 1
The graph of f

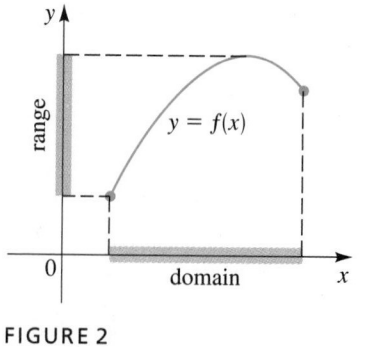

FIGURE 2
Domain and range of f

A function f of the form $f(x) = mx + b$ is called a **linear function** because its graph is the graph of the equation $y = mx + b$, which represents a line with slope m and y-intercept b. A special case of the linear function occurs when the

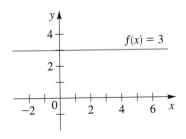

FIGURE 3

The constant function $f(x) = 3$

slope is $m = 0$. The function $f(x) = b$, where b is a given number, is called a **constant function** because all its values are the same number, namely, b. Its graph is the horizontal line $y = b$. Figure 3 shows the graph of the constant function $f(x) = 3$.

EXAMPLE 1 ■ **The Graph of a Linear Function**

Sketch the graph of the function $f(x) = 2x - 1$.

SOLUTION

To graph this linear function we graph the equation $y = 2x - 1$. We recognize this equation as being that of a line with slope 2 and y-intercept -1. This enables us to sketch the graph of f in Figure 4.

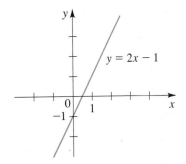

FIGURE 4

EXAMPLE 2 ■ **The Graph of the Squaring Function**

Sketch the graph of $f(x) = x^2$.

SOLUTION

To graph this function we graph the equation $y = x^2$. We draw it in Figure 5 by setting up a table of values and plotting points. Recall from Example 1 of Section 3.5 that the domain is $\mathbb{R}$ and the range is $[0, \infty)$. The graph is called a *parabola*.

x	$f(x) = x^2$
0	0
$\pm\frac{1}{2}$	$\frac{1}{4}$
± 1	1
± 2	4
± 3	9

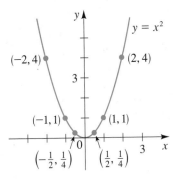

FIGURE 5

Donald Knuth was born in Milwaukee in 1938 and is now Professor of Computer Science at Stanford University. While still a graduate student at Caltech he started writing a monumental series of books entitled *The Art of Computer Programming.* President Carter awarded him the National Medal of Science in 1979. When Knuth was a high school student he became fascinated with graphs of functions and laboriously drew many hundreds of them because he wanted to see the behavior of a great variety of functions. (Now, with progress in computers and programming, it is far easier to use computers and graphing calculators to do this.) Knuth is also famous for his invention of $T_E X$, a system of computer-assisted typesetting. This system was used in the preparation of the manuscript for this textbook. He has written a novel entitled *Surreal Numbers: How Two Ex-Students Turned On to Pure Mathematics and Found Total Happiness.*

EXAMPLE 3 ■ Graphing a Function by Plotting Points

Sketch the graph of $f(x) = x^3$.

SOLUTION

We list some functional values and the corresponding points on the graph in the following table.

x	$f(x) = x^3$
0	0
$\frac{1}{2}$	$\frac{1}{8}$
1	1
2	8
$-\frac{1}{2}$	$-\frac{1}{8}$
-1	-1
-2	-8

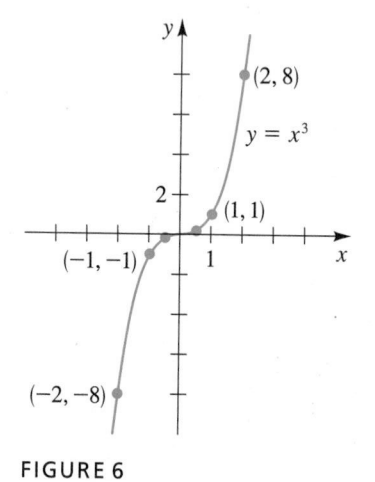

FIGURE 6

Then we plot these points and join them by a smooth curve to obtain the graph shown in Figure 6. ■

EXAMPLE 4 ■ Finding the Domain and Range from a Graph

(a) Sketch the graph of $f(x) = \sqrt{4 - x^2}$.
(b) Find the domain and range of f.

SOLUTION

(a) We must graph the equation $y = \sqrt{4 - x^2}$. Because we are taking the positive square root, we know that $y \geq 0$. Squaring each side of the equation, we get

$$y^2 = 4 - x^2$$

or

$$x^2 + y^2 = 4$$

which we recognize as the equation of a circle with center the origin and radius 2. But, since $y \geq 0$, the graph of f consists of just the upper half of this circle.

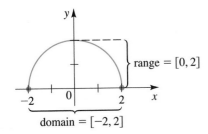

FIGURE 7

Graph of $f(x) = \sqrt{4 - x^2}$

(b) From the graph in Figure 7 we see that the domain is the closed interval $[-2, 2]$ and the range is $[0, 2]$.   ∎

The graph of a function is a curve in the xy-plane. But the question arises: Which curves in the xy-plane are graphs of functions? This is answered by the following test.

THE VERTICAL LINE TEST

A curve in the coordinate plane is the graph of a function if and only if no vertical line intersects the curve more than once.

We can see from Figure 8 why the Vertical Line Test is true. If each vertical line $x = a$ intersects a curve only once at (a, b), then exactly one functional value is defined by $f(a) = b$. But if a line $x = a$ intersects the curve twice, at (a, b) and at (a, c), then the curve cannot represent a function because a function cannot assign two different values to a.

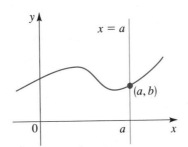
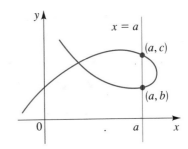

FIGURE 8

Vertical Line Test

Graph of a function   Not a graph of a function

EXAMPLE 5 ∎ **Using the Vertical Line Test**

Using the Vertical Line Test, we see that the curves in Figures 9(b) and (c) represent functions, whereas those in parts (a) and (d) do not.

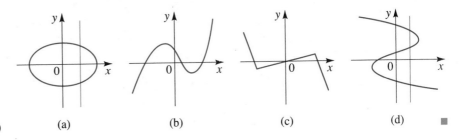

FIGURE 9   (a)   (b)   (c)   (d)   ∎

EXAMPLE 6 ■ Graphing the Square Root Function

Sketch the graph of $f(x) = \sqrt{x}$.

SOLUTION

First we note that the domain is $\{x \mid x \geq 0\} = [0, \infty)$. Then we plot the points given by the following table and use them to sketch the graph in Figure 10.

x	$f(x) = \sqrt{x}$
0	0
1	1
2	$\sqrt{2}$
3	$\sqrt{3}$
4	2
5	$\sqrt{5}$

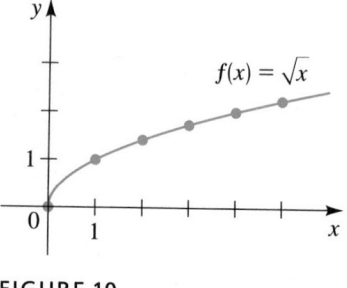

FIGURE 10   ■

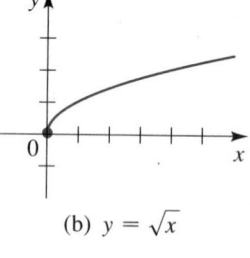

(b) $y = \sqrt{x}$

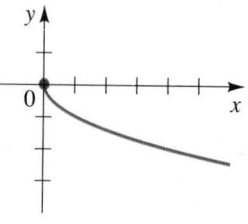

(c) $y = -\sqrt{x}$

FIGURE 11

NOTE: The graph of the square root function in Example 6 is the graph of the equation $y = \sqrt{x}$. If we square each side of this equation, we get $y^2 = x$, but we have to remember that $y \geq 0$. We saw in Section 3.2, Example 8, that the equation $x = y^2$ represents the parabola shown in Figure 11(a). By the Vertical line Test this parabola does not represent a function of x. But we can regard $y^2 = x$ as representing two functions of x; the upper and lower halves of this parabola are the graphs of the functions $f(x) = \sqrt{x}$ and $g(x) = -\sqrt{x}$, as in parts (b) and (c) of Figure 11.

It is very useful to know where the graph of a function rises and where it falls. The graph shown in Figure 12 rises, falls, then rises again as we move from left to right: it rises from A to B, falls from B to C, and rises again from C to D. The function f is said to be *increasing* when its graph rises and *decreasing* when its graph falls. We have the following definition.

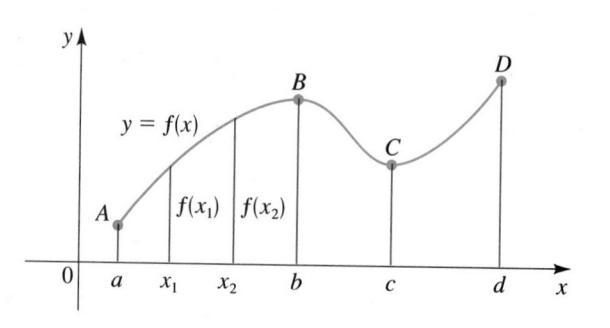

FIGURE 12

DEFINITION OF INCREASING AND DECREASING FUNCTIONS

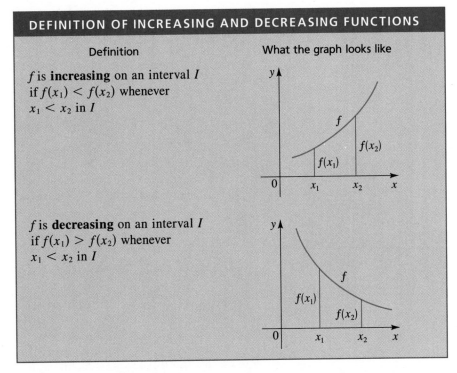

Definition

f is **increasing** on an interval I
if $f(x_1) < f(x_2)$ whenever
$x_1 < x_2$ in I

f is **decreasing** on an interval I
if $f(x_1) > f(x_2)$ whenever
$x_1 < x_2$ in I

What the graph looks like

For instance, the functions in Examples 1, 3, and 6 are all increasing on their domains. In Example 2, f is decreasing on $(-\infty, 0]$ and increasing on $[0, \infty)$. In Example 4, f is increasing on $[-2, 0]$ and decreasing on $[0, 2]$.

EXAMPLE 7 ■ Intervals on which a Function Increases or Decreases

State the intervals on which the function whose graph is shown in Figure 13 is increasing or decreasing.

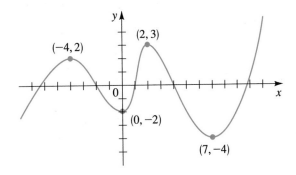

FIGURE 13

SOLUTION

The function is increasing on $(-\infty, -4]$, $[0, 2]$, and $[7, \infty)$. It is decreasing on $[-4, 0]$ and $[2, 7]$.   ■

Graphing calculators and computers are ideal devices to use for graphing functions because many graphing devices give exactly one value of y for each value of x, and this is the property that functions have. In Section 3.3 we observed that in order to graph an equation on a graphing calculator, we need to isolate y and write the equation as $y = $ *an expression involving only* x. We see now that such an equation actually represents a function $y = f(x)$.

EXAMPLE 8 ■ Using a Graph to Find Intervals where a Function Increases and Decreases

(a) Sketch the graph of the function $f(x) = x^{2/3}$.
(b) Find the domain and range of the function.
(c) Find the intervals on which f increases and decreases.

SOLUTION

(a) We use a graphing calculator to sketch the graph in Figure 14.

(b) From the graph, we observe that the domain of f is $\mathbb{R}$ and the range is $[0, \infty)$.

(c) From the graph, we see that f is decreasing on $(-\infty, 0]$ and increasing on $[0, \infty)$. ■

Some graphing calculators, such as the TI-81, do not evaluate $x^{2/3}$ [entered as $x^{\wedge}(2/3)$] for negative x. To graph a function like $f(x) = x^{2/3}$, we enter it as $y_1 = (x^{\wedge}(1/3))^{\wedge}2$ because these calculators evaluate correctly powers of the form $x^{\wedge}(1/n)$.

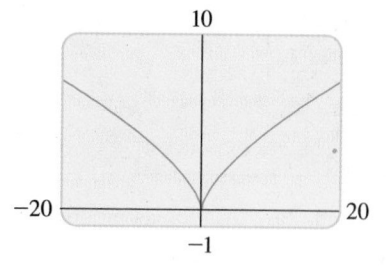

FIGURE 14
Graph of $f(x) = x^{2/3}$

To understand how the equation of a function relates to its graph, it is helpful to graph a **family of functions,** that is, a collection of functions whose equations are related. Usually, the equations differ only in a single constant called a *parameter.* In the next example we graph a family of **power functions** $f(x) = x^n$, where n is a positive integer. (In this case, n is the parameter.)

EXAMPLE 9 ■ A Family of Power Functions

(a) Graph the functions $f(x) = x^n$ for $n = 2, 4$, and 6 in the viewing rectangle $[-2, 2]$ by $[-1, 3]$.
(b) Graph the functions $f(x) = x^n$ for $n = 1, 3$, and 5 in the viewing rectangle $[-2, 2]$ by $[-2, 2]$.
(c) Graph all of the functions from parts (a) and (b) in the viewing rectangle $[-1, 3]$ by $[-1, 3]$.
(d) What conclusions can you draw from these graphs?

SOLUTION

The graphs for parts (a), (b), and (c) are shown in Figure 15.

(d) We see that the general shape of $f(x) = x^n$ depends on whether n is even or odd.

From part (a) we see that if n is even, then the graph of $f(x) = x^n$ is similar to the parabola $y = x^2$. These functions are decreasing on $(-\infty, 0]$ and increasing on $[0, \infty)$.

From part (b) we see that if n is odd, then the graph of $f(x) = x^n$ is similar to that of $y = x^3$. These functions are increasing on all of $\mathbb{R}$.

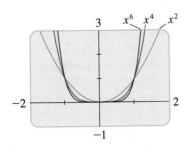

(a) Even powers of x

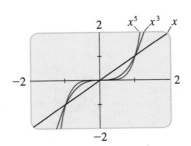

(b) Odd powers of x

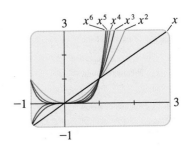

(c) Even and odd powers

FIGURE 15

Power functions $f(x) = x^n$

Notice from part (c) that as n increases the graph of $y = x^n$ becomes flatter near 0 and steeper when $x > 1$. When $0 < x < 1$, the lower powers of x are the "bigger" functions. But when $x > 1$, the higher powers of x are the dominant functions. ■

PIECEWISE-DEFINED FUNCTIONS

Each function we have graphed so far has been defined by just one equation. But the rule defining a function is often given by more than one equation. A function may be defined by different equations on different parts of its domain. Some examples from everyday experience are the cost of mailing a first-class letter as a function of its weight, the population of New York City as a function of time, and the cost of a taxi ride as a function of distance. The following examples show how to graph such functions.

EXAMPLE 10 ■ The Graph of a Piecewise-Defined Function

Sketch the graph of the function f defined by

$$f(x) = \begin{cases} 1 - x & \text{if } x \leq 1 \\ x^2 & \text{if } x > 1 \end{cases}$$

SOLUTION

Remember that a function is a rule. For this particular function the rule is the following: First look at the value of the input x. If it happens that $x \leq 1$, then the value of $f(x)$ is $1 - x$. On the other hand, if $x > 1$, then the value of $f(x)$ is x^2.

Since $0 \leq 1$, we have $f(0) = 1 - 0 = 1$.

Since $1 \leq 1$, we have $f(1) = 1 - 1 = 0$.

Since $2 > 1$, we have $f(2) = 2^2 = 4$.

Since $-3 \leq 1$, we have $f(-3) = 1 - (-3) = 4$.

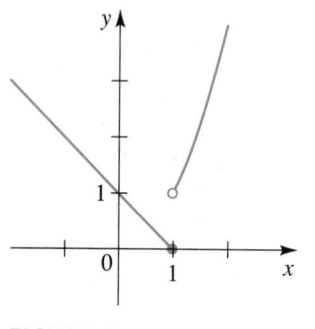

FIGURE 16

$$f(x) = \begin{cases} 1 - x & \text{if } x \leq 1 \\ x^2 & \text{if } x > 1 \end{cases}$$

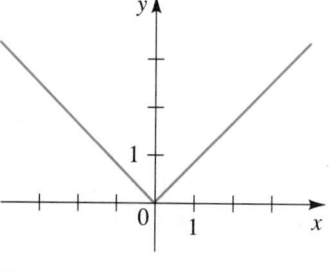

FIGURE 17

Graph of $f(x) = |x|$

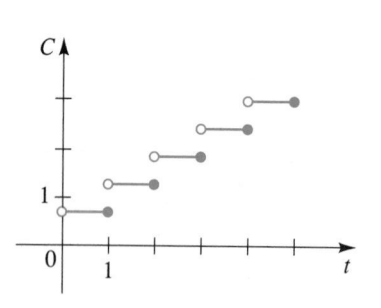

FIGURE 18

Cost of a long-distance call

How do we draw the graph of f? We observe that if $x \leq 1$, then $f(x) = 1 - x$, so the part of the graph of f that lies to the left of the vertical line $x = 1$ must coincide with the line $y = 1 - x$, which has slope -1 and y-intercept 1. If $x > 1$, then $f(x) = x^2$, so the part of the graph of f that lies to the right of the line $x = 1$ must coincide with the graph of $y = x^2$, which we sketched in Example 2. This enables us to sketch the graph in Figure 16. The solid dot at $(1, 0)$ indicates that this point is included on the graph; the open dot at $(1, 1)$ indicates that this point is excluded from the graph. ∎

EXAMPLE 11 ■ Graph of the Absolute Value Function

Sketch the graph of the absolute value function $f(x) = |x|$.

SOLUTION

Recall that

$$|x| = \begin{cases} x & \text{if } x \geq 0 \\ -x & \text{if } x < 0 \end{cases}$$

Using the same method as in Example 10, we note that the graph of f coincides with the line $y = x$ to the right of the y-axis and coincides with the line $y = -x$ to the left of the y-axis (see Figure 17). ∎

EXAMPLE 12 ■ The Cost Function for Long-Distance Phone Calls

The cost of a long-distance daytime phone call from Toronto to New York City is 69 cents for the first minute and 58 cents for each additional minute (or part of a minute). Draw the graph of the cost C (in dollars) of the phone call as a function of time t (in minutes).

SOLUTION

Let $C(t)$ be the cost for t minutes. Since $t > 0$, the domain of the function is $(0, \infty)$. From the given information, we have

$$C(t) = 0.69 \qquad\qquad\qquad \text{if } 0 < t \leq 1$$

$$C(t) = 0.69 + 0.58 = 1.27 \qquad \text{if } 1 < t \leq 2$$

$$C(t) = 0.69 + 2(0.58) = 1.85 \quad \text{if } 2 < t \leq 3$$

$$C(t) = 0.69 + 3(0.58) = 2.43 \quad \text{if } 3 < t \leq 4$$

and so on. The graph is shown in Figure 18. ∎

The following example shows how graphing calculators can be used to help draw piecewise-defined functions.

EXAMPLE 13 ■ Drawing a Piecewise-Defined Function Using a Graphing Device

Use a graphing device to draw the graph of the function

$$f(x) = \begin{cases} x^3 & \text{if } x \le 0 \\ x + 2 & \text{if } x > 0 \end{cases}$$

SOLUTION

First we use a calculator to draw the graphs of the functions

$$g(x) = x^3 \quad \text{and} \quad h(x) = x + 2$$

in the viewing rectangle $[-3, 3]$ by $[-5, 5]$, as shown in Figure 19(a). Then we draw the graph of f (by hand) in Figure 19(b) by taking the part of the graph of g to the left of $x = 0$ and combining it with the part of the graph of h to the right of $x = 0$. Note that $f(0) = 0^3 = 0$, so we place a solid dot at $(0, 0)$ and an open dot at $(0, 2)$.

On many graphing calculators the graph in Figure 19(b) can be produced by using the logical functions in the calculator. For example, on the TI-81 the following equation gives the required graph: $y = (x < 0)x^{\wedge}3 + (x > 0)(x + 2)$.

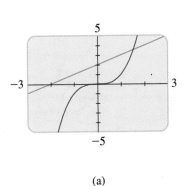

FIGURE 19

(a)

(b) ■

3.6 EXERCISES

1. The graph of a function f is given.
   (a) State the values of $f(-1)$, $f(0)$, $f(1)$, and $f(3)$.
   (b) State the domain and range of f.
   (c) State the intervals on which f is increasing and on which f is decreasing.

2. The graph of a function h is given.
   (a) State the values of $h(-2)$, $h(0)$, $h(2)$, and $h(3)$.
   (b) State the domain and range of h.
   (c) State the intervals on which h is increasing and on which h is decreasing.

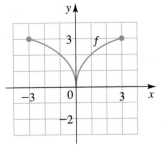

Graph of f

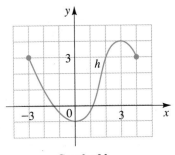

Graph of h

3. The graph of a function g is given.
   (a) State the values of $g(-4)$, $g(-2)$, $g(0)$, $g(2)$, and $g(4)$.
   (b) State the domain and range of g.
   (c) State the intervals on which g is increasing and on which g is decreasing.

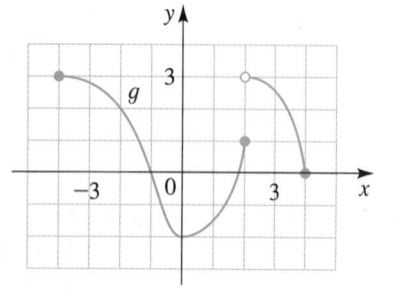

4. The graph of a function f is given.
   (a) State the values of $f(-3)$, $f(1)$, $f(2)$, and $f(3)$.
   (b) State the domain and range of f.

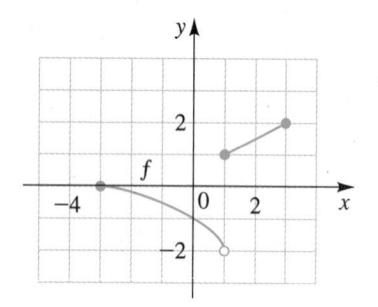
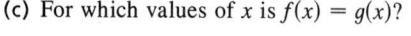

5. Graphs of the functions f and g are given.
   (a) Which is larger, $f(0)$ or $g(0)$?
   (b) Which is larger, $f(-3)$ or $g(-3)$?
   (c) For which values of x is $f(x) = g(x)$?

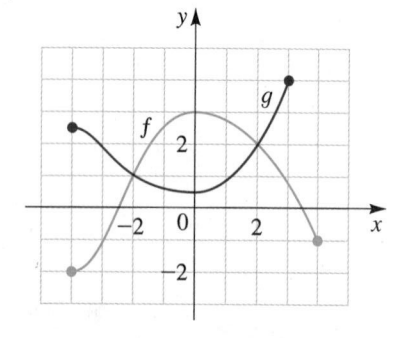

6. The graph of a function f is given.
   (a) Estimate $f(0.5)$ to the nearest tenth.
   (b) Estimate $f(3)$ to the nearest tenth.
   (c) Find all the numbers x in the domain of f so that $f(x) = 1$.

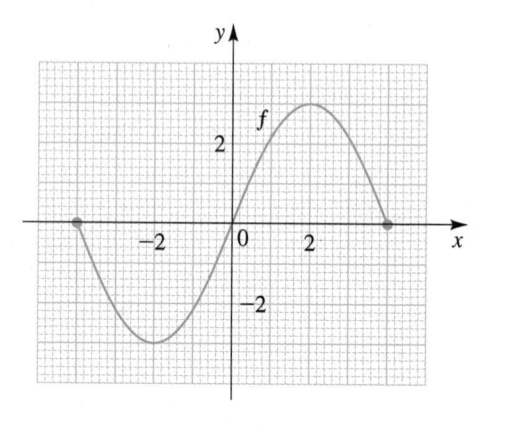

7–8 ■ Determine whether each curve is the graph of a function of x.

7. (a)   (b)

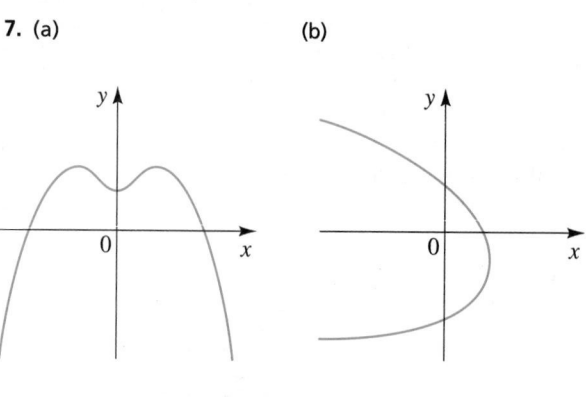

(c)   (d)

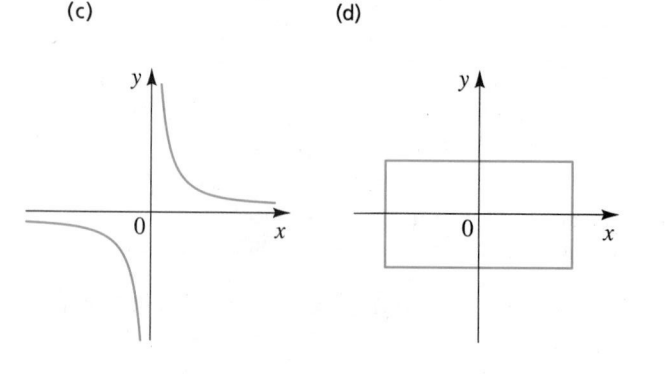

8. (a)

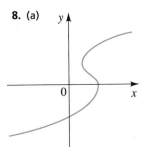

(b)

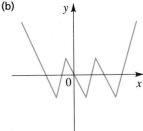

(c)

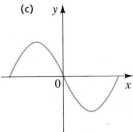

(d)

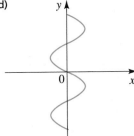

9–12 ■ Determine whether the curve is the graph of a function of x. If it is, state the domain and range of the function.

9.

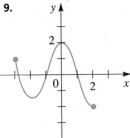

10.

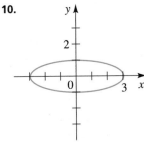

11.

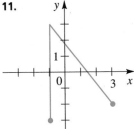

12.

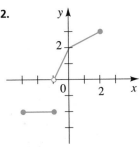

13–20 ■ A function f is given.
(a) Sketch the graph of f.
(b) Find the domain of f.

(c) Find the intervals on which f is increasing and on which f is decreasing.

13. $f(x) = 1 - x$

14. $f(x) = \frac{1}{2}(x + 1)$

15. $f(x) = x^2 - 4x$

16. $f(x) = x^2 - 4x + 4$

17. $f(x) = \sqrt{9 - x}$

18. $f(x) = \sqrt{2x + 6}$

19. $f(x) = \sqrt{16 - x^2}$

20. $f(x) = -\sqrt{25 - x^2}$

21–44 ■ Sketch the graph of the function.

21. $f(x) = 3$

22. $f(x) = -5$

23. $f(x) = 2x + 3$

24. $f(x) = 6 - 3x$

25. $f(x) = -x + 4, \quad -1 \leqslant x \leqslant 4$

26. $f(x) = \dfrac{x + 3}{2}, \quad -2 \leqslant x \leqslant 2$

27. $f(x) = -x^2$

28. $f(x) = x^2 - 4$

29. $f(x) = x^2 + 2x + 1$

30. $f(x) = x^2 + 6x - 7$

31. $g(x) = x^3 - 8$

32. $g(x) = 4x^2 - x^4$

33. $g(x) = \sqrt{-x}$

34. $g(x) = \sqrt{6 - 2x}$

35. $F(x) = \dfrac{1}{x}$

36. $F(x) = \dfrac{2}{x + 4}$

37. $H(x) = |2x|$

38. $H(x) = |x + 1|$

39. $G(x) = |x| + x$

40. $G(x) = |x| - x$

41. $f(x) = |2x - 2|$

42. $f(x) = \dfrac{x}{|x|}$

43. $f(x) = \dfrac{x^2 - 1}{x - 1}$

44. $f(x) = \dfrac{x^2 + 5x + 6}{x + 2}$

45–48 ■ A function f is given.
(a) Use a graphing device to draw the graph of f.
(b) State approximately the intervals on which f is increasing and on which f is decreasing.

45. $f(x) = x^{2/5}$

46. $f(x) = 4 - x^{2/3}$

47. $f(x) = x^3 + 2x^2 - x - 2$

48. $f(x) = x^4 - 4x^3 + 2x^2 + 4x - 3$

49. In this exercise we consider the family of root functions $f(x) = x^c$, where $0 < c \leqslant 1$.
(a) Draw the graphs of f for $c = \frac{1}{2}, \frac{1}{4}$, and $\frac{1}{6}$ on the same screen using the viewing rectangle $[-1, 4]$ by $[-1, 3]$.

(b) Draw the graphs of f for $c = 1$, $\frac{1}{3}$, and $\frac{1}{5}$ on the same screen using the viewing rectangle $[-3, 3]$ by $[-2, 2]$.

(c) Draw the graphs of f for $c = \frac{1}{2}, \frac{1}{3}, \frac{1}{4}$ and $\frac{1}{5}$ on the same screen using the viewing rectangle $[-1, 4]$ by $[-1, 3]$.

(d) What conclusions can you make from these graphs?

50. In this exercise we consider the family of functions $f(x) = 1/x^n$.

(a) Draw the graphs of f for $n = 1$ and $n = 3$ on the same screen using the viewing rectangle $[-3, 3]$ by $[-3, 3]$.

(b) Draw the graphs of f for $n = 2$ and $n = 4$ on the same screen using the same viewing rectangle.

(c) Draw the graphs of all the functions in parts (a) and (b) on the same screen using the viewing rectangle $[-1, 3]$ by $[-1, 3]$.

(d) What conclusions can you make from these graphs?

51. In this exercise we consider the family of functions $f(x) = x^2 + c$.

(a) Draw the graphs of f for $c = 0, 2, 4$, and 6 on the same screen using the viewing rectangle $[-5, 5]$ by $[-10, 10]$.

(b) Draw the graphs of f for $c = 0, -2, -4$, and -6 on the viewing rectangle $[-5, 5]$ by $[-10, 10]$.

(c) What conclusions can you make from these graphs?

52. In this example we consider the family of functions $f(x) = (x - c)^2$.

(a) Draw the graphs of f for $c = 0, 1, 2$, and 3 on the same screen using the viewing rectangle $[-5, 5]$ by $[-10, 10]$.

(b) Draw the graphs of f for $c = 0, -1, -2$, and -3 on the same screen using the viewing rectangle $[-5, 5]$ by $[-10, 10]$.

(c) What conclusions can you make from these graphs?

53. In this example we consider the family of functions $f(x) = (x - c)^3$.

(a) Draw the graphs of f for $c = 0, 2, 4$, and 6 on the same screen using the viewing rectangle $[-10, 10]$ by $[-10, 10]$.

(b) Draw the graphs of f for $c = 0, -2, -4$, and -6 on the same screen using the viewing rectangle $[-10, 10]$ by $[-10, 10]$.

(c) What conclusions can you make from these graphs?

54. In this example we consider the family of functions $f(x) = cx^2$.

(a) Draw the graphs of f for $c = 1, 2, 3$, and 4 on the same screen using the viewing rectangle $[-5, 5]$ by $[-10, 10]$.

(b) Draw the graphs of f for $c = 1, \frac{1}{2}, \frac{1}{4}$, and $\frac{1}{10}$ on the same screen using the viewing rectangle $[-5, 5]$ by $[-10, 10]$.

(c) Draw the graphs of f for $c = 1, -1, -\frac{1}{2}$, and -2 on the same screen using the viewing rectangle $[-5, 5]$ by $[-10, 10]$.

(d) What conclusions can you make from these graphs?

55. Use graphs to decide which of the functions $f(x) = 10x^2$ and $g(x) = \dfrac{x^3}{10}$ is eventually larger (that is, larger when x is very large).

56. Use graphs to decide which of the functions $f(x) = x^4 - 100x^3$ and $g(x) = x^3$ is eventually larger.

57. (a) Draw the graphs of the functions $f(x) = x^2 + x - 6$ and $g(x) = |x^2 + x - 6|$.

(b) How are the graphs of f and g related?

58. (a) Draw the graphs of the functions $f(x) = x^4 - 6x^2$ and $g(x) = |x^4 - 6x^2|$.

(b) How are the graphs of f and g related?

59–72 ■ Sketch the graph of the piecewise-defined function.

59. $f(x) = \begin{cases} 0 & \text{if } x < 2 \\ 1 & \text{if } x \geq 2 \end{cases}$

60. $f(x) = \begin{cases} 1 & \text{if } x \leq 1 \\ x + 1 & \text{if } x > 1 \end{cases}$

61. $f(x) = \begin{cases} 3 & \text{if } x < 2 \\ x - 1 & \text{if } x \geq 2 \end{cases}$

62. $f(x) = \begin{cases} 1 - x & \text{if } x < -2 \\ 5 & \text{if } x \geq -2 \end{cases}$

63. $f(x) = \begin{cases} x & \text{if } x \leq 0 \\ x + 1 & \text{if } x > 0 \end{cases}$

64. $f(x) = \begin{cases} 2x + 3 & \text{if } x < -1 \\ 3 - x & \text{if } x \geq -1 \end{cases}$

65. $f(x) = \begin{cases} x + 1 & \text{if } x \neq 1 \\ 1 & \text{if } x = 1 \end{cases}$

66. $f(x) = \begin{cases} x + 3 & \text{if } x \neq -2 \\ 4 & \text{if } x = -2 \end{cases}$

67. $f(x) = \begin{cases} -1 & \text{if } x < -1 \\ 1 & \text{if } -1 \leq x \leq 1 \\ -1 & \text{if } x > 1 \end{cases}$

68. $f(x) = \begin{cases} -1 & \text{if } x < -1 \\ x & \text{if } -1 \leq x \leq 1 \\ 1 & \text{if } x > 1 \end{cases}$

69. $f(x) = \begin{cases} 2 & \text{if } x \leq -1 \\ x^2 & \text{if } x > -1 \end{cases}$

70. $f(x) = \begin{cases} 1 - x^2 & \text{if } x \leq 2 \\ x & \text{if } x > 2 \end{cases}$

71. $f(x) = \begin{cases} 0 & \text{if } |x| \leq 2 \\ 3 & \text{if } |x| > 2 \end{cases}$

72. $f(x) = \begin{cases} x^2 & \text{if } |x| \leq 1 \\ 1 & \text{if } |x| > 1 \end{cases}$

73–76 ■ Use a graphing device to draw the graph of the piecewise-defined function.

73. $f(x) = \begin{cases} x + 2 & \text{if } x \leq -1 \\ x^2 & \text{if } x > -1 \end{cases}$

74. $f(x) = \begin{cases} 2x - x^2 & \text{if } x > 1 \\ (x - 1)^3 & \text{if } x \leq 1 \end{cases}$

75. $f(x) = \begin{cases} x^3 - 2x + 1 & \text{if } x \leq 0 \\ x - x^2 & \text{if } 0 < x < 1 \\ \sqrt[4]{x - 1} & \text{if } x \geq 1 \end{cases}$

76. $f(x) = \begin{cases} \sqrt{-x} & \text{if } x < 0 \\ \sqrt{2x - x^2} & \text{if } 0 \leq x \leq 2 \\ \sqrt{x - 2} & \text{if } x > 2 \end{cases}$

77. A taxi company charges \$2.00 for the first mile (or part of a mile) and 20 cents for each succeeding tenth of a mile (or part). Express the cost C (in dollars) of a ride as a function of the distance x traveled (in miles) for $0 < x < 2$, and sketch the graph of this function.

78. The domestic postage rate for first-class letters weighing 12 oz or less is 32 cents for a letter weighing one ounce or less and 23 cents for each additional ounce (or part of an ounce). Express the postage P as a function of the weight x of a letter, with $0 < x \leq 12$.

79–82 ■ Find a function whose graph is the given curve.

79. The line segment joining the points $(-2, 1)$ and $(4, -6)$

80. The line segment joining the points $(-3, -2)$ and $(6, 3)$

81. The bottom half of the parabola $x + (y - 1)^2 = 0$

82. The top half of the circle $(x - 1)^2 + y^2 = 1$

3.7 APPLIED FUNCTIONS

When scientists or applied mathematicians talk about a *mathematical model* for a real-world phenomenon, they mean a function that describes, at least approximately, the dependence of one physical quantity on another. For instance, the model may describe the population of an animal species as a function of time, or the pressure of a gas as a function of its volume. In this section we study such applied functions.

Our first example shows that even when a precise formula for functional dependence is not available, it is still possible to visualize the situation from a graph of the function.

EXAMPLE 1 ■ A Graph of Water Temperature

When you turn on a hot water faucet, the temperature T of the water depends on how long the water has been running. Draw a rough graph of T as a function of the time t that has elapsed since the faucet was turned on.

SOLUTION

The initial temperature of the water is close to room temperature because that water has been standing in the pipes. When the water from the hot water tank starts coming out, T increases quickly. In the next phase, T is constant at the temperature of the water in the tank. When the tank is drained, T decreases to the temperature of the water supply. This enables us to make the rough sketch of T as a function of t in Figure 1.

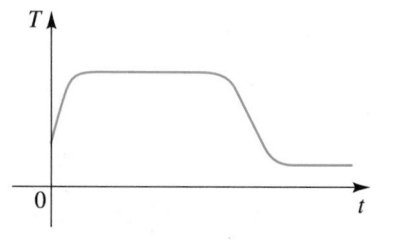

FIGURE 1

Graph of water temperature T as a function of time t

A more accurate graph of the function in Example 1 could be obtained by using a thermometer to measure the temperature of the water at 10-second intervals. In general, scientists collect experimental data and use them to sketch the graphs of functions; they then try to discover a formula that fits the graph. Once a defining formula for a function is found, scientists are then able to predict how a certain process will behave. In the next examples we find explicit formulas for applied functions. In discovering these functions we will rely on the problem-solving principles introduced on pages 61–63.

EXAMPLE 2 ■ Perimeter as a Function of Area

Express the perimeter of a square as a function of its area.

SOLUTION

Introduce notation

Let P, A, and s represent the perimeter, area, and side length of the square in Figure 2, respectively.

Draw a diagram

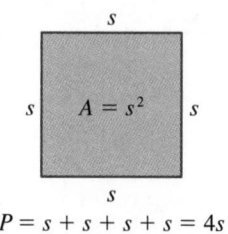

$$P = s + s + s + s = 4s$$

FIGURE 2

Our goal is to find the relationship between P and A. We start with the known formulas for perimeter and area:

$$P = 4s \qquad \text{and} \qquad A = s^2$$

Relate the quantities

The second formula gives $s = \sqrt{A}$. Substituting this into the first formula, we get

$$
\begin{aligned}
P &= 4s & &\text{Formula for perimeter} \\
&= 4\sqrt{A} & &\text{Substitute } \sqrt{A} \text{ for } s
\end{aligned}
$$

Eliminate s (in order to write P as a function of A)

So, the function that relates P and A is

$$P = 4\sqrt{A}$$

Since A must be positive, the domain is given by $A > 0$ or $(0, \infty)$. ■

The function $P = 4\sqrt{A}$ in Example 2 gives the precise relationship between the perimeter and the area of a square. The graph in Figure 3 allows us to visualize how fast the perimeter increases as the area increases.

In Example 2 we used the principles of problem solving. In the following box we adapt these principles to setting up applied functions.

FIGURE 3
Graph of $P = 4\sqrt{A}$

STEPS IN SETTING UP APPLIED FUNCTIONS

1. UNDERSTAND THE PROBLEM. Read the problem carefully until you understand it clearly. Ask yourself: *What is the unknown? What are the given quantities? How are the two related?*

2. DRAW A DIAGRAM. For most problems it is useful to draw a diagram and identify the given and required quantities on the diagram.

3. INTRODUCE NOTATION. If the problem asks for a certain quantity, assign a symbol to that quantity (let us call it Q for now). Also select symbols $(a, b, c, \ldots, x, y)$ for other unknown quantities or variables, and label the diagram with these symbols. It helps to use initials as suggestive symbols—for example, A for area, h for height, or t for time.

4. RELATE THE QUANTITIES. Express Q in terms of some of the other symbols from Step 3 using the given information.

5. ELIMINATE UNNEEDED VARIABLES. If Q has been expressed as a function of more than one variable in Step 4, then use the given information to find relationships (in the form of equations) among these variables. Use these equations to eliminate all but one of the variables in the expression for Q. Thus, Q will be expressed as a function of one variable.

EXAMPLE 3 ■ Surface Area as a Function of Radius

A can holds 1 L (liter) of oil. Express the surface area of the can as a function of its radius.

SOLUTION

| Introduce notation |

Let r be the radius and h be the height of the can (in centimeters) shown in Figure 4. Then the area of the top is πr^2, the area of the bottom is also πr^2, and the area of the sides is the circumference times the height, that is, $2\pi rh$.

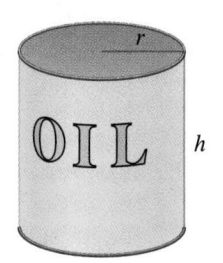

| Draw a diagram |

FIGURE 4

So the total surface area is

| Relate the quantities |

$$S = 2\pi r^2 + 2\pi rh$$

| Eliminate h |

To express S as a function of r we need to eliminate h, and we do this by using the fact that the volume is given as 1 L, which is eqivalent to 1000 cm³. Thus, the volume is

$$\pi r^2 h = 1000$$

The formula for the volume of a cylinder is given on the inside of the front cover of this textbook.

Substituting $h = 1000/(\pi r^2)$ into the expression for S, we have

$$S = 2\pi r^2 + 2\pi r\left(\frac{1000}{\pi r^2}\right) = 2\pi r^2 + \frac{2000}{r}$$

Therefore, the equation

$$S = 2\pi r^2 + \frac{2000}{r}, \quad r > 0$$

expresses S as a function of r. ■

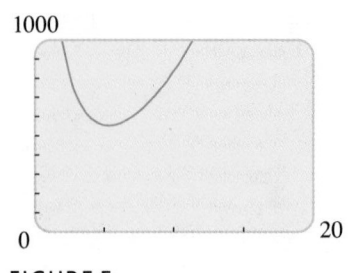

FIGURE 5

Graph of $S = 2\pi r^2 + \dfrac{2000}{r}$

  The graph of the function $S = 2\pi r^2 + \dfrac{2000}{r}$ for $r > 0$ gives us a picture of how the surface area of the can changes as the radius changes (see Figure 5): As r increases, the surface area S first decreases, then increases.

DIRECT AND INVERSE VARIATION

Two types of mathematical models occur so often in various sciences that they are given special names. The first is called *direct variation* and occurs when one quantity is a constant multiple of another.

> ### DIRECT VARIATION
>
> If the quantities x and y are related by an equation
>
> $$y = kx$$
>
> for some constant $k \neq 0$, we say that y **varies directly as** x, or y **is directly proportional to** x, or simply y **is proportional to** x. The constant k is called the **constant of proportionality.**

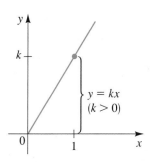

FIGURE 6
Direct variation

Recall that a linear function is a function of the form $f(x) = mx + b$ and its graph is a line with slope m and y-intercept b. So if y varies directly as x, then the equation

$$y = kx$$

or, equivalently, $f(x) = kx$, shows that y is a linear function of x. The graph of this function is a line with slope k, the constant of proportionality (see Figure 6). The y-intercept is $b = 0$, so this line passes through the origin.

EXAMPLE 4 ■ Direct Variation

During a thunderstorm you see the lightning before you hear the thunder because light travels much faster than sound. The distance between you and the center of the storm varies directly as the time interval between the lightning and the thunder.

(a) Suppose that the thunder from a storm whose center is 5400 ft away takes 5 seconds to reach you. Determine the constant of proportionality and write the equation for the variation.

(b) Sketch the graph of this equation. What does the constant of proportionality represent?

(c) If the time interval between the lightning and thunder is now 8 seconds, how far away is the center of the storm?

SOLUTION

Introduce notation

(a) Let d be the distance from you to the storm and let t be the length of the time interval. We are given that d varies directly as t, so

Relate d and t

$$d = kt$$

Eliminate k

where k is a constant. To find k, we use the fact that $t = 5$ when $d = 5400$. Substituting these values in the equation, we get

$$5400 = k(5)$$

Therefore,

$$k = \frac{5400}{5} = 1080$$

Substituting this value of k in the equation for d, we obtain

$$d = 1080t$$

as the equation for d as a function of t.

(b) The graph of the equation $d = 1080t$ is a line through the origin with slope 1080 and is shown in Figure 7. The constant $k = 1080$ is the approximate speed of sound (in ft/s).

(c) When $t = 8$, we have

$$d = 1080 \cdot 8 = 8640$$

So, the storm center is 8640 ft $\approx$ 1.6 mi away.   ∎

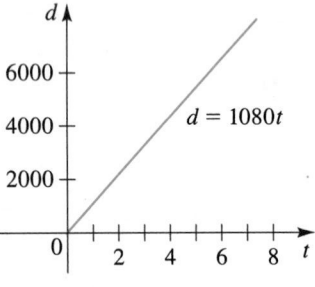

FIGURE 7

Another function that is used frequently in mathematical modeling is the function $f(x) = k/x$, where k is a constant.

INVERSE VARIATION

If the quantities x and y are related by the equation

$$y = \frac{k}{x}$$

for some constant $k \neq 0$, we say that y is **inversely proportional** to x, or y **varies inversely as** x.

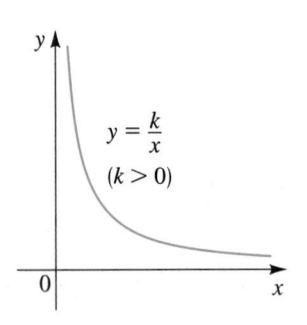

FIGURE 8
Inverse variation

The graph of the function $f(x) = k/x$ for $x > 0$ is shown in Figure 8 for the case $k > 0$. It gives a picture of what happens when y is inversely proportional to x.

EXAMPLE 5 ■ Inverse Variation

Boyle's law states that when a sample of gas is compressed at a constant temperature, the pressure of the gas is inversely proportional to the volume of the gas.

(a) Suppose that the pressure of a sample of air that occupies 0.106 m^3 at $25\,°C$ is 50 kPa. Find the constant of proportionality and write the equation that expresses the inverse proportionality.

(b) If the sample expands to a volume of 0.3 m^3, find the new pressure.

SOLUTION

Introduce notation

(a) Let P be the pressure of the sample of gas and let V be its volume. Then, by the definition of inverse proportionality, we have

Relate P and V

$$P = \frac{k}{V}$$

Eliminate k

where k is a constant. To find k we use the fact that $P = 50$ when $V = 0.106$. So

$$50 = \frac{k}{0.106}$$

Thus

$$k = (50)(0.106) = 5.3$$

Putting this value of k in the equation for P, we have

$$P = \frac{5.3}{V}$$

(b) When $V = 0.3$, we have

$$P = \frac{5.3}{0.3} \approx 17.7$$

So, the new pressure is about 17.7 kPa. ■

A physical quantity often depends on more than one other quantity. For instance, if the quantities x, y, and z are related by the equation

$$z = kxy$$

where k is a nonzero constant, then we say that z **varies jointly as** x and y, or z **is jointly proportional to** x and y. If

$$z = k\frac{x}{y}$$

we say that z **is proportional to** x and **inversely proportional to** y.

EXAMPLE 6 ■ Newton's Law of Gravitation

Newton's Law of Gravitation says that two objects with masses m_1 and m_2 attract each other with a force F that is jointly proportional to their masses and is inversely proportional to the square of the distance r between the objects. Express Newton's Law of Gravitation as an equation.

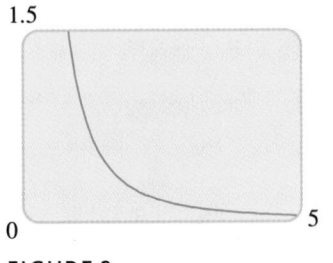

FIGURE 9

Graph of $F = \dfrac{1}{r^2}$

SOLUTION

Using the definitions of joint and inverse proportionality, and using the traditional notation G for the constant of proportionality, we have

$$F = G\frac{m_1 m_2}{r^2}$$

If m_1 and m_2 are fixed masses, then the gravitational force between them is $F = C/r^2$ (where $C = Gm_1 m_2$ is a constant). Figure 9 shows the graph of this function for $r > 0$ with $C = 1$. Observe how the gravitational attraction decreases with increasing distance. ∎

3.7 EXERCISES

1. The graph gives the weight of a certain person as a function of age. Describe in words how this person's weight has varied over time. What do you think happened when this person was 31 years old?

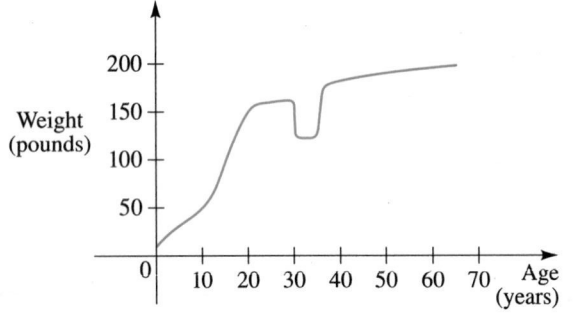

2. The graph gives a salesman's distance from his home as a function of time on a certain day. Describe in words what the graph indicates about his travels on this day.

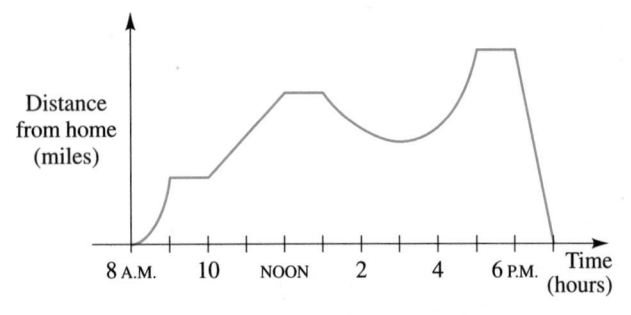

3. You put some ice cubes in a glass, fill the glass with cold water, and then let the glass sit on a table. Sketch a rough graph of the temperature of the water as a function of the elapsed time.

4. A home owner mows the lawn every Wednesday afternoon. Sketch a rough graph of the height of the grass as a function of time over the course of a four-week period beginning on a Sunday.

5. When a football is kicked from a tee, its height depends on the time elapsed since the kickoff. Sketch a rough graph of the height of the football as a function of time.

6. Sketch a rough graph of the number of hours of daylight as a function of the time of year in the Northern Hemisphere.

7. Sketch a rough graph of the outdoor temperature as a function of time during a typical spring day.

8. You place a frozen pie in an oven and bake it for an hour. Then you take it out and let it cool before eating it. Sketch a rough graph of the temperature of the pie as a function of time.

9–22 ■ Find a formula for the described function and state its domain.

9. A rectangle has a perimeter of 20 ft. Express the area A of the rectangle as a function of the length x of one of its sides.

10. A rectangle has an area of 16 m². Express the perimeter P of the rectangle as a function of the length x of one of its sides.

11. Express the area A of an equilateral triangle as a function of the length x of a side.

12. Express the surface area A of a cube as a function of its volume V.

13. Express the radius r of a circle as a function of its area A.

14. Express the area A of a circle as a function of the circumference C.

15. An open rectangular box with a volume of 12 ft^3 has a square base. Express the surface area A of the box as a function of the length x of a side of the base.

16. A woman 5 ft tall is standing near a street lamp that is 12 feet tall, as shown in the figure. Express the length L of her shadow as a function of her distance d from the base of the lamp.

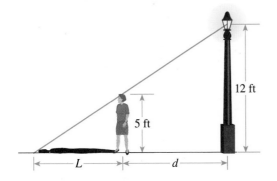

17. A Norman window has the shape of a rectangle surmounted by a semicircle as shown in the figure. If the perimeter of the window is 30 ft, express the area A of the window as a function of the width x of the window.

18. A box with an open top is to be constructed from a rectangular piece of cardboard with dimensions 12 in.

by 20 in. by cutting out equal squares of side x at each corner and then folding up the sides, as shown in the figure. Express the volume V of the box as a function of x.

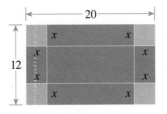

19. A farmer has 2400 ft of fencing and wants to fence off a rectangular field that borders a straight river, as shown in the figure. He needs no fence along the river. Express the area A of the field in terms of the width x of the field.

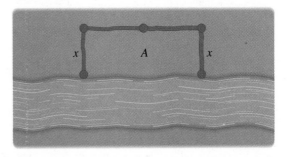

20. A rectangle is inscribed in a semicircle of radius r as shown in the figure. Express the area A of the rectangle as a function of the height h of the rectangle.

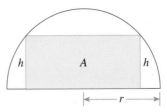

21. Two ships leave a port at the same time. One sails south at 15 mi/h and the other sails east at 20 mi/h. Express the distance d between the ships as a function of t, the time (in hours) elapsed since their departure.

22. A man is standing at point A on the bank of a straight river, 2 mi wide, and he wants to reach point B, 7 mi downstream on the opposite bank, by first rowing his boat to a point P on the opposite bank and then walking the remaining distance x to B. He can row at 2 mi/h and walk at 5 mi/h. Express the total time T that he takes to go from A to B as a function of x.

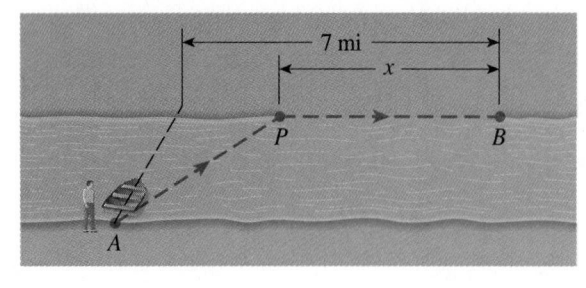

23. The weight w of an object at a height h above the surface of the earth is given by

$$w = \left(\frac{R}{R+h}\right)^2 w_0$$

where w_0 is the weight of the object at sea level and $R = 6400$ km is the radius of the earth. Suppose the weight of an object at sea level is 80 N. Using an appropriate viewing rectangle, sketch the graph of the weight of the object as a function of its height.

24. A highway engineer wants to estimate the maximum number of cars that can safely travel a particular highway at a given speed. She assumes that each car is 17 ft long, travels at a speed s, and follows the car in front of it at the "safe following distance" for that speed. The safe following distance at 20 mi/h is one car length. Using the fact that the distance required to stop is proportional to the square of the speed and the fact that 1 mi/h = 88 ft/s, she finds that the number N of cars per minute that pass a given point is a function of the speed given by

$$N(s) = \frac{88s}{17 + 17\left(\dfrac{s}{20}\right)^2}$$

Sketch a graph of this function in an appropriate viewing rectangle.

25. For a fish swimming at a speed v relative to the water, the energy expenditure per unit time is proportional to v^3. If the fish is swimming against a current with speed u miles per hour, where $u < v$, then the time required to travel a distance of L miles is $L/(v-u)$ and the total energy required to travel the distance is

$$E(v) = 2.73v^3 \frac{L}{v-u}$$

Suppose the speed of the current is $u = 5$ mi/h. Sketch

the graph of the energy $E(v)$ needed to swim a distance of 10 mi as a function of the speed v of the fish.

26. In the theory of relativity, the mass of an object changes as its speed changes. If m_0 is the rest mass of the object, then its mass m is a function of its speed v given by

$$m(v) = \frac{m_0}{\sqrt{1 - \dfrac{v^2}{c^2}}}$$

where $c = 3.0 \times 10^5$ km/s is the speed of light. Sketch the graph of the mass m as a function of the speed v for an object with rest mass 1000 kg.

27–34 ■ Write an equation that expresses the statement.

27. R varies directly as t.

28. P is directly proportional to u.

29. v is inversely proportional to z.

30. w is jointly proportional to m and n.

31. y is proportional to s and inversely proportional to t.

32. P varies inversely as T.

33. z is proportional to the square root of y.

34. A is proportional to the square of t and inversely proportional to the cube of x.

35–40 ■ Express the statement as a formula. Use the given information to find the constant of proportionality.

35. y is directly proportional to x. If $x = 4$, then $y = 72$.

36. z varies inversely as t. If $t = 3$, then $z = 5$.

37. M varies directly as x and inversely as y. If $x = 2$ and $y = 6$, then $M = 5$.

38. S varies jointly as p and q. If $p = 4$ and $q = 5$, then $S = 180$.

39. W is inversely proportional to the square of r. If $r = 6$, then $W = 10$.

40. t is directly proportional to x and y and is inversely proportional to r. If $x = 2$, $y = 3$, and $r = 12$, then $t = 25$.

41. Hooke's law states that the force needed to keep a spring stretched x units beyond its natural length is directly proportional to x. Here the constant of proportionality is called the **spring constant.**
(a) Write Hooke's law as an equation.

(b) If a spring has a natural length of 10 cm and a force of 40 N is required to maintain the spring stretched to a length of 15 cm, find the spring constant.

(c) What force is needed to keep the spring stretched to a length of 14 cm?

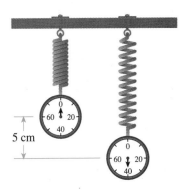

5 cm

42. The period of a pendulum (the time elapsed during one complete swing of the pendulum) varies directly with the square root of the length of the pendulum.

(a) Express this relationship by writing an equation.

(b) In order to double the period, how would we have to change the length?

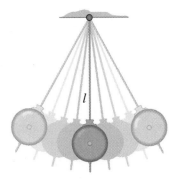

l

43. The cost of printing a magazine is jointly proportional to the number of pages in the magazine and the number of magazines printed.

(a) Write an equation for this joint variation if the printing cost is \$60,000 for 4000 copies of a 120-page magazine.

(b) How much would the printing cost be for 5000 copies of a 92-page magazine?

44. The pressure P of a sample of gas is directly proportional to the temperature T and inversely proportional to the volume V.

(a) Write an equation that expresses this fact if 100 L of gas exerts a pressure of 33.2 kPa at a temperature of 400 K (absolute temperature measured on the Kelvin scale).

(b) If the temperature is increased to 500 K and the volume is decreased to 80 L, what is the pressure of the gas?

45. The resistance R of a wire varies directly as its length L and inversely as the square of its diameter d.

(a) A wire 1.2 m long and 0.005 m in diameter has a resistance of 140 Ω. Write an equation for this variation and find the constant of proportionality.

(b) Find the resistance of a wire made of the same material that is 3 m long and has a diameter of 0.008 m.

46. Kepler's Third Law of planetary motion states that the square of the period T of a planet (the time it takes for the planet to make a complete revolution about the sun) is directly proportional to the cube of the average distance d from the sun.

(a) Express Kepler's Third Law as an equation.

(b) Find the constant of proportionality by using the fact that for our planet the period is about 365 days and the average distance is about 93 million miles.

(c) The planet Neptune is about 2.79×10^9 mi from the sun. Find the period of Neptune.

3.8 TRANSFORMATIONS OF FUNCTIONS

In this section we study how certain transformations of a function affect its graph. This will give us a better understanding of how to graph functions. The transformations we study are shifting, reflecting, and stretching.

EXAMPLE 1 ■ Vertical Shifts of Graphs

Sketch the graph of each function.

(a) $f(x) = x^2 + 3$ (b) $g(x) = x^2 - 2$

Recall that the graph of the function f is the same as the graph of the equation $y = f(x)$. It is often convenient to express a function in equation form, particularly when discussing its graph. For instance, we may refer to the function $f(x) = x^2$ by the equation $y = x^2$, as we do in Example 1.

SOLUTION

(a) We start with the graph of the function $y = x^2$ (from Example 2 in Section 3.6). The equation $y = x^2 + 3$ indicates that the y-coordinate of a point on the graph of f is 3 more than the y-coordinate of the corresponding point on the curve $y = x^2$. This means that we obtain the graph of $f(x) = x^2 + 3$ simply by shifting the graph of $y = x^2$ upward 3 units, as shown in Figure 1.

(b) Similarly, we get the graph of $g(x) = x^2 - 2$ by shifting the parabola $y = x^2$ downward 2 units (see Figure 1). ■

In general, suppose we know the graph of $y = f(x)$. How do we obtain from it the graphs of

$$y = f(x) + c \qquad \text{and} \qquad y = f(x) - c \qquad (c > 0)$$

The equation $y = f(x) + c$ tells us that the y-coordinate of each point on its graph is c units above the y-coordinate of the corresponding point on the graph of $y = f(x)$. So, we obtain the graph of $y = f(x) + c$ simply by shifting the graph of $y = f(x)$ upward c units. We summarize these observations in the following box.

FIGURE 1

(Figure: parabolas $y = x^2 + 3$, $y = x^2$, and $y = x^2 - 2$)

VERTICAL SHIFTS OF GRAPHS		
Equation	How to obtain the graph	What the graph looks like
$y = f(x) + c$ $(c > 0)$	Shift graph of $y = f(x)$ upward c units	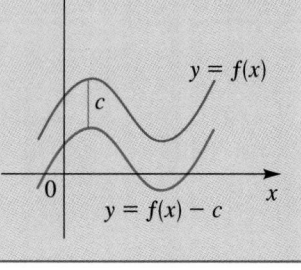
$y = f(x) - c$ $(c > 0)$	Shift graph of $y = f(x)$ downward c units	

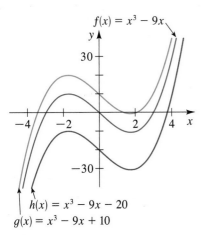

$f(x) = x^3 - 9x$

$h(x) = x^3 - 9x - 20$
$g(x) = x^3 - 9x + 10$

FIGURE 2

EXAMPLE 2 ■ Vertical Shifts of Graphs

Sketch the graph of each function.

(a) $f(x) = x^3 - 9x$ (b) $g(x) = x^3 - 9x + 10$

(c) $h(x) = x^3 - 9x - 20$

SOLUTION

(a) The graph of f was sketched in Example 9 in Section 3.2. It is sketched again in Figure 2.

(b) To obtain the graph of g, we shift the graph of f upward 10 units.

(c) To obtain the graph of h, we shift the graph of f downward 20 units.

The graphs of f, g, and h are sketched in Figure 2.   ■

In the next example we consider transformations that shift the graph of a function horizontally.

EXAMPLE 3 ■ Horizontal Shifts of Graphs

The graph of $y = f(x)$ is sketched in Figure 3. Sketch each of the following graphs.

(a) $y = f(x - 5)$ (b) $y = f(x + 8)$

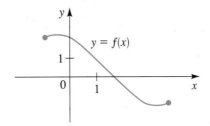

FIGURE 3

SOLUTION

(a) The value of $f(x - 5)$ at x is the same as the value of $f(x)$ at $x - 5$. So the graph of $y = f(x - 5)$ is just the graph of $y = f(x)$ shifted to the right 5 units (see Figure 4).

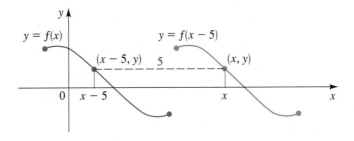

FIGURE 4

(b) Similar reasoning shows that the graph of $y = f(x + 8)$ is the graph of $y = f(x)$ shifted to the left 8 units (see Figure 5).

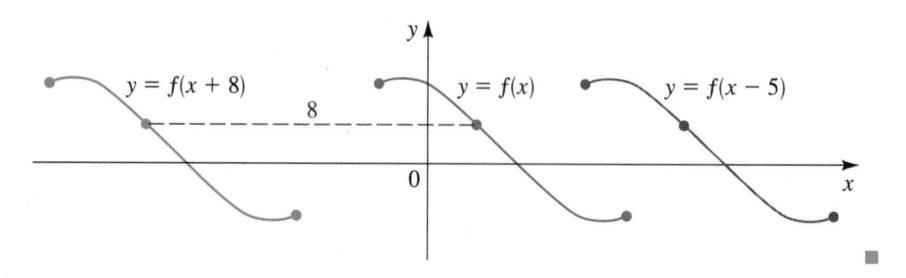

FIGURE 5

In general, suppose we know the graph of $y = f(x)$. How do we obtain from it the graphs of

$$y = f(x + c) \qquad \text{and} \qquad y = f(x - c) \qquad (c > 0)$$

The value of $f(x - c)$ at x is the same as the value of $f(x)$ at $x - c$. Since $x - c$ is c units to the left of x, it follows that the graph of $y = f(x - c)$ is just the graph of $y = f(x)$ shifted to the right c units. Similar reasoning shows that the graph of $y = f(x + c)$ is the graph of $y = f(x)$ shifted to the left c units. The following table summarizes these facts.

HORIZONTAL SHIFTS OF GRAPHS

Equation	How to obtain the graph	What the graph looks like
$y = f(x - c)$ $(c > 0)$	Shift graph of $y = f(x)$ to the right c units	
$y = f(x + c)$ $(c > 0)$	Shift graph of $y = f(x)$ to the left c units	

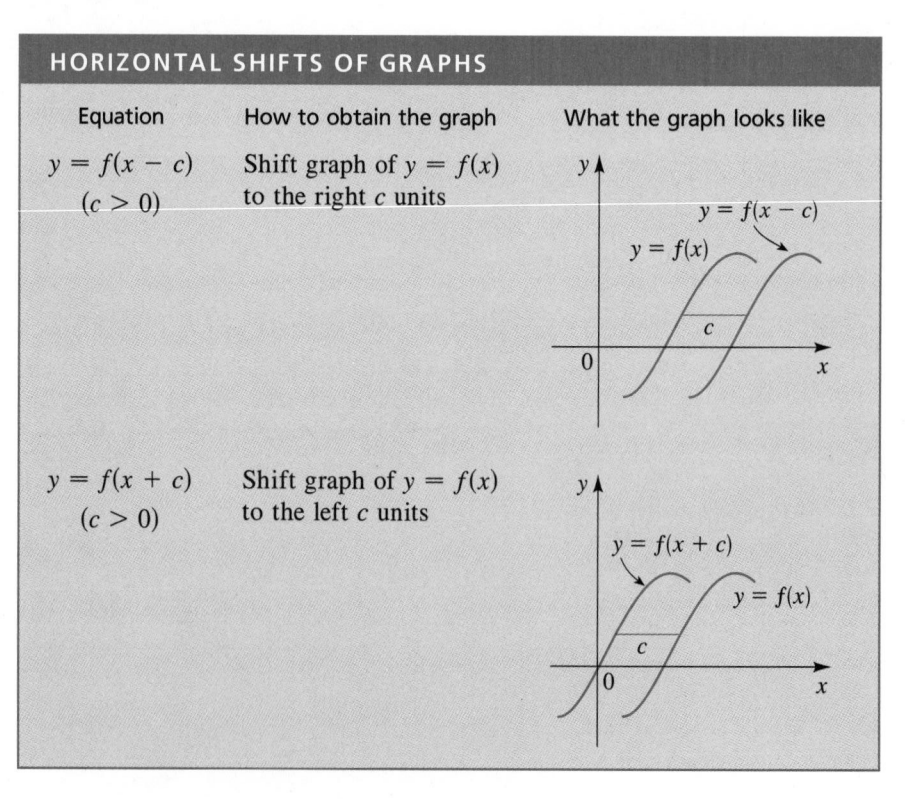

Sonya Kovalevsky (1850–1891) is considered the most important woman mathematician of the 19th century. She was born in Moscow to an aristocratic family. While a child, she was exposed to the principles of calculus in a very unusual fashion—her bedroom was temporarily wallpapered with the pages of a calculus book. She later wrote that she "spent many hours in front of that wall, trying to understand it." Since Russian law forbade women from studying in universities, she entered a marriage of convenience, which allowed her to travel to Germany and obtain a doctorate in mathematics from the University of Göttingen. She eventually was awarded a full professorship at the University of Stockholm, where she taught for eight years before dying in an influenza epidemic at the age of 41. Her research was instrumental in helping put the ideas and applications of functions and calculus on a sound and logical foundation. She received many accolades and prizes for her research work.

EXAMPLE 4 ■ Horizontal Shifts of Graphs

Sketch the graph of each function.

(a) $f(x) = (x + 4)^2$ (b) $g(x) = (x - 2)^2$

SOLUTION

We start with the graph of $y = x^2$, then move it to the left 4 units to get the graph of f and to the right 2 units to obtain the graph of g. See Figure 6.

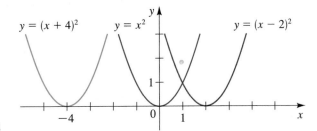

FIGURE 6

EXAMPLE 5 ■ Combining Horizontal and Vertical Shifts

Sketch the graph of the function $f(x) = \sqrt{x - 3} + 4$.

SOLUTION

We start with the graph of the square root function $y = \sqrt{x}$ (Example 6 in Section 3.6). We move it to the right 3 units to get the graph of $y = \sqrt{x - 3}$. Then we move it upward 4 units to obtain the graph of $f(x) = \sqrt{x - 3} + 4$ shown in Figure 7.

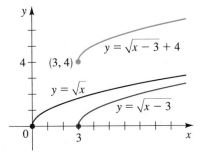

FIGURE 7

Suppose we know the graph of $y = f(x)$. How do we obtain from it the graphs of $y = -f(x)$ and $y = f(-x)$? The y-coordinate of each point on the graph of $y = -f(x)$ is simply the negative of the y-coordinate of the corresponding point on the graph of $y = f(x)$. So the desired graph is the reflection of the graph of $y = f(x)$ in the x-axis. The value of $y = f(-x)$ at x is the same as the value of $y = f(x)$ at $-x$. So the desired graph is the reflection of the graph of $y = f(x)$ in the y-axis. The following table summarizes these observations.

REFLECTING GRAPHS

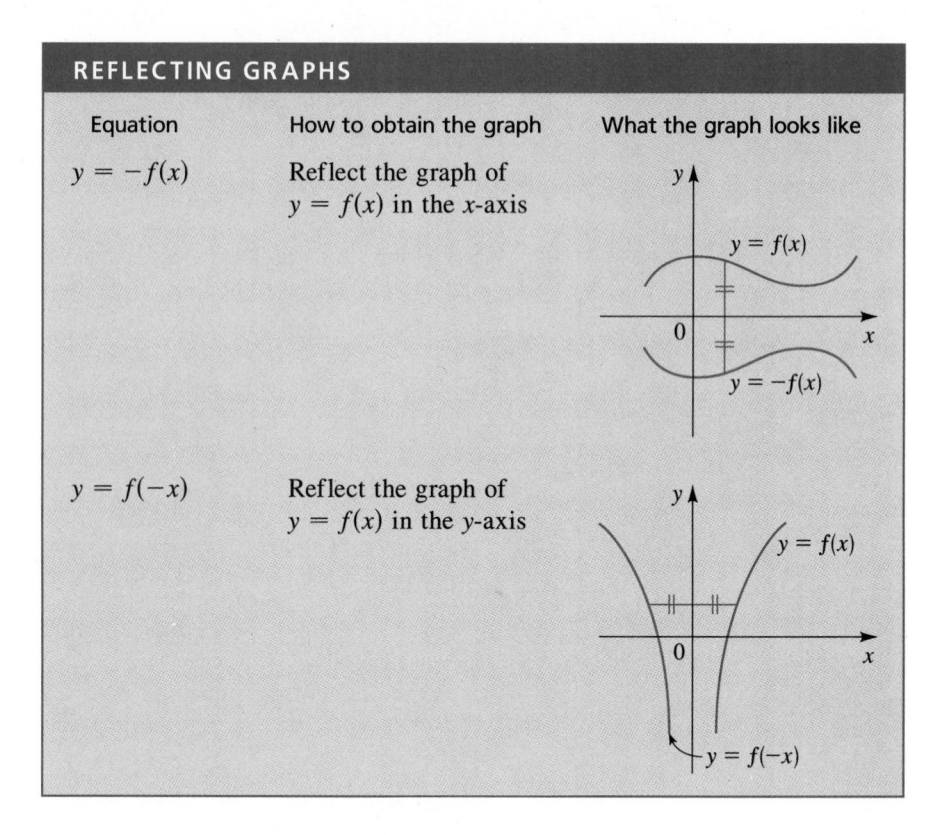

Equation	How to obtain the graph	What the graph looks like
$y = -f(x)$	Reflect the graph of $y = f(x)$ in the x-axis	
$y = f(-x)$	Reflect the graph of $y = f(x)$ in the y-axis	

EXAMPLE 6 ■ Reflecting Graphs

Sketch the graph of each function.

(a) $f(x) = -x^2$ (b) $g(x) = \sqrt{-x}$

SOLUTION

(a) We start with the graph of $y = x^2$. The graph of $f(x) = -x^2$ is the graph of $y = x^2$ reflected in the x-axis (see Figure 8).

(b) We start with the graph of $y = \sqrt{x}$ (Example 6 in Section 3.6). The graph of $g(x) = \sqrt{-x}$ is the graph of $y = \sqrt{x}$ reflected in the y-axis (see Figure 9). Note that the domain of the function $g(x) = \sqrt{-x}$ is $\{x \mid x \leq 0\}$.

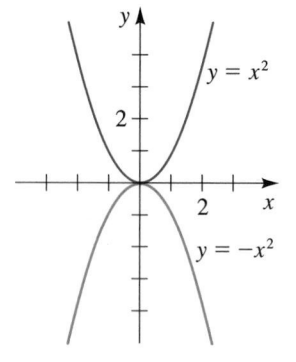

FIGURE 8

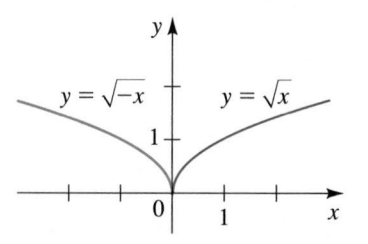

FIGURE 9

Suppose we know the graph of $y = f(x)$. How do we obtain from it the graph of $y = af(x)$? The y-coordinate of $y = af(x)$ at x is the same as the corresponding y-coordinate of $y = f(x)$ multiplied by a. Multiplying the y-coordinates by a has the effect of vertically stretching or shrinking (depending on whether $a > 1$ or $0 < a < 1$) the graph by a factor of a.

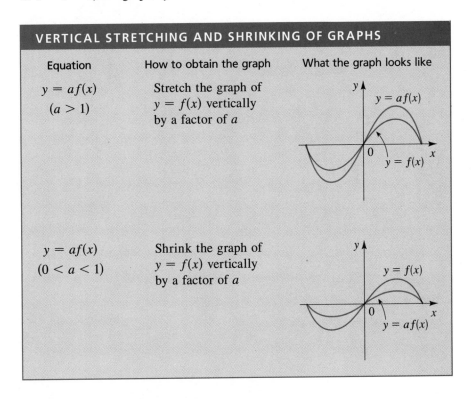

VERTICAL STRETCHING AND SHRINKING OF GRAPHS

Equation	How to obtain the graph	What the graph looks like
$y = af(x)$ $(a > 1)$	Stretch the graph of $y = f(x)$ vertically by a factor of a	
$y = af(x)$ $(0 < a < 1)$	Shrink the graph of $y = f(x)$ vertically by a factor of a	

Horizontal stretching and shrinking of graphs is discussed in Exercises 37 and 38.

EXAMPLE 7 ■ Vertical Stretching and Shrinking of Graphs

Sketch the graph of each function.

(a) $f(x) = 3x^2$ (b) $g(x) = \frac{1}{3}x^2$

SOLUTION

(a) We start with the parabola $y = x^2$. The graph of $f(x) = 3x^2$ is the graph of $y = x^2$ stretched vertically by a factor of 3. The result is a narrower parabola as shown in Figure 10(a) on page 210. To obtain the desired graph, we have multiplied the y-coordinate of each point on the graph of $y = x^2$ by 3.

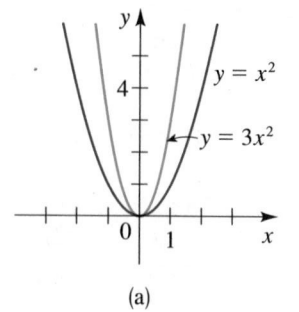

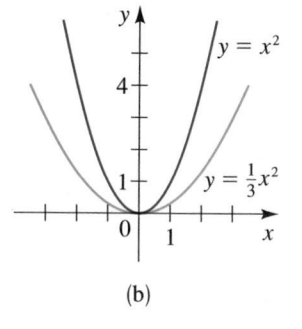

FIGURE 10   (a)   (b)

(b) The graph of $g(x) = \frac{1}{3}x^2$ is the graph of $y = x^2$ shrunk vertically by a factor of $\frac{1}{3}$. The result is a wider parabola as in Figure 10(b). To get the desired graph, we have multiplied the y-coordinate of each point on the graph of $y = x^2$ by $\frac{1}{3}$. ∎

EXAMPLE 8 ∎ Vertical Stretching, Shrinking, and Reflecting

Given the graph of f in Figure 11, draw the graphs of the following functions.

(a) $y = 2f(x)$ (b) $y = \frac{1}{2}f(x)$ (c) $y = -f(x)$

(d) $y = -2f(x)$ (e) $y = -\frac{1}{2}f(x)$

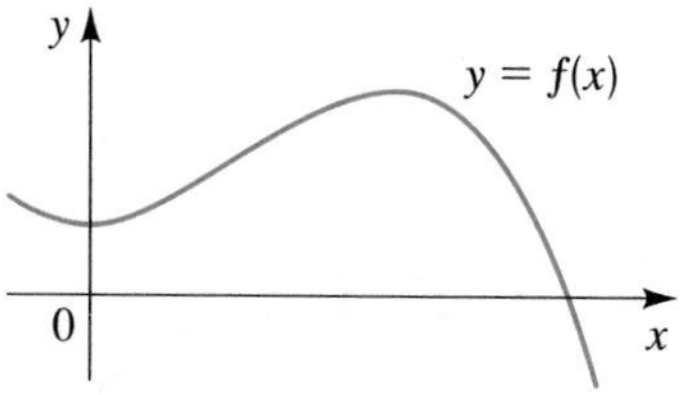

FIGURE 11

SOLUTION

The graphs are shown in Figure 12. Notice, for instance, that the graph of $y = -2f(x)$ is obtained by stretching the graph of f by a factor of 2 and then reflecting in the x-axis.

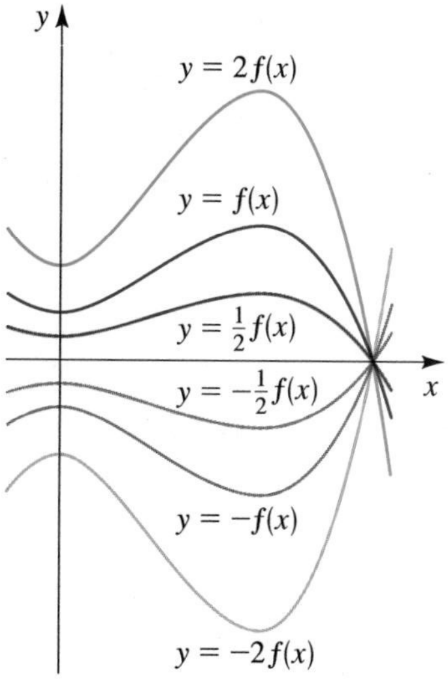

FIGURE 12

We illustrate the effect of combining shifts, reflections, and stretching in the following example.

EXAMPLE 9 ■ Combining Shifting, Stretching, and Reflecting

Sketch the graph of the function $f(x) = 1 - 2(x - 3)^2$.

SOLUTION

Starting with the graph of $y = x^2$, we first shift this graph to the right 3 units to get the graph of $y = (x - 3)^2$. Then we reflect in the x-axis and stretch by a factor of 2 to get the graph of $y = -2(x - 3)^2$. Finally, we shift upward 1 unit to get the graph of $f(x) = 1 - 2(x - 3)^2$ shown in Figure 13.

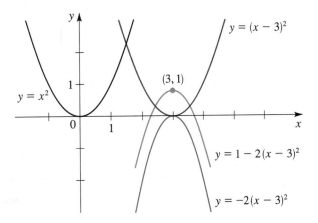

FIGURE 13

■

EVEN AND ODD FUNCTIONS

If a function f satisfies $f(-x) = f(x)$ for every number x in its domain, then f is called an **even function.** For instance, the function $f(x) = x^2$ is even because

$$f(-x) = (-x)^2 = (-1)^2 x^2 = x^2 = f(x)$$

The geometric significance of an even function is that its graph is symmetric with respect to the y-axis (see Figure 14). This means that if we have plotted the graph of f for $x \geq 0$, then we can obtain the entire graph simply by reflecting this portion in the y-axis.

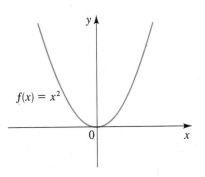

FIGURE 14

$f(x) = x^2$ is an even function.

If f satisfies $f(-x) = -f(x)$ for every number x in its domain, then f is called an **odd function.** For example, the function $f(x) = x^3$ is odd because

$$f(-x) = (-x)^3 = -x^3 = -f(x)$$

The graph of an odd function is symmetric about the origin (see Figure 15). If we have plotted the graph of f for $x \geq 0$, then we can obtain the entire graph by rotating this portion through $180°$ about the origin.

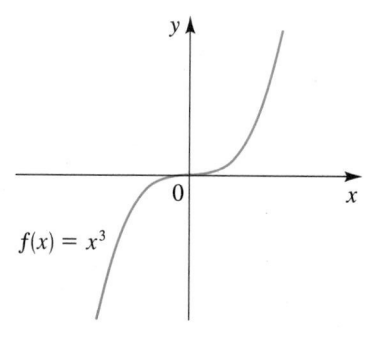

FIGURE 15
$f(x) = x^3$ is an odd function.

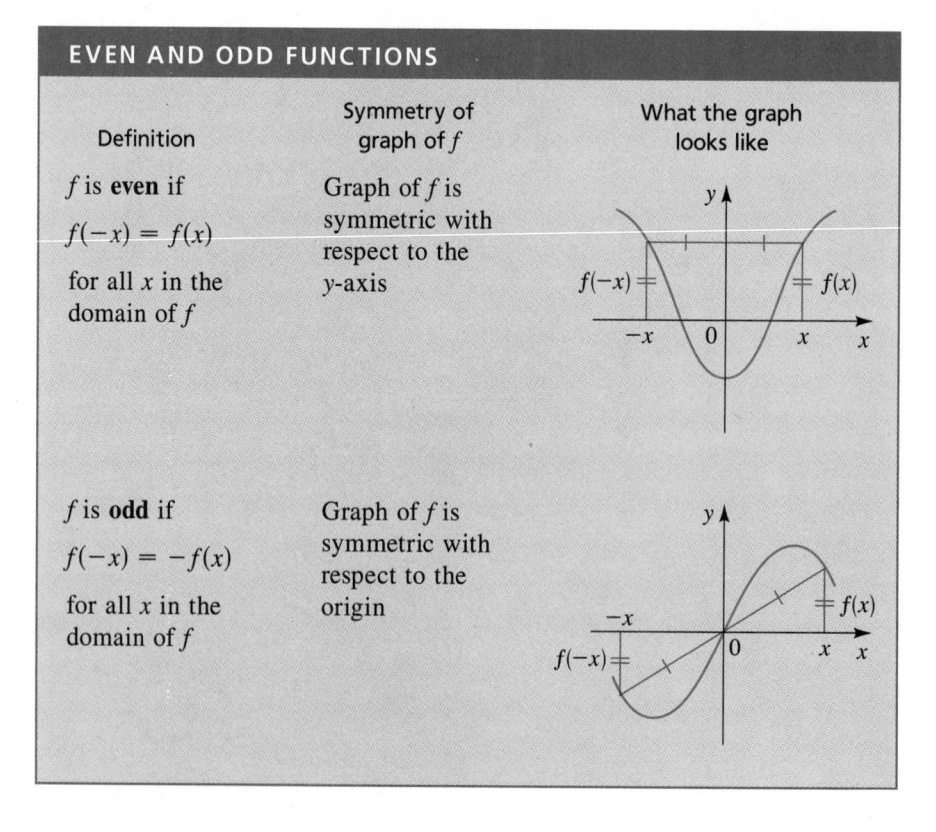

	EVEN AND ODD FUNCTIONS	
Definition	**Symmetry of graph of f**	**What the graph looks like**
f is **even** if $f(-x) = f(x)$ for all x in the domain of f	Graph of f is symmetric with respect to the y-axis	
f is **odd** if $f(-x) = -f(x)$ for all x in the domain of f	Graph of f is symmetric with respect to the origin	

EXAMPLE 10 ■ Even and Odd Functions

Determine whether each of the following functions is even, odd, or neither even nor odd.

(a) $f(x) = x^5 + x$ (b) $g(x) = 1 - x^4$

(c) $h(x) = 2x - x^2$

SOLUTION

(a) $f(-x) = (-x)^5 + (-x) = (-1)^5 x^5 + (-x)$

$= -x^5 - x = -(x^5 + x)$

$= -f(x)$

Therefore, f is an odd function.

(b) $g(-x) = 1 - (-x)^4 = 1 - x^4 = g(x)$

So g is even.

(c) $h(-x) = 2(-x) - (-x)^2 = -2x - x^2$

Since $h(-x) \neq h(x)$ and $h(-x) \neq -h(x)$, we conclude that h is neither even nor odd.   ■

The graphs of the functions in Example 10 are shown in Figure 16. The graph of f was drawn by plotting points for $x \geq 0$ and rotating $180°$ about the origin. The graph of g was drawn by plotting points for $x \geq 0$ and reflecting in the y-axis. Notice that the graph of h is symmetric neither about the y-axis nor about the origin.

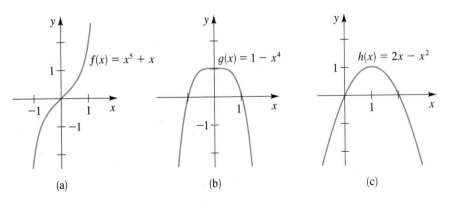

FIGURE 16   (a)   (b)   (c)

3.8 EXERCISES

1–8 ■ Suppose that the graph of f is given. Describe how the graph of each of the following functions can be obtained from the graph of f.

1. (a) $y = f(x) - 4$ (b) $y = f(x - 4)$

2. (a) $y = f(x + 5)$ (b) $y = f(x) + 5$

3. (a) $y = 3f(x)$ (b) $y = \frac{1}{3}f(x)$

4. (a) $y = -f(x)$ (b) $y = f(-x)$

5. (a) $y = -f(x) + 5$ (b) $y = f(-x) + 5$

6. (a) $y = -4f(x)$ (b) $y = -\frac{1}{4}f(x)$

7. (a) $y = f(x - 2) - 3$ (b) $y = 2f(x - 3)$

8. (a) $y = \frac{1}{2}f(x) + 10$ (b) $y = \frac{1}{2}f(x + 10)$

9. The graph of f is given. Draw the graphs of the following functions.

(a) $y = f(x - 2)$ (b) $y = f(x) - 2$
(c) $y = 2f(x)$ (d) $y = -f(x) + 3$
(e) $y = f(-x)$ (f) $y = \frac{1}{2}f(x - 1)$

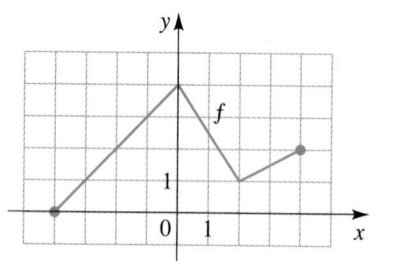

10. The graph of $y = f(x)$ is given. Match each equation with its graph.

(a) $y = f(x - 4)$ (b) $y = f(x) + 3$
(c) $y = \frac{1}{3}f(x)$ (d) $y = -f(x + 4)$
(e) $y = 2f(x + 6)$

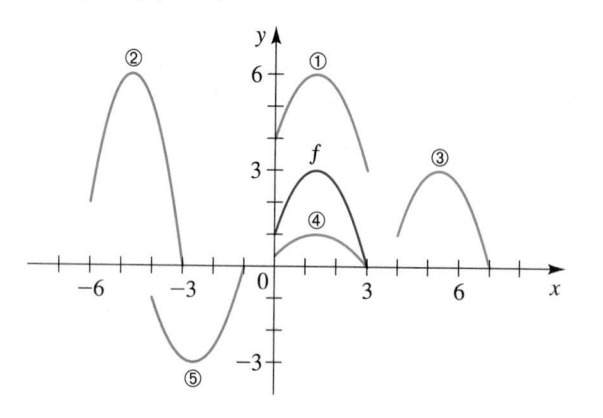

11. (a) Draw the graph of $f(x) = \dfrac{1}{x}$ by plotting points.

(b) Use the graph of f to draw the graph of each of the following functions.

(i) $y = -\dfrac{1}{x}$ (ii) $y = \dfrac{1}{x - 1}$

(iii) $y = \dfrac{2}{x + 2}$ (iv) $y = 1 + \dfrac{1}{x - 3}$

12. (a) Draw the graph of $g(x) = \sqrt[3]{x}$ by plotting points.

(b) Use the graph of g to draw the graph of each of the following functions.

(i) $y = \sqrt[3]{x - 2}$ (ii) $y = \sqrt[3]{x + 2} + 2$
(iii) $y = 1 - \sqrt[3]{x}$

13–28 ■ Sketch the graph of the function, not by plotting points, but by starting with the graph of a standard function and applying transformations.

13. $f(x) = (x - 2)^2$ **14.** $f(x) = (x + 7)^2$

15. $f(x) = -(x + 1)^2$ **16.** $f(x) = 1 - x^2$

17. $f(x) = x^3 + 2$ **18.** $f(x) = -x^3$

19. $y = 1 + \sqrt{x}$ **20.** $y = 2 - \sqrt{x + 1}$

21. $y = \frac{1}{2}\sqrt{x + 4} - 3$ **22.** $y = 3 - 2(x - 1)^2$

23. $y = 5 + (x + 3)^2$ **24.** $y = \frac{1}{3}x^3 - 1$

25. $y = |x| - 1$ **26.** $y = |x - 1|$

27. $y = |x + 2| + 2$ **28.** $y = 2 - |x|$

29–32 ■ Graph the functions on the same screen using the given viewing rectangle. How is each graph related to the graph in part (a)?

29. Viewing rectangle $[-8, 8]$ by $[-2, 8]$
(a) $y = \sqrt[4]{x}$ (b) $y = \sqrt[4]{x + 5}$
(c) $y = 2\sqrt[4]{x + 5}$ (d) $y = 4 + 2\sqrt[4]{x + 5}$

30. Viewing rectangle $[-8, 8]$ by $[-6, 6]$
(a) $y = |x|$ (b) $y = -|x|$
(c) $y = -3|x|$ (d) $y = -3|x - 5|$

31. Viewing rectangle $[-4, 6]$ by $[-4, 4]$
(a) $y = x^6$ (b) $y = \frac{1}{3}x^6$
(c) $y = -\frac{1}{3}x^6$ (d) $y = -\frac{1}{3}(x - 4)^6$

32. Viewing rectangle $[-6, 6]$ by $[-4, 4]$

(a) $y = \dfrac{1}{\sqrt{x}}$

(b) $y = \dfrac{1}{\sqrt{x + 3}}$

(c) $y = \dfrac{1}{2\sqrt{x + 3}}$

(d) $y = \dfrac{1}{2\sqrt{x + 3}} - 3$

33. The graphs of $f(x) = x^2 - 4$ and $g(x) = |x^2 - 4|$ are shown. Explain how the graph of g is obtained from the graph of f.

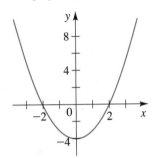

$f(x) = x^2 - 4$

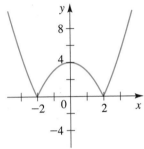

$g(x) = |x^2 - 4|$

34. The graph of $f(x) = x^4 - 4x^2$ is shown. Use this graph to sketch the graph of $g(x) = |x^4 - 4x^2|$.

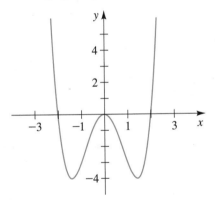

35–36 ■ Sketch the graph of each function.

35. (a) $f(x) = 4x - x^2$ (b) $g(x) = |4x - x^2|$

36. (a) $f(x) = x^3$ (b) $g(x) = |x^3|$

37. In this exercise we consider horizontal stretching and shrinking of graphs.
(a) The graph of f is given. Use it to graph the functions $y = f(2x)$ and $y = f(\tfrac{1}{2}x)$.
(b) If $a > 1$, how is the graph of $y = f(ax)$ obtained from the graph of f?
(c) If $0 < a < 1$, how is the graph of $y = f(ax)$ obtained from the graph of f?

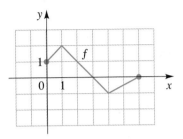

38. The graph of f is given. Use the conclusions of Exercise 37(b) and (c) to draw the graphs of the following functions.
(a) $y = f(2x)$ (b) $y = f(\tfrac{1}{2}x)$

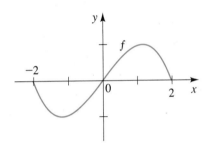

39. If $f(x) = \sqrt{2x - x^2}$, graph the following functions in the viewing rectangle $[-5, 5]$ by $[-4, 4]$. How is each graph related to the graph in part (a)?
(a) $y = f(x)$ (b) $y = f(2x)$
(c) $y = f(\tfrac{1}{2}x)$

40. If $f(x) = \sqrt{2x - x^2}$, graph the following functions in the viewing rectangle $[-5, 5]$ by $[-4, 4]$. How is each graph related to the graph in part (a)?
(a) $y = f(x)$ (b) $y = f(-x)$
(c) $y = -f(-x)$ (d) $y = f(-2x)$
(e) $y = f(-\tfrac{1}{2}x)$

41–48 ■ Determine whether the function f is even, odd, or neither. If f is even or odd, use symmetry to sketch its graph.

41. $f(x) = x^{-2}$ **42.** $f(x) = x^{-3}$

43. $f(x) = x^2 + x$ **44.** $f(x) = x^4 - 4x^2$

45. $f(x) = x^3 - x$ **46.** $f(x) = 3x^3 + 2x^2 + 1$

47. $f(x) = 1 - \sqrt[3]{x}$ **48.** $f(x) = x + \dfrac{1}{x}$

49–50 ■ Prove each statement.

49. (a) If f and g are even functions, then $f + g$ is an even function.

(b) If f and g are odd functions, then $f + g$ is an odd function.

50. (a) If f and g are even functions, then fg is an even function.

(b) If f and g are odd functions, then fg is an even function.

(c) If f is an even function and g is an odd function, then fg is an odd function.

3.9   EXTREME VALUES OF FUNCTIONS

An extreme value of a function is the largest or smallest value of the function on some interval. Finding points where functions achieve extreme values is important in applications. For example, if the function represents the profit in a business, we would be interested in the maximum values; if a function represents the amount of material to be used in a manufacturing process, we would be interested in the minimum values. We begin by giving a method of finding extreme values of quadratic functions and then use graphing devices to find extreme values of other functions.

EXTREME VALUES OF QUADRATIC FUNCTIONS

A **quadratic function** is a function f of the form

$$f(x) = ax^2 + bx + c$$

where a, b, and c are real numbers and $a \neq 0$.

In particular, if we take $a = 1$ and $b = c = 0$, we get the simple quadratic function $f(x) = x^2$ whose graph is the parabola that we drew in Example 2 of Section 3.6. In fact, the graph of any quadratic function is a **parabola**; it can be obtained from the graph of $y = x^2$ by the transformations given in Section 3.8.

EXAMPLE 1 ■ Graphing a Quadratic Function

Sketch the graph of the quadratic function $f(x) = -2x^2 + 3$.

SOLUTION

As in Section 3.8, we start with the graph of $y = x^2$. We stretch vertically by a factor of 2 to get the narrower parabola $y = 2x^2$. Then we reflect in the x-axis to get the graph of $y = -2x^2$ shown in Figure 1(a). Finally, we shift this graph upward 3 units to obtain the graph of $f(x) = -2x^2 + 3$ shown in Figure l(b). ■

If $f(x) = ax^2 + bx + c$ with $b \neq 0$, we first have to complete the square to put the function in a more convenient form for graphing.

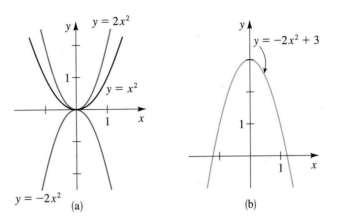

FIGURE 1

Steps in graphing $y = -2x^2 + 3$

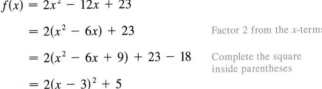

EXAMPLE 2 ■ **Graphing a Quadratic Function by Completing the Square**

Sketch the graph of the quadratic function $f(x) = 2x^2 - 12x + 23$.

SOLUTION

Since the coefficient of x^2 is not 1, we must factor this coefficient from the terms involving x before we complete the square:

$$f(x) = 2x^2 - 12x + 23$$

$$= 2(x^2 - 6x) + 23 \qquad \text{Factor 2 from the } x\text{-terms}$$

$$= 2(x^2 - 6x + 9) + 23 - 18 \qquad \text{Complete the square} \atop \text{inside parentheses}$$

$$= 2(x - 3)^2 + 5$$

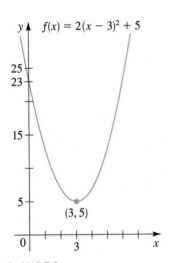

FIGURE 2

(When we add 9 inside the parentheses we actually add $2 \cdot 9 = 18$ because every term inside the parentheses is multiplied by 2. So, to preserve equality we subtract 18 outside the parentheses.) This form of the function tells us that we get the graph of f by taking the parabola $y = x^2$, shifting it to the right 3 units, stretching it by a factor of 2, and moving it upward 5 units. Notice that the vertex is at $(3, 5)$ and the parabola opens upward. We sketch the graph in Figure 2 after noting that the y-intercept is $f(0) = 23$. ■

In Figure 2 we can see that the graph of f has no x-intercept. Another way to see this is to compute the discriminant: $D = b^2 - 4ac = (-12)^2 - 4(2)(23) = -40$. From Section 2.3 we know that a negative discriminant means that the equation $2x^2 - 12x + 23 = 0$ has no real solution, so f has no x-intercept.

If we start with a general quadratic function $f(x) = ax^2 + bx + c$ and complete the square as in Example 2, we arrive at an expression in the standard form $f(x) = a(x - h)^2 + k$. We know from Section 3.8 that the graph of this function is obtained from the graph of $y = x^2$ by a horizontal shift, a stretch, and then a vertical shift. The vertex of the resulting parabola is the point (h, k). If $a > 0$, the parabola opens upward, as shown in Figure 3. But if $a < 0$, then a reflection in the x-axis is also involved, so the parabola opens downward as in Figure 4.

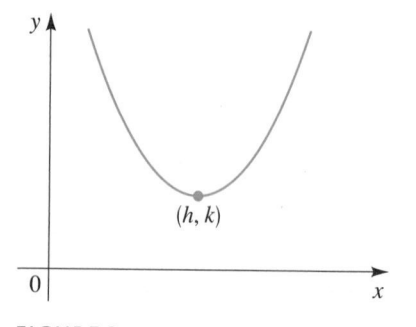

FIGURE 3
$f(x) = a(x - h)^2 + k,$
$a > 0, h > 0, k > 0$

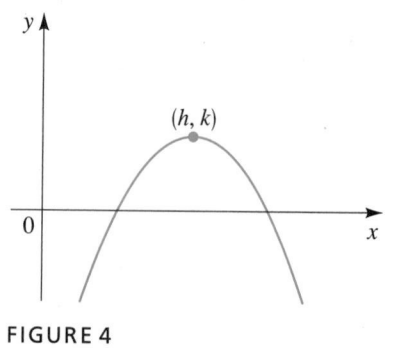

FIGURE 4
$f(x) = a(x - h)^2 + k,$
$a < 0, h > 0, k > 0$

Observe from Figure 3 that if $a > 0$, then the lowest point on the parabola is the vertex (h, k), so the minimum value of the function occurs when $x = h$ and this **minimum value** is $f(h) = k$. Even without the picture we could note that $(x - h)^2 \geq 0$ for all x, so $a(x - h)^2 \geq 0$ (since $a > 0$) and therefore

$$f(x) = a(x - h)^2 + k \geq k \quad \text{for all } x$$

and $f(h) = k$. Similarly, if $a < 0$, then the highest point on the parabola is (h, k), so the **maximum value** of f is $f(h) = k$. We summarize this discussion in the following box.

STANDARD FORM OF A QUADRATIC FUNCTION

A quadratic function $f(x) = ax^2 + bx + c$ can be expressed in the **standard form**

$$f(x) = a(x - h)^2 + k$$

by completing the square. The graph of f is a parabola with vertex (h, k); the parabola opens upward if $a > 0$ or opens downward if $a < 0$.

If $a > 0$, then the **minimum value** of f occurs at $x = h$ and this value is $f(h) = k$.

If $a < 0$, then the **maximum value** of f occurs at $x = h$ and this value is $f(h) = k$.

EXAMPLE 3 ■ Expressing a Quadratic Function in Standard Form

Consider the quadratic function $f(x) = 5x^2 - 30x + 49$.
(a) Express f in standard form.
(b) Sketch the graph of f.
(c) Find the minimum value of f.

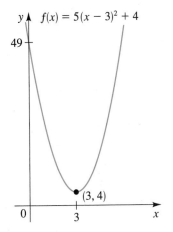

y $f(x) = 5(x - 3)^2 + 4$

49

(3, 4)

0

3

x

FIGURE 5

SOLUTION

(a) Completing the square, we have

$$f(x) = 5x^2 - 30x + 49$$

$$= 5(x^2 - 6x) + 49 \qquad \text{\small Factor 5 from the } x\text{-terms}$$

$$= 5(x^2 - 6x + 9) + 49 - 45 \qquad \text{\small Complete the square inside parentheses}$$

$$= 5(x - 3)^2 + 4$$

(When we add 9 inside the parentheses we actually add $5 \cdot 9 = 45$ because every term inside the parentheses is multiplied by 5. So, we subtract 45 outside the parentheses to preserve equality.)

(b) The graph is a parabola that has its vertex at $(3, 4)$ and opens upward, as sketched in Figure 5.

(c) Since the coefficient of x^2 is positive, f has a minimum value. The minimum value is $f(3) = 4$. (Note that the minimum value occurs at the vertex.) ■

EXAMPLE 4 ■ **Expressing a Quadratic Function in Standard Form**

Consider the quadratic function $f(x) = -x^2 + x + 1$.
(a) Express f in standard form.
(b) Sketch the graph of f.
(c) Find the maximum value of f.

SOLUTION

(a) To express this quadratic function in standard form, we complete the square:

$$y = -x^2 + x + 1$$

$$= -(x^2 - x) + 1 \qquad \text{\small Factor } -1 \text{ from the } x\text{-terms}$$

$$= -\left(x^2 - x + \tfrac{1}{4}\right) + 1 + \tfrac{1}{4} \qquad \text{\small Complete the square inside parentheses}$$

$$= -\left(x - \tfrac{1}{2}\right)^2 + \tfrac{5}{4}$$

[To complete the square of $x^2 - x$, we add $\left(-\tfrac{1}{2}\right)^2 = \tfrac{1}{4}$; but adding $\tfrac{1}{4}$ inside the parentheses is actually adding $-\tfrac{1}{4}$ because every term inside the parentheses is multiplied by -1. So, we must add $\tfrac{1}{4}$ outside the parentheses to preserve equality.]

(b) From the standard form we see that the graph is a parabola that opens downward and has vertex $\left(\tfrac{1}{2}, \tfrac{5}{4}\right)$. As an aid to sketching the graph, we find

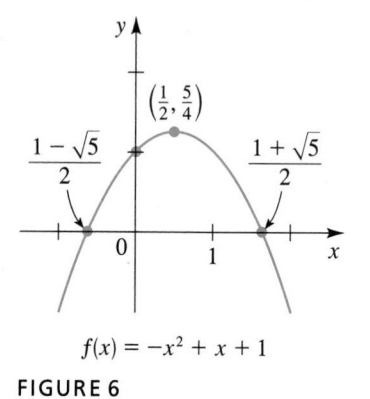

$f(x) = -x^2 + x + 1$

FIGURE 6

the intercepts. The y-intercept is $f(0) = 1$. To find the x-intercepts we use the quadratic formula to solve the equation $-x^2 + x + 1 = 0$:

$$x = \frac{-1 \pm \sqrt{1^2 - 4(-1)(1)}}{-2} = \frac{1 \pm \sqrt{5}}{2}$$

So the x-intercepts are $x \approx 1.62$ and $x \approx -0.62$. The graph of f is sketched in Figure 6.

(c) Since the coefficient of x^2 is negative, f has a maximum value. This maximum value is $f(\frac{1}{2}) = \frac{5}{4}$. (Note that the maximum value occurs at the vertex.) ∎

Expressing a quadratic function in standard form helps us sketch its graph as well as find its maximum or minimum value. If we are interested only in finding the maximum or minimum value, then a formula is available for doing so. This formula is obtained by completing the square for the general quadratic function as follows:

$$f(x) = ax^2 + bx + c$$

$$= a\left(x^2 + \frac{b}{a}x\right) + c \qquad \text{Factor } a \text{ from the } x\text{-terms}$$

$$= a\left(x^2 + \frac{b}{a}x + \frac{b^2}{4a^2}\right) + c - \frac{b^2}{4a} \qquad \text{Complete the square inside parentheses}$$

$$= a\left(x + \frac{b}{2a}\right)^2 + c - \frac{b^2}{4a}$$

This equation is in standard form with $h = -b/(2a)$ and $k = c - b^2/(4a)$. Since the maximum or minimum value occurs at $x = h$, we have the following result.

MAXIMUM OR MINIMUM VALUE OF A QUADRATIC FUNCTION

The maximum or minimum value of a quadratic function
$f(x) = ax^2 + bx + c$ occurs at

$$x = -\frac{b}{2a}$$

If $a > 0$, then the **minimum value** is $f\left(-\dfrac{b}{2a}\right)$.

If $a < 0$, then the **maximum value** is $f\left(-\dfrac{b}{2a}\right)$.

EXAMPLE 5 ■ Finding Maximum and Minimum Values of Quadratic Functions

Find the maximum or minimum value of each quadratic function.

(a) $f(x) = x^2 + 4x$ (b) $g(x) = -2x^2 + 4x - 5$

SOLUTION

(a) This is a quadratic function with $a = 1$ and $b = 4$. Thus, the maximum or minimum value occurs at

$$x = -\frac{b}{2a} = -\frac{4}{2 \cdot 1} = -2$$

Since $a > 0$, this quadratic function has the minimum value

$$f(-2) = (-2)^2 + 4(-2) = -4$$

(b) This is a quadratic function with $a = -2$ and $b = 4$. Thus, the maximum or minimum value occurs at

$$x = -\frac{b}{2a} = -\frac{4}{2 \cdot (-2)} = 1$$

Since $a < 0$, the function has the maximum value

$$f(1) = -2(1)^2 + 4(1) - 5 = -3$$ ■

APPLIED MAXIMUM AND MINIMUM PROBLEMS

In solving applied maximum and minimum problems, we use the steps for setting up applied problems that we discussed in Section 3.7.

EXAMPLE 6 ■ The Largest Product of Two Numbers with a Given Sum

Among all pairs of numbers whose sum is 100, find a pair whose product is as large as possible.

SOLUTION

| Introduce notation |

Let the numbers be x and y. We know that $x + y = 100$, so $y = 100 - x$. Thus their product is

| Relate x and y |

$$P = xy = x(100 - x) = -x^2 + 100x$$

| Eliminate y |

We must therefore find the maximum value of the function

$$P(x) = -x^2 + 100x$$

This is a quadratic function with $a = -1$ and $b = 100$, so the maximum occurs when

$$x = -\frac{b}{2a} = -\frac{100}{2(-1)} = 50$$

When $x = 50$, then $y = 100 - 50 = 50$, so the two numbers are 50 and 50. [Note that the maximum value of P is $P(50) = 50(100 - 50) = 2500$.]   ■

EXAMPLE 7 ■ Maximizing the Area of a Fenced Field

A farmer wants to enclose a rectangular field by a fence and divide it into two smaller rectangular fields by constructing another fence parallel to one side of the field. He has 3000 yards of fencing. Find the dimensions of the field so that the total enclosed area is a maximum.

SOLUTION

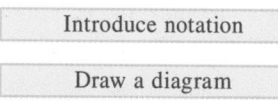

Let x and y be the dimensions of the field (in yards) and draw a diagram as in Figure 7. Then the area of the field is

$$A = xy$$

We eliminate y by using the fact that the total length of the fencing is 3000. Therefore,

$$x + x + x + y + y = 3000$$

$$3x + 2y = 3000$$

$$2y = 3000 - 3x$$

$$y = 1500 - \tfrac{3}{2}x$$

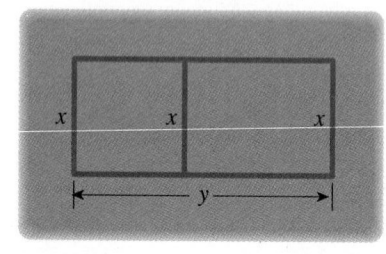

FIGURE 7

This enables us to express A in terms of x alone:

$$A = xy$$

$$= x\left(1500 - \tfrac{3}{2}x\right)$$

$$= 1500x - \tfrac{3}{2}x^2$$

This is a quadratic function with $a = -\tfrac{3}{2}$ and $b = 1500$, so its maximum value occurs when

$$x = -\frac{b}{2a} = -\frac{1500}{2\left(-\tfrac{3}{2}\right)} = 500$$

When $x = 500$, then $y = 1500 - \tfrac{3}{2}(500) = 750$. Thus the field should be 500 yd by 750 yd.   ■

EXAMPLE 8 ■ Maximizing Revenue from Ticket Sales

A hockey team plays in an arena with a seating capacity of 15,000 spectators. With the ticket price set at $12, average attendance at recent games has been 11,000. A market survey indicates that for each dollar that the ticket price is lowered, the average attendance will increase by 1000. What price should the owners of the team set as the ticket price to maximize their revenue from ticket sales?

SOLUTION

| Introduce notation |

Let x be the selling price of the ticket. Then $12 - x$ is the amount the ticket price has been lowered, so the number of tickets sold is

$$11,000 + 1000(12 - x) = 23,000 - 1000x$$

| Relate R and x |

The revenue is

$$R(x) = x(23,000 - 1000x)$$

$$= -1000x^2 + 23,000x$$

This is a quadratic function with $a = -1000$ and $b = 23,000$, so the maximum value occurs when

$$x = -\frac{b}{2a} = -\frac{23,000}{2(-1000)} = 11.5$$

This shows that the revenue is maximized when $x = 11.5$. The owners should set the ticket price at $11.50. ■

 ## USING GRAPHING DEVICES TO FIND EXTREME VALUES

The methods we have discussed apply to finding extreme values of quadratic functions only. We now show how to locate extreme values of any function that can be graphed with a calculator or computer.

EXAMPLE 9 ■ Peaks and Valleys of a Graph

Draw the graph of the function $f(x) = x^3 - 8x + 1$ in the viewing rectangle $[-5, 5]$ by $[-20, 20]$. Comment on the peaks and valleys of the graph.

SOLUTION

The graph is shown in Figure 8. A high point of the graph occurs for a value of x between -3 and 0. In fact, if we restrict our attention to the viewing rectangle $[-3, 0]$ by $[0, 15]$, as in Figure 9(a), we see that there is a *highest* point on the restricted graph. Similarly, when we confine our attention to the

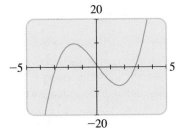

FIGURE 8
Graph of $f(x) = x^3 - 8x + 1$

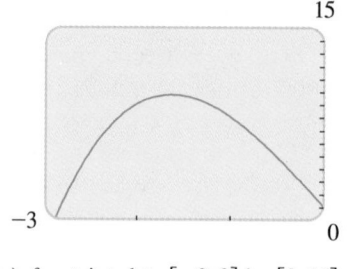
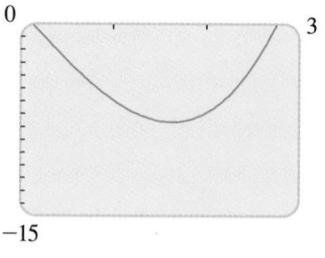

FIGURE 9   (a) f restricted to $[-3, 0]$ by $[0, 15]$ (b) f restricted to $[0, 3]$ by $[-15, 0]$

viewing rectangle $[0, 3]$ by $[-15, 0]$ in Figure 9(b), we see that there is also a *lowest* point on the restricted graph.   ∎

In general, if there is a viewing rectangle such that the point $(a, f(a))$ is the highest point on the graph of f *within* the viewing rectangle (not on the edge), then the number $f(a)$ is called a **local maximum value** of f (see Figure 10). Notice that $f(a) \geq f(x)$ for all numbers x that are close to a.

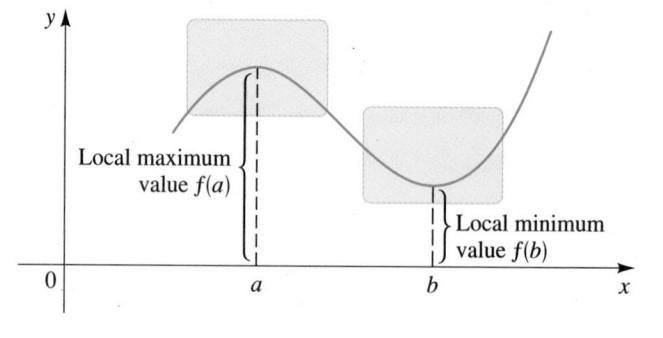

FIGURE 10

Similarly, if there is a viewing rectangle such that the point $(b, f(b))$ is the lowest point on the graph of f within the viewing rectangle, then the number $f(b)$ is called a **local minimum value** of f. In this case, $f(b) \leq f(x)$ for all numbers x that are close to b.

EXAMPLE 10 ■ Finding Local Maxima and Minima from a Graph

Find the approximate local maximum and minimum values of the function $f(x) = x^3 - 8x + 1$ of Example 9.

SOLUTION

Let us first find the local maximum value. Figure 9(a) shows that the maximum point lies in the rectangle $[-2, -1]$ by $[9, 10]$, so we draw the graph restricted to this viewing rectangle in Figure 11(a). By moving the cursor close to the maximum point, we see that the y-coordinates do not change very much in the vicinity of the maximum. The maximum value is about 9.7, and since the distance between the scale marks is 0.1, this estimate for the maximum value is accurate to within 0.1.

If we want an even more accurate estimate, we can zoom in to a smaller rectangle containing the maximum point. Figure 11(b) shows the graph restricted to the viewing rectangle $[-1.7, -1.6]$ by $[9.6, 9.8]$. The curve in Figure 11(b) looks quite flat. It is easier to pinpoint the maximum value if we use a viewing rectangle that is much wider than tall. So we use the viewing rectangle $[-1.7, -1.6]$ by $[9.7, 9.71]$ in Figure 11(c). By moving the cursor along the curve and observing how the y-coordinates change, we can estimate the maximum value quite accurately. It appears that the local maximum value is about 9.709, and this value occurs when x is about -1.633.

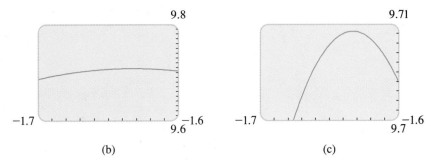

(a)

(b)

(c)

FIGURE 11

Steps in finding the local maximum of $f(x) = x^3 - 8x + 1$

We locate the local minimum in a similar fashion. By zooming in to the rectangles shown in Figure 12, we find that the local minimum value is about -7.709, and this value occurs when $x \approx 1.633$.

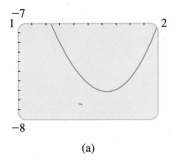

(a)

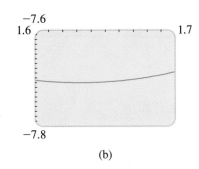

(b)

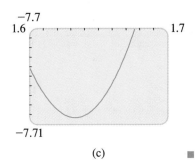

(c)

FIGURE 12

Steps in finding the local minimum of $f(x) = x^3 - 8x + 1$

EXAMPLE 11 ■ Minimizing Cost

A can is to be made to hold 1 L of oil. Find the value of the radius that minimizes the cost of the metal to manufacture the can.

SOLUTION

In order to minimize the cost of the metal, we minimize the total surface area of the can. In Example 3 in Section 3.7 we found the surface area to be

$$S = 2\pi r^2 + \frac{2000}{r}, \qquad r > 0$$

where r is the radius of the can (in centimeters). The graph of S as a function of r is shown in Figure 13(a). We see that the minimum point occurs in the rectangle $[5, 6]$ by $[500, 600]$, so we draw the graph in that viewing rectangle in Figure 13(b). But that graph is very flat, so we change the view to $[5, 6]$ by $[550, 560]$ in Figure 13(c). Moving the cursor near the minimum point and observing the y-coordinates (or S-coordinates in this case), we see that the minimum value of S is about 554, and this value occurs when the radius is about 5.4 cm.

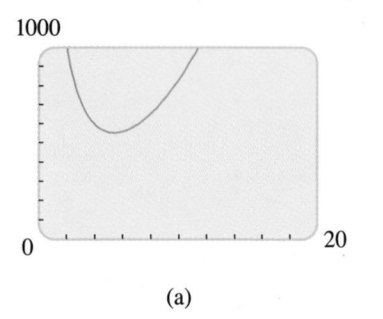
(a)

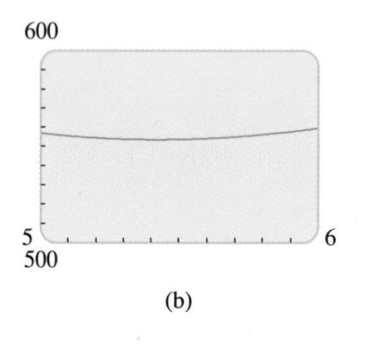
(b)

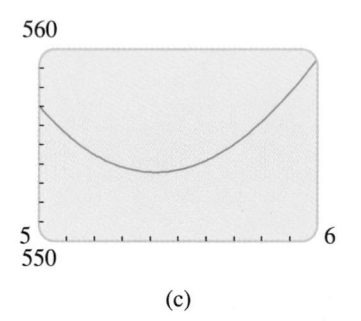
(c)

FIGURE 13

3.9 EXERCISES

1–12 ■ Sketch the graph of the given parabola and state the coordinates of its vertex and its intercepts.

1. $y = x^2 - 8$

2. $y = 4 - x^2$

3. $y = -x^2 - 2$

4. $y = x^2 + 6x$

5. $y = 2x^2 - 6x$

6. $y = x^2 - 2x + 2$

7. $y = x^2 + 6x + 8$

8. $y = -x^2 - 4x + 4$

9. $y = 2x^2 + 4x + 3$

10. $y = -3x^2 + 6x - 2$

11. $y = 2x^2 - 20x + 57$

12. $y = 2x^2 + x - 6$

13–22 ■ A quadratic function is given.
(a) Express the quadratic function in standard form.
(b) Sketch its graph.
(c) Find its maximum or minimum value.

13. $f(x) = 2x - x^2$

14. $f(x) = x + x^2$

15. $f(x) = x^2 + 2x - 1$

16. $f(x) = x^2 - 8x + 8$

17. $f(x) = -x^2 - 3x + 3$

18. $f(x) = 1 - 6x - x^2$

19. $g(x) = 3x^2 - 12x + 13$

20. $g(x) = 2x^2 + 8x + 11$

21. $h(x) = 1 - x - x^2$

22. $h(x) = 3 - 4x - 4x^2$

23–30 ■ Find the maximum or minimum value of the function.

23. $f(x) = x^2 + x + 1$

24. $f(x) = 1 + 3x - x^2$

25. $f(t) = 100 - 49t - 7t^2$

26. $f(t) = 10t^2 + 40t + 113$

27. $f(s) = s^2 - 1.2s + 16$

28. $g(x) = 100x^2 - 1500x$

29. $h(x) = \frac{1}{2}x^2 + 2x - 6$

30. $f(x) = -\dfrac{x^2}{3} + 2x + 7$

31. Find a function whose graph is a parabola with vertex $(1, -2)$ and that passes through the point $(4, 16)$.

32. Find a function whose graph is a parabola with vertex $(3, 4)$ and that passes through the point $(1, -8)$.

33–34 ■ Find the domain and range of the function.

33. $f(x) = -x^2 + 4x - 3$

34. $f(x) = x^2 - 2x - 3$

35. If a ball is thrown directly upward into the air with a velocity of 40 ft/s, its height (in feet) after t seconds is given by $y = 40t - 16t^2$. What is the maximum height attained by the ball?

36. The effectiveness of a television commercial depends on how many times a viewer watches it. After some experiments, an advertising agency found that if the effectiveness E is measured on a scale of 0 to 10, then

$$E(n) = \tfrac{2}{3}n - \tfrac{1}{90}n^2$$

where n is the number of times a viewer watches a given commercial. For a commercial to have maximum effectiveness, how many times should a viewer watch it?

37. Find two numbers whose difference is 100 and whose product is as small as possible.

38. Find two positive numbers whose sum is 100 and the sum of whose squares is a minimum.

39. Find two numbers whose sum is -24 and whose product is a maximum.

40. Among all rectangles that have a perimeter of 20 ft, find the dimensions of the one with the largest area.

41. A farmer has 2400 ft of fencing and wants to fence off a rectangular field that borders a straight river. He needs no fence along the river. (See the figure for Exercise 19 Section 3.7.) What are the dimensions of the field that has the largest area?

42. Find the area of the largest rectangle that can be inscribed in a right triangle with legs of lengths 3 cm and 4 cm if two sides of the rectangle lie along the legs as in the figure.

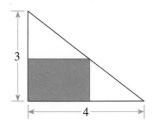

43. A farmer with 750 ft of fencing wants to enclose a rectangular area and then divide it into four pens with fencing parallel to one side of the rectangle (see the figure). What is the largest possible total area of the four pens?

44. A student makes and sells necklaces at the beach during the summer months. The material for each necklace costs her $6 and she has been selling about 20 per day at $10 each. She has been wondering whether or not to raise the price, so she takes a survey and finds that for every dollar increase she would lose two sales a day. What price should she set for the necklaces to maximize her profit?

 45–46 ■ A quadratic function is given.
(a) Use a graphing calculator or computer to find the maximum or minimum value of the quadratic function f, correct to two decimal places.
(b) Find the exact maximum or minimum value of f, and compare with your answer to part (a).

45. $f(x) = x^2 + 1.79x - 3.21$

46. $f(x) = 1 + x - \sqrt{2}x^2$

 47–54 ■ Find the local maximum and minimum values of the function and the value of x at which each occurs. State each answer correct to two decimal places.

47. $f(x) = x^3 - x$

48. $f(x) = 3 + x + x^2 - x^3$

49. $g(x) = x^4 - 2x^3 - 11x^2$

50. $g(x) = x^5 - 8x^3 + 20x$

51. $U(x) = x\sqrt{6 - x}$ **52.** $U(x) = x\sqrt{x - x^2}$

53. $V(x) = \dfrac{1 - x^2}{x^3}$ **54.** $V(x) = \dfrac{1}{x^2 + x + 1}$

 55. For a fish swimming at a speed v relative to the water, the energy expenditure per unit of time is proportional to v^3. Exercise 25 in Section 3.7 stated that if a fish is swimming against a current of 5 mi/h, then the total energy required to travel a distance of 10 mi is

$$E(v) = 2.73v^3 \frac{10}{v - 5}$$

Biologists believe that migrating fish try to minimize the total energy required to swim a fixed distance. Find the value of v that minimizes energy required.

NOTE: This result has been verified; migrating fish swim against a current at a speed 50% greater than the speed of the current.

 56. A highway engineer wants to estimate the maximum number of cars that can safely travel a particular highway at a given speed. She assumes that each car is

17 ft long, travels at a speed s, and follows the car in front of it at the "safe following distance" for that speed. We saw in Exercise 24 in Section 3.7 that the number N of cars per minute that pass a given point is a function of the speed given by

$$N(s) = \frac{88s}{17 + 17\left(\dfrac{s}{20}\right)^2}$$

At what speed can the greatest number of cars travel the highway safely?

57. An open box with a square base must have a volume of 12 ft^3. Find the dimensions of the box that minimize the amount of material used. (See Exercise 15 in Section 3.7.)

58. An open box is to be constructed from a rectangular piece of cardboard with dimensions 12 in. by 20 in. by cutting out equal squares of side x at each corner and then folding up the sides. (See the figure in Exercise 18 in Section 3.7.) Find the largest volume that such a box can have.

59. A Norman window has the shape of a rectangle surmounted by a semicircle. (See the figure in Exercise 17 in Section 3.7.) If the perimeter of the window is 30 ft, find the dimensions of the window so that the greatest possible amount of light is admitted.

60. Find the area of the largest rectangle that can be inscribed in a semicircle of radius 1 ft. (See the figure in Exercise 20 in Section 3.7.)

61. Find the dimensions of the rectangle of largest area that has its base on the x-axis and its other two vertices above the x-axis, lying on the parabola $y = 8 - x^2$.

62. A man is standing at a point A on the bank of a straight river, 2 mi wide, and he wants to reach point B, 7 mi downstream on the opposite bank, by first rowing his boat to a point P on the opposite bank and then walking the remaining distance x to B. (See the figure in Exercise 22 in Section 3.7.) He can row at a speed of 2 mi/h and walk at a speed of 5 mi/h. Where should he land in order to reach B as soon as possible?

63. A kite frame is to be made from six pieces of wood. The four pieces that form the border of the kite have been cut to the lengths indicated in the figure. Let x be as in the figure.
(a) Show that the area of the kite is given by the function $A(x) = x(\sqrt{25 - x^2} + \sqrt{144 - x^2})$.
(b) How long should each of the two crosspieces be to maximize the area of the kite?

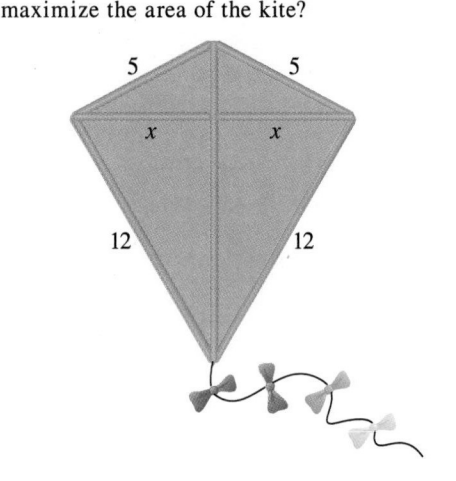

64. Find the maximum value of the function $f(x) = 3 + 4x^2 - x^4$. [*Hint:* Let $t = x^2$.]

65. Find a function whose graph is a parabola that passes through the points $(1, -1)$, $(-1, -3)$, and $(3, 9)$.

3.10 COMBINING FUNCTIONS

Two functions f and g can be combined to form new functions $f + g$, $f - g$, fg, and f/g in a manner similar to the way we add, subtract, multiply, and divide real numbers. For example, we define the function $f + g$ by

$$(f + g)(x) = f(x) + g(x)$$

The new function $f + g$ is called the **sum** of the functions f and g; its value at x is $f(x) + g(x)$. Of course, the sum on the right-hand side makes sense only if

The sum of f and g is defined by

$$(f + g)(x) = f(x) + g(x)$$

The name of the new function is "$f + g$". So this $+$ sign stands for the operation of addition of *functions*. The $+$ sign on the right side, however, stands for addition of the *numbers* $f(x)$ and $g(x)$.

both $f(x)$ and $g(x)$ are defined, that is, if x belongs to the domain of f and also to the domain of g. So, if the domain of f is A and the domain of g is B, then the domain of $f + g$ is the intersection of these domains, that is, $A \cap B$. Similarly, we can define the **difference** $f - g$, the **product** fg, and the **quotient** f/g of the functions f and g. Their domains are $A \cap B$, but in the case of the quotient we must remember not to divide by 0.

ALGEBRA OF FUNCTIONS

Let f and g be functions with domains A and B. Then the functions $f + g, f - g, fg, f/g$ are defined as follows.

$$(f + g)(x) = f(x) + g(x) \qquad \text{Domain } A \cap B$$

$$(f - g)(x) = f(x) - g(x) \qquad \text{Domain } A \cap B$$

$$(fg)(x) = f(x)g(x) \qquad \text{Domain } A \cap B$$

$$\left(\frac{f}{g}\right)(x) = \frac{f(x)}{g(x)} \qquad \text{Domain } \{x \in A \cap B \mid g(x) \neq 0\}$$

EXAMPLE 1 ■ Combining Functions

Let $f(x) = x^3$ and $g(x) = \sqrt{x}$.
(a) Find the functions $f + g, f - g, fg$, and f/g and their domains.
(b) Find $(f + g)(4), (f - g)(4), (fg)(3)$, and $(f/g)(1)$.

SOLUTION

(a) The domain of f is $\mathbb{R}$ and the domain of g is $\{x \mid x \geq 0\}$. We have

$$(f + g)(x) = f(x) + g(x) = x^3 + \sqrt{x} \qquad \text{Domain } \{x \mid x \geq 0\}$$

$$(f - g)(x) = f(x) - g(x) = x^3 - \sqrt{x} \qquad \text{Domain } \{x \mid x \geq 0\}$$

$$(fg)(x) = f(x)g(x) = x^3\sqrt{x} \qquad \text{Domain } \{x \mid x \geq 0\}$$

$$\left(\frac{f}{g}\right)(x) = \frac{f(x)}{g(x)} = \frac{x^3}{\sqrt{x}} \qquad \text{Domain } \{x \mid x > 0\}$$

Note that in the domain of f/g we exclude 0 because $g(0) = 0$.

(b) $(f + g)(4) = f(4) + g(4) = 4^3 + \sqrt{4} = 66$

$(f - g)(4) = f(4) - g(4) = 4^3 - \sqrt{4} = 62$

$(fg)(3) = f(3)g(3) = 3^3\sqrt{3} = 27\sqrt{3}$

$\left(\frac{f}{g}\right)(1) = \frac{f(1)}{g(1)} = \frac{1^3}{\sqrt{1}} = 1$

■

EXAMPLE 2 ■ **Finding Domains of Combinations of Functions**

If $f(x) = \sqrt{x}$ and $g(x) = \sqrt{4 - x^2}$, find the functions $f + g, f - g, fg$, and f/g and their domains.

SOLUTION

The domain of $f(x) = \sqrt{x}$ is $[0, \infty)$. The domain of $g(x) = \sqrt{4 - x^2}$ is the interval $[-2, 2]$ (from Example 4 in Section 3.6). The intersection of the domains of f and g is

$$[0, \infty) \cap [-2, 2] = [0, 2]$$

Thus, we have

$$(f + g)(x) = \sqrt{x} + \sqrt{4 - x^2} \qquad \text{Domain } \{x \mid 0 \le x \le 2\}$$

$$(f - g)(x) = \sqrt{x} - \sqrt{4 - x^2} \qquad \text{Domain } \{x \mid 0 \le x \le 2\}$$

$$(fg)(x) = \sqrt{x}\,\sqrt{4 - x^2} = \sqrt{4x - x^3} \qquad \text{Domain } \{x \mid 0 \le x \le 2\}$$

$$\left(\frac{f}{g}\right)(x) = \frac{\sqrt{x}}{\sqrt{4 - x^2}} = \sqrt{\frac{x}{4 - x^2}} \qquad \text{Domain } \{x \mid 0 \le x < 2\}$$

Notice that the domain of f/g is the interval $[0, 2)$ because we must exclude the points where $g(x) = 0$, that is, $x = \pm2$. ■

The graph of the function $f + g$ is obtained from the graphs of f and g by **graphical addition.** This means that we add corresponding y-coordinates as in Figure 1.

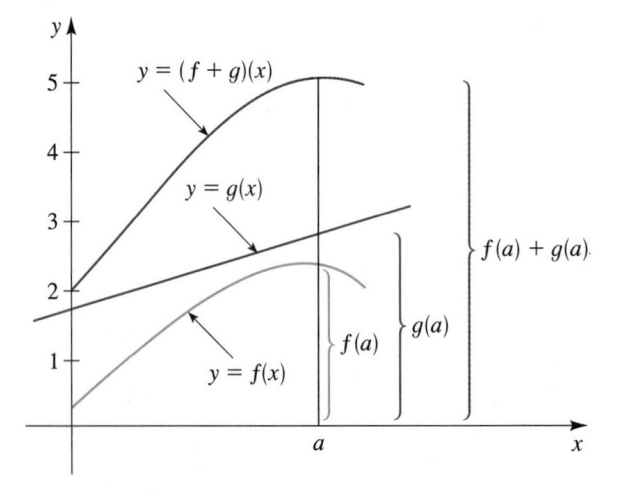

FIGURE 1
Graphical addition

EXAMPLE 3 ■ Using Graphical Addition

Use graphical addition to graph the function $f + g$ from Example 2.

SOLUTION

To graph the function $(f + g)(x) = \sqrt{x} + \sqrt{4 - x^2}$ (from Example 2), we first graph the function $f(x) = \sqrt{x}$ and the function $g(x) = \sqrt{4 - x^2}$ (the top half of the circle $x^2 + y^2 = 4$) in Figure 2. To obtain the graph of $f + g$ we add the y-coordinates of the points on the graphs of f and g. Given the points $Q(x, f(x))$ and $R(x, g(x))$, the point S has coordinates $(x, f(x) + g(x))$. Thus $d(P, Q) = d(R, S)$.

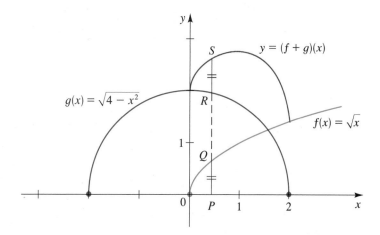

FIGURE 2

■

COMPOSITION OF FUNCTIONS

We now consider a very important way of combining two functions to get a new function. Suppose $f(x) = \sqrt{x}$ and $g(x) = x^2 + 1$. We may define a function h as

$$h(x) = f(g(x)) = f(x^2 + 1) = \sqrt{x^2 + 1}$$

The function h is made up of the functions f and g in an interesting way: Given a number x we first apply to it the function g, then apply f to the result. In this case f is the rule "take the square root," g is the rule "square, then add 1," and h is the rule "square, then add 1, then take the square root." In other words, we get the rule h by applying the rule g and then the rule f.

In general, given any two functions f and g, we start with a number x in the domain of g and find its image $g(x)$. If this number $g(x)$ is in the domain of f, then we can calculate the value of $f(g(x))$. The result is a new function $h(x) = f(g(x))$ obtained by substituting g into f. It is called the *composition* (or *composite*) of f and g and is denoted by $f \circ g$ ("f circle g").

COMPOSITION OF FUNCTIONS

Given two functions f and g, the **composite function** $f \circ g$ (also called the **composition** of f and g) is defined by

$$(f \circ g)(x) = f(g(x))$$

The domain of $f \circ g$ is the set of all x in the domain of g such that $g(x)$ is in the domain of f. In other words, $(f \circ g)(x)$ is defined whenever both $g(x)$ and $f(g(x))$ are defined. The best way to picture $f \circ g$ is by a machine diagram (Figure 3) or an arrow diagram (Figure 4).

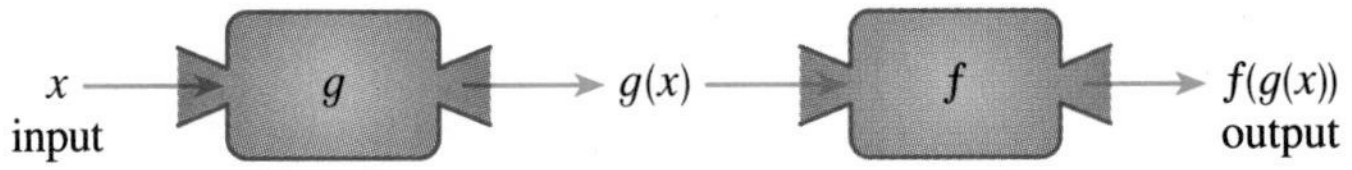

FIGURE 3

The $f \circ g$ machine is composed of the g machine (first) and then the f machine.

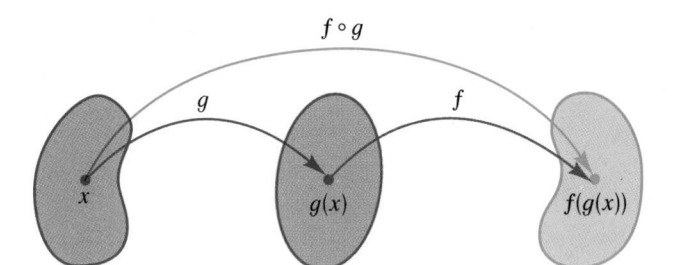

FIGURE 4

Arrow diagram for $f \circ g$

EXAMPLE 4 ■ Finding the Composition of Functions

Let $f(x) = x^2$ and $g(x) = x - 3$.

(a) Find the functions $f \circ g$ and $g \circ f$ and their domains.

(b) Find $(f \circ g)(5)$ and $(g \circ f)(7)$.

SOLUTION

(a) We have

$$\begin{aligned}(f \circ g)(x) &= f(g(x)) && \text{Definition of } f \circ g\\ &= f(x-3) && \text{Definition of } g\\ &= (x-3)^2 && \text{Definition of } f\end{aligned}$$

and

$$\begin{aligned}(g \circ f)(x) &= g(f(x)) && \text{Definition of } g \circ f\\ &= g(x^2) && \text{Definition of } f\\ &= x^2 - 3 && \text{Definition of } g\end{aligned}$$

The domains of both $f \circ g$ and $g \circ f$ are $\mathbb{R}$.

In Example 4, f is the rule "square" and g is the rule "subtract 3." The function $f \circ g$ *first* subtracts 3 and *then* squares; the function $g \circ f$ *first* squares and *then* subtracts 3.

(b) We have

$$(f \circ g)(5) = f(g(5)) = f(2) = (2)^2 = 4$$

$$(g \circ f)(7) = g(f(7)) = g(49) = 49 - 3 = 46$$

An alternative method is to use the results from part (a). Since $(f \circ g)(x) = (x - 3)^2$, we have

$$(f \circ g)(5) = (5 - 3)^2 = 4$$

Also, since $(g \circ f)(x) = x^2 - 3$, we have

$$(g \circ f)(7) = 7^2 - 3 = 46$$ ∎

The graphs of f and g of Example 5, as well as $f \circ g$, $g \circ f$, $f \circ f$, and $g \circ g$ are shown. These graphs indicate that the operation of composition can produce functions quite different from the original functions.

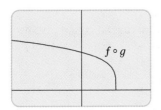

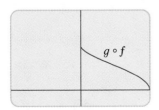

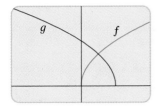

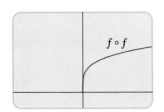

You can see from Example 4 that, in general, $f \circ g \neq g \circ f$. Remember that the notation $f \circ g$ means that the function g is applied first and then f is applied second.

EXAMPLE 5 ■ Finding the Composition of Functions

If $f(x) = \sqrt{x}$ and $g(x) = \sqrt{2 - x}$, find the following functions and their domains.

(a) $f \circ g$ (b) $g \circ f$ (c) $f \circ f$ (d) $g \circ g$

SOLUTION

(a) $(f \circ g)(x) = f(g(x))$ Definition of $f \circ g$

$$= f(\sqrt{2 - x})$$ Definition of g

$$= \sqrt{\sqrt{2 - x}}$$ Definition of f

$$= \sqrt[4]{2 - x}$$

The domain of $f \circ g$ is $\{x \mid 2 - x \geq 0\} = \{x \mid x \leq 2\} = (-\infty, 2]$.

(b) $(g \circ f)(x) = g(f(x))$ Definition of $g \circ f$

$$= g(\sqrt{x})$$ Definition of f

$$= \sqrt{2 - \sqrt{x}}$$ Definition of g

For $\sqrt{x}$ to be defined, we must have $x \geq 0$. For $\sqrt{2 - \sqrt{x}}$ to be defined, we must have $2 - \sqrt{x} \geq 0$, that is, $\sqrt{x} \leq 2$, or $x \leq 4$. Thus, we have $0 \leq x \leq 4$, so the domain of $g \circ f$ is the closed interval $[0, 4]$.

(c)
$$(f \circ f)(x) = f(f(x)) \qquad \text{Definition of } f \circ f$$
$$= f(\sqrt{x}) \qquad\qquad \text{Definition of } f$$
$$= \sqrt{\sqrt{x}} \qquad\qquad \text{Definition of } f$$
$$= \sqrt[4]{x}$$

The domain of $f \circ f$ is $[0, \infty)$.

(d)
$$(g \circ g)(x) = g(g(x)) \qquad \text{Definition of } g \circ g$$
$$= g(\sqrt{2 - x}) \qquad\quad \text{Definition of } g$$
$$= \sqrt{2 - \sqrt{2 - x}} \quad\; \text{Definition of } g$$

This expression is defined when both $2 - x \geq 0$ and $2 - \sqrt{2 - x} \geq 0$. The first inequality means $x \leq 2$, and the second is equivalent to $\sqrt{2 - x} \leq 2$, or $2 - x \leq 4$, that is, $x \geq -2$. Thus $-2 \leq x \leq 2$, so the domain of $g \circ g$ is $[-2, 2]$. ∎

It is possible to take the composition of three or more functions. For instance, the composite function $f \circ g \circ h$ is found by first applying h, then g, and then f as follows:

$$(f \circ g \circ h)(x) = f(g(h(x)))$$

EXAMPLE 6 ■ A Composition of Three Functions

Find $f \circ g \circ h$ if $f(x) = x/(x + 1)$, $g(x) = x^{10}$, and $h(x) = x + 3$.

SOLUTION

$$(f \circ g \circ h)(x) = f(g(h(x))) \qquad \text{Definition of } f \circ g \circ h$$
$$= f(g(x + 3)) \qquad\quad \text{Definition of } h$$
$$= f((x + 3)^{10}) \qquad\; \text{Definition of } g$$
$$= \frac{(x + 3)^{10}}{(x + 3)^{10} + 1} \qquad \text{Definition of } f$$
∎

So far we have used composition to build complicated functions from simpler ones. But in calculus it is useful to be able to "decompose" a complicated function into simpler ones, as shown in the following example.

EXAMPLE 7 ■ Recognizing a Composition of Functions

Given $F(x) = \sqrt[4]{x + 9}$, find functions f and g such that $F = f \circ g$.

SOLUTION

Since the formula for F says to first add 9 and then take the fourth root, we let

$$g(x) = x + 9 \qquad \text{and} \qquad f(x) = \sqrt[4]{x}$$

Then

$$(f \circ g)(x) = f(g(x))$$
$$= f(x + 9)$$
$$= \sqrt[4]{x + 9}$$
$$= F(x) \qquad \blacksquare$$

EXAMPLE 8 ■ An Application of Composition of Functions

A ship is traveling at 20 mi/h parallel to a straight shoreline. The ship is 5 mi from shore. It passes a lighthouse at noon.

(a) Express the distance s between the lighthouse and the ship as a function of d, the distance the ship has traveled since noon; that is, find f so that $s = f(d)$.

(b) Express d as a function of t, the time elapsed since noon; that is, find g so that $d = g(t)$.

(c) Find $f \circ g$. What does this function represent?

SOLUTION

We draw a diagram as in Figure 5.

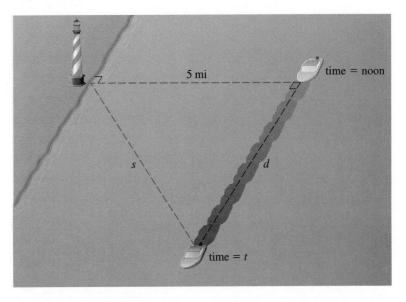

FIGURE 5

(a) We can relate the distances s and d by the Pythagorean Theorem. Thus, s can be expressed as a function of d by

$$s = f(d) = \sqrt{25 + d^2}$$

(b) Since the ship is traveling at 20 mi/h, the distance d it has traveled is a function of t as follows:

distance = rate × time

$$d = g(t) = 20t$$

(c) We have

$$(f \circ g)(t) = f(g(t)) \qquad \text{Definition of } f \circ g$$

$$= f(20t) \qquad \text{Definition of } g$$

$$= \sqrt{25 + (20t)^2} \qquad \text{Definition of } f$$

The function $f \circ g$ gives the distance of the ship from the lighthouse as a function of time. ∎

3.10   EXERCISES

1–6 ■ Find $f + g$, $f - g$, fg, and f/g and their domains.

1. $f(x) = x^2 - x$, $g(x) = x + 5$

2. $f(x) = x^3 + 2x^2$, $g(x) = 3x^2 - 1$

3. $f(x) = \sqrt{1 + x}$, $g(x) = \sqrt{1 - x}$

4. $f(x) = \sqrt{9 - x^2}$, $g(x) = \sqrt{x^2 - 1}$

5. $f(x) = \dfrac{2}{x}$, $g(x) = -\dfrac{2}{x + 4}$

6. $f(x) = \dfrac{1}{x + 1}$, $g(x) = \dfrac{x}{x + 1}$

7–8 ■ Find the domain of the function.

7. $F(x) = \dfrac{\sqrt{4 - x} + \sqrt{3 + x}}{x^2 - 1}$

8. $F(x) = \sqrt{1 - x} + \sqrt{x - 2}$

9–10 ■ Use graphical addition to sketch the graph of $f + g$.

9.

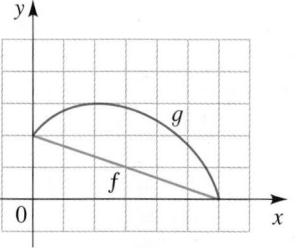

10.

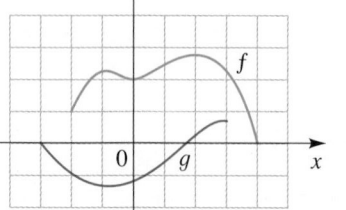

11–14 ■ Draw the graphs of f, g, and $f + g$ on a common screen to illustrate graphical addition.

11. $f(x) = \sqrt{1 + x}$, $g(x) = \sqrt{1 - x}$

12. $f(x) = x^2$, $g(x) = \sqrt{x}$

13. $f(x) = x^2$, $g(x) = x^3$

14. $f(x) = \sqrt[4]{1 - x}$, $g(x) = \sqrt{1 - \dfrac{x^2}{9}}$

15–20 ■ Use $f(x) = 3x - 5$ and $g(x) = 2 - x^2$ to evaluate the expression.

15. (a) $f(g(0))$ (b) $g(f(0))$

16. (a) $f(f(4))$ (b) $g(g(3))$

17. (a) $(f \circ g)(-2)$ (b) $(g \circ f)(-2)$

18. (a) $(f \circ f)(-1)$ (b) $(g \circ g)(2)$

19. (a) $(f \circ g)(x)$ (b) $(g \circ f)(x)$

20. (a) $(f \circ f)(x)$ (b) $(g \circ g)(x)$

21–26 ■ Use the given graphs of f and g to evaluate the expression.

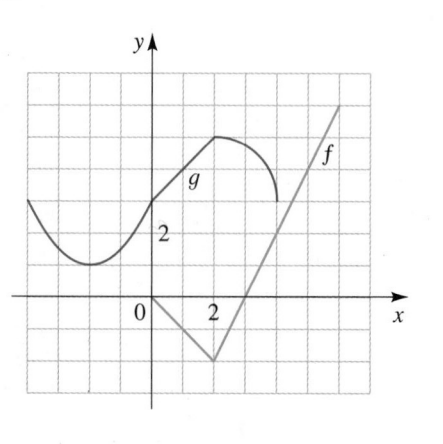

21. $f(g(2))$

22. $g(f(0))$

23. $(g \circ f)(4)$

24. $(f \circ g)(0)$

25. $(g \circ g)(-2)$

26. $(f \circ f)(4)$

27–38 ■ Find the functions $f \circ g$, $g \circ f$, $f \circ f$, and $g \circ g$ and their domains.

27. $f(x) = 2x + 3$, $g(x) = 4x - 1$

28. $f(x) = 6x - 5$, $g(x) = \dfrac{x}{2}$

29. $f(x) = 2x^2 - x$, $g(x) = 3x + 2$

30. $f(x) = \dfrac{1}{x}$, $g(x) = x^3 + 2x$

31. $f(x) = \sqrt{x - 1}$, $g(x) = x^2$

32. $f(x) = x + 2$, $g(x) = \dfrac{1}{x - 2}$

33. $f(x) = \dfrac{1}{x - 1}$, $g(x) = \dfrac{x - 1}{x + 1}$

34. $f(x) = \dfrac{2}{x}$, $g(x) = \dfrac{x}{x + 2}$

35. $f(x) = \sqrt[3]{x}$, $g(x) = 1 - \sqrt{x}$

36. $f(x) = \sqrt{x^2 - 1}$, $g(x) = \sqrt{1 - x}$

37. $f(x) = \dfrac{x + 2}{2x + 1}$, $g(x) = \dfrac{x}{x - 2}$

38. $f(x) = \dfrac{1}{\sqrt{x}}$, $g(x) = x^2 - 4x$

39–42 ■ Find $f \circ g \circ h$.

39. $f(x) = x - 1$, $g(x) = \sqrt{x}$, $h(x) = x - 1$

40. $f(x) = \dfrac{1}{x}$, $g(x) = x^3$, $h(x) = x^2 + 2$

41. $f(x) = x^4 + 1$, $g(x) = x - 5$, $h(x) = \sqrt{x}$

42. $f(x) = \sqrt{x}$, $g(x) = \dfrac{x}{x - 1}$, $h(x) = \sqrt[3]{x}$

43–48 ■ Express the function in the form $f \circ g$.

43. $F(x) = (x - 9)^5$

44. $F(x) = \sqrt{x} + 1$

45. $G(x) = \dfrac{x^2}{x^2 + 4}$

46. $G(x) = \dfrac{1}{x + 3}$

47. $H(x) = |1 - x^3|$

48. $H(x) = \sqrt{1 + \sqrt{x}}$

49–52 ■ Express the function in the form $f \circ g \circ h$.

49. $F(x) = \dfrac{1}{x^2 + 1}$

50. $F(x) = \sqrt[3]{\sqrt{x} - 1}$

51. $G(x) = \left(4 + \sqrt[3]{x}\right)^9$

52. $G(x) = \dfrac{2}{\left(3 + \sqrt{x}\right)^2}$

53. A stone is dropped into a lake, creating a circular ripple that travels outward at a speed of 60 cm/s. Express the area of this circle as a function of time t (in seconds).

54. A spherical balloon is being inflated. If the radius of the balloon is increasing at a rate of 1 cm/s, express the volume of the balloon as a function of time t (in seconds).

55. An airplane is flying at a speed of 350 mi/h at an altitude of one mile. The plane passes directly above a radar station at time $t = 0$.
  (a) Express the horizontal distance d (in miles) that the plane has flown as a function of t (in hours).
  (b) Express the distance s between the plane and the radar station as a function of d.
  (c) Use composition to express s as a function of t.

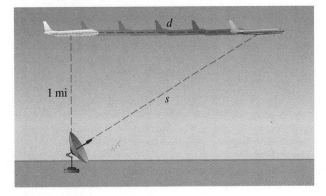

56. A savings account earns 5% interest compounded annually. If you invest x dollars in such an account, then the amount of the investment after one year is $A(x) = x + 0.05x = 1.05x$. Find $A \circ A$, $A \circ A \circ A$, and $A \circ A \circ A \circ A$. What do these compositions represent?

57. If $f(x) = 3x + 5$ and $h(x) = 3x^2 + 3x + 2$, find a function g such that $f \circ g = h$.

58. If $f(x) = x + 4$ and $h(x) = 4x - 1$, find a function g such that $g \circ f = h$.

59. Suppose g is an even function and let $h = f \circ g$. Is h always an even function?

60. Suppose g is an odd function and let $h = f \circ g$. Is h always an odd function? What if f is odd? What if f is even?

3.11 ONE-TO-ONE FUNCTIONS AND THEIR INVERSES

Let us compare the functions f and g whose arrow diagrams are shown in Figure 1. Note that f never takes on the same value (any two numbers in A have different images), whereas g does take on the same value twice (both 2 and 3 have the same image, 4). In symbols, $g(2) = g(3)$ but $f(x_1) \neq f(x_2)$ whenever $x_1 \neq x_2$. Functions that have this latter property are called *one-to-one*.

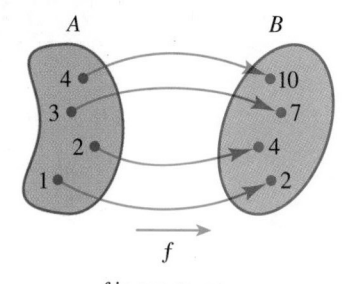
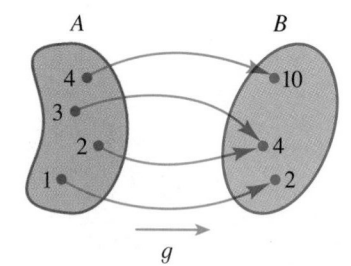

FIGURE 1

f is one-to-one g is not one-to-one

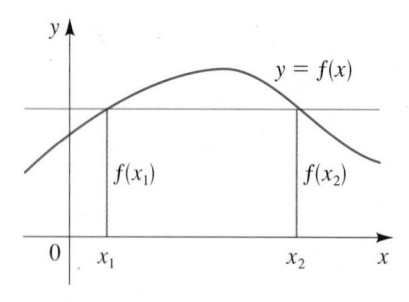

FIGURE 2

This function is not one-to-one because $f(x_1) = f(x_2)$.

DEFINITION OF A ONE-TO-ONE FUNCTION

A function with domain A is called a **one-to-one function** if no two elements of A have the same image, that is,

$$f(x_1) \neq f(x_2) \qquad \text{whenever } x_1 \neq x_2$$

An equivalent way of writing the condition for a one-to-one function is this:

$$\text{If } f(x_1) = f(x_2), \text{ then } x_1 = x_2.$$

If a horizontal line intersects the graph of f in more than one point, then we see from Figure 2 that there are numbers $x_1 \neq x_2$ such that $f(x_1) = f(x_2)$. This means that f is not one-to-one. Therefore, we have the following geometric method for determining whether a function is one-to-one.

HORIZONTAL LINE TEST

A function is one-to-one if and only if no horizontal line intersects its graph more than once.

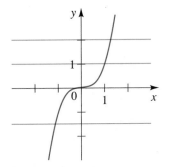

FIGURE 3
$f(x) = x^3$ is one-to-one.

EXAMPLE 1 ■ Deciding Whether a Function is One-to-One

Is the function $f(x) = x^3$ one-to-one?

SOLUTION 1

If $x_1 \neq x_2$, then $x_1^3 \neq x_2^3$ (two different numbers cannot have the same cube). Therefore, $f(x) = x^3$ is one-to-one.

SOLUTION 2

From Figure 3 we see that no horizontal line intersects the graph of $f(x) = x^3$ more than once. Therefore, by the Horizontal Line Test, f is one-to-one. ■

Notice that the function f of Example 1 is increasing and is also one-to-one. In fact it can be proved that every increasing function and every decreasing function is one-to-one (see Exercise 49).

EXAMPLE 2 ■ Deciding Whether a Function is One-to-One

Is the function $g(x) = x^2$ one-to-one?

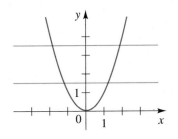

FIGURE 4
$g(x) = x^2$ is not one-to-one.

SOLUTION 1

This function is not one-to-one because, for instance,

$$g(1) = 1 = g(-1)$$

and so 1 and -1 have the same image.

SOLUTION 2

From Figure 4 we see that there are horizontal lines that intersect the graph of g more than once. Therefore, by the Horizontal Line Test, g is not one-to-one. ■

Although the function g in Example 2 is not one-to-one, it is possible to restrict its domain so that the resulting function is one-to-one. In fact, if we define

$$h(x) = x^2, \qquad x \geq 0$$

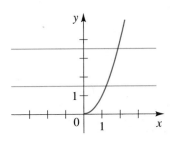

FIGURE 5
$h(x) = x^2 \ (x \geq 0)$ is one-to-one.

then h is one-to-one, as you can see from Figure 5 and the Horizontal Line Test.

EXAMPLE 3 ■ Showing that a Function Is One-to-One

Show that the function $f(x) = 3x + 4$ is one-to-one.

SOLUTION

Suppose there are numbers x_1 and x_2 such that $f(x_1) = f(x_2)$. Then

$$3x_1 + 4 = 3x_2 + 4$$

$$3x_1 = 3x_2$$

$$x_1 = x_2$$

Therefore, f is one-to-one.   ■

INVERSE FUNCTIONS

One-to-one functions are important because they are precisely the functions that possess inverse functions according to the following definition.

Do not mistake the -1 in f^{-1} for an exponent.

$$f^{-1} \quad does \ not \ mean \quad \frac{1}{f(x)}$$

The reciprocal $1/f(x)$ is written as $[f(x)]^{-1}$.

> **DEFINITION OF INVERSE FUNCTION**
>
> Let f be a one-to-one function with domain A and range B. Then its **inverse function** f^{-1} has domain B and range A and is defined by
>
> $$f^{-1}(y) = x \quad \Longleftrightarrow \quad f(x) = y$$
>
> for any y in B.

This definition says that if f takes x into y, then f^{-1} takes y back into x. (If f were not one-to-one, then f^{-1} would not be defined uniquely.) The arrow diagram in Figure 6 indicates that f^{-1} reverses the effect of f. From the definition, we have

$$\text{domain of } f^{-1} = \text{range of } f$$

$$\text{range of } f^{-1} = \text{domain of } f$$

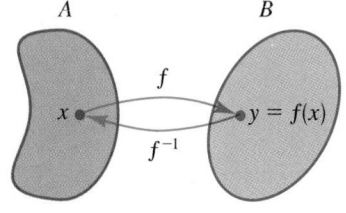

FIGURE 6

EXAMPLE 4 ■ Finding f^{-1} for Specific Values

If $f(1) = 5$, $f(3) = 7$, and $f(8) = -10$, find $f^{-1}(5)$, $f^{-1}(7)$, and $f^{-1}(-10)$.

SOLUTION

From the definition of f^{-1}, we have

$$f^{-1}(5) = 1 \qquad \text{because} \qquad f(1) = 5$$

$$f^{-1}(7) = 3 \qquad \text{because} \qquad f(3) = 7$$

$$f^{-1}(-10) = 8 \qquad \text{because} \qquad f(8) = -10$$

The diagram in Figure 7 makes it clear how f^{-1} reverses the effect of f in this case.

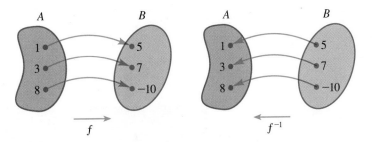

FIGURE 7

By its definition, the inverse function f^{-1} undoes what f does: If we start with x, apply f, and then apply f^{-1}, we arrive back at x, where we started. Similarly, f undoes what f^{-1} does. In general, any function that reverses the effect of f in this way must be the inverse of f. These observations are expressed precisely as follows.

PROPERTY OF INVERSE FUNCTIONS

Let f be a one-to-one function with domain A and range B. The inverse function f^{-1} satisfies the following cancellation properties.

$$f^{-1}(f(x)) = x \qquad \text{for every } x \text{ in } A$$

$$f(f^{-1}(x)) = x \qquad \text{for every } x \text{ in } B$$

Conversely, any function f^{-1} satisfying these equations is the inverse of f.

These properties indicate that f is the inverse function of f^{-1}, so we often say that f and f^{-1} are *inverses of each other*.

EXAMPLE 5 ■ Verifying that Two Functions Are Inverses

Show that $f(x) = x^3$ and $g(x) = x^{1/3}$ are inverses of each other.

SOLUTION

Note that the domain and range of both f and g is $\mathbb{R}$. We have

$$g(f(x)) = g(x^3) = (x^3)^{1/3} = x$$

$$f(g(x)) = f(x^{1/3}) = (x^{1/3})^3 = x$$

This shows that f and g are inverses of each other. These equations say simply that the cube function and the cube root function, when composed, cancel each other out.

We now explain how to compute inverse functions. We first observe from the definition of f^{-1} that

$$y = f(x) \iff f^{-1}(y) = x$$

So, if $y = f(x)$ and if we are able to solve this equation for x in terms of y, then we must have $x = f^{-1}(y)$. If we then interchange x and y, we have $y = f^{-1}(x)$, which is the desired equation.

HOW TO FIND THE INVERSE FUNCTION OF A ONE-TO-ONE FUNCTION

1. Write $y = f(x)$.

2. Solve this equation for x in terms of y (if possible).

3. Interchange x and y. The resulting equation is $y = f^{-1}(x)$.

Note that Steps 2 and 3 can be reversed. In other words, it is possible to interchange x and y first and then solve for y in terms of x.

EXAMPLE 6 ■ Finding an Inverse Function

In Example 6, note how f^{-1} reverses the effect of f. The function f is the rule "multiply by 3, then subtract 2," whereas f^{-1} is the rule "add 2, then divide by 3."

Find the inverse function of $f(x) = 3x - 2$.

SOLUTION

First we write $y = f(x)$:

$$y = 3x - 2$$

Then we solve this equation for x:

$$3x = y + 2$$

$$x = \frac{y + 2}{3}$$

Finally, we interchange x and y:

$$y = \frac{x + 2}{3}$$

Therefore, the inverse function is $f^{-1}(x) = \dfrac{x + 2}{3}$.

CHECK YOUR ANSWER

We can check our answer by using the Property of Inverse Functions:

$$f^{-1}(f(x)) = f^{-1}(3x - 2)$$
$$= \frac{(3x - 2) + 2}{3}$$
$$= \frac{3x}{3} = x$$

$$f(f^{-1}(x)) = f\left(\frac{x + 2}{3}\right)$$
$$= 3\left(\frac{x + 2}{3}\right) - 2$$
$$= x + 2 - 2 = x \quad \checkmark$$

■

In Example 7, note how f^{-1} reverses the effect of f. The function f is the rule "take the 5th power, subtract 3, then divide by 2," whereas f^{-1} is the rule "multiply by 2, add 3, then take the 5th root."

CHECK YOUR ANSWER

We can check our answer by using the Property of Inverse Functions:

$$f^{-1}(f(x)) = f^{-1}\left(\frac{x^5 - 3}{2}\right)$$

$$= \left[2\left(\frac{x^5 - 3}{2}\right) + 3\right]^{1/5}$$

$$= (x^5 - 3 + 3)^{1/5}$$

$$= (x^5)^{1/5} = x$$

$$f(f^{-1}(x)) = f((2x + 3)^{1/5})$$

$$= \frac{[(2x + 3)^{1/5}]^5 - 3}{2}$$

$$= \frac{2x + 3 - 3}{2} = \frac{2x}{2} = x \quad \checkmark$$

EXAMPLE 7 ■ Finding an Inverse Function

Find the inverse function of $f(x) = \dfrac{x^5 - 3}{2}$.

SOLUTION

We first write $y = (x^5 - 3)/2$ and solve for x:

$$y = \frac{x^5 - 3}{2}$$

$$2y = x^5 - 3$$

$$x^5 = 2y + 3$$

$$x = (2y + 3)^{1/5}$$

Then we interchange x and y to get $y = (2x + 3)^{1/5}$. Therefore, the inverse function is $f^{-1}(x) = (2x + 3)^{1/5}$. ∎

The principle of interchanging x and y to find the inverse function also gives us the method for obtaining the graph of f^{-1} from the graph of f. If $f(a) = b$, then $f^{-1}(b) = a$. Thus, the point (a, b) is on the graph of f if and only if the point (b, a) is on the graph of f^{-1}. But we get the point (b, a) from the point (a, b) by reflecting in the line $y = x$ (see Figure 8).

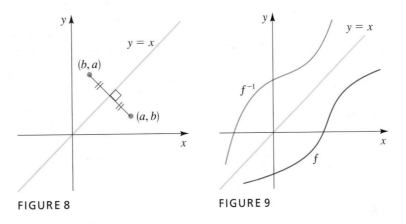

FIGURE 8   **FIGURE 9**

Therefore, as illustrated in Figure 9:

The graph of f^{-1} is obtained by reflecting the graph of f in the line $y = x$.

EXAMPLE 8 ■ Finding an Inverse Function

(a) Sketch the graph of $f(x) = \sqrt{x - 2}$.
(b) Use the graph of f to sketch the graph of f^{-1}.
(c) Find an equation for f^{-1}.

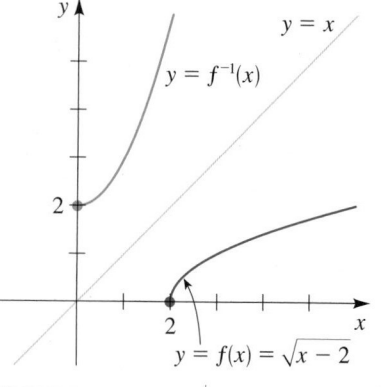

FIGURE 10

SOLUTION

(a) Using the transformations from Section 3.8, we sketch the graph of $y = \sqrt{x - 2}$ by plotting the graph of the function $y = \sqrt{x}$ (Example 6 in Section 3.6) and moving it to the right 2 units.

(b) The graph of f^{-1} is obtained from the graph of f in part (a) by reflecting it in the line $y = x$, as shown in Figure 10.

(c) Solve $y = \sqrt{x - 2}$ for x:

$$\sqrt{x - 2} = y \qquad \text{Note that } y \geqslant 0$$

$$x - 2 = y^2$$

$$x = y^2 + 2, \qquad y \geqslant 0$$

Interchange x and y:

$$y = x^2 + 2, \qquad x \geqslant 0$$

Thus

$$f^{-1}(x) = x^2 + 2, \qquad x \geqslant 0$$

This expression shows that the graph of f^{-1} is the right half of the parabola $y = x^2 + 2$ and, from the graph shown in Figure 10, this seems reasonable. ■

In Example 8, note how f^{-1} reverses the effect of f. The function f is the rule "subtract 2, then take the square root"; f^{-1} is the rule "square, then add 2."

3.11   EXERCISES

1–6 ■ The graph of a function f is given. Determine whether f is one-to-one.

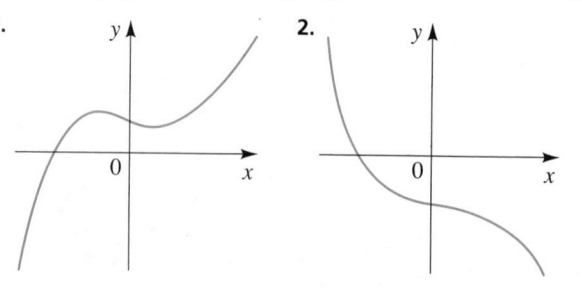

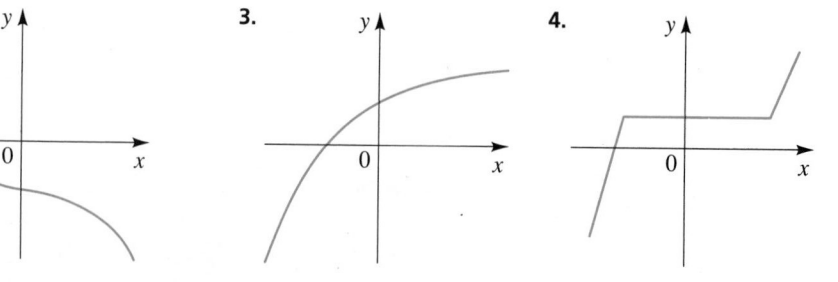

5.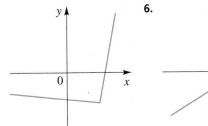

6.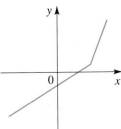

7–12 ■ Determine whether the function is one-to-one.

7. $f(x) = 7x - 3$

8. $f(x) = x^2 - 2x + 5$

9. $g(x) = \sqrt{x}$

10. $g(x) = |x|$

11. $h(x) = x^4 + 5$

12. $h(x) = x^4 + 5, \quad 0 \leq x \leq 2$

13–16 ■ Assume f is a one-to-one function.

13. (a) If $f(2) = 7$, find $f^{-1}(7)$.
   (b) If $f^{-1}(3) = -1$, find $f(-1)$.

14. (a) If $f(5) = 18$, find $f^{-1}(18)$.
   (b) If $f^{-1}(4) = 2$, find $f(2)$.

15. If $f(x) = 5 - 2x$, find $f^{-1}(3)$.

16. If $g(x) = x^2 + 4x$ with $x \geq -2$, find $g^{-1}(5)$.

17–22 ■ Use the Property of Inverse Functions to show that f and g are inverses of each other.

17. $f(x) = 2x - 5; \quad g(x) = \dfrac{x + 5}{2}$

18. $f(x) = \dfrac{3 - x}{4}; \quad g(x) = 3 - 4x$

19. $f(x) = x^2 - 4, \quad x \geq 0;$
   $g(x) = \sqrt{x + 4}, \quad x \geq -4$

20. $f(x) = x^3 + 1; \quad g(x) = (x - 1)^{1/3}$

21. $f(x) = \dfrac{1}{x - 1}, \quad x \neq 1;$
   $g(x) = \dfrac{1}{x} + 1, \quad x \neq 0$

22. $f(x) = \sqrt{4 - x^2}, \quad 0 \leq x \leq 2;$
   $g(x) = \sqrt{4 - x^2}, \quad 0 \leq x \leq 2$

23–40 ■ Find the inverse function of f.

23. $f(x) = 2x + 1$

24. $f(x) = 6 - x$

25. $f(x) = 4x + 7$

26. $f(x) = 3 - 5x$

27. $f(x) = \dfrac{1}{x + 2}, \quad x > -2$

28. $f(x) = \dfrac{x - 2}{x + 2}$

29. $f(x) = \dfrac{1 + 3x}{5 - 2x}$

30. $f(x) = 5 - 4x^3$

31. $f(x) = \sqrt{2 + 5x}$

32. $f(x) = x^2 + x, \quad x \geq -\frac{1}{2}$

33. $f(x) = 4 - x^2, \quad x \geq 0$

34. $f(x) = \sqrt{2x - 1}$

35. $f(x) = 4 + \sqrt[3]{x}$

36. $f(x) = (2 - x^3)^5$

37. $f(x) = 1 + \sqrt{1 + x}$

38. $f(x) = \sqrt{9 - x^2}, \quad 0 \leq x \leq 3$

39. $f(x) = x^4, \quad x \geq 0$

40. $f(x) = 1 - x^3$

41–44 ■ A function f is given.
(a) Sketch the graph of f.
(b) Use the graph of f to sketch the graph of f^{-1}.
(c) Find f^{-1}.

41. $f(x) = 3x - 6$

42. $f(x) = 16 - x^2, \quad x \geq 0$

43. $f(x) = \sqrt{x + 1}$

44. $f(x) = x^3 - 1$

45–48 ■ Draw the graph of f and use it to determine whether the function is one-to-one.

45. $f(x) = x^3 - x$

46. $f(x) = x^3 + x$

47. $f(x) = \dfrac{x + 12}{x - 6}$

48. $f(x) = \sqrt{x^3 - 4x + 1}$

49. Prove that if f is an increasing function on its entire domain, then f is one-to-one.

50. Show that it is possible to restrict the domain of the function $f(x) = x^2 + x + 1$ in such a way that the resulting function is one-to-one.

51. For what values of the number m is the linear function $f(x) = mx + b$ one-to-one? For those values of m, find f^{-1}.

52. Under what condition on the constants a, b, c, and d is the following function one-to-one?

$$f(x) = \frac{ax + b}{cx + d}$$

3   REVIEW

KEY TOPICS ■ Define, state, or discuss each of the following.

1. Coordinate plane
2. Distance Formula
3. Midpoint Formula
4. Graph of an equation
5. x- and y-intercepts
6. Equation of a circle
7. Symmetry with respect to the x-axis, the y-axis, and the origin
8. Slope of a line
9. Point-slope equation for a line
10. Slope-intercept equation for a line
11. General equation for a line (linear equation)
12. Parallel and perpendicular lines
13. Function
14. Domain and range of a function
15. Independent and dependent variables
16. Graph of a function
17. Linear function
18. Constant function

19. Vertical Line Test
20. Increasing and decreasing functions
21. Direct, inverse, and joint proportionality
22. Vertical and horizontal shifting of graphs
23. Vertical stretching and shrinking of graphs
24. Vertical and horizontal reflection of graphs
25. Even and odd functions
26. Quadratic function
27. Standard form of a quadratic function
28. Maximum and minimum values of a function
29. Sum, difference, product, and quotient of functions
30. Composition of functions
31. One-to-one function
32. Horizontal Line Test
33. Inverse function
34. Procedure for finding an inverse function
35. Graph of an inverse function

EXERCISES

1–4 ■ Two points P and Q are given.
(a) Plot P and Q on a coordinate plane.
(b) Find the distance from P to Q.
(c) Find the midpoint of the segment PQ.
(d) Sketch the line determined by P and Q and find its equation in slope-intercept form.
(e) Sketch the circle that passes through Q and has center P, and find the equation of this circle.

1. $P(0, 0)$, $Q(4, 3)$ **2.** $P(2, 0)$, $Q(-5, 12)$
3. $P(7, -1)$, $Q(2, -11)$ **4.** $P(8, 0)$, $Q(0, -6)$

5–6 ■ Sketch the region in the plane.
5. $\{(x, y) \mid |x| < 4 \text{ and } |y| < 2\}$
6. $\{(x, y) \mid |x| \geqslant 4 \text{ or } |y| \geqslant 2\}$

7. Which of the points $A(4, 4)$ or $B(5, 3)$ is closer to the point $C(-1, -3)$?

8. Show that the points $A(1, 2)$, $B(2, 6)$, and $C(9, 0)$ are the vertices of a right triangle. Find the area of the triangle.

9. Find an equation of the circle that has center $(2, -5)$ and radius $\sqrt{2}$.

10. Find an equation of the circle that has center $(-5, -1)$ and passes through the origin.

11. Find an equation of the circle that contains the points $P(2, 3)$ and $Q(-1, 8)$ and has the midpoint of the segment PQ as its center.

12. Find an equation of the circle that contains the points $P(1, -3)$ and $Q(-1, 2)$ as the endpoints of a diameter.

13–16 ■ Determine whether the equation represents a circle, a point, or has no graph. If the equation is that of a circle, find its center and radius.

13. $x^2 + y^2 + 2x - 6y + 9 = 0$

14. $2x^2 + 2y^2 - 2x + 8y = \frac{1}{2}$

15. $x^2 + y^2 + 72 = 12x$

16. $x^2 + y^2 - 6x - 10y + 34 = 0$

17. Find an equation for the line that passes through the points $(-1, -6)$ and $(2, -4)$.

18. Find an equation for the line that passes through the point $(6, -3)$ and has slope $-\frac{1}{2}$.

19. Find an equation for the line that has x-intercept 4 and y-intercept 12.

20. Find an equation for the line that passes through the point $(1, 7)$ and is perpendicular to the line $x - 3y + 16 = 0$.

21. Find an equation for the line that passes through the origin and is parallel to the line $3x + 15y = 22$.

22. Find an equation for the line that passes through the point $(5, 2)$ and is parallel to the line passing through $(-1, -3)$ and $(3, 2)$.

23–24 ■ Find equations for the circle and the line in the figure.

23.

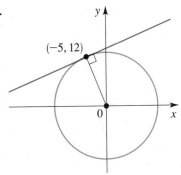

24.

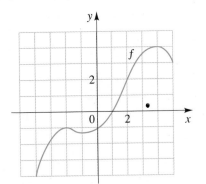

25–38 ■ Test the equation for symmetry and sketch its graph.

25. $y = 2 - 3x$

26. $2x - y + 1 = 0$

27. $x + 3y = 21$

28. $x = 2y + 12$

29. $\dfrac{x}{2} - \dfrac{y}{7} = 1$

30. $\dfrac{x}{4} + \dfrac{y}{5} = 0$

31. $y = 16 - x^2$

32. $8x + y^2 = 0$

33. $x = \sqrt{y}$

34. $y = -\sqrt{1 - x^2}$

35. $2x^2 + 2y^2 = 5$

36. $xy = 4$

37. $y = x^3 - 4x$

38. $x = y^4 - 4y^2$

39. If $f(x) = x^2 - x + 1$, find $f(0), f(2), f(-2), f(a), f(-a), f(x + 1), f(2x),$ and $2f(x) - 2$.

40. If $f(x) = 1 + \sqrt{x - 1}$, find $f(5), f(9), f(a + 1), f(-x), f(x^2),$ and $[f(x)]^2$.

41. The graph of a function is given.

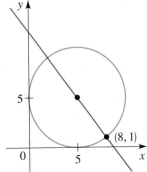

(a) Find $f(-2)$ and $f(2)$.
(b) Find the domain of f.
(c) Find the range of f.

(d) On what intervals is f increasing? On what intervals is f decreasing?

(e) Is f one-to-one?

42. Which of the following figures are graphs of functions? Which of the functions are one-to-one?

(a)

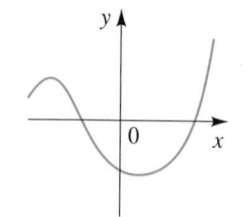

(b)

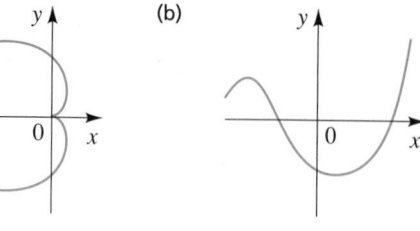

(c)

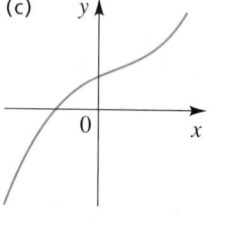

(d)

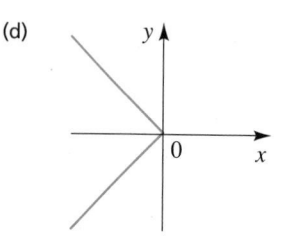

43–44 ■ Find the domain and range of the function.

43. $f(x) = \sqrt{x + 3}$

44. $F(t) = t^2 + 2t + 5$

45–52 ■ Find the domain of the function.

45. $f(x) = 7x + 15$

46. $f(x) = \dfrac{2x + 1}{2x - 1}$

47. $f(x) = \sqrt{x + 4}$

48. $f(x) = 3x - \dfrac{2}{\sqrt{x + 1}}$

49. $f(x) = \dfrac{1}{x} + \dfrac{1}{x + 1} + \dfrac{1}{x + 2}$

50. $g(x) = \dfrac{2x^2 + 5x + 3}{2x^2 - 5x - 3}$

51. $h(x) = \sqrt{4 - x} + \sqrt{x^2 - 1}$

52. $f(x) = \dfrac{\sqrt[3]{2x + 1}}{\sqrt[3]{2x + 2}}$

53–70 ■ Sketch the graph of the function.

53. $f(x) = 1 - 2x$

54. $f(x) = \frac{1}{3}(x - 5), \quad 2 \leqslant x \leqslant 8$

55. $f(t) = 1 - \frac{1}{2}t^2$

56. $g(t) = t^2 - 2t$

57. $f(x) = x^2 - 6x + 6$

58. $f(x) = 3 - 8x - 2x^2$

59. $y = 1 - \sqrt{x}$

60. $y = -|x|$

61. $y = \frac{1}{2}x^3$

62. $y = \sqrt{x + 3}$

63. $h(x) = \sqrt[3]{x}$

64. $H(x) = x^3 - 3x^2$

65. $g(x) = \dfrac{1}{x^2}$

66. $G(x) = \dfrac{1}{(x - 3)^2}$

67. $f(x) = \begin{cases} 1 - x & \text{if } x < 0 \\ 1 & \text{if } x \geqslant 0 \end{cases}$

68. $f(x) = \begin{cases} 1 - 2x & \text{if } x \leqslant 0 \\ 2x - 1 & \text{if } x > 0 \end{cases}$

69. $f(x) = \begin{cases} x + 6 & \text{if } x < -2 \\ x^2 & \text{if } x \geqslant -2 \end{cases}$

70. $f(x) = \begin{cases} -x & \text{if } x < 0 \\ x^2 & \text{if } 0 \leqslant x < 2 \\ 1 & \text{if } x \geqslant 2 \end{cases}$

71. Determine which of the following viewing rectangles produces the most appropriate graph of the function $f(x) = 6x^3 - 15x^2 + 4x - 1$.

(i) $[-2, 2]$ by $[-2, 2]$

(ii) $[-8, 8]$ by $[-8, 8]$

(iii) $[-4, 4]$ by $[-12, 12]$

(iv) $[-100, 100]$ by $[-100, 100]$

72. Determine which of the following viewing rectangles produces the most appropriate graph of the function $f(x) = \sqrt{100 - x^3}$.

(i) $[-4, 4]$ by $[-4, 4]$

(ii) $[-10, 10]$ by $[-10, 10]$

(iii) $[-10, 10]$ by $[-10, 40]$

(iv) $[-100, 100]$ by $[-100, 100]$

73–76 ■ Draw the graph of the function in an appropriate viewing rectangle.

73. $f(x) = x^2 + 25x + 173$

74. $f(x) = 1.1x^3 - 9.6x^2 - 1.4x + 3.2$

75. $y = \dfrac{x}{\sqrt{x^2 + 16}}$

76. $y = |x(x + 2)(x + 4)|$

77–80 ■ Use a graphing device to solve the equation or inequality correct to two decimals.

77. $x^3 - 3x - 1 = 0$

78. $x^2 = \sqrt{x + 4}$

79. $0.1x + 1 \geqslant x^3$

80. $4 - x^2 \leqslant 3 - \sqrt{x + 3}$

 81. Find, approximately, the domain of the function
$f(x) = \sqrt{x^3 - 4x + 1}$.

 82. Find, approximately, the range of the function
$f(x) = x^4 - x^3 + x^2 + 3x - 6$.

83. Suppose that the graph of f is given. Describe how the graph of each of the following functions can be obtained from the graph of f.

(a) $y = f(x) + 8$ (b) $y = f(x + 8)$
(c) $y = 1 + 2f(x)$ (d) $y = f(x - 2) - 2$
(e) $y = f(-x)$ (f) $y = -f(-x)$
(g) $y = -f(x)$ (h) $y = f^{-1}(x)$

84. The graph of f is given. Draw the graph of each of the following functions.

(a) $y = f(x - 2)$ (b) $y = -f(x)$
(c) $y = 3 - f(x)$ (d) $y = \frac{1}{2}f(x) - 1$
(e) $y = f^{-1}(x)$ (f) $y = f(-x)$

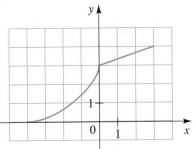

85. Determine whether f is even, odd, or neither even nor odd.

(a) $f(x) = 2x^5 - 3x^2 + 2$ (b) $f(x) = x^3 - x^7$

(c) $f(x) = \dfrac{1 - x^2}{1 + x^2}$ (d) $f(x) = \dfrac{1}{x + 2}$

86. Determine whether the function in the figure is even, odd, or neither.

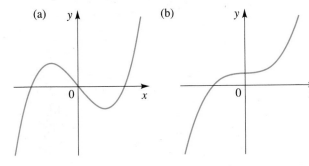

(a)   (b)

(c)

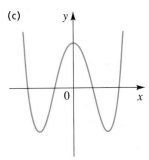

87. Express the quadratic function $f(x) = x^2 + 4x + 1$ in standard form.

88. Express the quadratic function $f(x) = -2x^2 + 12x + 12$ in standard form.

89. Find the minimum value of the function
$g(x) = 2x^2 + 4x - 5$.

90. Find the maximum value of the function
$f(x) = 1 - x - x^2$.

91–92 ■ Find the local maximum and minimum values of the function and the values of x at which they occur. State each answer correct to two decimal places.

91. $f(x) = 3.3 + 1.6x - 2.5x^3$

92. $f(x) = x^{2/3}(6 - x)^{1/3}$

93. If $f(x) = x^2 - 3x + 2$ and $g(x) = 4 - 3x$, find each of the following functions.

(a) $f + g$ (b) $f - g$ (c) fg
(d) f/g (e) $f \circ g$ (f) $g \circ f$

94. If $f(x) = 1 + x^2$ and $g(x) = \sqrt{x - 1}$, find each of the following functions.

(a) $f \circ g$ (b) $g \circ f$
(c) $(f \circ g)(2)$ (d) $(f \circ f)(2)$
(e) $f \circ g \circ f$ (f) $g \circ f \circ g$

95–96 ■ Find the functions $f \circ g$, $g \circ f$, $f \circ f$, and $g \circ g$ and their domains.

95. $f(x) = 3x - 1$, $g(x) = 2x - x^2$

96. $f(x) = \sqrt{x}$, $g(x) = \dfrac{2}{x - 4}$

97. Find $f \circ g \circ h$, where $f(x) = \sqrt{1 - x}$, $g(x) = 1 - x^2$, and $h(x) = 1 + \sqrt{x}$.

98. If $T(x) = \dfrac{1}{\sqrt{1 + \sqrt{x}}}$, find functions f, g, and h such that $f \circ g \circ h = T$.

99–104 ■ Determine whether the function is one-to-one.

99. $f(x) = 3 + x^3$

100. $g(x) = 2 - 2x + x^2$

101. $h(x) = \dfrac{1}{x^4}$

102. $r(x) = 2 + \sqrt{x + 3}$

103. $p(x) = 3.3 + 1.6x - 2.5x^3$

104. $q(x) = 3.3 + 1.6x + 2.5x^3$

105. Show that $f(x) = 3x - 2$ is a one-to-one function and find its inverse function.

106. If $f(x) = 1 + \sqrt[5]{x - 2}$, find f^{-1}.

107. (a) Sketch the graph of the function

$$f(x) = x^2 - 4, \quad x \geq 0$$

  (b) Use part (a) to sketch the graph of f^{-1}.
  (c) Find an equation for f^{-1}.

108. (a) Show that the function $f(x) = 1 + \sqrt[4]{x}$ is one-to-one.
  (b) Sketch the graph of f.
  (c) Use part (b) to sketch the graph of f^{-1}.
  (d) Find an equation for f^{-1}.

109. Suppose that M varies directly as z, and $M = 120$ when $z = 15$. Write an equation that expresses this variation.

110. Suppose that z is inversely proportional to y. If $y = 16$, then $z = 12$. Use this information to write an equation that expresses z in terms of y.

111. The intensity of illumination I from a light varies inversely as the square of the distance d from the light.
  (a) Write this statement as an equation.
  (b) Determine the constant of proportionality if it is known that a lamp has an intensity of 1000 candles at a distance of 8 m.
  (c) What is the intensity of this lamp at a distance of 20 m?

112. The number of Christmas cards sold by a greeting-card store depends on the time of year. Sketch a rough graph of the number of Christmas cards sold as a function of the time of year.

113. An isosceles triangle has a perimeter of 8 cm. Express the area A of the triangle as a function of the length b of the base of the triangle.

114. A rectangle is inscribed in an equilateral triangle with a perimeter of 30 cm as in the figure.
  (a) Express the area A of the rectangle as a function of the width x of the rectangle.
  (b) Find the dimensions of the rectangle with the largest area.

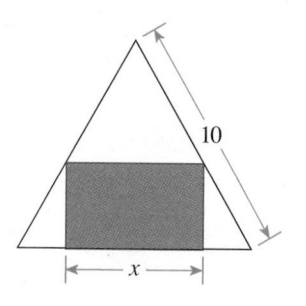

115. A piece of wire 10 m long is cut into two pieces. One piece, of length x, is bent into the shape of a square. The other piece is bent into the shape of an equilateral triangle.
  (a) Express the total area enclosed as a function of x.
  (b) For what value of x is this total area a minimum?

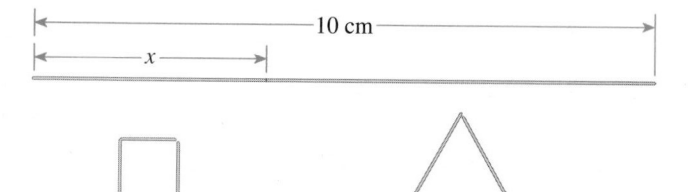

116. A baseball team plays in a stadium that holds 55,000 spectators. With the ticket price set at $10, the average attendance at recent games has been 27,000. A market survey indicates that for every dollar that the ticket price is lowered, attendance will increase by 3000.

How should the ticket price be set to maximize revenue?

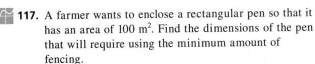

117. A farmer wants to enclose a rectangular pen so that it has an area of 100 m². Find the dimensions of the pen that will require using the minimum amount of fencing.

118. A bird is released from point A on an island, 5 mi from the nearest point B on a straight shoreline. The bird flies to a point C on the shoreline, and then flies along the shoreline to its nesting area D (see the figure at the right). Suppose the bird requires 10 kcal/mi of energy to fly over land and 14 kcal/mi to fly over water (see Example 7 in Section 2.5). If the bird instinctively chooses a path that minimizes its energy expenditure, to what point does it fly?

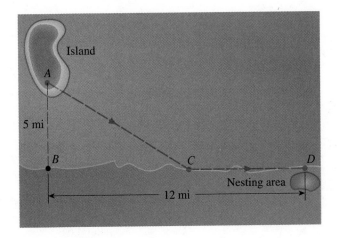

1. Let $A(-7, 4)$ and $B(5, -12)$ be points in the plane.
   (a) Find the length of the segment AB.
   (b) Find the midpoint of the segment AB.
   (c) Find an equation of the line that passes through the points A and B.
   (d) Find the equation of the perpendicular bisector of AB.
   (e) Find an equation of the circle for which AB is a diameter.

2. Sketch the graph of each equation.
   (a) $x^2 + y^2 - 6x + 10y + 9 = 0$
   (b) $2x^2 + 2y^2 + 6x + 10y + 17 = 0$

3. Find an equation for the line with the given property.
   (a) Passing through $(-2, 3)$ and parallel to the line $2x + 6y = 17$
   (b) Having x-intercept -3 and y-intercept 12

4. Sketch the graph of each equation.
   (a) $-3x + 4y = 24$
   (b) $x = y^2 - 1$

5. Which of the following are graphs of functions? If the graph is that of a function, is it one-to-one?

   (a)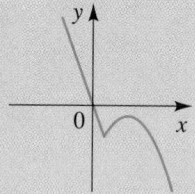

   (b)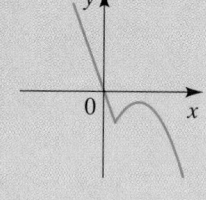

   (c)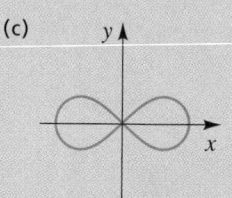

   (d)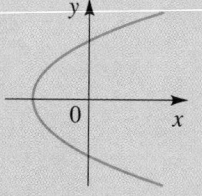

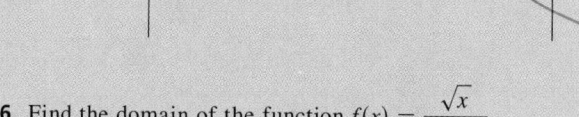
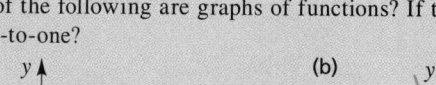

6. Find the domain of the function $f(x) = \dfrac{\sqrt{x}}{x - 1}$.

7. (a) Sketch the graph of the function $f(x) = x^3$.
   (b) Use part (a) to graph the function $g(x) = (x - 1)^3 - 2$.

8. (a) How is the graph of $y = 2 - f(x + 3)$ obtained from the graph of f?
   (b) How is the graph of $y = -f(-x)$ obtained from the graph of f?

9. (a) Sketch the graph of the function $f(x) = 2x^2 - 8x + 13$.
   (b) What is the minimum value of f?

10. Let $f(x) = \begin{cases} 1 - x^2 & \text{if } x \leq 0 \\ 2x + 1 & \text{if } x > 0 \end{cases}$

   (a) Evaluate $f(-2)$ and $f(1)$.
   (b) Sketch the graph of f.

11. If $f(x) = x^2 + 2x - 1$ and $g(x) = 2x - 3$, find each of the following.
   (a) $f \circ g$ (b) $g \circ f$
   (c) $f(g(2))$ (d) $g(f(2))$
   (e) $g \circ g \circ g$

12. (a) If $f(x) = \sqrt{3 - x}$, find the inverse function f^{-1}.
   (b) Sketch the graphs of f and f^{-1} on the same coordinate axes.

13. (a) If 800 ft of fencing is available to build three adjacent pens, as shown in the diagram, express the total area of the pens as a function of x.
   (b) What value of x will maximize the total area?

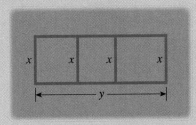

14. Let $f(x) = 3x^4 - 14x^2 + 5x - 3$.
   (a) Draw the graph of f in an appropriate viewing rectangle.
   (b) Is f one-to-one?
   (c) Find the local maximum and minimum values of f and the values of x at which they occur. State each answer correct to two decimal places.
   (d) Find the approximate range of f.
   (e) Find the approximate intervals on which f is increasing and on which f is decreasing.

FOCUS ON PROBLEM SOLVING

One of the most important principles of problem solving is the **recognition of patterns.** If we can discern that a numerical pattern or a geometrical pattern occurs in a problem, then we may be able to guess what the continuing pattern is and thereby solve the problem.

This principle is often combined with the principle of **analogy.** If we are able to solve the simpler problems obtained by taking special cases or by using smaller numbers, we can often see that they lead to a pattern, which, in turn, gives us the answer to the original problem.

PROBLEM 1 ■ Finding Patterns in Powers of an Integer

Find the final digit in the number 3^{459}.

SOLUTION

Analogy

First notice that 3^{459} is a very large number—far too large for a calculator. Therefore, we attack this problem by first looking at analogous problems. A similar, but simpler, problem would be to find the final digit in 3^9 or 3^{59}. In fact, let us start with the exponents 1, 2, 3, ... and see what happens.

Number	Final Digit
3^1	3
3^2	9
3^3	7
3^4	1
3^5	3
3^6	9
3^7	7
3^8	1

Pattern

By now you can see a pattern. The final digits repeat in a cycle with length 4: 3, 9, 7, 1, 3, 9, 7, 1, 3, 9, 7, 1, ... Which number occurs in the 459th position? If we divide 459 by 4, the remainder is 3. So the final digit is the third number in the cycle, namely, 7.

The final digit in the number 3^{459} is 7. ■

PROBLEM 2 ■ Finding Patterns in Compositions of a Function

If $f_0(x) = 1/(1 - x)$ and $f_{n+1} = f_0 \circ f_n$ for $n = 0, 1, 2, \ldots$, find $f_{1000}(1000)$.

Analogy

Instead of computing $f_{1000}(x)$ directly, we start with the easier problem of computing $f_1(x)$, $f_2(x)$, and so on. Putting $n = 0$ in the equation

$$f_{n+1}(x) = f_0(f_n(x))$$

we get

$$f_1(x) = f_0(f_0(x)) = f_0\left(\frac{1}{1-x}\right)$$

$$= \frac{1}{1 - \dfrac{1}{1-x}} = \frac{1}{\dfrac{1-x-1}{1-x}} = \frac{1}{\dfrac{-x}{1-x}}$$

$$= \frac{1-x}{-x} = 1 - \frac{1}{x}$$

Next we use $n = 1$:

$$f_2(x) = f_0(f_1(x)) = f_0\left(1 - \frac{1}{x}\right)$$

$$= \frac{1}{1 - \left(1 - \dfrac{1}{x}\right)} = \frac{1}{\dfrac{1}{x}} = x$$

Now let $n = 2$:

$$f_3(x) = f_0(f_2(x)) = f_0(x) = \frac{1}{1-x}$$

Pattern

Thus, we have returned to the function we started with, and we see a pattern:

$$f_0(x) = \frac{1}{1-x} \qquad f_1(x) = 1 - \frac{1}{x} \qquad f_2(x) = x$$

$$f_3(x) = \frac{1}{1-x} \qquad f_4(x) = 1 - \frac{1}{x} \qquad f_5(x) = x$$

The pattern is that successive functions repeat in a cycle of length 3.

Where does $f_{1000}(x)$ occur in the cycle? When we divide 1000 by 3, we get a remainder of 1, so

$$f_{1000}(x) = f_1(x) = 1 - \frac{1}{x}$$

Therefore

$$f_{1000}(1000) = 1 - \frac{1}{1000} = \frac{999}{1000}$$

PROBLEMS

1. Find the final digit in the number 947^{362}.

2. How many digits does the number $8^{15} \cdot 5^{37}$ have?

3. If $f_0(x) = x^2$ and $f_{n+1}(x) = f_0(f_n(x))$ for $n = 0, 1, 2, \ldots$, find a formula for $f_n(x)$.

4. If $f_0(x) = \dfrac{1}{2-x}$ and $f_{n+1} = f_0 \circ f_n$ for $n = 0, 1, 2, \ldots$, find $f_{100}(3)$.

5. Use the techniques of solving a simpler problem and looking for a pattern to evaluate the number

$$3999999999999^2$$

6. Find the domain of the function

$$f(x) = \sqrt{1 - \sqrt{2 - \sqrt{3-x}}}$$

7. Sketch the graph of the function $f(x) = \left| x^2 - 4|x| + 3 \right|$.

8. Sketch the graph of the function $g(x) = \left| x^2 - 1 \right| - \left| x^2 - 4 \right|$.

9. The **greatest integer function** is defined by $[\![x]\!]$ = the largest integer that is less than or equal to x. (For instance, $[\![4]\!] = 4$, $[\![4.8]\!] = 4$, $[\![\pi]\!] = 3$, $[\![-\frac{1}{2}]\!] = -1$.) Sketch the graph of the greatest integer function.

10. Sketch the graph of the function $f(x) = x - [\![x]\!]$, where $[\![x]\!]$ is the greatest integer function defined in Problem 9.

11. Sketch the region in the plane defined by the equation

$$[\![x]\!]^2 + [\![y]\!]^2 = 1$$

where $[\![x]\!]$ denotes the greatest integer function defined in Problem 9.

12. The positive integers are written in order starting with 1:

$$1234567891011121314151617181920 21 \ldots$$

What digit is in the 300,000th position? [*Hint:* Using analogy, start with the simpler problem of finding the digit in the 80th position. Then find the digit in the 800th position. Armed with the clues from solving these simpler problems, tackle the given problem.]

Srinivasa Ramanujan (1887–1920) was born into a poor family in the small town of Kumbakonam in India. He was self-taught in mathematics and worked in virtual isolation from other mathematicians. At the age of 25 he wrote a letter to G. H. Hardy, the leading British mathematician at the time, listing some of his discoveries. Hardy immediately recognized Ramanujan's genius and for the next six years the two worked together in London until Ramanujan fell ill and went back to his hometown in India, where he died a year later. Ramanujan

(continued)

was a genius with phenomenal ability to see hidden patterns in the properties of numbers. Most of his discoveries were written as complicated infinite series, the importance of which was not recognized until many years after his death. In the last year of his life he wrote 130 pages of mysterious formulas, many of which still defy proof. Hardy tells the story that when he visited Ramanujan in a hospital and arrived in a taxi, he remarked to Ramanujan that the cab's number, 1729, was uninteresting. Ramanujan replied "No, it is a very interesting number. It is the smallest number expressible as the sum of two cubes in two different ways." (See Problem 13.)

13. The number 1729 is the smallest positive integer that can be represented in two different ways as the sum of two cubes. What are the two different ways?

14. Suppose that each point in the coordinate plane is colored either red or blue. Show that there must always be two points of the same color that are exactly one unit apart.

15. Suppose that each point (x, y) in the plane, both of whose coordinates are rational numbers, represents a tree. If you are standing at the point $(0, 0)$, how far could you see in this forest?

16. A thousand points are graphed in the coordinate plane. Explain why it is possible to draw a straight line in the plane so that half of the points are on one side of the line and half are on the other.   [*Hint:* Consider the slopes of the lines determined by each *pair* of points.]

17. A piece of wire is bent as shown in the figure. You can see that one cut through the wire produces four pieces and two parallel cuts produce seven pieces. How many pieces will be produced by 142 parallel cuts? Write a formula for the number of pieces produced by n parallel cuts.

18. Draw the graph of the equation $|x| + |y| = 1 + |xy|$.

19. Sketch the region in the plane consisting of all points (x, y) such that
$$|x| + |y| \leq 1$$

20. Sketch the region in the plane consisting of all points (x, y) such that
$$|x - y| + |x| - |y| \leq 2$$

21. Draw the graph of the equation
$$x^2 y - y^3 - 5x^2 + 5y^2 = 0$$

4

POLYNOMIALS AND RATIONAL FUNCTIONS

In applications of mathematics, many situations can be modeled by polynomial functions. For example, the volume of a silo of fixed height is a polynomial function of its radius.

Each problem that I solved became a rule which served afterwards to solve other problems.

RENÉ DESCARTES

We have previously studied constant, linear, and quadratic functions, which have the equations $f(x) = c$, $f(x) = mx + b$, and $f(x) = ax^2 + bx + c$, respectively. All these functions are special cases of an important class of functions called polynomials. A **polynomial P of degree** n is a function of the form

$$P(x) = a_nx^n + a_{n-1}x^{n-1} + \cdots + a_1x + a_0$$

where $a_n \neq 0$. Polynomials are constructed using the operations of addition, subtraction, and multiplication. If we introduce division as well, we obtain the set of rational functions. A **rational function** r is a function of the form

$$r(x) = \frac{P(x)}{Q(x)}$$

where P and Q are polynomials.

Virtually all the functions used in mathematics and the sciences are evaluated numerically by using polynomial approximations. In this chapter we study this important class of functions by first learning about the graphs of polynomial functions. We then learn how to find rational, irrational, and imaginary solutions of polynomial equations. Finally, we study rational functions and their graphs.

Working with polynomials and rational functions has numerical, algebraic, and graphical aspects, and we use all three approaches to give us a deeper understanding of these functions.

4.1 POLYNOMIAL FUNCTIONS AND THEIR GRAPHS

Before we learn to work with polynomials, we must agree on some terminology.

POLYNOMIAL FUNCTIONS

The function

$$P(x) = a_nx^n + a_{n-1}x^{n-1} + \cdots + a_1x + a_0$$

where $a_n \neq 0$, is a **polynomial of degree** n. The numbers $a_0, a_1, a_2, \ldots, a_n$ are called the **coefficients** of the polynomial. The number a_0 is the **constant coefficient** and the number a_n, the coefficient of the highest power, is the **leading coefficient.**

If a polynomial consists of just a single term, then it is called a **monomial.** For example, $P(x) = x^3$ and $Q(x) = -6x^5$ are monomials.

The graphs of polynomials of degree 0 or 1 are lines, and the graphs of polynomials of degree 2 are parabolas. Accurate graphing techniques for these polynomials were considered in Chapter 3. The greater the degree of a polynomial, the more complicated its graph; to draw accurately the graph of a polynomial function of degree 3 or more requires the techniques of calculus or the use of a graphing calculator or computer (see Section 3.3). In this section we study the general features of polynomial graphs. We will see that the graph of every polynomial is a smooth, continuous curve with no gap or jump. We begin by graphing the simplest polynomial functions, those of the form $y = x^n$.

EXAMPLE 1 ■ Graphs of Simple Monomials

Graph each of the following functions.

(a) $y = x$ (b) $y = x^2$ (c) $y = x^3$ (d) $y = x^4$ (e) $y = x^5$

SOLUTION

We are already familiar from our previous work with some of these graphs but we include them for completeness.

x	$y = x$	$y = x^2$	$y = x^3$	$y = x^4$	$y = x^5$
0.1	0.1	0.01	0.001	0.0001	0.00001
0.2	0.2	0.04	0.008	0.0016	0.00032
0.5	0.5	0.25	0.125	0.0625	0.03125
0.7	0.7	0.49	0.343	0.2401	0.16807
1.0	1.0	1.0	1.0	1.0	1.0
1.2	1.2	1.44	1.728	2.0736	2.48832
1.5	1.5	2.25	3.375	5.0625	7.59375
2.0	2.0	4.0	8.0	16.0	32.0

If n is odd, then $(-x)^n = -x^n$, so $y = x^n$ is an odd function. If n is even, then $(-x)^n = x^n$, so $y = x^n$ is an even function (see Section 3.8). We use these facts to find the values for negative x from the table. Plotting the points leads to the graphs in Figure 1.

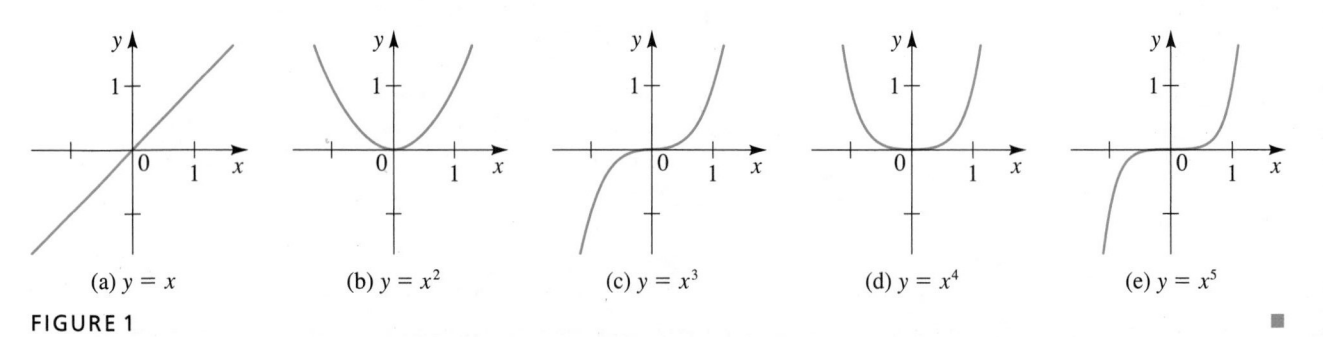

(a) $y = x$ (b) $y = x^2$ (c) $y = x^3$ (d) $y = x^4$ (e) $y = x^5$

FIGURE 1   ■

As Example 1 suggests, when n is odd, the graph of $y = x^n$ has the same general shape as $y = x^3$, and when n is even, the graph of $y = x^n$ has more or less the same U-shape as $y = x^2$. However, note that as the degree n becomes larger, the graphs become flatter around the origin and steeper elsewhere.

EXAMPLE 2 ■ Transformations of Monomials

Sketch the graph of each of the following functions.

(a) $y = -x^3$ (b) $y = (x - 2)^4$ (c) $y = -2x^5 + 4$

SOLUTION

We use the graphs in Example 1 and transform them using the techniques of Section 3.8.

(a) The function $y = -x^3$ is the negative of $y = x^3$, so we simply reflect the graph in Figure l(c) in the x-axis to obtain Figure 2.

(b) The graph of the function $y = (x - 2)^4$ has the same shape as $y = x^4$. Replacing x by $x - 2$ shifts the graph of $y = x^4$ in Figure l(d) to the right 2 units (see Figure 3).

(c) We begin with the graph of $y = x^5$ in Figure l(e). Multiplying the function by 2 stretches the graph vertically. The negative sign reflects the graph about the x-axis, so at this stage we have the graph in Figure 4. Finally, adding 4 to the function shifts the graph upward 4 units (see Figure 5). Since $-2x^5 + 4 = 0$ when $x^5 = 2$, the graph crosses the x-axis at $x = \sqrt[5]{2}$.

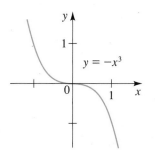

FIGURE 2

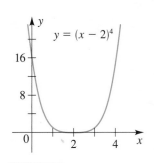

FIGURE 3

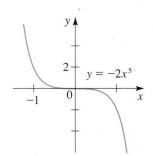

FIGURE 4

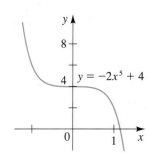

FIGURE 5

If $y = P(x)$ is a polynomial function and if c is a number such that $P(c) = 0$, then we say that c is a **zero** of P. In other words, the zeros of P are the solutions of the polynomial equation $P(x) = 0$. In the next example we use the zeros of the polynomial to help us sketch its graph. Note that if $P(c) = 0$, then the graph of $y = P(x)$ touches the x-axis where $x = c$; that is, c is an x-intercept of the graph. Between any two successive zeros, the values of the polynomial are either all positive or all negative. Thus, the graph will lie entirely above or entirely below the x-axis between any two successive zeros.

Sir Isaac Newton (1642–1727) continues to be universally regarded, more than 265 years after his death, as one of the giants of physics and mathematics. He is most well known for discovering the laws of motion and gravity and for inventing the calculus, but his contributions also include the binomial theorem, the laws of optics, and methods for solving polynomial equations to any desired accuracy. He was born on Christmas Day, a sickly infant whose father had died a few months before. After an unhappy childhood, he entered Cambridge University, where he learned mathematics by studying the writings of Euclid and Descartes.

During the plague years of 1665 and 1666 the university was closed, and Newton spent his time thinking and writing about his ideas that were later to revolutionize the sciences. Because of a pathological fear of criticism, he did not publish these writings until much later, partly as a result of the encouragement of Edmund Halley (who discovered the now-famous comet). After his
(continued)

EXAMPLE 3 ■ Graph of a Third-Degree Polynomial

Sketch the graph of the function $P(x) = x^3 - x^2 - 4x + 4$.

SOLUTION

We first find the zeros of P by factoring.

$$
\begin{aligned}
P(x) &= x^3 - x^2 - 4x + 4 \\
&= x^2(x - 1) - 4(x - 1) && \text{Group terms and factor} \\
&= (x^2 - 4)(x - 1) && \text{Factor out } x - 1 \\
&= (x + 2)(x - 2)(x - 1) && \text{Difference of squares}
\end{aligned}
$$

Thus, $P(x) = 0$ when $x + 2 = 0$, $x - 2 = 0$, and when $x - 1 = 0$. It follows that the zeros of P are -2, 2, and 1, so these are the x-intercepts of its graph. Since $P(0) = 4$, the y-intercept is 4.

To sketch the graph, we must calculate $P(x)$ for other values of x as well. We choose values between (and to the right and left of) successive zeros to determine whether $P(x)$ is positive or negative on each of the intervals determined by the zeros. If $P(x)$ is positive at any x between successive zeros, then the graph of $y = P(x)$ lies above the x-axis on the interval between those zeros, and if $P(x)$ is negative on such an interval, then the graph lies below the x-axis on that interval.

x	$P(x)$
-3	-20
-2	0
-1	6
0	4
1	0
$\frac{3}{2}$	$-\frac{7}{8}$
2	0
3	10

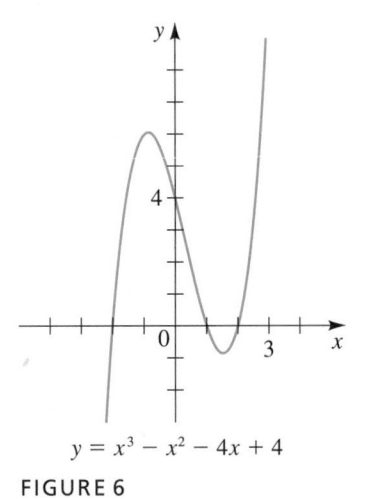

$$y = x^3 - x^2 - 4x + 4$$

FIGURE 6

Plotting the points given in the table and completing the sketch gives us the graph in Figure 6.   ■

The graph in Example 3 has two **turning points** (or **local extrema**), which are points at which a particle moving along the graph would change from moving upward to moving downward, or vice versa. The turning points are thus "peaks"

work was published, it brought Newton enormous fame and prestige in his own lifetime. Even the poets were moved to express praise for Newton. Alexander Pope wrote:

Nature and Nature's Laws
   lay hid in Night,
God said, "Let Newton be"
   and all was Light.

Newton himself was far more modest about his accomplishments. He said, "I seem to have been only like a boy playing on the seashore...while the great ocean of truth lay all undiscovered before me." Newton was knighted by Queen Anne in 1705 and was buried with great honor in Westminster Abbey.

and "valleys" on the graph. Without using calculus we cannot find the exact locations of such points. We cannot even prove that we have in fact found all the turning points that a graph may have, but it can be shown (again, using calculus) that a polynomial function of degree n can have at most $n - 1$ turning points (although it may have fewer than $n - 1$). But by plotting sufficiently many points, we can locate them approximately, as we did in Example 3. In Section 4.2 we use a graphing calculator to find the approximate locations of the turning points of a polynomial.

EXAMPLE 4 ■ Graph of a Fourth-Degree Polynomial

Sketch the graph of the function $P(x) = -2x^4 - x^3 + 3x^2$.

SOLUTION

We factor to obtain

$$P(x) = -x^2(2x^2 + x - 3)$$
$$= -x^2(2x + 3)(x - 1)$$

so the zeros of P are 0, $-\frac{3}{2}$, and 1. The y-intercept is $P(0) = 0$. In the following table we give the values of P at a number of other points. (Such values are most easily obtained using a programmable calculator.) Remember that we must choose points between successive zeros to determine whether the polynomial is positive or negative between these zeros.

x	$P(x)$
-2	-12
-1.5	0
-1	2
-0.5	0.75
0	0
0.5	0.5
1	0
1.5	-6.75

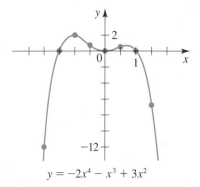

$$y = -2x^4 - x^3 + 3x^2$$

FIGURE 7

Plotting these points and completing the graph, we obtain the curve shown in Figure 7. The graph cannot have any more turning points than the three we have found, since the degree of P is 4.

Note that although $x = 0$ is a zero of the polynomial, the graph does not cross the x-axis at the x-intercept 0. It just touches the x-axis there and then moves upward again. We say that the graph is *tangent* to the x-axis at $x = 0$. ■

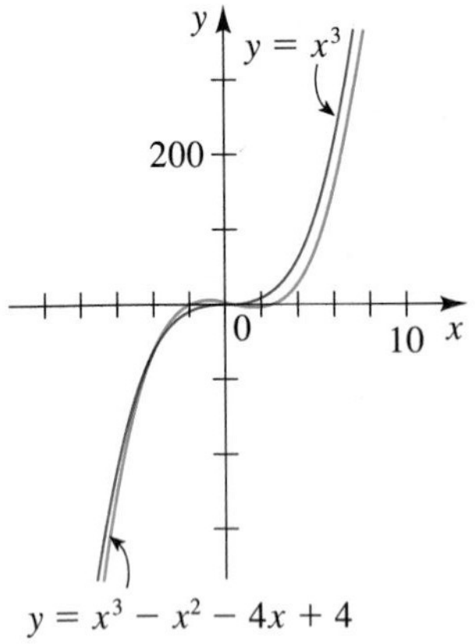

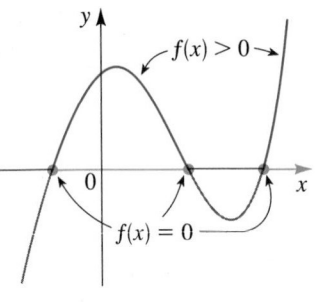

$y = x^3 - x^2 - 4x + 4$

FIGURE 8

As a further aid in graphing polynomials, note that for large $|x|$, the values of

$$a_n x^n + a_{n-1} x^{n-1} + \cdots + a_1 x + a_0$$

are close to $a_n x^n$ because

$$a_n x^n + a_{n-1} x^{n-1} + \cdots + a_1 x + a_0 = a_n x^n \left[1 + \frac{a_{n-1}}{a_n x} + \frac{a_{n-2}}{a_n x^2} + \cdots + \frac{a_0}{a_n x^n} \right]$$

When $|x|$ is large the quantity inside the brackets is close to 1 in value because the reciprocals of large numbers are close to 0. For example, for large $|x|$ the graph of $P(x) = x^3 - x^2 - 4x + 4$ (sketched in Figure 6) looks very much like the graph of $y = x^3$ (see Figure 8). Of course, when $|x|$ is small (say, $-5 < x < 5$), the two graphs are quite different, since $y = P(x)$ crosses the x-axis three times, at $x = -2, 1,$ and 2, whereas $y = x^3$ crosses the x-axis only at the origin.

Figure 7 in Section 4.2 shows the general features of the graphs of polynomials of even and odd degree.

INEQUALITIES

In Chapter 2 we learned how to solve equations and inequalities *algebraically* by using factoring. But any equation of the form $f(x) = 0$ or any inequality, such as $f(x) > 0$, can be solved by analyzing the graph of $y = f(x)$. As we see in Figure 9, the solution of the *equation* $f(x) = 0$ is the set of x-intercepts of the graph of f. The solution of the *inequality* $f(x) > 0$ is the set of all values of x for which $f(x)$ is positive, that is, the set of all x for which the point $(x, f(x))$ lies above the x-axis. The same idea can also be used to solve inequalities of the forms $f(x) \geq 0, f(x) < 0,$ and $f(x) \leq 0$. In the next example, we apply this principle to a polynomial inequality.

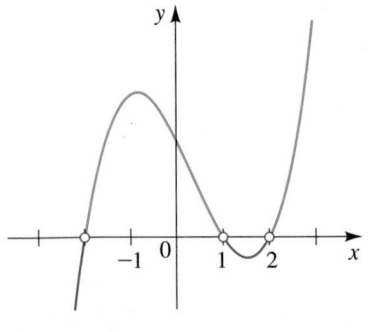

FIGURE 9
$y = f(x)$

$y = x^3 - x^2 - 4x + 4 > 0$

FIGURE 10
$x^3 - x^2 - 4x + 4 > 0$

EXAMPLE 5 ■ Solving an Inequality Graphically

Solve the inequality $x^3 + 4 > x^2 + 4x$.

SOLUTION

After moving all terms in the inequality to the left side, we get

$$x^3 - x^2 - 4x + 4 > 0$$

The polynomial in this inequality is the function of Example 3, which was graphed in Figure 6. The solution consists of all values x at which the corresponding point on the graph of the polynomial lies above the x-axis. We see from Figure 10 that this consists of the set

$$(-2, 1) \cup (2, \infty)$$ ■

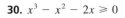

4.1 EXERCISES

1–8 ■ Sketch the graph of the function by transforming the graph of an appropriate function of the form $y = x^n$. Indicate all x- and y-intercepts on each graph.

1. $y = x^3 - 8$

2. $y = (x - 2)^3$

3. $y = -x^4 + 16$

4. $y = -2(x - 1)^3$

5. $y = -(x - 1)^4 + 1$

6. $y = 3x^5 - 9$

7. $y = 4(x - 2)^5 - 4$

8. $y = 3x^4 - 27$

9–26 ■ Sketch the graph of the function by first plotting all x-intercepts, the y-intercept, and sufficiently many other points to detect the shape of the curve, and then filling in the rest of the graph.

9. $y = (x - 3)(x + 1)$

10. $y = (x - 1)(x + 1)(x - 2)$

11. $y = x(x - 2)(x + 1)$

12. $y = \frac{1}{5}x(x - 5)^2$

13. $y = (x - 1)^2(x - 3)$

14. $y = \frac{1}{4}(x + 1)^3(x - 3)$

15. $y = \frac{1}{12}(x + 2)^2(x - 3)^2$

16. $y = x^3 + x^2 - x - 1$

17. $y = x^3 + 3x^2 - 4x - 12$

18. $y = 2x^3 - x^2 - 18x + 9$

19. $y = x^3 - x^2 - 6x$

20. $y = (x^2 - 2x - 3)^2$

21. $y = \frac{1}{8}(2x^4 + 3x^3 - 16x - 24)$

22. $y = x^4 - 3x^2 - 4$

23. $y = x^4 - 2x^3 - 8x + 16$

24. $y = x^4 - 2x^3 + 8x - 16$

25. $y = x^5 - 9x^3$

26. $y = x^6 - 2x^3 + 1$

27–30 ■ The polynomial that appears in the inequality is graphed in the figure. Use the graph to solve the inequality.

27. $x^3 - 4x > 0$

28. $x^4 - 16x^2 < 0$

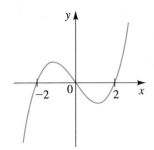

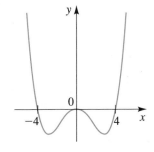

29. $x^3 + 3x^2 \le 0$

30. $x^3 - x^2 - 2x \ge 0$

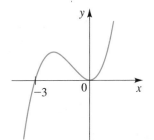

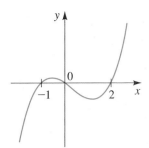

31. (a) On the same coordinate axes, sketch graphs (as accurately as possible) of the functions

$$y = x^3 - 2x^2 - x + 2 \quad \text{and} \quad y = -x^2 + 5x + 2$$

(b) Based on your sketch in part (a), at how many points do the two graphs appear to intersect?

(c) Find the coordinates of all the intersection points.

32. Portions of the graphs of $y = x^2$, $y = x^3$, $y = x^4$, $y = x^5$, and $y = x^6$ are plotted in the figures. Determine which function belongs to each graph.

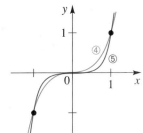

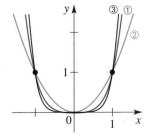

33. Recall that a function f is *odd* if $f(-x) = -f(x)$ or *even* if $f(-x) = f(x)$ for all real x.

(a) Show that if f and g are both odd, then so is the function $f + g$.

(b) Show that if f and g are both even, then so is the function $f + g$.

(c) Show that if f is odd and g is even, and neither has constant value 0, then the function $f + g$ is neither even nor odd.

(d) Show that a polynomial $P(x)$ that contains only odd powers of x is an odd function.

(e) Show that a polynomial $P(x)$ that contains only even powers of x is an even function.

(f) Show that if a polynomial $P(x)$ contains both odd and even powers of x, then it is neither an odd nor an even function.

(g) Express the function

$$P(x) = x^5 + 6x^3 - x^2 - 2x + 5$$

as the sum of an odd function and an even function.

34. On a cold winter night, the temperature in Parksville was given by the formula $T(x) = x^3 - 8x^2 + 15x$, where $T(x)$ is the temperature in °F at time x (measured in hours from midnight) with $0 \leq x \leq 6$.

(a) At what times between midnight and 6:00 A.M. (inclusive) was the temperature exactly $0\,°F$?

(b) At what times in this interval was the temperature below $0\,°F$?

(c) Sketch a graph of the temperature between midnight and 6:00 A.M.

35. A cardboard box has a square base, with each of the four edges of the base having length x inches, as shown in the figure. The total length of all 12 edges of the box is 144 inches.

(a) Express the volume V of the box as a function of x.

(b) Sketch the graph of the function V.

(c) Since both x and V represent positive quantities (length and volume, respectively), what is the domain of V?

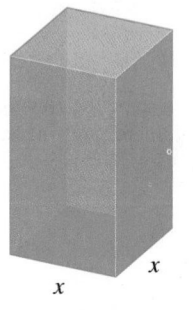

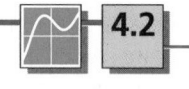 **4.2 USING GRAPHING DEVICES TO GRAPH POLYNOMIALS**

In the preceding section we studied some basic properties of the graphs of polynomial functions. In this section we use graphing devices to explore more features of their graphs.

Recall from Section 3.9 that if the point $(a, f(a))$ is the highest point on the graph of f within some viewing rectangle, then $f(a)$ is a local maximum value of f, and if $(b, f(b))$ is the lowest point on the graph of f within a viewing rectangle, then $f(b)$ is a local minimum value (see Figure 1). We say that such a point $(a, f(a))$ is a **local maximum point** on the graph and that $(b, f(b))$ is a **local minimum point.** The set of all local maximum and minimum points on the graph of a function are called its **local extrema.**

FIGURE 1

EXAMPLE 1 ■ Graphing a Third-Degree Polynomial

Graph the polynomial function

$$P(x) = x^3 - 3x^2 - x + 15$$

Determine the *x*- and *y*-intercepts and the coordinates of the local extrema of this graph.

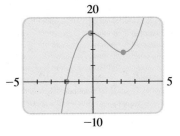

FIGURE 2
$P(x) = x^3 - 3x^2 - x + 15$

SOLUTION

By setting $x = 0$, we see that the *y*-intercept of the graph is 15, so we must choose a viewing rectangle that extends at least this far upward. The graph of *P* in the viewing rectangle $[-5, 5]$ by $[-10, 20]$ is shown in Figure 2. The graph has one *x*-intercept, $x \approx -1.86$, and two extrema, a local maximum and a local minimum. By zooming in on each of these and tracing along the graph with the cursor (as described in Section 3.7), we find that the local maximum point is $(-0.16, 15.08)$ and the local minimum point is $(2.15, 8.92)$, rounded to two decimal places. ■

EXAMPLE 2 ■ Polynomials of Fourth and Fifth Degree

Using the viewing rectangle $[-5, 5]$ by $[-100, 100]$, graph each of the following polynomials, and determine how many local extrema each function has.

(a) $P_1(x) = x^4 + x^3 - 16x^2 - 4x + 48$

(b) $P_2(x) = x^5 + 3x^4 - 5x^3 - 15x^2 + 4x - 15$

SOLUTION

The graphs are shown in Figure 3. We see that P_1 has two local minimum points and one local maximum point, for a total of three local extrema. The polynomial P_2 has two local minimum and two local maximum points—a total of four local extrema.

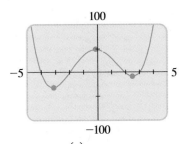
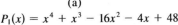
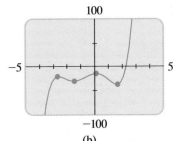

(a)   (b)

FIGURE 3 $P_1(x) = x^4 + x^3 - 16x^2 - 4x + 48$ $P_2(x) = x^5 + 3x^4 - 5x^3 - 15x^2 + 4x - 15$

■

The polynomial P in Example 1 is of degree 3 and its graph has 2 local extrema. In Example 2, P_1 is a fourth-degree polynomial and has 3 local extrema, whereas P_2 is of degree 5 and has 4 local extrema. The number of extrema in each case is one less than the degree. This is not a coincidence, as the principle in the following box indicates. (A proof of this principle requires the methods of calculus.)

LOCAL EXTREMA OF POLYNOMIALS

If $P(x) = a_nx^n + a_{n-1}x^{n-1} + \cdots + a_1x + a_0$ is a polynomial of degree n, then the graph of P has at most $n - 1$ local extrema.

A polynomial of degree n may in fact have less than $n - 1$ local extrema, as we see in the following example. The preceding principle tells us only that a polynomial of degree n can have *no more* than $n - 1$ local extrema.

EXAMPLE 3 ■ Local Extrema of Polynomials

Graph each of the following polynomials using a graphing device, and determine the approximate coordinates of all local extrema.

(a) $F(x) = -2x^3 + 3x^2 - 5x + 2$

(b) $G(x) = 7x^4 + 3x^2 - 10x$

SOLUTION

(a) Using the viewing rectangle $[-5, 5]$ by $[-100, 100]$ gives us the graph of F shown in Figure 4(a). As the graph is plotted, we see that the values of $F(x)$ appear to decline continuously, so the function has no local maximum or minimum. If we select other viewing rectangles, we observe the same behavior, so the function has no local extrema.

(b) In the viewing rectangle $[-5, 5]$ by $[-100, 100]$, the graph of G has the shape shown in Figure 4(b). It appears to have just a single local minimum in quadrant IV. To confirm this, we zoom in on the portion of the graph

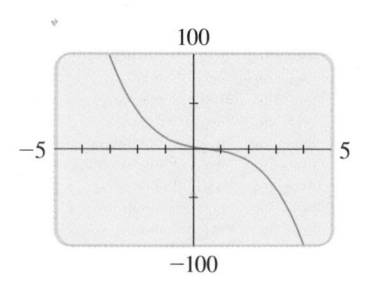

(a) $F(x) = -2x^3 + 3x^2 - 5x + 2$

FIGURE 4

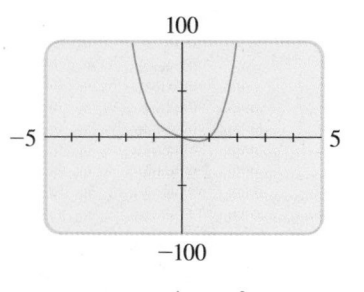

(b) $G(x) = 7x^4 + 3x^2 - 10x$

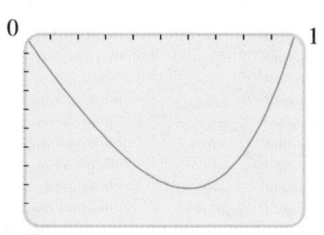
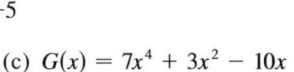

(c) $G(x) = 7x^4 + 3x^2 - 10x$

near this minimum point. The viewing rectangle $[0, 1]$ by $[-5, 0]$ gives us the graph in Figure 4(c). Using the cursor, we locate the local minimum point at $(0.61, -4.01)$, correct to two decimals.   ■

END BEHAVIOR

All the polynomial functions that we have graphed so far have the property that, if the viewing rectangle is wide enough, the graph of the function eventually disappears off the top or bottom of the screen. This means that $|y|$ can be made as large as we wish on the graph of the polynomial $y = P(x)$ if we choose $|x|$ sufficiently large. By the **end behavior** of a function, we mean a description of what happens to the values of the function as $|x|$ becomes large. In the next example we show how end behavior is described for polynomials.

EXAMPLE 4 ■ End Behavior of a Third-Degree Polynomial

Graph the polynomials

$$P(x) = 3x^5 - 5x^3 + 2x \qquad \text{and} \qquad Q(x) = 3x^5$$

on the same screen, first using the viewing rectangle $[-2, 2]$ by $[-2, 2]$ and then changing the rectangle to $[-10, 10]$ by $[-10,000, 10,000]$. Describe and compare the end behavior of the two functions.

SOLUTION

Figure 5(a) shows that the two functions look quite different on the smaller viewing rectangle. The polynomial P has four local extrema, but Q has none. On the large viewing rectangle shown in Figure 5(b), however, the two functions look almost the same. This means that P and Q have the same end behavior. As x becomes large in the positive direction, the y-coordinates of points on the graphs of both P and Q become large in the positive direction as well, and as x becomes large in the negative direction, so does y. In symbols, we write this as follows:

$$y \to \infty \quad \text{as} \quad x \to \infty \qquad \text{and} \qquad y \to -\infty \quad \text{as} \quad x \to -\infty$$

FIGURE 5
$P(x) = 3x^5 - 5x^3 + 2x$
$Q(x) = 3x^5$

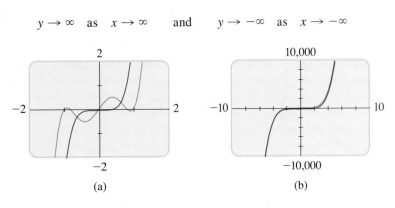

(a)   (b)

■

In Example 4 we saw that the end behavior of a polynomial P is determined by the term containing the highest power of the variable (in this case, the term $3x^5$). This is because when $|x|$ is large, most of the value of $P(x)$ is determined by the highest-power term. For example, substituting $x = 10$ into the polynomials of Example 4 gives

$$P(10) = 3(10)^5 - 5(10)^3 + 2(10) = 300{,}000 - 5000 + 20 = 295{,}020$$

$$Q(10) = 3(10)^5 = 300{,}000$$

The two values are, relatively speaking, very close—they are less than 0.5% apart.

EXAMPLE 5 ■ End Behavior of a Fourth-Degree Polynomial

Describe the end behavior of the polynomial

$$P(x) = -x^4 + 3x^3 - 6x + 100$$

SOLUTION

We graph P on a large viewing rectangle, since we are interested only in end behavior. From Figure 6, we see that

$$y \to -\infty \quad \text{as} \quad x \to \infty \qquad \text{and} \qquad y \to -\infty \quad \text{as} \quad x \to -\infty \qquad ■$$

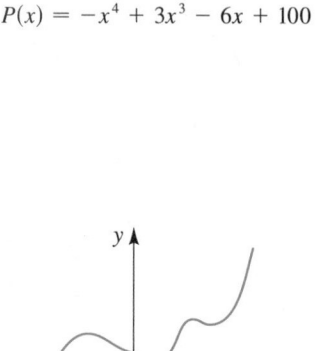

10,000

−20   20

−10,000

FIGURE 6
$P(x) = -x^4 + 3x^3 - 6x + 100$

As Examples 4 and 5 suggest, if $y = P(x)$ is a polynomial function of *odd* degree, the values of y go in opposite directions as $x \to \infty$ and as $x \to -\infty$, but if P is of *even* degree, then the values of y go in the same direction as $x \to \infty$ and as $x \to -\infty$ (see Figure 7).

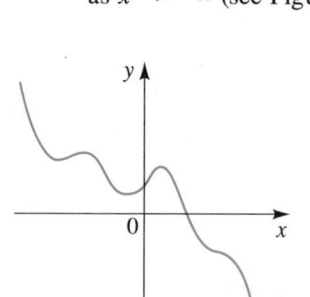
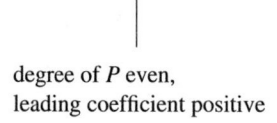
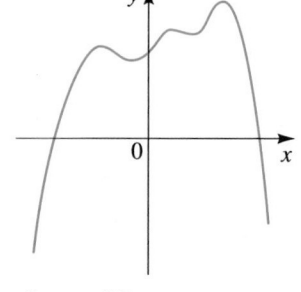

degree of P odd, leading coefficient positive	degree of P odd, leading coefficient negative	degree of P even, leading coefficient positive	degree of P even, leading coefficient negative

FIGURE 7

SOLVING INEQUALITIES

In the next example, we solve an inequality using the graph of a polynomial.

EXAMPLE 6 ■ Solving an Inequality Using a Graphing Calculator

Solve the inequality

$$2x^4 - 6x^3 - 5x^2 - 3x - 3 \leqslant 0$$

SOLUTION

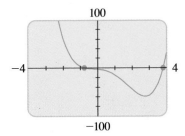

First we graph the polynomial $P(x) = 2x^4 - 6x^3 - 5x^2 - 3x - 3$, as in Figure 8. To solve the inequality $P(x) \leqslant 0$ we must find all values of x for which the graph of the function lies on or below the x-axis. From the graph we see that the solution is the interval that lies between the two x-intercepts. By zooming in we find that the x-intercepts are -0.79 and 3.79 (correct to two decimal places), so the approximate solution of the inequality is the interval $[-0.79, 3.79]$. ■

FIGURE 8
$P(x) = 2x^4 - 6x^3 - 5x^2 - 3x - 3$

FAMILIES OF POLYNOMIALS

A graphing calculator enables us to quickly draw the graphs of many functions at once, on the same viewing screen. This enables us to see how changing a value in the definition of the functions affects the shape of its graph. In the next example we apply this principle to a family of third-degree polynomials.

EXAMPLE 7 ■ A Family of Polynomials

Sketch the family of polynomials $P(x) = x^3 - cx^2$ for $c = 0, 1, 2,$ and 3. How does changing the value of c affect the graph?

SOLUTION

The polynomials

$$P_0(x) = x^3 \qquad\qquad P_1(x) = x^3 - x^2$$

$$P_2(x) = x^3 - 2x^2 \qquad P_3(x) = x^3 - 3x^2$$

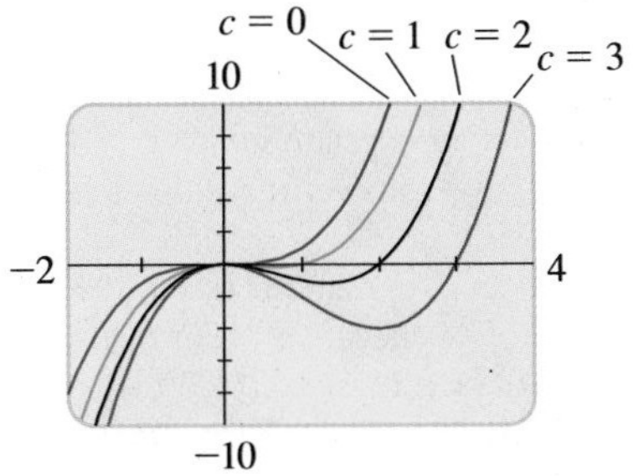

FIGURE 9
$P(x) = x^3 - cx^2$

are graphed in Figure 9. We see that increasing the value of c causes the graph to develop an increasingly deep "valley" to the right of the y-axis, creating a local maximum at the origin and a local minimum at a point in quadrant IV. This local minimum moves lower and further to the right as c increases. ■

 **4.2** **EXERCISES**

1–8 ■ Graph the polynomials in the given viewing rectangle. Find the x-intercept(s), if any, the y-intercept, and the coordinates of all local extrema. State each answer correct to two decimals.

1. $y = -x^2 + 8x$, $[-4, 12]$ by $[-50, 30]$

2. $y = x^3 - 3x^2$, $[-2, 5]$ by $[-10, 10]$

3. $y = x^3 - 12x + 9$, $[-5, 5]$ by $[-30, 30]$

4. $y = 2x^3 - 3x^2 - 12x - 32$, $[-5, 5]$ by $[-60, 30]$

5. $y = x^4 + 4x^3$, $[-5, 5]$ by $[-30, 30]$

6. $y = x^4 - 18x^2 + 32$, $[-5, 5]$ by $[-100, 100]$

7. $y = 3x^5 - 5x^3 + 3$, $[-3, 3]$ by $[-5, 10]$

8. $y = x^5 - 5x^2 + 6$, $[-3, 3]$ by $[-5, 10]$

9–20 ■ Find all local extrema of the polynomial, correct to two decimals.

9. $y = -2x^2 + 3x + 5$

10. $y = x^3 + 12x$

11. $y = x^3 - x^2 - x$

12. $y = 6x^3 + 3x + 1$

13. $y = x^4 - 5x^2 + 4$

14. $y = 1.2x^5 + 3.75x^4 - 7x^3 - 15x^2 + 18x$

15. $y = (x - 2)^5 + 32$

16. $y = (x^2 - 2)^3$

17. $y = (x - 4)(2x + 1)^2$

18. $y = (x + 2)^2(x - 3)^2$

19. $y = x^8 - 3x^4 + x$

20. $y = \frac{1}{3}x^7 - 17x^2 + 7$

21–28 ■ Graph the polynomial and describe its end behavior.

21. $y = 3x^3 - x^2 + 5x + 1$

22. $y = -\frac{1}{8}x^3 + \frac{1}{4}x^2 + 12x$

23. $y = x^4 - 7x^2 + 5x + 5$

24. $y = (1 - x)^5$

25. $y = x^{11} - 9x^9$

26. $y = 2x^2 - x^{12}$

27. $y = 200x^2 - 0.001x^5$

28. $y = x^{20}$

29–36 ■ Solve the inequality and express the solution in interval notation.

29. $3x^2 + 6x - 14 < 0$

30. $x^3 + x^2 + x + 1 \geqslant 0$

31. $x^3 - 2x^2 - 5x + 6 \geqslant 0$

32. $2x^3 + x^2 - 8x - 4 \leqslant 0$

33. $2x^3 - 3x + 1 < 0$

34. $x^4 - 4x^3 + 8x > 0$

35. $5x^4 < 8x^3$

36. $x^5 + x^3 \geqslant x^2 + 6x$

37–42 ■ Graph the family of polynomials in the same viewing rectangle, using the given values of c. Explain how changing the value of c affects the graph.

37. $P(x) = cx^3$; $c = 0, 2, 5, \frac{1}{2}$

38. $P(x) = (x - c)^4$; $c = -1, 0, 1, 2$

39. $P(x) = x^4 + c$; $c = -1, 0, 1, 2$

40. $P(x) = x^3 + cx$; $c = 2, 0, -2, -4$

41. $P(x) = x^4 - cx$; $c = 0, 1, 8, 27$

42. $P(x) = x^c$; $c = 1, 3, 5, 7$

43. A market analyst working for a small-appliance manufacturer finds that if the firm produces and sells x blenders annually, the total profit (in dollars) is

$$P(x) = 8x + 0.3x^2 - 0.0013x^3 - 372$$

Graph the function P in an appropriate viewing rectangle and use the graph to answer the following questions.

(a) When just a few blenders are manufactured, the firm loses money (profit is negative). [For example, $P(10) = -263.3$, so the firm loses \$263.30 if it produces and sells only 10 blenders.] How many blenders must the firm produce to break even?

(b) Does profit increase indefinitely as more blenders are produced and sold? If not, what is the largest possible profit the firm could have?

44. The rabbit population on a small island is observed to be given by the function

$$P(t) = 120t - 0.4t^4 + 1000$$

where t is the time (in months) since observations of the island began.

(a) When is the maximum population attained, and what is that maximum population?

(b) When does the rabbit population disappear from the island?

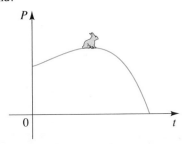

45. (a) Graph the function $P(x) = (x - 1)(x - 3)(x - 4)$ and find all local extrema, correct to the nearest tenth.

(b) Graph the function

$$Q(x) = (x - 1)(x - 3)(x - 4) + 5$$

and use your answers to part (a) to find all local extrema, correct to the nearest tenth.

(c) If $a < b < c$, explain why the function

$$P(x) = (x - a)(x - b)(x - c)$$

must have two local extrema.

(d) If $a < b < c$ and d is any real number, explain why the function $Q(x) = (x - a)(x - b)(x - c) + d$ must have two local extrema.

46. Give an example of a polynomial that has six local extrema.

47. (a) How many x-intercepts and how many local extrema does the polynomial $P(x) = x^3 - 4x$ have?

(b) How many x-intercepts and how many local extrema does the polynomial $Q(x) = x^3 + 4x$ have?

(c) If $a > 0$, how many x-intercepts and how many local extrema does each of the polynomials $P(x) = x^3 - ax$ and $Q(x) = x^3 + ax$ have? Explain your answer.

48. Is it possible for a third-degree polynomial to have exactly one local extremum? Explain your answer. [*Hint:* Think about the end behavior of such a polynomial.]

4.3 DIVIDING POLYNOMIALS

So far in this chapter we have been studying polynomial functions *graphically*. In this section we begin to study polynomials *algebraically*. Most of our work will be concerned with finding the solutions of polynomial equations, but first we consider division of polynomials.

Long division for polynomials is very much like the familiar process of long division for numbers. For example, to divide $6x^2 - 26x + 12$ (the **dividend**) by $x - 4$ (the **divisor**), we arrange our work as follows.

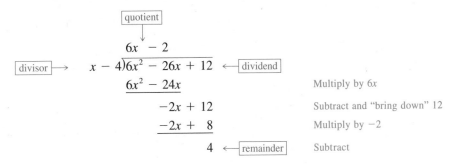

The **quotient** at the top of the division table is obtained by first dividing the initial term of the dividend by the initial term of the divisor ($6x^2/x = 6x$), which gives the first term of the quotient. The process then continues as explained beside each line of the division table. The process stops if, after the subtraction has been performed, the result has degree *smaller* than the degree of the divisor and nothing is left in the original dividend to "bring down." The last line of the table contains the **remainder,** and the result of the division can be interpreted in either of two ways:

$$\frac{6x^2 - 26x + 12}{x - 4} = 6x - 2 + \frac{4}{x - 4}$$

or

$$6x^2 - 26x + 12 = (x - 4)(6x - 2) + 4$$

We summarize what happens in this or any such long division problem with the following theorem.

DIVISION ALGORITHM

If $P(x)$ and $D(x)$ are polynomials, with $D(x) \neq 0$, then there exist unique polynomials $Q(x)$ and $R(x)$ such that

$$P(x) = D(x) \cdot Q(x) + R(x)$$

where $R(x)$ is either 0 or of degree less than the degree of $D(x)$. The polynomials $P(x)$ and $D(x)$ are called the **dividend** and **divisor,** respectively, $Q(x)$ is the **quotient,** and $R(x)$ is the **remainder.**

EXAMPLE 1 ■ Long Division of Polynomials

Let $P(x) = 8x^4 + 6x^2 - 3x + 1$ and $D(x) = 2x^2 - x + 2$. Find polynomials $Q(x)$ and $R(x)$ such that $P(x) = D(x) \cdot Q(x) + R(x)$.

SOLUTION

We use long division after first inserting the term $0x^3$ into the dividend to ensure that the columns will line up correctly in the long division process.

$$
\begin{array}{r}
4x^2 + 2x \phantom{{}+{}} \\
2x^2 - x + 2 \overline{\smash{)}\,8x^4 + 0x^3 + 6x^2 - 3x + 1} \\
\underline{8x^4 - 4x^3 + 8x^2} \phantom{{}- 3x + 1} \\
4x^3 - 2x^2 - 3x \phantom{{}+ 1} \\
\underline{4x^3 - 2x^2 + 4x} \phantom{{}+ 1} \\
-7x + 1
\end{array}
$$

The process is completed at this point because $-7x + 1$ is of lesser degree than the divisor $2x^2 - x + 2$. From the long division table, we see that $Q(x) = 4x^2 + 2x$ and $R(x) = -7x + 1$, so

$$8x^4 + 6x^2 - 3x + 1 = (2x^2 - x + 2)(4x^2 + 2x) + (-7x + 1) \qquad \blacksquare$$

THE REMAINDER THEOREM

If the divisor in the Division Algorithm is of the form $x - c$ for some real number c, then the remainder must be a constant (since the degree of the remainder is less than the degree of the divisor). If we call this constant r, then

$$P(x) = (x - c) \cdot Q(x) + r$$

Setting $x = c$ in this equation, we get $P(c) = (c - c) \cdot Q(x) + r = 0 + r = r$. This proves the next theorem.

REMAINDER THEOREM

If the polynomial $P(x)$ is divided by $x - c$, then the remainder is the value $P(c)$.

$$
\begin{array}{r}
x^2 - 2x - 1 \\
x - 2\overline{)x^3 - 4x^2 + 3x + 5} \\
\underline{x^3 - 2x^2} \\
-2x^2 + 3x \\
\underline{-2x^2 + 4x} \\
-x + 5 \\
\underline{-x + 2} \\
3
\end{array}
$$

EXAMPLE 2 ■ Using the Remainder Theorem to Find the Value of A Polynomial

Let $P(x) = x^3 - 4x^2 + 3x + 5$. If we divide $P(x)$ by $x - 2$ using long division, we obtain a quotient of $x^2 - 2x - 1$ and a remainder of 3. Thus, by the Remainder Theorem, the value of $P(2)$ should be 3. To verify this, we calculate

$$P(2) = 2^3 - 4 \cdot 2^2 + 3 \cdot 2 + 5 = 8 - 16 + 6 + 5 = 3 \qquad \blacksquare$$

FACTORS AND ZEROS OF POLYNOMIALS

If $R(x) = 0$ in the Division Algorithm, then $P(x) = D(x) \cdot Q(x)$ and we say that $D(x)$ and $Q(x)$ are **factors** of $P(x)$. If $D(x) = x - c$ is a factor of $P(x)$, then obviously $P(c) = (c - c) \cdot Q(c) = 0$. Conversely, if $P(c) = 0$, then by the Remainder Theorem

$$P(x) = (x - c) \cdot Q(x) + 0 = (x - c) \cdot Q(x)$$

so $x - c$ is a factor of $P(x)$. Thus, we have proved the following theorem.

FACTOR THEOREM

$P(c) = 0$ if and only if $x - c$ is a factor of $P(x)$.

EXAMPLE 3 ■ Factoring a Polynomial Using the Factor Theorem

Let $P(x) = x^3 - 7x + 6$. Show that $P(1) = 0$, and use this fact to factor $P(x)$ completely.

SOLUTION

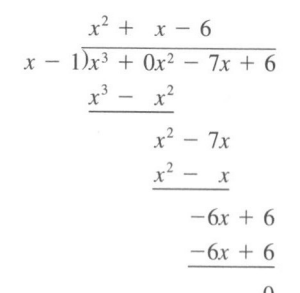

Substituting, we see that $P(1) = 1^3 - 7 \cdot 1 + 6 = 0$. By the Factor Theorem, this means that $x - 1$ is a factor of $P(x)$. Using long division, we see that

$$P(x) = (x - 1)(x^2 + x - 6)$$

$$= (x - 1)(x - 2)(x + 3) \qquad \text{Factor the quadratic } x^2 + x - 6 \qquad ■$$

Recall that if $P(c) = 0$, then the number c is a **zero** of the polynomial P. We also express this by saying that c is a **root** of the polynomial equation $P(x) = 0$. Thus the zeros of the polynomial of Example 3 are 1, 2, and -3. With this terminology, the Factor Theorem tells us that $x - c$ is a factor of $P(x)$ if and only if c is a zero of $P(x)$.

EXAMPLE 4 ■ Finding a Polynomial with Specified Zeros

Find a polynomial $F(x)$ of degree 4 that has zeros -3, 0, 1, and 5.

SOLUTION

By the Factor Theorem, $x - (-3)$, $x - 0$, $x - 1$, and $x - 5$ must all be factors of the desired polynomial, so let

$$F(x) = (x + 3)(x - 0)(x - 1)(x - 5) = x^4 - 3x^3 - 13x^2 + 15x$$

Since $F(x)$ is to have degree 4, any other solution of the problem must be a constant multiple of the polynomial we have chosen (because multiplication by any polynomial other than a constant will increase the degree).   ■

In Section 1.4 we developed some techniques and formulas for factoring certain special kinds of polynomials. We have seen in Example 3 that the Factor Theorem can help us factor polynomials for which those techniques and formulas do not work. We will exploit this use of the Factor Theorem more fully in Section 4.4.

SYNTHETIC DIVISION

The Remainder Theorem does not seem at first glance to provide a particularly useful way to evaluate polynomials. After all, to find the remainder when $x - c$

Gerolamo Cardano (1501–1576) is certainly one of the most colorful figures in the history of mathematics. He was the most renowned physician in Europe in his day, yet he was plagued throughout his life by numerous maladies, including ruptures, hemorrhoids, and irrational fears of encountering rabid dogs. His beloved sons broke his heart— his favorite was eventually beheaded for murdering his wife. He was a compulsive gambler, but he made the best of this vice by writing the *Book on Games of Chance,* the first study of probability from a sound mathematical point of view.

His major mathematical work was the *Ars Magna,* in which he detailed the solution of the general third- and fourth-degree polynomial equations. At the time of its publication, mathematicians were uncomfortable even with negative numbers, but Cardano's formulas paved the way for the acceptance of not just negative but also imaginary numbers, because they occurred naturally in solving polynomial equations. For example, one of his formulas gives the solution

$$x = \sqrt[3]{2 + \sqrt{-121}}$$
$$- \sqrt[3]{-2 + \sqrt{-121}}$$

for the cubic equation

$$x^3 - 15x - 4 = 0$$

This value for x actually turns out to be the *integer* 4, yet to find it Cardano had to use the imaginary number $\sqrt{-121} = 11i$.

is divided into a polynomial requires division, and the long division process appears much more complicated than simply substituting c in place of x and evaluating the polynomial. However, it turns out that if the divisor is of the form $x - c$, the long division process can be substantially simplified. For example, let us divide the polynomial $2x^3 - 7x^2 + 5$ by $x - 3$:

$$
\begin{array}{r}
2x^2 - x - 3 \\
x - 3 \overline{\smash{)}2x^3 - 7x^2 + 0x + 5} \\
\underline{2x^3 - 6x^2} \\
-x^2 + 0x \\
\underline{-x^2 + 3x} \\
-3x + 5 \\
\underline{-3x + 9} \\
-4
\end{array}
$$

Notice that the powers of x act simply as place-holders in this format, and we can certainly omit writing them down. Moreover, all the coefficients in the lines underneath the dividend, except those in boxes, are repeated either in the dividend or in the quotient. (For example, the 0 and the 5 are brought down from the dividend, and the 2, -1, and -3 are repeated in the quotient line, in the same column as they appear below the dividend.) So, we can omit these as well.

If we change the sign of the numbers shown in boxes, then we can *add* instead of *subtract* to obtain the coefficients underneath the boxed numbers. If we now write the quotient line, together with the remainder, at the bottom instead of the top of the division table, the long division format simplifies to

$$
\begin{array}{r|rrrr}
3 & 2 & -7 & 0 & 5 \\
 & & 6 & -3 & -9 \\
\hline
 & 2 & -1 & -3 & -4
\end{array}
$$

Compare this carefully with the long division. The top left corner shows the *negative* of the constant term in the dividend, and the rest of the top line contains the coefficients of the divisor. The bottom line displays the coefficients of the quotient and the remainder. Note that each entry in the bottom line is the sum of the two numbers above it. In the middle line, the 6 is the product $3 \cdot 2$, the -3 is the product $3 \cdot (-1)$, and the -9 is the product $3 \cdot (-3)$. This multiplication process is indicated by the arrows. We summarize what we have done in the following box.

SYNTHETIC DIVISION

To divide $a_n x^n + a_{n-1}x^{n-1} + \cdots + a_1 x + a_0$ by $x - c$, we proceed as follows:

$$
\begin{array}{c|ccccccc}
c & a_n & a_{n-1} & a_{n-2} & a_{n-3} & \cdots & a_2 & a_1 & a_0 \\
  &   & cb_{n-1} & cb_{n-2} & cb_{n-3} & \cdots & cb_2 & cb_1 & cb_0 \\
\hline
  & b_{n-1} & b_{n-2} & b_{n-3} & b_{n-4} & \cdots & b_1 & b_0 & r
\end{array}
$$

Each number in the bottom row is obtained by adding the numbers above it. In particular, $b_{n-1} = a_n$. The quotient is

$$b_{n-1}x^{n-1} + b_{n-2}x^{n-2} + \cdots + b_1 x + b_0$$

and the remainder is r.

EXAMPLE 5 ■ Synthetic Division

Find the quotient and remainder when $3x^5 + 5x^4 - 4x^3 + 7x + 3$ is divided by $x + 2$.

SOLUTION

Since $x + 2 = x - (-2)$, the synthetic division table for this problem takes the following form.

$$
\begin{array}{c|cccccc}
-2 & 3 & 5 & -4 & 0 & 7 & 3 \\
   &   & -6 & 2 & 4 & -8 & 2 \\
\hline
   & 3 & -1 & -2 & 4 & -1 & 5
\end{array}
$$

The quotient is $3x^4 - x^3 - 2x^2 + 4x - 1$ and the remainder is 5. Thus

$$3x^5 + 5x^4 - 4x^3 + 7x + 3 = (x + 2)(3x^4 - x^3 - 2x^2 + 4x - 1) + 5$$

or $$\frac{3x^5 + 5x^4 - 4x^3 + 7x + 3}{x + 2} = 3x^4 - x^3 - 2x^2 + 4x - 1 + \frac{5}{x + 2}$$ ■

EXAMPLE 6 ■ Using Synthetic Division to Find the Value of a Polynomial

Let $P(x) = x^5 - 5x^4 + 20x^2 - 10x + 10$. Find $P(4)$.

SOLUTION

Since $P(4)$ is the remainder when $P(x)$ is divided by $x - 4$ (by the Remainder Theorem), we can use synthetic division.

$$
\begin{array}{r|rrrrrr}
4 & 1 & -5 & 0 & 20 & -10 & 10 \\
  &   & 4 & -4 & -16 & 16 & 24 \\
\hline
  & 1 & -1 & -4 & 4 & 6 & 34
\end{array}
$$

The remainder is 34, so $P(4) = 34$. ∎

In Example 6 we could also have evaluated $P(4)$ by direct substitution:

$$P(4) = (4)^5 - 5(4)^4 + 20(4)^2 - 10(4) + 10$$

$$= 1024 - 1280 + 320 - 40 + 10 = 34$$

When evaluating polynomials without the aid of a calculator, synthetic division usually involves less work than direct substitution.

4.3 EXERCISES

1–25 ■ Find the quotient and remainder.

1. $\dfrac{x^2 - 5x + 4}{x - 3}$

2. $\dfrac{x^2 - 5x + 4}{x - 1}$

3. $\dfrac{x^3 + 2x^2 + 2x + 1}{x + 2}$

4. $\dfrac{3x^3 - 12x^2 - 9x + 1}{x - 5}$

5. $\dfrac{x^3 - 8x + 2}{x + 3}$

6. $\dfrac{x^4 - x^3 + x^2 - x + 2}{x - 2}$

7. $\dfrac{x^5 + 3x^3 - 6}{x - 1}$

8. $\dfrac{x^3 - 9x^2 + 27x - 27}{x - 3}$

9. $\dfrac{x^3 + 6x + 3}{x^2 - 2x + 2}$

10. $\dfrac{3x^4 - 5x^3 - 20x - 5}{x^2 + x + 3}$

11. $\dfrac{6x^3 + 2x^2 + 22x}{2x^2 + 5}$

12. $\dfrac{9x^2 - x + 5}{3x^2 - 7x}$

13. $\dfrac{x^5 + x^4 + x^3 + x^2 + x + 1}{x^2 + x + 1}$

14. $\dfrac{x^3 - 27}{x - 3}$

15. $\dfrac{x^3 - 4x + 7}{x}$

16. $\dfrac{x^6 + x^4 + x^2 + 1}{x^2 + 1}$

17. $\dfrac{x^5 - 1}{x^2 - 1}$

18. $\dfrac{2x^5 - 7x^4 - 13}{4x^2 - 6x + 8}$

19. $\dfrac{2x^3 + 3x^2 - 2x + 1}{x - \frac{1}{2}}$

20. $\dfrac{6x^4 + 10x^3 + 5x^2 + x + 1}{x + \frac{2}{3}}$

21. $\dfrac{x^{101} - 1}{x - 1}$

22. $\dfrac{x^{100} - 1}{x + 1}$

[*Hint:* In Exercises 21 and 22 you are obviously not being asked to carry out more than 100 steps of a division process. Think! Use the Remainder Theorem, and look for a pattern when finding the quotient.]

23. $\dfrac{x^2 - x - 3}{2x - 4}$

$$\left[\text{*Hint:* Note that } \frac{x^2 - x - 3}{2x - 4} = \frac{1}{2}\left(\frac{x^2 - x - 3}{x - 2} \right). \right]$$

24. $\dfrac{x^3 + 3x^2 + 4x + 3}{3x + 6}$ **25.** $\dfrac{x^4 - 2x^3 + x + 2}{2x - 1}$

26–39 ■ Use synthetic division and the Remainder Theorem to evaluate $P(c)$.

26. $P(x) = x^2 - 3x + 6$, $c = 1$

27. $P(x) = 4x^2 + 12x + 5$, $c = -1$

28. $P(x) = 2x^2 + 9x + 1$, $c = \frac{1}{2}$

29. $P(x) = 2x^2 + 9x + 1$, $c = 0.1$

30. $P(x) = x^3 + 3x^2 - 7x + 6$, $c = 2$

31. $P(x) = 2x^3 - 21x^2 + 9x - 200$, $c = 11$

32. $P(x) = 5x^4 + 30x^3 - 40x^2 + 36x + 14$, $c = -7$

33. $P(x) = 6x^5 + 10x^3 + x + 1$, $c = -2$

34. $P(x) = x^7 - 3x^2 - 1$, $c = 3$

35. $P(x) = -2x^6 + 7x^5 + 40x^4 - 7x^2 + 10x + 112$, $c = -3$

36. $P(x) = 3x^3 + 4x^2 - 2x + 1$, $c = \frac{2}{3}$

37. $P(x) = x^3 - x + 1$, $c = \frac{1}{4}$

38. $P(x) = x^3 + 2x^2 - 3x - 8$, $c = \sqrt{3}$

39. $P(x) = -2x^3 + 3x^2 + 4x + 6$, $c = 1 + \sqrt{2}$

40. Let

$$P(x) = 6x^7 - 40x^6 + 16x^5 - 200x^4$$
$$- 60x^3 - 69x^2 + 13x - 139$$

Calculate $P(7)$ by (a) using synthetic division and (b) substituting $x = 7$ into the polynomial and evaluating directly.

41–44 ■ Use the Factor Theorem to show that $x - c$ is a factor of $P(x)$ for the given value(s) of c.

41. $P(x) = x^3 - 3x^2 + 3x - 1$, $c = 1$

42. $P(x) = x^3 + 2x^2 - 3x - 10$, $c = 2$

43. $P(x) = 2x^3 + 7x^2 + 6x - 5$, $c = \frac{1}{2}$

44. $P(x) = x^4 + 3x^3 - 16x^2 - 27x + 63$, $c = 3, -3$

45–46 ■ Show that the given value(s) of c are zeros of $P(x)$, and find all other zeros of $P(x)$.

45. $P(x) = x^3 - x^2 - 11x + 15$, $c = 3$

46. $P(x) = 3x^4 - x^3 - 21x^2 - 11x + 6$, $c = \frac{1}{3}, -2$

47. Find a polynomial of degree 3 that has zeros 1, -2, and 3, and in which the coefficient of x^2 is 3.

48. Find a polynomial of degree 4 that has integer coefficients and zeros 1, -1, 2, and $\frac{1}{2}$.

49. Find a polynomial of degree 4 that has integer coefficients, constant coefficient 24, and zeros 3, 4, and -2.

50. Find the remainder when the polynomial $6x^{1000} - 17x^{562} + 12x + 26$ is divided by $x + 1$.

51. Is $x - 1$ a factor of $x^{567} - 3x^{400} + x^9 + 2$?

52. If we divide the polynomial $P(x) = x^4 + kx^2 - kx + 2$ by $x + 2$, the remainder is 36. What must the value of k be?

53. For what values of k will $x - 3$ be a factor of $P(x) = k^2x^2 + 2kx - 12$?

54. The quadratic equation $x^2 - 2kx + 12 = 0$ has two positive roots, one of which is three times the other. What is the value of k, and what are the two roots of the equation?

4.4 REAL ZEROS OF POLYNOMIALS

In the preceding section we studied the Factor Theorem, which tells us that if $x = c$ is a root or solution of the polynomial equation $P(x) = 0$ (or, equivalently, a zero of the polynomial P), then $x - c$ is a factor of the polynomial. This means that solving a polynomial equation is really the same thing as factoring it into linear factors. In this section we study some algebraic methods that help us to find the real zeros of a polynomial, and thereby factor the polynomial. The first theorem in this section helps us find the *rational* zeros of a polynomial.

We begin with an illustration. By expanding the left side of the following equation, we see that

$$(x - 2)(x - 3)(x + 4) = x^3 - x^2 - 14x + 24$$

Thus, the zeros of the polynomial $P(x) = x^3 - x^2 - 14x + 24$ are 2, 3, and -4, since those are the values that, in turn, make each factor of P equal 0. Notice that when expanding the left side of the displayed equation, the constant 24 on the right is obtained by multiplying $(-2) \times (-3) \times 4$. This means that the zeros of the polynomial all must divide its constant term evenly. The next theorem states that for all polynomials with integer coefficients, we can always list all possible rational zeros of a polynomial using a generalization of this principle.

RATIONAL ROOTS THEOREM

If p/q is a rational root in lowest terms of the polynomial equation with integer coefficients

$$a_n x^n + a_{n-1} x^{n-1} + \cdots + a_1 x + a_0 = 0$$

(where $a_n \neq 0$ and $a_0 \neq 0$), then p divides a_0 evenly and q divides a_n evenly.

A *rational number* is one that can be expressed as a quotient (or ratio) of integers; for example, $\dfrac{2}{3}, \dfrac{-1}{5}, \dfrac{3}{1}$.

If $a_n = 1$ in the Rational Roots Theorem, then q must be 1 or -1, so in this case any rational root p/q is in fact an integer that divides a_0. A proof of the Rational Roots Theorem is given at the end of this section. In the following examples we show how to use the theorem to find the zeros of polynomials.

EXAMPLE 1 ■ Using the Rational Roots Theorem to Factor a Polynomial

Factor the polynomial $P(x) = 2x^3 + x^2 - 13x + 6$.

SOLUTION

By the Rational Roots Theorem, if p/q is a zero of $P(x)$, then p divides 6 and q divides 2, so p/q is of the form

$$\frac{\text{divisor of 6}}{\text{divisor of 2}}$$

The divisors of 6 are ± 1, ± 2, ± 3, ± 6, and the divisors of 2 are ± 1, ± 2. Thus the possible values of p/q are

$$\pm\frac{1}{1}, \quad \pm\frac{2}{1}, \quad \pm\frac{3}{1}, \quad \pm\frac{6}{1}, \quad \pm\frac{1}{2}, \quad \pm\frac{2}{2}, \quad \pm\frac{3}{2}, \quad \pm\frac{6}{2}$$

Simplifying the fractions and eliminating duplicates, we get the following list of possible values of p/q:

$$\pm 1, \quad \pm 2, \quad \pm 3, \quad \pm 6, \quad \pm \tfrac{1}{2}, \quad \pm \tfrac{3}{2}$$

Now we check to see which of these *possible* zeros actually *are* zeros by substituting them, one at a time, into the polynomial P until we find one that makes $P(x) = 0$. We have

$$P(1) = 2(1)^3 + (1)^2 - 13(1) + 6 = -4 \qquad \text{1 is \textit{not} a zero of } P$$

$$P(2) = 2(2)^3 + (2)^2 - 13(2) + 6 = 0 \qquad \text{2 \textit{is} a zero of } P$$

Since $x = 2$ is a zero of P, it follows that $x - 2$ divides $P(x)$. Using synthetic division (as shown in the margin), we obtain the following factorization:

$$
\begin{array}{c|cccc}
2 & 2 & 1 & -13 & 6 \\
  &   & 4 & 10 & -6 \\
\hline
  & 2 & 5 & -3 & 0
\end{array}
$$

$$
\begin{aligned}
P(x) &= 2x^3 + x^2 - 13x + 6 \\
     &= (x - 2)(2x^2 + 5x - 3) \\
     &= (x - 2)(2x - 1)(x + 3) \qquad \text{Factor } 2x^2 + 5x - 3 \qquad \blacksquare
\end{aligned}
$$

In the remaining examples in this section, we use synthetic division (described in Section 4.3) both for evaluating and for dividing polynomials. For those who prefer not to use synthetic division, simple substitution can be used to evaluate polynomials and long division can be used to divide them.

EXAMPLE 2 ■ Using the Rational Roots Theorem and the Quadratic Formula

Find all rational roots of the following equation, and then solve it completely.

$$x^3 - 17x + 4 = 0$$

$$
\begin{array}{c|cccc}
1 & 1 & 0 & -17 & 4 \\
  &   & 1 & 1   & -16 \\
\hline
  & 1 & 1 & -16 & -12
\end{array}
$$

SOLUTION

Let $P(x) = x^3 - 17x + 4$. Since the leading coefficient of P is 1, all the rational roots are integer divisors of 4, so the possible candidates for rational roots are ± 1, ± 2, and ± 4. Using synthetic division, we check these candidates one at a time, starting with the positive ones (see the margin). We find that 1 and 2 are not roots, but that 4 is a root because $P(4) = 0$. Thus

$$
\begin{array}{c|cccc}
2 & 1 & 0 & -17 & 4 \\
  &   & 2 & 4   & -26 \\
\hline
  & 1 & 2 & -13 & -22
\end{array}
$$

$$x^3 - 17x + 4 = (x - 4)(x^2 + 4x - 1)$$

The quadratic formula now gives us the zeros of the quotient $x^2 + 4x - 1$:

$$x = \frac{-4 \pm \sqrt{4^2 - 4 \cdot 1 \cdot (-1)}}{2} = -2 \pm \sqrt{5}$$

$$
\begin{array}{c|cccc}
4 & 1 & 0 & -17 & 4 \\
  &   & 4 & 16 & -4 \\
\hline
  & 1 & 4 & -1 & 0 \\
  &   &   &   & \uparrow \\
  &   &   & \text{remainder is 0}
\end{array}
$$

So the only rational root is 4, and the remaining roots are $-2 + \sqrt{5}$ and $-2 - \sqrt{5}$, which are irrational. $\qquad \blacksquare$

From now on, whenever an example involves using synthetic division several times with the same dividend, we will use an abbreviated version of the synthetic division table: We omit the middle line and write the top line only once, with the result of each division written below this. For example, the three divisions we performed in Example 2 are combined as follows.

$$
\begin{array}{r|rrrr}
 & 1 & 0 & -17 & 4 \\
\hline
1 & 1 & 1 & -16 & -12 \\
2 & 1 & 2 & -13 & -22 \\
4 & 1 & 4 & -1 & 0 \\
\end{array}
$$

DESCARTES' RULE OF SIGNS

In some cases the following rule, discovered by the French philosopher and mathematician René Descartes around 1637 (see page 150), is helpful in eliminating candidates from lengthy lists of possible rational roots. Before we state the rule (which we do not prove), we must first explain what is meant by a *variation in sign*. If $P(x)$ is a polynomial with real coefficients, written with descending powers of x (and omitting powers with coefficient 0), then a **variation in sign** occurs whenever adjacent coefficients have opposite signs. For example,

$$P(x) = 5x^7 - 3x^5 - x^4 + 2x^2 + x - 3$$

has three variations in sign.

DESCARTES' RULE OF SIGNS

If $P(x)$ is a polynomial with real coefficients, then

1. The number of positive real zeros of $P(x)$ is either equal to the number of variations in sign in $P(x)$ or is less than that by an even whole number.

2. The number of negative real zeros of $P(x)$ is either equal to the number of variations in sign in $P(-x)$ or is less than that by an even whole number.

EXAMPLE 3 ■ Using Descartes' Rule

Use Descartes' Rule of Signs to determine the possible number of positive and negative real zeros of the polynomial

$$P(x) = 3x^6 + 4x^5 + 3x^3 - x - 3$$

SOLUTION

The polynomial has one variation in sign and so it has one positive zero. Now

$$P(-x) = 3(-x)^6 + 4(-x)^5 + 3(-x)^3 - (-x) - 3$$

$$= 3x^6 - 4x^5 - 3x^3 + x - 3$$

So, $P(-x)$ has three variations in sign. Thus, $P(x)$ has either three or one negative root(s), making a total of either two or four real roots. The remaining roots are imaginary numbers, which we will study in Section 4.7. ∎

UPPER AND LOWER BOUNDS FOR ROOTS

We say that a is a **lower bound** and b is an **upper bound** for the roots of a polynomial equation if every real root c of the equation satisfies $a \leq c \leq b$. The next theorem helps us find such bounds for any polynomial equation.

THE UPPER AND LOWER BOUNDS THEOREM

Let $P(x)$ be a polynomial with real coefficients.

1. If we divide $P(x)$ by $x - b$ (with $b > 0$) using synthetic division, and if the row that contains the quotient and remainder has no negative entry, then b is an upper bound for the real roots of $P(x) = 0$.

2. If we divide $P(x)$ by $x - a$ (with $a < 0$) using synthetic division, and if the row that contains the quotient and remainder has entries that are alternately nonpositive and nonnegative, then a is a lower bound for the real roots of $P(x) = 0$.

A proof of this theorem is suggested in Exercises 63 and 64. The phrase "alternatively nonpositive and nonnegative" simply means that the signs of the numbers alternate, with 0 considered to be positive or negative as required.

EXAMPLE 4 ∎ Upper and Lower Bounds for Zeros of a Polynomial

Show that all the real roots of the equation $x^4 - 3x^2 + 2x - 5 = 0$ lie between -3 and 2.

SOLUTION

We divide the polynomial by $x - 2$ and $x + 3$ using synthetic division.

$$
\begin{array}{r|rrrrr}
 & 1 & 0 & -3 & 2 & -5 \\
\hline
2 & 1 & 2 & 1 & 4 & 3 \quad \leftarrow \text{all entries positive} \\
-3 & 1 & -3 & 6 & -16 & 43 \quad \leftarrow \text{entries alternate in sign}
\end{array}
$$

By the Upper and Lower Bounds Theorem, -3 is a lower bound and 2 is an upper bound for the roots. Since neither -3 nor 2 is a root (the remainders are not 0 in the division table), all the real roots lie between them. ∎

EXAMPLE 5 ■ Factoring a Fifth-Degree Polynomial

Factor completely the polynomial

$$P(x) = 2x^5 + 5x^4 - 8x^3 - 14x^2 + 6x + 9$$

SOLUTION

The possible rational zeros of $P(x)$ are $\pm \frac{1}{2}$, ± 1, $\pm \frac{3}{2}$, ± 3, $\pm \frac{9}{2}$, and ± 9. We check the positive candidates first, beginning with the smallest.

$$
\begin{array}{r|rrrrrr}
 & 2 & 5 & -8 & -14 & 6 & 9 \\
\hline
\frac{1}{2} & 2 & 6 & -5 & -\frac{33}{2} & -\frac{9}{4} & \frac{63}{8} \\
1 & 2 & 7 & -1 & -15 & -9 & 0 \quad \leftarrow P(1) = 0
\end{array}
$$

So 1 is a zero, and $P(x)$ factors as $(x - 1)(2x^4 + 7x^3 - x^2 - 15x - 9)$. We continue by factoring the quotient.

$$
\begin{array}{r|rrrrr}
 & 2 & 7 & -1 & -15 & -9 \\
\hline
1 & 2 & 9 & 8 & -7 & -16 \\
\frac{3}{2} & 2 & 10 & 14 & 6 & 0 \quad \leftarrow \text{all entries nonnegative}
\end{array}
$$

We see that $\frac{3}{2}$ is both a zero and an upper bound for the zeros of $P(x)$, so we do not need to check any further for positive zeros, because all the remaining candidates are greater than $\frac{3}{2}$.

$$P(x) = (x - 1)\left(x - \tfrac{3}{2}\right)(2x^3 + 10x^2 + 14x + 6)$$

$$= (x - 1)(2x - 3)(x^3 + 5x^2 + 7x + 3)$$

By Descartes' Rule of Signs, $x^3 + 5x^2 + 7x + 3$ has no positive zero, so the only possible rational zeros are -1 and -3.

$$
\begin{array}{r|rrrr}
 & 1 & 5 & 7 & 3 \\
\hline
-1 & 1 & 4 & 3 & 0
\end{array}
$$

Therefore

$$P(x) = (x - 1)(2x - 3)(x + 1)(x^2 + 4x + 3)$$

$$= (x - 1)(2x - 3)(x + 1)^2(x + 3)$$

This means that the zeros of P are 1, $\frac{3}{2}$, -1, and -3.   ■

To summarize, the following procedure gives a systematic, efficient method for finding all rational roots of a polynomial equation.

FINDING THE RATIONAL ZEROS OF A POLYNOMIAL

Step 1 List all possible rational roots using the Rational Roots Theorem.

Step 2 Use Descartes' Rule of Signs to see how many positive and negative real roots the equation may have. (In some cases the possible positive or negative rational candidates may be eliminated completely by this step.)

Step 3 Check (from smallest to largest, in absolute value) the positive and the negative candidates for rational roots provided by Step 1. Stop when you find a root, reach an upper or lower bound, or have found the maximum number of positive or negative roots predicted by Descartes' Rule of Signs.

Step 4 If you find a root, repeat the process with the quotient. (Remember that you do not need to check possible roots that have not worked at a previous stage.) Once you reach a quotient that is quadratic or is in some other way easily factorable, use the quadratic formula or other techniques you have learned to find the remaining roots.

PROOF OF THE RATIONAL ROOTS THEOREM

Let $P(x) = a_n x^n + a_{n-1} x^{n-1} + \cdots + a_1 x + a_0$ be a polynomial with integer coefficients (with $a_n \neq 0$ and $a_0 \neq 0$), and suppose that p/q is a rational number in lowest terms for which

$$a_n \left(\frac{p}{q} \right)^n + a_{n-1} \left(\frac{p}{q} \right)^{n-1} + \cdots + a_1 \left(\frac{p}{q} \right) + a_0 = 0$$

so that p/q is a rational zero of $P(x)$. We are going to show that p divides a_0 evenly and that q divides a_n evenly.

First we multiply both sides of the equation by q^n, giving

(1) $a_n p^n + a_{n-1} p^{n-1} q + a_{n-2} p^{n-2} q^2 + \cdots + a_1 p q^{n-1} + a_0 q^n = 0$

Subtracting $a_0 q^n$ from each side of the equation and factoring p from the left side, we get

$$p(a_n p^{n-1} + a_{n-1} p^{n-2} q + a_{n-2} p^{n-3} q^2 + \cdots + a_1 q^{n-1}) = -a_0 q^n$$

Now we can see that p divides the left side of the equation evenly, so it must divide the right side as well. But since p/q is in lowest terms, p and q have no integer factor in common, and so neither do p and q^n. Thus, all the integer factors of p must go evenly into a_0, which means that p divides a_0.

If we take Equation 1 and subtract $a_n p^n$ from each side, and then factor q from the left side, we get

$$q(a_{n-1}p^{n-1} + a_{n-2}p^{n-2}q + \cdots + a_1 pq^{n-2} + a_0 q^{n-1}) = -a_n p^n$$

Using the same reasoning as before, we can now show that q must divide a_n. This proves the Rational Roots Theorem.   ☐

4.4 EXERCISES

1–4 ■ List all possible rational zeros given by the Rational Roots Theorem (but do not check to see which values actually are zeros).

1. $P(x) = x^3 - 2x^2 - 5x + 3$

2. $Q(x) = x^3 - 4x^2 + 6x + 20$

3. $R(x) = 2x^4 - 3x^3 - x + 6$

4. $S(x) = 6x^4 - x^3 + x^2 - 12$

5–32 ■ Find all rational roots of the equation, and then find the irrational roots, if any.

5. $2x^2 + x - 6 = 0$

6. $3x^2 - 13x + 4 = 0$

7. $x^3 - x^2 + x - 1 = 0$

8. $x^3 - 2x^2 - x + 2 = 0$

9. $x^3 - 3x^2 - 4x + 12 = 0$

10. $x^3 - x^2 - 8x + 12 = 0$

11. $x^3 - 4x^2 + x + 6 = 0$

12. $x^3 + 3x^2 - 4 = 0$

13. $x^3 - 7x^2 + 14x - 8 = 0$

14. $x^3 - 4x^2 - 7x + 10 = 0$

15. $x^3 + 4x^2 - 11x + 6 = 0$

16. $x^3 + 7x^2 + 8x - 16 = 0$

17. $x^3 + 3x^2 + 6x + 4 = 0$

18. $x^3 - 2x^2 - 2x - 3 = 0$

19. $x^4 + 3x^2 - 4 = 0$

20. $x^4 - 5x^2 + 4 = 0$

21. $x^4 - 2x^3 - 3x^2 + 8x - 4 = 0$

22. $x^4 + 3x^3 - 7x^2 - 15x + 18 = 0$

23. $x^4 + 6x^3 + 7x^2 - 6x - 8 = 0$

24. $x^4 - x^3 - 23x^2 - 3x + 90 = 0$

25. $4x^4 - 25x^2 + 36 = 0$

26. $x^4 - x^3 - 5x^2 + 3x + 6 = 0$

27. $4x^3 - 7x + 3 = 0$

28. $2x^3 - 3x^2 - 2x + 3 = 0$

29. $4x^3 + 4x^2 - x - 1 = 0$

30. $8x^3 + 10x^2 - x - 3 = 0$

31. $x^8 - 1 = 0$

32. $x^9 - 1 = 0$

33–40 ■ Use Descartes' Rule of Signs to determine how many positive and how many negative real zeros the polynomial can have. Then determine the possible total number of real zeros.

33. $x^3 - x^2 - x - 3$

34. $2x^3 - x^2 + 4x - 7$

35. $2x^6 + 5x^4 - x^3 - 5x - 1$

36. $x^4 + x^3 + x^2 + x + 12$

37. $x^5 + 4x^3 - x^2 + 6x$

38. $125x^{12} + 76x^5 - 12x^2 - 1$

39. $4x^7 - 3x^5 + 5x^4 + x^3 - 3x^2 + 2x - 5$

40. $x^8 - x^5 + x^4 - x^3 + x^2 - x + 1$

41–44 ■ Show that the given values for a and b are lower and upper bounds, respectively, for the real roots of the equation.

41. $2x^3 + 5x^2 + x - 2 = 0;\quad a = -3, b = 1$

42. $x^4 - 2x^3 - 9x^2 + 2x + 8 = 0;\quad a = -3, b = 5$

43. $8x^3 + 10x^2 - 39x + 9 = 0;\quad a = -3, b = 2$

44. $3x^4 - 17x^3 + 24x^2 - 9x + 1 = 0;\quad a = 0, b = 6$

45–48 ■ Find integers that are upper and lower bounds for the real roots of the equation.

45. $x^3 - 3x^2 + 4 = 0$

46. $2x^3 - 3x^2 - 8x + 12 = 0$

47. $x^4 - 2x^3 + x^2 - 9x + 2 = 0$

48. $x^5 - x^4 + 1 = 0$

49–56 ■ Find all rational roots of the equation, and then find the irrational roots, if any. Whenever appropriate, use the Rational Roots Theorem, the Upper and Lower Bounds Theorem, Descartes' Rule of Signs, the quadratic formula, or other factoring techniques.

49. $2x^4 + 3x^3 - 4x^2 - 3x + 2 = 0$

50. $2x^4 + 15x^3 + 31x^2 + 20x + 4 = 0$

51. $4x^4 - 21x^2 + 5 = 0$

52. $6x^4 - 7x^3 - 8x^2 + 5x = 0$

53. $x^5 - 7x^4 + 9x^3 + 23x^2 - 50x + 24 = 0$

54. $8x^5 - 14x^4 - 22x^3 + 57x^2 - 35x + 6 = 0$

55. $x^4 - \frac{11}{4}x^3 - \frac{11}{2}x^2 + \frac{5}{4}x + 3 = 0$

56. $x^5 - \frac{1}{5}x^4 - 5x^3 + x^2 + 4x - \frac{4}{5} = 0$

57–60 ■ Show that the polynomial does not have any rational zeros.

57. $x^3 - x - 2$

58. $2x^4 - x^3 + x + 2$

59. $3x^3 - x^2 - 6x + 12$

60. $x^{50} - 5x^{25} + x^2 - 1$

61. A city planner counts the number of cars per minute that travel through a certain intersection at 1:00, 2:00, 3:00, 4:00, and 5:00 P.M., and obtains the following data.

t (time)	1	2	3	4	5
$C(t)$ (cars per minute)	30	16	36	‘12	41

(a) Plot these data points on a coordinate plane.

(b) Is it possible to find a quadratic or third-degree polynomial whose graph passes through all these points? Why or why not?

(c) If C is a polynomial function of t, what is the smallest possible degree that C can have, and still be consistent with the given data?

62. A box with a square base has length plus girth of 108 inches. (Girth is the distance "around" the box.) What is the length of the box if its volume is 2200 in^3?

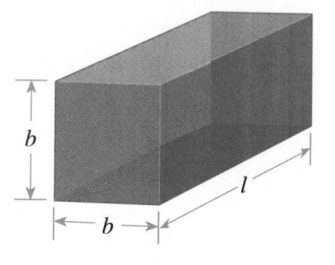

63. Let $P(x)$ be a polynomial with real coefficients and let $b > 0$. Use the Division Algorithm to write

$$P(x) = (x - b) \cdot Q(x) + r$$

Suppose that $r \geq 0$ and that all the coefficients in $Q(x)$ are nonnegative. Let $z > b$.

(a) Show that $P(z) > 0$.

(b) Prove the first part of the Upper and Lower Bounds Theorem.

64. Use the first part of the Upper and Lower Bounds Theorem to prove the second part. [*Hint:* Show that if $P(x)$ satisfies the second part of the theorem, then $P(-x)$ satisfies the first part.]

65. Show that the equation

$$x^5 - x^4 - x^3 - 5x^2 - 12x - 6 = 0$$

has exactly one rational root, and then prove that it must have either two or four irrational roots.

66. Consider the equation $x^3 + kx + 4 = 0$, where k is an integer.

(a) For which values of k does this equation have at least one rational root?

(b) For which values of k does this equation have one rational root and two irrational roots?

(c) What is the largest number of rational roots this equation can have?

67. For what integer values of k does the equation $x^3 - kx^2 + k^2x - 3 = 0$ have a rational root?

4.5 APPROXIMATING IRRATIONAL ZEROS OF POLYNOMIALS

For linear and quadratic equations we have formulas for the exact roots, including the irrational ones. In the preceding section we described a method for finding all the rational roots for equations of higher degree. Although there are formulas analogous to the quadratic formula for finding the roots of the general third- and fourth-degree equations, they are complicated, difficult to remember, and cumbersome to use. For example, the formula for one of the three roots of the general cubic equation $ax^3 + bx^2 + cx + d = 0$ is

$$x = \frac{-b}{3a} + \frac{1}{a}\sqrt[3]{\frac{-b^3}{27} + \frac{abc}{6} - \frac{a^2d}{2} + a\sqrt{\frac{a^2d^2}{4} - \frac{b^2c^2}{108} + \frac{b^3d}{27} + \frac{ac^3}{27} - \frac{abcd}{6}}}$$

$$+ \frac{1}{a}\sqrt[3]{\frac{-b^3}{27} + \frac{abc}{6} - \frac{a^2d}{2} - a\sqrt{\frac{a^2d^2}{4} - \frac{b^2c^2}{108} + \frac{b^3d}{27} + \frac{ac^3}{27} - \frac{abcd}{6}}}$$

For equations of degree 5 and higher, no such formulas exist. In fact, the French mathematician Evariste Galois proved in 1832 (shortly before his death at the age of 20) that it is *impossible* to construct formulas (involving radicals and the usual algebraic operations) for the roots of the general nth-degree polynomial equation for $n \geq 5$.

In some special cases the exact values of the irrational zeros of a higher-degree polynomial can be found by factoring and then using the quadratic formula, as we saw in Example 2 in Section 4.4. In this section we will demonstrate an effective numerical method for finding decimal approximations to irrational roots of polynomial equations to any desired accuracy. The technique depends on the following special case of the Intermediate Value Theorem, which is studied more fully in calculus courses.

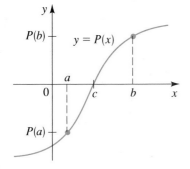

FIGURE 1

INTERMEDIATE VALUE THEOREM FOR POLYNOMIALS

If P is a polynomial with real coefficients and if $P(a)$ and $P(b)$ have opposite signs, then there is at least one value c between a and b for which $P(c) = 0$.

Although we will not prove this theorem, Figure 1 shows why it is intuitively plausible. A part of the graph of $y = P(x)$ is shown, with the case $P(a) < 0$ and $P(b) > 0$ illustrated.

EXAMPLE 1 ■ Using the Intermediate Value Theorem to Approximate a Root

(a) Show that the polynomial $P(x) = x^3 + 8x - 30$ has a zero between 2 and 3.
(b) Find an approximate value for this zero, correct to the nearest tenth.

Evariste Galois (1811–1832) is one of the very few mathematicians to have an entire theory named in his honor. Though he died while not yet 21, he completely settled the central problem in the theory of equations. He did so by describing a criterion that reveals whether any given equation can be solved by algebraic operations. Although Galois was one of the greatest mathematicians in the world at that time, no one knew it but him. He repeatedly sent his work to the eminent mathematicians Cauchy and Poisson, who either lost his letters or did nor understand his ideas. Galois wrote in a terse style and included few details, which probably played a role in his failure to pass the entrance exams at the Ecole Polytechnique in Paris. Galois was a political radical and spent several months in prison for his revolutionary activities. His brief life came to a tragic end when he was killed in a duel over a love affair. The night before his duel, fearing that he would be killed, Galois wrote

(continued)

SOLUTION

(a) Since $P(2) = -6$ and $P(3) = 21$, the Intermediate Value Theorem says that there is a number c between 2 and 3 for which $P(c) = 0$.

(b) We evaluate P at successive tenths between 2 and 3 until we find where P changes sign. This will "trap" the zero between successive tenths.

x	$P(x)$
2.1	-3.94
2.2	-1.75 ⎫
2.3	0.57 ⎬ opposite in sign

This means that the number c for which $P(c) = 0$ lies between 2.2 and 2.3. To see whether it is closer to 2.2 or 2.3, we check the value of P halfway between these two numbers:

$$P(2.25) \approx -0.61$$

Since the value of $P(x)$ is still negative when $x = 2.25$, the zero we are looking for is closer to 2.3 than to 2.2, as illustrated in Figure 2. Correct to the nearest tenth, the zero is 2.3.

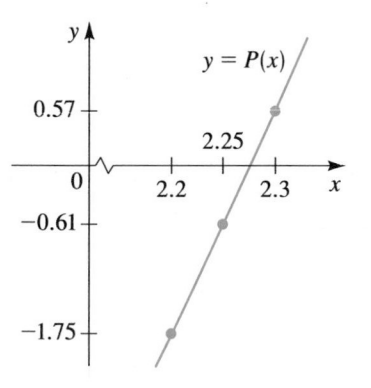

FIGURE 2

EXAMPLE 2 ■ Approximating a Root to Two Decimals

Show that the following equation has exactly one positive irrational root, and find the decimal value of this root correct to the nearest hundredth.

$$P(x) = x^3 + x^2 - 2x - 3 = 0$$

SOLUTION

By Descartes' Rule of Signs, the equation has exactly one positive real root. The only possible positive rational roots are 1 and 3, but $P(1) = -3$ and $P(3) = 27$, so neither actually turns out to be a root. So the positive real root must be irrational and, moreover, it must lie between 1 and 3, because $P(1)$ and $P(3)$ are opposite in sign. We calculate $P(x)$ for enough values of x between 1 and 3 to locate this root between successive tenths. To help us decide where to start this process, we observe that $P(1)$ is much closer to 0 than is $P(3)$. This leads us to guess that the zero we are looking for is much closer to 1 than to 3 (see Figure 3), so we begin our search at $x = 1.3$.

x	$P(x)$	
1.3	−1.713	
1.4	−1.096	
1.5	−0.375	opposite in sign
1.6	0.456	

This means that the root lies somewhere between 1.5 and 1.6. We now attempt to locate it between successive hundredths.

x	$P(x)$	
1.53	−0.138	
1.54	−0.056	opposite in sign
1.55	0.026	

So the root lies between 1.54 and 1.55. To see which of these is closer to the actual value, we calculate the value of P at 1.545, halfway between the two possibilities. We find that

$$P(1.545) \approx -0.015$$

Since $P(1.545)$ and $P(1.55)$ are opposite in sign, the root lies between 1.545 and 1.55 and thus is closer to 1.55 than to 1.54. So the positive irrational root is 1.55, correct to the nearest hundredth. ■

The process described in Example 2 can be continued to obtain the value of the root to any number of decimal places. The calculations involved are long and tedious, however. More efficient methods for approximating the roots of polynomial equations are studied in calculus courses.

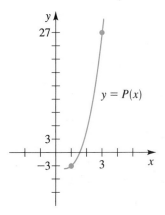

FIGURE 3

Part of the graph of
$P(x) = x^3 + x^2 - 2x - 3$

4.5 EXERCISES

1–8 ■ Show that the polynomial has a zero between the given integers, and then find that zero correct to the indicated number of decimal places.

1. $P(x) = x^3 + x - 1$;   between 0 and 1;
to the nearest tenth

2. $P(x) = x^3 + x - 7$;   between 1 and 2;   to one decimal

3. $P(x) = x^3 + x^2 + x - 2$;   between 0 and 1;
to one decimal

4. $P(x) = x^3 - x^2 - 5$;   between 2 and 3;
to two decimals

5. $P(x) = x^3 - 4x^2 + 2$;   between 0 and 1;
to two decimal places

6. $P(x) = 2x^4 - 4x^2 + 1$;   between 1 and 2;
to the nearest hundredth

7. $P(x) = 2x^4 - 4x^2 + 1$;   between −1 and 0;
to the nearest hundredth

8. $P(x) = x^5 - x^3 + 1$;   between −2 and −1;
to two decimal places

9–14 ■ Find the indicated irrational zero or root, correct to two decimal places.

9. The only real zero of $2x^3 - x^2 - 6$

10. The positive zero of $x^3 - 4x - 2$

11. The smaller positive zero of $x^3 - 3x + 1$

12. The negative root of $x^3 - 2x^2 - x + 3 = 0$

13. The positive root of $x^4 + 2x^3 + x^2 - 1 = 0$

14. The negative root of $x^4 + 2x^3 + x^2 - 1 = 0$

15–24 ■ Find all rational and irrational roots of the equation. Find the exact values of irrational roots using the quadratic formula whenever possible (as in Example 2 of Section 4.4). Otherwise, find approximate values, correct to two decimal places.

15. $x^3 + 2x^2 - 6x - 4 = 0$

16. $3x^3 + 10x^2 + 6x + 1 = 0$

17. $2x^5 + 3x^4 + x^2 + x - 1 = 0$

18. $x^4 - 4x^3 + 2x^2 - 5x + 10 = 0$

19. $x^3 - 3x^2 + 3 = 0$

20. $x^3 + 2x^2 - 4x - 7 = 0$

21. $x^5 + x^4 - 4x^3 - x^2 + 5x - 2 = 0$

22. $x^5 + 11x^4 + 43x^3 + 73x^2 + 52x + 12 = 0$

23. $2x^4 - 7x^3 + 9x^2 + 5x - 4 = 0$

24. $5x^4 + 13x^3 + 9x^2 - 16x + 4 = 0$

25. A rectangle with an area of 10 ft² has a diagonal that is 2 ft longer than one of its sides. Find the dimensions of the rectangle, correct to the nearest thousandth of a foot.

26. An open box with a volume of 1500 cm³ is to be constructed by taking a 20-cm-by-40-cm piece of cardboard, cutting squares of side length x cm from each corner, and folding up the sides. Show that this can be done in two different ways, and find the exact dimensions of the box in each case.

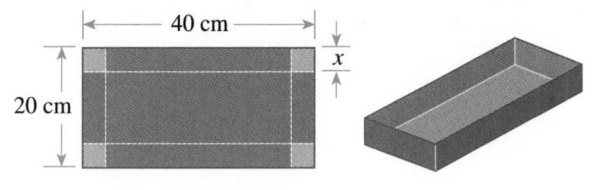

27. A rocket consists of a right circular cylinder of height 20 m surmounted by a cone whose height and diameter are equal and whose radius is the same as that of the cylindrical section. What should this radius be (correct to two decimals) if the total volume is to be $500\pi/3$ m³?

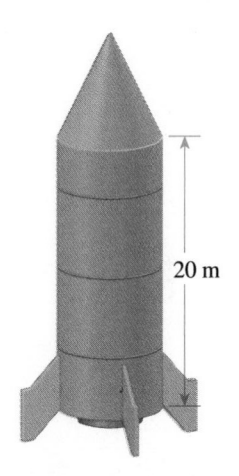

28. A rectangular box with a volume of $2\sqrt{2}$ ft³ has a square base. The diagonal of the box (between a pair of opposite corners) is 1 ft longer than each side of the base.

(a) If the base has sides of length x feet, show that

$$x^6 - 2x^5 - x^4 + 8 = 0$$

(b) Show that there are two different boxes that satisfy the given conditions. Find the dimensions in each case, correct to the nearest hundredth of a foot.

 4.6 USING ALGEBRA AND GRAPHING DEVICES TO SOLVE POLYNOMIAL EQUATIONS

In Section 4.2 we used graphing calculators and computer graphics programs to help graph polynomials. In this section we use the graphs that such devices produce, together with the theorems of Sections 4.4 and 4.5, to find the zeros of polynomials.

Recall that the *x*-intercepts of a function are the *x*-coordinates of the points at which the graph of the function intersects the *x*-axis (see Figure 1). Since the *x*-axis consists of all points in the coordinate plane whose *y*-coordinates are 0, this means that the *x*-intercepts of the function $y = f(x)$ are the solutions of the equation

$$0 = f(x)$$

So, to solve equations of this form, we can use the *x*-intercepts of the graph of $y = f(x)$, as we have seen in Section 3.3. In this section, we apply this procedure to polynomial equations.

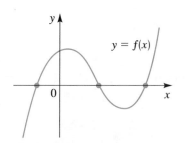

FIGURE 1
x-intercepts of $y = f(x)$

EXAMPLE 1 ■ Solving a Third-Degree Equation

Find all solutions of the equation $2x^3 - 15x^2 + 22x + 15 = 0$.

SOLUTION

We begin by graphing the function $y = 2x^3 - 15x^2 + 22x + 15$ using a graphing device. Since we want to be sure to see all the *x*-intercepts of the curve on our graph, we initially choose a wide viewing rectangle. By the Rational Roots Theorem, 15 and -15 are the largest and smallest possible rational roots of the equation, so let's choose the viewing rectangle $[-15, 15]$ by $[-50, 50]$. In this rectangle, the graphing device gives the graph in Figure 2. (We have used a scale of 1 unit along the *x*-axis and 10 units along the *y*-axis.)

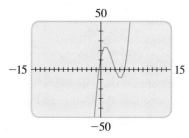

FIGURE 2
$P(x) = 2x^3 - 15x^2 + 22x + 15$

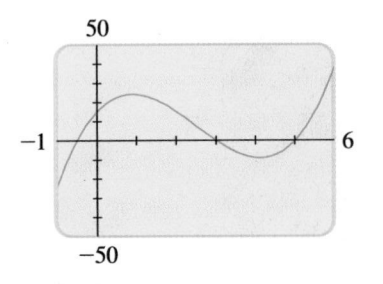

FIGURE 3

$P(x) = 2x^3 - 15x^2 + 22x + 15$

It appears that the graph has three x-intercepts, lying somewhere between -1 and 6. Since the x-intercepts are the only parts of the graph we are interested in, we change the viewing rectangle to $[-1, 6]$ by $[-50, 50]$. This gives the graph in Figure 3. The x-intercepts seem to be at $x = -\frac{1}{2}$, 3, and 5. To confirm this, we could use the zoom feature on the graphing device to get a closer look at each of the three x-intercepts. Or, to prove that these are in fact the solutions of the equation, we can substitute each of them into the equation:

$$2\left(-\tfrac{1}{2}\right)^3 - 15\left(-\tfrac{1}{2}\right)^2 + 22\left(-\tfrac{1}{2}\right) + 15 = -\tfrac{1}{4} - \tfrac{15}{4} - 11 + 15 = 0$$

$$2(3)^3 - 15(3)^2 + 22(3) + 15 = 54 - 135 + 66 + 15 = 0$$

$$2(5)^3 - 15(5)^2 + 22(5) + 15 = 250 - 375 + 110 + 15 = 0$$

The solutions of the equation are $-\frac{1}{2}$, 3, and 5. ∎

How can we be sure that the equation in Example 1 has no solutions other than the ones we found? Remember that the Factor Theorem says that if c is a solution of the polynomial equation $P(x) = 0$, then $x - c$ is a factor of $P(x)$. If the equation in Example 1 had *four* different solutions (call them c_1, c_2, c_3, and c_4), then the polynomial $2x^3 - 15x^2 + 22x + 15$ would have to be evenly divisible by $x - c_1$, $x - c_2$, $x - c_3$, and $x - c_4$, and hence by the product

$$(x - c_1)(x - c_2)(x - c_3)(x - c_4)$$

as well. But if this product is multiplied out, it contains the term x^4, so it is impossible for this to divide the third-degree polynomial $2x^3 - 15x^2 + 22x + 15$ evenly. This illustrates the general principle given in the following box.

THE NUMBER OF ROOTS OF A POLYNOMIAL EQUATION

If $P(x) = 0$ is a polynomial of degree n, then the equation

$$P(x) = 0$$

can have at most n different real roots.

In Examples 2 and 3 we can see that the number of real roots of a polynomial equation of degree n can in fact be less than n. The preceding principle just says that there *cannot* be *any more* roots than the degree of the equation. We will consider this fact in more detail in Section 4.7, when we study the Complete Factorization Theorem.

Another way to ensure that we are finding all the real solutions of a polynomial equation is to use the Upper and Lower Bounds Theorem, as in the next example.

EXAMPLE 2 ■ Solving a Fourth-Degree Equation

Find all real solutions of the following equation, correct to the nearest tenth.

$$3x^4 + 4x^3 - 7x^2 - 2x - 3 = 0$$

SOLUTION

We use the Upper and Lower Bounds Theorem to see where the roots can be found.

This equation can have at most four real roots because it is of degree 4. First we use the Upper and Lower Bounds Theorem to find two numbers between which all the solutions must lie. This will allow us to choose a viewing rectangle that is certain to contain all the x-intercepts of the polynomial function. We use synthetic division and proceed by trial and error.

		3	4	−7	−2	−3
1		3	7	0	−2	−5
2		3	10	13	24	45

← all entries positive

Thus 2 is an upper bound for the roots of the equation. Now we look for a lower bound.

		3	4	−7	−2	−3
−1		3	1	−8	6	−9
−2		3	−2	−3	4	−11
−3		3	−5	8	−26	75

← entries alternate in sign

Thus −3 is a lower bound for the solutions. This means that if we use the viewing rectangle $[-3, 2]$ by $[-20, 20]$ we are certain to see all the x-intercepts of the function $y = 3x^4 + 4x^3 - 7x^2 - 2x - 3$. With this rectangle we obtain the graph in Figure 4. The graph has two x-intercepts, one between −3 and −2, and the other between 1 and 2. To determine the values of these intercepts to the nearest tenth, we change the scale along the x-axis to 0.1, and then zoom in to examine each of them more closely. From the resulting graphs in Figures 5 and 6, we see that the negative x-intercept is close to −2.3 and the positive one is close to 1.3.

The solutions of the equation, to the nearest tenth, are −2.3 and 1.3.

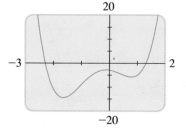

FIGURE 4

$y = 3x^4 + 4x^3 - 7x^2 - 2x - 3$

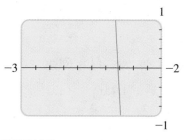

FIGURE 5

$y = 3x^4 + 4x^3 - 7x^2 - 2x - 3$

FIGURE 6

$y = 3x^4 + 4x^3 - 7x^2 - 2x - 3$ ■

In the next example, we see that by zooming in closer and closer to an x-intercept, we can find the solution of an equation to any desired degree of accuracy (within the limits allowed by the graphing device we are using).

EXAMPLE 3 ■ Solving a Third-Degree Equation

Find all solutions of the following equation, correct to three decimal places.

$$-3x^3 + 4x^2 + 5 = 0$$

SOLUTION

We use Descartes' Rule of Signs and see that there is just one root, and it is positive.

Since the polynomial in the equation has one sign change, it has one positive zero (by Descartes' Rule of Signs). Changing x to $-x$ in the polynomial gives

$$-3(-x)^3 + 4(-x)^2 + 5 = 3x^3 + 4x^2 + 5$$

which has no sign change, so there are no negative solutions to the equation. Thus, the equation has only one solution and it is positive. Graphing the function $y = -3x^3 + 4x^2 + 5$ in the viewing rectangle $[0, 3]$ by $[-10, 10]$ gives the graph in Figure 7. The solution of the equation is slightly less than 2. To find its value correct to three decimal places, we change the scale on the x-axis to 0.001 and zoom in several times to get the graph in Figure 8. Using the cursor (or by inspection), we see that the x-intercept is 1.831, correct to the nearest thousandth. The only real solution of the equation is $x \approx 1.831$.

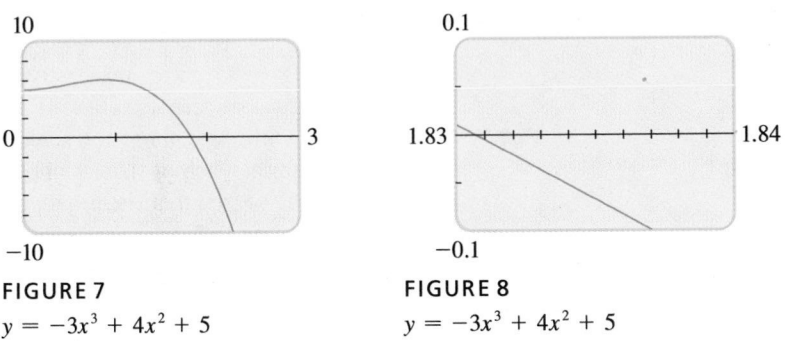

FIGURE 7
$y = -3x^3 + 4x^2 + 5$

FIGURE 8
$y = -3x^3 + 4x^2 + 5$

EXAMPLE 4 ■ Solving a Fifth-Degree Equation

Find all real solutions of the following equation, correct to two decimals.

$$x^5 - 1.45x^4 - 3.19x^3 + 1.31x^2 + 1.50x - 0.62 = 0$$

SOLUTION

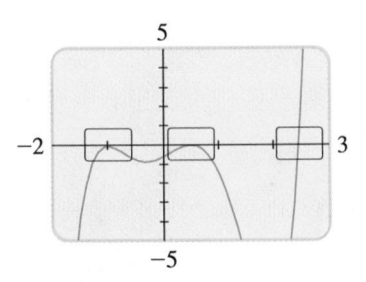

FIGURE 9

Using the viewing rectangle $[-2, 3]$ by $[-5, 5]$, we obtain the graph shown in Figure 9. The graph appears to touch the x-axis near -1 and $\frac{1}{2}$ and to cross the x-axis between 2 and 3. However, if we zoom in to each of these three locations, we see that the graph does not actually touch the x-axis near $x = -1$, so no solution occurs here (Figure 10). At $x = \frac{1}{2}$ the graph does touch the x-axis, so even though the graph doesn't *cross* the axis, $\frac{1}{2}$ is still an x-intercept (Figure 11). You can verify by substitution that $x = 0.5$ is in fact an *exact* solution of the equation (although the calculations involve a lot of work!). Zooming in several times on the remaining x-intercept shows that its value is 2.45, correct to the nearest hundredth (Figure 12). The solutions of the equation are

$$x = 0.50 \qquad \text{and} \qquad x \approx 2.45$$

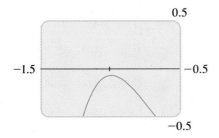

FIGURE 10

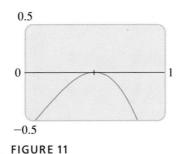

FIGURE 11

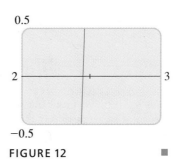

FIGURE 12   ■

EXAMPLE 5 ■ Determining the Size of a Fuel Tank

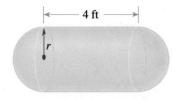

FIGURE 13

A fuel tank consists of a cylindrical center section that is 4 ft long and two hemispherical end sections, as shown in Figure 13. If the tank has a volume of 100 ft³, what is the radius r shown in the figure, correct to the nearest hundredth of a foot?

SOLUTION

Volume of a cylinder: $V = \pi r^2 h$

Using the volume formula listed on the inside of the front cover of this book, we see that the volume of the cylindrical section of the tank is

$$\pi \cdot r^2 \cdot 4$$

The two hemispherical parts together form a complete sphere whose volume is

Volume of a sphere: $V = \frac{4}{3}\pi r^3$

$$\frac{4}{3}\pi r^3$$

Because the total volume of the tank is 100 ft³, we get the following equation:

$$\frac{4}{3}\pi r^3 + 4\pi r^2 = 100$$

A negative solution for r would be meaningless in this physical situation, and by substitution we can verify that $r = 3$ leads to a tank that is over 226 ft³ in volume, much larger than the required 100 ft³. Thus, we know the correct radius lies some where between 0 and 3 ft, and so we use a viewing rectangle

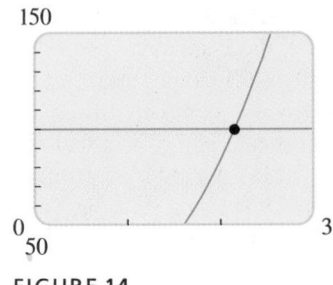

FIGURE 14
$y = \frac{4}{3}\pi x^3 + 4\pi x^2$ and $y = 100$

of $[0, 3]$ by $[50, 150]$ to graph the function $y = \frac{4}{3}\pi x^3 + 4\pi x^2$, as shown in Figure 14. Since we want the value of this function to be 100, we also graph the horizontal line $y = 100$ in the same viewing rectangle. The correct radius will be the x-coordinate of the point of intersection of the curve and the line. Using the cursor and zooming in, we see that at the point of intersection $x \approx 2.15$, correct to two decimals. Thus the tank has a radius of about 2.15 ft. ∎

Note that we also could have solved the equation in Example 5 by first writing it as

$$\frac{4}{3}\pi r^3 + 4\pi r^2 - 100 = 0$$

and then finding the x-intercept of the function $y = \frac{4}{3}\pi x^3 + 4\pi x^2 - 100$.

4.6 EXERCISES

1–10 ■ All the real solutions of the given equation are rational. List all possible rational roots using the Rational Roots Theorem, and then graph the polynomial in the given viewing rectangle to determine which values are actually the solutions of the equation. (All solutions can be seen in the given viewing rectangle.)

1. $x^3 - 3x^2 - 4x + 12 = 0$; $[-4, 4]$ by $[-15, 15]$

2. $x^4 - 5x^2 + 4 = 0$; $[-4, 4]$ by $[-30, 30]$

3. $2x^3 + x^2 + 6x + 3 = 0$; $[-2, 2]$ by $[-10, 10]$

4. $4x^3 - 5x + 6 = 0$; $[-3, 3]$ by $[-30, 30]$

5. $2x^4 - 5x^3 - 14x^2 + 5x + 12 = 0$;
   $[-2, 5]$ by $[-40, 40]$

6. $2x^4 - 5x^3 - 8x^2 + 17x - 6 = 0$; $[-3, 5]$ by $[-60, 60]$

7. $2x^3 - x^2 = 8x + 5$; $[-2, 3]$ by $[-30, 30]$

8. $x^4 + 12x + 36 = 2x^3 + 11x^2$; $[-4, 4]$ by $[-50, 50]$

9. $3x^3 + 8x^2 + 5x + 2 = 0$; $[-3, 3]$ by $[-10, 10]$

10. $4x^4 + 4x^3 + 7x^2 = x + 2$; $[-2, 2]$ by $[-40, 40]$

11–24 ■ Find all real solutions of the equation, correct to two decimals.

11. $x^3 - 5x^2 - 4 = 0$

12. $x^4 - x - 4 = 0$

13. $3x^3 + x^2 + x - 2 = 0$

14. $2x^3 - 8x^2 + 9x - 9 = 0$

15. $10x^4 - 9x^3 - 11x^2 + 5x - 3 = 0$

16. $3x^4 + 8x^3 + 2x^2 + 5x + 2 = 0$

17. $x^3 + 6 = 6x^2$

18. $x^4 + x^3 = 4$

19. $x^4 + 8x + 16 = 2x^3 + 8x^2$

20. $2x^5 + 9x^4 - 5x^2 = 21x^3 + 11x - 2$

21. $4.00x^4 + 4.00x^3 - 10.96x^2 - 5.88x + 9.09 = 0$

22. $x^5 + 2.00x^4 + 0.96x^3 + 5.00x^2 + 10.00x + 4.80 = 0$

23. $x^5 - 3x^4 - 4x^3 + 12x^2 + 4x - 12 = 0$

24. $3x^6 + 3x^5 = 7x^3 + x^2 - 2$

25. A grain silo consists of a cylindrical main section and a hemispherical roof. If the total volume of the silo (including the part inside the roof section) is 15,000 ft^3, and the cylindrical part is 30 ft tall, what is the radius of the silo, correct to the nearest tenth of a foot?

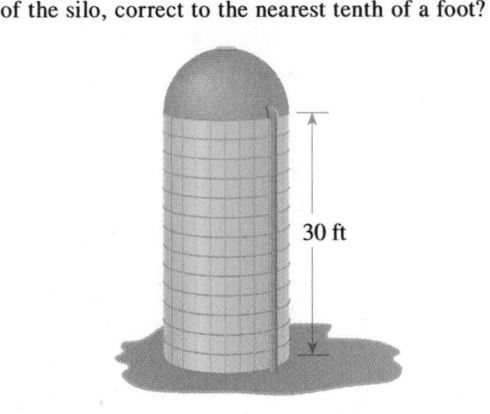

30 ft

26. Suppose a silo like the one described in Exercise 25 has a volume of 20,000 ft^3, and the total height of the silo at the top of the roof is 40 ft. What is the approximate radius of the silo in this case?

27. A piece of sheet metal measuring 18 in. square is to be made into a box with an open top by cutting equal squares from each corner and then folding up and soldering the sides. The resulting box is to have a volume of 400 in^3. Show that this can be done in two different ways, and find the dimensions of the box, correct to the nearest tenth of an inch, for each of the two cases.

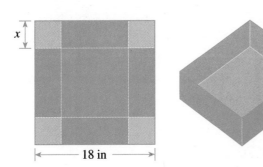

28. A rectangular parcel of land has an area of 5000 ft^2. A diagonal between opposite corners is measured to be

10 ft longer than one side of the parcel. What are the dimensions of the land, correct to the nearest foot?

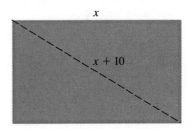

29. Snow began falling at noon on Sunday. The amount of snow on the ground at a certain location at time t was given by the function

$$h(t) = 11.60t - 12.41t^2 + 6.20t^3$$
$$- 1.58t^4 + 0.20t^5 - 0.01t^6$$

where t is measured in days from the start of the snowfall and $h(t)$ is the depth of snow in inches. Sketch a graph of this function and use your graph to answer the following questions.
(a) What happened shortly after noon on Tuesday?
(b) Was there ever more than 5 inches of snow on the ground? If so, on what day(s)?
(c) On what day and at what time (to the nearest hour) did the snow disappear completely?

4.7

COMPLEX ROOTS AND THE FUNDAMENTAL THEOREM OF ALGEBRA

We have already seen that the quadratic equation $ax^2 + bx + c = 0$, with $a \neq 0$, has the solutions

$$x = \frac{-b \pm \sqrt{b^2 - 4ac}}{2a}$$

If $b^2 - 4ac < 0$, then the equation has no real solution. But in the complex number system, this equation will always have solutions, because negative numbers have square roots in this expanded setting.

EXAMPLE 1 ■ Using the Quadratic Formula to Find Complex Roots

Solve each of the following equations.

(a) $x^2 + 9 = 0$

(b) $x^2 + 4x + 5 = 0$

SOLUTION

(a) $x^2 + 9 = 0$ means $x^2 = -9$, so $x = \pm\sqrt{-9} = \pm i\sqrt{9} = \pm 3i$. The solutions are $3i$ and $-3i$.

(b) By the quadratic formula,

$$x = \frac{-4 \pm \sqrt{4^2 - 4 \cdot 5}}{2}$$

$$= \frac{-4 \pm \sqrt{-4}}{2}$$

$$= \frac{-4 \pm 2i}{2} = -2 \pm i$$

so the solutions are $-2 + i$ and $-2 - i$.   ■

EXAMPLE 2 ■ Factoring and Using the Quadratic Formula to Find Complex Roots

Find all the roots of the equation $x^6 - 64 = 0$.

SOLUTION

$$x^6 - 64 = (x^3)^2 - 8^2$$

$$= (x^3 - 8)(x^3 + 8)$$

$$= (x - 2)(x^2 + 2x + 4)(x + 2)(x^2 - 2x + 4)$$

Difference of squares formula
Difference of cubes and sum of cubes formulas

This expression will equal 0 when any factor is 0, so we find the solutions as follows.

$x - 2 = 0$ means $x = 2$

$x^2 + 2x + 4 = 0$ means $x = \dfrac{-2 \pm \sqrt{2^2 - 4 \cdot 4}}{2}$

$$= \frac{-2 \pm \sqrt{-12}}{2} = -1 \pm i\sqrt{3}$$

$x + 2 = 0$ means $x = -2$

$x^2 - 2x + 4 = 0$ means $x = \dfrac{2 \pm \sqrt{(-2)^2 - 4 \cdot 4}}{2}$

$$= \frac{2 \pm \sqrt{-12}}{2} = 1 \pm i\sqrt{3}$$

The roots of the equation are

$$2, \quad -2, \quad -1 + i\sqrt{3}, \quad -1 - i\sqrt{3}, \quad 1 + i\sqrt{3}, \quad \text{and} \quad 1 - i\sqrt{3} \qquad ■$$

THE FUNDAMENTAL THEOREM OF ALGEBRA

It is a remarkable fact that adding just the number $\sqrt{-1}$ and its real multiples to the real number system is sufficient to provide a number system in which *every* polynomial equation has a root. Although we will not prove this fact (a proof requires mathematical expertise well beyond the scope of this book), it nevertheless forms the basis for much of our work in solving polynomial equations. This theorem was proved by the German mathematician C. F. Gauss in 1799.

FUNDAMENTAL THEOREM OF ALGEBRA

Every polynomial

$$P(x) = a_n x^n + a_{n-1} x^{n-1} + \cdots + a_1 x + a_0 \qquad (n \geq 1, a_n \neq 0)$$

with complex coefficients has at least one complex zero.

Because any real number is also a complex number, the theorem applies to polynomials with real coefficients as well.

Since every zero c of a polynomial corresponds to a factor of the form $x - c$ (by the Factor Theorem), the Fundamental Theorem of Algebra ensures that we can factor any polynomial $P(x)$ of degree n as follows:

$$P(x) = (x - c_1) \cdot Q_1(x)$$

where $Q_1(x)$ is of degree $n - 1$ and c_1 is a zero of $P(x)$. But now applying the Fundamental Theorem to the quotient $Q_1(x)$ gives us the factorization

$$P(x) = (x - c_1) \cdot (x - c_2) \cdot Q_2(x)$$

where $Q_2(x)$ is of degree $n - 2$ and c_2 is a zero of $Q_1(x)$. Continuing this process for n steps, we will get a final quotient $Q_n(x)$ of degree 0, which is therefore a nonzero constant that we will call a. This proves the following corollary of the Fundamental Theorem of Algebra.

COMPLETE FACTORIZATION THEOREM

If $P(x)$ is a polynomial of degree $n > 0$, then there exist complex numbers $a, c_1, c_2, \ldots, c_n$ (with $a \neq 0$) such that

$$P(x) = a(x - c_1)(x - c_2) \cdots (x - c_n)$$

The number a is clearly the coefficient of x^n in $P(x)$. The numbers $c_1, c_2, \ldots, c_n$ are the zeros of $P(x)$ (by the Factor Theorem). These need not all be different: If the factor $x - c$ appears k times in the complete factorization of $P(x)$, then we say that c is a zero of **multiplicity** k.

$P(x)$ can have no zero other than $c_1, c_2, \ldots, c_n$, because if

$$P(c) = a(c - c_1)(c - c_2)\cdots(c - c_n) = 0$$

then at least one of the factors $c - c_i$ must be zero, so $c = c_i$ for some $i \in \{1, 2, \ldots, n\}$. We have thus proved the following theorem.

ZEROS THEOREM

Every polynomial of degree $n \geq 1$ has exactly n zeros, provided that a zero with multiplicity k is counted k times.

In Example 2, for instance, the polynomial $x^6 - 64$ is of degree 6, and it has exactly 6 zeros.

EXAMPLE 3 ■ Factoring a Polynomial with Complex Zeros

Find the complete factorization and all five zeros of the polynomial

$$P(x) = 3x^5 + 24x^3 + 48x$$

SOLUTION

The terms of P have $3x$ as a common factor, so we get the following factorization:

$$P(x) = 3x(x^4 + 8x^2 + 16)$$

$$= 3x(x^2 + 4)^2$$

To factor $x^2 + 4$, note that $2i$ and $-2i$ are zeros of this polynomial. Thus, $x^2 + 4 = (x - 2i)(x + 2i)$, and so

$$P(x) = 3x[(x - 2i)(x + 2i)]^2$$

$$= 3x(x - 2i)(x - 2i)(x + 2i)(x + 2i)$$

Setting each factor equal to zero in turn, we see that the zeros of P are 0, $2i$, and $-2i$. However, $2i$ and $-2i$ are each counted twice, since the factor of P that corresponds to each occurs twice in the factorization of P. Each of these is a zero of multiplicity 2 (or a *double zero*).   ■

EXAMPLE 4 ■ Finding Polynomials with Specified Zeros

Find a polynomial that satisfies the given description.

(a) A polynomial $P(x)$ of degree 3, with zeros 1, 2, and -4 and constant coefficient 16.
(b) A polynomial $Q(x)$ of degree 4, with zeros i, $-i$, 2, and -2 and with $Q(3) = 25$.
(c) A polynomial $R(x)$ of degree 4, with zeros -2 and 0, where -2 is a zero of multiplicity three.

SOLUTION

(a) From the given description, we see that $P(x)$ has the complete factorization $a(x - 1)(x - 2)(x - (-4))$ for some a. Thus

$$P(x) = a(x - 1)(x - 2)(x + 4)$$

$$= a(x^2 - 3x + 2)(x + 4)$$

$$= a(x^3 + x^2 - 10x + 8)$$

$$= ax^3 + ax^2 - 10ax + 8a$$

Since the constant coefficient is 16, we see that $a = 2$, so

$$P(x) = 2x^3 + 2x^2 - 20x + 16$$

(b) The required polynomial has the form

$$Q(x) = a(x - i)(x - (-i))(x - 2)(x - (-2))$$

$$= a(x^2 + 1)(x^2 - 4)$$

$$= a(x^4 - 3x^2 - 4)$$

We know that $Q(3) = a(3^4 - 3 \cdot 3^2 - 4) = 50a = 25$, so $a = \frac{1}{2}$, and

$$Q(x) = \tfrac{1}{2}x^4 - \tfrac{3}{2}x^2 - 2$$

(c) We require

$$R(x) = a(x - (-2))^3(x - 0)$$

$$= a(x + 2)^3 x$$

$$= a(x^3 + 6x^2 + 12x + 8)x \qquad \text{Special product formula 4 (Section 1.4)}$$

$$= a(x^4 + 6x^3 + 12x^2 + 8x)$$

Carl Friedrich Gauss
(1777–1855) is considered the
greatest mathematician of modern
times. He was referred to by his
contemporaries as "The Prince of
Mathematics." Gauss was born
into a poor family; his father
made a living as a mason. As a
very small child he found a
calculation error in his father's
accounts. This was the first of
many incidents that gave evidence
of his mathematical precocity.
(See also page 609.) At the age of
19 Gauss demonstrated that the
regular 17-sided polygon can be
constructed with straightedge and
compass alone. This was remarkable because, since the time of
Euclid, it was thought that the
only regular polygons construct-
ible in this way were the triangle
and pentagon. Because of this
discovery Gauss decided to
pursue a career in mathematics
instead of languages, his other
passion. In his doctoral disserta-
tion, written at the age of 22,
Gauss proved the Fundamental
Theorem of Algebra: A polynomial of degree n with complex
coefficients has n roots. His other
accomplishments range over
every branch of mathematics, as
well as physics and astronomy.

Since we are given no information about R other than its zeros and their
multiplicity, we can choose any number for a. If we use $a = 1$, we get

$$R(x) = x^4 + 6x^3 + 12x^2 + 8x$$

EXAMPLE 5 ■ Finding All the Zeros of a Polynomial

Find all five zeros of $P(x) = 3x^5 - 2x^4 - x^3 - 12x^2 - 4x$.

SOLUTION

By Descartes' Rule of Signs, $P(x)$ has one positive real zero and either three or
one negative real zero(s). Since $x = 0$ is obviously a zero (but is neither positive
nor negative), it follows that there are either five real zeros and no imaginary
zero, or three real zeros and two imaginary zeros. Checking through the list of
possible rational zeros of $P(x)/x$, we see that $P(2) = 0$ and $P(-\frac{1}{3}) = 0$, so by
the Factor Theorem, x, $x - 2$, and $x + \frac{1}{3}$ are factors. We divide $P(x)$ by each of
these factors, one after another (using synthetic or long division), to find that

$$P(x) = 3x^5 - 2x^4 - x^3 - 12x^2 - 4x$$

$$= x(3x^4 - 2x^3 - x^2 - 12x - 4) \qquad \text{Factor } x$$

$$= x(x - 2)(3x^3 + 4x^2 + 7x + 2) \qquad \text{Factor } x - 2$$

$$= x(x - 2)\left(x + \tfrac{1}{3}\right)(3x^2 + 3x + 6) \qquad \text{Factor } x + \tfrac{1}{3}$$

$$= x(x - 2)\left(x + \tfrac{1}{3}\right)(3)(x^2 + x + 2) \qquad \text{Factor } 3$$

$$= x(x - 2)(3x + 1)(x^2 + x + 2) \qquad \text{Multiply}$$

The zeros of the quadratic factor are

$$x = \frac{-1 \pm \sqrt{1 - 8}}{2} \qquad \text{Quadratic formula}$$

$$= -\frac{1}{2} \pm i\frac{\sqrt{7}}{2}$$

so the zeros of $P(x)$ are

$$0, \quad 2, \quad -\frac{1}{3}, \quad -\frac{1}{2} + i\frac{\sqrt{7}}{2}, \quad \text{and} \quad -\frac{1}{2} - i\frac{\sqrt{7}}{2}$$

As you may have noticed from the examples so far, the imaginary roots of
polynomial equations with real coefficients come in pairs. Whenever $a + bi$ is a
root, so is its complex conjugate $a - bi$. This is always the case, as the following
theorem states.

> ### CONJUGATE ROOTS THEOREM
>
> If the polynomial $P(x)$ of degree $n > 0$ has real coefficients, and if the complex number z is a root of the equation $P(x) = 0$, then so is its complex conjugate $\bar{z}$.

The proof of this theorem is given at the end of this section.

EXAMPLE 6 ■ A Polynomial with Specified Imaginary Zeros

Find a polynomial $P(x)$ of degree 5 that has integer coefficients, and zeros $\frac{1}{2}$, $3 - i$, and $2i$.

SOLUTION

Since $3 - i$ and $2i$ are zeros, then so are $3 + i$ and $-2i$ by the Conjugate Roots Theorem. This means that $P(x)$ has all the factors in the following product:

$$\left(x - \tfrac{1}{2}\right)[x - (3 - i)][x - (3 + i)](x - 2i)(x + 2i)$$

$$= \left(x - \tfrac{1}{2}\right)[(x - 3) + i][(x - 3) - i](x^2 + 4)$$

$$= \left(x - \tfrac{1}{2}\right)[(x - 3)^2 - i^2](x^2 + 4)$$

$$= \left(x - \tfrac{1}{2}\right)(x^2 - 6x + 10)(x^2 + 4)$$

$$= x^5 - \tfrac{13}{2}x^4 + 17x^3 - 31x^2 + 52x - 20$$

Multiplying by 2 to make all coefficients integers, we get

$$P(x) = 2x^5 - 13x^4 + 34x^3 - 62x^2 + 104x - 40$$

Any other polynomial that satisfies the given requirements must be an integer multiple of this one. ■

EXAMPLE 7 ■ Finding All Roots of an Equation, Given One Imaginary Root

Find all roots of the equation $x^4 - 12x^3 + 56x^2 - 120x + 96 = 0$, given that one root is $3 + i\sqrt{3}$.

SOLUTION

The complex conjugate of $3 + i\sqrt{3}$ is also a root, so both $x - \left(3 + i\sqrt{3}\right)$ and $x - \left(3 - i\sqrt{3}\right)$, and hence their product, must divide the polynomial in the equation:

$$\left[x - \left(3 + i\sqrt{3}\right)\right] \cdot \left[x - \left(3 - i\sqrt{3}\right)\right] = x^2 - 6x + 12$$

We divide the original polynomial by $x^2 - 6x + 12$:

$$
\begin{array}{r}
x^2 - 6x + 8 \\
x^2 - 6x + 12\overline{\smash{\big)}\,x^4 - 12x^3 + 56x^2 - 120x + 96} \\
\underline{x^4 - 6x^3 + 12x^2} \\
-6x^3 + 44x^2 - 120x \\
\underline{-6x^3 + 36x^2 - 72x} \\
8x^2 - 48x + 96 \\
\underline{8x^2 - 48x + 96} \\
0
\end{array}
$$

The quotient factors as

$$x^2 - 6x + 8 = (x - 2)(x - 4)$$

so the complete factorization of the original polynomial is

$$x^4 - 12x^3 + 56x^2 - 120x + 96$$
$$= \left[x - (3 + i\sqrt{3}\,)\right]\left[x - (3 - i\sqrt{3}\,)\right](x - 2)(x - 4)$$

and the roots of the equation are $3 + i\sqrt{3}$, $3 - i\sqrt{3}$, 2, and 4. ∎

PROOF OF THE CONJUGATE ROOTS THEOREM

Let

$$P(x) = a_n x^n + a_{n-1}x^{n-1} + \cdots + a_1 x + a_0$$

where each coefficient is real. Suppose that $P(z) = 0$. To prove the Conjugate Roots Theorem, we must prove that $\bar{z}$ is also a zero of P. We use the facts that the complex conjugate of a sum of two complex numbers is the sum of the conjugates and that the conjugate of a product is the product of the conjugates (see Exercises 59 and 60 of Section 2.4).

$$P(\bar{z}) = a_n(\bar{z})^n + a_{n-1}(\bar{z})^{n-1} + \cdots + a_1\bar{z} + a_0$$

$$= \overline{a_n z^n} + \overline{a_{n-1}z^{n-1}} + \cdots + \overline{a_1 z} + \overline{a_0} \qquad \text{Because the coefficients are real}$$

$$= \overline{a_n z^n} + \overline{a_{n-1}z^{n-1}} + \cdots + \overline{a_1 z} + \overline{a_0}$$

$$= \overline{a_n z^n + a_{n-1}z^{n-1} + \cdots + a_1 z + a_0}$$

$$= \overline{P(z)} = \bar{0} = 0$$

This derivation shows that the conjugate $\bar{z}$ is also a zero of $P(x)$, and we have proved the theorem. □

4.7 EXERCISES

1–14 ■ Find all solutions of the equation.

1. $x^2 + 4 = 0$

2. $25x^2 + 9 = 0$

3. $x^2 - x + 1 = 0$

4. $x^2 + 2x + 2 = 0$

5. $x^2 + 4x + 8 = 0$

6. $2x^2 + 2x + 1 = 0$

7. $3x^2 - 5x + 4 = 0$

8. $2x^2 - 3x + 2 = 0$

9. $x^2 - 8x + 17 = 0$

10. $3x^2 - 4x + 2 = 0$

11. $t + 3 + \dfrac{3}{t} = 0$

12. $\theta^3 + \theta^2 + \theta = 0$

13. $z^2 - iz = 0$

14. $2z^3 + iz^2 = 0$

15–20 ■ Find a polynomial with integer coefficients that satisfies the given conditions.

15. $P(x)$ has degree 3, zeros 2 and i, and leading coefficient 1.

16. $Q(x)$ has degree 4, zeros $1 + i$ and 1, with 1 a zero of multiplicity 2, and $Q(0) = 4$.

17. $S(x)$ has degree 4, zeros $1 + i$ and $3 - 4i$, and the coefficient of x^2 is 39.

18. $R(x)$ has degree 3, with 2 a zero of multiplicity 3.

19. $T(x)$ has constant coefficient 8 and is of the smallest possible degree consistent with having $1 - i$ as a zero of multiplicity 2.

20. $U(x)$ has degree 3, with $U(i) = 0$, $U(2) = 0$, and $U(1) = -10$.

21–24 ■ Show that the indicated value of x is a solution of the equation, and then find all solutions.

21. $x^3 - 2x^2 + 4x - 8 = 0$, $x = 2i$

22. $x^3 + 5x^2 + 8x + 6 = 0$, $x = -1 + i$

23. $2x^4 + 9x^2 + 4 = 0$, $x = 2i$

24. $x^5 + 5x^3 + 4x = 0$, $x = i$

25–42 ■ Find all solutions of the equation.

25. $x^4 - 1 = 0$

26. $x^3 - 64 = 0$

27. $x^3 + 8 = 0$

28. $x^4 + 4 = 0$

29. $x^4 - 16 = 0$

30. $16x^4 - 81 = 0$

31. $x^6 - 729 = 0$

32. $x^4 + 2x^2 + 1 = 0$

33. $x^4 + 10x^2 + 25 = 0$

34. $x^6 + 7x^3 - 8 = 0$

35. $x^3 + 2x^2 + 4x + 8 = 0$

36. $x^3 - 7x^2 + 17x - 15 = 0$

37. $x^3 - 2x^2 + 2x - 1 = 0$

38. $x^3 + 7x^2 + 18x + 18 = 0$

39. $x^3 - 3x^2 + 3x - 2 = 0$

40. $2x^3 - 8x^2 + 9x - 9 = 0$

41. $x^4 + x^3 + 7x^2 + 9x - 18 = 0$

42. $x^5 + x^3 + 8x^2 + 8 = 0$ [*Hint:* Factor by grouping.]

43–48 ■ Find the complete factorization of the polynomial.

43. $x^3 + 27$

44. $x^4 - 625$

45. $x^6 - 64$

46. $x^5 + 3x^3 + 2x$

47. $2x^3 + 7x^2 + 12x + 9$

48. $x^4 - x^3 + 7x^2 - 9x - 18$

49. Show that every polynomial with real coefficients and odd degree has at least one real root. [*Hint:* Use the Conjugate Roots Theorem.]

50–51 ■ Find the value of the polynomial at the given number.

50. $2x^3 + ix^2 - 3ix - (50 + 5i)$, 3

51. $3x^3 - x^2 + x - 4$, $2i$

52. (a) Show that $2i$ and $1 - i$ are both solutions of the equation
$$x^2 - (1 + i)x + (2 + 2i) = 0$$
but that their complex conjugates $-2i$ and $1 + i$ are not.

(b) Explain why the result of part (a) does not violate the Conjugate Roots Theorem.

53. (a) Find the polynomial with *real* coefficients of the smallest possible degree for which i and $1 + i$ are zeros and in which the coefficient of the highest power is 1.

(b) Find the polynomial with *complex* coefficients of the smallest possible degree for which i and $1 + i$ are zeros and in which the coefficient of the highest power is 1.

54. The steps in this problem provide a proof of the fact that it is impossible for the graphs of two different polynomials, each of degree $\leq n$, to intersect at more than n points.

(a) Let

$$P(x) = a_n x^n + a_{n-1} x^{n-1} + \cdots + a_1 x + a_0$$

and

$$Q(x) = b_n x^n + b_{n-1} x^{n-1} + \cdots + b_1 x + b_0$$

Suppose that the graphs of P and Q intersect in the $n + 1$ points (x_1, y_1), (x_2, y_2), ..., (x_n, y_n), and (x_{n+1}, y_{n+1}). Let $F(x) = P(x) - Q(x)$. Show that $F(x)$ has at least $n + 1$ zeros.

(b) Show that $F(x) = 0$ for all x.   [*Hint:* Use the degree of F and the Zeros Theorem.]

(c) Conclude that P and Q are the same polynomial.

55. By the Zeros Theorem, every nth-degree polynomial equation has exactly n solutions (including possibly some that are repeated). Some of these may be real and some may be imaginary. Use a graphing calculator to determine how many real and how many imaginary solutions each of the following equations has.

(a) $x^4 - 2x^3 - 11x^2 + 12x = 0$

(b) $x^4 - 2x^3 - 11x^2 + 12x - 5 = 0$

(c) $x^4 - 2x^3 - 11x^2 + 12x + 40 = 0$

4.8   RATIONAL FUNCTIONS

A **rational function** is a function of the form

$$r(x) = \frac{P(x)}{Q(x)}$$

where P and Q are polynomials. We assume that $P(x)$ and $Q(x)$ have no factor in common. Although polynomial functions are defined for all real values of x, rational functions are not defined for those values of x for which the denominator $Q(x)$ is 0. The x-intercepts (if any) of r are the zeros of the numerator $P(x)$, since a fraction is 0 only when its numerator is 0.

EXAMPLE 1 ■ The Domain and Intercepts of a Rational Function

Find the domain, the x-intercepts, and the y-intercept of the function

$$r(x) = \frac{x^2 - 2x - 3}{2x^2 - x}$$

SOLUTION

We factor the numerator and denominator to write

$$r(x) = \frac{(x + 1)(x - 3)}{x(2x - 1)}$$

A fraction is 0 if any only if its numerator is 0.

The function is defined for all x except those for which the denominator is 0, so the domain of $r(x)$ consists of all real numbers except 0 and $\frac{1}{2}$.

The x-intercepts are the zeros of the numerator, $x = -1$ and $x = 3$. The y-intercept is the value of the function when $x = 0$. Since the given function $r(x)$ is not defined for $x = 0$, the graph has no y-intercept.   ■

The most important feature that distinguishes the graphs of rational functions is the presence of *asymptotes*. Before we formulate a precise definition of this word, we consider an example to illustrate the concept.

EXAMPLE 2 ■ A Rational Function with One Vertical and One Horizontal Asymptote

Sketch a graph of the function $y = \dfrac{x - 3}{x - 2}$.

SOLUTION

The function is not defined for $x = 2$, so we first examine the nature of the function for values of x near 2.

We see from the first table that as x approaches 2 from the left (that is, gets progressively closer to 2 while remaining smaller than 2), the values of y increase without bound. In fact, by taking x sufficiently close to 2 on the left, we can make y larger than any given number. We describe this situation by saying "y approaches infinity as x approaches 2 from the left" and we write this phrase using the notation

$$y \to \infty \quad \text{as} \quad x \to 2^-$$

Similarly, the second table shows that as x approaches 2 from the right, the values of y decrease without bound, and by taking x sufficiently close to 2 on the right, we can make y smaller than any given negative number. In this situation we say that y approaches negative infinity as x approaches 2 from the right, and we write

$$y \to -\infty \quad \text{as} \quad x \to 2^+$$

The graph of $y = r(x)$ therefore has the shape near $x = 2$ shown in Figure 1.

x	y
1	2
1.5	3
1.9	11
1.95	21
1.99	101
1.999	1001

x	y
3	0
2.5	−1
2.1	−9
2.05	−19
2.01	−99
2.001	−999

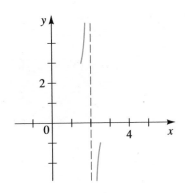

FIGURE 1

Now we examine the behavior of the function as x becomes progressively larger in absolute value (for both negative and positive x).

x	y
10	0.8750
100	0.9898
1000	0.9990
10,000	0.9999

x	y
-10	1.0833
-100	1.0098
-1000	1.0010
$-10,000$	1.0001

As $|x|$ becomes larger and larger, the values of y get progressively closer to 1. This means that the graph of $y = r(x)$ will approach the horizontal line $y = 1$ as x increases or decreases without bound. We express this by saying "y approaches 1 as x approaches infinity or negative infinity" and we write

$$y \to 1 \quad \text{as} \quad x \to \infty \quad \text{and} \quad y \to 1 \quad \text{as} \quad x \to -\infty$$

This means we can complete the graph of $y = r(x)$ as shown in Figure 2.   ■

In Example 2 we introduced the notation described in the following box.

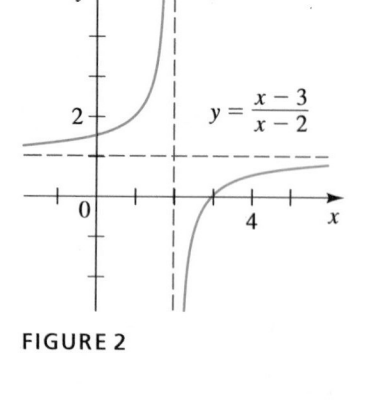

$$y = \frac{x-3}{x-2}$$

FIGURE 2

ARROW NOTATION	
Symbol	**Meaning**
$x \to a^{+}$	x approaches a from the right
$x \to a^{-}$	x approaches a from the left
$x \to \infty$	x goes to infinity; that is, x increases indefinitely
$x \to -\infty$	x goes to negative infinity; that is, x decreases indefinitely

The line $x = 2$ is called a *vertical asymptote* of the graph in Figure 2 and the line $y = 1$ is a *horizontal asymptote*. Informally speaking, an asymptote of a function is a line that the graph of the function gets closer and closer to as one travels along that line in either direction. More formally, we make the following definitions.

VERTICAL ASYMPTOTES

The line $x = a$ is a **vertical asymptote** of the function $y = f(x)$ if $y \to \infty$ or $y \to -\infty$ as $x \to a^{+}$ or $x \to a^{-}$.

If the function $y = f(x)$ is a rational function, then its vertical asymptotes are the lines $x = a$, where a is a zero of the denominator of the function; this is because only when the denominator is 0 does the function fail to have a real, finite value.

> ## HORIZONTAL ASYMPTOTES
>
> The line $y = b$ is a **horizontal asymptote** of the function $y = f(x)$ if $y \to b$ as $x \to \infty$ or as $x \to -\infty$.

In the next example, we show the most efficient method for finding the horizontal asymptote of a rational function.

EXAMPLE 3 ■ Graphing a Rational Function

Find all vertical and horizontal asymptotes and the x- and y-intercepts of the function $r(x)$. Use this information to graph the function.

$$r(x) = \frac{x^2 - x - 6}{2x^2 + 5x - 3}$$

SOLUTION

Factoring the numerator and denominator, we see that

Factor

$$r(x) = \frac{(x - 3)(x + 2)}{(2x - 1)(x + 3)}$$

Intercepts

The x-intercepts of the graph are the zeros of the numerator: $x = 3$ and $x = -2$. To find the y-intercept, we substitute $x = 0$ into the original form of the function to find

$$r(0) = \frac{0^2 - 0 - 6}{2 \cdot 0^2 + 5 \cdot 0 - 3} = \frac{-6}{-3} = 2$$

Horizontal asymptote

The horizontal asymptote (if it exists) will be the value that y approaches as $x \to \pm\infty$. To help us find this value, let us begin by dividing both the numerator and the denominator of $r(x)$ by x^2 (the highest power of x that appears in the denominator).

$$y = \frac{x^2 - x - 6}{2x^2 + 5x - 3} \cdot \frac{1/x^2}{1/x^2} = \frac{1 - \dfrac{1}{x} - \dfrac{6}{x^2}}{2 + \dfrac{5}{x} - \dfrac{3}{x^2}}$$

Any function of the form c/x^n approaches 0 as $x \to \pm\infty$ (if n is a positive integer). For example, in the following table we examine the values of $6/x^2$

(which appears in the preceding quotient) as x increases.

x	$6/x^2$
10	0.06
100	0.0006
1000	0.000006
10,000	0.00000006

$\leftarrow$ approaching 0

This means that as $x \to \pm\infty$,

$$y \to \frac{1 - 0 - 0}{2 + 0 - 0} = \frac{1}{2}$$

so $y = \frac{1}{2}$ is the horizontal asymptote.

Vertical asymptotes

The vertical asymptotes occur where the function is undefined (or, in other words, where the denominator is 0). So, the vertical asymptotes here are $x = \frac{1}{2}$ and $x = -3$. To be able to graph the function, we need to know whether $y \to \infty$ or $y \to -\infty$ on each side of these vertical lines. Thus, we need to determine the sign of y for values of x near the vertical asymptotes. As $x \to \frac{1}{2}^+$, the values of x are slightly larger than $\frac{1}{2}$, so

$$x - 3 < 0$$

$$x + 2 > 0$$

$$2x - 1 > 2 \cdot \tfrac{1}{2} - 1 = 0$$

and

$$x + 3 > 0$$

This means that as $x \to \frac{1}{2}^+$, the value of y is the quotient of one negative and three positive factors and therefore must be negative. So $y \to -\infty$ as $x \to \frac{1}{2}^+$. We can represent what happens here and at other sides of the vertical asymptotes schematically as in the following table:

As $x \to$	$\frac{1}{2}^+$	$\frac{1}{2}^-$	-3^+	-3^-
the sign of $y = \dfrac{(x-3)(x+2)}{(2x-1)(x+3)}$ is	$\dfrac{(-)(+)}{(+)(+)}$	$\dfrac{(-)(+)}{(-)(+)}$	$\dfrac{(-)(-)}{(-)(+)}$	$\dfrac{(-)(-)}{(-)(-)}$
so $y \to$	$-\infty$	∞	$-\infty$	∞

Using all this information about intercepts, asymptotes, and the behavior of the function near asymptotes, we obtain the partial graph in Figure 3, where the asymptotes have been plotted as broken lines.

Sketch the asymptotes and intercepts

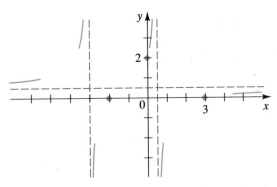

FIGURE 3

All we need to do now is to plot a few more points and fill in the rest of the graph, which is shown in Figure 4.

Complete the graph

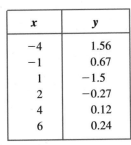

x	y
−4	1.56
−1	0.67
1	−1.5
2	−0.27
4	0.12
6	0.24

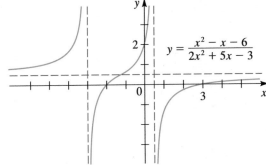

$$y = \frac{x^2 - x - 6}{2x^2 + 5x - 3}$$

FIGURE 4

We summarize the procedure to be followed in graphing rational functions by the following sequence of steps.

SKETCHING GRAPHS OF RATIONAL FUNCTIONS

1. FACTOR. Factor the numerator and denominator.

2. INTERCEPTS. Find the x-intercepts by determining the zeros of the numerator, and the y-intercept from the value of the function at $x = 0$.

3. VERTICAL ASYMPTOTES. Find the vertical asymptotes by determining the zeros of the denominator, and then see if $y \to \infty$ or $y \to -\infty$ on each side of every vertical asymptote.

4. HORIZONTAL ASYMPTOTE. Find the horizontal asymptote (if any) by dividing both numerator and denominator by the highest power of x that appears in the denominator, and then letting $x \to \pm\infty$.

5. SKETCH THE GRAPH. Sketch a partial graph using the information provided by the first four steps of this procedure. Then plot as many additional points as needed to fill in the rest of the graph of the function.

THE THREE FORMS OF A RATIONAL FUNCTION

To find the *y*-intercept of a rational function, we use the *original form* of the function.

To find the *x*-intercepts and **vertical asymptotes,** we use the *factored form* of the function.

To find the **horizontal asymptote,** we use a *compound-fraction form* of the function.

Notice that we use three forms of the equation defining the rational function when we perform this analysis. In Example 3, the *original form*

$$r(x) = \frac{x^2 - x - 6}{2x^2 + 5x - 3}$$

easily gave us the *y*-intercept $r(0) = (-6)/(-3) = 2$. The *factored form*

$$r(x) = \frac{(x-3)(x+2)}{(2x-1)(x+3)}$$

gave us the *x*-intercepts 3 and -2, and the vertical asymptotes $x = \frac{1}{2}$ and $x = -3$. Finally, the *compound-fraction form*

$$r(x) = \frac{1 - \dfrac{1}{x} - \dfrac{6}{x^2}}{2 + \dfrac{5}{x} - \dfrac{3}{x^2}}$$

tells us that the horizontal asymptote is $y = \frac{1}{2}$.

EXAMPLE 4 ■ Graphing a Rational Function with Two Vertical Asymptotes

Sketch a graph of the rational function

$$r(x) = \frac{x-2}{x^2 - 1}$$

SOLUTION

x-intercept: $\quad x - 2 = 0 \quad$ or $\quad x = 2$

y-intercept: $\quad r(0) = \dfrac{0-2}{0^2 - 1} = 2$

Vertical asymptotes: $\quad y = \dfrac{x-2}{(x-1)(x+1)}$, so vertical asymptotes are $x = 1$ and $x = -1$.

Horizontal asymptote: $\quad y = \dfrac{\dfrac{x}{x^2} - \dfrac{2}{x^2}}{\dfrac{x^2}{x^2} - \dfrac{1}{x^2}} = \dfrac{\dfrac{1}{x} - \dfrac{2}{x^2}}{1 - \dfrac{1}{x^2}} \rightarrow \dfrac{0-0}{1-0} = \dfrac{0}{1} = 0$ as

$x \rightarrow \infty$, so the horizontal asymptote is $y = 0$.

Behavior near vertical asymptotes:

As $x \rightarrow$	1^+	1^-	-1^+	-1^-
the sign of $y = \dfrac{x-2}{(x-1)(x+1)}$ is	$\dfrac{(-)}{(+)(+)}$	$\dfrac{(-)}{(-)(+)}$	$\dfrac{(-)}{(-)(+)}$	$\dfrac{(-)}{(-)(-)}$
so $y \rightarrow$	$-\infty$	∞	∞	$-\infty$

Table of additional values: **Graph:**

x	y
-2	-1.33
-0.5	3.33
0.5	2
1.5	-0.4
3	0.125
4	0.133
5	0.125

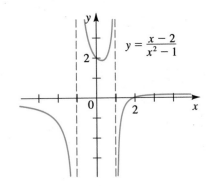

FIGURE 5

 From the graph in Figure 5, we see that, contrary to common misconception, *a graph may cross a horizontal asymptote.*

We can determine whether a rational function has a horizontal asymptote by considering the degrees of the numerator and denominator of the rational function. To see how this works, let

$$r(x) = \frac{P(x)}{Q(x)} = \frac{a_n x^n + a_{n-1} x^{n-1} + \cdots + a_1 x + a_0}{b_m x^m + b_{m-1} x^{m-1} + \cdots + b_1 x + b_0}$$

be a rational function. If the degrees of P and Q are the same (so that $n = m$), then we can see by dividing both numerator and denominator by x^n that $y = a_n/b_m$ is the horizontal asymptote of $y = r(x)$. This was the case in Example 3. If the degree of P is less than the degree of Q (that is, $n < m$), then dividing numerator and denominator by x^m shows that $y = 0$ is the horizontal asymptote, as in Example 4. Finally, if $n > m$ the same procedure shows that the function has no horizontal asymptote.

 SLANT ASYMPTOTES

If $r(x) = P(x)/Q(x)$ is a rational function in which the degree of the numerator is one more than the degree of the denominator, we can use the Division Algorithm to express the function in the form

$$r(x) = ax + b + \frac{R(x)}{Q(x)}$$

where the degree of R is less than the degree of Q and $a \neq 0$. This means that as $x \to \pm\infty$, $R(x)/Q(x) \to 0$, so for large values of $|x|$, the graph of $y = r(x)$ approaches the graph of the line $y = ax + b$. In this situation we say that $y = ax + b$ is a **slant asymptote,** or an **oblique asymptote.**

EXAMPLE 5 ■ A Rational Function with a Slant Asymptote

Graph the function

$$r(x) = \frac{x^2 - 4x - 5}{x - 3}$$

SOLUTION

Since the degree of the numerator is one more than the degree of the denominator, the function has a slant asymptote. By dividing $x - 3$ into $x^2 - 4x - 5$, we obtain

$$r(x) = x - 1 - \frac{8}{x - 3}$$

so the slant asymptote is the line $y = x - 1$. The line $x = 3$ is a vertical asymptote, and it is easy to see that $r(x) \to -\infty$ as $x \to 3^+$ and $r(x) \to \infty$ as $x \to 3^-$.

Factoring the numerator in the original form for r gives

$$r(x) = \frac{(x + 1)(x - 5)}{x - 3}$$

so the x-intercepts are -1 and 5, and the y-intercept is $\frac{5}{3}$. Plotting the asymptotes, intercepts, and the additional points listed in the table, we can complete the graph as shown in Figure 6.

x	y
-2	-1.4
1	4
2	9
4	-5
6	2.33

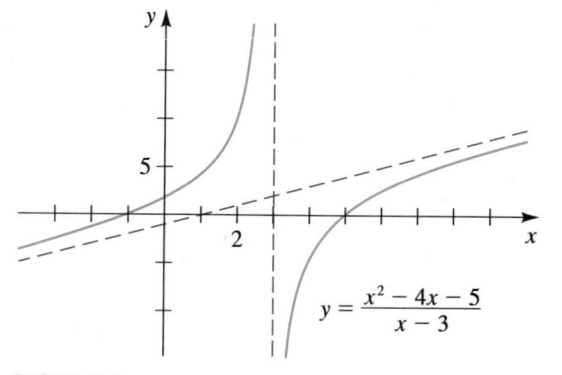

FIGURE 6

When graphing a rational function in which the degree of the numerator is one plus the degree of the denominator, we must modify Step 4 in the procedure given on page 313. Instead of finding the horizontal asymptote (which does not exist in this case), we find the slant asymptote using the method described in Example 5.

The following box summarizes what we have observed about horizontal and slant asymptotes.

HORIZONTAL AND SLANT ASYMPTOTES

Let

$$r(x) = \frac{P(x)}{Q(x)}$$

be a rational function with $P(x)$ of degree n:

$$P(x) = a_n x^n + a_{n-1} x^{n-1} + \cdots + a_1 x + a_0$$

and with $Q(x)$ of degree m:

$$Q(x) = b_m x^m + b_{m-1} x^{m-1} + \cdots + b_1 x + b_0$$

1. If $n < m$, then r has horizontal asymptote $y = 0$.

2. If $n = m$, then r has horizontal asymptote $y = \dfrac{a_n}{b_m}$.

3. If $n = m + 1$, then r has a slant asymptote.

4. If $n > m + 1$, then r has no horizontal or slant asymptote.

4.8 EXERCISES

1–6 ■ Find the x- and y-intercepts of the function.

1. $y = \dfrac{x - 6}{x + 1}$

2. $y = \dfrac{2}{x - 2}$

3. $y = \dfrac{x}{x^2 - 2x - 15}$

4. $y = \dfrac{x^2 - 2x - 15}{x}$

5. $y = \dfrac{x^2 + 10}{2x}$

6. $y = \dfrac{x^2 - 9}{x^3 - 1}$

7–16 ■ Find all asymptotes (including vertical, horizontal, and slant).

7. $y = \dfrac{5}{x + 3}$

8. $y = \dfrac{3x + 3}{x - 3}$

9. $y = \dfrac{x^2}{x^2 - x - 6}$

10. $y = \dfrac{2x - 4}{x^2 + 2x + 1}$

11. $y = \dfrac{6}{x^2 + 2}$

12. $y = \dfrac{(x - 1)(x - 2)}{(x - 3)(x - 4)}$

13. $y = \dfrac{x^2 + 2}{x - 1}$

14. $y = \dfrac{x^3 + 3x^2}{x^2 - 4}$

15. $y = \dfrac{2x^3 - x^2 - 8x + 4}{x + 3}$

16. $y = \dfrac{6x^4}{x^2 - 3}$

17–48 ■ Find the intercepts and asymptotes, and then graph the rational function.

17. $y = \dfrac{4}{x - 2}$

18. $y = \dfrac{9}{x + 3}$

19. $y = \dfrac{x - 1}{x - 2}$

20. $y = \dfrac{x + 9}{x - 3}$

21. $y = \dfrac{4x - 4}{x + 2}$

22. $y = \dfrac{2x + 6}{-6x + 3}$

23. $y = \dfrac{2x - 4}{x}$

24. $y = \dfrac{x}{2x - 4}$

25. $y = \dfrac{18}{(x - 3)^2}$

26. $y = \dfrac{x - 2}{(x + 1)^2}$

27. $y = \dfrac{4x + 8}{(x - 4)(x + 1)}$

28. $y = \dfrac{x - 9}{(x + 3)(x - 1)}$

29. $y = \dfrac{(x - 1)(x + 2)}{(x + 1)(x - 3)}$

30. $y = \dfrac{2x(x + 4)}{(x - 1)(x - 2)}$

31. $y = \dfrac{x^2 - 2x + 1}{x^2 + 2x + 1}$

32. $y = \dfrac{4x^2}{x^2 - 2x - 3}$

33. $y = \dfrac{2x^2 + 10x - 12}{x^2 + x - 6}$

34. $y = \dfrac{2x^2 + 2x - 4}{x^2 + x}$

35. $y = \dfrac{x^2 - x - 6}{x^2 + 3x}$

36. $y = \dfrac{x^2 + 3x}{x^2 - x - 6}$

37. $y = \dfrac{3x^2 + 6}{x^2 - 2x - 3}$

38. $y = \dfrac{5x^2 + 5}{x^2 + 4x + 4}$

39. $y = \dfrac{x^2}{x - 2}$

40. $y = \dfrac{x^2 + 2x}{x - 1}$

41. $y = \dfrac{x^2 - 2x - 8}{x}$

42. $y = \dfrac{3x - x^2}{2x - 2}$

43. $y = \dfrac{x^2 + 5x + 4}{x - 3}$

44. $y = \dfrac{x^2 + 4}{2x^2 + x - 1}$

45. $y = \dfrac{x^3 + x^2}{x^2 - 4}$

46. $y = \dfrac{2x^3 + 2x}{x^2 - 1}$

47. $y = \dfrac{x^3}{x - 2}$

48. $y = \dfrac{x^4 - 16}{x^2 - 1}$

49–53 ■ In this chapter we have adopted the convention that in rational functions, the numerator and denominator do not share a common factor. Here we study the graphs of rational functions that do not satisfy this rule.

49. Show that the graph of

$$r(x) = \frac{3x^2 - 3x - 6}{x - 2}$$

is the line $y = 3x + 3$ with the point $(2, 9)$ removed. [*Hint:* Divide. What is the domain of r?]

50–53 ■ Graph the rational function.

50. $y = \dfrac{x^2 + x - 20}{x + 5}$

51. $y = \dfrac{2x^2 - x - 1}{x - 1}$

52. $y = \dfrac{x^2 - 3x + 2}{x^2 - 4x + 4}$

53. $y = \dfrac{2x^2 - 5x - 3}{x^2 - 2x - 3}$

54–57 ■ Construct a rational function $y = P(x)/Q(x)$ that has the indicated properties and in which the degrees of P and Q are as small as possible.

54. The function has vertical asymptote $x = 3$, horizontal asymptote $y = 0$, and y-intercept $-\frac{1}{3}$, and it never crosses the x-axis.

55. The function has vertical asymptotes $x = 1$ and $x = -4$, horizontal asymptote $y = 1$, and x-intercepts 2 and 3.

56. The function has horizontal asymptote $y = 2$ but no vertical asymptote. The origin is the only x-intercept, and i is a zero of $Q(x)$.

57. The function has slant asymptote $y = 3x - 6$ and vertical asymptote $x = \frac{1}{2}$, and its graph passes through the origin.

58. Show that the function

$$y = \frac{x^6 + 10}{x^4 + 8x^2 + 15}$$

has no x-intercept and no horizontal, vertical, or slant asymptote.

59. The rabbit population on Mr. Jenkins' farm follows the formula

$$p(t) = \frac{3000t}{t + 1}$$

where $t \geq 0$ is the time (in months) since the beginning of the year.
(a) Sketch a graph of the rabbit population.
(b) What eventually happens to the rabbit population?

60. After a certain drug is injected into a patient, the concentration c of the drug in the blood is monitored.

At time $t \geq 0$ (in minutes since the injection), the concentration (in mg/L) is given by $c(t) = \dfrac{30t}{t^2 + 2}$.

(a) Sketch a graph of the drug concentration.
(b) What eventually happens to the concentration of drug in the blood?

4.9 USING GRAPHING DEVICES TO GRAPH RATIONAL FUNCTIONS

In the preceding section we learned how to find horizontal and vertical asymptotes for rational functions. Since the graph of a rational function is very steep near a vertical asymptote, some graphing calculators and computer graphing programs do not properly graph functions with vertical asymptotes (see Example 5 in Section 3.3). A sudden, nearly vertical jump in a graph produced by a graphing calculator often signals the presence of a vertical asymptote.

EXAMPLE 1 ■ A Rational Function, Viewed from Up Close and Far Away

Graph the function

$$f(x) = \frac{2x + 1}{x - 3}$$

in the viewing rectangles $[-10, 10]$ by $[-10, 10]$ and $[-1000, 1000]$ by $[-10, 10]$. Interpret the result in each case.

SOLUTION

The graph in the smaller viewing rectangle is shown in Figure 1(a). The graph consists of two parts—because f is undefined when $x = 3$, the line $x = 3$ is a vertical asymptote. (On some graphing calculators, a vertical line may actually be shown on the screen. This is not part of the graph of the function and may be interpreted simply as an asymptote.)

In the very wide viewing rectangle shown in Figure 1(b), the graph of f looks almost like the horizontal line $y = 2$. Because the graphing device does not plot enough points to discern the vertical asymptote, it disappears from the

FIGURE 1
$$f(x) = \frac{2x + 1}{x - 3}$$

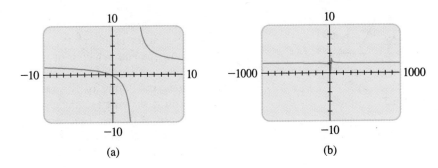

(a)

(b)

screen. This happens because the calculator connects points plotted on each side of the vertical asymptote. We might make the same mistake ourselves if we were to graph the function simply by plotting points and joining them with a smooth curve, without analyzing the behavior of the function. However, this wide viewing rectangle shows the horizontal asymptote well, since a horizontal asymptote is a line that the graph approaches when $|x|$ is large. The horizontal asymptote $y = 2$ represents the end behavior of the function. ∎

In the preceding section we considered only horizontal and slant asymptotes as end behaviors for rational functions. In the next example we graph a function that behaves like a parabola for large values of $|x|$.

EXAMPLE 2 ∎ End Behavior of a Rational Function

Graph the function

$$f(x) = \frac{x^3 - 2x^2 + 3}{x - 2}$$

in appropriate viewing rectangles to show the vertical asymptote and to determine its end behavior.

SOLUTION

First we graph the function in a narrow viewing rectangle to see the vertical asymptote. The function is undefined when $x = 2$, so we choose the viewing rectangle $[-4, 4]$ by $[-20, 20]$ and obtain the graph in Figure 2(a). The function has x-intercept -1, vertical asymptote $x = 2$, and a local minimum point with approximate coordinates $(2.74, 11.56)$. To determine the end behavior, we try a larger viewing rectangle—in this case $[-30, 30]$ by $[-200, 200]$. In the graph in Figure 2(b) the vertical asymptote has all but disappeared, and the graph looks like a parabola. To see why this is the case, we divide the denominator of f into the numerator and write the result in quotient-remainder form:

$$f(x) = x^2 + \frac{3}{x - 2}$$

When $|x|$ is large, $3/(x - 2)$ is small; that is, $3/(x - 2) \to 0$ as $x \to \pm\infty$. This means that for large $|x|$, the graph of f will be close to the graph of $y = x^2$. Thus, the end behavior of the function f is like that of the parabola $y = x^2$.

In Figure 3 the graphs of $y = (x^3 - 2x^2 + 3)/(x - 2)$ and $y = x^2$ are displayed in the viewing rectangle $[-8, 8]$ by $[-5, 20]$. From the figure we see that the graphs of the two functions are very close to each other everywhere except near the vertical asymptote. ∎

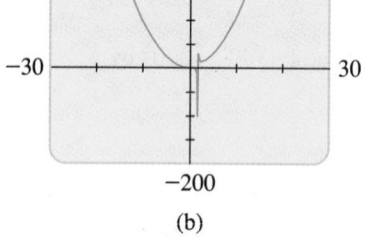

(a)

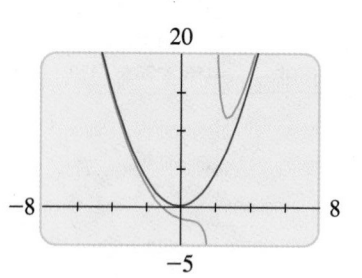

(b)

FIGURE 2
$$f(x) = \frac{x^3 - 2x^2 + 3}{x - 2}$$

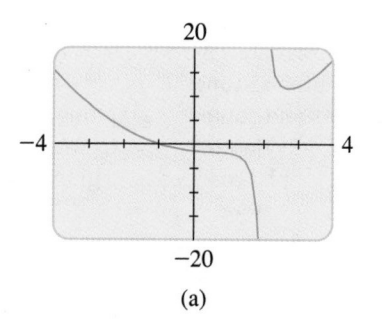

FIGURE 3
$$y = \frac{x^3 - 2x^2 + 3}{x - 2} \quad \text{and} \quad y = x^2$$

Rational functions occur frequently in the applications of algebra to the sciences. In the next example we analyze the graph of a function from the theory of electricity.

EXAMPLE 3 ■ Electrical Resistance

When two resistors with resistances R_1 and R_2 are connected in parallel, their combined resistance R is given by the formula

$$R = \frac{R_1 R_2}{R_1 + R_2}$$

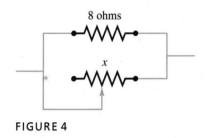

8 ohms

x

FIGURE 4

Suppose that a fixed 8-ohm resistor is connected in parallel with a variable resistor, as shown in Figure 4. If the resistance of the variable resistor is denoted by x, then the combined resistance R is a function of x. Graph R and give a physical interpretation of the graph.

SOLUTION

Substituting $R_1 = 8$ and $R_2 = x$ into the formula gives the function

$$R(x) = \frac{8x}{8 + x}$$

Since resistance cannot be negative, this function has physical meaning only when $x > 0$. The function is graphed in Figure 5(a) using the viewing rectangle $[0, 20]$ by $[0, 10]$. The function has no vertical asymptote when x is restricted to positive values. The combined resistance R increases as the variable resistance x increases. If we widen the viewing rectangle to $[0, 100]$ by $[0, 10]$, we obtain the graph in Figure 5(b). For large x, the combined resistance R levels off, getting closer and closer to the horizontal asymptote $R = 8$. No matter how large the variable resistance x, the combined resistance is never greater than 8 ohms.

FIGURE 5

$$R(x) = \frac{8x}{8 + x}$$

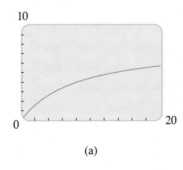

(a)

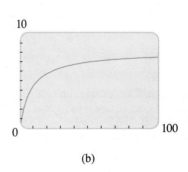

(b)

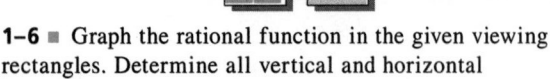

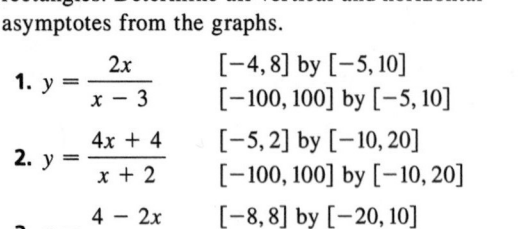 **4.9**   **EXERCISES**

1–6 ■ Graph the rational function in the given viewing rectangles. Determine all vertical and horizontal asymptotes from the graphs.

1. $y = \dfrac{2x}{x-3}$ $[-4, 8]$ by $[-5, 10]$
                  $[-100, 100]$ by $[-5, 10]$

2. $y = \dfrac{4x+4}{x+2}$ $[-5, 2]$ by $[-10, 20]$
                  $[-100, 100]$ by $[-10, 20]$

3. $y = \dfrac{4-2x}{x}$ $[-8, 8]$ by $[-20, 10]$
                  $[-100, 100]$ by $[-20, 10]$

4. $y = \dfrac{5x-10}{x^2}$ $[-10, 10]$ by $[-10, 3]$
                  $[-100, 100]$ by $[-10, 3]$

5. $y = \dfrac{3x^2+1}{x^2-9}$ $[-10, 10]$ by $[-10, 10]$
                  $[-100, 100]$ by $[-10, 10]$

6. $y = \dfrac{x^2-6}{x^2-3x}$ $[-5, 10]$ by $[-5, 5]$
                  $[-100, 100]$ by $[-5, 5]$

7–18 ■ Graph the rational function in an appropriate viewing rectangle and determine its vertical and horizontal asymptotes, its x- and y-intercepts, and all local extrema, correct to two decimals.

7. $y = \dfrac{x+1}{x-1}$

8. $y = \dfrac{4}{x-2}$

9. $y = \dfrac{7x-14}{x}$

10. $y = \dfrac{2-12x}{4+3x}$

11. $y = \dfrac{4x}{x^2-4}$

12. $y = \dfrac{x^2+9}{x^2-9}$

13. $y = \dfrac{6x^2-6}{x^2+2}$

14. $y = \dfrac{x^2+1}{x^3-27}$

15. $y = \dfrac{4}{(x-1)^2}$

16. $y = \dfrac{x^2-4x}{(x+1)^2}$

17. $y = \dfrac{2x^2+3x-2}{x^2}$

18. $y = \dfrac{1}{x^3-2x^2}$

19–22 ■ Graph the rational function f and determine all vertical asymptotes from your graph. Then graph f and g in a sufficiently large viewing rectangle to show that they have the same end behavior.

19. $f(x) = \dfrac{2x^2+6x+6}{x+3}$, $g(x) = 2x$

20. $f(x) = \dfrac{-x^3+6x^2-5}{x^2-2x}$, $g(x) = -x+4$

21. $f(x) = \dfrac{x^3-2x^2+16}{x-2}$, $g(x) = x^2$

22. $f(x) = \dfrac{-x^4+2x^3-2x}{(x-1)^2}$, $g(x) = 1-x^2$

23–26 ■ Graph the rational function and find all vertical asymptotes, x- and y-intercepts, and local extrema, correct to the nearest decimal. Then use long division to find **a** polynomial that has the same end behavior as the rational function, and graph both functions in a sufficiently large viewing rectangle to verify that the end behaviors of the polynomial and the rational function are the same.

23. $y = \dfrac{2x^2-5x}{2x+3}$

24. $y = \dfrac{x^4-3x^3+x^2-3x+3}{x^2-3x}$

25. $y = \dfrac{x^5}{x^3-1}$

26. $y = \dfrac{x^4}{x^2-2}$

27. A drug is administered to a patient and the concentration of the drug in the bloodstream is monitored. At time $t \geq 0$ (in hours since giving the drug) the concentration (in mg/L) is given by

$$c(t) = \frac{5t}{t^2+1}$$

Graph the function c with a graphing device.
(a) What is the highest concentration of drug that is reached in the patient's bloodstream?
(b) What happens to the drug concentration after a long period of time?
(c) How long does it take for the concentration to drop below 0.3 mg/L?

28. Suppose a rocket is fired upward from the surface of the earth with an initial velocity v (measured in m/s). Then the maximum height h (in meters) reached by the rocket is given by the function

$$h(v) = \frac{Rv^2}{2gR - v^2}$$

where $R = 6.4 \times 10^6$ m is the radius of the earth and $g = 9.8$ m/s^2 is the acceleration due to gravity. Use a graphing device to sketch a graph of the function h.

(Note that h and v must both be positive, so the viewing rectangle need not contain negative values.) What does the vertical asymptote represent physically?

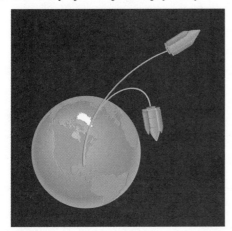

29. As a train moves towards an observer (see the figure), the pitch of its whistle sounds higher to the observer than it would if the train were at rest, because the crests of the sound waves are compressed closer together. This phenomenon is called the *Doppler effect*. The observed pitch P is a function of the speed v of the train and is given by

$$P(v) = P_0\left(\frac{s_0}{s_0 - v}\right)$$

where P_0 is the actual pitch of the whistle at the source and $s_0 = 332$ m/s is the speed of sound in air. Suppose that a train has a whistle pitched at $P_0 = 440$ Hz. Graph the function $y = P(v)$ using a graphing device. How can the vertical asymptote of this function be interpreted physically?

30. Give an example of a rational function that has horizontal asymptote $y = 2$, vertical asymptotes $x = 1$ and $x = 4$, and whose graph passes through the origin. Graph the function using a graphing device to confirm that it has the required properties.

31. Give an example of a rational function that has vertical asymptotes $x = \pm 1$, whose end behavior is the same as the end behavior of $y = x^2$, and whose graph passes through the origin. Graph the function using a graphing device to confirm that it has the required properties.

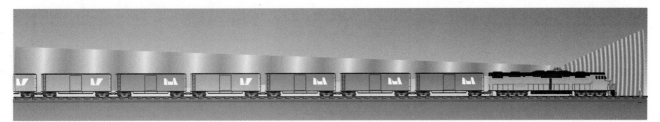

---

4 REVIEW

KEY TOPICS ■ Define, state, or discuss each of the following.

1. Polynomial of degree n

2. The graph of $y = x^n$ for n a positive integer

3. The graph of a polynomial

4. Turning points; local maxima and minima of polynomials

5. End behavior of a polynomial

6. Dividend, divisor, quotient, and remainder

7. Division Algorithm

8. Remainder Theorem

9. Synthetic division

10. Factor Theorem

11. Zero of a polynomial

12. Root of a polynomial equation

13. Rational Roots Theorem

14. Descartes' Rule of Signs

15. Upper and lower bounds for roots

16. Upper and Lower Bounds Theorem

17. Intermediate Value Theorem for Polynomials

18. Finding approximate values for irrational zeros of polynomials

19. Fundamental Theorem of Algebra

20. Complete Factorization Theorem

21. Multiplicity of a zero

22. Zeros Theorem

23. Conjugate Roots Theorem

24. Rational function

25. Vertical, horizontal, and slant asymptotes

EXERCISES

1–6 ■ Graph the polynomial. Show clearly all x- and y-intercepts.

1. $y = (x - 2)^3 + 8$

2. $y = 32 - 2x^4$

3. $y = x^3 - 9x$

4. $y = x^3 - 5x^2 - 6x$

5. $y = x^3 - 5x^2 - 4x + 20$

6. $y = x^4 - 9x^2$

7–10 ■ Use a graphing device to graph the polynomial. Find the x- and y-intercepts and the coordinates of all local extrema, correct to the nearest decimal. Describe the end behavior of the function.

7. $y = 2x^3 + x^2 - 18x - 9$

8. $y = x^4 - 8x^2 + 16$

9. $y = x^5 + x^2 - 5$

10. $y = 3x^5 + x^4 - 4x$

11–18 ■ Find the quotient and remainder.

11. $\dfrac{x^3 - x^2 + x - 11}{x - 3}$

12. $\dfrac{x^4 + 30x + 12}{2x + 6}$

13. $\dfrac{x^3 - x^2 - 11x + 6}{x^2 + 2x - 5}$

14. $\dfrac{x^5 - 3x^4 + 3x^3 + 20x - 6}{x^2 + 2x - 6}$

15. $\dfrac{x^4 - 25x^2 + 4x + 15}{x + 5}$

16. $\dfrac{2x^3 - x^2 - 5}{x - \frac{3}{2}}$

17. $\dfrac{x^4 + x^3 - 2x^2 - 3x - 1}{x - \sqrt{3}}$

18. $\dfrac{15x - 7}{5x + 12}$

19–20 ■ Find the indicated value of the polynomial using the Remainder Theorem.

19. $P(x) = 2x^3 - 9x^2 - 7x + 13;$ find $P(5)$

20. $Q(x) = x^4 + 4x^3 + 7x^2 + 10x + 15;$ find $Q(-3)$

21. Show that $\frac{1}{2}$ is a zero of the polynomial

$$2x^4 + x^3 - 5x^2 + 10x - 4$$

22. Use the Factor Theorem to show that $x + 4$ is a factor of the polynomial

$$x^5 + 4x^4 - 7x^3 - 23x^2 + 23x + 12$$

23. What is the remainder when the polynomial $x^{500} + 6x^{201} - x^2 - 2x + 4$ is divided by $x - 1$?

24. What is the remainder when $x^{101} - x^4 + 2$ is divided by $x + 1$?

25–26 ■ List all possible rational roots (without testing to see if they actually are roots), and then determine the possible number of positive and negative real roots using Descartes' Rule of Signs.

25. $x^5 - 6x^3 - x^2 + 2x + 18 = 0$

26. $6x^4 + 3x^3 + x^2 + 3x + 4 = 0$

27–30 ■ Show that the equation has a root between the given two integers, and then find that root correct to the indicated number of decimal places.

27. $x^3 - 2x - 1 = 0;$
between 1 and 2;   to the nearest tenth

28. $3x^3 + x^2 + 4 = 0;$
between -2 and -1;   to one decimal place

29. $x^6 - 5x^4 + 10 = 0$;
between 1 and 2; to two decimal places

30. $x^6 - 5x^4 + 10 = 0$;
between 2 and 3; to the nearest hundredth

31. Find a polynomial of degree 3 with constant coefficient 12 and zeros $-\frac{1}{2}$, 2, and 3.

32. Find a polynomial of degree 4 having integer coefficients and zeros $3i$ and 4, with 4 a double zero.

33. Does there exist a polynomial of degree 4 with integer coefficients that has zeros i, $2i$, $3i$, and $4i$? If so, find it. If not, explain why.

34. Prove that the equation $3x^4 + 5x^2 + 2 = 0$ has no real root.

35–44 ■ Find all rational, irrational, and imaginary roots (and state their multiplicities). Use Descartes' Rule of Signs, the Upper and Lower Bounds Theorem, the quadratic formula, or other factoring techniques to help you whenever possible.

35. $x^3 - 3x^2 - 13x + 15 = 0$

36. $2x^3 + 5x^2 - 6x - 9 = 0$

37. $x^4 + 6x^3 + 17x^2 + 28x + 20 = 0$

38. $x^4 + 7x^3 + 9x^2 - 17x - 20 = 0$

39. $x^5 - 3x^4 - x^3 + 11x^2 - 12x + 4 = 0$

40. $x^4 = 81$

41. $x^6 = 64$

42. $18x^3 + 3x^2 - 4x - 1 = 0$

43. $6x^4 - 18x^3 + 6x^2 - 30x + 36 = 0$

44. $x^4 + 15x^2 + 54 = 0$

45. Show that $2 - i\sqrt{2}$ is a root of
$$x^4 - 5x^3 + 8x^2 + 2x - 12 = 0$$
and find all other roots of the equation.

46. Show that i and $-1 + i\sqrt{6}$ are roots of
$$x^6 + x^5 + 7x^4 - 4x^3 + 13x^2 - 5x + 7 = 0$$
and find all other roots of the equation.

47. The polynomial $x^4 - x^2 - x - 2 = 0$ has one positive real root. Show that the root is irrational, and find a decimal approximation for it, correct to two decimal places.

48. (a) Show that the polynomial $x^4 + 2x^2 - x - 1$ has exactly two real zeros and that neither zero is rational.

(b) Find decimal approximations of each of the two real zeros, correct to the nearest hundredth.

49–52 ■ Use a graphing device to find all real solutions of the equation.

49. $2x^2 = 5x + 3$

50. $x^3 + x^2 - 14x - 24 = 0$

51. $x^4 - 3x^3 - 3x^2 - 9x - 2 = 0$

52. $x^5 = x + 3$

53–58 ■ Graph the rational function. Show clearly all x- and y-intercepts and asymptotes.

53. $y = \dfrac{3x - 12}{x + 1}$

54. $y = \dfrac{1}{(x + 2)^2}$

55. $y = \dfrac{x - 2}{x^2 - 2x - 8}$

56. $y = \dfrac{2x^2 - 6x - 7}{x - 4}$

57. $y = \dfrac{x^2 - 9}{2x^2 + 1}$

58. $y = \dfrac{x^3 + 27}{x + 4}$

59–62 ■ Use a graphing device to analyze the graph of the rational function. Find all x- and y-intercepts, vertical, horizontal, and slant asymptotes, and the coordinates of local extrema. If the function has no horizontal or slant asymptote, find a polynomial that has the same end behavior as the rational function

59. $y = \dfrac{x - 3}{2x + 6}$

60. $y = \dfrac{2x - 7}{x^2 + 9}$

61. $y = \dfrac{x^3 + 8}{x^2 - x - 2}$

62. $y = \dfrac{2x^3 - x^2}{x + 1}$

63. (a) Show that -1 is a root of the equation
$$2x^4 + 5x^3 + x + 4 = 0$$

(b) Use the information from part (a) to show that $2x^3 + 3x^2 - 3x + 4 = 0$ has no positive real root. [*Hint:* Compare the coefficients of this polynomial to your synthetic division table from part (a).]

64. Find the coordinates of all points of intersection of the graphs of
$$y = x^4 + x^2 + 24x \quad \text{and} \quad y = 6x^3 + 20$$

1. Graph the function $f(x) = x^3 - x^2 - 9x + 9$, showing clearly all x- and y-intercepts.

2. Let $P(x) = 2x^4 - 17x^3 + 53x^2 - 72x + 36$.
   (a) List all possible rational zeros of P given by the Rational Roots Theorem.
   (b) Use Descartes' Rule of Signs to determine how many positive and how many negative real roots the equation $P(x) = 0$ may have.
   (c) Show that 9 is an upper bound for the real roots of $P(x) = 0$ but is not itself a root.
   (d) Use the information from your answers to parts (a), (b), and (c) to construct a new, shorter list of possible rational zeros of P.
   (e) Determine which of the possible rational zeros listed in part (d) actually are zeros of P.
   (f) Factor P in the form given in the Complete Factorization Theorem.

3. Find a fifth-degree polynomial with integer coefficients that has zeros $1 + 2i$ and -1, with -1 a zero of multiplicity 3.

4. Let

$$P(x) = x^{23} - 5x^{12} + 8x - 1 \qquad Q(x) = 3x^4 + x^2 - x - 15$$
$$R(x) = 4x^6 + x^4 + 2x^2 + 16$$

   (a) Explain why an even integer could not possibly be a zero of any of these three polynomials.
   (b) Does R have any real zeros? Why or why not?
   (c) How many real zeros does Q have? Why?
   (d) Show that P has no rational zero.

5. Find a decimal approximation for the positive root of the equation $x^4 + 2x^2 - x - 4 = 0$, correct to the nearest tenth.

6. Find all roots of the equation $2x^4 - 17x^3 + 42x^2 - 25x + 4 = 0$.

7. Find all roots of the equation $x^4 + x^3 + 5x^2 + 4x + 4 = 0$ given that $-2i$ is one of its roots.

8. Consider the following four rational functions:

$$r(x) = \frac{2x - 1}{x^2 - x - 2} \qquad s(x) = \frac{x^3 + 27}{x^2 + 4} \qquad t(x) = \frac{x^3 - 9x}{x + 2} \qquad u(x) = \frac{x^2 + x - 6}{x^2 - 25}$$

   (a) Which of these functions has a horizontal asymptote?
   (b) Which of these functions has a slant asymptote?
   (c) Which of these functions has no vertical asymptote?
   (d) Graph $y = u(x)$, showing clearly any asymptotes and x- and y-intercepts the function may have.

9. Choose an appropriate viewing rectangle to graph the following function, and find all its x-intercepts and local extrema, correct to two decimals:

$$P(x) = x^4 - 4x^3 + 8x$$

FOCUS ON PROBLEM SOLVING

It is often necessary to combine different problem-solving principles. The principles of **drawing a diagram** and **introducing something extra** are used in the two problems that follow. Euler's solution of the "Königsberg bridge problem" is so brilliant that it is the basis for a whole field of mathematics called *network theory*, which has practical applications to electric circuits and economics.

PROBLEM 1 ■ The Königsberg Bridge Problem

The old city of Königsberg, now called Kaliningrad, is situated on both banks of the Pregel River and on two islands in the river. Seven bridges connect the parts of the city, as shown in Figure 1. After the last bridge was built, the Königsberg townspeople posed the problem of finding a route through the city that would cross all seven bridges in a continuous walk, without recrossing any bridge. No one was able to find such a route. When Euler heard of the problem he realized that an interesting mathematical principle must be at work. To discover the principle Euler first stripped away the nonessential parts of the problem by drawing a simpler diagram of the city so that the land masses were represented by points and the bridges by lines connecting the points (Figure 2).

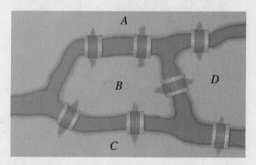

FIGURE 1

FIGURE 2

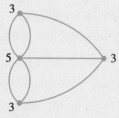

FIGURE 3

The problem now was to draw the diagram without lifting the pencil and without retracing any line. To show that this is impossible, Euler introduced something extra by labeling each point with the number of line segments that end at the point (Figure 3). He then argued that every pass through a point accounts for two line ends (one entering and one leaving). Thus each point, except possibly the starting point and the ending point, must be labeled by an even number in order for the walk to be possible. Since more than two points are labeled by odd numbers, the walk is impossible!

The beauty of Euler's solution is that it is completely general. In modern language a diagram consisting of points and connecting lines is called a *network*, each point is a *vertex*, each line an *edge*, and the number of edges ending at a vertex is the *order* of the vertex. A network is *traversable* if it can be drawn in one stroke without retracing any edge. Euler's reasoning about the Königsberg bridge problem proved the following principle: *If a network has more than two vertices with odd order, then it is not traversable.* ■

Here is another clever use of the principle of **introducing something extra.**

PROBLEM 2 ■ The Impossible Museum Tour

The floor plan of a museum is in the shape of a square with 6 square rooms to a side. Each room is connected to each adjacent room by a door. The museum entrance and exit are at diagonally opposite corners, as shown in Figure 4.

A visitor making a tour of this museum would like to visit each room exactly once and then exit. Can you find such a path? Here are examples of attempts that failed:

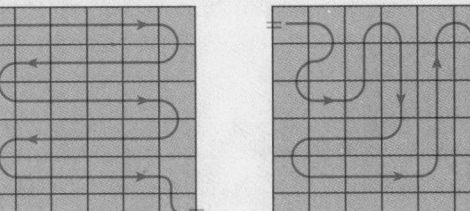

Trial and error will make it plausible to you that such a path is in fact not possible. But trial and error does not *prove* that our proposed tour of the museum is impossible. How do we give a convincing argument for this? Here is a clever idea: Imagine that the rooms are colored alternately black and white, like a checkerboard, as shown in Figure 5. So there are 18 black rooms and 18 white rooms. By coloring the rooms we have introduced something extra that at first does not seem relevant to the problem. Indeed, the tour can or cannot be done, regardless of the colors of the rooms. But notice how coloring the rooms in this fashion allows us to give a convincing argument that explains why this tour is impossible. We reason as follows:

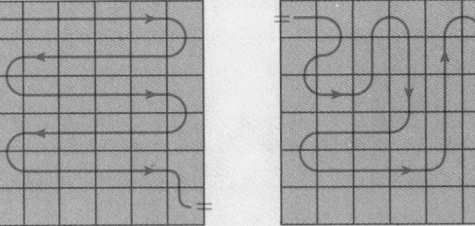

Entrance

Exit

FIGURE 4

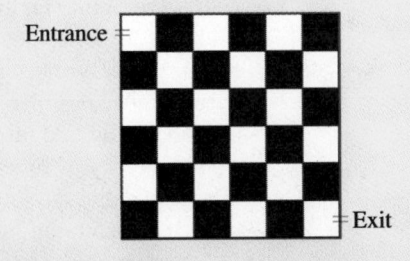

Entrance

Exit

FIGURE 5

If our visitor is in a white room, he must next go to a black room, and if he is in a black room, he must next go to a white room. The entrance is in a white room, so our visitor's path, in terms of colors of rooms, is

$$W \quad B \quad W \quad B \quad W \quad B \quad \dots$$

Since the museum contains an even number of rooms and the visitor starts in a white room, he must end his tour in a black room. But the exit is in a white room, so the proposed tour is impossible! ■

PROBLEMS

1–6 ■ Determine if each network is traversable. If it is, find a path that traverses it.

1.
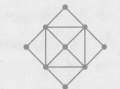

2.

3.

4.

5.
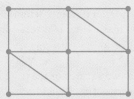

6.

7. Is the museum tour possible if there are 5 rooms to a side instead of 6? Find a general result about when the tour is possible for an $n \times n$ museum.

8. A Tibetan monk leaves the monastery at 7:00 A.M. and walks along his usual path to the top of the mountain, arriving at 7:00 P.M. The following morning, he starts at 7:00 A.M. at the top and takes the same path back, arriving at the monastery at 7:00 P.M. Show that there is exactly one point on the path that the monk will cross at exactly the same time of day on both trips.

9. A curve that starts and ends at the same point and does not intersect itself is called a *Jordan curve*. Such a curve always has an "inside" and an "outside." Is the point A in the figure inside or outside this Jordan curve? Find an easy method of determining whether a point is inside or outside a Jordan curve. [*Hint:* The point B is clearly outside the curve. Add a line segment connecting A and B.]

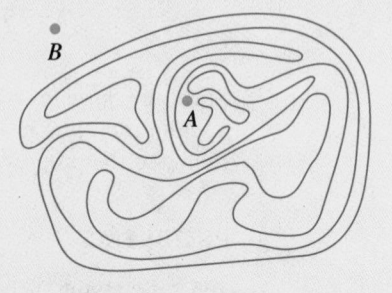

10. Consider the following sequence of polynomials:

$$P_1(x) = (x - a)$$
$$P_2(x) = (x - a)(x - b)$$
$$P_3(x) = (x - a)(x - b)(x - c)$$
$$\vdots \qquad \vdots \qquad \vdots$$

(a) Expand the polynomials P_2 and P_3.

(b) Using the pattern you observe from part (a), write the expansions of P_4 and P_5 without actually multiplying out these polynomials. The patterns you have found are called *Viète's relations* (see page 91).

(c) For a polynomial of degree n, express the coefficient of x^{n-1} and the constant coefficient in terms of the zeros of the polynomial.

11. If the equation $x^4 + ax^2 + bx + c = 0$ has roots 1, 2, and 3, find c.

12. Find a polynomial of degree 3 with integer coefficients that has $2 - \sqrt[3]{2}$ as one of its zeros. How many other real zeros does the polynomial have?

13. You have an 8×8 grid with squares removed from two diagonally opposite corners, as shown in the figure. You also have a set of dominoes, each of which covers exactly two squares of the grid. Is it possible to cover all the remaining squares of the grid with dominoes, without overlapping any dominoes? [*Hint:* Color the grid as a checkerboard.]

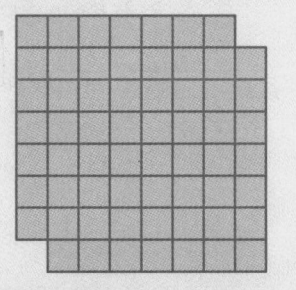

14. A rectangle is divided into smaller rectangles in such a way that each of the smaller rectangles has at least one side of integer length. Show that one of the sides of the original, large rectangle must have integer length. [*Hint:* Imagine the rectangle lying in the coordinate plane, which has been tiled into a checkerboard pattern of black and white squares, each $\frac{1}{2}$ unit by $\frac{1}{2}$ unit. Observe that a rectangle has at least one integer side if and only if its area is composed of equal amounts of black and white.]

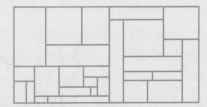

15. (a) Show that if you multiply four consecutive integers and then add 1 to the result, you get a perfect square.

(b) Show that if you multiply three consecutive integers and then add the middle integer to the result, you get a perfect cube.

16. The equation $x^3 = 1$ has only one real solution, so it must have two imaginary solutions. Let ω be one of the imaginary solutions. Show that the expression

$$(1 - \omega + \omega^2)(1 + \omega - \omega^2)$$

is a real number, and find its value.

17. The Indian mathematician Bhaskara sketched the two figures shown here and wrote below them "Behold!" Explain how his sketches prove the Pythagorean Theorem.

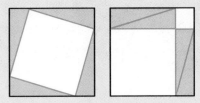

18. If the lengths of the sides of a right triangle, in increasing order, are a, b, and c, show that $a^3 + b^3 < c^3$.

5

EXPONENTIAL AND LOGARITHMIC FUNCTIONS

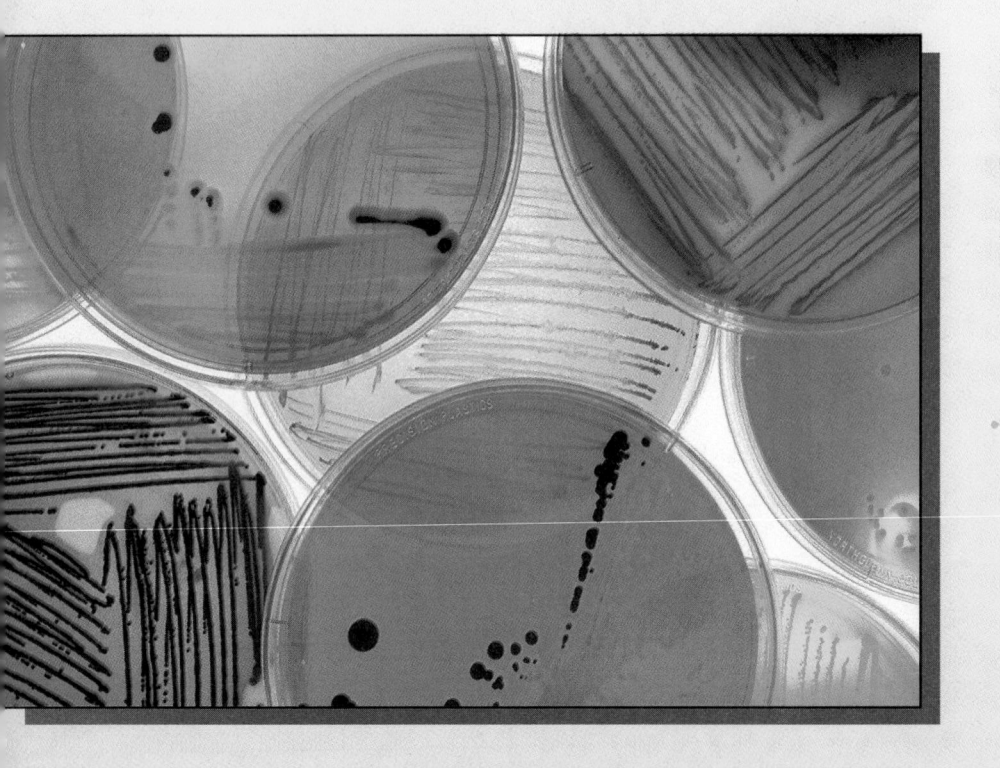

Population growth—such as the growth of a bacteria sample in an experiment—is modeled by exponential functions. These same functions describe compound interest, radioactive decay, and temperature changes.

Mathematics compares the most diverse phenomena and discovers the secret analogies which unite them.

JOSEPH FOURIER

So far we have studied relatively simple functions such as polynomials and rational functions. We now turn our attention to two of the most important functions in mathematics, the exponential function and its inverse function, the logarithmic function. We use these functions to describe exponential growth in biology and economics and radioactive decay in physics and chemistry.

5.1 EXPONENTIAL FUNCTIONS

In Section 1.3 we defined a^x if $a > 0$ and x is a rational number, but we have not yet defined irrational powers. For instance, what is meant by $2^{\sqrt{3}}$ and 5^π? To help us answer these questions we first look at the graph of the function $f(x) = 2^x$, where x is rational. A very crude representation of this graph is shown in Figure 1. Holes occur throughout the graph, and we want to fill in these holes with a smooth curve. To do this we enlarge the domain of $f(x) = 2^x$ to all of $\mathbb{R}$ by defining irrational powers of 2 appropriately. For example, to define $2^{\sqrt{3}}$ we use rational approximations of $\sqrt{3}$. Since

$$\sqrt{3} \approx 1.73205\dots$$

we successively approximate $2^{\sqrt{3}}$ by the following rational powers:

$$2^{1.7}, \quad 2^{1.73}, \quad 2^{1.732}, \quad 2^{1.7320}, \quad 2^{1.73205}, \quad \dots$$

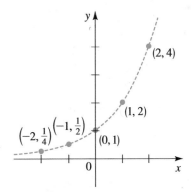

FIGURE 1
Representation of $f(x) = 2^x$,
x rational

Using advanced mathematics it can be shown that there is exactly one number that these powers approach. We define $2^{\sqrt{3}}$ to be this number. Intuitively, these rational powers of 2 are getting closer and closer to $2^{\sqrt{3}}$. By this approximation process we can compute $2^{\sqrt{3}}$ correct to 6 decimal places:

$$2^{\sqrt{3}} \approx 3.321997$$

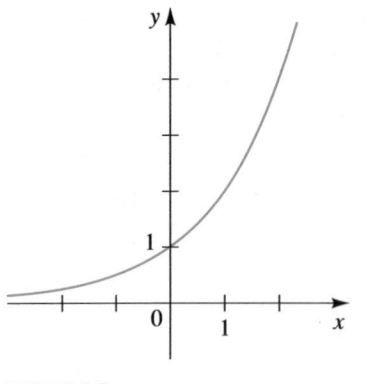

FIGURE 2
$f(x) = 2^x$, x real

To demonstrate just how quickly $f(x) = 2^x$ increases, let us perform the following thought experiment. Suppose we start with a piece of paper a thousandth of an inch thick, and we fold it in half 50 times. Each time we fold the paper, the thickness of the paper stack doubles, so the thickness of the resulting stack would be $2^{50}/1000$ inches. How thick do you think that is? It works out to be more than 17 million miles!

Similarly, we can define 2^x (or a^x if $a > 0$) where x is any irrational number. It can be proved that *the Laws of Exponents are still true when the exponents are real numbers.*

Using this definition of irrational powers, we can graph $f(x) = 2^x$ on all of $\mathbb{R}$, as shown in Figure 2. It is an increasing function and, in fact, it increases very rapidly when $x > 0$ (see the margin note).

EXAMPLE 1 ■ Graphing Exponential Functions by Plotting Points

Draw the graph of each of the following functions.

(a) $f(x) = 3^x$ (b) $g(x) = \left(\dfrac{1}{3}\right)^x$

SOLUTION

We calculate values of $f(x)$ and $g(x)$ and plot points to sketch the graphs in Figure 3.

x	$f(x) = 3^x$	$g(x) = (\frac{1}{3})^x$
-3	$\frac{1}{27}$	27
-2	$\frac{1}{9}$	9
-1	$\frac{1}{3}$	3
0	1	1
1	3	$\frac{1}{3}$
2	9	$\frac{1}{9}$
3	27	$\frac{1}{27}$

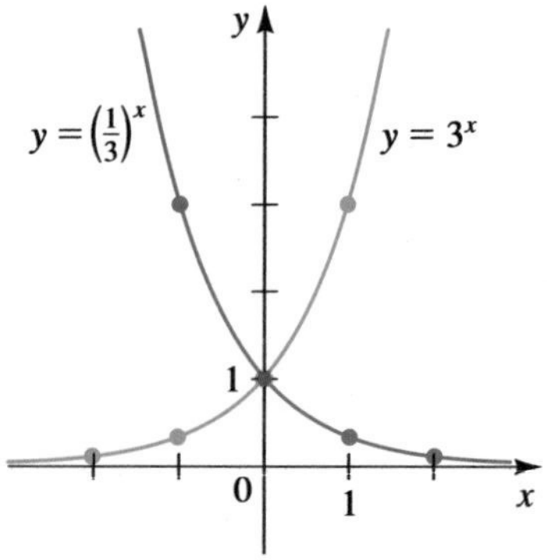

FIGURE 3

Notice that

$$g(x) = \left(\frac{1}{3}\right)^x = \frac{1}{3^x} = 3^{-x} = f(-x)$$

and so we could have obtained the graph of g from the graph of f by reflecting in the y-axis. ■

Figure 4 shows the graphs of the exponential function $f(x) = a^x$ for various values of the base a. All of these graphs pass through the same point $(0, 1)$ because $a^0 = 1$ for $a \neq 0$. You can see from Figure 4 that there are three kinds of exponential functions $y = a^x$: If $0 < a < 1$, the exponential function decreases rapidly. If $a = 1$, it is constant. If $a > 1$, the function increases rapidly, and the larger the base, the more rapid the increase.

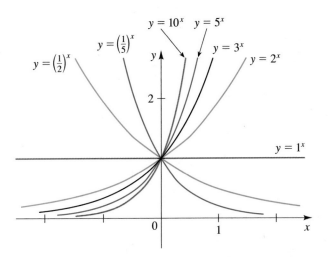

$y = \left(\dfrac{1}{2}\right)^x$ $y = \left(\dfrac{1}{5}\right)^x$ $y = 10^x$ $y = 5^x$ $y = 3^x$ $y = 2^x$ $y = 1^x$

FIGURE 4

If $a > 1$, then the graph of $y = a^x$ approaches zero as x decreases through negative values and so the x-axis is a horizontal asymptote. If $0 < a < 1$, the graph approaches zero as x increases indefinitely and, again, the x-axis is a horizontal asymptote. In either case the graph never touches the x-axis because $a^x > 0$ for all x. Thus, for $a \neq 1$, the exponential function $f(x) = a^x$ has domain $\mathbb{R}$ and range $(0, \infty)$. We summarize the preceding discussion.

EXPONENTIAL FUNCTIONS

If $a > 0$, the **exponential function with base a** is defined by

$$f(x) = a^x$$

For $a \neq 1$, the domain of f is $\mathbb{R}$, the range of f is $(0, \infty)$, and the graph of f has one of the following shapes:

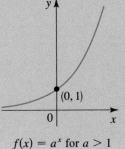

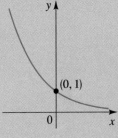

$f(x) = a^x$ for $a > 1$ $f(x) = a^x$ for $0 < a < 1$

EXAMPLE 2 ■ Identifying the Graphs of Exponential Functions

Find the exponential function $f(x) = a^x$ whose graph is given.

(a)

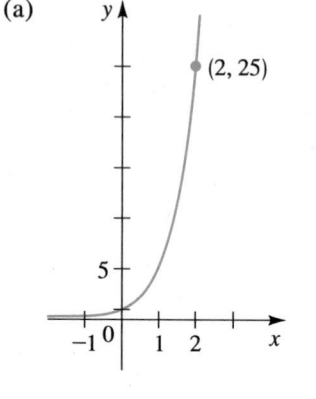

(b)

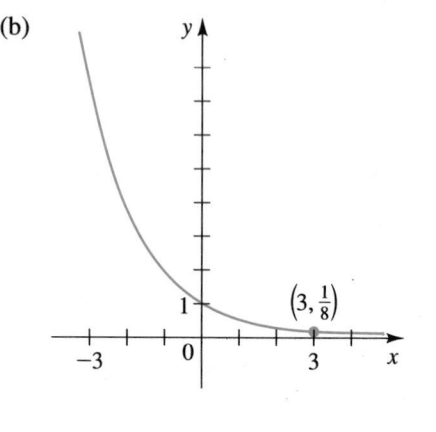

SOLUTION

(a) Since $f(2) = a^2 = 25$, we see by inspection that the base is $a = 5$. So this is the graph of the exponential function $f(x) = 5^x$.

(b) Since $f(3) = a^3 = \frac{1}{8}$, we see by inspection that the base is $a = \frac{1}{2}$. So this is the graph of the exponential function $f(x) = \left(\frac{1}{2}\right)^x$.   ■

In the next two examples we see how to graph certain functions, not by plotting points but by taking the basic graphs of the exponential functions in Figure 4 and applying the shifting and reflecting transformations of Section 3.8.

EXAMPLE 3 ■ Transformations of Exponential Functions

Vertical shifting and reflecting of graphs is explained in Section 3.8.

Use the graph of $f(x) = 2^x$ to sketch the graph of each of the following functions.

(a) $g(x) = 3 + 2^x$

(b) $h(x) = -2^x$

SOLUTION

(a) The graph of $g(x) = 3 + 2^x$ is obtained by starting with the graph of $f(x) = 2^x$ in Figure 5(a) and shifting it upward 3 units. Notice from Figure 5(b) that the line $y = 3$ is now a horizontal asymptote.

(b) Again we start with the graph of $f(x) = 2^x$, but here we reflect in the x-axis to get the graph of $h(x) = -2^x$ shown in Figure 5(c).

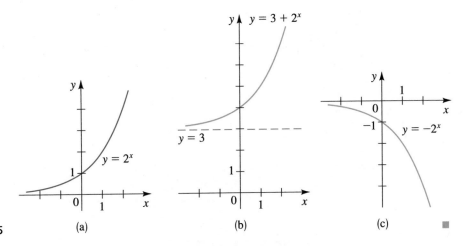

FIGURE 5   (a)   (b)   (c)

EXAMPLE 4 ■ **Transformations of Exponential Functions**

Vertical and horizontal shifting of graphs is explained in Section 3.8.

(a) Use the graph of $f(x) = 10^x$ to sketch the graph of $g(x) = 10^{x-1} - 2$.

(b) State the asymptote, the domain, and the range of this function.

SOLUTION

(a) Recall from Section 3.8 that we get the graph of $y = f(x - 1)$ from the graph of $y = f(x)$ by shifting to the right 1 unit. Thus, we get the graph of $y = 10^{x-1} - 2$ by shifting the graph of $y = 10^x$ to the right 1 unit and downward 2 units, as shown in Figure 6.

(b) The horizontal asymptote is $y = -2$, the domain is $\mathbb{R}$, and the range is

$$\{y \mid y > -2\} = (-2, \infty)$$

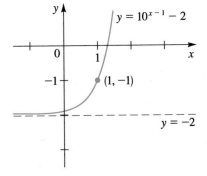

FIGURE 6

EXAMPLE 5 ■ **Comparing Exponential and Power Functions**

Compare the rates of growth of the exponential function $f(x) = 2^x$ and the power function $g(x) = x^2$ by drawing the graphs of both functions in each of the following viewing rectangles.

(a) $[0, 3]$ by $[0, 8]$ (b) $[0, 6]$ by $[0, 25]$ (c) $[0, 20]$ by $[0, 1000]$

SOLUTION

(a) Figure 7(a) on the next page shows that the graph of $g(x) = x^2$ catches up with, and becomes higher than, the graph of $f(x) = 2^x$ at $x = 2$.

(b) The larger viewing rectangle in Figure 7(b) on the next page shows that the graph of $f(x) = 2^x$ overtakes that of $g(x) = x^2$ when $x = 4$.

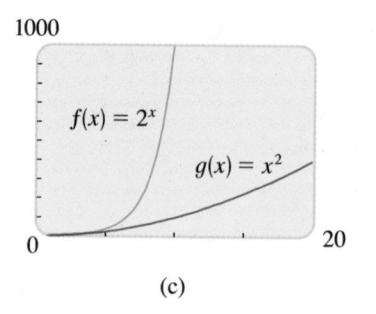

(a)

(b)

(c)

FIGURE 7

(c) Figure 7(c) gives a more global view and shows that, when x is large, $f(x) = 2^x$ is much larger than $g(x) = x^2$. ∎

5.1 EXERCISES

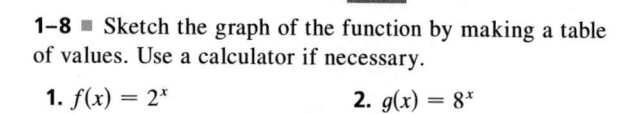

1–8 ■ Sketch the graph of the function by making a table of values. Use a calculator if necessary.

1. $f(x) = 2^x$ **2.** $g(x) = 8^x$

3. $h(x) = 6^x$ **4.** $h(x) = (0.8)^x$

5. $f(x) = \left(\frac{1}{3}\right)^x$

6. $h(x) = (1.1)^x$

7. $g(x) = \left(\frac{1}{4}\right)^x$

8. $f(x) = \left(\frac{3}{2}\right)^x$

9–10 ■ Graph both functions on one set of axes.

9. $y = 4^x$ and $y = 7^x$

10. $y = \left(\frac{2}{3}\right)^x$ and $y = \left(\frac{4}{3}\right)^x$

11–14 ■ Find the exponential function $f(x) = a^x$ whose graph is given.

11.

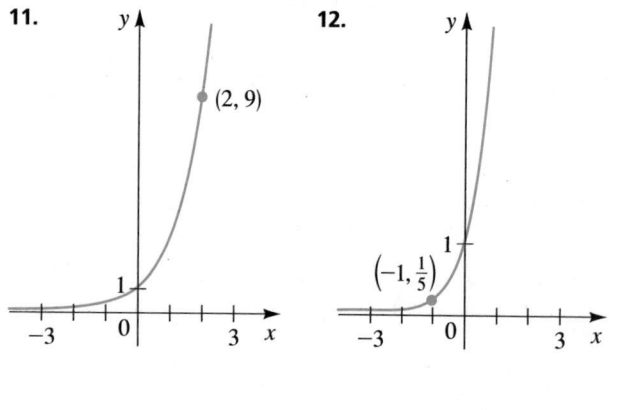

12.

13.

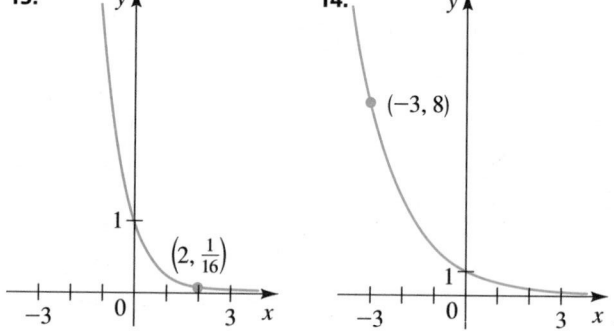

14.

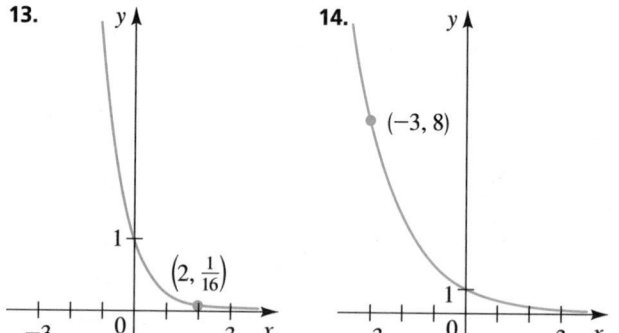

15–20 ■ Match the exponential function with one of the graphs I–VI.

15. $f(x) = 5^x$ **16.** $f(x) = -5^x$

17. $f(x) = 5^{-x}$ **18.** $f(x) = 5^x + 3$

19. $f(x) = 5^{x-3}$ **20.** $f(x) = 5^{x+1} - 4$

I

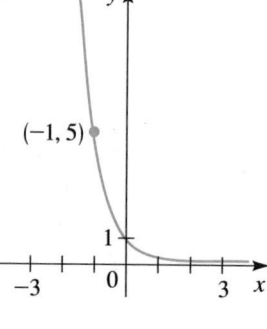

II

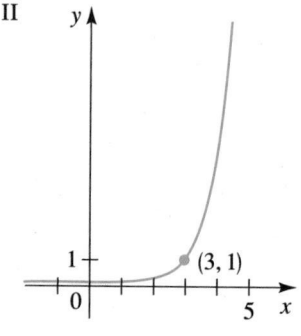

III

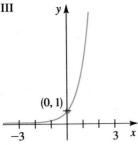

(0, 1)

IV

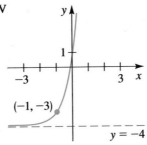

(−1, −3)

y = −4

V

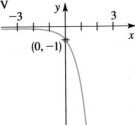

(0, −1)

VI

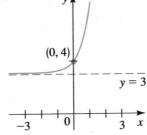

(0, 4)

y = 3

21–36 ■ Graph the function, not by plotting points, but by starting from the graphs in Figure 4. State the domain, range, and asymptote.

21. $f(x) = -3^x$

22. $f(x) = 10^{-x}$

23. $g(x) = 2^x - 3$

24. $g(x) = 2^{x-3}$

25. $h(x) = 4 + \left(\frac{1}{2}\right)^x$

26. $h(x) = 6 - 3^x$

27. $f(x) = 10^{x+3}$

28. $f(x) = -\left(\frac{1}{5}\right)^x$

29. $f(x) = -3^{-x}$

30. $f(x) = 10^{-x} - 4$

31. $y = 5^{-2x}$

32. $y = 1 + 2^{x+1}$

33. $f(x) = 5 - 2^{x-1}$

34. $f(x) = 1 - 2^{-x}$

35. $y = 2^{|x|}$

36. $y = 2^{-|x|}$

37–38 ■ Find the function of the form $f(x) = Ca^x$ whose graph is given.

37.

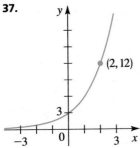

(2, 12)

38.

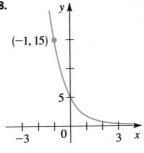

(−1, 15)

39. (a) Sketch the graphs of $f(x) = 2^x$ and $g(x) = 3(2^x)$.
   (b) How are the graphs related?

40. (a) Sketch the graphs of $f(x) = 9^{x/2}$ and $g(x) = 3^x$.
   (b) Use the Laws of Exponents to explain the relationship between these graphs.

41. If $f(x) = 10^x$, show that

$$\frac{f(x + h) - f(x)}{h} = 10^x \left(\frac{10^h - 1}{h}\right)$$

42. Compare the functions $f(x) = x^3$ and $g(x) = 3^x$ by evaluating both of them for $x = 0, 1, 2, 3, 4, 5, 6, 7, 8, 9, 10, 15,$ and 20. Then draw the graphs of f and g on the same set of axes.

43. Suppose you are offered a job that lasts one month, and you are to be very well paid. Which of the following methods of payment is more profitable for you?
   (i) One million dollars at the end of the month.
   (ii) Two cents on the first day of the month, 4 cents on the second day, 8 cents on the third day, and, in general, 2^n cents on the nth day.

44. Your mathematics teacher asks you to sketch a graph of the exponential function $f(x) = 2^x$ for x between 0 and 40, using a scale of 10 units to one inch. What are the dimensions of the sheet of paper you will need to sketch this graph?

45. (a) Compare the rates of growth of the functions $f(x) = 2^x$ and $g(x) = x^5$ by drawing the graphs of both functions in each of the following viewing rectangles.
   (i) $[0, 5]$ by $[0, 20]$
   (ii) $[0, 25]$ by $[0, 10^7]$
   (iii) $[0, 50]$ by $[0, 10^8]$
   (b) Find the solutions of the equation $2^x = x^5$, correct to one decimal place.

46. (a) Compare the rates of growth of the functions $f(x) = 3^x$ and $g(x) = x^4$ by drawing the graphs of both functions in each of the the following viewing rectangles:
   (i) $[-4, 4]$ by $[0, 20]$
   (ii) $[0, 10]$ by $[0, 5000]$
   (iii) $[0, 20]$ by $[0, 10^5]$
   (b) Find the solutions of the equation $3^x = x^4$, correct to two decimal places.

47–48 ■ Draw graphs of the given family of functions for $c = 0.25, 0.5, 1, 2, 4$. How are the graphs related?

47. $f(x) = c2^x$ **48.** $f(x) = 2^{cx}$

49–50 ■ Graph the function and comment on vertical and horizontal asymptotes.

49. $y = 2^{1/x}$ **50.** $y = \dfrac{2^x}{x}$

51–52 ■ Find the local maximum and minimum values of the function and the value of x at which each occurs. State each answer correct to two decimal places.

51. $g(x) = x^x$ $(x > 0)$ **52.** $g(x) = \sqrt[x]{x}$ $(x > 0)$

53–54 ■ Find, correct to two decimals, (a) the intervals on which the function is increasing or decreasing and (b) the range of the function.

53. $y = 10^{x - x^2}$ **54.** $y = x2^x$

5.2 THE NATURAL EXPONENTIAL FUNCTION

The notation e for the base of the natural exponential function was chosen by the Swiss mathematician Leonhard Euler (see page 100), probably because it is the first letter of the word *exponential*.

Any positive number can be used as the base for an exponential function, but some bases are used more frequently than others. We will see in the remaining sections of this chapter that the bases 2 and 10 are convenient for certain applications. But the most important base is the number denoted by the letter e.

The number e is defined as the value that $(1 + 1/n)^n$ approaches as n becomes large. (In calculus this idea is made more precise through the concept of a limit. See Exercise 41.) The following table shows the values of the expression $(1 + 1/n)^n$ for increasingly large values of n.

n	$\left(1 + \dfrac{1}{n}\right)^n$
1	2.00000
5	2.48832
10	2.59374
100	2.70481
1000	2.71692
10,000	2.71815
100,000	2.71827
1,000,000	2.71828
10,000,000	2.71828

It appears that, correct to five decimal places, $e \approx 2.71828$; in fact, the approximate value to 20 decimal places is

$$e \approx 2.71828182845904523536$$

It can be shown that e is an irrational number, so we cannot write its exact value.

Why use such a strange base for an exponential function? It may seem at first that a base such as 10 is easier to work with. We will see, however, that in certain applications the number e is the best possible. In this section we study how e occurs in the description of compound interest and population growth.

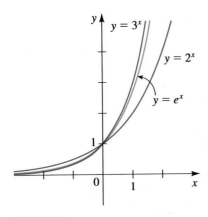

FIGURE 1

Graph of the natural exponential
function

THE NATURAL EXPONENTIAL FUNCTION

The **natural exponential function** is the exponential function

$$f(x) = e^x$$

with base e. It is often referred to as *the* exponential function.

Since $2 < e < 3$, the graph of the natural exponential function lies between the graphs of $y = 2^x$ and $y = 3^x$, as shown in Figure 1.

Scientific calculators have a special key for the function $f(x) = e^x$. We use this key in the next example.

EXAMPLE 1 ■ Evaluating the Exponential Function

Evaluate each expression correct to five decimal places.

(a) e^3 (b) $2e^{-0.53}$ (c) $e^{4.8}$

SOLUTION

We use the $\boxed{e^x}$ key on a calculator to evaluate the exponential function.

(a) $e^3 \approx 20.08554$ (b) $2e^{-0.53} \approx 1.17721$ (c) $e^{4.8} \approx 121.51042$ ■

EXAMPLE 2 ■ Transformations of the Exponential Function

Sketch the graph of each function.

(a) $f(x) = e^{-x}$ (b) $g(x) = 3e^{0.5x}$

SOLUTION

Reflecting graphs is explained in
Section 3.8.

(a) We start with the graph of $y = e^x$ and reflect in the y-axis to obtain the graph of $y = e^{-x}$ as in Figure 2.

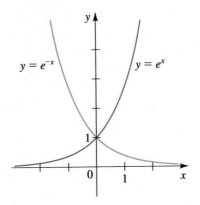

FIGURE 2

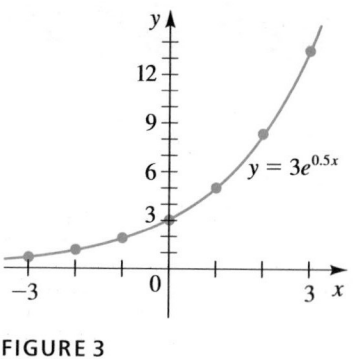

FIGURE 3

(b) We calculate several values, plot the resulting points, then connect the points with a smooth curve. The graph is shown in Figure 3.

x	$f(x) = 3e^{0.5x}$
-3	0.67
-2	1.10
-1	1.82
0	3
1	4.95
2	8.15
3	13.45

■

COMPOUND INTEREST

The natural exponential function occurs in calculating *continuously* compounded interest. We begin by explaining compound interest.

If an amount of money P, called the **principal,** is invested at a simple interest rate r, then after one time period the interest is Pr and the amount A of money is

$$A = P + Pr = P(1 + r)$$

If the interest is reinvested, then the new principal is now $P(1 + r)$ and the amount after another time period is $A = P(1 + r)(1 + r) = P(1 + r)^2$. Similarly, after a third time period the amount is $A = P(1 + r)^3$. In general, after k periods the amount is

$$A = P(1 + r)^k$$

Notice that this is an exponential function with base $1 + r$.

If interest is compounded n times per year, then in each time period the interest rate is r/n and there are nt time periods in t years. This leads to the following formula for the amount after t years.

COMPOUND INTEREST

Compound interest is calculated by the formula

$$A = P\left(1 + \frac{r}{n}\right)^{nt}$$

where P = principal

r = interest rate

n = number of times interest is compounded per year

t = number of years

A = amount after t years

EXAMPLE 3 ■ Calculating Compound Interest

A sum of $1000 is invested at an interest rate of 12% per year. Find the amount in the account after three years for each compounding method.

(a) annual   (b) semiannual   (c) quarterly   (d) monthly   (e) daily

SOLUTION

We use the compound interest formula with $P = \$1000$, $r = 0.12$, and $t = 3$.

(a) With annual compounding, $n = 1$.

$$A = 1000\left(1 + \frac{0.12}{1}\right)^{1(3)} = 1000(1.12)^3 = \$1404.93$$

(b) With semiannual compounding, $n = 2$.

$$A = 1000\left(1 + \frac{0.12}{2}\right)^{2(3)} = 1000(1.06)^6 = \$1418.52$$

(c) With quarterly compounding, $n = 4$.

$$A = 1000\left(1 + \frac{0.12}{4}\right)^{4(3)} = 1000(1.03)^{12} = \$1425.76$$

(d) With monthly compounding, $n = 12$.

$$A = 1000\left(1 + \frac{0.12}{12}\right)^{12(3)} = 1000(1.01)^{36} = \$1430.77$$

(e) With daily compounding, $n = 365$.

$$A = 1000\left(1 + \frac{0.12}{365}\right)^{365(3)} = \$1433.24$$

■

We see from Example 3 that the interest paid increases as the number of compounding periods (n) increases. Let us see what happens as n increases indefinitely. If we let $m = n/r$, then

$$A = P\left(1 + \frac{r}{n}\right)^{nt} = P\left[\left(1 + \frac{r}{n}\right)^{n/r}\right]^{rt} = P\left[\left(1 + \frac{1}{m}\right)^{m}\right]^{rt}$$

Recall that as m becomes large, the quantity $(1 + 1/m)^m$ approaches the number e. Thus, the amount approaches $A = Pe^{rt}$. This expression gives the amount when the interest is compounded at "every instant."

CONTINUOUSLY COMPOUNDED INTEREST

Continuously compounded interest is calculated by the formula

$$A = Pe^{rt}$$

where P = principal

r = interest rate

t = number of years

A = amount after t years

EXAMPLE 4 ■ Calculating Continuously Compounded Interest

Find the amount after three years if $1000 is invested at an interest rate of 12% per year, compounded continuously.

SOLUTION

We use the formula for continuously compounded interest with P = $1000, $r = 0.12$, and $t = 3$ to get

$$A = 1000e^{(0.12)3} = 1000e^{0.36} = \$1433.33$$

Compare this amount with the amounts in Example 3.   ■

EXAMPLE 5 ■ Comparing Compounded and Continuously Compounded Interest

Find the amount after five years if $10,000 is invested at an interest rate of 8% per year and the interest is compounded (a) quarterly or (b) continuously.

SOLUTION

(a) Using the formula for compound interest with P = $10,000, $r = 0.08$, $n = 4$, and $t = 5$, we have

$$A = 10{,}000\left(1 + \frac{0.08}{4}\right)^{(4)5} = 10{,}000(1.02)^{20} = \$14{,}859.47$$

(b) Using the formula for continuously compounded interest with P = $10,000, $r = 0.08$, and $t = 5$, we have

$$A = 10{,}000e^{(0.08)5} = 10{,}000e^{0.4} = \$14{,}918.25$$

Notice that continuous compounding gives a larger return than quarterly compounding, but not much larger for this relatively short time period.   ■

EXPONENTIAL GROWTH

Biologists have observed that the population of a species always doubles its size in a fixed period of time. For example, under ideal conditions a certain bacteria population doubles in size every three hours. If the culture is started with 1000 bacteria, then after three hours there will be 2000 bacteria, after another three hours there will be 4000, and so on. If we let $n = n(t)$ be the number of bacteria after t hours, then

$$n(0) = 1000$$

$$n(3) = 1000 \cdot 2$$

$$n(6) = (1000 \cdot 2) \cdot 2 = 1000 \cdot 2^2$$

$$n(9) = (1000 \cdot 2^2) \cdot 2 = 1000 \cdot 2^3$$

$$n(12) = (1000 \cdot 2^3) \cdot 2 = 1000 \cdot 2^4$$

$$n(15) = (1000 \cdot 2^4) \cdot 2 = 1000 \cdot 2^5$$

The **Gateway Arch** in St. Louis, Missouri, is shaped in the form of the graph of a combination of exponential functions (*not* a parabola, as it might first appear). Specifically, it is a **catenary,** which is the graph of an equation of the form

$$y = a(e^{bx} + e^{-bx})$$

(see Exercise 43). This shape was chosen because it is optimal for distributing the internal structural forces of the arch. Chains and cables suspended between two points (for example, the stretches of cable between two telephone poles) hang in the shape of a catenary.

From the pattern in the table, it appears that the number of bacteria after t hours is

$$n(t) = 1000 \cdot 2^{t/3}$$

In general, suppose that the initial size of a population is n_0 and the doubling period is a. Then the size of the population at time t is

$$n(t) = n_0 2^{ct}$$

where $c = 1/a$. If we knew the tripling time b, then the formula would be $n(t) = n_0 3^{ct}$ where $c = 1/b$. These formulas indicate that the growth of the bacteria is exponential. But what base should we use in our formula? The answer is e, because then it can be shown (using calculus) that the formula is

$$n(t) = n_0 e^{rt}$$

where r is the *relative rate of growth of population, expressed as a proportion of the population at any time.* For instance, if $r = 0.02$ then at any time t the growth is 2% of the population at time t.

Notice that the formula for population growth is the same as that for continuously compounded interest. In fact, the same principle is at work in both cases: The growth of a population (or an investment) per time period is proportional to the size of the population (or the amount of the investment). A population of 1,000,000 will increase more in one year than a population of 1000; in exactly the same way, an investment of $1,000,000 will increase more in one year than an investment of $1000.

EXPONENTIAL GROWTH

A population that experiences **exponential growth** increases according to the formula

$$n(t) = n_0 e^{rt}$$

where n_0 = initial size of the population

r = relative rate of growth (expressed as a proportion of the population)

t = time

$n(t)$ = population at time t

EXAMPLE 6 ■ Predicting the Size of a Population

The initial bacteria count in a culture is 500. A biologist later makes a sample count of bacteria in the culture and finds that the relative rate of growth is 40% per hour.

(a) Find a formula for the number of bacteria $n(t)$ after t hours.
(b) What is the estimated count after 10 hours?
(c) Sketch the graph of the function $n(t)$.

SOLUTION

(a) We use the formula for exponential population growth with $n_0 = 500$ and $r = 0.4$ to get

$$n(t) = 500 e^{0.4t}$$

where t is measured in hours.

(b) Using the formula in part (a) we find that the bacteria count after 10 hours is

$$n(10) = 500 e^{0.4(10)} = 500 e^4 \approx 27{,}300$$

(c) We plot several points and sketch the graph in Figure 4.   ■

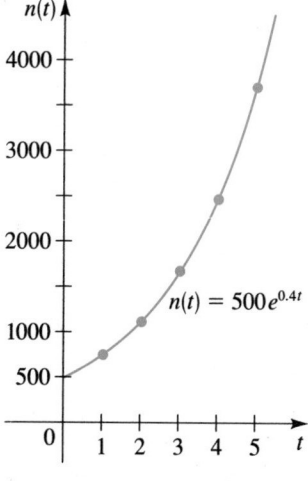

FIGURE 4

The curve is labeled $n(t) = 500 e^{0.4t}$

EXAMPLE 7 ■ Comparing Effects of Different Rates of Population Growth

In 1995 the population of the world was 5.8 billion and the relative rate of growth was 2% per year. It is claimed that a rate of 1.6% per year would make a significant difference in the total population in just a few decades. Test this claim by estimating the population of the world in the year 2035 if the relative rate of growth is (a) 2% per year or (b) 1.6% per year.

Graph the population functions for the next 100 years for the two relative growth rates in the same viewing rectangle.

SOLUTION

(a) By the formula for exponential population growth, we have

$$n(t) = 5.8e^{0.02t}$$

where $n(t)$ is measured in billions and t is measured in years since 1995. Because the year 2035 is 40 years after 1995, we find

$$n(40) = 5.8e^{0.02(40)} = 5.8e^{0.8} \approx 12.9$$

Thus, the estimated population in the year 2035 is 13 billion.

(b) We use the formula

$$n(t) = 5.8e^{0.016t}$$

and find $n(40) = 5.8e^{0.016(40)} = 5.8e^{0.64} \approx 11.0$

So, the estimated population in the year 2035 is 11 billion.

The graphs in Figure 5 show that a small change in the relative rate of growth can, over time, make a large difference in population size.   ■

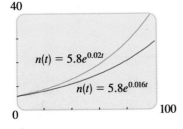

40

$n(t) = 5.8e^{0.02t}$

$n(t) = 5.8e^{0.016t}$

0   100

FIGURE 5

EXAMPLE 8 ■ Finding the Initial Population

A certain breed of rabbit was introduced onto a small island about eight years ago. The current rabbit population on the island is now estimated to be 4100 and the estimated relative rate of growth is 55% per year.

(a) What was the initial size of the rabbit population?
(b) What is the population estimated to be 12 years from now?

SOLUTION

(a) From the formula for exponential population growth, we have

$$n(t) = n_0 e^{0.55t}$$

and we know that the population at time $t = 8$ is $n(8) = 4100$. We substitute what we know into the equation and solve for n_0:

$$4100 = n_0 e^{0.55(8)}$$

$$4100 \approx n_0\, 81.45$$

$$n_0 \approx \frac{4100}{81.45} \approx 50$$

Thus, we estimate that 50 rabbits were introduced onto the island.

(b) Now that we know n_0, we can write a formula for population growth:

$$n(t) = 50e^{0.55t}$$

Twelve years from now, $t = 20$ and

$$n(20) = 50e^{0.55(20)} \approx 2{,}993{,}707$$

So, the rabbit population on the island 12 years from now is estimated to be about 3 million. ∎

Can the rabbit population in Example 8(b) actually reach such a high number? In reality, as the island becomes overpopulated with rabbits, the rabbit population growth will be slowed due to food shortage and other factors. One model, or formula, that takes into account such factors is the *logistic growth formula* described in Exercises 31 and 32.

5.2　EXERCISES

1–2 ■ Complete the table and use it to graph the function.

1. $f(x) = 3e^x$

x	$f(x) = 3e^x$
-2	
-1.5	
-1	
-0.5	
0	
0.5	
1	
1.5	
2	

2. $g(x) = 2e^{-0.5x}$

x	$g(x) = 2e^{-0.5x}$
-4	
-3	
-2	
-1	
0	
1	
2	
3	
4	

3–8 ■ Graph the function, not by plotting points, but by starting from the graph of $y = e^x$ in Figure 1. State the domain, range, and asymptote.

3. $y = -e^x$

4. $y = 1 - e^x$

5. $y = e^{-x} - 1$

6. $y = -e^{-x}$

7. $y = e^{x-2}$

8. $y = e^{x-3} + 4$

9. If \$10,000 is invested at an interest rate of 10% per year, compounded semiannually, find the value of the investment after the given number of years.
(a) 5 years　　(b) 10 years　　(c) 15 years

10. If \$4000 is borrowed at a rate of 16% interest per year, compounded quarterly, find the amount due at the end of the given number of years.
(a) 4 years　　(b) 6 years　　(c) 8 years

11. If \$3000 is invested at an interest rate of 9% per year, find the amount of the investment at the end of five years for each of the following compounding methods.
(a) annual　　(b) semiannual　　(c) monthly
(d) weekly　　(e) daily　　(f) hourly
(g) continuously

12. If \$4000 is invested in an account for which interest is compounded quarterly, find the amount of the investment at the end of five years for each of the following interest rates.
(a) 6%   (b) $6\frac{1}{2}$%   (c) 7%   (d) 8%

13. Which of the given interest rates and compounding interest periods would provide the better investment?
(i) $8\frac{1}{2}$% per year, compounded semiannually
(ii) $8\frac{1}{4}$% per year, compounded quarterly
(iii) 8% per year, compounded continuously

14. Which of the given interest rates and compounding periods would provide the better investment?
(i) $9\frac{1}{4}$% per year, compounded semiannually
(ii) 9% per year, compounded continuously

15–16 ■ The **present value** of a sum of money is the amount that must be invested now, at a given rate of interest, to produce the desired sum at a later date.

15. Find the present value of \$10,000 if interest is paid at a rate of 9% per year, compounded semiannually, for three years.

16. Find the present value of \$100,000 if interest is paid at a rate of 8% per year, compounded monthly, for five years.

17. The number of bacteria in a culture is given by the formula

$$n(t) = 500e^{0.45t}$$

where t is measured in hours.
(a) What is the relative rate of growth of this bacteria population? Express your answer as a percentage.
(b) What is the initial population of the culture (at $t = 0$)?
(c) How many bacteria will the culture contain at time $t = 5$?

18. The rat population in New York City is given by the formula

$$n(t) = 54e^{0.12t}$$

where t is measured in years since 1990 and $n(t)$ is measured in millions.
(a) What is the relative rate of growth of the rat population? Express your answer as a percentage.

(b) What was the rat population in 1990?
(c) What is the rat population expected to be in the year 2000?
(d) Sketch a graph of the population function.

19. The fox population in a certain region has a relative growth rate of 8% per year. It is estimated that the population in 1992 was 18,000.
(a) Find a formula for the population t years after 1992.
(b) Use the formula from part (a) to estimate the fox population in the year 2000.
(c) Sketch the graph of the fox population function for the years 1992–2000.

20. The population of a certain city has a relative growth rate of 5% per year. The population in 1988 was 421,000. Find the projected population of the city for each of the given years.
(a) 2000   (b) 2030

21. The population of a country has relative growth rate of 3% per year. The government is trying to reduce the growth rate to 2%. The population in 1995 was approximately 110 million. Find the projected population for the year 2020 for each of the following conditions.
(a) The relative growth rate remains at 3% per year.
(b) The relative growth rate is reduced to 2% per year.

22. The population of a certain city was 680,000 in 1992 and is growing at the relative growth rate of 12% per year.
(a) Find a formula for the population of this city t years after 1992.
(b) Estimate the population in the year 2000.

23. The population of the world in 1987 was 5 billion and the relative growth rate was estimated at 2% per year. Assuming that the world population follows an exponential growth model, find the projected world population in 1995. (Compare this with the actual world population of 5.8 billion in 1995.)

24. The relative growth rate for a certain bacteria population is 80% per hour. A small culture is formed, and three hours later a count shows approximately 21,500 bacteria in the culture.
(a) Find the initial number of bacteria in the culture.
(b) Estimate the number of bacteria five hours from the time the culture was started.

25. Under ideal conditions, a certain type of bacteria has a relative growth rate of 220% per hour. A number of these bacteria are introduced accidentally into a food product. Two hours after contamination, a bacteria count shows that there are about 40,000 bacteria in the food.

(a) Find the initial number of bacteria introduced into the food.

(b) Find the estimated number of bacteria in the food three hours after contamination.

26. A radioactive substance decays in such a way that the amount of mass remaining after t days is given by the function

$$m(t) = 13e^{-0.015t}$$

where $m(t)$ is measured in kilograms.

(a) Find the mass at time $t = 0$.

(b) How much of the mass remains after 45 days?

27. Radioactive iodine is used by doctors as a tracer in diagnosing certain thyroid gland disorders. This type of iodine decays in such a way that the mass remaining after t days is given by the function

$$m(t) = 6e^{-0.087t}$$

where $m(t)$ is measured in grams.

(a) Find the mass at time $t = 0$.

(b) How much of the mass remains after 20 days?

28. A sky diver is dropped from a reasonable height above the ground. The air resistance she experiences is proportional to her velocity, and the constant of proportionality is 0.2. It can be shown that the velocity of the sky diver at time t is given by

$$v(t) = 80(e^{-0.2t} - 1)$$

where t is measured in seconds and $v(t)$ is measured in feet per second (ft/s).

(a) Find the initial velocity of the sky diver.

(b) Find the velocity after 5 s and after 10 s.

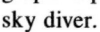 (c) Draw a graph of the velocity function $v(t)$.

(d) The maximum velocity of a falling object with wind resistance is called its *terminal velocity*. From the graph in part (c), find the terminal velocity of this sky diver.

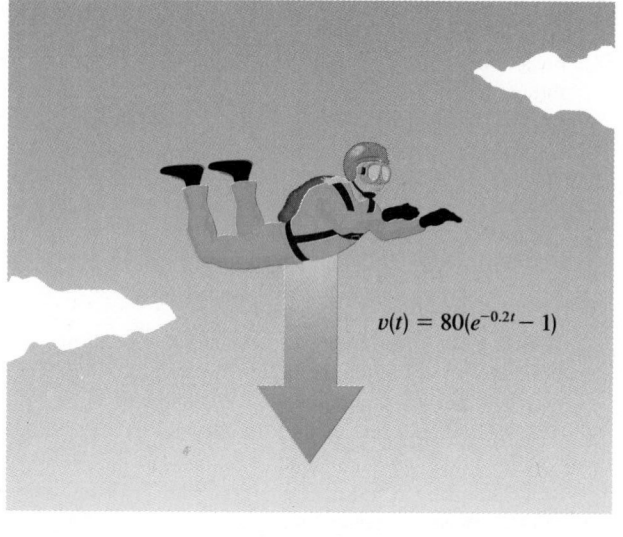

$$v(t) = 80(e^{-0.2t} - 1)$$

29. A 50-gallon barrel is filled completely with pure water. Salt water with a concentration of 0.3 lb/gal is then pumped into the barrel, and the resulting mixture overflows at the same rate. The amount of salt in the barrel at time t is given by

$$Q(t) = 15(1 - e^{-0.04t})$$

where t is measured in minutes and $Q(t)$ is measured in pounds.

(a) How much salt is in the barrel after five minutes?

(b) How much salt is in the barrel after ten minutes?

(c) Draw a graph of the function $Q(t)$.

(d) From the graph in part (c), what value does the amount of salt in the barrel approach as t becomes large? Is this what you would expect?

$$Q(t) = 15(1 - e^{-0.04t})$$

30. A sum of $5000 is invested at an interest rate of 9% per year, compounded semiannually.
(a) Find the value $A(t)$ of the investment after t years.
(b) Draw the graph of $A(t)$.
(c) Use the graph of $A(t)$ to determine when this investment will amount to $25,000.

 31. Assume that the rabbit population in Example 8 behaves according to the *logistic growth formula*

$$n(t) = \frac{300}{0.05 + \left(\dfrac{300}{n_0} - 0.05\right)e^{-0.55t}}$$

where n_0 is the initial rabbit population.
(a) If the initial population is 50 rabbits, what will the population be after 12 years?
(b) Draw the graphs of the function $n(t)$ for $n_0 = 50$, 500, 2000, 8000, and 12,000 in the viewing rectangle $[0, 15]$ by $[0, 12,000]$.
(c) From the graphs in part (b), observe that, regardless of the initial population, the rabbit population seems to approach a certain number as time goes on. What is that number? (This is the number of rabbits that the island can support.)

32. The population of a certain species of bird is limited by the type of habitat required for nesting. The population appears to be fairly accurately modeled by the *logistic growth formula*

$$n(t) = \frac{5600}{0.5 + 27.5e^{-0.044t}}$$

where t is measured in years.
(a) Find the initial bird population.
(b) Draw a graph of the function $n(t)$.
(c) What size does the population approach as time goes on?

33. Use a calculator to help graph the function $f(x) = e^{-x^2}$ for $x \geq 0$. Then use the fact that f is an even function to draw the rest of the graph.

34. (a) Compare the functions $f(x) = e^x$ and $g(x) = x^3$ by drawing their graphs in each of the following viewing rectangles:
(i) $[0, 3]$ by $[0, 15]$
(ii) $[0, 6]$ by $[0, 120]$
(iii) $[0, 20]$ by $[0, 10,000]$
(b) Find the solutions of the equation $e^x = x^3$, correct to two decimal places.

35. The *hyperbolic cosine function* is defined by

$$\cosh(x) = \frac{e^x + e^{-x}}{2}$$

Sketch the graphs of the functions $y = \frac{1}{2}e^x$ and $y = \frac{1}{2}e^{-x}$ on the same axes and use graphical addition (see Section 3.10) to draw the graph of $y = \cosh(x)$.

36. The *hyperbolic sine function* is defined by

$$\sinh(x) = \frac{e^x - e^{-x}}{2}$$

Graph this function using graphical addition as in Exercise 35.

37–40 ■ Use the definitions in Exercises 35 and 36 to prove the identity.

37. $\cosh(-x) = \cosh(x)$

38. $\sinh(-x) = -\sinh(x)$

39. $[\cosh(x)]^2 - [\sinh(x)]^2 = 1$

40. $\sinh(x + y) = \sinh(x)\cosh(y) + \cosh(x)\sinh(y)$

 41. Illustrate the definition of the number e by graphing the curve $y = (1 + 1/x)^x$ and the line $y = e$ on the same screen using the viewing rectangle $[0, 40]$ by $[0, 4]$.

42. Investigate the behavior of the function

$$f(x) = \left(1 - \frac{1}{x}\right)^x$$

as $x \to \infty$ by drawing its graph and the line $y = 1/e$ on the same screen using the viewing rectangle $[0, 20]$ by $[0, 1]$.

43. (a) Draw the graphs of the family of functions

$$f(x) = \frac{a}{2}(e^{x/a} + e^{-x/a})$$

for $a = 0.5$, 1, 1.5, and 2.
(b) How does a larger value of a affect the graph?

44. Graph the function $y = e^x/x$ and comment on vertical and horizontal asymptotes.

45–46 ■ Find the local maximum and minimum values of the function and the value of x at which each occurs. State each answer correct to two decimal places.

45. $f(x) = xe^{-x}$

46. $f(x) = e^x + e^{-3x}$

5.3 LOGARITHMIC FUNCTIONS

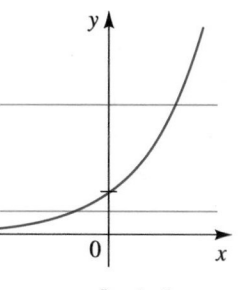

$y = a^x$, $a > 1$

FIGURE 1
$f(x) = a^x$ is one-to-one

We read $\log_a x = y$ as "log base a of x is y."

By tradition, the name of the logarithmic function is $\log_a$, not just a single letter. Also, we usually omit the parentheses in the function notation and write

$$\log_a(x) = \log_a x$$

Every exponential function $f(x) = a^x$, with $a > 0$ and $a \neq 1$, is a one-to-one function by the Horizontal Line Test (see Figure 1 for the case $a > 1$) and, therefore, has an inverse function. The inverse function f^{-1} is called the *logarithmic function with base a* and is denoted by $\log_a$. Recall from Section 3.11 that f^{-1} is defined by

$$f^{-1}(x) = y \iff f(y) = x$$

This leads to the following definition of the logarithmic function.

DEFINITION OF THE LOGARITHMIC FUNCTION

Let a be a positive number with $a \neq 1$. The **logarithmic function with base a,** denoted by **$\log_a$,** is defined by

$$\log_a x = y \iff a^y = x$$

In words, this says that

$\log_a x$ is the exponent to which the base a must be raised to give x.

When we use the definition of logarithms to switch back and forth between the **logarithmic form** $\log_a x = y$ and the **exponential form** $a^y = x$, it is helpful to notice that, in both forms, the base is the same:

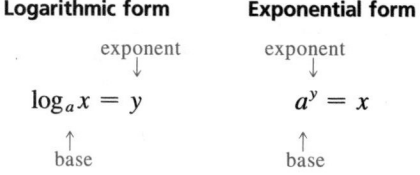

The following examples illustrate how to change an equation from one of these forms to the other.

EXAMPLE 1 ■ Logarithmic and Exponential Forms

The logarithmic and exponential forms are equivalent equations—if one is true, then so is the other. So, we can switch from one form to the other as in the following illustrations:

Logarithmic form	Exponential form
$\log_{10} 100{,}000 = 5$	$10^5 = 100{,}000$
$\log_2 8 = 3$	$2^3 = 8$
$\log_2\left(\frac{1}{2}\right) = -1$	$2^{-1} = \frac{1}{2}$
$\log_3 z = t$	$3^t = z$
$\log_2\left(\frac{1}{8}\right) = -3$	$2^{-3} = \frac{1}{8}$
$\log_5 s = r$	$5^r = s$

x	$\log_{10} x$
10	1
10^2	2
10^3	3
10^4	4
10^{-1}	-1
10^{-2}	-2
10^{-3}	-3
10^{-4}	-4

It is important to understand that $\log_a x$ is an *exponent*. For example, the numbers in the right column of the table (in the margin) are the logarithms (base 10) of the numbers in the left column. This is the case for all bases, as the following example illustrates.

EXAMPLE 2 ■ Evaluating Logarithms

(a) $\log_{10} 1000 = 3$ because $10^3 = 1000$

(b) $\log_2 32 = 5$ because $2^5 = 32$

(c) $\log_3 81 = 4$ because $3^4 = 81$

(d) $\log_{10} 0.1 = -1$ because $10^{-1} = 0.1$

(e) $\log_{16} 4 = \frac{1}{2}$ because $16^{1/2} = 4$

(f) $\log_5 125 = 3$ because $5^3 = 125$

GRAPHS OF LOGARITHMIC FUNCTIONS

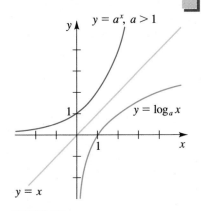

FIGURE 2

Graph of the logarithmic function $f(x) = \log_a x$

Recall that if a one-to-one function f has domain A and range B, then its inverse function f^{-1} has domain B and range A. Since the exponential function $f(x) = a^x$ with $a \neq 1$ has domain $\mathbb{R}$ and range $(0, \infty)$, we conclude that its inverse function, $f^{-1}(x) = \log_a x$, has domain $(0, \infty)$ and range $\mathbb{R}$.

The graph of $f^{-1}(x) = \log_a x$ is obtained by reflecting the graph of $f(x) = a^x$ in the line $y = x$. Figure 2 shows the case $a > 1$. (Most of the important logarithmic functions have base $a > 1$.) The fact that $y = a^x$ (for $a > 1$) is a very rapidly increasing function for $x > 0$ implies the fact that $y = \log_a x$ is a very slowly increasing function for $x > 1$ (see Exercise 71).

Notice that since $a^0 = 1$, we have

$$\log_a 1 = 0$$

and so the x-intercept of the function $y = \log_a x$ is 1. Notice also that because the x-axis is a horizontal asymptote of $y = a^x$, the y-axis is a vertical asymptote of

$y = \log_a x$. In fact, using the notation of Section 4.8, we can write

$$\log_a x \to -\infty \quad \text{as} \quad x \to 0^+$$

for the case $a > 1$.

EXAMPLE 3 ■ Graphing a Logarithmic Function by Plotting Points

Sketch the graph of $f(x) = \log_2 x$.

SOLUTION

To make a table of values, we choose the x-values to be powers of 2 so that we can easily find their logarithms. We plot these points and connect them with a smooth curve as in Figure 3.

x	$\log_2 x$
1	0
2	1
2^2	2
2^3	3
2^4	4
2^{-1}	-1
2^{-2}	-2
2^{-3}	-3
2^{-4}	-4

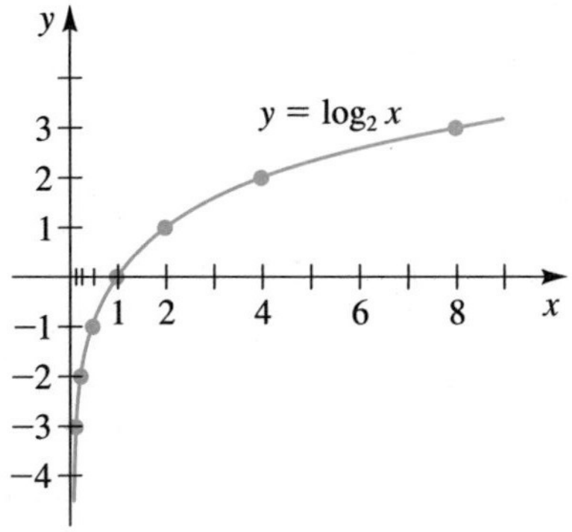

FIGURE 3

Figure 4 shows the relationship among the graphs of the logarithmic functions with bases 2, 3, 5, and 10. These graphs are drawn by reflecting the graphs of $y = 2^x$, $y = 3^x$, $y = 5^x$, and $y = 10^x$ (see Figure 4 in Section 5.1) in the line $y = x$. We can also plot points as an aid to sketching these graphs, as illustrated in Example 3.

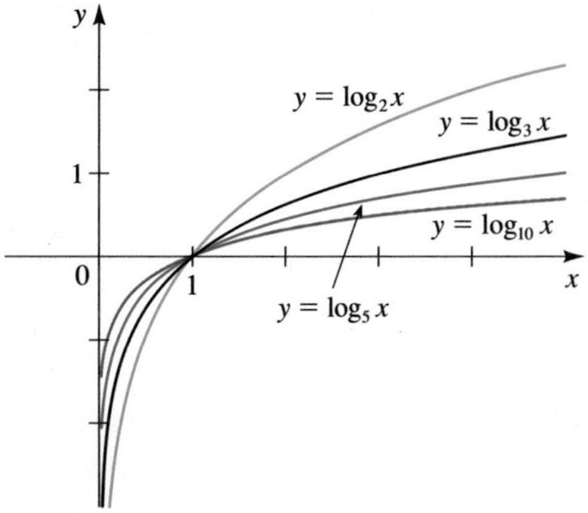

FIGURE 4
Logarithmic functions with
various bases

In the next two examples we graph logarithmic functions by starting with the basic graphs in Figure 4 and using the transformations of Section 3.8.

EXAMPLE 4 ■ **Reflecting Graphs of Logarithmic Functions**

Sketch the graph of each function.

(a) $y = -\log_2 x$ (b) $y = \log_2(-x)$

SOLUTION

Reflecting graphs is explained in Section 3.8.

(a) We start with the graph of $y = \log_2 x$ in Figure 5(a) and reflect in the x-axis to get the graph of $y = -\log_2 x$ in Figure 5(b).

(b) To obtain the graph of $y = \log_2(-x)$, we reflect the graph of $y = \log_2 x$ in the y-axis. See Figure 5(c).

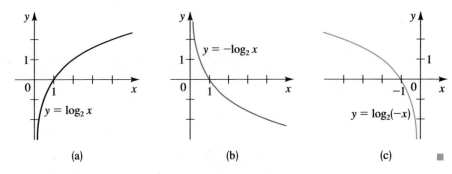

FIGURE 5 (a) (b) (c) ■

EXAMPLE 5 ■ **Shifting Graphs of Logarithmic Functions**

Find the domain of each function, and sketch the graph.

(a) $f(x) = 2 + \log_5 x$ (b) $g(x) = \log_{10}(x - 3)$

SOLUTION

(a) The graph of f is obtained from the graph of $y = \log_5 x$ (Figure 4) by shifting upward 2 units (see Figure 6). The domain of f is $(0, \infty)$.

(b) The graph of g is obtained from the graph of $y = \log_{10} x$ (Figure 4) by shifting to the right 3 units (see Figure 7). The line $x = 3$ is a vertical asymptote.

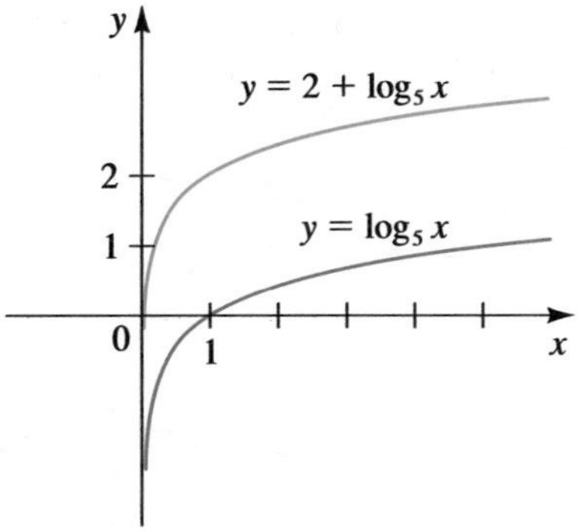

FIGURE 6

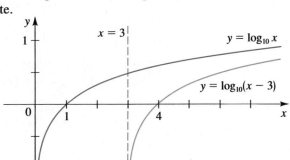

FIGURE 7

The domain of $y = \log_a x$ is the interval $(0, \infty)$, so $\log_{10} x$ is defined only when $x > 0$. Therefore, the domain of $g(x) = \log_{10}(x - 3)$ is

$$\{x \mid x - 3 > 0\} = \{x \mid x > 3\} = (3, \infty)$$ ■

In Section 3.11 we saw that a function f and its inverse function f^{-1} satisfy the equations

$$f^{-1}(f(x)) = x \qquad \text{for } x \text{ in the domain of } f$$
$$f(f^{-1}(x)) = x \qquad \text{for } x \text{ in the domain of } f^{-1}$$

When applied to $f(x) = a^x$ and $f^{-1}(x) = \log_a x$, these equations become

$$\log_a(a^x) = x \qquad x \in \mathbb{R}$$
$$a^{\log_a x} = x \qquad x > 0$$

We list these and other properties of logarithms discussed in this section.

PROPERTIES OF LOGARITHMS

Property	Reason
$\log_a 1 = 0$	We must raise a to the power 0 to get 1.
$\log_a a = 1$	We must raise a to the power 1 to get a.
$\log_a a^x = x$	We must raise a to the power x to get a^x.
$a^{\log_a x} = x$	$\log_a x$ is the power to which a must be raised to get x.

EXAMPLE 6 ■ Applying Properties of Logarithms

We illustrate the properties of logarithms when the base is 5.

(a) $\log_5 1 = 0$ (b) $\log_5 5 = 1$

(c) $\log_5 5^8 = 8$ (d) $5^{\log_5 12} = 12$ ■

COMMON AND NATURAL LOGARITHMS

We now use a calculator to evaluate logarithms with base 10 (common logarithms) and base e (natural logarithms). In the next section we will see that we can use a calculator to find logarithms to *any* base.

COMMON LOGARITHM

The logarithm with base 10 is called the **common logarithm** and is denoted by omitting the base:

$$\log x = \log_{10} x$$

From the definition of logarithms we can easily find that

$$\log 10 = 1 \quad \text{and} \quad \log 100 = 2$$

But how can we find $\log 50$? We need to find the exponent y such that $10^y = 50$. Clearly, 1 is too small and 2 is too large. So

$$1 < \log 50 < 2$$

To find a better approximation we can experiment to find a power of 10 closer to 50. Fortunately, scientific calculators are equipped with a $\boxed{\log}$ key that directly gives the values of common logarithms.

EXAMPLE 7 ■ Evaluating Common Logarithms

Use a calculator to find appropriate values of $f(x) = \log x$ and use the values to sketch the graph.

SOLUTION

We make a table of values, using a calculator to evaluate the function at those values of x that are not powers of 10. We plot those points and connect them by a smooth curve as in Figure 8.

x	$\log x$
0.01	-2
0.1	-1
0.5	-0.301
1	0
4	0.602
5	0.699
10	1
15	1.176

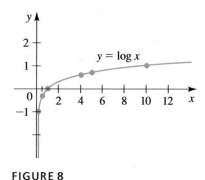

FIGURE 8 ■

Of all possible bases a for logarithms, it turns out that the most convenient choice for the purposes of calculus is the number e, which we defined in Section 5.2.

The notation ln is an abbreviation for *logarithmus naturalis.*

NATURAL LOGARITHM

The logarithm with base e is called the **natural logarithm** and is denoted by ln:

$$\ln x = \log_e x$$

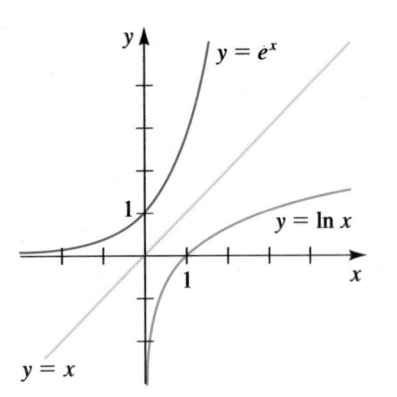

FIGURE 9

Graph of the natural logarithmic function

The natural logarithmic function $y = \ln x$ is the inverse function of the exponential function $y = e^x$. Both functions are graphed in Figure 9. By the definition of inverse functions we have

$$\ln x = y \iff e^y = x$$

If we put $a = e$ and write "ln" for "$\log_e$" in the properties of logarithms mentioned earlier, we obtain the following properties of natural logarithms.

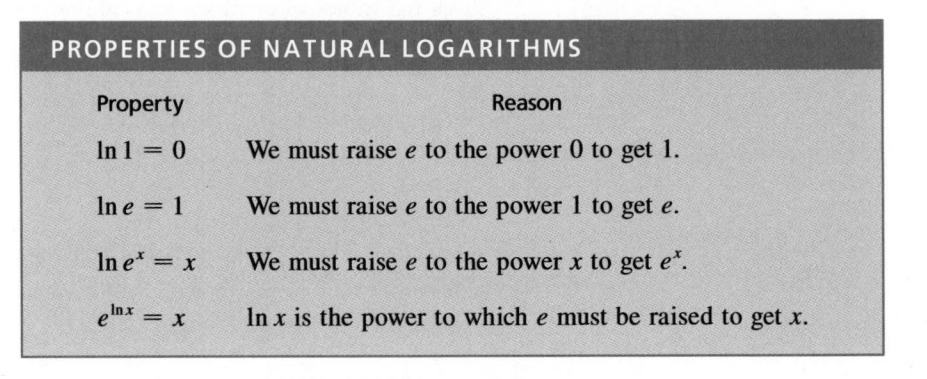

PROPERTIES OF NATURAL LOGARITHMS

Property	Reason
$\ln 1 = 0$	We must raise e to the power 0 to get 1.
$\ln e = 1$	We must raise e to the power 1 to get e.
$\ln e^x = x$	We must raise e to the power x to get e^x.
$e^{\ln x} = x$	$\ln x$ is the power to which e must be raised to get x.

Calculators are equipped with an $\boxed{\text{ln}}$ key that directly gives the values of natural logarithms.

EXAMPLE 8 ■ Evaluating the Natural Logarithm Function

(a) $\ln e^8 = 8$ Definition of natural logarithm

(b) $\ln\left(\dfrac{1}{e^2}\right) = \ln e^{-2} = -2$ Definition of natural logarithm

(c) $\ln 5 \approx 1.609$ Use the $\boxed{\text{ln}}$ key on a calculator   ■

EXAMPLE 9 ■ Finding the Domain of a Logarithmic Function

Find the domain of the function $f(x) = \ln(4 - x^2)$.

SOLUTION

As with any logarithmic function, $\ln x$ is defined when $x > 0$. Thus, the domain of f is

$$\{x \mid 4 - x^2 > 0\} = \{x \mid x^2 < 4\} = \{x \mid |x| < 2\}$$

$$= \{x \mid -2 < x < 2\} = (-2, 2)$$ ■

EXAMPLE 10 ■ Drawing the Graph of a Logarithmic Function

Draw the graph of the function $y = x\ln(4 - x^2)$ and use it to find the asymptotes and local maximum and minimum values.

SOLUTION

As in Example 9, the domain of this function is the interval $(-2, 2)$, so we choose the viewing rectangle $[-3, 3]$ by $[-3, 3]$. The graph is shown in Figure 10, and from it we see that the lines $x = -2$ and $x = 2$ are vertical asymptotes.

The function has a local maximum point to the right of $x = 1$ and a local minimum point to the left of $x = -1$. By zooming in and tracing along the graph with the cursor, we find that the local maximum value is approximately 1.13 and this occurs when $x \approx 1.15$. Similarly (or by noticing that the function is odd), we find that the local minimum value is about -1.13, and it occurs when $x \approx -1.15$. ■

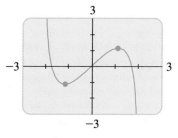

FIGURE 10
$y = x\ln(4 - x^2)$

| 5.3 | **EXERCISES** |

1–6 ■ Express the equation in exponential form.

1. (a) $\log_2 32 = 5$ (b) $\log_5 1 = 0$

2. (a) $\log_{10} 0.1 = -1$ (b) $\log_8 512 = 3$

3. (a) $\log_4 2 = \frac{1}{2}$ (b) $\log_2\left(\frac{1}{16}\right) = -4$

4. (a) $\log_3 81 = 4$ (b) $\log_8 4 = \frac{2}{3}$

5. (a) $\ln 5 = x$ (b) $\ln y = 5$

6. (a) $\ln(x + 1) = 2$ (b) $\ln(x - 1) = 4$

7–12 ■ Express the equation in logarithmic form.

7. (a) $2^3 = 8$ (b) $10^{-3} = 0.001$

8. (a) $10^4 = 10,000$ (b) $81^{1/2} = 9$

9. (a) $4^{-3/2} = 0.125$ (b) $7^3 = 343$

10. (a) $8^{-1} = \frac{1}{8}$ (b) $10^m = n$

11. (a) $e^x = 2$ (b) $e^3 = y$

12. (a) $e^{x+1} = 0.5$ (b) $e^{0.5x} = t$

13–22 ■ Evaluate the expression.

13. (a) $\log_5 5^4$ (b) $\log_4 64$ (c) $\log_9 9$

14. (a) $\log_3 3$ (b) $\log_3 1$ (c) $\log_3 3^2$

15. (a) $\log_8 64$ (b) $\log_7 49$ (c) $\log_7 7^{10}$

16. (a) $\log_2 32$ (b) $\log_8 8^{17}$ (c) $\log_6 1$

17. (a) $\log_3\left(\frac{1}{27}\right)$ (b) $\log_{10} \sqrt{10}$ (c) $\log_5 0.2$

18. (a) $\log_5 125$ (b) $\log_{49} 7$ (c) $\log_9 \sqrt{3}$

19. (a) $2^{\log_2 37}$ (b) $3^{\log_3 8}$ (c) $e^{\ln \sqrt{5}}$

20. (a) $e^{\ln \pi}$ (b) $10^{\log 5}$ (c) $10^{\log 87}$

21. (a) $\log_8 0.25$ (b) $\ln e^4$ (c) $\ln(1/e)$

22. (a) $\log_4 \sqrt{2}$ (b) $\log_4\left(\frac{1}{2}\right)$ (c) $\log_4 8$

23–28 ■ Use the definition of the logarithmic function to find x.

23. (a) $\log_2 x = 5$ (b) $\log_2 16 = x$

24. (a) $\log_5 x = 4$ (b) $\log_{10} 0.1 = x$

25. (a) $\log_{10} x = 2$ (b) $\log_5 x = 2$

26. (a) $\log_x 1000 = 3$ (b) $\log_x 25 = 2$

27. (a) $\log_x 16 = 4$ (b) $\log_x 8 = \frac{3}{2}$

28. (a) $\log_x 6 = \frac{1}{2}$ (b) $\log_x 3 = \frac{1}{3}$

29–32 ■ Use a calculator to evaluate the expression, correct to four decimal places.

29. (a) $\log 2$ (b) $\log 35.2$ (c) $\log\left(\frac{2}{3}\right)$

30. (a) $\log 50$ (b) $\log \sqrt{2}$ (c) $\log(3\sqrt{2})$

31. (a) $\ln 5$ (b) $\ln 25.3$ (c) $\ln(1 + \sqrt{3})$

32. (a) $\ln 27$ (b) $\ln 7.39$ (c) $\ln 54.6$

33–36 ■ Find the function of the form $y = \log_a x$ whose graph is given.

33.

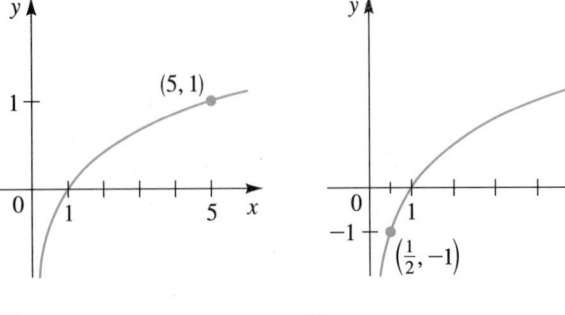

(5, 1)

34.

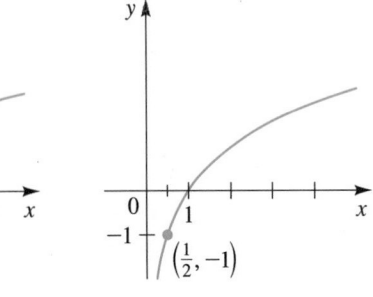

$(\frac{1}{2}, -1)$

35.

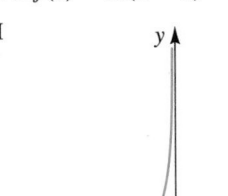

$(3, \frac{1}{2})$

36.

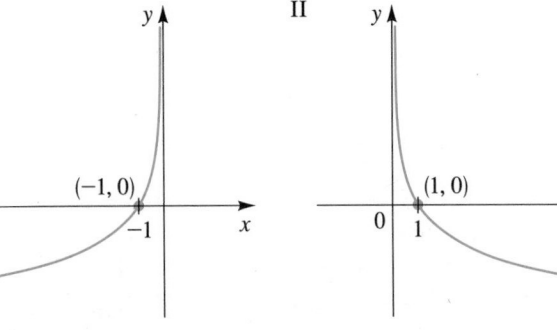

(9, 2)

37–42 ■ Match the logarithmic function with one of the graphs I–VI.

37. $f(x) = -\ln x$

38. $f(x) = \ln(x - 2)$

39. $f(x) = 2 + \ln x$

40. $f(x) = \ln(-x)$

41. $f(x) = \ln(2 - x)$

42. $f(x) = -\ln(-x)$

I

(-1, 0)
-1

II

(1, 0)

III

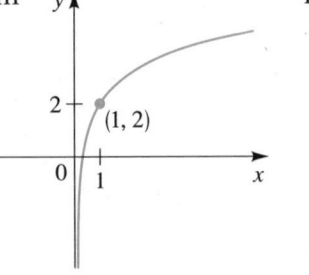

(1, 2)

IV

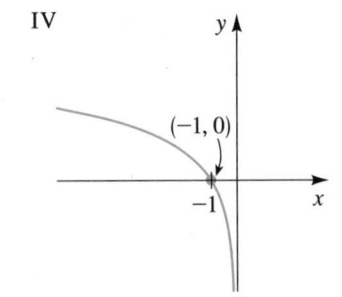

(-1, 0)

V

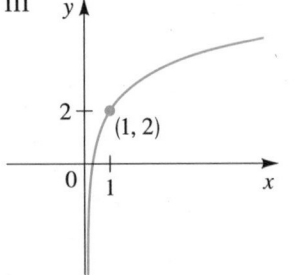

$x = 2$
(3, 0)

VI

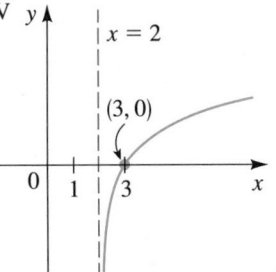

$x = 2$
(1, 0)

43. Draw the graph of $y = 4^x$, then use it to draw the graph of $y = \log_4 x$.

44. Draw the graph of $y = 3^x$, then use it to draw the graph of $y = \log_3 x$.

45–54 ■ Graph the function, not by plotting points, but by starting from the graphs in Figures 4 and 9. State the domain, range, and asymptote.

45. $f(x) = \log_2(x - 4)$ **46.** $f(x) = -\log_{10} x$

47. $g(x) = \log_5(-x)$ **48.** $g(x) = \ln(x + 2)$

49. $y = 2 + \log_3 x$ **50.** $y = \log_3(x - 1) - 2$

51. $y = 1 - \log_{10} x$ **52.** $y = 1 + \ln(-x)$

53. $y = |\ln x|$ **54.** $y = \ln|x|$

55–60 ■ Find the domain of the function.

55. $f(x) = \log_{10}(2 + 5x)$ **56.** $f(x) = \log_2(10 - 3x)$

57. $g(x) = \log_3(x^2 - 1)$ **58.** $g(x) = \ln(x - x^2)$

59. $h(x) = \ln x + \ln(2 - x)$

60. $h(x) = \sqrt{x - 2} - \log_5(10 - x)$

61–66 ■ Draw the graph of the function in a suitable viewing rectangle and use it to find the domain, the asymptotes, and the local maximum and minimum values.

61. $y = \log_{10}(1 - x^2)$ **62.** $y = \ln(x^2 - x)$

63. $y = x + \ln x$ **64.** $y = x(\ln x)^2$

65. $y = \dfrac{\ln x}{x}$ **66.** $y = x \log_{10}(x + 10)$

 67. Compare the rates of growth of the functions
$f(x) = \ln x$ and $g(x) = \sqrt{x}$ by drawing their graphs on
a common screen using the viewing rectangle $[-1, 30]$
by $[-1, 6]$.

 68. (a) By drawing the graphs of the functions

$$f(x) = 1 + \ln(1 + x) \qquad \text{and} \qquad g(x) = \sqrt{x}$$

in a suitable viewing rectangle, show that even when
a logarithmic function starts out higher than a root
function, it is ultimately overtaken by the root
function.

(b) Find, correct to two decimal places, the solutions of
the equation $\sqrt{x} = 1 + \ln(1 + x)$

 69–70 ■ A family of functions is given.
(a) Draw graphs of the family for $c = 1, 2, 3,$ and 4.

(b) How are the graphs in part (a) related?

69. $f(x) = \log(cx)$ **70.** $f(x) = c \log x$

71. Suppose that the graph of $y = 2^x$ is drawn on a
coordinate plane where the unit of measurement is
an inch.
(a) Show that at a distance 2 ft to the right of the origin
the height of the graph is about 265 mi.
(b) If the graph of $y = \log_2 x$ is drawn on the same set
of axes, how far to the right of the origin do we have
to go before the height of the curve reaches 2 ft?

72. Which is larger, $\log_4 17$ or $\log_5 24$?

73–74 ■ A function $f(x)$ is given.
(a) Find the domain of the function f.
(b) Find the inverse function of f.

73. $f(x) = \log_2(\log_{10} x)$ **74.** $f(x) = \ln(\ln(\ln x))$

75. (a) Find the inverse of the function $f(x) = \dfrac{2^x}{1 + 2^x}$.

(b) What is the domain of the inverse function?

5.4 LAWS OF LOGARITHMS

Since logarithms are exponents, the Laws of Exponents give rise to the Laws of
Logarithms. These properties give logarithmic functions a wide range of applications, as we will see in Section 5.6.

LAWS OF LOGARITHMS

Let a be a positive number, with $a \neq 1$. Let $x > 0$, $y > 0$, and r be any
real number.

Law	Description
1. $\log_a(xy) = \log_a x + \log_a y$	The logarithm of a product of numbers is the sum of the logarithms of the numbers.
2. $\log_a\left(\dfrac{x}{y}\right) = \log_a x - \log_a y$	The logarithm of a quotient of numbers is the difference of the logarithms of the numbers.
3. $\log_a(x^r) = r \log_a x$	The logarithm of a power of a number is the exponent times the logarithm of the number.

John Napier (1550–1617) was a Scottish landowner for whom mathematics was a hobby. Today he is best known as the inventor of logarithms. He published his invention in 1614 under the title *A Description of the Marvelous Rule of Logarithms*. In Napier's time logarithms were used exclusively for simplifying complicated calculations. For example, to multiply two large numbers we write them as powers of 10. The exponents are simply the logarithms of the numbers. For example,

$$4532 \times 57783 \approx 10^{3.65629} \times 10^{4.76180}$$
$$= 10^{8.41809}$$
$$\approx 261{,}872{,}564$$

The idea is that multiplying powers of 10 is easy (we simply add exponents). Napier produced extensive tables giving the logarithms (or exponents) of numbers. Since the advent of calculators and computers, logarithms are no longer used for this purpose. The logarithmic functions, however, have found many applications, some of which are described in this chapter. Napier wrote on

(continued)

■ **Proof** We make use of the property $\log_a(a^x) = x$ from Section 5.3.

1. Let
$$\log_a x = b \quad \text{and} \quad \log_a y = c$$

When written in exponential form, these equations become

$$a^b = x \quad \text{and} \quad a^c = y$$

Thus
$$\log_a(xy) = \log_a(a^b a^c)$$
$$= \log_a(a^{b+c})$$
$$= b + c$$
$$= \log_a x + \log_a y$$

2. Using Law 1, we have

$$\log_a x = \log_a\left[\left(\frac{x}{y}\right)y\right] = \log_a\left(\frac{x}{y}\right) + \log_a y$$

so
$$\log_a\left(\frac{x}{y}\right) = \log_a x - \log_a y$$

3. Let $\log_a x = b$. Then $a^b = x$, so

$$\log_a(x^r) = \log_a(a^b)^r = \log_a(a^{rb}) = rb = r\log_a x \qquad \square$$

As the following examples illustrate, these laws are used in both directions. Since the domain of any logarithmic function is the interval $(0, \infty)$, we assume that all quantities whose logarithms occur are positive.

EXAMPLE 1 ■ **Using the Laws of Logarithms to Expand Expressions**

Use the Laws of Logarithms to rewrite each expression.

(a) $\log_2(6x)$

(b) $\log \sqrt{5}$

(c) $\log_5(x^3 y^6)$

(d) $\ln\left(\dfrac{ab}{\sqrt[3]{c}}\right)$

SOLUTION

(a) $\log_2(6x) = \log_2 6 + \log_2 x$ Law 1

(b) $\log \sqrt{5} = \log 5^{1/2}$
$$= \tfrac{1}{2}\log 5 \qquad \text{Law 3}$$

(c) $\log_5(x^3 y^6) = \log_5 x^3 + \log_5 y^6$ Law 1
$$= 3\log_5 x + 6\log_5 y \qquad \text{Law 3}$$

many other topics. One of his most colorful works is a book entitled *A Plaine Discovery of the Whole Revelation of Saint John*, in which he predicted that the world would end in the year 1700.

(d) $\ln\left(\dfrac{ab}{\sqrt[3]{c}}\right) = \ln(ab) - \ln\sqrt[3]{c}$ Law 2

$= \ln a + \ln b - \ln c^{1/3}$ Law 1

$= \ln a + \ln b - \tfrac{1}{3}\ln c$ Law 3 ■

EXAMPLE 2 ■ **Using the Laws of Logarithms to Evaluate Expressions**

Evaluate each expression.

(a) $\log_4 2 + \log_4 32$ (b) $\log_2 80 - \log_2 5$ (c) $-\tfrac{1}{3}\log 8$

SOLUTION

(a) $\log_4 2 + \log_4 32 = \log_4(2 \cdot 32)$ Law 1

$= \log_4 64 = 3$ Because $4^3 = 64$

(b) $\log_2 80 - \log_2 5 = \log_2\left(\tfrac{80}{5}\right)$ Law 2

$= \log_2 16 = 4$ Because $2^4 = 16$

(c) $-\tfrac{1}{3}\log 8 = \log 8^{-1/3}$ Law 3

$= \log\left(\tfrac{1}{2}\right)$ Property of exponents

≈ -0.301 Use a calculator ■

EXAMPLE 3 ■ **Writing an Expression as a Single Logarithm**

Express $3\log x + \tfrac{1}{2}\log(x + 1)$ as a single logarithm.

SOLUTION

$3\log x + \tfrac{1}{2}\log(x + 1) = \log x^3 + \log(x + 1)^{1/2}$ Law 3

$= \log(x^3(x + 1)^{1/2})$ Law 1 ■

EXAMPLE 4 ■ **Writing an Expression as a Single Logarithm**

Express $3\ln s + \tfrac{1}{2}\ln t - 4\ln(t^2 + 1)$ as a single logarithm.

SOLUTION

$3\ln s + \tfrac{1}{2}\ln t - 4\ln(t^2 + 1) = \ln s^3 + \ln t^{1/2} - \ln(t^2 + 1)^4$ Law 3

$= \ln(s^3 t^{1/2}) - \ln(t^2 + 1)^4$ Law 1

$= \ln\left(\dfrac{s^3\sqrt{t}}{(t^2 + 1)^4}\right)$ Law 2 ■

WARNING Although the Laws of Logarithms tell us how to compute the logarithm of a product or a quotient, *there is no corresponding rule for the logarithm of a sum or a difference*. For instance,

⊘
$$\log_a(x + y) \neq \log_a x + \log_a y$$

In fact, we know that the right side is equal to $\log_a(xy)$.

Also, don't improperly simplify quotients or powers of logarithms. For instance,

⊘
$$\frac{\log 6}{\log 2} \neq \log\left(\frac{6}{2}\right)$$

⊘
$$(\log_2 x)^3 \neq 3\log_2 x$$

CHANGE OF BASE

For some purposes it is useful to be able to change from logarithms in one base to logarithms in another base. Suppose that we are given $\log_a x$ and want to find $\log_b x$. Let

$$y = \log_b x$$

We write this in exponential form and take the logarithm, with base a, of each side.

$$b^y = x \qquad \text{Exponential form}$$

$$\log_a(b^y) = \log_a x \qquad \text{Take } \log_a \text{ of each side}$$

$$y\log_a b = \log_a x \qquad \text{Law 3}$$

$$y = \frac{\log_a x}{\log_a b} \qquad \text{Divide by } \log_a b$$

This proves the following formula.

We may write the change of base formula as

$$\log_b x = \left(\frac{1}{\log_a b}\right)\log_a x$$

So, $\log_b x$ is just a constant multiple of $\log_a x$; the constant is $\dfrac{1}{\log_a b}$.

CHANGE OF BASE FORMULA

$$\log_b x = \frac{\log_a x}{\log_a b}$$

In particular if we put $x = a$, then $\log_a a = 1$ and this formula becomes

$$\log_b a = \frac{1}{\log_a b}$$

We can now evaluate a logarithm to *any* base by using the change of base formula to express the logarithm in terms of common logarithms or natural logarithms and then using a calculator.

EXAMPLE 5 ■ Using the Change of Base Formula to Evaluate Logarithms

Use the change of base formula and common logarithms to evaluate each logarithm, correct to five decimal places.

(a) $\log_8 5$ (b) $\log_4 35$

SOLUTION

(a) We use the change of base formula with $b = 8$ and $a = 10$ to convert to common logarithms:

$$\log_8 5 = \frac{\log_{10} 5}{\log_{10} 8} \approx 0.77398$$

(b) We use the change of base formula with $b = 4$ and $a = 10$ to convert to common logarithms:

$$\log_4 35 = \frac{\log_{10} 35}{\log_{10} 4} \approx 2.56464$$

■

EXAMPLE 6 ■ Using the Change of Base Formula to Evaluate Logarithms

Use the change of base formula and natural logarithms to evaluate each logarithm, correct to five decimal places.

(a) $\log_9 20$ (b) $\log_5 3$

SOLUTION

(a) We use the change of base formula with $b = 9$ and $a = e$ to convert to natural logarithms:

$$\log_9 20 = \frac{\ln 20}{\ln 9} \approx 1.36342$$

(b) We use the change of base formula with $b = 5$ and $a = e$ to convert to natural logarithms:

$$\log_5 3 = \frac{\ln 3}{\ln 5} \approx 0.68261$$

■

EXAMPLE 7 ■ Using the Change of Base Formula to Graph a Logarithmic Function

Use a graphing calculator to graph $f(x) = \log_6 x$.

SOLUTION

Calculators do not have a key for $\log_6$, so we use the change of base formula to write

$$f(x) = \log_6 x = \frac{\ln x}{\ln 6}$$

Since calculators do have an $\boxed{\ln}$ key, we can enter this new form of the function and graph it. The graph is shown in Figure 1. ■

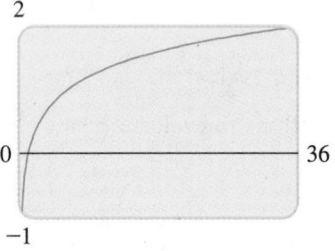

FIGURE 1

$$f(x) = \log_6 x = \frac{\ln x}{\ln 6}$$

5.4   EXERCISES

1–24 ■ Use the Laws of Logarithms to rewrite the expression in a form with no logarithm of a product, quotient, or power.

1. $\log_2(x(x - 1))$

2. $\log_5\left(\dfrac{x}{2}\right)$

3. $\log 7^{23}$

4. $\ln(\pi x)$

5. $\log_2(AB^2)$

6. $\log_6 \sqrt[4]{17}$

7. $\log_3(x\sqrt{y}\,)$

8. $\log_2(xy)^{10}$

9. $\log_5 \sqrt[3]{x^2 + 1}$

10. $\log_a\left(\dfrac{x^2}{yz^3}\right)$

11. $\ln\sqrt{ab}$

12. $\ln\sqrt[3]{3r^2s}$

13. $\log\left(\dfrac{x^3 y^4}{z^6}\right)$

14. $\log\left(\dfrac{a^2}{b^4 \sqrt{c}}\right)$

15. $\log_2\left(\dfrac{x(x^2 + 1)}{\sqrt{x^2 - 1}}\right)$

16. $\log_5 \sqrt{\dfrac{x - 1}{x + 1}}$

17. $\ln\left(x\sqrt{\dfrac{y}{z}}\right)$

18. $\ln\dfrac{3x^2}{(x + 1)^{10}}$

19. $\log\sqrt[4]{x^2 + y^2}$

20. $\log\left(\dfrac{x}{\sqrt[3]{1 - x}}\right)$

21. $\log\sqrt{\dfrac{x^2 + 4}{(x^2 + 1)(x^3 - 7)^2}}$

22. $\log\sqrt{x\sqrt{y\sqrt{z}}}$

23. $\ln\left(\dfrac{z^4 \sqrt{x}}{\sqrt[3]{y^2 + 6y + 17}}\right)$

24. $\log\left(\dfrac{10^x}{x(x^2 + 1)(x^4 + 2)}\right)$

25–34 ■ Evaluate the expression.

25. $\log_5 \sqrt{125}$

26. $\log_2 112 - \log_2 7$

27. $\log 2 + \log 5$

28. $\log \sqrt{0.1}$

29. $\log_4 192 - \log_4 3$

30. $\log_{12} 9 + \log_{12} 16$

31. $\ln 6 - \ln 15 + \ln 20$

32. $e^{3\ln 5}$

33. $10^{2\log 4}$

34. $\log_2 8^{33}$

35–44 ■ Rewrite the expression as a single logarithm.

35. $\log_3 5 + 5\log_3 2$

36. $\log 12 + \tfrac{1}{2}\log 7 - \log 2$

37. $\log_2 A + \log_2 B - 2\log_2 C$

38. $\log_5(x^2 - 1) - \log_5(x - 1)$

39. $4\log x - \tfrac{1}{3}\log(x^2 + 1) + 2\log(x - 1)$

40. $\ln(a + b) + \ln(a - b) - 2\ln c$

41. $\ln 5 + 2\ln x + 3\ln(x^2 + 5)$

42. $2[\log_5 x + 2\log_5 y - 3\log_5 z]$

43. $\tfrac{1}{3}\log(2x + 1) + \tfrac{1}{2}[\log(x - 4) - \log(x^4 - x^2 - 1)]$

44. $\log_a b + c\log_a d - r\log_a s$

45–54 ■ State whether the equation is an identity.

45. $\log_2(x - y) = \log_2 x - \log_2 y$

46. $\log_5\left(\dfrac{a}{b^2}\right) = \log_5 a - 2\log_5 b$

47. $\log 2^z = z \log 2$

48. $(\log P)(\log Q) = \log P + \log Q$

49. $\dfrac{\log a}{\log b} = \log a - \log b$

50. $(\log_2 7)^x = x \log_2 7$

51. $\log_a a^a = a$

52. $\log(x - y) = \dfrac{\log x}{\log y}$

53. $-\ln\left(\dfrac{1}{A}\right) = \ln A$

54. $r \ln s = \ln(s^r)$

55–62 ■ Use the change of base formula and a calculator to evaluate the logarithm, correct to six decimal places. Use either natural or common logarithms.

55. $\log_2 7$

56. $\log_5 2$

57. $\log_3 11$

58. $\log_6 92$

59. $\log_7 3.58$

60. $\log_6 532$

61. $\log_4 322$

62. $\log_{12} 2.5$

 63. Use the change of base formula to show that

$$\log_3 x = \frac{\ln x}{\ln 3}$$

Then use this fact to draw the graph of the function $f(x) = \log_3 x$.

 64. Use the method of Exercise 63 to draw the graphs of the functions $y = \log_2 x$, $y = \ln x$, $y = \log_5 x$, and $y = \log_{10} x$ on the same screen using the viewing rectangle $[0, 5]$ by $[-3, 3]$. How are these graphs related?

65. Use the change of base formula to show that

$$\log e = \frac{1}{\ln 10}$$

66. Simplify: $(\log_2 5)(\log_5 7)$

67. Show that $-\ln\left(x - \sqrt{x^2 - 1}\right) = \ln\left(x + \sqrt{x^2 - 1}\right)$.

68. Find the error: $\log 0.1 < 2 \log 0.1$

$$= \log(0.1)^2$$

$$= \log 0.01$$

$$\log 0.1 < \log 0.01$$

$$0.1 < 0.01$$

5.5 EXPONENTIAL AND LOGARITHMIC EQUATIONS

In this section we solve equations that involve exponential or logarithmic functions. The techniques we develop here will be used in the next section for solving applied problems.

EXPONENTIAL EQUATIONS

An exponential equation is one in which the variable occurs in the exponent. For example,

$$2^x = 7$$

The variable x presents a difficulty because it is in the exponent. To deal with this difficulty we take the logarithm of each side and then use the Laws of Logarithms to "bring down x" from the exponent.

$$2^x = 7$$

$$\ln 2^x = \ln 7 \qquad \text{Take ln of each side}$$

$$x \ln 2 = \ln 7 \qquad \text{Law 3 (bring down the exponent)}$$

$$x = \frac{\ln 7}{\ln 2} \qquad \text{Solve for } x$$

$$\approx 2.807 \qquad \text{Use a calculator}$$

Recall that Law 3 of the Laws of Logarithms says that $\log_a x^r = r \log_a x$.

The method we used to solve $2^x = 7$ is typical of the methods we use to solve all exponential equations, and it can be summarized in the following three steps.

GUIDELINES FOR SOLVING EXPONENTIAL EQUATIONS

1. Isolate the exponential expression on one side of the equation.

2. Take the logarithm of each side, then use the Laws of Logarithms to "bring down the exponent."

3. Solve for the variable.

These techniques are illustrated in the following examples.

EXAMPLE 1 ■ Solving an Exponential Equation

Solve the equation $8e^{2x} = 20$.

SOLUTION

We first divide by 8 in order to isolate the exponential term on one side of the equation.

$$8e^{2x} = 20$$

$$e^{2x} = \frac{20}{8} \qquad \text{Divide by 8}$$

$$\ln e^{2x} = \ln 2.5 \qquad \text{Take ln of each side}$$

$$2x = \ln 2.5 \qquad \text{Property of ln}$$

$$x = \frac{\ln 2.5}{2} \qquad \text{Divide by 2}$$

$$\approx 0.458 \qquad \text{Use a calculator}$$

CHECK YOUR ANSWER

Substituting $x = 0.458$ into the original equation and using a calculator, we get $8e^{2(0.458)} \approx 20$ ✓

EXAMPLE 2 ■ Solving an Exponential Equation

Find the solution of the equation $3^{x+2} = 7$, correct to six decimal places.

SOLUTION

We take the common logarithm of each side and use Law 3.

$$3^{x+2} = 7$$

$$\log(3^{x+2}) = \log 7 \qquad \text{Take log of each side}$$

$$(x + 2)\log 3 = \log 7 \qquad \text{Law 3 (bring down the exponent)}$$

$$x + 2 = \frac{\log 7}{\log 3} \qquad \text{Divide by } \log 3$$

$$x = \frac{\log 7}{\log 3} - 2 \qquad \text{Subtract 2}$$

$$\approx -0.228756 \qquad \text{Use a calculator}$$

We could have used natural logarithms instead of common logarithms. In fact, using the same steps, we get

$$x = \frac{\ln 7}{\ln 3} - 2 \approx -0.228756$$

CHECK YOUR ANSWER ■ Substituting $x = -0.228756$ into the original equation and using a calculator, we get

$$3^{(-0.228756)+2} \approx 7 \qquad ✓$$

An alternative method is to solve graphically (see Section 3.3). For the equation in Example 3 we draw the graphs of $y = e^{3-2x}$ and $y = 4$ in the same viewing rectangle.

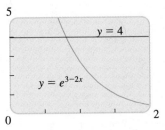

Zooming in on the point of intersection of the two graphs, we see that the solution is $x \approx 0.807$.

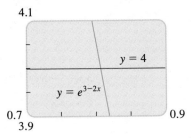

EXAMPLE 3 ■ Solving an Exponential Equation

Solve the equation $e^{3-2x} = 4$.

SOLUTION

Since the base of the exponential term is e, we use natural logarithms to solve this equation.

$$\ln(e^{3-2x}) = \ln 4 \qquad \text{Take ln of each side}$$

$$3 - 2x = \ln 4 \qquad \text{Property of ln}$$

$$2x = 3 - \ln 4$$

$$x = \tfrac{1}{2}(3 - \ln 4) \approx 0.807$$

Check that this answer satisfies the original equation. ■

EXAMPLE 4 ■ Solving an Exponential Equation

Solve the equation $e^{2x} - e^x - 6 = 0$.

SOLUTION

To isolate the exponential term, we factor.

$$e^{2x} - e^x - 6 = 0$$

$$(e^x)^2 - e^x - 6 = 0 \qquad \text{Property of exponents}$$

$$(e^x - 3)(e^x + 2) = 0 \qquad \text{Factor (a quadratic in } e^x)$$

$$e^x - 3 = 0 \quad \text{or} \quad e^x + 2 = 0 \qquad \text{Zero product property}$$

$$e^x = 3 \qquad\qquad e^x = -2$$

The equation $e^x = 3$ leads to $x = \ln 3$. But the equation $e^x = -2$ has no solution because $e^x > 0$ for all x. Thus, $x = \ln 3 \approx 1.0986$ is the only solution. Check that this answer satisfies the original equation. ■

EXAMPLE 5 ■ Solving an Exponential Equation

Solve the equation $3x^2 e^x + x^3 e^x = 0$.

SOLUTION

First we factor the left side of the equation.

$$3x^2 e^x + x^3 e^x = 0$$

$$(3x^2 + x^3)e^x = 0 \qquad \text{Factor out } e^x$$

$$x^2(3 + x)e^x = 0 \qquad \text{Factor out } x^2$$

$$x^2 = 0 \quad \text{or} \quad 3 + x = 0 \quad \text{or} \quad e^x = 0 \qquad \text{Zero product property}$$

$$x = 0 \qquad\qquad x = -3 \qquad\qquad e^x = 0$$

CHECK YOUR ANSWERS

If $x = 0$, we get

$$3(0)^2 e^0 + 0^3 e^0 = 0 \quad \checkmark$$

If $x = -3$, we get

$$3(-3)^2 e^{-3} + (-3)^3 e^{-3}$$
$$= 27e^{-3} - 27e^{-3} = 0 \quad \checkmark$$

The equation $e^x = 0$ has no solution because $e^x > 0$ for all x. Thus, $x = 0$ and $x = -3$ are the only solutions. ■

LOGARITHMIC EQUATIONS

A logarithmic equation is one in which a logarithm of the variable occurs. For example,

$$\log_2(x + 2) = 5$$

Radiocarbon dating is a method archeologists use to determine the age of ancient objects. The carbon dioxide in the atmosphere always contains a fixed fraction of radioactive carbon, carbon-14 (C^{14}), with a half-life of about 5730 years. Plants absorb carbon dioxide from the atmosphere, which then makes its way to animals through the food chain. Thus, all living creatures contain the same fixed proportions of C^{14} to nonradioactive C^{12} as the atmosphere.

After an organism dies, it stops assimilating C^{14}, and the amount of C^{14} in it begins to decay exponentially. We can then determine the time elapsed since the death of the organism by measuring the amount of C^{14} left in it.

For example, if a donkey bone contains 73% as much C^{14} as a living donkey and it died t years ago, then by the formula for radioactive decay (Section 5.6),

$$0.73 = (1.00)e^{-(t \ln 2)/5730}$$

We solve this to find $t \approx 2600$, so the bone is about 2600 years old.

To solve for x we write the equation in exponential form.

$$x + 2 = 2^5 \qquad \text{Exponential form}$$

$$x = 32 - 2 = 30 \qquad \text{Solve for } x$$

Another way of looking at the first step is to raise the base, 2, to each side of the equation.

$$2^{\log_2(x+2)} = 2^5 \qquad \text{Raise 2 to each side}$$

$$x + 2 = 2^5 \qquad \text{Property of logarithms}$$

$$x = 32 - 2 = 30 \qquad \text{Solve for } x$$

The methods used to solve this simple problem are typical. We summarize the steps as follows.

> **GUIDELINES FOR SOLVING LOGARITHMIC EQUATIONS**
>
> 1. Isolate the logarithmic term on one side of the equation; you may need to first combine the logarithmic terms.
>
> 2. Write the equation in exponential form (or raise the base to each side of the equation).
>
> 3. Solve for the variable.

In the remaining examples in the section we illustrate the procedure.

EXAMPLE 6 ■ Solving Logarithmic Equations

Solve each of the following equations for x.

(a) $\ln x = 8$ (b) $\log_2(25 - x) = 3$

SOLUTION

(a) $$\ln x = 8$$

$$x = e^8 \qquad \text{Exponential form}$$

Therefore, $x = e^8 \approx 2981$.

Another way to solve this problem is as follows.

$$\ln x = 8$$

$$e^{\ln x} = e^8 \qquad \text{Raise } e \text{ to each side}$$

$$x = e^8 \qquad \text{Property of ln}$$

(b) The first step is to rewrite the equation in exponential form.

$$\log_2(25 - x) = 3$$
$$25 - x = 2^3 \qquad \text{Exponential form (or raise 2 to each side)}$$
$$25 - x = 8$$
$$x = 25 - 8 = 17$$

∎

EXAMPLE 7 ■ Solving Logarithmic Equations

Solve the equation $4 + 3\log(2x) = 16$

SOLUTION

We first isolate the logarithmic term. This will allow us to write the equation in exponential form.

$$4 + 3\log(2x) = 16$$
$$3\log(2x) = 12 \qquad \text{Subtract 4}$$
$$\log(2x) = 4 \qquad \text{Divide by 3}$$
$$2x = 10^4 \qquad \text{Exponential form (or raise 10 to each side)}$$
$$x = 5000 \qquad \text{Divide by 2}$$

∎

EXAMPLE 8 ■ Solving Logarithmic Equations

Solve the equation $\log(x + 2) + \log(x - 1) = 1$.

SOLUTION

We first combine the logarithmic terms using the Laws of Logarithms.

$$\log[(x + 2)(x - 1)] = 1 \qquad \text{Law 1}$$
$$(x + 2)(x - 1) = 10 \qquad \text{Exponential form (or raise 10 to each side)}$$
$$x^2 + x - 2 = 10 \qquad \text{Expand left side}$$
$$x^2 + x - 12 = 0 \qquad \text{Subtract 10}$$
$$(x + 4)(x - 3) = 0 \qquad \text{Factor}$$
$$x = -4 \quad \text{or} \quad x = 3$$

Let us check to see if these values satisfy the original equation. If $x = -4$, we have

$$\log(x + 2) = \log(-4 + 2) = \log(-2)$$

which is undefined, so $x = -4$ is not a solution. Thus the only solution is $x = 3$, as you can verify. (See *Check Your Answer*.)

∎

EXAMPLE 9 ■ The Length of a Stalactite

A volcanic eruption in the north Atlantic in 1963 created a new island, now called Surtsey. This gave scientists the unusual opportunity to observe stalactites and stalagmites from the beginning of their growth. It appears that they grow fast at first, then their growth slows down. It seems possible that their growth is logarithmic. The length of a particular stalactite might be given by

$$L(t) = 24 \log(0.177t + 1)$$

where t is measured in years and L is measured in inches.

(a) What is the length of the stalactite at time $t = 0$?
(b) What is the length after 20 years?
(c) When will the length be 5 ft?

SOLUTION

(a) At time $t = 0$ the length is

$$L(0) = 24 \log(0.177(0) + 1) = 24 \log 1 = 0$$

As expected, the formula tells us that the stalactite has length 0 at the start.

(b) At time $t = 20$ we have

$$L(20) = 24 \log(0.177(20) + 1) = 24 \log 4.54 \approx 15.769$$

Thus, the stalactite is approximately 15.8 in. long.

(c) We require that $L(t) = 60$ in. and we need to solve for t.

$$60 = 24 \log(0.177t + 1)$$

$\log(0.177t + 1) = 2.5$	Divide by 24
$0.177t + 1 = 10^{2.5}$	Exponential form
$0.177t = 10^{2.5} - 1$	Subtract 1
$t = \dfrac{10^{2.5} - 1}{0.177}$	Divide by 0.177
≈ 1780.9	Use a calculator

Thus, it would take almost 1800 years for the stalactite to reach a length of 5 ft.   ■

5.5 EXERCISES

1–24 ■ Find the solution of the exponential equation, correct to four decimal places.

1. $5^x = 16$

2. $10^{-x} = 2$

3. $2^{1-x} = 3$

4. $3^{2x-1} = 5$

5. $3e^x = 10$

6. $2e^{12x} = 17$

7. $e^{1-4x} = 2$

8. $4(1 + 10^{5x}) = 9$

9. $4 + 3^{5x} = 8$

10. $2^{3x} = 34$

11. $8^{0.4x} = 5$

12. $3^{x/14} = 0.1$

13. $5^{-x/100} = 2$

14. $e^{3-5x} = 16$

15. $e^{2x+1} = 200$

16. $\left(\frac{1}{4}\right)^x = 75$

17. $5^x = 4^{x+1}$

18. $10^{1-x} = 6^x$

19. $2^{3x+1} = 3^{x-2}$

20. $7^{x/2} = 5^{1-x}$

21. $\dfrac{50}{1 + e^{-x}} = 4$

22. $\dfrac{10}{1 + e^{-x}} = 2$

23. $100(1.04)^{2t} = 300$

24. $(1.00625)^{12t} = 2$

25–32 ■ Solve the equation.

25. $x^2 2^x - 2^x = 0$

26. $x^2 10^x - x10^x = 2(10^x)$

27. $4x^3 e^{-3x} - 3x^4 e^{-3x} = 0$

28. $x^2 e^x + x e^x - e^x = 0$

29. $e^{2x} - 3e^x + 2 = 0$

30. $e^{2x} - e^x - 6 = 0$

31. $e^{4x} + 4e^{2x} - 21 = 0$

32. $e^x - 12e^{-x} - 1 = 0$

33–48 ■ Solve the logarithmic equation for x.

33. $\ln x = 10$

34. $\ln(2 + x) = 1$

35. $\log x = -2$

36. $\log(x - 4) = 3$

37. $\log(3x + 5) = 2$

38. $\log_3(2 - x) = 3$

39. $2 - \ln(3 - x) = 0$

40. $\log_2(x^2 - x - 2) = 2$

41. $\log_2 3 + \log_2 x = \log_2 5 + \log_2(x - 2)$

42. $2 \log x = \log 2 + \log(3x - 4)$

43. $\log x + \log(x - 1) = \log(4x)$

44. $\log_5 x + \log_5(x + 1) = \log_5 20$

45. $\log_5(x + 1) - \log_5(x - 1) = 2$

46. $\log x + \log(x - 3) = 1$

47. $\log_9(x - 5) + \log_9(x + 3) = 1$

48. $\ln(x - 1) + \ln(x + 2) = 1$

49. For what value of x is it true that $\log(x + 3) = \log x + \log 3$?

50. For what value of x is it true that $(\log x)^3 = 3 \log x$?

51. Solve for x: $2^{2/\log_5 x} = \frac{1}{16}$

52. Solve for x: $\log_2(\log_3 x) = 4$

53. A 15-g sample of radioactive iodine decays in such a way that the mass remaining after t days is given by $m(t) = 15e^{-0.087t}$ where $m(t)$ is measured in grams. After how many days is there only 5 g remaining?

54. The velocity of a sky diver t seconds after jumping is given by $v(t) = 80(e^{-0.2t} - 1)$. After how many seconds is the velocity 70 ft/s?

55. The figure shows an electric circuit containing a battery producing a voltage of 60 volts, a resistor with a resistance of 13 ohms, and an inductor with an inductance of 5 henrys. Using calculus it can be shown that the current $I = I(t)$ (in amperes) t seconds after the switch is closed is $I = \frac{60}{13}(1 - e^{-13t/5})$.
   (a) Use this equation to express the time t as a function of the current I.
   (b) After how many seconds is the current 2 amperes?

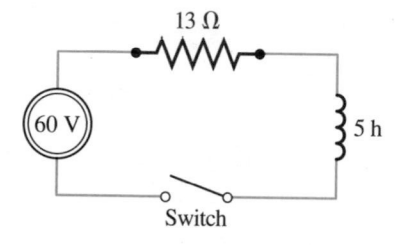

56. A *learning curve* is a graph of a function $P(t)$ that measures the performance of someone learning a skill as a function of the training time t. At first, the rate of learning is rapid. Then, as performance increases and approaches a maximal value M, the rate of learning decreases. It has been found that the function $P(t) = M - Ce^{-kt}$, where k and C are positive constants and $C < M$, is a reasonable model for learning.
   (a) Express the learning time t as a function of the performance level P.
   (b) For a pole-vaulter in training, the learning curve is given by $P(t) = 20 - 14e^{-0.024t}$, where $P(t)$ is the

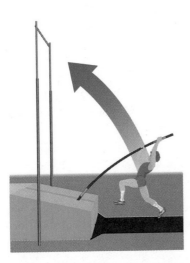

height he is able to pole-vault after t months. After how many months of training is he able to vault 12 ft?

 (c) Draw a graph of the learning curve in part (b).

 57–64 ■ Use a graphing calculator to find all solutions of the equation, correct to two decimal places.

57. $\ln x = 3 - x$

58. $\log_{10} x = x^2 - 2$

59. $x^3 - x = \log_{10}(x + 1)$

60. $x = \ln(4 - x^2)$

61. $e^x = -x$

62. $2^{-x} = x - 1$

63. $4^{-x} = \sqrt{x}$

64. $e^{x^2} - 2 = x^3 - x$

65. Solve the inequality $\log(x - 2) + \log(9 - x) < 1$.

66. Solve the inequality $3 \leq \log_2 x \leq 4$.

67. Solve the inequality $2 < 10^x < 5$.

68. Solve the inequality $x^2 e^x - 2e^x < 0$.

69. Solve the equation $(x - 1)^{\log(x-1)} = 100(x - 1)$.

70. Solve the equation $\log_2 x + \log_4 x + \log_8 x = 11$.

71. Solve the equation $4^x - 2^{x+1} = 3$.
[*Hint:* First write the equation as a quadratic equation in 2^x.]

APPLICATIONS OF EXPONENTIAL AND LOGARITHMIC FUNCTIONS

5.6

Logarithms were invented by John Napier (1550–1617) to eliminate the tedious calculations involved in multiplying, dividing, and taking powers and roots of the large numbers that occur in astronomy and other sciences. With the advent of computers and calculators, logarithms are no longer important for such calculations. However, logarithms arise in problems of exponential growth and decay because they are inverses of exponential functions. Because of the Laws of Logarithms, they also turn out to be useful in the measurement of the loudness of sounds, the intensity of earthquakes, and many other phenomena. In this section we study some of these applications.

COMPOUND INTEREST

Recall the formulas for compound interest that we found in Section 5.2.

> If a principal P is invested at an interest rate r for a period of t years, then the amount A of the investment is given by
>
> $$A = P\left(1 + \frac{r}{n}\right)^{nt} \qquad \text{interest compounded } n \text{ times per year}$$
>
> $$A = Pe^{rt} \qquad \text{interest compounded continuously}$$

We can use logarithms to determine the time it takes for the principal to increase to a given amount.

EXAMPLE 1 ■ Finding the Time for an Investment to Double

A sum of $5000 is invested at an interest rate of 9% per year. Find the time required for the money to double if the interest is compounded according to the following method.

(a) semiannually   (b) continuously

SOLUTION

(a) We use the formula for compound interest with $P = \$5000$, $A = \$10,000$, $r = 0.09$, $n = 2$, and solve the resulting exponential equation for t.

$$5000\left(1 + \frac{0.09}{2}\right)^{2t} = 10,000$$

$$(1.045)^{2t} = 2 \qquad \text{Divide by 5000}$$

$$\log 1.045^{2t} = \log 2 \qquad \text{Take log of each side}$$

$$2t \log 1.045 = \log 2 \qquad \text{Law 3}$$

$$t = \frac{\log 2}{2 \log 1.045} \qquad \text{Divide by } 2 \log 1.045$$

$$\approx 7.9 \qquad \text{Use a calculator}$$

Thus, the money will double in 7.9 years.

(b) We use the formula for continuously compounding interest with $P = \$5000$, $A = \$10,000$, $r = 0.09$, and solve the resulting exponential equation.

$$5000e^{0.09t} = 10,000$$

$$e^{0.09t} = 2 \qquad \text{Divide by 5000}$$

$$\ln e^{0.09t} = \ln 2 \qquad \text{Take ln of each side}$$

$$0.09t = \ln 2 \qquad \text{Property of ln}$$

$$t = \frac{\ln 2}{0.09} \qquad \text{Divide by 0.09}$$

$$t \approx 7.702 \qquad \text{Use a calculator}$$

Thus, the money will double in 7.7 years.   ■

EXPONENTIAL GROWTH

In Section 5.2 we studied the formula for exponential growth, which describes the growth of an animal or bacteria population.

> If n_0 is the initial size of the population, then the population $n(t)$ at time t is given by
>
> $$n(t) = n_0 e^{rt}$$
>
> where r is the relative rate of growth expressed as a fraction of the population.

Now that we are equipped with logarithms, we can answer questions concerning the time at which the population reaches a certain size.

EXAMPLE 2 ■ World Population Projections

The population of the world in 1995 was 5.8 billion and the estimated relative growth rate is 2% per year. If the population continues to grow at this rate, when will it reach 58 billion?

SOLUTION

We use the formula for population growth with $n_0 = 5.8$ billion, $r = 0.02$, and $n(t) = 58$ billion. This leads to an exponential equation, which we solve for t.

$$5.8e^{0.02t} = 58$$

$$e^{0.02t} = 10 \qquad \text{Divide by 5.8}$$

$$\ln e^{0.02t} = \ln 10 \qquad \text{Take ln of each side}$$

$$0.02t = \ln 10 \qquad \text{Property of ln}$$

$$t = \frac{\ln 10}{0.02} \qquad \text{Divide by 0.02}$$

$$t \approx 115.129 \qquad \text{Use a calculator}$$

Thus, the population will reach 58 billion in approximately 115 years; that is, in the year $1995 + 115 = 2110$. ■

EXAMPLE 3 ■ The Number of Bacteria in a Culture

A bacteria culture starts with 10,000 bacteria and the number doubles every 40 minutes.

(a) Find a formula for the number of bacteria at time t.
(b) Find the number of bacteria after one hour.

STANDING ROOM ONLY

The population of the world was about 5.8 billion in 1995. The recently observed rate of increase is 2% per year. Using the exponential model for population growth, and assuming that each person occupies an average of 4 ft^2 of the surface of the earth, we find that by the year 2557 there will be standing room only! (The total land surface area of the world is about 1.8×10^{15} ft^2.)

(c) After how many minutes will there be 50,000 bacteria?

(d) Sketch the graph of the number of bacteria at time t.

SOLUTION

(a) In order to find the formula for population growth, we need to find the rate r. To do this we use the formula with $n_0 = 10{,}000$, $t = 40$, and $n(t) = 20{,}000$, and then solve for r.

$$10{,}000 \cdot e^{r(40)} = 20{,}000$$

$$e^{40r} = 2 \qquad \text{Divide by 10,000}$$

$$\ln e^{40r} = \ln 2 \qquad \text{Take ln of each side}$$

$$40r = \ln 2 \qquad \text{Property of ln}$$

$$r = \frac{\ln 2}{40} \qquad \text{Divide by 40}$$

$$r \approx 0.01733 \qquad \text{Use a calculator}$$

Now that we know $r \approx 0.01733$ we can write the formula for the population growth.

$$n(t) = 10{,}000 \cdot e^{0.01733t}$$

(b) Using the formula we found in part (a) with $t = 60$ minutes (one hour), we get

$$n(60) = 10{,}000 \cdot e^{0.01733(60)} \approx 28{,}287$$

Thus, the number of bacteria after one hour is approximately 28,000.

(c) We use the formula we found in part (a) with $n(t) = 50{,}000$ and solve the resulting exponential equation for t.

$$10{,}000 \cdot e^{0.01733t} = 50{,}000$$

$$e^{0.01733t} = 5 \qquad \text{Divide by 10,000}$$

$$\ln e^{0.01733t} = \ln 5 \qquad \text{Take ln of each side}$$

$$0.01733t = \ln 5 \qquad \text{Property of ln}$$

$$t = \frac{\ln 5}{0.01733} \qquad \text{Divide by 0.01733}$$

$$t \approx 92.9 \qquad \text{Use a calculator}$$

Thus, the bacteria count will reach 50,000 in approximately 93 minutes.

(d) We sketch a graph of the function $n(t) = 10{,}000 \cdot e^{0.01733t}$ in Figure 1.

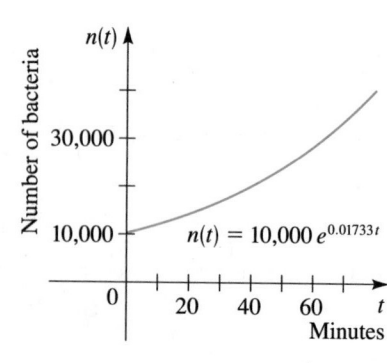

$n(t) = 10{,}000\, e^{0.01733t}$

FIGURE 1

RADIOACTIVE DECAY

Radioactive substances decay by spontaneously emitting radiation. The rate of decay is directly proportional to the mass of the substance. This is analogous to population growth, except that the mass of radioactive material *decreases*. It can be shown that the formula for the mass $m(t)$ remaining at time t is given by

$$m(t) = m_0 e^{-rt}$$

where r is the rate of decay expressed as a proportion of the mass and m_0 is the initial mass. Physicists express the rate of decay in terms of **half-life,** the time required for half the mass to decay. We can obtain the rate r from this as follows. If h is the half-life, then a mass of one unit will become $\frac{1}{2}$ unit when $t = h$. Substituting this into the formula, we get

$$\frac{1}{2} = 1 \cdot e^{-rh}$$

$$\ln\left(\tfrac{1}{2}\right) = -rh \qquad \text{Take ln of each side}$$

$$r = -\frac{1}{h} \ln(2^{-1}) \qquad \text{Solve for } r$$

$$r = \frac{\ln 2}{h} \qquad \ln 2^{-1} = -\ln 2 \text{ by Law 3}$$

This last equation allows us to find the rate r from the half-life h.

If m_0 is the initial mass of a radioactive substance with half-life h, then the mass $m(t)$ remaining at time t is given by

$$m(t) = m_0 e^{-rt}$$

where $r = \dfrac{\ln 2}{h}$.

EXAMPLE 4 ■ Radioactive Decay

Polonium-210 has a half-life of 140 days. Suppose a sample has a mass of 300 mg.

(a) Find a formula for the amount of the sample remaining at time t.
(b) Find the mass remaining after one year.
(c) How long will it take for the sample to decay to a mass of 200 mg?
(d) Draw the graph of the sample mass as a function of time.

SOLUTION

(a) Using the formula for radioactive decay with $m_0 = 300$ and

$$r = \frac{\ln 2}{140} \approx 0.00495, \text{ we have}$$

$$m(t) = 300 e^{-0.00495t}$$

(b) We use the formula we found in part (a) with $t = 365$ (one year).

$$m(365) = 300e^{-0.00495(365)} \approx 49.256$$

Thus, approximately 50 mg remain after one year.

(c) We use the formula we found in part (a) with $m(t) = 200$ and solve the resulting exponential equation for t.

$$300e^{-0.00495t} = 200$$

$$e^{-0.00495t} = \tfrac{2}{3} \qquad \text{Divide by 300}$$

$$\ln e^{-0.00495t} = \ln\!\left(\tfrac{2}{3}\right) \qquad \text{Take ln of each side}$$

$$-0.00495t = \ln\!\left(\tfrac{2}{3}\right) \qquad \text{Property of ln}$$

$$t = -\frac{\ln\!\left(\tfrac{2}{3}\right)}{0.00495} \qquad \text{Divide by } -0.00495$$

$$t \approx 81.9 \qquad \text{Use a calculator}$$

The time required is about 82 days.

(d) A graph of the function $m(t) = 300e^{-0.00495t}$ is shown in Figure 2. ∎

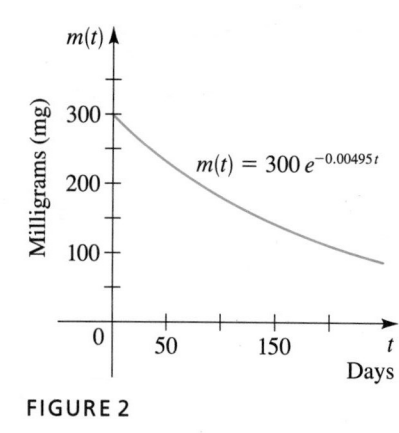

$m(t) = 300\,e^{-0.00495t}$

FIGURE 2

NEWTON'S LAW OF COOLING

Newton's Law of Cooling states that the rate of cooling of an object is proportional to the temperature difference between the object and its surroundings, provided that the temperature difference is not too large. Using calculus, the following formula can be deduced from this law.

> If D_0 is the initial temperature difference between an object and its surroundings, and its surroundings have temperature T_s, then the temperature of the object at time t is given by
>
> $$T(t) = T_s + D_0 e^{-kt}$$
>
> where k is a positive constant that depends on the type of object.

EXAMPLE 5 ■ Newton's Law of Cooling

A cup of coffee has a temperature of 200 °F and is placed in a room that has a temperature of 70 °F. After 10 minutes the temperature of the coffee is 150 °F.

(a) Find a formula for the temperature of the coffee at time t.
(b) Find the temperature of the coffee after 15 minutes.
(c) When will the coffee have cooled to 100 °F?
(d) Illustrate by drawing the graph of the temperature function.

Half-lives of **radioactive elements** vary from very short to very long. Here are some examples.

Element	Half-life
Thorium-232	14.5 billion years
Uranium-235	4.5 billion years
Thorium-230	80,000 years
Plutonium-239	24,360 years
Carbon-14	5,730 years
Radium-226	1,600 years
Cesium-137	30 years
Strontium-90	28 years
Polonium-210	140 days
Thorium-234	25 days
Iodine-135	8 days
Radon-222	3.8 days
Lead-211	3.6 minutes
Krypton-91	10 seconds

SOLUTION

(a) The temperature of the room is $T_s = 70\,°F$ and the initial temperature difference is

$$D_0 = 200 - 70 = 130\,°F$$

so, by Newton's Law of Cooling, the temperature after t minutes is

$$T(t) = 70 + 130e^{-kt}$$

We need to find the constant k associated with this cup of coffee. To do this we use the fact that when $t = 10$, the temperature is $T(10) = 150$. So, we have

$$70 + 130e^{-10k} = 150$$

$$130e^{-10k} = 80 \qquad \text{Subtract 70}$$

$$e^{-10k} = \tfrac{8}{13} \qquad \text{Divide by 130}$$

$$-10k = \ln\!\left(\tfrac{8}{13}\right) \qquad \text{Take ln of each side}$$

$$k = -\tfrac{1}{10}\ln\!\left(\tfrac{8}{13}\right) \qquad \text{Divide by } -10$$

$$k \approx 0.04855 \qquad \text{Use a calculator}$$

Putting this value of k into the expression for $T(t)$, we get

$$T(t) = 70 + 130e^{-0.04855t}$$

(b) We use the fomula we found in part (a) with $t = 15$.

$$T(15) = 70 + 130e^{-0.04855(15)} \approx 133\,°F$$

(c) We use the formula we found in part (a) with $T(t) = 100$ and solve the resulting exponential equation for t.

$$70 + 130e^{-0.04855t} = 100$$

$$130e^{-0.04855t} = 30 \qquad \text{Subtract 70}$$

$$e^{-0.04855t} = \tfrac{3}{13} \qquad \text{Divide by 130}$$

$$-0.04855t = \ln\!\left(\tfrac{3}{13}\right) \qquad \text{Take ln of each side}$$

$$t = -\frac{\ln\!\left(\tfrac{3}{13}\right)}{0.04855} \qquad \text{Divide by } -0.04855$$

$$t \approx 30.2 \qquad \text{Use a calculator}$$

The coffee will have cooled to $100\,°F$ after about half an hour.

(d) The graph of the temperature function is sketched in Figure 3. Notice that the line $T = 70$ is a horizontal asymptote. (Why?)

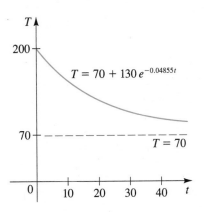

FIGURE 3

Temperature of coffee after t minutes

LOGARITHMIC SCALES

When physical quantities can vary over very large ranges, it is often convenient to take their logarithms in order to have a more manageable set of numbers. We discuss three such situations: the pH scale in chemistry; the Richter scale, which measures the intensity of earthquakes; and the decibel scale, which measures the loudness of sounds. Other quantities that are measured on logarithmic scales are light intensity, information capacity, and radiation.

THE pH SCALE   Chemists measured the acidity of a solution by giving its hydrogen ion concentration until Sorensen, in 1909, proposed a more convenient measure. He defined

$$\text{pH} = -\log[\text{H}^+]$$

where $[\text{H}^+]$ is the concentration of hydrogen ions measured in moles per liter (M). He did this to avoid very small numbers and negative exponents. For instance,

$$\text{if} \quad [\text{H}^+] = 10^{-4}\,\text{M}, \quad \text{then} \quad \text{pH} = -\log_{10}(10^{-4}) = -(-4) = 4$$

Solutions with a pH of 7 are defined as *neutral*; those with pH < 7 are *acidic*; and those with pH > 7 are *basic*. Notice that when the pH increases by one unit, $[\text{H}^+]$ decreases by a factor of 10.

EXAMPLE 6 ■ pH Scale and Hydrogen Ion Concentration

(a) The hydrogen ion concentration of a sample of human blood was measured to be $[\text{H}^+] = 3.16 \times 10^{-8}$ M. Find the pH and classify the blood as acidic or basic.

(b) The most acidic rainfall ever measured was in Scotland in 1974, when the pH was 2.4. Find the hydrogen ion concentration.

SOLUTION

(a) A calculator gives

$$\text{pH} = -\log[\text{H}^+] = -\log(3.16 \times 10^{-8}) \approx 7.5$$

Since this is greater than 7, the blood is basic.

(b) To find the hydrogen ion concentration, we need to solve for $[\text{H}^+]$ in the logarithmic equation

$$\log[\text{H}^+] = -\text{pH}$$

So, we write it in exponential form.

$$[\text{H}^+] = 10^{-\text{pH}}$$

In this case pH $= 2.4$, so

$$[\text{H}^+] = 10^{-2.4} \approx 4.0 \times 10^{-3} \text{ M}$$ ∎

THE RICHTER SCALE In 1935 the American geologist Charles Richter (1900–1984) defined the magnitude of an earthquake to be

$$M = \log \frac{I}{S}$$

where I is the intensity of the earthquake (measured by the amplitude of a seismograph reading, located 100 km from the epicenter of the earthquake) and S is the intensity of a "standard" earthquake (whose amplitude is 1 micron $= 10^{-4}$ cm). The magnitude of a standard earthquake is

$$M = \log \frac{S}{S} = \log 1 = 0$$

Richter studied many earthquakes that occurred between 1900 and 1950. The largest had magnitude 8.9 on the Richter scale, and the smallest had magnitude 0. This corresponds to a ratio of intensities of 800,000,000, so the Richter scale provides more manageable numbers to work with. For instance, an earthquake of magnitude 6 is ten times stronger than an earthquake of magnitude 5.

EXAMPLE 7 ■ **Magnitude of Earthquakes**

The 1906 earthquake in San Francisco had an estimated magnitude of 8.3 on the Richter scale. In the same year the strongest earthquake ever recorded occurred on the Colombia-Ecuador border and was four times as intense. What was the magnitude of the Colombia-Ecuador earthquake on the Richter scale?

SOLUTION

If I is the intensity of the San Francisco earthquake, then from the definition of magnitude we have

$$\log \frac{I}{S} = 8.3$$

The intensity of the Colombia-Ecuador earthquake was $4I$, so its magnitude was

$$\log \frac{4I}{S} = \log 4 + \log \frac{I}{S} = \log 4 + 8.3 \approx 8.9$$ ∎

EXAMPLE 8 ■ Intensity of Earthquakes

The 1989 Loma Prieta earthquake that shook San Francisco had a magnitude of 7.1 on the Richter scale. How many times more intense was the 1906 earthquake (see Example 7) than the 1989 event?

SOLUTION

If I_1 and I_2 are the intensities of the 1906 and 1989 earthquakes, then we are required to find I_1/I_2. To relate this to the definition of magnitude, we divide numerator and denominator by S and we first find the common logarithm of I_1/I_2.

$$\log \frac{I_1}{I_2} = \log \frac{I_1/S}{I_2/S}$$

$$= \log \frac{I_1}{S} - \log \frac{I_2}{S}$$

$$= 8.3 - 7.1 = 1.2$$

Therefore

$$\frac{I_1}{I_2} = 10^{\log(I_1/I_2)} = 10^{1.2} \approx 16$$

The 1906 earthquake was about 16 times as intense as the 1989 earthquake. ■

THE DECIBEL SCALE   The ear is sensitive to an extremely wide range of sound intensities. We take as a reference intensity $I_0 = 10^{-12}$ watts/m^2 at a frequency of 1000 hertz, which measures a sound that is just barely audible (the threshold of hearing). The psychological sensation of loudness varies with the logarithm of the intensity (the Weber-Fechner Law) and so the **intensity level** β, measured in decibels (dB), is defined as

$$\beta = 10 \log \frac{I}{I_0}$$

The intensity level of the barely audible reference sound is

$$\beta = 10 \log \frac{I_0}{I_0} = 10 \log 1 = 0 \text{ dB}$$

EXAMPLE 9 ■ Sound Intensity of a Jet Takeoff

Find the decibel intensity level of a jet plane engine during takeoff if the intensity was measured at 100 watts/m^2.

SOLUTION

From the definition of intensity level we see that

$$\beta = 10 \log \frac{I}{I_0} = 10 \log \frac{10^2}{10^{-12}} = 10 \log 10^{14} = 140 \text{ dB}$$

Thus, the intensity level is 140 dB. ∎

The following table lists decibel intensity levels for some common sounds ranging from the threshold of hearing to the jet takeoff of Example 9. The threshold of pain is about 120 dB.

Source of Sound	β (dB)
Jet takeoff (40 m away)	140
Jackhammer	130
Rock concert (2 m from speakers)	120
Subway	100
Heavy traffic	80
Ordinary traffic	70
Normal conversation	50
Whisper	30
Rustling leaves	10–20
Threshold of hearing	0

5.6 EXERCISES

1. A man invests $10,000 in an account that pays 8.5% per year, compounded quarterly.
   (a) Find the amount after three years.
   (b) How long will it take for the investment to double?

2. A man invests $6500 in an account that pays 6% per year, compounded continuously.
   (a) What is the amount after two years?
   (b) How long will it take for the amount to be $8000?

3. Find the time required for an investment of $5000 to grow to $8000 at an interest rate of 9.5% per year, compounded quarterly.

4. Nancy wants to invest $4000 in saving certificates that bear an interest rate of 9.75% per year, compounded semiannually. How long a time period should she choose in order to save an amount of $5000?

5. How long will it take for an investment to double in value if the interest rate is 8.5% per year, compounded continuously?

6. A sum of $1000 was invested for four years and the interest was compounded semiannually. If this sum amounted to $1435.77 in the given time, what was the interest rate?

7. The number of bacteria in a culture is given by the formula

$$n(t) = 500e^{0.45t}$$

where t is measured in hours.
   (a) What is the initial number of bacteria?
   (b) What is the relative rate of growth of this bacteria population? Express your answer as a percentage.

(c) How many bacteria are in the culture after three hours?

(d) After how many hours will the number of bacteria reach 10,000?

8. The number of fish of a certain species is given by the formula $n(t) = 12e^{0.012t}$ where t is measured in years and $n(t)$ is measured in millions.

(a) What is the relative rate of growth of the fish population? Express your answer as a percentage.

(b) What will the fish population be after five years?

(c) After how many years will the number of fish reach 30 million?

(d) Sketch a graph of the fish population function $n(t)$.

9. The population of a certain city was 112,000 in 1994 and the observed relative growth rate is 4% per year.

(a) Find a formula for the population $n(t)$ after t years.

(b) Find the projected population in the year 2000.

(c) In what year will the population reach 200,000?

10. The frog population in a small pond grows exponentially. The current population is 85 frogs and the relative growth rate is 18% per year.

(a) Find a formula for the population $n(t)$ after t years.

(b) Find the projected population after three years.

(c) Find the number of years required for the frog population to reach 600.

11. The graph shows the deer population in a Pennsylvania county between 1990 and 1994. Assume that the population grows exponentially.

(a) What was the deer population in 1990?

(b) Find a formula for the deer population t years after 1990.

(c) What is the projected deer population in 1998?

(d) In what year will the deer population reach 100,000?

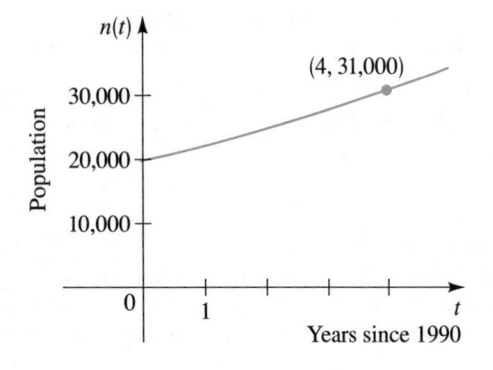

12. The graph shows the estimated beaver population in northern Manitoba between 1991 and 1994. Assume that the population grows exponentially.

(a) Find a formula for the beaver population t years after 1991.

(b) What is the projected beaver population in 1997?

(c) In what year will the beaver population reach 300,000?

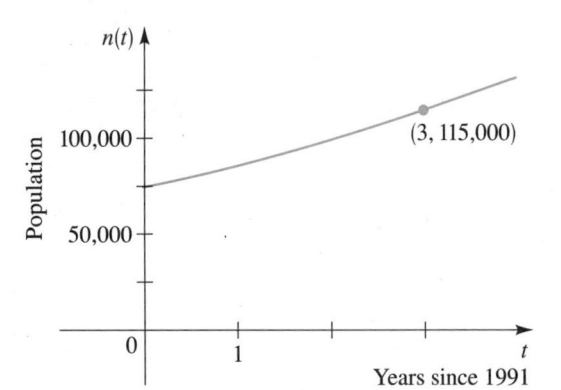

13. A bacteria culture contains 1500 bacteria initially and doubles every 30 minutes.

(a) Find a formula for the number of bacteria $n(t)$ after t minutes.

(b) Find the number of bacteria after two hours.

(c) After how many minutes will the culture contain 4000 bacteria?

14. If a bacteria culture starts with 8000 bacteria and doubles every 20 minutes, when will the population reach 30,000?

15. A bacteria culture starts with 8600 bacteria. After one hour the count is 10,000.

(a) Find a formula for the number of bacteria $n(t)$ after t hours.

(b) Find the number of bacteria after two hours.

(c) After how many hours will the number of bacteria double?

16. The count in a bacteria culture was 400 after two hours and 25,600 after six hours.

(a) What is the relative rate of growth of the bacteria population? Express your answer as a percentage.

(b) What was the initial size of the culture?

(c) Find a formula for the number of bacteria $n(t)$ after t hours.

(d) Find the number of bacteria after 4.5 hours.

(e) When will the number of bacteria be 50,000?

17. The population of the world was 5.8 billion in 1995 and the observed relative growth rate was 2% per year.
(a) By what year will the population have doubled?
(b) By what year will the population have tripled?

18. The population of California was 10,586,223 in 1950 and 23,668,562 in 1980. Assume the population grows exponentially.
(a) Find a formula for the population t years after 1950.
(b) Find the time required for the population to double.
(c) Use the data to predict the population of California in the year 2000.

19. An infectious strain of bacteria increases in number at a relative growth rate of 200% per hour. When a certain critical number of bacteria are present in the bloodstream, a person becomes ill. If a single bacterium infects a person, the critical level is reached in 24 hours. How long will it take for the critical level to be reached if the same person is infected with 10 bacteria?

20. The half-life of radium-226 is 1600 years. Suppose we have a 22-mg sample.
(a) Find a formula for the mass remaining after t years.
(b) How much of the sample remains after 4000 years?
(c) After how long will only 18 mg of the sample remain?

21. The half-life of cesium-137 is 30 years. Suppose we have a 10-g sample.
(a) Find a formula for the mass remaining after t years.
(b) How much of the sample remains after 80 years?
(c) After how long will only 2 g of the sample remain?

22. The mass $m(t)$ remaining after t days from a 40-g sample of thorium-234 is given by

$$m(t) = 40e^{-0.0277t}$$

(a) How much of the sample remains after 60 days?
(b) After how long will only 10 g of the sample remain?
(c) Find the half-life of thorium-234.

23. The half-life of strontium-90 is 25 years. How long will it take a 50-mg sample to decay to a mass of 32 mg?

24. Radium-221 has a half-life of 30 s. How long will it take for 95% of a sample to decompose?

25. If 250 mg of a radioactive element decays to 200 mg in 48 hours, find the half-life of the element.

26. After three days a sample of radon-222 has decayed to 58% of its original amount.
(a) What is the half-life of radon-222?
(b) How long will it take the sample to decay to 20% of its original amount?

27. A wooden artifact from an ancient tomb contains 65% of the carbon-14 that is present in living trees. How long ago was the artifact made? (The half-life of carbon-14 is 5730 years.)

28. The burial cloth of an Egyptian mummy is estimated to contain 59% of the carbon-14 it contained originally. How long ago was the mummy buried? (The half-life of carbon-14 is 5730 years.)

29. A hot bowl of soup is served at at a dinner party. It starts to cool according to Newton's Law of Cooling so that its temperature at time t is given by

$$T(t) = 65 + 145e^{-0.05t}$$

where t is measured in minutes and T is measured in °F.
(a) What is the initial temperature of the soup?
(b) What is the temperature after 10 minutes?
(c) After how long will the temperature be 100 °F?

30. Newton's Law of Cooling is used in homicide investigations to determine the time of death. The normal body temperature is 98.6 °F. Immediately following death the body begins to cool. It has been determined experimentally that the constant in Newton's Law of Cooling is approximately $k = 0.1947$. If the temperature of the surroundings is 60 °F and the temperature of the body is 72 °F, how long ago was the time of death?

31. A roasted turkey is taken from an oven when its temperature has reached 185 °F and is placed on a table in a room where the temperature is 75 °F.
(a) If the temperature of the turkey is 150 °F after half an hour, what is its temperature after 45 minutes?
(b) When will the turkey cool to 100 °F?

32. A kettle full of water is brought to a boil in a room with temperature 20 °C. After 15 minutes the temperature of the water has decreased from 100 °C to 75 °C. Find the temperature after another 10 minutes. Illustrate by sketching the graph of the temperature function.

33. The hydrogen ion concentration of a sample of each substance is given. Calculate the pH of the substance.
(a) lemon juice: $[H^+] = 5.0 \times 10^{-3}$ M

(b) tomato juice: $[H^+] = 3.2 \times 10^{-4}$ M
(c) seawater: $[H^+] = 5.0 \times 10^{-9}$ M

34. An unknown substance has a hydrogen ion concentration of $[H^+] = 3.1 \times 10^{-8}$ M. Find the pH and classify the substance as acidic or basic.

35. The pH reading of a sample of each substance is given. Calculate the hydrogen ion concentration of the substance.
(a) vinegar: pH $= 3.0$
(b) milk: pH $= 6.5$

36. The pH reading of a glass of liquid is given. Find the hydrogen ion concentration of the liquid.
(a) beer: pH $= 4.6$
(b) water: pH $= 7.3$

37. The hydrogen ion concentrations in cheeses range from 4.0×10^{-7} M to 1.6×10^{-5} M. Find the corresponding range of pH readings.

38. The pH readings for wines vary from 2.8 to 3.8. Find the corresponding range of hydrogen ion concentrations.

39. If one earthquake is 20 times as intense as another, how much larger is its magnitude on the Richter scale?

40. The 1906 earthquake in San Francisco had a magnitude of 8.3 on the Richter scale. At the same time in Japan there was an earthquake with magnitude 4.9 that caused only minor damage. How many times more intense was the San Francisco earthquake than the Japanese earthquake?

41. The Alaska earthquake of 1964 had a magnitude of 8.6 on the Richter scale. How many times more intense was this than the 1906 San Francisco earthquake?

42. The Northridge, California, earthquake of 1994 had a magnitude of 6.8 on the Richter scale. A year later, a 7.2-magnitude earthquake struck Kobe, Japan. How many times more intense was the Kobe earthquake than the Northridge earthquake?

43. The 1985 Mexico City earthquake had a magnitude of 8.1 on the Richter scale. The 1976 earthquake in Tangshan, China, was 1.26 times as intense. What was the magnitude of the Tangshan earthquake?

44. The intensity of the sound of traffic at a busy intersection was measured at 2.0×10^{-5} watts/m². Find the intensity level in decibels.

45. The intensity level of the sound of a subway train was measured at 98 dB. Find the intensity in watts/m².

46. The noise from a power mower was measured at 106 dB. The noise level at a rock concert was measured at 120 dB. Find the ratio of the intensity of the rock music to that of the power mower.

47. It is a law of physics that the intensity of sound is inversely proportional to the square of the distance d from the source:

$$I = \frac{k}{d^2}$$

(a) Use this and the equation $\beta = 10 \log \dfrac{I}{I_0}$ (described in this section) to show that the decibel levels β_1 and β_2 at distances d_1 and d_2 from a sound source are related by the equation

$$\beta_2 = \beta_1 + 20 \log \frac{d_1}{d_2}$$

(b) The intensity level at a rock concert is 120 dB at a distance 2 m from the speakers. Find the intensity level at a distance of 10 m.

48. The table shows the mean distances d of the planets from the sun (taking the unit of measurement to be the distance from the earth to the sun) and their periods T (time of revolution in years). Try to discover a relationship between T and d. [*Hint:* Consider their logarithms.]

Planet	d	T
Mercury	0.387	0.241
Venus	0.723	0.615
Earth	1.000	1.000
Mars	1.523	1.881
Jupiter	5.203	11.861
Saturn	9.541	29.457
Uranus	19.190	84.008
Neptune	30.086	164.784
Pluto	39.507	248.35

5 | REVIEW

KEY IDEAS ■ Define, state, or discuss each of the following.

1. The exponential function with base a
2. The number e
3. The natural exponential function
4. Graphs of exponential functions
5. Compound interest
6. Continuous compounding of interest
7. Exponential growth
8. The logarithmic function with base a
9. Natural logarithms

10. Common logarithms
11. Laws of logarithms
12. Change of base formula
13. Radioactive decay
14. Half-life
15. Newton's Law of Cooling
16. The pH scale
17. The Richter scale
18. The decibel scale

EXERCISES

1–12 ■ Sketch the graph of the function. State the domain, range, and asymptote.

1. $f(x) = \dfrac{1}{2^x}$

2. $g(x) = 3^{x-2}$

3. $y = 5 - 10^x$

4. $y = 1 + 5^{-x}$

5. $f(x) = \log_3(x - 1)$

6. $g(x) = \log(-x)$

7. $y = 2 - \log_2 x$

8. $y = 3 + \log_5(x + 4)$

9. $F(x) = e^x - 1$

10. $G(x) = \frac{1}{2} e^{x-1}$

11. $y = 2 \ln x$

12. $y = \ln(x^2)$

13–14 ■ Find the domain of the function.

13. $f(x) = 10^{x^2} + \log(1 - 2x)$ **14.** $g(x) = \ln(2 + x - x^2)$

15–18 ■ Write the equation in exponential form.

15. $\log_2 1024 = 10$ **16.** $\log_6 37 = x$

17. $\log x = y$ **18.** $\ln c = 17$

19–22 ■ Write the equation in logarithmic form.

19. $2^6 = 64$ **20.** $49^{-1/2} = \frac{1}{7}$

21. $10^x = 74$ **22.** $e^k = m$

23–38 ■ Evaluate the expression without using a calculator.

23. $\log_2 128$ **24.** $\log_8 1$

25. $10^{\log 45}$ **26.** $\log 0.000001$

27. $\ln(e^6)$ **28.** $\log_4 8$

29. $\log_3\left(\frac{1}{27}\right)$ **30.** $2^{\log_2 13}$

31. $\log_5 \sqrt{5}$ **32.** $e^{2\ln 7}$

33. $\log 25 + \log 4$ **34.** $\log_3 \sqrt{243}$

35. $\log_2 16^{23}$ **36.** $\log_5 250 - \log_5 2$

37. $\log_8 6 - \log_8 3 + \log_8 2$ **38.** $\log \log 10^{100}$

39–44 ■ Rewrite the expression in a form with no logarithms of products, quotients, or powers.

39. $\log(AB^2C^3)$ **40.** $\log_2(x\sqrt{x^2 + 1}\,)$

41. $\ln \sqrt{\dfrac{x^2 - 1}{x^2 + 1}}$ **42.** $\log\left(\dfrac{4x^3}{y^2(x - 1)^5}\right)$

43. $\log_5\left(\dfrac{x^2(1 - 5x)^{3/2}}{\sqrt{x^3 - x}}\right)$ **44.** $\ln\left(\dfrac{\sqrt[3]{x^4 + 12}}{(x + 16)\sqrt{x - 3}}\right)$

45–50 ■ Rewrite the expression as a single logarithm.

45. $\log 6 + 4 \log 2$

46. $\log x + \log(x^2 y) + 3 \log y$

47. $\frac{3}{2} \log_2(x - y) - 2 \log_2(x^2 + y^2)$

48. $\log_5 2 + \log_5(x + 1) - \frac{1}{3} \log_5(3x + 7)$

49. $\log(x - 2) + \log(x + 2) - \frac{1}{2}\log(x^2 + 4)$

50. $\frac{1}{2}[\ln(x - 4) + 5\ln(x^2 + 4x)]$

51–60 ■ Use a calculator to find the solution of the equation, correct to two decimal places.

51. $\log_2(1 - x) = 4$ **52.** $2^{3x-5} = 7$

53. $5^{5-3x} = 26$ **54.** $\ln(2x - 3) = 14$

55. $e^{3x/4} = 10$ **56.** $2^{1-x} = 3^{2x+5}$

57. $\log x + \log(x + 1) = \log 12$

58. $\log_8(x + 5) - \log_8(x - 2) = 1$

59. $x^2 e^{2x} + 2xe^{2x} = 8e^{2x}$ **60.** $2^{3^x} = 5$

61–64 ■ Use a calculator to find the solution of the equation, correct to six decimal places.

61. $5^{-2x/3} = 0.63$ **62.** $2^{3x-5} = 7$

63. $5^{2x+1} = 3^{4x-1}$ **64.** $e^{-15k} = 10,000$

65–68 ■ Draw the graph of the function and use it to determine the asymptotes and the local maximum and minimum values.

65. $y = e^{x/(x+2)}$ **66.** $y = 2x^2 - \ln x$

67. $y = \log(x^3 - x)$ **68.** $y = 10^x - 5^x$

69–70 ■ Find the solutions of the equations, correct to two decimal places.

69. $3\log x = 6 - 2x$ **70.** $4 - x^2 = e^{-2x}$

71–72 ■ Solve the inequality graphically.

71. $\ln x > x - 2$ **72.** $e^x < 4x^2$

73. Use a graph of $f(x) = e^x - 3e^{-x} - 4x$ to find, approximately, the intervals on which f is increasing and on which f is decreasing.

74. Find an equation of the line shown in the figure.

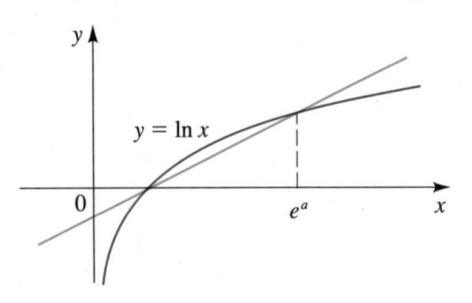

75. Evaluate $\log_4 15$, correct to six decimal places.

76. Solve the inequality: $0.2 \leq \log x < 2$

77. Which is larger, $\log_4 258$ or $\log_5 620$?

78. Find the inverse function of the function $f(x) = 2^{3^x}$ and state its domain and range.

79. If \$12,000 is invested at an interest rate of 10% per year, find the amount of the investment at the end of three years for each compounding method.
(a) semiannual   (b) monthly
(c) daily   (d) continuously

80. A sum of \$5000 is invested at an interest rate of $8\frac{1}{2}\%$ per year, compounded semiannually.
(a) Find the amount of the investment after $1\frac{1}{2}$ years.
(b) After what period of time will the investment amount to \$7000?

81. The stray-cat population in a small town grows exponentially. In 1994, the town had 30 stray cats and the relative growth rate was 15% per year.
(a) Find a formula for the stray-cat population $n(t)$ after t years.
(b) Find the projected population after 4 years.
(c) Find the number of years required for the stray-cat population to reach 500.

82. A bacteria culture contains 10,000 bacteria initially. After an hour the bacteria count is 25,000.
(a) Find the doubling period.
(b) Find the population after three hours.

83. Uranium-234 has a half-life of 2.7×10^5 years.
(a) Find the amount remaining from a 10-mg sample after a thousand years.
(b) How long will it take this sample to decompose until its mass is 7 mg?

84. A sample of bismuth-210 decayed to 33% of its original mass after eight days.
(a) Find the half-life of this element.
(b) Find the mass remaining after 12 days.

85. The half-life of radium-226 is 1590 years.
(a) If a sample has a mass of 150 mg, find a formula for the mass that remains after t years.
(b) Find the mass that remains after 1000 years.
(c) After how many years will only 50 mg remain?

86. The half-life of palladium-100 is four days. After 20 days a sample has been reduced to a mass of 0.375 g.
(a) What was the initial mass of the sample?
(b) Find a formula for the mass remaining after t days.
(c) What is the mass after three days?
(d) After how many days will only 0.15 g remain?

87. The graph shows the population of a rare species of bird, where t represents years since 1988 and $n(t)$ is measured in thousands.

(a) Find a formula for the bird population at time t in the form $n(t) = n_0 e^{rt}$.

(b) What is the bird population expected to be in the year 1999?

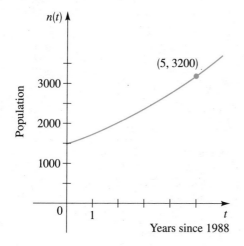

88. A car engine runs at a temperature of $190\,°F$. When the engine is turned off, it cools according to Newton's Law of Cooling with constant $k = 0.341$. Find the time needed for the engine to cool to $90\,°F$ if the surrounding temperature is $60\,°F$.

89. The hydrogen ion concentration of fresh egg whites was measured as

$$[H^+] = 1.3 \times 10^{-8}\ M$$

Find the pH, and classify the substance as acidic or basic.

90. The pH of lime juice is 1.9. Find the hydrogen ion concentration.

91. If one earthquake has magnitude 6.5 on the Richter scale, what is the magnitude of another quake that is 35 times as intense?

92. The noise from a jackhammer was measured at $132\,dB$. The sound of whispering was measured at $28\,dB$. Find the ratio of the intensity of the jackhammer to that of the whispering.

1. Graph the functions $y = 4^x$ and $y = \log_4 x$ on the same axes.

2. Sketch the graph of the function $f(x) = \log(x + 2)$ and state the domain, range, and asymptote.

3. Evaluate each logarithmic expression.
   (a) $\log_3 \sqrt{27}$ (b) $\log_2 56 - \log_2 7$
   (c) $\log_8 4$ (d) $\log_6 4 + \log_6 9$

4. Use the Laws of Logarithms to rewrite the expression

$$\log \sqrt{\frac{x^2 - 1}{x^3(y^2 + 1)^5}}$$

   without logarithms of products, quotients, powers, or roots.

5. Write as a single logarithm: $\ln x - 2\ln(x^2 + 1) + \frac{1}{2}\ln(3 - x^4)$

6. Find the solution of the equation, correct to two decimal places.
   (a) $2^{x-1} = 10$ (b) $5\ln(3 - x) = 4$
   (c) $10^{x+3} = 6^{2x}$ (d) $\log_2(x + 2) + \log_2(x - 1) = 2$

7. The initial size of a bacteria culture is 1000. After one hour the bacteria count is 8000.
   (a) Find a formula for the population after t hours.
   (b) Find the population after 1.5 hours.
   (c) When will the population reach 15,000?
   (d) Sketch the graph of the population function.

8. How long will it take for an investment to double in value if the interest rate is 8.5% per year, compounded semiannually?

9. Find the domain of the function $f(x) = \log(x + 4) + \log(8 - 5x)$.

10. Let $f(x) = \dfrac{e^x}{x^3}$.
    (a) Graph f in an appropriate viewing rectangle.
    (b) State the asymptotes of f.
    (c) Find, correct to two decimal places, the local minimum value of f and the value of x at which it occurs.
    (d) Find the range of f.
    (e) Solve the equation $\dfrac{e^x}{x^3} = 2x + 1$. State each solution correct to two decimal places.

FOCUS ON PROBLEM SOLVING

One way to show that a statement is true is to show that its opposite leads to a contradiction. This type of **indirect reasoning** is an important problem-solving tool. The two examples we give here are problems that were solved more than 2000 years ago but continue to have a profound influence in mathematics.

PROBLEM 1 ■ Irrational Numbers Exist

The Pythagoreans were fascinated by the beauty of the natural numbers: $1, 2, 3, \ldots$. They hoped that the length of every interval constructed in geometry would be a ratio of two natural numbers, and so the natural numbers would describe all of geometry.

But what is the length of the diagonal of a square of side 1? The Pythagoreans knew (by the Pythagorean Theorem) that this length is $\sqrt{2}$. The question was: Is $\sqrt{2}$ the ratio of two natural numbers? Hippasus, one of the Pythagoreans, is reputed to have proved that $\sqrt{2}$ is in fact *not* rational while traveling onboard a ship. His fellow Pythagoreans were so upset by this discovery that they hurled him overboard. The proof, nevertheless, is a classic use of indirect reasoning.

Suppose $\sqrt{2}$ is rational, so that

$$\sqrt{2} = \frac{a}{b}$$

where a and b are natural numbers with no factor in common. Then

$$a = \sqrt{2}\, b$$

$$a^2 = 2b^2$$

This means that a^2 is an even number and so a is an even number, say $a = 2m$. So, from the preceding equation we get

$$(2m)^2 = 2b^2$$

$$4m^2 = 2b^2$$

$$2m^2 = b^2$$

Thus, b is also even. So a and b have 2 as a common factor, and this contradicts our assumption that a and b have no factor in common. Thus, the assumption that $\sqrt{2}$ is rational leads to a contradiction, and so $\sqrt{2}$ must be irrational.

Eratosthenes (circa 276–195 B.C.) was a renowned Greek geographer, mathematician, and astronomer. He accurately calculated the circumference of the earth by an ingenious method. He is most famous, however, for his method for finding primes, now called the *sieve of Eratosthenes*. The method consists of listing the integers, beginning with 2 (the first prime), and then crossing out all the multiples of 2, which are not prime. The next number remaining on the list is 3 (the second prime), so we again cross out all multiples of it. The next remaining number is 5 (the third prime number), and we cross out all multiples of it, and so on. In this way all numbers that are not prime are crossed out and the remaining numbers are the primes.

A prime number is one that has no factor other than 1 and itself. The first primes are

$$2, \ 3, \ 5, \ 7, \ 11, \ 13, \ 17, \ 19, \ 23, \ 29, \ 31, \ \ldots$$

How do we find the next prime? No formula is known that will produce the primes and no one has found a pattern for the location of the primes. One thing that is known is that there are infinitely many primes. Euclid gave a proof for this fact over 2000 years ago by a brilliant use of indirect reasoning. Here is his famous proof:

Suppose that there is only a finite number of primes. We list them as

$$p_1, \ p_2, \ p_3, \ \ldots, \ p_n$$

Then the number

$$N = p_1 \cdot p_2 \cdot p_3 \cdot \cdots \cdot p_n + 1$$

is not divisible by any prime (Why?) and so is itself prime. But N is not in our list of primes. (Why?) This contradicts our assumption that the list contained all the primes. Thus, there are infinitely many primes. ■

PROBLEMS

1. (a) Prove that $\sqrt{6}$ is an irrational number.
   (b) Prove that $\sqrt{2} + \sqrt{3}$ is an irrational number.

2. Prove that $\log_2 5$ is irrrational.

3. (a) What is the smallest positive integer by which 12 can be multiplied to obtain a perfect cube?
   (b) What is the smallest positive integer by which 15 can be multiplied to obtain a perfect cube?
   (c) Can you find a rule for doing problems like parts (a) and (b) for any number n?

4. Evaluate: $(\log_2 3)\,(\log_3 4)\,(\log_4 5)\cdots(\log_{31} 32)$

5. Show that if $x > 0$ and $x \neq 1$, then

$$\frac{1}{\log_2 x} + \frac{1}{\log_3 x} + \frac{1}{\log_5 x} = \frac{1}{\log_{30} x}$$

6. Solve for x: $(\log_a x)(\log_5 x) = \log_a 5$

7. Solve the inequality: $\log(x^2 - 2x - 2) \leqslant 0$

8. Solve the inequality: $\log_{1/2}(1 + x) + \log_2\left(1 + \dfrac{1}{x}\right) \geqslant 1$

9. (a) Show that the number of digits in any positive integer n is $[\![\log n]\!] + 1$.
   (*The greatest integer function* $[\![\ ]\!]$ is explained in Problem 9 on page 256.)
   (b) How many digits does the number 2^{500} have?

10. Prove that at any party there are two people who know the same number of people. (Assume that if person A knows person B, then B knows A. Assume also that everyone knows himself or herself.)

11. Three tangent circles of radius 10 cm are drawn. All centers lie on the line AB. The tangent AC to the right-hand circle is drawn, intersecting the middle circle at D and E. Find the length of the segment DE.

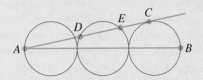

12. Augustus DeMorgan, the famous 19th-century logician, once stated that he was x years old in the year x^2. He died at age 65. In what year did he die?

6

SYSTEMS OF EQUATIONS AND INEQUALITIES

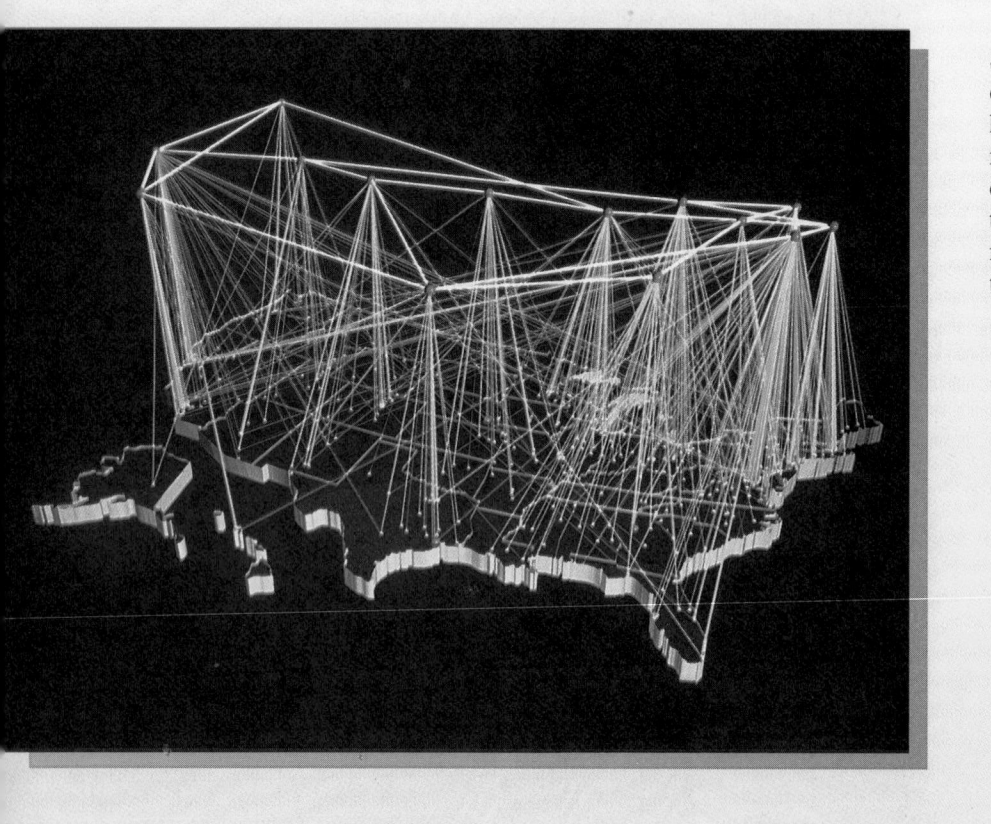

Systems of equations and inequalities are used to determine how resources can be allocated most effectively. Long-distance companies, for example, use these techniques to find the most efficient routing for a telephone call.

Mathematics is the key and door to the sciences.

GALILEO GALILEI

Many of the problems to which we can apply the techniques of algebra give rise to sets of equations with several unknowns, rather than to just a single equation in a single variable. A set of equations with common variables is called a *system of equations,* and in this chapter we develop techniques for finding simultaneous solutions of systems. We first consider pairs of linear equations with two unknowns, the simplest case of this situation. To help us solve linear equations in an arbitrary number of variables, we study the algebra of matrices and determinants. We also study systems of inequalities and linear programming, which is an optimization technique used widely in business and the social sciences.

6.1 PAIRS OF LINES

In Section 3.4, we saw that the graph of any equation of the form

$$Ax + By = C$$

is a line. Let us consider a **system** of two such equations:

$$\begin{cases} ax + by = c \\ dx + ey = f \end{cases}$$

A **solution** of this system is an ordered pair of numbers (x_0, y_0) that simultaneously makes both equations true statements when x is replaced by x_0 and y by y_0. This means that the point (x_0, y_0) lies on both of the lines in the system, and so it must be a point at which they intersect. For example, $(2, 6)$ is a solution of the system

$$\begin{cases} 3x - y = 0 \\ 5x + 2y = 22 \end{cases}$$

because

$$3(2) - (6) = 0$$

and

$$5(2) + 2(6) = 22$$

Graphing the lines given by these equations, we see in Figure 1 that $(2, 6)$ is their point of intersection. The graph also shows that there can be no other solution of the system because the lines do not intersect anywhere else.

In general, there are three situations that can occur when we graph two linear equations. The graphs may intersect at a single point (Figure 2), they may be parallel with no intersection point (Figure 3), or the two equations may just be different equations for the same line (Figure 4). This means that the system can have one solution, no solution, or infinitely many solutions.

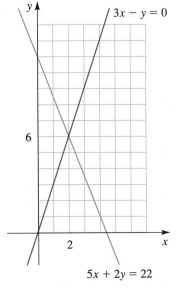

FIGURE 1

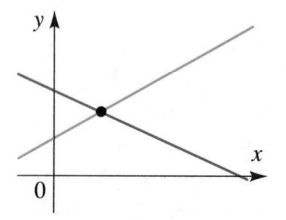

FIGURE 2
Linear system with one solution.
Lines intersect at a single point.

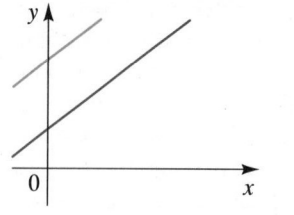

FIGURE 3
Linear system with no solution.
Lines are parallel—they do not
intersect.

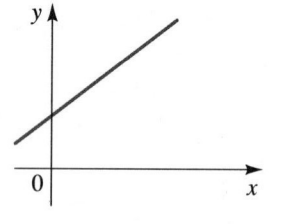

FIGURE 4
Linear system with infinitely
many solutions.
Equations are for the same line.

There are two basic methods for solving systems of two linear equations. The first, the **substitution method,** is perhaps the more obvious, and we use it in Example 1. In this method, we solve one of the equations for one variable in terms of the other, and then we substitute this expression into the remaining equation and solve it. The second method, the *elimination method,* is easier to extend to situations where we have more equations and more variables. We use it in the remaining examples.

EXAMPLE 1 ■ The Substitution Method

Solve the following system and graph the lines.

$$\begin{cases} 4x - 3y = 11 \\ 6x + 2y = -3 \end{cases}$$

SOLUTION

Solving the second equation for y in terms of x, we get

$$6x + 2y = -3$$

$$2y = -6x - 3$$

$$y = -3x - \tfrac{3}{2}$$

Now we can substitute this expression for y into the first equation, which gives us an equation that involves only the variable x:

$$4x - 3\left(-3x - \tfrac{3}{2}\right) = 11$$

$$13x + \tfrac{9}{2} = 11$$

$$13x = \tfrac{13}{2}$$

$$x = \tfrac{1}{2}$$

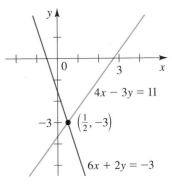

FIGURE 5

We now substitute this value for x back into the original expression for y:

$$y = -3\left(\tfrac{1}{2}\right) - \tfrac{3}{2} = -3$$

The solution of the system is $\left(\tfrac{1}{2}, -3\right)$. This is the intersection point of the lines in the system (see Figure 5). ■

Another way of solving pairs of linear equations is to eliminate either x or y from the equations by adding a suitable multiple of one to the other. This is called the **elimination method,** and we illustrate it in the next example.

EXAMPLE 2 ■ **The Elimination Method**

Solve the system

$$\begin{cases} x - 3y = 6 \\ -2x + 5y = -5 \end{cases}$$

SOLUTION

If we multiply each side of the first equation by 2, then the coefficients of x in the two equations will be negatives of each other:

$$\begin{cases} 2x - 6y = 12 \\ -2x + 5y = -5 \end{cases}$$

Adding corresponding sides of the two equations eliminates the variable x, and we can solve for y:

$$-y = 7$$

$$y = -7$$

At this point we could substitute this value for y into either of the original equations and solve for x. We use the first one, because it looks a little easier.

$$x - 3(-7) = 6 \qquad \text{Substitute } y = -7$$

$$x + 21 = 6 \qquad \text{Simplify}$$

$$x = -15 \qquad \text{Subtract 21}$$

The solution of the system is $x = -15$, $y = -7$. To check our answer, we make sure that these values do indeed satisfy both equations.

First equation: $\quad x - 3y = (-15) - 3(-7) = -15 + 21 = 6$

Second equation: $\quad -2x + 5y = -2(-15) + 5(-7) = 30 - 35 = -5$

Both equations are verified, so the solution $(-15, -7)$ is correct. ■

EXAMPLE 3 ■ A Pair of Linear Equations with No Solution

Solve the following system.

$$\begin{cases} 8x - 2y = 5 \\ -12x + 3y = 7 \end{cases}$$

SOLUTION

This time we try to find a suitable combination of the two equations to eliminate the variable y. Multiplying the first equation by 3 and the second by 2 gives

$$\begin{cases} 24x - 6y = 15 \\ -24x + 6y = 14 \end{cases}$$

Adding the two equations eliminates *both x and y* in this case, and we end up with $0 = 29$, which is obviously false. No matter what values we assign to x and y, we cannot make this statement true, so the system has *no solution*. ■

A system that has no solution, like the one in Example 3, is said to be **inconsistent.** In slope-intercept form the equations in the system are

$$y = 4x - \tfrac{5}{2} \qquad \text{and} \qquad y = 4x + \tfrac{7}{3}$$

These lines are parallel, with different y-intercepts (see Figure 6), so they have no intersection point.

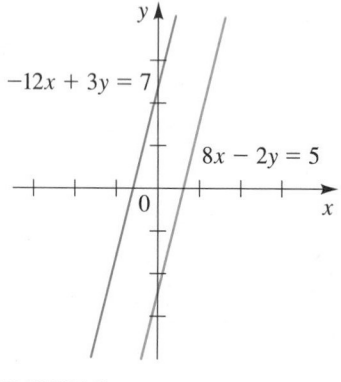

FIGURE 6

EXAMPLE 4 ■ A Pair of Linear Equations with Infinitely Many Solutions

Solve the following system.

$$\begin{cases} 3x - 6y = 12 \\ 4x - 8y = 16 \end{cases}$$

SOLUTION

We multiply the first equation by 4 and the second by 3 to prepare for subtracting the equations to eliminate x. The new equations are

$$\begin{cases} 12x - 24y = 48 \\ 12x - 24y = 48 \end{cases}$$

We see that the two equations in the original system are just different ways of expressing the equation of one single line. The coordinates of any point on this line give a solution of the system. Writing the equation in slope-intercept form, we have $y = \tfrac{1}{2}x - 2$, so any pair of the form

$$\left(x, \tfrac{1}{2}x - 2\right)$$

where x can be any real number, is a solution of the system. The system has infinitely many solutions.   ■

APPLIED LINEAR SYSTEMS

Frequently, when we use equations to solve problems in the sciences or in other areas, we obtain systems like the ones we have been considering. The next two examples illustrate such situations.

EXAMPLE 5 ■ A Distance-Speed-Time Problem

A woman rows a boat upstream from one point on a river to another point 4 mi away in $1\frac{1}{2}$ h. The return trip, traveling with the current, takes only 45 min. How fast does she row relative to the water, and at what speed is the current flowing?

SOLUTION

In this and any other problem that involves distance, time, and speed, we make use of the fundamental relationships between these quantities that are shown in the margin. We use these equations to translate the English sentences of the problem into mathematical form. Since we are asked to find the rowing speed and the speed of the current, we give names to these quantities. Let

$$x = \text{rowing speed (in mi/hr)}$$

$$y = \text{speed of the current (in mi/hr)}$$

$$\text{speed} = \frac{\text{distance}}{\text{time}}$$

$$\text{distance} = \text{speed} \times \text{time}$$

$$\text{time} = \frac{\text{distance}}{\text{speed}}$$

When she is traveling with the current (downstream), she will be moving at a total of $x + y$ miles per hour, but upstream she moves at $x - y$ miles per hour, since the current decreases her net speed. The distance both upstream and downstream is 4 mi, so using the fact that distance = speed × time for both parts of the trip, we get the equations

$$4 = (x - y) \cdot \tfrac{3}{2} \quad \text{and} \quad 4 = (x + y) \cdot \tfrac{3}{4}$$

(Note: All times have been converted to hours, since we are expressing the speeds in miles per *hour.*) If we multiply the equations by 2 and 4, respectively, to clear the denominators, we get the system

$$\begin{cases} 3x - 3y = 8 \\ 3x + 3y = 16 \end{cases}$$

Adding the equations eliminates the variable y:

$$6x = 24$$

$$x = 4$$

CHECK YOUR ANSWER

Speed upstream is

$$\frac{\text{distance}}{\text{time}} = \frac{4 \text{ mi}}{1\frac{1}{2}\text{ h}} = 2\frac{2}{3}\text{ mi/h}$$

and this should equal

rowing speed − current flow
$$= 4 \text{ mi/h} - \tfrac{4}{3} \text{ mi/h}$$
$$= 2\tfrac{2}{3} \text{ mi/h} \qquad \checkmark$$

Speed downstream is

$$\frac{\text{distance}}{\text{time}} = \frac{4 \text{ mi}}{\frac{3}{4}\text{ h}} = 5\frac{1}{3}\text{ mi/h}$$

and this should equal

rowing speed + current flow
$$= 4 \text{ mi/h} + \tfrac{4}{3} \text{ mi/h} = 5\tfrac{1}{3} \text{ mi/h} \quad \checkmark$$

Substituting this value of x into the first equation in the system (the second works just as well) and solving for y gives

$$3(4) - 3y = 8$$

$$-3y = 8 - 12$$

$$y = \tfrac{4}{3}$$

The woman rows at 4 mi/h and the current flows at $1\frac{1}{3}$ mi/h.   ■

Many students find that the hardest part of solving a word problem is getting started. The following steps, adapted from the problem-solving principles on pages 61–63, should help you find your way to the equations that are needed to solve the problem.

SOLVING APPLIED PROBLEMS

1. GIVE NAMES TO THE VARIABLES. Assign letters to denote the variable quantities in the problem. Usually the last sentence of the problem tells you what is being asked for, so this is what the variable names will represent.

2. ORGANIZE THE GIVEN INFORMATION. If possible, draw a diagram or create a table that helps you see the relationship between the quantities involved in the problem.

3. TRANSLATE THE GIVEN INFORMATION INTO EQUATIONS. Translate the information about the variables given in the problem—whether or not you have been able to organize it into a table or diagram—into mathematical equations. Remember that *an equation is just a sentence written using the symbols of mathematics.*

4. SOLVE THE EQUATIONS AND INTERPRET THE RESULTS. Solve the equations given by Step 3, and state in English what the solutions mean in terms of the original meanings of the variables.

EXAMPLE 6 ■ A Mixture Problem

A vintner wishes to fortify wine that contains 10% alcohol by adding to it some 70% alcohol solution. The resulting mixture is to have an alcoholic strength of 16% and is to fill 1000 one-liter bottles. How many liters (L) of the wine and of the alcohol solution should he use?

SOLUTION

Let

x = number of liters of wine to be used

y = number of liters of alcohol solution to be used

Name the variables

To help us translate the information in the problem into equations, we organize the given data in a table.

Organize the information into a table

	Wine	Alcohol solution	Resulting mixture
Volume	x	y	1000
Percent alcohol	10%	70%	16%
Amount of alcohol	$(0.10)x$	$(0.70)y$	$(0.16)1000$

The volume of the mixture must be the total of the two volumes the vintner is adding together, so

$$x + y = 1000$$

Similarly, the amount of alcohol in the mixture must be the total of the alcohol contributed by the wine and by the alcohol solution, which means that

Translate the information into equations

$$(0.10)x + (0.70)y = (0.16)1000$$

$$(0.10)x + (0.70)y = 160$$

$$x + 7y = 1600 \qquad \text{Multiply by 10 to clear decimals}$$

Thus, we must solve the system

$$\begin{cases} x + y = 1000 \\ x + 7y = 1600 \end{cases}$$

Subtracting the first equation from the second eliminates the variable x, and we get

Solve the equations

$$6y = 600$$

$$y = 100$$

We now substitute this value into the first equation and solve for x:

$$x + 100 = 1000$$

$$x = 900$$

The vintner should use 900 L of wine and 100 L of the alcohol solution. ■

$$\boxed{6.1}\quad\textbf{EXERCISES}$$

1–6 ■ Graph each pair of lines on a single set of axes. Determine whether the lines are parallel, and if they are not parallel, estimate the coordinates of their point of intersection from your graph.

1. $\begin{cases} x + y = 3 \\ 2x - y = 0 \end{cases}$

2. $\begin{cases} 3x + 2y = 3 \\ -x + 5y = 16 \end{cases}$

3. $\begin{cases} 2x + 3y = 12 \\ x - y = 1 \end{cases}$

4. $\begin{cases} 3x + 5y = 15 \\ x + \frac{5}{3}y = 5 \end{cases}$

5. $\begin{cases} 2x + 5y = 15 \\ 4x + 8y = 22 \end{cases}$

6. $\begin{cases} -4x + 14y = 28 \\ 10x - 35y = 70 \end{cases}$

7–12 ■ Solve the system using the substitution method.

7. $\begin{cases} -x + y = 2 \\ 4x - 3y = -3 \end{cases}$

8. $\begin{cases} 4x - 3y = 28 \\ 9x - y = -6 \end{cases}$

9. $\begin{cases} x + 2y = 7 \\ 5x - y = 2 \end{cases}$

10. $\begin{cases} -4x + 12y = 0 \\ 12x + 4y = 160 \end{cases}$

11. $\begin{cases} \frac{1}{2}x + \frac{1}{3}y = 2 \\ \frac{1}{5}x - \frac{2}{3}y = 8 \end{cases}$

12. $\begin{cases} 0.2x - 0.2y = -1.8 \\ -0.3x + 0.5y = 3.3 \end{cases}$

13–26 ■ Solve the system using the elimination method. If a system has infinitely many solutions, express them in the form given in Example 4.

13. $\begin{cases} 3x + 2y = 8 \\ x - 2y = 0 \end{cases}$

14. $\begin{cases} 4x + 2y = 16 \\ x - 5y = 70 \end{cases}$

15. $\begin{cases} 2x - 6y = 10 \\ -3x + 9y = -15 \end{cases}$

16. $\begin{cases} 18x + y = 30 \\ 12x - 3y = -24 \end{cases}$

17. $\begin{cases} 3x + 5y = 17 \\ 7x + 9y = 29 \end{cases}$

18. $\begin{cases} 2x - 3y = -8 \\ 14x - 21y = 3 \end{cases}$

19. $\begin{cases} 8s - 3t = -3 \\ 5s - 2t = -1 \end{cases}$

20. $\begin{cases} u - 30v = -5 \\ -3u + 80v = 5 \end{cases}$

21. $\begin{cases} \frac{1}{2}x + \frac{3}{5}y = 3 \\ \frac{1}{3}x + 2y = -6 \end{cases}$

22. $\begin{cases} \frac{3}{2}x - \frac{1}{3}y = \frac{1}{2} \\ 2x - \frac{1}{2}y = -\frac{1}{2} \end{cases}$

23. $\begin{cases} 0.2r + 0.3s = 0.16 \\ -1.2r + 4s = 1.36 \end{cases}$

24. $\begin{cases} 4.8x - 1.6y = 8 \\ 3.6x - 1.2y = 6 \end{cases}$

25. $\begin{cases} \sqrt{3}\,x + \sqrt{2}\,y = 5 \\ 2\sqrt{6}\,x + 4y = \sqrt{5} \end{cases}$

26. $\begin{cases} \sqrt{10}\,w - \sqrt{2}\,z = -2 + 5\sqrt{2} \\ \sqrt{2}\,w + 2\sqrt{5}\,z = 3\sqrt{10} \end{cases}$

27–30 ■ Solve the system by first making a substitution that will turn the equations into equations of lines and then using either of the methods discussed in the text. The appropriate substitutions are given in Exercises 27 and 28, but you must determine them for yourself in Exercises 29 and 30.

27. $\begin{cases} \dfrac{2}{u} + \dfrac{1}{v} = 1 \\ \dfrac{3}{u} - \dfrac{2}{v} = 1 \end{cases}$
$\left[\text{Let } x = \dfrac{1}{u},\, y = \dfrac{1}{v}.\right]$

28. $\begin{cases} 2r^2 + 3s^2 = 11 \\ 6r^2 - s^2 = 23 \end{cases}$
$[\text{Let } x = r^2,\, y = s^2.]$

29. $\begin{cases} 2z^3 + \frac{1}{2}w^3 = 2 \\ -3z^3 + \frac{3}{2}w^3 = 15 \end{cases}$

30. $\begin{cases} \dfrac{2}{x} - \dfrac{4}{y^2} = 8 \\ \dfrac{1}{x} - \dfrac{3}{y^2} = 6 \end{cases}$

31–34 ■ Solve the system.

31. $\begin{cases} \dfrac{2x - 5}{3} + \dfrac{y - 1}{6} = \dfrac{1}{2} \\ \dfrac{x}{5} + \dfrac{3y - 6}{12} = 1 \end{cases}$

32. $\begin{cases} x - 3y = 4x - 6y - 10 \\ 2x = 12y + 10 \end{cases}$

33. $x - 2y = 2x + 2y = 1$

34. $x = 2x + y = 2y + 1$

35–38 ■ Find x and y in terms of a and b.

35. $\begin{cases} x + y = 0 \\ x + ay = 1 \end{cases}$ $(a \ne 1)$

36. $\begin{cases} ax + by = 0 \\ x + y = 1 \end{cases}$ $(a \ne b)$

37. $\begin{cases} ax + by = 1 \\ bx + ay = 1 \end{cases}$ $(a^2 - b^2 \ne 0)$

38. $\begin{cases} ax + by = 0 \\ a^2x + b^2y = 1 \end{cases}$ $(a \ne 0, b \ne 0, a \ne b)$

39. Find two numbers whose sum is 34 and whose difference is 10.

40. The sum of two numbers is twice their difference. The larger number is 6 more than twice the smaller. Find the numbers.

41. A man has 14 coins in his pocket, all of which are dimes and quarters. If the total value of his change is $2.75, how many dimes and how many quarters does he have?

42. The admission fee at an amusement park is $1.50 for children and $4.00 for adults. On a certain day, 2200 people entered the park and the admission fees collected totaled $5050. How many children and how many adults were admitted?

43. A man flies a small airplane from Fargo to Bismarck, North Dakota—a distance of 180 mi. Because he is flying into a head wind, the trip takes him 2 h. On the way back, the wind is still blowing at the same speed, so the return trip takes only 1 h 12 min. What is his speed in still air, and how fast is the wind blowing?

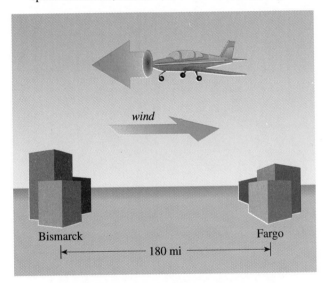

wind

Bismarck   Fargo

$\longleftarrow$ 180 mi $\longrightarrow$

44. A boat on a river travels downstream between two points, 20 mi apart, in 1 h. The return trip against the current takes $2\frac{1}{2}$ h. What is the boat's speed, and how fast does the current in the river flow?

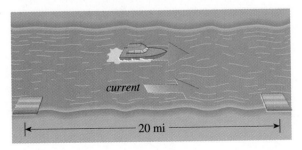

current

$\longleftarrow$ 20 mi $\longrightarrow$

45. A woman keeps fit by bicycling and running every day. On Monday she spends $\frac{1}{2}$ h at each activity, covering a total of $12\frac{1}{2}$ mi. On Tuesday, she runs for 12 min and cycles for 45 min, covering a total of 16 mi. Assuming her running and cycling speeds do not change from day to day, find these speeds.

46. A biologist has two brine solutions, one containing 5% salt and another containing 20% salt. How many milliliters of each solution should he mix to obtain 1 L of a solution that contains 14% salt?

47. A researcher performs an experiment to test a hypothesis that involves the nutrients niacin and retinol. She wishes to feed one of her groups of laboratory rats a diet that contains precisely 32 units of niacin and 22,000 units of retinol per day. She has two types of commercial pellet foods available. Food A contains 0.12 unit of niacin and 100 units of retinol per gram. Food B contains 0.20 unit of niacin and 50 units of retinol per gram. How many grams of each food should she feed this group of rats each day?

48. A customer in a coffee shop wishes to purchase a blend of two coffees: Kenyan, costing $3.50 a pound, and Sri Lankan, costing $5.60 a pound. He ends up buying 3 lb of such a blend, which costs him $11.55. How many pounds of each kind went into the mixture?

49. A chemist has two large containers of sulfuric acid solution, with different concentrations of acid in each container. Blending 300 mL of the first solution and 600 mL of the second gives a mixture that is 15% acid, whereas 100 mL of the first mixed with 500 mL of the second gives a $12\frac{1}{2}$% acid mixture. What is the concentration of sulfuric acid in each of the original containers?

50. John and Mary leave their house at the same time and drive off in opposite directions. John drives at 60 mi/h and travels 35 mi farther than Mary, who drives at 40 mi/h. Mary's trip takes 15 min longer than John's. For what length of time does each of them drive?

51. A business executive normally leaves her office at 5:00 P.M. to drive home. On Monday, she is able to leave the office at 4:45 P.M. She is able to drive at 30 mi/h and arrives home 19 min earlier than usual. On Tuesday she leaves the office at 5:20 P.M. Because the traffic is so heavy, she drives at an average speed of only 15 mi/h and arrives home 36 min later than usual. What is her usual speed on the way home, and at what time does she usually arrive?

52. The sum of the digits of a two-digit number is 7. When the digits are reversed, the number is increased by 27. Find the number.

53. The sum of the digits of a two-digit number is 9. When the digits are reversed, the value of the number is decreased to $\frac{3}{8}$ of its original value. What is the number?

54. Find the area of the triangle that lies in the first quadrant (with its base on the x-axis) and that is bounded by the lines $y = 2x - 4$ and $y = -4x + 20$.

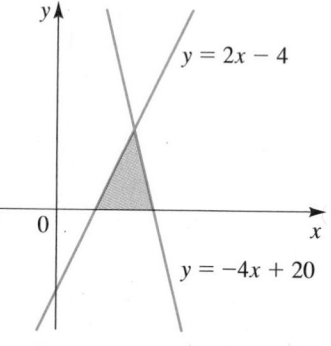

55. Find the equation of the parabola that passes through the origin and through the points $(1, 12)$ and $(3, 6)$. [*Hint:* Recall that the general equation of a parabola of this type is $y = ax^2 + bx + c$.]

56. Find the equation of the parabola that passes through the point $(-1, 8)$, that has y-intercept 14, and whose vertex has x-coordinate -2.   [*Hint:* Recall that the equation of a parabola with vertex (h, k) is of the form $y = a(x - h)^2 + k$.]

57–62 ■ Use a graphing device to graph both lines in the same viewing rectangle. (Note that you must solve for y in terms of x before graphing if you are using a graphing calculator.) Solve the system by zooming in to the point of intersection and then using the cursor to find its coordinates, correct to two decimals.

57. $\begin{cases} 0.21x + 3.17y = 9.51 \\ 2.35x - 1.17y = 5.89 \end{cases}$

58. $\begin{cases} 18.72x - 14.91y = 12.33 \\ 6.21x - 12.92y = 17.82 \end{cases}$

59. $\begin{cases} 2371x - 6552y = 13{,}591 \\ 9815x + 992y = 618{,}555 \end{cases}$

60. $\begin{cases} -435x + 912y = 0 \\ 132x + 455y = 994 \end{cases}$

61. $\begin{cases} \sqrt{3}\,x + \sqrt{5}\,y = \sqrt{7} \\ -\sqrt{2}\,x + \sqrt{11}\,y = -\sqrt{13} \end{cases}$

62. $\begin{cases} \frac{1}{23}x - \frac{1}{42}y = 3 \\ ex + e^2 y = 2.71 \end{cases}$

6.2 SYSTEMS OF LINEAR EQUATIONS

A **linear equation in n variables** is an equation which can be put in the form

$$a_1 x_1 + a_2 x_2 + \cdots + a_n x_n = c$$

Linear equations

$6x_1 - 3x_2 + \sqrt{5}\,x_3 = 1000$

$x + y + z = 2w - \frac{1}{2}$

Nonlinear equations

$x^2 + 3y - \sqrt{z} = 5$

$x_1 x_2 + 6x_3 = -6$

where $a_1, a_2, \ldots, a_n$ and c are real numbers, and $x_1, x_2, \ldots, x_n$ are the variables. If the number of variables is no more than three or four, we generally use x, y, z, and w instead of x_1, x_2, x_3, and x_4. Such equations are called *linear* because if we have just two variables, the equation is

$$a_1 x + a_2 y = c$$

which is the equation of a line. Each term of a linear equation is either a constant or a constant multiple of one of the variables.

We are going to adapt the elimination method introduced in Section 6.1 to solve systems of linear equations in any number of variables. We begin with an example to show how the method works before formally describing the technique.

EXAMPLE 1 ■ Solving a System Using the Elimination Method

Solve the following system.

$$\begin{cases} x + 2y + 4z = 7 \\ -x + y + 2z = 5 \\ 2x + 3y + 3z = 7 \end{cases}$$

SOLUTION

Our goal in solving this system is to eliminate all but one of the variables from one equation in the system, and all but two variables from another equation. More specifically, let's try to use elimination to end up with a system whose last equation involves only the variable z, and the next-to-last involves just y and z. As we shall see, this allows us to work backward to find first z, then y, and finally x.

We begin by eliminating the x terms from the second and third equations. If we add the first equation to the second, we get

$$\begin{array}{r} x + 2y + 4z = 7 \\ -x + y + 2z = 5 \\ \hline 3y + 6z = 12 \end{array}$$

The resulting equation does not contain the variable x. Similarly, if we add -2 times the first equation to the third, we get

$$\begin{array}{r} -2x - 4y - 8z = -14 \\ 2x + 3y + 3z = 7 \\ \hline -y - 5z = -7 \end{array}$$

This equation also does not contain x. This means that the original system is equivalent to the simpler system

$$\begin{cases} x + 2y + 4z = 7 \\ 3y + 6z = 12 \\ -y - 5z = -7 \end{cases}$$

Since each term in the second equation has a common factor of 3, we multiply both sides by $\frac{1}{3}$.

$$\begin{cases} x + 2y + 4z = 7 \\ y + 2z = 4 \\ -y - 5z = -7 \end{cases}$$

We now eliminate the variable y from the last equation by adding the second equation to it. This gives

$$\begin{cases} x + 2y + 4z = & 7 \\ y + 2z = & 4 \\ -3z = & -3 \end{cases}$$

Finally, multiplying the last equation by $-\frac{1}{3}$, we get

$$\begin{cases} x + 2y + 4z = 7 \\ y + 2z = 4 \\ z = 1 \end{cases}$$

We now know from the third equation that $z = 1$. Substituting this back into the second equation allows us to solve for y:

$$y + 2(1) = 4$$

$$y = 4 - 2 = 2$$

Substituting these values for y and z back into the first equation now gives us x:

$$x + 2(2) + 4(1) = 7$$

$$x = 7 - 4 - 4 = -1$$

Thus, the simultaneous solution to the system is the ordered triple $(-1, 2, 1)$, that is, $x = -1$, $y = 2$, and $z = 1$. ∎

The technique we have been using here is called **Gaussian elimination** in honor of the German mathematician C. F. Gauss (see page 304). The method consists of using algebraic operations to change the linear system we are solving into an equivalent system in *triangular form*. A system of three equations in three variables x, y, z is in **triangular form** if the second equation does not contain an x term and the third has neither x nor y. The last system in the solution to Example 1 is in triangular form. We solve a system in this form by using **back substitution**: We substitute the value of z obtained from the last equation back into the equation that involves just y and z, and solve for y. Then we substitute y and z back into the first equation and solve for z.

The algebraic operations we are permitted to use to change the systems to triangular form are called the **elementary row operations**. They are listed in the following box.

ELEMENTARY ROW OPERATIONS

1. Add a multiple of one equation to another.

2. Multiply an equation by a nonzero constant.

3. Rearrange the order of the equations.

None of these operations changes the solution of a system of equations, so after performing them we always end up with an equivalent system, that is, one that has the same solution.

If we examine the solution to Example 1, we see that the variables x, y, and z act simply as place-holders in our computations. It is only the coefficients of the variables and the constants that actually enter into the calculations. We will use this fact to simplify our notation for the Gaussian elimination process. Instead of writing out the equations of a system in full, we write only the coefficients and constants in a rectangular array, called the **matrix form** of the system. The matrix form of the system of Example 1 is as follows.

$$
\begin{array}{cc}
\text{System of linear equations} & \text{Matrix form} \\
\begin{cases} x + 2y + 4z = 7 \\ -x + y + 2z = 5 \\ 2x + 3y + 3z = 7 \end{cases} &
\begin{bmatrix} 1 & 2 & 4 & 7 \\ -1 & 1 & 2 & 5 \\ 2 & 3 & 3 & 7 \end{bmatrix}
\end{array}
$$

The **rows** of the matrix are the horizontal lists of numbers in the array. For example, the first row of the matrix in the preceding display is $[1 \quad 2 \quad 4 \quad 7]$. Each row of the matrix form of a system represents an equation. To further simplify the notation, we use the following symbols to represent the elementary row operations.

NOTATION FOR THE ELEMENTARY ROW OPERATIONS

Symbol	Description
$R_i + kR_j \rightarrow R_i$	Change the ith row by adding k times row j to it, and then put the result back in row i.
kR_i	Multiply the ith row by k.
$R_i \leftrightarrow R_j$	Interchange the ith and jth rows.

In the next example, we compare the two ways of writing systems of linear equations.

EXAMPLE 2 ■ Using Gaussian Elimination to Solve a System

Solve the following system, using Gaussian elimination.

$$
\begin{array}{cc}
\text{System} & \text{Matrix form} \\
\begin{cases} x - y + 3z = 4 \\ x + 2y - 2z = 10 \\ 3x - y + 5z = 14 \end{cases} &
\begin{bmatrix} 1 & -1 & 3 & 4 \\ 1 & 2 & -2 & 10 \\ 3 & -1 & 5 & 14 \end{bmatrix}
\end{array}
$$

SOLUTION

Our goal is to put the system into triangular form. In each step we write both the full form of the system and the shorthand matrix form. First we eliminate the x term from the second and third equations, and then the y term from the third.

	System		Matrix form

$$\begin{cases} x - y + 3z = 4 \\ x + 2y - 2z = 10 \\ 3x - y + 5z = 14 \end{cases} \qquad \begin{bmatrix} 1 & -1 & 3 & 4 \\ 1 & 2 & -2 & 10 \\ 3 & -1 & 5 & 14 \end{bmatrix}$$

Subtract the first equation from the second.
Subtract 3 times the first from the third.

$$\begin{cases} x - y + 3z = 4 \\ \ 3y - 5z = 6 \\ \ 2y - 4z = 2 \end{cases} \xrightarrow[\begin{subarray}{c} R_2 - R_1 \to R_2 \\ R_3 - 3R_1 \to R_3 \end{subarray}]{} \begin{bmatrix} 1 & -1 & 3 & 4 \\ 0 & 3 & -5 & 6 \\ 0 & 2 & -4 & 2 \end{bmatrix}$$

Multiply the third equation by $\frac{1}{2}$.

$$\begin{cases} x - y + 3z = 4 \\ \ 3y - 5z = 6 \\ \ y - 2z = 1 \end{cases} \xrightarrow[\ ]{\frac{1}{2}R_3} \begin{bmatrix} 1 & -1 & 3 & 4 \\ 0 & 3 & -5 & 6 \\ 0 & 1 & -2 & 1 \end{bmatrix}$$

Subtract 3 times the third equation from the second (to eliminate y from the second equation).

$$\begin{cases} x - y + 3z = 4 \\ \ z = 3 \\ \ y - 2z = 1 \end{cases} \xrightarrow[\ ]{R_2 - 3R_3 \to R_2} \begin{bmatrix} 1 & -1 & 3 & 4 \\ 0 & 0 & 1 & 3 \\ 0 & 1 & -2 & 1 \end{bmatrix}$$

Interchange the second and third equations (to put the system into triangular form).

$$\begin{cases} x - y + 3z = 4 \\ \ y - 2z = 1 \\ \ z = 3 \end{cases} \xrightarrow[\ ]{R_2 \leftrightarrow R_3} \begin{bmatrix} 1 & -1 & 3 & 4 \\ 0 & 1 & -2 & 1 \\ 0 & 0 & 1 & 3 \end{bmatrix}$$

Using $z = 3$, we now substitute back into the second equation in the last system to get

$$y - 2(3) = 1$$

$$y = 7$$

We then substitute $y = 7$ and $z = 3$ back into the first equation to get x:

$$x - (7) + 3(3) = 4$$

$$x = 2$$

The solution of the system is $(2, 7, 3)$. ∎

CHECK YOUR ANSWER
$x = 2, y = 7, z = 3$:
$$\begin{cases} (2) - (7) + 3(3) = 4 \\ (2) + 2(7) - 2(3) = 10 \\ 3(2) - (7) + 5(3) = 14 \end{cases} ✓$$

A rectangular array of numbers like the ones we have been using is called a **matrix.** We will study the algebra of matrices in Sections 6.4 and 6.5, but for now we just use them as a shorthand device when solving systems of equations.

Note that the goal in Gaussian elimination is to arrive at a system in triangular form, one that can be solved easily. Each elementary row operation we perform should take us one step closer to that goal. At each stage in the process we will have several possible operations to choose from, so there is no single "right" way to solve such a problem. Of course, no matter what route we decide to take to arrive at the triangular form, the final answer will be the same.

EXAMPLE 3 ■ A System with Four Equations and Four Variables

Solve the following system.

$$\begin{cases} 3x + y + 4z + w = 6 \\ 2x \quad\;\; + 3z + 4w = 13 \\ \quad\;\; y - 2z - w = 0 \\ x - y + z + w = 3 \end{cases}$$

SOLUTION

In matrix form the system is

$$\begin{bmatrix} 3 & 1 & 4 & 1 & 6 \\ 2 & 0 & 3 & 4 & 13 \\ 0 & 1 & -2 & -1 & 0 \\ 1 & -1 & 1 & 1 & 3 \end{bmatrix}$$

Since it is advantageous to have a 1 in the upper left corner, we first rearrange the order of the rows and then create as many zeros as possible in the first column.

$$\begin{bmatrix} 1 & -1 & 1 & 1 & 3 \\ 0 & 1 & -2 & -1 & 0 \\ 3 & 1 & 4 & 1 & 6 \\ 2 & 0 & 3 & 4 & 13 \end{bmatrix} \xrightarrow[\;R_4 - 2R_1 \to R_4\;]{R_3 - 3R_1 \to R_3} \begin{bmatrix} 1 & -1 & 1 & 1 & 3 \\ 0 & 1 & -2 & -1 & 0 \\ 0 & 4 & 1 & -2 & -3 \\ 0 & 2 & 1 & 2 & 7 \end{bmatrix}$$

To get this matrix into triangular form, we must change the 4 and the 2 in the second column to zeros, so we continue as follows:

$$\xrightarrow[\;R_4 - 2R_2 \to R_4\;]{R_3 - 4R_2 \to R_3} \begin{bmatrix} 1 & -1 & 1 & 1 & 3 \\ 0 & 1 & -2 & -1 & 0 \\ 0 & 0 & 9 & 2 & -3 \\ 0 & 0 & 5 & 4 & 7 \end{bmatrix}$$

We would now like to change the 5 in the last row to a 0. Although we might be tempted to do this by subtracting five times row 1 from row 4, this approach will not work because it would eliminate the first two zeros in that row, which we have worked so hard to get. So, instead, we perform the following operations on the last two rows:

$$
\xrightarrow{R_3 - 2R_4 \to R_3}
\begin{bmatrix}
1 & -1 & 1 & 1 & 3 \\
0 & 1 & -2 & -1 & 0 \\
0 & 0 & -1 & -6 & -17 \\
0 & 0 & 5 & 4 & 7
\end{bmatrix}
\xrightarrow{R_4 + 5R_3 \to R_4}
$$

$$
\begin{bmatrix}
1 & -1 & 1 & 1 & 3 \\
0 & 1 & -2 & -1 & 0 \\
0 & 0 & -1 & -6 & -17 \\
0 & 0 & 0 & -26 & -78
\end{bmatrix}
\xrightarrow[-\frac{1}{26}R_4]{-R_3}
\begin{bmatrix}
1 & -1 & 1 & 1 & 3 \\
0 & 1 & -2 & -1 & 0 \\
0 & 0 & 1 & 6 & 17 \\
0 & 0 & 0 & 1 & 3
\end{bmatrix}
$$

The last equation tells us that $w = 3$, so working backward through the equations as before, we get

$$z + 6w = 17 \qquad\qquad y - 2z - w = 0 \qquad\qquad x - y + z + w = 3$$

$$z + 6(3) = 17 \qquad\quad y - 2(-1) - 3 = 0 \qquad\quad x - 1 + (-1) + 3 = 3$$

$$z = -1 \qquad\qquad\qquad y = 1 \qquad\qquad\qquad\qquad x = 2$$

The solution is $(2, 1, -1, 3)$. ∎

Linear equations, often containing hundreds or even thousands of variables, occur frequently in the applications of algebra to the sciences and to other fields. For now, we consider an example that involves only three variables.

EXAMPLE 4 ■ Nutritional Analysis Using a System of Linear Equations

A nutritionist is performing an experiment on student volunteers. He wishes to feed one of his subjects a daily diet that consists of a combination of three commercial diet foods: MiniCal, SloStarve, and SlimQuick. For the experiment it is important that the subject consume exactly 500 mg of potassium, 75 g of protein, and 1150 units of vitamin D every day. The amounts of these nutrients in one ounce of each food are given in the following table.

	MiniCal	SloStarve	SlimQuick
Potassium (mg)	50	75	10
Protein (g)	5	10	3
Vitamin D (units)	90	100	50

How many ounces of each food should the subject eat every day to satisfy the nutrient requirements exactly?

SOLUTION

Let x, y, and z represent the number of ounces of MiniCal, SloStarve, and SlimQuick, respectively, that the subject should eat every day. This means that he will get $50x$ mg of potassium from MiniCal, $75y$ mg from SloStarve, and $10z$ mg from SlimQuick, for a total of $50x + 75y + 10z$ mg potassium in all. Since the potassium requirement is 500 mg, we get the equation

$$50x + 75y + 10z = 500$$

Similar reasoning for the protein and vitamin D requirements leads to the equations

$$5x + 10y + 3z = 75$$

and
$$90x + 100y + 50z = 1150$$

Dividing the first equation by 5 and the third by 10 gives the system

$$\begin{cases} 10x + 15y + 2z = 100 \\ 5x + 10y + 3z = 75 \\ 9x + 10y + 5z = 115 \end{cases}$$

We solve this system using Gaussian elimination.

$$\begin{bmatrix} 10 & 15 & 2 & 100 \\ 5 & 10 & 3 & 75 \\ 9 & 10 & 5 & 115 \end{bmatrix} \xrightarrow{R_1 - R_3 \to R_1} \begin{bmatrix} 1 & 5 & -3 & -15 \\ 5 & 10 & 3 & 75 \\ 9 & 10 & 5 & 115 \end{bmatrix}$$

$$\xrightarrow[R_3 - 9R_1 \to R_3]{R_2 - 5R_1 \to R_2} \begin{bmatrix} 1 & 5 & -3 & -15 \\ 0 & -15 & 18 & 150 \\ 0 & -35 & 32 & 250 \end{bmatrix} \xrightarrow{-\frac{1}{3}R_2} \begin{bmatrix} 1 & 5 & -3 & -15 \\ 0 & 5 & -6 & -50 \\ 0 & -35 & 32 & 250 \end{bmatrix}$$

$$\xrightarrow{R_3 + 7R_2 \to R_3} \begin{bmatrix} 1 & 5 & -3 & -15 \\ 0 & 5 & -6 & -50 \\ 0 & 0 & -10 & -100 \end{bmatrix} \xrightarrow{-\frac{1}{10}R_3} \begin{bmatrix} 1 & 5 & -3 & -15 \\ 0 & 5 & -6 & -50 \\ 0 & 0 & 1 & 10 \end{bmatrix}$$

Now we work backward through the equations to get $z = 10$, $y = 2$, and $x = 5$. The subject should be fed 5 oz of MiniCal, 2 oz of SloStarve, and 10 oz of SlimQuick every day. ∎

A more practical application might involve dozens of foods and nutrients, rather than just three. As you may imagine, such a problem would be almost impossible to solve without the assistance of a computer. Many graphing calculators, including the TI-82 and TI-85, are capable of performing elementary row operations on matrices, so you can use them to solve more complicated systems.

CHECK YOUR ANSWER

$x = 5$, $y = 2$, $z = 10$:

$$\begin{cases} 10(5) + 15(2) + 2(10) = 100 \\ 5(5) + 10(2) + 3(10) = 75 \\ 9(5) + 10(2) + 5(10) = 115 \end{cases} \checkmark$$

$$\boxed{6.2}\quad \textbf{EXERCISES}$$

1–6 ■ State whether the equation or system of equations is linear.

1. $6x - 3y + 1000z - w = \sqrt{13}$

2. $6xy - 3yz + 15zx = 0$

3. $x_1^2 + x_2^2 + x_3^2 = 36$

4. $e^2 x_1 + \pi x_2 - \sqrt{5} = x_3 - \frac{1}{2}x_4$

5. $\begin{cases} x - 3xy + 5y = 0 \\ 12x + 321y = 123 \end{cases}$

6. $\begin{cases} x - 3y = 15z + \dfrac{1}{\sqrt{3}} \\ x = 3z \\ x - z = \dfrac{x}{\sqrt{47}} \end{cases}$

7–10 ■ Write a system of equations that corresponds to the given matrix.

7. $\begin{bmatrix} 2 & 3 & 1 \\ 4 & 2 & 3 \end{bmatrix}$

8. $\begin{bmatrix} 1 & 2 & 4 & 6 \\ 3 & -1 & 2 & 4 \\ -1 & -1 & 0 & 7 \end{bmatrix}$

9. $\begin{bmatrix} 0 & 1 & 0 & 0 \\ 1 & 0 & 1 & 0 \\ 0 & -2 & 2 & 7 \end{bmatrix}$

10. $\begin{bmatrix} 1 & 2 & 3 & 4 & 5 \\ -1 & 0 & 1 & 0 & 6 \\ 2 & 3 & 5 & 0 & 0 \\ 0 & 1 & 1 & 0 & -2 \end{bmatrix}$

11–32 ■ Use Gaussian elimination to solve the system.

11. $\begin{cases} x - 2y + z = 1 \\ y + 2z = 5 \\ x + y + 3z = 8 \end{cases}$

12. $\begin{cases} x + y + 6z = 3 \\ x + y + 3z = 3 \\ x + 2y + 4z = 7 \end{cases}$

13. $\begin{cases} x + y + z = 2 \\ 2x - 3y + 2z = 4 \\ 4x + y - 3z = 1 \end{cases}$

14. $\begin{cases} x + y + z = 4 \\ -x + 2y + 3z = 17 \\ 2x - y = -7 \end{cases}$

15. $\begin{cases} x + 2y - z = -2 \\ x + z = 0 \\ 2x - y - z = -3 \end{cases}$

16. $\begin{cases} 2y + z = 4 \\ x + y = 4 \\ 3x + 3y - z = 10 \end{cases}$

17. $\begin{cases} x_1 + 2x_2 - x_3 = 9 \\ 2x_1 - x_3 = -2 \\ 3x_1 + 5x_2 + 2x_3 = 22 \end{cases}$

18. $\begin{cases} 2x_1 + x_2 = 7 \\ 2x_1 - x_2 + x_3 = 6 \\ 3x_1 - 2x_2 + 4x_3 = 11 \end{cases}$

19. $\begin{cases} 2x - 3y - z = 13 \\ -x + 2y - 5z = 6 \\ 5x - y - z = 49 \end{cases}$

20. $\begin{cases} 10x + 10y - 20z = 60 \\ 15x + 20y + 30z = -25 \\ -5x + 30y - 10z = 45 \end{cases}$

21. $\begin{cases} 0.1x + y - 0.2z = 0.6 \\ -0.2x + 1.1y + 0.6z = -1.6 \\ 0.3x + 0.2y + z = -1.4 \end{cases}$

22. $\begin{cases} \frac{1}{2}x + \frac{1}{3}y - \frac{1}{6}z = 8 \\ x - \frac{2}{3}y + \frac{1}{3}z = -4 \\ -\frac{1}{3}x + \frac{1}{2}y - z = 10 \end{cases}$

23. $\begin{cases} 3x + y + z = \frac{3}{2} \\ 3x + 12z = -5 \\ 2y - 4z = 4 \end{cases}$

24. $\begin{cases} x - y + 2z = 1.9 \\ 5x - 6y + z = 8.0 \\ 7x + y - 2z = -1.1 \end{cases}$

25. $\begin{cases} x + y = -2 \\ 2y + z = 1 \\ x - 3z = -20 \end{cases}$

26. $\begin{cases} 3x + 5z = 56 \\ 4y - 2z = 14 \\ 7x + 4y = 77 \end{cases}$

27. $\begin{cases} x_1 + 7x_3 = -20 \\ 2x_1 - 5x_2 = 7 \\ -3x_2 + x_3 = 0 \end{cases}$

28. $\begin{cases} x + y - z - w = 6 \\ 2x + z - 3w = 8 \\ x - y + 4w = -10 \\ 3x + 5y - z - w = 20 \end{cases}$

29. $\begin{cases} -x + 2y + z - 3w = 3 \\ 3x - 4y + z + w = 9 \\ -x - y + z + w = 0 \\ 2x + y + 4z - 2w = 3 \end{cases}$

30. $\begin{cases} x_1 + x_2 - x_3 = 0 \\ x_1 + 3x_4 = 13 \\ 3x_2 - 2x_3 = 0 \\ 2x_1 + 5x_3 = 17 \end{cases}$

31. $\begin{cases} x_1 - x_2 + x_3 + 2x_4 + 3x_5 = 0 \\ -x_1 - 2x_2 + x_3 - 2x_4 + x_5 = 7 \\ -x_1 + x_2 + x_4 - x_5 = -4 \\ 2x_1 - 2x_2 + 3x_3 - x_4 = 12 \\ x_1 + x_3 - x_4 - 5x_5 = 5 \end{cases}$

32. $\begin{cases} x + y + z + w + u + v = 12 \\ y - z + u - v = -1 \\ 2x - 2z + 4w - 4v = -6 \\ 3y - z + v = 4 \\ x - y + z - w + u - v = 0 \\ -x - y + z + w = 2 \end{cases}$

33. A doctor recommends that a patient take 50 mg each of niacin, riboflavin, and thiamin daily to alleviate a vitamin deficiency. Looking into his medicine chest at home, the patient finds three brands of vitamin pills. The amounts of the relevant vitamins per pill are given in the table.

	VitaMax	Vitron	VitaPlus
Niacin (mg)	5	10	15
Riboflavin (mg)	15	20	0
Thiamin (mg)	10	10	10

How many pills of each type should he take every day to fulfill the prescription?

34. A chemist has three containers of acid solution at various concentrations. The first is 10% acid, the second is 20%, and the third is 40%. How many milliliters of each should he mix together to make 100 mL of acid at 18% concentration, if he has to use four times as much of the 10% solution as the 40% solution?

35. The drawer of a cash register contains 30 coins: pennies, nickels, dimes, and quarters. The total value of the coins is $3.31. The total number of pennies and

nickels combined is the same as the total number of dimes and quarters combined. The total value of the quarters is five times the total value of the dimes. How many coins of each type are in the drawer?

36. A small school has 100 students who occupy three classrooms: rooms A, B, and C. After the first period of the school day, half the students in room A move to room B, one-fifth of the students in room B move to room C, and one-third of the students in room C move to room A. Nevertheless, the total number of students in each room is the same for each period. How many students occupy each room?

37. A hotel offers three classes of accommodation: standard, deluxe, and first-class. A group of ten employees of a manufacturing company attend a trade convention and stay in this hotel. If six of them take standard rooms, two take deluxe, and two take first-class, the total hotel bill will be $530 per day. If five stay in standard rooms, four in deluxe, and only one in a first-class room, the bill will decrease to $510 per day. If they splurge and have three stay in standard rooms, three in deluxe, and four in first-class rooms, the bill will be $645 per day. How much is the daily rate for each type of room?

38. Amanda, Bryce, and Corey enter a race in which they have to run, swim, and cycle over a marked course. Amanda runs at an average speed of 10 mi/h, swims at 4 mi/h, and cycles at 20 mi/h during this race. Bryce runs at $7\frac{1}{2}$ mi/h, swims at 6 mi/h, and cycles at 15 mi/h. Corey runs at 15 mi/h, swims at 3 mi/h, and cycles at 40 mi/h. Corey finishes first with a total time of 1 h 45 min. Amanda comes in second with a time of 2 h 30 min. Bryce finishes last with a time of 3 h. Find the distance (in miles) for each part of the race.

39. Determine a, b, and c so that the graph of the parabola $y = ax^2 + bx + c$ passes through the points $(-2, 24)$, $(1, 3)$, and $(3, 9)$.

40. Determine a, b, c, and d so that the points $(1, 1)$, $(2, 45)$, $(-1, -3)$, and $(-3, 225)$ all lie on the graph of the function

$$f(x) = ax^4 + bx^2 + cx + d$$

41–42 ■ Solve the system.

41. $5x + 2y = 4x - z = 4y + 3z = 1$

42. $6x + 2y = 2z - 2y = 3w - 7y = x + z = 2$

6.3 INCONSISTENT AND DEPENDENT SYSTEMS

All the systems of linear equations that we considered in the last section had one unique solution for each of the unknowns. But as we saw in Section 6.1, a system of two linear equations in two variables can have one solution, no solution, or infinitely many solutions. The same cases arise when we study linear systems with more equations and more variables. To solve a general system of linear equations we use Gaussian elimination to reduce the matrix that represents the system to a special form, which we now describe.

ECHELON FORM OF A MATRIX

A matrix is in **echelon form** if it has the following properties:

1. The first nonzero number in each row (reading from left to right) is 1. This is called the **leading entry** of the row.

2. The leading entry in each row is to the right of the leading entry in the row immediately above it.

3. Rows that consist entirely of 0's are at the bottom of the matrix.

For example, the matrix on the left below is in echelon form. However, the matrix on the right is not, because the leading entries of successive rows do not step down toward the right.

$$\begin{bmatrix} 1 & 3 & -6 & 10 & 0 \\ 0 & 0 & 1 & 4 & -3 \\ 0 & 0 & 0 & 1 & \frac{1}{2} \\ 0 & 0 & 0 & 0 & 0 \end{bmatrix} \qquad \begin{bmatrix} 0 & 1 & -\frac{1}{2} & 0 & 7 \\ 1 & 0 & 3 & 4 & -5 \\ 0 & 0 & 0 & 1 & 0.4 \\ 0 & 1 & 1 & 0 & 0 \end{bmatrix}$$

If the matrix form of a system of equations is in echelon form, we call the variables that correspond to the leading entries the **leading variables** of the system. For example, the matrix in echelon form on the left above corresponds to the system

$$\begin{cases} x + 3y - 6z + 10w = & 0 \\ z + 4w = & -3 \\ w = & \frac{1}{2} \\ 0 = & 0 \end{cases}$$

In this system, x, z, and w are leading variables, but y is not a leading variable.

In Example 1, we see how to determine that a system has no solution. In this example we use the following procedure, which always produces an echelon form for a matrix. First we use elementary row operations to change all entries below the *first* leading entry to zeros. Then we change all the entries below the *next* leading entry to zeros, and so on.

EXAMPLE 1 ■ A System with No Solution

Solve the following system.

$$\begin{cases} x - 3y + 2z = 12 \\ 2x - 5y + 5z = 14 \\ x - 2y + 3z = 20 \end{cases}$$

SOLUTION

$$\begin{bmatrix} 1 & -3 & 2 & 12 \\ 2 & -5 & 5 & 14 \\ 1 & -2 & 3 & 20 \end{bmatrix} \xrightarrow[\substack{R_3 - R_1 \to R_3}]{R_2 - 2R_1 \to R_2} \begin{bmatrix} 1 & -3 & 2 & 12 \\ 0 & 1 & 1 & -10 \\ 0 & 1 & 1 & 8 \end{bmatrix}$$

$$\xrightarrow{R_3 - R_2 \to R_3} \begin{bmatrix} 1 & -3 & 2 & 12 \\ 0 & 1 & 1 & -10 \\ 0 & 0 & 0 & 18 \end{bmatrix} \xrightarrow{\frac{1}{18}R_3} \begin{bmatrix} 1 & -3 & 2 & 12 \\ 0 & 1 & 1 & -10 \\ 0 & 0 & 0 & 1 \end{bmatrix}$$

This is in echelon form, so we may stop the Gaussian elimination process. Now if we translate the last row back into equation form, we get $0x + 0y + 0z = 1$, or $0 = 1$, which is false. No matter what values we pick for x, y, and z, the last equation will never be a true statement. This means the system *has no solution.*

■

A system that has no solution is said to be **inconsistent.** The procedure we used to show that the system in Example 1 is inconsistent works in general. If we use Gaussian elimination to change a system to echelon form, and if one of the equations we arrive at is false, then the system is inconsistent. (The false equation will always have the form $0 = c$, where c is nonzero.)

The next example shows what happens when we apply Gaussian elimination to a system with infinitely many solutions.

EXAMPLE 2 ■ A System with Infinitely Many Solutions

Find the complete solution of the following system.

$$\begin{cases} -3x - 5y + 36z = 10 \\ -x + 7z = 5 \\ x + y - 10z = -4 \end{cases}$$

SOLUTION

$$\begin{bmatrix} -3 & -5 & 36 & 10 \\ -1 & 0 & 7 & 5 \\ 1 & 1 & -10 & -4 \end{bmatrix} \xrightarrow{R_1 \leftrightarrow R_3} \begin{bmatrix} 1 & 1 & -10 & -4 \\ -1 & 0 & 7 & 5 \\ -3 & -5 & 36 & 10 \end{bmatrix}$$

$$\xrightarrow[\substack{R_3 + 3R_1 \to R_3}]{R_2 + R_1 \to R_2} \begin{bmatrix} 1 & 1 & -10 & -4 \\ 0 & 1 & -3 & 1 \\ 0 & -2 & 6 & -2 \end{bmatrix} \xrightarrow{R_3 + 2R_2 \to R_3} \begin{bmatrix} 1 & 1 & -10 & -4 \\ 0 & 1 & -3 & 1 \\ 0 & 0 & 0 & 0 \end{bmatrix}$$

When you study calculus or linear algebra, you will learn that the graph of a linear equation in three variables is a *plane* in a three-dimensional coordinate system. For a system of three equations in three variables, the following situations arise:

The three planes intersect in a single point.
The system has a unique solution.

The three planes intersect in more than one point.
The system has infinitely many solutions.

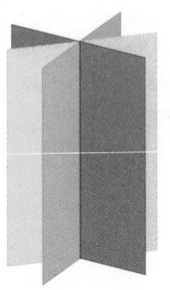

The three planes have no point in common.
The system has no solution.

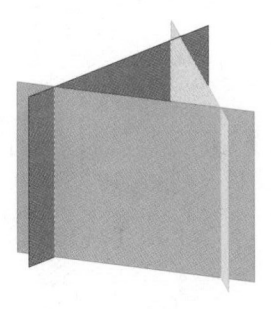

The system is now in echelon form, so we stop using Gaussian elimination. Translating the last row back into an equation, we get

$$0x + 0y + 0z = 0$$

or

$$0 = 0$$

This equation is always true, no matter what values are used for x, y, and z. Since the equation adds no new information about the variables, we can drop it from the system, which we now write in the form

$$\begin{bmatrix} 1 & 1 & -10 & -4 \\ 0 & 1 & -3 & 1 \end{bmatrix}$$

This corresponds to the system

$$\begin{cases} x + y - 10z = -4 \\ y - 3z = 1 \end{cases}$$

Neither of these equations determines a value for z, but we can use them to express the leading variables x and y in terms of z. From the last equation we get

$$y = 3z + 1$$

Substituting this value for y into the first equation gives us

$$x + (3z + 1) - 10z = -4$$

$$x - 7z + 1 = -4$$

$$x = 7z - 5$$

Since no value is determined for z, we can get a solution to the system by letting z be any real number and then using the equations we have found to calculate x and y. The system has infinitely many solutions because z can be given any value. We write the complete solution as follows:

$$x = 7z - 5$$

$$y = 3z + 1$$

$$z = \text{any real number}$$

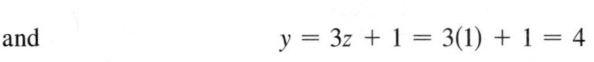

To get specific solutions, we give a specific value to z. For example, if $z = 1$, then

$$x = 7z - 5 = 7(1) - 5 = 2$$

and

$$y = 3z + 1 = 3(1) + 1 = 4$$

Thus $(2, 4, 1)$ is a solution to the system. We would get a different solution if we let $z = 2$, because then we have

$$x = 7z - 5 = 7(2) - 5 = 9$$

and

$$y = 3z + 1 = 3(2) + 1 = 7$$

So $(9, 7, 2)$ is also a solution.

A system that has infinitely many solutions is called **dependent.** In the complete solution to such a system, the variables that are *not* leading variables will be arbitrary, and the leading variables will *depend* on the arbitrary one(s). In Example 2, the leading variables x and y depend on (that is, are expressed in terms of) the nonleading variable z. If we use Gaussian elimination to convert a dependent system to echelon form and then discard any equations of the form $0 = 0$, we arrive at a system that has fewer equations than variables. Example 2 ended up with only two equations in the three variables x, y, and z. In general, if we arrive at n equations in m variables ($m > n$) after this process, the complete solution will have $m - n$ arbitrary (nonleading) variables, and the leading variables will be expressed in terms of these.

The following box summarizes what we have learned about systems with no solution, one solution, or infinitely many solutions.

SOLVING A SYSTEM IN ECHELON FORM

Suppose the matrix that represents a system of linear equations has been transformed by Gaussian elimination into echelon form.

1. If the echelon form contains a row that represents the equation $0 = c$, where c is nonzero, then the system has *no solution.*

2. If each of the variables in the echelon form of the system is a leading variable, then the system has *exactly one solution,* which we find using back substitution.

3. If not all of the variables in the echelon form are leading variables, then the system has *infinitely many solutions.* To solve the system, we use back substitution to express the leading variables in terms of the arbitrary variables.

Sometimes it is convenient to continue using Gaussian elimination on a matrix in echelon form to change each of the numbers *above* each leading entry to 0. The matrix is then said to be in *reduced echelon form.*

> ## REDUCED ECHELON FORM OF A MATRIX
>
> A matrix is in **reduced echelon form** if it is in echelon form and every number above and below each leading entry is 0.

The matrix on the left below is in reduced echelon form, but the one on the right is not, because the leading entries in the second and third rows do not have only 0's above them.

$$\begin{bmatrix} 1 & 3 & 0 & 0 & 0 \\ 0 & 0 & 1 & 0 & -3 \\ 0 & 0 & 0 & 1 & \frac{1}{2} \\ 0 & 0 & 0 & 0 & 0 \end{bmatrix} \qquad \begin{bmatrix} 1 & 3 & -\frac{1}{2} & 2 & 0 \\ 0 & 0 & 1 & 0 & -3 \\ 0 & 0 & 0 & 1 & \frac{1}{2} \\ 0 & 0 & 0 & 0 & 0 \end{bmatrix}$$

In the next example, we use the reduced echelon form to solve a system.

EXAMPLE 3 ■ **Using Reduced Echelon Form to Solve a Dependent System**

Find the complete solution of the following system.

$$\begin{cases} x + 2y - 3z - 4w = 10 \\ x + 3y - 3z - 4w = 15 \\ 2x + 2y - 6z - 8w = 10 \end{cases}$$

SOLUTION

$$\begin{bmatrix} 1 & 2 & -3 & -4 & 10 \\ 1 & 3 & -3 & -4 & 15 \\ 2 & 2 & -6 & -8 & 10 \end{bmatrix} \xrightarrow[\;R_3 - 2R_1 \to R_3\;]{R_2 - R_1 \to R_2} \begin{bmatrix} 1 & 2 & -3 & -4 & 10 \\ 0 & 1 & 0 & 0 & 5 \\ 0 & -2 & 0 & 0 & -10 \end{bmatrix}$$

$$\xrightarrow{R_3 + 2R_2 \to R_3} \begin{bmatrix} 1 & 2 & -3 & -4 & 10 \\ 0 & 1 & 0 & 0 & 5 \\ 0 & 0 & 0 & 0 & 0 \end{bmatrix} \xrightarrow{R_1 - 2R_2 \to R_1} \begin{bmatrix} 1 & 0 & -3 & -4 & 0 \\ 0 & 1 & 0 & 0 & 5 \\ 0 & 0 & 0 & 0 & 0 \end{bmatrix}$$

This is in reduced echelon form. Since the last row represents the equation $0 = 0$, we may discard it, and we then have the system

$$\begin{bmatrix} 1 & 0 & -3 & -4 & 0 \\ 0 & 1 & 0 & 0 & 5 \end{bmatrix}$$

At this stage we have two equations in four unknowns, so the system is dependent. The leading variables are x and y, and the arbitrary variables are z

and w. From the second equation we get $y = 5$, and from the first,

$$x - 3z - 4w = 0$$

$$x = 3z + 4w$$

The complete solution is therefore

$$x = 3z + 4w$$

$$y = 5$$

$$z = \text{any real number}$$

$$w = \text{any real number} \qquad \blacksquare$$

Note that we did not have to use back substitution in Example 3. If a system is in reduced form, the solution can be determined directly, without back substitution.

 Note also that z and w do *not* necessarily have to be the *same* real number in the solution for Example 3. We can choose arbitrary values for each if we wish to construct a specific solution to the system. For example, if we let $z = 1$ and we let $w = 2$, we get the solution $(11, 5, 1, 2)$. You should check that this does indeed satisfy all three of the original equations in Example 3.

Many of the linear systems that arise in practical problems are inconsistent or dependent. Both situations occur in the next two examples.

EXAMPLE 4 ■ An Application of an Inconsistent System

A biologist is performing an experiment on the effects of various combinations of vitamins. She wishes to feed each of her laboratory rabbits a diet that contains exactly 9 mg of niacin, 14 mg of thiamin, and 32 mg of riboflavin. She has available three different types of commercial rabbit pellets. The vitamin content per ounce is given in the following table for each type.

	Type A	Type B	Type C
Niacin (mg)	2	3	1
Thiamin (mg)	3	1	3
Riboflavin (mg)	8	5	7

How many ounces of each type of food should each rabbit be given daily to satisfy the experiment requirements?

SOLUTION

If we let x represent the amount of type A to be fed to each rabbit, y the amount of type B, and z the amount of type C, then the daily requirements the biologist has established lead to the linear equations

$$\begin{cases} 2x + 3y + z = 9 & \text{Niacin requirement} \\ 3x + y + 3z = 14 & \text{Thiamin requirement} \\ 8x + 5y + 7z = 32 & \text{Riboflavin requirement} \end{cases}$$

We solve this system as follows.

$$\begin{bmatrix} 2 & 3 & 1 & 9 \\ 3 & 1 & 3 & 14 \\ 8 & 5 & 7 & 32 \end{bmatrix} \xrightarrow{R_2 - R_1 \to R_2} \begin{bmatrix} 2 & 3 & 1 & 9 \\ 1 & -2 & 2 & 5 \\ 8 & 5 & 7 & 32 \end{bmatrix}$$

$$\xrightarrow[R_3 - 8R_2 \to R_3]{R_1 - 2R_2 \to R_1} \begin{bmatrix} 0 & 7 & -3 & -1 \\ 1 & -2 & 2 & 5 \\ 0 & 21 & -9 & -8 \end{bmatrix} \xrightarrow[R_1 \longleftrightarrow R_2]{R_3 - 3R_1 \to R_3} \begin{bmatrix} 1 & -2 & 2 & 5 \\ 0 & 7 & -3 & -1 \\ 0 & 0 & 0 & -5 \end{bmatrix}$$

Since the last row translates into the equation $0 = -5$, which is false, we need go no further. The system has no solution, so no combination of the three food types will satisfy the vitamin requirements.   ■

EXAMPLE 5 ■ An Application of a Dependent System

Suppose that the biologist in Example 4 had specified 37 mg instead of 32 mg as the riboflavin requirement, but all the other aspects of the experiment remained unchanged. Would there now be a combination of the three foods that would satisfy the vitamin requirements?

SOLUTION

The only change we need to make in the solution to Example 4 is to replace the 32 in the original system of equations by 37. If we then carry out the same row operations as before, we arrive at the matrix

$$\begin{bmatrix} 1 & -2 & 2 & 5 \\ 0 & 7 & -3 & -1 \\ 0 & 0 & 0 & 0 \end{bmatrix}$$

We now continue using Gaussian elimination to put this matrix into reduced echelon form.

$$\xrightarrow{\frac{1}{7}R_2} \begin{bmatrix} 1 & -2 & 2 & 5 \\ 0 & 1 & -\frac{3}{7} & -\frac{1}{7} \\ 0 & 0 & 0 & 0 \end{bmatrix} \xrightarrow{R_1 + 2R_2 \to R_1} \begin{bmatrix} 1 & 0 & \frac{8}{7} & \frac{33}{7} \\ 0 & 1 & -\frac{3}{7} & -\frac{1}{7} \\ 0 & 0 & 0 & 0 \end{bmatrix}$$

The last equation now simply states that $0 = 0$ and so can be eliminated. The system has infinitely many solutions, with the leading variables x and y depending on the arbitrary variable z. The solution is

$$\begin{cases} x = -\frac{8}{7}z + \frac{33}{7} \\ y = \frac{3}{7}z - \frac{1}{7} \\ z = \text{any real number} \end{cases}$$

Because an amount of food cannot be negative, not every solution to the system provides a practical solution to the problem.

Since $y \geq 0$,   Since $x \geq 0$,

$y = \frac{3}{7}z - \frac{1}{7} \geq 0$ $x = -\frac{8}{7}z + \frac{33}{7} \geq 0$

$\frac{3}{7}z \geq \frac{1}{7}$ $-\frac{8}{7}z \geq -\frac{33}{7}$

$z \geq \frac{1}{3}$ $z \leq \frac{33}{8}$

This means that the solution to the problem would have to include the condition that the amount z of type C rabbit food used should be between $\frac{1}{3}$ oz and $\frac{33}{8}$ oz. ∎

6.3 EXERCISES

1–6 ■ Determine whether the matrix is in echelon form. If it is, determine whether it is in reduced echelon form.

1. $\begin{bmatrix} 1 & 0 & -3 \\ 0 & 1 & 5 \end{bmatrix}$

2. $\begin{bmatrix} 1 & 3 & -3 \\ 0 & 1 & 5 \end{bmatrix}$

3. $\begin{bmatrix} 2 & 0 & 8 & 0 \\ 0 & 1 & 3 & 2 \\ 0 & 0 & 0 & 0 \end{bmatrix}$

4. $\begin{bmatrix} 1 & 0 & -7 & 0 \\ 0 & 1 & 3 & 0 \\ 0 & 0 & 0 & 1 \end{bmatrix}$

5. $\begin{bmatrix} 1 & 0 & 0 & 0 \\ 0 & 0 & 0 & 0 \\ 0 & 1 & 5 & 1 \end{bmatrix}$

6. $\begin{bmatrix} 1 & 0 & 0 & 1 \\ 0 & 1 & 0 & 2 \\ 0 & 0 & 1 & 3 \end{bmatrix}$

7–28 ■ Find the complete solution to each system of equations, or show that none exists.

7. $\begin{cases} x + y + z = 2 \\ y - 3z = 1 \\ 2x + y + 5z = 0 \end{cases}$

8. $\begin{cases} x \quad\quad + 3z = 3 \\ 2x + y - 2z = 5 \\ -y + 8z = 1 \end{cases}$

9. $\begin{cases} x - y + 3z = 3 \\ 4x - 8y + 32z = 24 \\ 2x - 3y + 11z = 4 \end{cases}$

10. $\begin{cases} -2x + 6y - 2z = -12 \\ x - 3y + 2z = 10 \\ -x + 3y + 2z = 6 \end{cases}$

11. $\begin{cases} x + 5y = 12 \\ 3x - 7y = 14 \\ 2x - 4y = 10 \end{cases}$

12. $\begin{cases} 12x - 7y = 11 \\ -3x - 14y = 12 \\ 15x + 8y = 13 \end{cases}$

13. $\begin{cases} x - 3y = 1 \\ 3x - y = 5 \\ 4x - 8y = 3 \end{cases}$

14. $\begin{cases} x + 2y + 3z = 7 \\ 3x + 2y + z = 21 \end{cases}$

15. $\begin{cases} x - y - z = 0 \\ 4x - 3y + 8z = 12 \end{cases}$

16. $\begin{cases} 3x - 6y - 12z = 0 \\ -4x + 8y + 16z = 0 \end{cases}$

17. $\begin{cases} 2x - y + 5z = 12 \\ x + 4y - 2z = -3 \\ 8x + 5y + 11z = 30 \end{cases}$

18. $\begin{cases} 3r + 2s - 3t = 10 \\ r - s - t = -5 \\ r + 4s - t = 20 \end{cases}$

19. $\begin{cases} 2x + y - 2z = 12 \\ -x - \frac{1}{2}y + z = -6 \\ 3x + \frac{3}{2}y - 3z = 18 \end{cases}$

20. $\begin{cases} y - 5z = 7 \\ 3x + 2y = 12 \\ 3x + 10z = 80 \end{cases}$

21. $\begin{cases} x + y + z + w = 8 \\ y - w = 0 \\ 3x + 2y + z = 12 \\ -3x - 2y + z + 4w = 0 \end{cases}$

22. $\begin{cases} y - z + 2w = 0 \\ 3x + 2y \phantom{{}-z} + w = 0 \\ 2x \phantom{{}+2y-z} + 4w = 12 \\ -2x \phantom{{}+2y} - 2z + 5w = 6 \end{cases}$

23. $\begin{cases} 2x - y + 2z + w = 5 \\ -x + y + 4z - w = 3 \\ 3x - 2y - z \phantom{{}+w} = 0 \end{cases}$

24. $\begin{cases} 3t - u + v + 2w = 5 \\ t + u - v - w = 7 \\ 4t - 4u + 4v + 6w = 3 \end{cases}$

25. $\begin{cases} x - y \phantom{{}-z} + w = 0 \\ 3x \phantom{{}-y} - z + 2w = 0 \\ x - 4y + z + 2w = 0 \end{cases}$

26. $\begin{cases} 3x_1 - 2x_2 + 4x_3 = -2 \\ x_1 - 2x_2 + x_3 = 0 \\ 4x_1 - 4x_2 + 5x_3 = -2 \\ - 4x_2 - x_3 = 2 \end{cases}$

27. $\begin{cases} 2x - y + z = 5 \\ 3x - 4y - 2z = 1 \\ x - 2y + 4z = 9 \\ 2x - 3y + 5z = 0 \end{cases}$

28. $\begin{cases} a + b + c + d + e = 2 \\ a \phantom{{}+b} - c \phantom{{}+d} + e = 2 \\ -2a + b \phantom{{}+c} - d \phantom{{}+e} = 0 \\ 2b \phantom{{}+c+d} + 2e = 4 \end{cases}$

29. A nutritionist wishes to make a milk substitute by combining soya powder, ground millet, and nonfat dried milk powder with enough water to make one quart. She wants the mixture to contain 1.1 mg of thiamin, 3.1 mg of riboflavin, and 3.5 mg of niacin. The vitamin content per ounce is given in the following table for each ingredient.

	Soya powder	Ground millet	Dried milk
Thiamin (mg)	0.2	0.5	0.4
Riboflavin (mg)	0.2	2.0	1.4
Niacin (mg)	1.0	1.0	1.0

How many ounces of each ingredient should she combine to satisfy the vitamin requirements? (Give all possible combinations.)

30. If the nutritionist of Exercise 29 decides she wants the milk substitute to have 1.2 mg of thiamin instead of 1.1 mg (without changing the other requirements), what combination of the three ingredients could she use?

31. A furniture factory makes wooden tables, chairs, and armoires. Each piece of furniture requires three production steps: cutting the wood, assembling, and finishing. The number of hours (h) of each operation required to make a piece of furniture is given in the following table.

	Table	Chair	Armoire
Cutting (h)	$\frac{1}{2}$	1	1
Assembling (h)	$\frac{1}{2}$	$1\frac{1}{2}$	1
Finishing (h)	1	$1\frac{1}{2}$	2

The workers in the factory can provide 300 h of cutting, 400 h of assembling, and 590 h of finishing each work week. How many tables, chairs, and armoires should be produced so that all available labor-hours are used? Or is this impossible?

32. Rework Exercise 31 assuming that one worker has been laid off, so that only 550 h of finishing labor are available each week.

33. I have some pennies, nickels, and dimes in my pocket. The total value of the coins is 72 cents, and the number of dimes is one-third the total number of nickels and pennies. How many coins of each denomination do I have? [*Hint:* The number of each type of coin must be a nonnegative integer.]

34. A diagram of a section of the street network in a city is shown in the figure, where the arrows indicate one-way streets. The numbers on the diagram show how many cars enter or leave this section of the city via the indicated street in a certain one-hour period. The variables x, y, z, and w represent the number of cars that travel along the portions of First, Second, Avocado, and Birch Streets shown in the figure during this period. Find x, y, z, and w, assuming that none of the cars involved in this problem stop or park on any of the streets shown in the diagram.

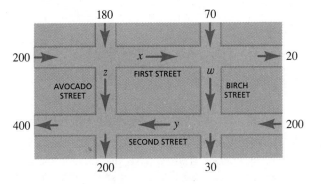

35. (a) Suppose that (x_0, y_0, z_0) and (x_1, y_1, z_1) are solutions of the system

$$\begin{cases} a_1x + b_1y + c_1z = d_1 \\ a_2x + b_2y + c_2z = d_2 \\ a_3x + b_3y + c_3z = d_3 \end{cases}$$

Show that $\left(\dfrac{x_0 + x_1}{2}, \dfrac{y_0 + y_1}{2}, \dfrac{z_0 + z_1}{2} \right)$ is also a solution.

(b) Use the result of part (a) to prove that if the system has two different solutions, then it has infinitely many solutions.

6.4 THE ALGEBRA OF MATRICES

Up to this point we have been using matrices simply as a notational convenience, to make our work in solving linear equations a little easier. Matrices have many other uses in mathematics and the sciences, and for most of these applications a knowledge of matrix algebra is essential. Like numbers, matrices can be added, subtracted, multiplied, and divided under certain circumstances, and in this section we learn how to perform these algebraic operations on matrices.

Recall that a matrix is simply a rectangular array of numbers enclosed between brackets. For example, let A be the matrix

$$A = \begin{bmatrix} -1 & 4 & 7 & 0 \\ 0 & 2 & 13 & 14 \\ \frac{1}{2} & 22 & 8 & -2 \end{bmatrix}$$

The **dimension** of a matrix is a pair of numbers that indicates how many rows and columns a matrix has. The matrix A is a 3×4 matrix because it has 3 horizontal rows and 4 vertical columns. The individual numbers that make up a matrix are called its **entries,** and they are specified by their row and column position. In the matrix A, the number 13 is the $(2, 3)$ entry, since it is in the second row and the third column. If the name of a matrix is A, we will often use the symbol a_{ij} to denote the (i, j) entry of the matrix. Thus, for the preceding matrix, we have $a_{24} = 14$ and $a_{32} = 22$.

Two matrices are **equal** if they have the same dimension and their corresponding entries are equal. So

> $A = B$ if and only if both A and B have dimension $m \times n$ and $a_{ij} = b_{ij}$ for $i = 1, 2, \ldots, m$ and $j = 1, 2, \ldots, n$.

For example,

$$\begin{bmatrix} \sqrt{4} & 2^2 & 0 & e^0 \\ 0.5 & 1 & 0 & 1-1 \end{bmatrix} = \begin{bmatrix} 2 & 4 & 0 & 1 \\ \frac{1}{2} & \frac{2}{2} & 0 & 0 \end{bmatrix}$$

but

$$\begin{bmatrix} 1 & 2 \\ 3 & 4 \\ 5 & 6 \end{bmatrix} \neq \begin{bmatrix} 1 & 3 & 5 \\ 2 & 4 & 6 \end{bmatrix}$$

ALGEBRAIC OPERATIONS ON MATRICES

Two matrices can be added or subtracted if they have the same dimension. (Otherwise, their sum or difference is undefined.) We add or subtract the matrices by adding or subtracting corresponding entries. To multiply a matrix by a number, we multiply every element of the matrix by that number. This is called the *scalar product.*

SUM, DIFFERENCE, AND SCALAR PRODUCT OF MATRICES

If A and B are matrices of the same dimension and if k is any real number, then

1. The **sum** $A + B$ is the matrix of the same dimension as A and B, and its (i, j) entry is $a_{ij} + b_{ij}$.

2. The **difference** $A - B$ is the matrix of the same dimension as A and B, and its (i, j) entry is $a_{ij} - b_{ij}$.

3. The **scalar product** kA is the matrix of the same dimension as A, and its (i, j) entry is ka_{ij}.

EXAMPLE 1 ■ Performing Algebraic Operations on Matrices

Let

$$A = \begin{bmatrix} 2 & -3 \\ 0 & 5 \\ 7 & -\frac{1}{2} \end{bmatrix} \qquad B = \begin{bmatrix} 1 & 0 \\ -3 & 1 \\ 2 & 2 \end{bmatrix}$$

$$C = \begin{bmatrix} 7 & -3 & 0 \\ 0 & 1 & 5 \end{bmatrix} \qquad D = \begin{bmatrix} 6 & 0 & -6 \\ 8 & 1 & 9 \end{bmatrix}$$

Carry out each indicated operation, or explain why it cannot be performed.

(a) $A + B$ (b) $C - D$ (c) $C + A$ (d) $5A$

SOLUTION

(a) $A + B = \begin{bmatrix} 2 & -3 \\ 0 & 5 \\ 7 & -\frac{1}{2} \end{bmatrix} + \begin{bmatrix} 1 & 0 \\ -3 & 1 \\ 2 & 2 \end{bmatrix} = \begin{bmatrix} 3 & -3 \\ -3 & 6 \\ 9 & \frac{3}{2} \end{bmatrix}$

(b) $C - D = \begin{bmatrix} 7 & -3 & 0 \\ 0 & 1 & 5 \end{bmatrix} - \begin{bmatrix} 6 & 0 & -6 \\ 8 & 1 & 9 \end{bmatrix} = \begin{bmatrix} 1 & -3 & 6 \\ -8 & 0 & -4 \end{bmatrix}$

(c) $C + A$ is undefined because we can't add matrices of different dimensions.

(d) $5A = 5 \begin{bmatrix} 2 & -3 \\ 0 & 5 \\ 7 & -\frac{1}{2} \end{bmatrix} = \begin{bmatrix} 10 & -15 \\ 0 & 25 \\ 35 & -\frac{5}{2} \end{bmatrix}$ ∎

MATRIX MULTIPLICATION

Multiplication of two matrices is more difficult to describe. We will see in later examples why taking the matrix product involves the following rather complex procedure.

First, the product AB (or $A \cdot B$) of two matrices A and B is defined only when the number of columns in A is equal to the number of rows in B. This means that if we write their dimensions side by side, the two inner numbers must match:

matrices	A	B
dimensions	$m \times n$	$n \times k$
	↑	↑
	columns in A	rows in B

If the dimensions of A and B match in this fashion, then the product AB will have dimension $m \times k$. Before describing the procedure for obtaining the elements of AB, we define the *inner product* of a row of A and a column of B.

If $\begin{bmatrix} a_1 & a_2 & \cdots & a_n \end{bmatrix}$ is a row of A, and if $\begin{bmatrix} b_1 \\ b_2 \\ \cdot \\ \cdot \\ \cdot \\ b_n \end{bmatrix}$ is a column of B, then their **inner product** is the number $a_1b_1 + a_2b_2 + \cdots + a_nb_n$.

For example,

$$[2 \quad -1 \quad 0 \quad 4] \cdot \begin{bmatrix} 5 \\ 4 \\ -3 \\ \frac{1}{2} \end{bmatrix} = 2 \cdot 5 + (-1) \cdot 4 + 0 \cdot (-3) + 4 \cdot \frac{1}{2} = 8$$

We now define the **product** AB of two matrices.

THE PRODUCT OF TWO MATRICES

Suppose that A is an $m \times n$ matrix and B an $n \times k$ matrix. Then $C = AB$ is an $m \times k$ matrix, where c_{ij} is the inner product of the ith row of A and the jth column of B.

$$\begin{array}{cc} A & \cdot & B \\ \downarrow & & \downarrow \\ 2 \times 2 & & 2 \times 3 \end{array}$$

Inner numbers match, so product is defined.
Outer numbers give dimension of product: 2×3.

EXAMPLE 2 ■ **Multiplying Matrices**

Let

$$A = \begin{bmatrix} 1 & 3 \\ -1 & 0 \end{bmatrix} \quad \text{and} \quad B = \begin{bmatrix} -1 & 5 & 2 \\ 0 & 4 & 7 \end{bmatrix}$$

Calculate, if possible, the products AB and BA.

SOLUTION

Since A has dimension 2×2 and B has dimension 2×3, the product AB will have dimension 2×3. We can thus write

$$AB = \begin{bmatrix} 1 & 3 \\ -1 & 0 \end{bmatrix} \begin{bmatrix} -1 & 5 & 2 \\ 0 & 4 & 7 \end{bmatrix} = \begin{bmatrix} ? & ? & ? \\ ? & ? & ? \end{bmatrix}$$

where the question marks must be filled in using the rule defining the entries of a matrix product. The $(1, 1)$ entry will be the inner product of the first row of A and the first column of B:

$$\begin{bmatrix} 1 & 3 \\ -1 & 0 \end{bmatrix} \begin{bmatrix} -1 & 5 & 2 \\ 0 & 4 & 7 \end{bmatrix} \qquad 1 \cdot (-1) + 3 \cdot 0 = -1$$

Similarly, we calculate the remaining entries as follows.

Entry	Inner Product of:	Value	Product matrix
$(1, 2)$	$\begin{bmatrix} 1 & 3 \\ -1 & 0 \end{bmatrix} \begin{bmatrix} -1 & 5 & 2 \\ 0 & 4 & 7 \end{bmatrix}$	$1 \cdot 5 + 3 \cdot 4 = 17$	$\begin{bmatrix} -1 & 17 & \\ & & \end{bmatrix}$
$(1, 3)$	$\begin{bmatrix} 1 & 3 \\ -1 & 0 \end{bmatrix} \begin{bmatrix} -1 & 5 & 2 \\ 0 & 4 & 7 \end{bmatrix}$	$1 \cdot 2 + 3 \cdot 7 = 23$	$\begin{bmatrix} -1 & 17 & 23 \\ & & \end{bmatrix}$
$(2, 1)$	$\begin{bmatrix} 1 & 3 \\ -1 & 0 \end{bmatrix} \begin{bmatrix} -1 & 5 & 2 \\ 0 & 4 & 7 \end{bmatrix}$	$(-1) \cdot (-1) + 0 \cdot 0 = 1$	$\begin{bmatrix} -1 & 17 & 23 \\ 1 & & \end{bmatrix}$
$(2, 2)$	$\begin{bmatrix} 1 & 3 \\ -1 & 0 \end{bmatrix} \begin{bmatrix} -1 & 5 & 2 \\ 0 & 4 & 7 \end{bmatrix}$	$(-1) \cdot 5 + 0 \cdot 4 = -5$	$\begin{bmatrix} -1 & 17 & 23 \\ 1 & -5 & \end{bmatrix}$
$(2, 3)$	$\begin{bmatrix} 1 & 3 \\ -1 & 0 \end{bmatrix} \begin{bmatrix} -1 & 5 & 2 \\ 0 & 4 & 7 \end{bmatrix}$	$(-1) \cdot 2 + 0 \cdot 7 = -2$	$\begin{bmatrix} -1 & 17 & 23 \\ 1 & -5 & -2 \end{bmatrix}$

Thus, we have

$$AB = \begin{bmatrix} -1 & 17 & 23 \\ 1 & -5 & -2 \end{bmatrix}$$

The product BA is not defined, however, because the dimensions are

$$2 \times 3 \quad \text{and} \quad 2 \times 2$$

The inner two numbers are not the same, so the rows and columns will not match up when we try to calculate the product. ■

The next example shows that even when both AB and BA are defined, they are not necessarily equal. This result will prove that matrix multiplication is *not* commutative.

EXAMPLE 3 ■ **Matrix Multiplication Is Not Commutative**

Let

$$A = \begin{bmatrix} 5 & 7 \\ -3 & 0 \end{bmatrix} \quad \text{and} \quad B = \begin{bmatrix} 1 & 2 \\ 9 & -1 \end{bmatrix}$$

Calculate the products AB and BA.

SOLUTION

Since both matrices A and B have dimension 2×2, both products AB and BA are defined, and each product is also a 2×2 matrix.

$$AB = \begin{bmatrix} 5 & 7 \\ -3 & 0 \end{bmatrix} \begin{bmatrix} 1 & 2 \\ 9 & -1 \end{bmatrix} = \begin{bmatrix} 5 \cdot 1 + 7 \cdot 9 & 5 \cdot 2 + 7 \cdot (-1) \\ (-3) \cdot 1 + 0 \cdot 9 & (-3) \cdot 2 + 0 \cdot (-1) \end{bmatrix}$$

$$= \begin{bmatrix} 68 & 3 \\ -3 & -6 \end{bmatrix}$$

$$BA = \begin{bmatrix} 1 & 2 \\ 9 & -1 \end{bmatrix} \begin{bmatrix} 5 & 7 \\ -3 & 0 \end{bmatrix} = \begin{bmatrix} 1 \cdot 5 + 2 \cdot (-3) & 1 \cdot 7 + 2 \cdot 0 \\ 9 \cdot 5 + (-1) \cdot (-3) & 9 \cdot 7 + (-1) \cdot 0 \end{bmatrix}$$

$$= \begin{bmatrix} -1 & 7 \\ 48 & 63 \end{bmatrix}$$

This shows that, in general, $AB \neq BA$. In fact, in this example, AB and BA do not even have any entry in common. ∎

Although matrix multiplication is not commutative, it does obey the Associative and Distributive Properties.

PROPERTIES OF MATRIX MULTIPLICATION

Let A, B, C, and D be matrices for which the following products are defined. Then

$$A(BC) = (AB)C \qquad \text{Associative Property}$$

$$A(B + C) = AB + AC$$
$$(B + C)D = BD + CD \qquad \text{Distributive Property}$$

The next two examples give some indication of why mathematicians chose to define the matrix product in such an apparently bizarre fashion.

EXAMPLE 4 ■ **Writing a System of Equations as a Matrix Equation**

Show that the matrix equation

$$\begin{bmatrix} 1 & 2 & 4 \\ -1 & 1 & 2 \\ 2 & 3 & 3 \end{bmatrix} \begin{bmatrix} x \\ y \\ z \end{bmatrix} = \begin{bmatrix} 7 \\ 5 \\ 7 \end{bmatrix}$$

is equivalent to the system of equations in Example 1 of Section 6.2.

SOLUTION

If we perform the matrix multiplication on the left side of the given equation, we get

$$\begin{bmatrix} x + 2y + 4z \\ -x + y + 2z \\ 2x + 3y + 3z \end{bmatrix} = \begin{bmatrix} 7 \\ 5 \\ 7 \end{bmatrix}$$

Because two matrices are equal if their corresponding entries are equal, this matrix equation means that

$$\begin{cases} x + 2y + 4z = 7 \\ -x + y + 2z = 5 \\ 2x + 3y + 3z = 7 \end{cases}$$

This is exactly the system of equations we had in Example 1 of Section 6.2. ■

The preceding example shows that our definition of matrix product allows us to express a system of linear equations as a single matrix equation in a natural way.

EXAMPLE 5 ■ Representing Demographic Data in Terms of Matrices

In a certain city the proportion of voters in each age group who are registered as Democrats, Republicans, or Independents is given by the following matrix.

	Age		
	18–30	31–50	Over 50
Democrat	0.30	0.60	0.50
Republican	0.50	0.35	0.25
Independent	0.20	0.05	0.25

$= A$

The next matrix gives the distribution, by age and sex, of the voting population of this city.

		Male	Female
Age	18–30	5,000	6,000
	31–50	10,000	12,000
	Over 50	12,000	15,000

$= B$

For the purpose of this problem, let us make the (highly unrealistic) assumption that within each age group, political preference is not related to gender.

Olga Taussky-Todd (b. 1906) is one of the world's leaders in developing applications of Matrix Theory. She has been described as "in love with anything matrices can do." She has successfully applied matrices to the study of aerodynamics, a field used in the design of airplanes and rockets. Taussky-Todd is also famous for her work in Number Theory, a subject which deals with prime numbers and divisibility. Although Number Theory has often been called the least applicable branch of mathematics, it is now used in significant ways in the computer industry.

Taussky-Todd studied mathematics at a time when it was very uncommon for a young woman to want to be a mathematician. She says "When I entered university I had no idea what it meant to study mathematics." But she became one of the most respected mathematicians of her time. She was for many years a professor of mathematics at the California Institute of Technology in Pasadena.

That is, the percentage of Democrat males in the 18–30 group, for example, is the same as the percentage of Democrat females in this group.

(a) Calculate the product AB.

(b) How many males are registered as Democrats in this city?

(c) How many females are registered as Republicans?

SOLUTION

(a) $$AB = \begin{bmatrix} 0.30 & 0.60 & 0.50 \\ 0.50 & 0.35 & 0.25 \\ 0.20 & 0.05 & 0.25 \end{bmatrix} \begin{bmatrix} 5{,}000 & 6{,}000 \\ 10{,}000 & 12{,}000 \\ 12{,}000 & 15{,}000 \end{bmatrix} = \begin{bmatrix} 13{,}500 & 16{,}500 \\ 9{,}000 & 10{,}950 \\ 4{,}500 & 5{,}550 \end{bmatrix}$$

(b) When we take the inner product of a row from A with a column from B, we are adding the number of people in each of the three age groups who belong to the category in question. For example, the $(2, 1)$ entry of AB (the 9,000) was obtained by taking the inner product of the Republican row from A with the Male column from B. This number is therefore the total number of male Republicans in this city. We can label the rows and columns of AB as follows.

$$\begin{array}{c} & \text{Male} \quad \text{Female} \\ \begin{matrix} \text{Democrat} \\ \text{Republican} \\ \text{Independent} \end{matrix} & \begin{bmatrix} 13{,}500 & 16{,}500 \\ 9{,}000 & 10{,}950 \\ 4{,}500 & 5{,}550 \end{bmatrix} = AB \end{array}$$

There are 13,500 males registered as Democrats in this city.

(c) There are 10,950 females registered as Republicans. ∎

If we add the entries in the columns of matrix A in Example 5, we see that in each case the sum is 1. (Can you see why this has to be true, given what the matrix describes?) A matrix with this property is called **stochastic.** Stochastic matrices are studied extensively in statistics, where they arise frequently in situations like the one described in Example 5.

6.4 EXERCISES

1–21 ■ The matrices A, B, C, D, E, F, and G are defined as follows.

$$A = \begin{bmatrix} 2 & -5 \\ 0 & 7 \end{bmatrix} \qquad B = \begin{bmatrix} 3 & \frac{1}{2} & 5 \\ 1 & -1 & 3 \end{bmatrix} \qquad C = \begin{bmatrix} 2 & -\frac{5}{2} & 0 \\ 0 & 2 & -3 \end{bmatrix} \qquad D = \begin{bmatrix} 7 & 3 \end{bmatrix}$$

$$E = \begin{bmatrix} 1 \\ 2 \\ 0 \end{bmatrix} \qquad F = \begin{bmatrix} 1 & 0 & 0 \\ 0 & 1 & 0 \\ 0 & 0 & 1 \end{bmatrix} \qquad G = \begin{bmatrix} 5 & -3 & 10 \\ 6 & 1 & 0 \\ -5 & 2 & 2 \end{bmatrix}$$

Carry out the indicated algebraic operation, or explain why it cannot be performed.

1. $B + C$ **2.** $B + F$ **3.** $C - B$

4. $5A$ **5.** $3B + 2C$ **6.** $C - 5A$

7. $2C - 6B$ **8.** DA **9.** AD

10. BC **11.** BF **12.** GF

13. $(DA)B$ **14.** $D(AB)$ **15.** GE

16. A^2 **17.** A^3 **18.** $DB + DC$

19. B^2 **20.** F^2 **21.** $BF + FE$

22. What must be true about the dimensions of the matrices A and B if both products AB and BA are defined?

23–26 ■ Write the system of equations as a matrix equation (see Example 4).

23. $\begin{cases} 2x - 5y = 7 \\ 3x + 2y = 4 \end{cases}$

24. $\begin{cases} 6x - y + z = 12 \\ 2x \quad\;\; + z = 7 \\ \quad\;\; y - 2z = 4 \end{cases}$

25. $\begin{cases} 3x_1 + 2x_2 - x_3 + x_4 = 0 \\ x_1 \quad\;\; - x_3 \quad\;\; = 5 \\ 3x_2 + x_3 - x_4 = 4 \end{cases}$

26. $\begin{cases} x - y + z = 2 \\ 4x - 2y - z = 2 \\ x + y + 5z = 2 \\ -x - y - z = 2 \end{cases}$

27–28 ■ Solve for x and y.

27. $\begin{bmatrix} 6 & x \\ 1 & 0 \end{bmatrix} \begin{bmatrix} y & 2 \\ -1 & 2 \end{bmatrix} = \begin{bmatrix} 4 & 16 \\ 1 & 2 \end{bmatrix}$

28. $\begin{bmatrix} 2 & 5 \\ 1 & 3 \end{bmatrix} \begin{bmatrix} x \\ y \end{bmatrix} = \begin{bmatrix} -2 \\ -3 \end{bmatrix}$

29–32 ■ Solve the matrix equation for the unknown matrix X, or explain why there is no solution. Here

$$A = \begin{bmatrix} 4 & 6 \\ 1 & 3 \end{bmatrix} \qquad B = \begin{bmatrix} 2 & 5 \\ 3 & 7 \end{bmatrix}$$

$$C = \begin{bmatrix} 2 & 3 \\ 1 & 0 \\ 0 & 2 \end{bmatrix} \qquad D = \begin{bmatrix} 10 & 20 \\ 30 & 20 \\ 10 & 0 \end{bmatrix}$$

29. $2X - A = B$ **30.** $5(X - C) = D$

31. $3X + B = C$ **32.** $A + D = 3X$

33. Let O represent the 2×2 **zero matrix**:

$$O = \begin{bmatrix} 0 & 0 \\ 0 & 0 \end{bmatrix}$$

If A and B are 2×2 matrices with $AB = O$, is it necessarily true that $A = O$ or $B = O$?

34. Let the matrix O be as in Exercise 33. Find a matrix $A \neq O$ such that $A^2 = O$.

35. Prove that if A and B are 2×2 matrices, then

$$(A + B)^2 = A^2 + AB + BA + B^2$$

36. If A and B are 2×2 matrices, is it necessarily true that

$$(A + B)^2 \overset{?}{=} A^2 + 2AB + B^2$$

37. Let

$$A = \begin{bmatrix} 1 & 1 \\ 0 & 1 \end{bmatrix}$$

(a) Calculate A^2, A^3, and A^4.
(b) Find a general formula for A^n.

38. Let

$$A = \begin{bmatrix} 1 & 1 \\ 1 & 1 \end{bmatrix}$$

(a) Calculate A^2, A^3, and A^4.
(b) Find a general formula for A^n.

39. A small fast-food chain has restaurants in Santa Monica, Long Beach, and Anaheim. Only hamburgers, hot dogs, and milk shakes are sold by this chain. On a certain day, sales were distributed according to the following matrix.

Number of items sold

	Santa Monica	Long Beach	Anaheim
Hamburgers	4000	1000	3500
Hot dogs	400	300	200
Milk shakes	700	500	9000

$= A$

The price of each item is given by the following matrix.

Hamburger	Hot dog	Milk shake
[$0.90	$0.80	$1.10]

$= B$

(a) Calculate the product BA.
(b) Interpret the entries in the product matrix BA.

40. A specialty car manufacturer has plants in Auburn, Biloxi, and Chattanooga. Three models are produced, with daily production given in the following matrix.

Cars produced each day

	Model K	Model R	Model W	
Auburn	12	10	0	
Biloxi	4	4	20	= A
Chattanooga	8	9	12	

Because of a wage increase, February profits are less than January profits. The profit per car is tabulated by model in the following matrix.

	January	February	
Model K	$1000	$500	
Model R	$2000	$1200	= B
Model W	$1500	$1000	

(a) Calculate AB.

(b) Assuming all cars produced were sold, what was the daily profit in January from the Biloxi plant?

(c) What was the total daily profit (from all three plants) in February?

41. Let

$$A = \begin{bmatrix} 1 & 0 & 6 & -1 \\ 2 & \frac{1}{2} & 4 & 0 \end{bmatrix} \qquad C = \begin{bmatrix} 1 \\ 0 \\ -1 \\ -2 \end{bmatrix}$$

$$B = \begin{bmatrix} 1 & 7 & -9 & 2 \end{bmatrix}$$

Determine which of the following products are defined, and calculate the ones that are:

$$ABC \qquad ACB \qquad BAC$$

$$BCA \qquad CAB \qquad CBA$$

6.5   INVERSES OF MATRICES AND MATRIX EQUATIONS

We have seen in the preceding section that matrices can, when the dimensions are appropriate, be added, subtracted, and multiplied. In this section we investigate division of matrices. With this operation we can solve equations that involve matrices.

First, we define *identity matrices,* which play the same role for matrix multiplication that the number 1 does for ordinary multiplication of numbers; that is, $1 \cdot a = a \cdot 1 = a$ for all numbers a. In the following definition the term **main diagonal** refers to the entries of a square matrix whose row and column numbers are the same. (Note that these entries stretch diagonally down the matrix, from top left to bottom right.)

> The **identity matrix** I_n is the $n \times n$ matrix for which each main diagonal entry is a 1 and for which all other entries are 0.

Thus, the 2×2, 3×3, and 4×4 identity matrices are, respectively,

$$I_2 = \begin{bmatrix} 1 & 0 \\ 0 & 1 \end{bmatrix} \qquad I_3 = \begin{bmatrix} 1 & 0 & 0 \\ 0 & 1 & 0 \\ 0 & 0 & 1 \end{bmatrix} \qquad I_4 = \begin{bmatrix} 1 & 0 & 0 & 0 \\ 0 & 1 & 0 & 0 \\ 0 & 0 & 1 & 0 \\ 0 & 0 & 0 & 1 \end{bmatrix}$$

Identity matrices behave like the number 1 in the sense that

$$A \cdot I_n = A \qquad \text{and} \qquad I_n \cdot B = B$$

whenever these products are defined. Thus, multiplication by an identity of the appropriate size leaves a matrix unchanged. For example, we can verify by direct calculation that

$$\begin{bmatrix} 1 & 0 \\ 0 & 1 \end{bmatrix} \begin{bmatrix} 3 & 5 & 6 \\ -1 & 2 & 7 \end{bmatrix} = \begin{bmatrix} 3 & 5 & 6 \\ -1 & 2 & 7 \end{bmatrix}$$

or that

$$\begin{bmatrix} -1 & 7 & \frac{1}{2} \\ 12 & 1 & 3 \\ -2 & 0 & 7 \end{bmatrix} \begin{bmatrix} 1 & 0 & 0 \\ 0 & 1 & 0 \\ 0 & 0 & 1 \end{bmatrix} = \begin{bmatrix} -1 & 7 & \frac{1}{2} \\ 12 & 1 & 3 \\ -2 & 0 & 7 \end{bmatrix}$$

If A and B are $n \times n$ matrices, and if $AB = BA = I_n$, then we say that B is the *inverse* of A, and we write $B = A^{-1}$. The concept of the inverse of a matrix is analogous to that of the reciprocal of a real number.

INVERSE OF A MATRIX

Let A be a square $n \times n$ matrix. If there exists an $n \times n$ matrix A^{-1} with the property that

$$AA^{-1} = A^{-1}A = I_n$$

then we say that A^{-1} is the **inverse** of A.

Not every square matrix has an inverse. The following rule provides a simple way for calculating the inverse of a 2×2 matrix, when it exists. For larger matrices, there is a more general procedure for finding inverses, which we consider later in this section.

INVERSE OF A 2 × 2 MATRIX

If $A = \begin{bmatrix} a & b \\ c & d \end{bmatrix}$ then $A^{-1} = \dfrac{1}{ad - bc} \begin{bmatrix} d & -b \\ -c & a \end{bmatrix}.$

EXAMPLE 1 ■ Finding the Inverse of a 2 × 2 Matrix

Let

$$A = \begin{bmatrix} 4 & 5 \\ 2 & 3 \end{bmatrix}$$

Find A^{-1} and verify that $AA^{-1} = A^{-1}A = I_2$.

SOLUTION

Using the rule, we get

$$A^{-1} = \frac{1}{4 \cdot 3 - 5 \cdot 2} \begin{bmatrix} 3 & -5 \\ -2 & 4 \end{bmatrix} = \frac{1}{2} \begin{bmatrix} 3 & -5 \\ -2 & 4 \end{bmatrix} = \begin{bmatrix} \frac{3}{2} & -\frac{5}{2} \\ -1 & 2 \end{bmatrix}$$

To verify that this is indeed the inverse of A, we calculate AA^{-1} and $A^{-1}A$:

$$AA^{-1} = \begin{bmatrix} 4 & 5 \\ 2 & 3 \end{bmatrix} \begin{bmatrix} \frac{3}{2} & -\frac{5}{2} \\ -1 & 2 \end{bmatrix} = \begin{bmatrix} 4 \cdot \frac{3}{2} + 5(-1) & 4(-\frac{5}{2}) + 5 \cdot 2 \\ 2 \cdot \frac{3}{2} + 3(-1) & 2(-\frac{5}{2}) + 3 \cdot 2 \end{bmatrix} = \begin{bmatrix} 1 & 0 \\ 0 & 1 \end{bmatrix}$$

$$A^{-1}A = \begin{bmatrix} \frac{3}{2} & -\frac{5}{2} \\ -1 & 2 \end{bmatrix} \begin{bmatrix} 4 & 5 \\ 2 & 3 \end{bmatrix} = \begin{bmatrix} \frac{3}{2} \cdot 4 + (-\frac{5}{2})2 & \frac{3}{2} \cdot 5 + (-\frac{5}{2})3 \\ (-1)4 + 2 \cdot 2 & (-1)5 + 2 \cdot 3 \end{bmatrix} = \begin{bmatrix} 1 & 0 \\ 0 & 1 \end{bmatrix}$$

■

The quantity $ad - bc$ that appears in the rule for calculating the inverse is called the **determinant** of the matrix. If the determinant is 0, then the matrix does not have an inverse (since we cannot divide by 0). In the next section we will learn how to calculate the determinant of a square matrix of any size, and how to use determinants to solve systems of equations.

INVERSES OF $n \times n$ MATRICES

For 3×3 and larger square matrices, the following technique provides the most efficient way to calculate the inverse. If A is an $n \times n$ matrix, we begin by constructing the $n \times 2n$ matrix that has the entries of A on the left and of the identity matrix I_n on the right:

$$\begin{bmatrix} a_{11} & a_{12} & \cdots & a_{1n} & 1 & 0 & \cdots & 0 \\ a_{21} & a_{22} & \cdots & a_{2n} & 0 & 1 & \cdots & 0 \\ \vdots & \vdots & & \vdots & \vdots & \vdots & & \vdots \\ a_{n1} & a_{n2} & \cdots & a_{nn} & 0 & 0 & \cdots & 1 \end{bmatrix}$$

We then use the elementary row operations on this new large matrix to change the left side into the identity matrix. The right side will be transformed automatically into A^{-1}. (We omit the proof of this fact.)

EXAMPLE 2 ■ Finding the Inverse of a 3 × 3 Matrix

Find the inverse of the matrix A, and verify that $AA^{-1} = A^{-1}A = I_3$.

$$A = \begin{bmatrix} 1 & -2 & -4 \\ 2 & -3 & -6 \\ -3 & 6 & 15 \end{bmatrix}$$

SOLUTION

We begin with the 3×6 matrix whose left half is A and whose right half is the identity matrix.

$$\begin{bmatrix} 1 & -2 & -4 & 1 & 0 & 0 \\ 2 & -3 & -6 & 0 & 1 & 0 \\ -3 & 6 & 15 & 0 & 0 & 1 \end{bmatrix}$$

We then transform the left half of this new matrix into the identity matrix by performing the following sequence of elementary row operations on the *entire* new matrix:

$$\xrightarrow[\begin{array}{c} R_2 - 2R_1 \to R_2 \\ R_3 + 3R_1 \to R_3 \end{array}]{} \begin{bmatrix} 1 & -2 & -4 & 1 & 0 & 0 \\ 0 & 1 & 2 & -2 & 1 & 0 \\ 0 & 0 & 3 & 3 & 0 & 1 \end{bmatrix}$$

$$\xrightarrow{\frac{1}{3}R_3} \begin{bmatrix} 1 & -2 & -4 & 1 & 0 & 0 \\ 0 & 1 & 2 & -2 & 1 & 0 \\ 0 & 0 & 1 & 1 & 0 & \frac{1}{3} \end{bmatrix}$$

$$\xrightarrow{R_1 + 2R_2 \to R_1} \begin{bmatrix} 1 & 0 & 0 & -3 & 2 & 0 \\ 0 & 1 & 2 & -2 & 1 & 0 \\ 0 & 0 & 1 & 1 & 0 & \frac{1}{3} \end{bmatrix}$$

$$\xrightarrow{R_2 - 2R_3 \to R_2} \begin{bmatrix} 1 & 0 & 0 & -3 & 2 & 0 \\ 0 & 1 & 0 & -4 & 1 & -\frac{2}{3} \\ 0 & 0 & 1 & 1 & 0 & \frac{1}{3} \end{bmatrix}$$

We have now transformed the left half of this matrix into the identity matrix. (This means we have put the entire matrix into reduced echelon form.) Note that to do this in as systematic a fashion as possible, we first changed the elements below the main diagonal to zeros, just as we would if we were using

Arthur Cayley (1821–1895) was an English mathematician who invented matrices and developed Matrix Theory. He practiced law until the age of 42, but his primary interest from adolescence was mathematics, and he published almost 200 papers on the subject in his spare time. In 1863 he accepted the offer of a professorship in mathematics at Cambridge, where he taught until his death. Cayley's work on matrices was of purely theoretical interest in his day, but in the 20th century many of his results have found applications in physics, the social sciences, business, and other fields.

Gaussian elimination. We then changed each main diagonal element to a one by multiplying by the appropriate constant(s). Finally, we completed the process by changing the remaining entries on the left side to zeros. The right half is now A^{-1}.

$$A^{-1} = \begin{bmatrix} -3 & 2 & 0 \\ -4 & 1 & -\frac{2}{3} \\ 1 & 0 & \frac{1}{3} \end{bmatrix}$$

To verify this, we find the products AA^{-1} and $A^{-1}A$:

$$AA^{-1} = \begin{bmatrix} 1 & -2 & -4 \\ 2 & -3 & -6 \\ -3 & 6 & 15 \end{bmatrix} \begin{bmatrix} -3 & 2 & 0 \\ -4 & 1 & -\frac{2}{3} \\ 1 & 0 & \frac{1}{3} \end{bmatrix} = \begin{bmatrix} 1 & 0 & 0 \\ 0 & 1 & 0 \\ 0 & 0 & 1 \end{bmatrix}$$

$$A^{-1}A = \begin{bmatrix} -3 & 2 & 0 \\ -4 & 1 & -\frac{2}{3} \\ 1 & 0 & \frac{1}{3} \end{bmatrix} \begin{bmatrix} 1 & -2 & -4 \\ 2 & -3 & -6 \\ -3 & 6 & 15 \end{bmatrix} = \begin{bmatrix} 1 & 0 & 0 \\ 0 & 1 & 0 \\ 0 & 0 & 1 \end{bmatrix}$$

The next example shows that not every square matrix has an inverse.

EXAMPLE 3 ■ A Matrix that Does Not Have an Inverse

Try to find the inverse of the matrix

$$\begin{bmatrix} 2 & -3 & -7 \\ 1 & 2 & 7 \\ 1 & 1 & 4 \end{bmatrix}$$

SOLUTION

We proceed as follows.

$$\begin{bmatrix} 2 & -3 & -7 & 1 & 0 & 0 \\ 1 & 2 & 7 & 0 & 1 & 0 \\ 1 & 1 & 4 & 0 & 0 & 1 \end{bmatrix} \xrightarrow{R_1 \leftrightarrow R_2} \begin{bmatrix} 1 & 2 & 7 & 0 & 1 & 0 \\ 2 & -3 & -7 & 1 & 0 & 0 \\ 1 & 1 & 4 & 0 & 0 & 1 \end{bmatrix}$$

$$\xrightarrow[R_3 - R_1 \to R_3]{R_2 - 2R_1 \to R_2} \begin{bmatrix} 1 & 2 & 7 & 0 & 1 & 0 \\ 0 & -7 & -21 & 1 & -2 & 0 \\ 0 & -1 & -3 & 0 & -1 & 1 \end{bmatrix}$$

$$\xrightarrow{-\frac{1}{7}R_2} \begin{bmatrix} 1 & 2 & 7 & 0 & 1 & 0 \\ 0 & 1 & 3 & -\frac{1}{7} & \frac{2}{7} & 0 \\ 0 & -1 & -3 & 0 & -1 & 1 \end{bmatrix}$$

$$\xrightarrow[R_1 - 2R_2 \to R_1]{R_3 + R_2 \to R_3} \begin{bmatrix} 1 & 0 & 1 & \frac{2}{7} & \frac{3}{7} & 0 \\ 0 & 1 & 3 & -\frac{1}{7} & \frac{2}{7} & 0 \\ 0 & 0 & 0 & -\frac{1}{7} & -\frac{5}{7} & 1 \end{bmatrix}$$

At this point we would like to change the 0 in the $(3, 3)$ position of this matrix to a 1, without changing the zeros in the $(3, 1)$ and $(3, 2)$ positions. But there is no way to accomplish this, because no matter what multiple of rows 1 and/or 2 we add to row 3, we cannot change the third zero in row 3 without changing the first or second as well. Thus, we cannot change the left half to the identity matrix. The original matrix does not have an inverse.   ∎

If we encounter a row of zeros on the left when trying to find an inverse, as in Example 3, then the original matrix does not have an inverse.

MATRIX EQUATIONS

We saw in Section 6.4 that a system of linear equations can be written as a single matrix equation. For example, the system

$$\begin{cases} x + 2y + 4z = 7 \\ -x + \ \ y + 2z = 5 \\ 2x + 3y + 3z = 7 \end{cases}$$

is equivalent to the matrix equation

$$\begin{bmatrix} 1 & 2 & 4 \\ -1 & 1 & 2 \\ 2 & 3 & 3 \end{bmatrix} \begin{bmatrix} x \\ y \\ z \end{bmatrix} = \begin{bmatrix} 7 \\ 5 \\ 7 \end{bmatrix}$$

(See Example 4, Section 6.4.)

If we let

$$A = \begin{bmatrix} 1 & 2 & 4 \\ -1 & 1 & 2 \\ 2 & 3 & 3 \end{bmatrix} \qquad X = \begin{bmatrix} x \\ y \\ z \end{bmatrix} \qquad B = \begin{bmatrix} 7 \\ 5 \\ 7 \end{bmatrix}$$

then this matrix equation can be written as

$$AX = B$$

Solving the matrix equation $AX = B$ is very similar to solving the simple real number equation

$$3x = 12$$

We solve this latter equation by multiplying each side by the reciprocal (or inverse) of 3:

$$\tfrac{1}{3}(3x) = \tfrac{1}{3}(12)$$

$$x = 4$$

We solve this matrix equation by multiplying each side by the inverse of A (provided this inverse exists):

$$AX = B$$

$$A^{-1}(AX) = A^{-1}B$$

$$(A^{-1}A)X = A^{-1}B$$

$$I_3 X = A^{-1}B$$

$$X = A^{-1}B$$

In this example,

$$A^{-1} = \frac{1}{9}\begin{bmatrix} 3 & -6 & 0 \\ -7 & 5 & 6 \\ 5 & -1 & -3 \end{bmatrix}$$

(Verify!), so that from $X = A^{-1}B$ we have

$$\begin{bmatrix} x \\ y \\ z \end{bmatrix} = \frac{1}{9}\begin{bmatrix} 3 & -6 & 0 \\ -7 & 5 & 6 \\ 5 & -1 & -3 \end{bmatrix}\begin{bmatrix} 7 \\ 5 \\ 7 \end{bmatrix}$$

$$= \frac{1}{9}\begin{bmatrix} -9 \\ 18 \\ 9 \end{bmatrix} = \begin{bmatrix} -1 \\ 2 \\ 1 \end{bmatrix}$$

Thus $x = -1$, $y = 2$, and $z = 1$ is the solution to the original system. (Compare this with the solution to Example 1 in Section 6.2.)

SOLVING A MATRIX EQUATION

If A is a square $n \times n$ matrix that has an inverse A^{-1}, and if X is a variable matrix and B a known matrix, both with n rows, then the solution of the matrix equation $AX = B$ is given by

$$X = A^{-1}B$$

EXAMPLE 4 ■ Solving a System Using the Matrix Inverse

Solve the following system of equations.

$$\begin{cases} 2x - 5y = 15 \\ 3x - 6y = 36 \end{cases}$$

SOLUTION

We first convert this to a matrix equation of the form $AX = B$:

$$\begin{bmatrix} 2 & -5 \\ 3 & -6 \end{bmatrix}\begin{bmatrix} x \\ y \end{bmatrix} = \begin{bmatrix} 15 \\ 36 \end{bmatrix}$$

Using the rule for calculating the inverse of a 2×2 matrix, we get

Find A^{-1}.

$$\begin{bmatrix} 2 & -5 \\ 3 & -6 \end{bmatrix}^{-1} = \frac{1}{2(-6) - (-5)3}\begin{bmatrix} -6 & -(-5) \\ -3 & 2 \end{bmatrix} = \frac{1}{3}\begin{bmatrix} -6 & 5 \\ -3 & 2 \end{bmatrix}$$

Multiplying each side of the matrix equation by this inverse matrix, we get

$X = A^{-1}B$

$$\begin{bmatrix} x \\ y \end{bmatrix} = \frac{1}{3}\begin{bmatrix} -6 & 5 \\ -3 & 2 \end{bmatrix}\begin{bmatrix} 15 \\ 36 \end{bmatrix} = \begin{bmatrix} 30 \\ 9 \end{bmatrix}$$

So $x = 30$ and $y = 9$.   ■

EXAMPLE 5 ■ An Application of Matrix Equations

A pet store owner feeds his hamsters and gerbils different mixtures of three types of rodent food pellets, which we will call brands A, B, and C. He wishes to feed his animals the correct amount of each brand to satisfy exactly their daily requirements for protein, fat, and carbohydrates. Suppose that hamsters require 340 mg of protein, 280 mg of fat, and 440 mg of carbohydrates, and gerbils need 480 mg of protein, 360 mg of fat, and 680 mg of carbohydrates each day. The amount of each nutrient in one gram of each brand of food is given in the following table. How many grams of each food should the storekeeper feed his hamsters and gerbils daily to satisfy their nutrient requirements?

	Brand A	Brand B	Brand C
Protein (mg)	10	0	20
Fat (mg)	10	20	10
Carbohydrates (mg)	5	10	30

SOLUTION

If we let x_1, x_2, and x_3 be the grams of brands A, B, and C, respectively, that the hamsters should eat, and if we let y_1, y_2, and y_3 be the corresponding amounts for the gerbils, then we want to solve the matrix equations

$$\begin{bmatrix} 10 & 0 & 20 \\ 10 & 20 & 10 \\ 5 & 10 & 30 \end{bmatrix}\begin{bmatrix} x_1 \\ x_2 \\ x_3 \end{bmatrix} = \begin{bmatrix} 340 \\ 280 \\ 440 \end{bmatrix} \qquad \text{Hamster equation}$$

and

$$\begin{bmatrix} 10 & 0 & 20 \\ 10 & 20 & 10 \\ 5 & 10 & 30 \end{bmatrix}\begin{bmatrix} y_1 \\ y_2 \\ y_3 \end{bmatrix} = \begin{bmatrix} 480 \\ 360 \\ 680 \end{bmatrix} \qquad \text{Gerbil equation}$$

Since the coefficient matrix on the left is the same in both of these equations, we can solve each one by multiplying each side by the inverse of this matrix.

We therefore begin by finding this inverse.

$$\left[\begin{array}{cccccc} 10 & 0 & 20 & 1 & 0 & 0 \\ 10 & 20 & 10 & 0 & 1 & 0 \\ 5 & 10 & 30 & 0 & 0 & 1 \end{array}\right] \xrightarrow{2R_3} \left[\begin{array}{cccccc} 10 & 0 & 20 & 1 & 0 & 0 \\ 10 & 20 & 10 & 0 & 1 & 0 \\ 10 & 20 & 60 & 0 & 0 & 2 \end{array}\right]$$

$$\xrightarrow[R_3 - R_1 \to R_3]{R_2 - R_1 \to R_2} \left[\begin{array}{cccccc} 10 & 0 & 20 & 1 & 0 & 0 \\ 0 & 20 & -10 & -1 & 1 & 0 \\ 0 & 20 & 40 & -1 & 0 & 2 \end{array}\right]$$

$$\xrightarrow{R_3 - R_2 \to R_3} \left[\begin{array}{cccccc} 10 & 0 & 20 & 1 & 0 & 0 \\ 0 & 20 & -10 & -1 & 1 & 0 \\ 0 & 0 & 50 & 0 & -1 & 2 \end{array}\right]$$

$$\xrightarrow{\frac{1}{5}R_3} \left[\begin{array}{cccccc} 10 & 0 & 20 & 1 & 0 & 0 \\ 0 & 20 & -10 & -1 & 1 & 0 \\ 0 & 0 & 10 & 0 & -\frac{1}{5} & \frac{2}{5} \end{array}\right]$$

$$\xrightarrow[R_1 - 2R_3 \to R_1]{R_2 + R_3 \to R_2} \left[\begin{array}{cccccc} 10 & 0 & 0 & 1 & \frac{2}{5} & -\frac{4}{5} \\ 0 & 20 & 0 & -1 & \frac{4}{5} & \frac{2}{5} \\ 0 & 0 & 10 & 0 & -\frac{1}{5} & \frac{2}{5} \end{array}\right]$$

$$\xrightarrow{\frac{1}{10}R_1, \frac{1}{20}R_2, \frac{1}{10}R_3} \left[\begin{array}{cccccc} 1 & 0 & 0 & 0.10 & 0.04 & -0.08 \\ 0 & 1 & 0 & -0.05 & 0.04 & 0.02 \\ 0 & 0 & 1 & 0 & -0.02 & 0.04 \end{array}\right]$$

So

$$\left[\begin{array}{ccc} 10 & 0 & 20 \\ 10 & 20 & 10 \\ 5 & 10 & 30 \end{array}\right]^{-1} = \frac{1}{100} \left[\begin{array}{ccc} 10 & 4 & -8 \\ -5 & 4 & 2 \\ 0 & -2 & 4 \end{array}\right]$$

and if we now multiply each side of our matrix equations by this inverse matrix, we get

$$\left[\begin{array}{c} x_1 \\ x_2 \\ x_3 \end{array}\right] = \frac{1}{100} \left[\begin{array}{ccc} 10 & 4 & -8 \\ -5 & 4 & 2 \\ 0 & -2 & 4 \end{array}\right] \left[\begin{array}{c} 340 \\ 280 \\ 440 \end{array}\right] = \left[\begin{array}{c} 10 \\ 3 \\ 12 \end{array}\right]$$

and

$$\left[\begin{array}{c} y_1 \\ y_2 \\ y_3 \end{array}\right] = \frac{1}{100} \left[\begin{array}{ccc} 10 & 4 & -8 \\ -5 & 4 & 2 \\ 0 & -2 & 4 \end{array}\right] \left[\begin{array}{c} 480 \\ 360 \\ 680 \end{array}\right] = \left[\begin{array}{c} 8 \\ 4 \\ 20 \end{array}\right]$$

We interpret these solution matrices as follows: Each hamster should be fed 10 g of brand A, 3 g of brand B, and 12 g of brand C, and each gerbil should be fed 8 g of brand A, 4 g of brand B, and 20 g of brand C daily.   ∎

Since a lot of work is usually involved in finding the inverse of a 3×3 or larger matrix, the method used in Example 5 is really useful only when we are solving several systems of equations with the same coefficient matrix. However, if we have access to a calculator or computer program that calculates matrix inverses, then this becomes the preferred method in all circumstances. Most graphing calculators can compute matrix inverses.

6.5 EXERCISES

1–2 ■ Find the inverse of the matrix and verify that $A^{-1}A = AA^{-1} = I_2$ and $B^{-1}B = BB^{-1} = I_3$.

1. $A = \begin{bmatrix} 7 & 4 \\ 3 & 2 \end{bmatrix}$

2. $B = \begin{bmatrix} 1 & 3 & 2 \\ 0 & 2 & 2 \\ -2 & -1 & 0 \end{bmatrix}$

3–18 ■ Find the inverse of the matrix if it exists.

3. $\begin{bmatrix} 5 & 3 \\ 3 & 2 \end{bmatrix}$

4. $\begin{bmatrix} 3 & 4 \\ 7 & 9 \end{bmatrix}$

5. $\begin{bmatrix} 2 & 5 \\ -5 & -13 \end{bmatrix}$

6. $\begin{bmatrix} -7 & 4 \\ 8 & -5 \end{bmatrix}$

7. $\begin{bmatrix} 6 & -3 \\ -8 & 4 \end{bmatrix}$

8. $\begin{bmatrix} \frac{1}{2} & \frac{1}{3} \\ 5 & 4 \end{bmatrix}$

9. $\begin{bmatrix} 0.4 & -1.2 \\ 0.3 & 0.6 \end{bmatrix}$

10. $\begin{bmatrix} 4 & 2 & 3 \\ 3 & 3 & 2 \\ 1 & 0 & 1 \end{bmatrix}$

11. $\begin{bmatrix} 2 & 4 & 1 \\ -1 & 1 & -1 \\ 1 & 4 & 0 \end{bmatrix}$

12. $\begin{bmatrix} 5 & 7 & 4 \\ 3 & -1 & 3 \\ 6 & 7 & 5 \end{bmatrix}$

13. $\begin{bmatrix} 1 & 2 & 3 \\ 4 & 5 & -1 \\ 1 & -1 & -10 \end{bmatrix}$

14. $\begin{bmatrix} 2 & 1 & 0 \\ 1 & 1 & 4 \\ 2 & 1 & 2 \end{bmatrix}$

15. $\begin{bmatrix} 0 & -2 & 2 \\ 3 & 1 & 3 \\ 1 & -2 & 3 \end{bmatrix}$

16. $\begin{bmatrix} 3 & -2 & 0 \\ 5 & 1 & 1 \\ 2 & -2 & 0 \end{bmatrix}$

17. $\begin{bmatrix} 1 & 2 & 0 & 3 \\ 0 & 1 & 1 & 1 \\ 0 & 1 & 0 & 1 \\ 1 & 2 & 0 & 2 \end{bmatrix}$

18. $\begin{bmatrix} 1 & 0 & 1 & 0 \\ 0 & 1 & 0 & 1 \\ 1 & 1 & 1 & 0 \\ 1 & 1 & 1 & 1 \end{bmatrix}$

19–26 ■ Solve the system of equations by converting to a matrix equation and using the inverse of the coefficient matrix, as in Example 4. Use the inverses from Exercises 3–6, 11, 12, 16, and 17.

19. $\begin{cases} 5x + 3y = 4 \\ 3x + 2y = 0 \end{cases}$

20. $\begin{cases} 3x + 4y = 10 \\ 7x + 9y = 20 \end{cases}$

21. $\begin{cases} 2x + 5y = 2 \\ -5x - 13y = 20 \end{cases}$

22. $\begin{cases} -7x + 4y = 0 \\ 8x - 5y = 100 \end{cases}$

23. $\begin{cases} 2x + 4y + z = 7 \\ -x + y - z = 0 \\ x + 4y = -2 \end{cases}$

24. $\begin{cases} 5x + 7y + 4z = 1 \\ 3x - y + 3z = 1 \\ 6x + 7y + 5z = 1 \end{cases}$

25. $\begin{cases} 3x - 2y = 6 \\ 5x + y + z = 12 \\ 2x - 2y = 18 \end{cases}$

26. $\begin{cases} x + 2y + 3w = 0 \\ y + z + w = 1 \\ y + w = 2 \\ x + 2y + 2w = 3 \end{cases}$

27–28 ■ Solve the matrix equation by multiplying each side by the appropriate inverse matrix.

27. $\begin{bmatrix} 3 & -2 \\ -4 & 3 \end{bmatrix} \begin{bmatrix} x & y & z \\ u & v & w \end{bmatrix} = \begin{bmatrix} 1 & 0 & -1 \\ 2 & 1 & 3 \end{bmatrix}$

28. $\begin{bmatrix} 0 & -2 & 2 \\ 3 & 1 & 3 \\ 1 & -2 & 3 \end{bmatrix} \begin{bmatrix} x & u \\ y & v \\ z & w \end{bmatrix} = \begin{bmatrix} 3 & 6 \\ 6 & 12 \\ 0 & 0 \end{bmatrix}$

29. A nutritionist is studying the effects of the nutrients folic acid, choline, and inositol. He has three types of food available, and each type contains the following amounts of these nutrients per ounce:

	Type A	Type B	Type C
Folic acid (mg)	3	1	3
Choline (mg)	4	2	4
Inositol (mg)	3	2	4

(a) Find the inverse of the matrix

$$\begin{bmatrix} 3 & 1 & 3 \\ 4 & 2 & 4 \\ 3 & 2 & 4 \end{bmatrix}$$

and use it to solve the remaining parts of this problem.

(b) How many ounces of each food should the nutritionist feed his laboratory rats if he wants their daily diet to contain 10 mg of folic acid, 14 mg of choline, and 13 mg of inositol?

(c) How much of each food should be given to supply 9 mg of folic acid, 12 mg of choline, and 10 mg of inositol?

(d) Will any combination of these foods supply 2 mg of folic acid, 4 mg of choline, and 11 mg of inositol?

30. Refer to Exercise 29. Suppose it is found that food type C has been improperly labeled, and actually contains 4 mg of folic acid, 6 mg of choline, and 5 mg of inositol per ounce. Would it still be possible to use matrix inversion to solve parts (b), (c), and (d) of Exercise 29? Why or why not?

31. An encyclopedia salesman works for a company that offers three different grades of bindings for its encyclopedias: standard, deluxe, and leather. For each set that he sells he earns a commission that is based on the set's binding grade. One week he sells one standard, one deluxe, and two leather sets and makes $675 in commission. The next week he sells two standard, one deluxe, and one leather set for a $600 commission. The third week he sells one standard, two deluxe, and one leather set, earning $625 in commission.

(a) Let x, y, and z represent the commission he earns on standard, deluxe, and leather sets, respectively. Translate the given information into a system of equations in x, y, and z.

(b) Express the system of equations you found in part (a) as a matrix equation of the form $AX = B$.

(c) Find the inverse of the coefficient matrix A and use it to solve the matrix equation in part (b). How much commission does the salesman earn on a set of encyclopedias in each grade of binding?

32–35 ■ Find the inverse of the matrix. For what value(s) of x, if any, does the matrix have no inverse?

32. $\begin{bmatrix} 2x^2 & x \\ 3x & 1 \end{bmatrix}$

33. $\begin{bmatrix} x & 1 \\ -1 & 1/x \end{bmatrix}$

34. $\begin{bmatrix} e^x & -e^{2x} \\ e^{2x} & e^{3x} \end{bmatrix}$

35. $\begin{bmatrix} 1 & e^x & 0 \\ e^x & -e^{2x} & 0 \\ 0 & 0 & 2 \end{bmatrix}$

36. A matrix that has an inverse is called **invertible.** Find two 2×2 invertible matrices whose sum is not invertible.

37. Find the inverse of the following matrix, where $abcd \neq 0$:

$$\begin{bmatrix} a & 0 & 0 & 0 \\ 0 & b & 0 & 0 \\ 0 & 0 & c & 0 \\ 0 & 0 & 0 & d \end{bmatrix}$$

6.6 DETERMINANTS AND CRAMER'S RULE

If a matrix is **square** (that is, if it has the same number of rows as columns), then we can assign to it a number called its **determinant.** Determinants can be used to solve matrix equations, as we will see later in this section. They are also useful in determining whether a matrix has an inverse, without our actually going through the process of trying to find its inverse.

We denote the determinant of a square matrix A by the symbol $|A|$, and we begin by defining $|A|$ for the simplest case. If A is a 1×1 matrix, then it has only one entry, and we define its determinant to be the value of that entry; that is, if $A = [a]$, then $|A| = a$. If A is a 2×2 matrix, then

$$A = \begin{bmatrix} a & b \\ c & d \end{bmatrix}$$

and we define the determinant of A to be

$$|A| = \begin{vmatrix} a & b \\ c & d \end{vmatrix} = ad - bc$$

EXAMPLE 1 ■ Determinant of a 2 × 2 Matrix

Evaluate $|A|$ for $A = \begin{bmatrix} 6 & -3 \\ 2 & 3 \end{bmatrix}$.

SOLUTION

$$\begin{vmatrix} 6 & -3 \\ 2 & 3 \end{vmatrix} = 6 \cdot 3 - (-3)2 = 18 - (-6) = 24 \qquad ■$$

We can think of the evaluation of a 2×2 determinant as a "cross-product" operation. We take the product of the diagonal from top left to bottom right, and subtract the product from top right to bottom left.

To define the concept of determinant for an arbitrary $n \times n$ matrix, we must first introduce the following terminology.

Let A be an $n \times n$ matrix.

1. The **minor** M_{ij} of the element a_{ij} is the determinant of the matrix obtained by deleting the ith row and jth column of A.

2. The **cofactor** A_{ij} of the element a_{ij} is

$$A_{ij} = (-1)^{i+j} M_{ij}$$

For example, if A is the matrix

$$\begin{bmatrix} 2 & 3 & -1 \\ 0 & 2 & 4 \\ -2 & 5 & 6 \end{bmatrix}$$

then M_{12} is the determinant of the matrix obtained by deleting the first row and second column from A.

Thus

$$M_{12} = \begin{vmatrix} 2 & 3 & -1 \\ 0 & 2 & 4 \\ -2 & 5 & 6 \end{vmatrix} = \begin{vmatrix} 0 & 4 \\ -2 & 6 \end{vmatrix} = 0(6) - 4(-2) = 8$$

so
$$A_{12} = (-1)^{1+2} M_{12} = -8$$

Similarly,

$$M_{33} = \begin{vmatrix} 2 & 3 & -1 \\ 0 & 2 & 4 \\ -2 & 5 & 6 \end{vmatrix} = \begin{vmatrix} 2 & 3 \\ 0 & 2 \end{vmatrix} = 2 \cdot 2 - 3 \cdot 0 = 4$$

so
$$A_{33} = (-1)^{3+3} M_{33} = 4$$

Note that the cofactor of a_{ij} is just the minor of a_{ij} multiplied by either 1 or -1, depending on whether $i + j$ is even or odd. Thus, in a 3×3 matrix we obtain the cofactor of any element by prefixing its minor with the sign obtained from the following checkerboard pattern:

$$\begin{bmatrix} + & - & + \\ - & + & - \\ + & - & + \end{bmatrix}$$

We are now ready to define the determinant of any square matrix.

THE DETERMINANT OF A SQUARE MATRIX

If A is an $n \times n$ matrix, then the **determinant** of A is obtained by multiplying each element of the first row by its cofactor, and then adding the results. In symbols,

$$|A| = \begin{vmatrix} a_{11} & a_{12} & \cdots & a_{1n} \\ a_{21} & a_{22} & \cdots & a_{2n} \\ \cdot & \cdot & \cdot & \cdot \\ \cdot & \cdot & \cdot & \cdot \\ \cdot & \cdot & \cdot & \cdot \\ a_{n1} & a_{n2} & \cdots & a_{nn} \end{vmatrix} = a_{11}A_{11} + a_{12}A_{12} + \cdots + a_{1n}A_{1n}$$

EXAMPLE 2 ■ Determinant of a 3 × 3 Matrix

Evaluate the determinant of the matrix

$$A = \begin{bmatrix} 2 & 3 & -1 \\ 0 & 2 & 4 \\ -2 & 5 & 6 \end{bmatrix}$$

Emmy Noether (1882–1935) was one of the foremost mathematicians of the early 20th century. Her groundbreaking work in abstract algebra provided much of the foundation for this field, and her work in Invariant Theory was essential in the development of Einstein's theory of general relativity. Although women were not allowed to study at German universities at the time, she audited courses unofficially and went on to receive a doctorate at Erlangen *summa cum laude*, despite the opposition of the academic senate, which declared that women students would "overthrow all academic order." She subsequently taught mathematics at Göttingen, Moscow, and Frankfurt. In 1933 she left Germany to escape Nazi persecution, accepting a position at Bryn Mawr College in suburban Philadelphia. She lectured there and at the Institute for Advanced Study in Princeton, New Jersey, until her untimely death in 1935.

SOLUTION

$$|A| = 2\begin{vmatrix} 2 & 4 \\ 5 & 6 \end{vmatrix} - 3\begin{vmatrix} 0 & 4 \\ -2 & 6 \end{vmatrix} + (-1)\begin{vmatrix} 0 & 2 \\ -2 & 5 \end{vmatrix}$$

$$= 2(2 \cdot 6 - 4 \cdot 5) - 3[0 \cdot 6 - 4(-2)] - [0 \cdot 5 - 2(-2)]$$

$$= -16 - 24 - 4$$

$$= -44$$ ∎

In our definition of the determinant, we used the cofactors of elements in the first row only. This is called **expanding the determinant by the first row.** In fact, *we can expand the determinant by any row or column in the same way, and obtain the same result.* Although we will not prove this, the next example illustrates this principle.

EXAMPLE 3 ∎ **Expanding a Determinant about a Row and a Column**

Expand the determinant of the matrix A in Example 2 by the second row and by the third column, and show that the value obtained is the same in each case.

SOLUTION

Expanding by the second row, we get

$$|A| = \begin{vmatrix} 2 & 3 & -1 \\ 0 & 2 & 4 \\ -2 & 5 & 6 \end{vmatrix} = -0\begin{vmatrix} 3 & -1 \\ 5 & 6 \end{vmatrix} + 2\begin{vmatrix} 2 & -1 \\ -2 & 6 \end{vmatrix} - 4\begin{vmatrix} 2 & 3 \\ -2 & 5 \end{vmatrix}$$

$$= 0 + 2[2 \cdot 6 - (-1)(-2)] - 4[2 \cdot 5 - 3(-2)]$$

$$= 0 + 20 - 64$$

$$= -44$$

Expanding by the third column gives

$$|A| = -1\begin{vmatrix} 0 & 2 \\ -2 & 5 \end{vmatrix} - 4\begin{vmatrix} 2 & 3 \\ -2 & 5 \end{vmatrix} + 6\begin{vmatrix} 2 & 3 \\ 0 & 2 \end{vmatrix}$$

$$= -[0 \cdot 5 - 2(-2)] - 4[2 \cdot 5 - 3(-2)] + 6(2 \cdot 2 - 3 \cdot 0)$$

$$= -4 - 64 + 24$$

$$= -44$$

In both cases we obtain the same value for the determinant as when we expanded by the first row in Example 2. ∎

The following principle allows us to determine whether a square matrix has an inverse without actually calculating the inverse. This is one of the most important uses of the determinant in matrix algebra, and it is the reason for the name *determinant*.

INVERTIBILITY CRITERION

If A is a square matrix, then A has an inverse if and only if $|A| \neq 0$.

Although we will not prove this fact, we have already seen (in the preceding section) why it is true in the case of 2×2 matrices.

EXAMPLE 4 ■ **Using the Determinant to Show that a Matrix Is Not Invertible**

Show that the matrix A has no inverse.

$$
A = \begin{bmatrix} 1 & 2 & 0 & 4 \\ 0 & 0 & 0 & 3 \\ 5 & 6 & 2 & 6 \\ 2 & 4 & 0 & 9 \end{bmatrix}
$$

SOLUTION

We begin by calculating the determinant of A. Since all but one of the elements of the second row is zero, we expand the determinant by the second row. If we do this, we see from the following equation that only the cofactor A_{24} will have to be calculated.

$$
|A| = -0 \cdot A_{21} + 0 \cdot A_{22} - 0 \cdot A_{23} + 3 \cdot A_{24} = 3A_{24}
$$

$$
= 3 \begin{vmatrix} 1 & 2 & 0 \\ 5 & 6 & 2 \\ 2 & 4 & 0 \end{vmatrix} \quad \text{Expand this by the third column}
$$

$$
= 3(-2) \begin{vmatrix} 1 & 2 \\ 2 & 4 \end{vmatrix}
$$

$$
= 3(-2)(1 \cdot 4 - 2 \cdot 2) = 0
$$

Since the determinant of A is zero, A cannot have an inverse, by the Invertibility Criterion. ■

The preceding example shows that if we expand a determinant about a row or column that contains many zeros, our work is reduced considerably because we do not have to evaluate the cofactors of the elements that are zero. The following principle enables us in many cases to simplify the process of finding a determinant by introducing zeros into it without changing its value.

> ### ROW AND COLUMN TRANSFORMATIONS OF A DETERMINANT
>
> If A is a square matrix, and if the matrix B is obtained from A by adding a multiple of one row to another, or a multiple of one column to another, then $|A| = |B|$.

EXAMPLE 5 ■ **Using Row and Column Transformations to Calculate a Determinant**

Find the determinant of the matrix A. Does it have an inverse?

$$A = \begin{bmatrix} 8 & 2 & -1 & -4 \\ 3 & 5 & -3 & 11 \\ 24 & 6 & 1 & -12 \\ 2 & 2 & 7 & -1 \end{bmatrix}$$

SOLUTION

If we subtract three times row 1 from row 3, we will change all but one of the elements of row 3 to zeros:

$$\begin{bmatrix} 8 & 2 & -1 & -4 \\ 3 & 5 & -3 & 11 \\ 0 & 0 & 4 & 0 \\ 2 & 2 & 7 & -1 \end{bmatrix}$$

This new matrix has the same determinant as A, and if we expand its determinant by the third row, we get

$$|A| = 4\begin{vmatrix} 8 & 2 & -4 \\ 3 & 5 & 11 \\ 2 & 2 & -1 \end{vmatrix}$$

Now, adding two times column 3 to column 1 in this determinant gives us

$$|A| = 4\begin{vmatrix} 0 & 2 & -4 \\ 25 & 5 & 11 \\ 0 & 2 & -1 \end{vmatrix} \qquad \text{Expand this by the first column}$$

$$= 4(-25)\begin{vmatrix} 2 & -4 \\ 2 & -1 \end{vmatrix}$$

$$= 4(-25)[2(-1) - (-4)2] = -600$$

Since the determinant of A is not zero, A does have an inverse.   ■

CRAMER'S RULE

The solutions of linear equations can sometimes be expressed using determinants. To illustrate, let's try to solve the following pair of linear equations for the variable x:

$$\begin{cases} ax + by = r \\ cx + dy = s \end{cases}$$

If $d \neq 0$, then we can eliminate the variable y from the first equation by multiplying the second equation by b/d and then subtracting it from the first, which gives

$$ax - \left(\frac{b}{d}\right)cx = r - \left(\frac{b}{d}\right)s$$

If we now multiply each side of this equation by d and factor x from the left, we get

$$(ad - bc)x = rd - bs$$

Assuming that $ad - bc \neq 0$, we can now solve this equation for x, obtaining

$$x = \frac{rd - bs}{ad - bc}$$

The numerator and denominator of this fraction look like the determinants of 2×2 matrices. In fact, we can write the solution for x as

$$x = \frac{\begin{vmatrix} r & b \\ s & d \end{vmatrix}}{\begin{vmatrix} a & b \\ c & d \end{vmatrix}}$$

Using the same sort of technique, we can solve the original pair of equations for y, to get

$$y = \frac{\begin{vmatrix} a & r \\ c & s \end{vmatrix}}{\begin{vmatrix} a & b \\ c & d \end{vmatrix}}$$

Notice that the denominator in each case is the determinant of the coefficient matrix, which we will call D. The numerator in the solution for x is the determinant of the matrix obtained from D by replacing the first column, the coefficients of x, by r and s, respectively. Similarly, in the solution for y the numerator is the determinant of the matrix obtained from D by replacing the second column, the coefficients of y, by r and s. Thus, if we define

$$D = \begin{bmatrix} a & b \\ c & d \end{bmatrix} \qquad D_x = \begin{bmatrix} r & b \\ s & d \end{bmatrix} \qquad D_y = \begin{bmatrix} a & r \\ c & s \end{bmatrix}$$

then we can write the solution of the system as

$$x = \frac{|D_x|}{|D|} \quad \text{and} \quad y = \frac{|D_y|}{|D|}$$

This pair of formulas is known as **Cramer's Rule,** and it can be used to solve any pair of linear equations in two unknowns in which the determinant of the coefficient matrix is not zero.

EXAMPLE 6 ■ **Using Cramer's Rule to Solve a System with Two Variables**

Use Cramer's Rule to solve the following system.

$$\begin{cases} 2x + 6y = -1 \\ x + 8y = 2 \end{cases}$$

SOLUTION

For this system, we have

$$|D| = \begin{vmatrix} 2 & 6 \\ 1 & 8 \end{vmatrix} = 2 \cdot 8 - 6 \cdot 1 = 10$$

$$|D_x| = \begin{vmatrix} -1 & 6 \\ 2 & 8 \end{vmatrix} = (-1)8 - 6 \cdot 2 = -20$$

and $\qquad |D_y| = \begin{vmatrix} 2 & -1 \\ 1 & 2 \end{vmatrix} = 2 \cdot 2 - (-1)1 = 5$

The solution is

$$x = \frac{|D_x|}{|D|} = \frac{-20}{10} = -2$$

and $\qquad y = \dfrac{|D_y|}{|D|} = \dfrac{5}{10} = \dfrac{1}{2}$ ■

Cramer's Rule can be extended to apply to any system of n linear equations in n variables in which the determinant of the coefficient matrix is not zero. As we saw in the preceding section, any such system can be written in matrix form as

$$\begin{bmatrix} a_{11} & a_{12} & \cdots & a_{1n} \\ a_{21} & a_{22} & \cdots & a_{2n} \\ \vdots & \vdots & \ddots & \vdots \\ a_{n1} & a_{n2} & \cdots & a_{nn} \end{bmatrix} \begin{bmatrix} x_1 \\ x_2 \\ \vdots \\ x_n \end{bmatrix} = \begin{bmatrix} b_1 \\ b_2 \\ \vdots \\ b_n \end{bmatrix}$$

By analogy with what we did in the case of two equations in two unknowns, we let D be the coefficient matrix in the above system, and we let D_{x_i} be the matrix obtained by replacing the ith column of D by the numbers $b_1, b_2, \ldots, b_n$ that appear to the right of the equal sign in the system. The solution of the system is then given by the following rule.

CRAMER'S RULE

If a system of n linear equations in the n variables $x_1, x_2, \ldots, x_n$ is equivalent to the matrix equation $DX = B$, and if $|D| \neq 0$, then its solutions are

$$x_1 = \frac{|D_{x_1}|}{|D|}, \quad x_2 = \frac{|D_{x_2}|}{|D|}, \quad \ldots, \quad x_n = \frac{|D_{x_n}|}{|D|}$$

where D_{x_i} is the matrix obtained by replacing the ith column of D by the $n \times 1$ matrix B.

EXAMPLE 7 ■ **Using Cramer's Rule to Solve a System with Three Variables**

Use Cramer's Rule to solve the following system.

$$\begin{cases} 2x - 3y + 4z = 1 \\ x \quad\quad\; + 6z = 0 \\ 3x - 2y \quad\quad = 5 \end{cases}$$

SOLUTION

First we evaluate the determinants that appear in Cramer's Rule.

$$|D| = \begin{vmatrix} 2 & -3 & 4 \\ 1 & 0 & 6 \\ 3 & -2 & 0 \end{vmatrix} = -38 \qquad |D_x| = \begin{vmatrix} 1 & -3 & 4 \\ 0 & 0 & 6 \\ 5 & -2 & 0 \end{vmatrix} = -78$$

$$|D_y| = \begin{vmatrix} 2 & 1 & 4 \\ 1 & 0 & 6 \\ 3 & 5 & 0 \end{vmatrix} = -22 \qquad |D_z| = \begin{vmatrix} 2 & -3 & 1 \\ 1 & 0 & 0 \\ 3 & -2 & 5 \end{vmatrix} = 13$$

Now we use Cramer's Rule to get the solution:

$$x = \frac{|D_x|}{|D|} = \frac{-78}{-38} = \frac{39}{19}$$

$$y = \frac{|D_y|}{|D|} = \frac{-22}{-38} = \frac{11}{19}$$

$$z = \frac{|D_z|}{|D|} = \frac{13}{-38} = -\frac{13}{38}$$ ∎

Solving the system in Example 7 using Gaussian elimination would involve matrices whose elements are fractions with fairly large denominators. Thus, in cases like Examples 6 and 7, Cramer's Rule provides an efficient method for solving systems of linear equations. But in systems with more than three equations, evaluating the various determinants involved is usually a long and tedious task. Moreover, the rule does not apply if $|D| = 0$ or if D is not a square matrix. So, Cramer's Rule is a useful alternative to Gaussian elimination, but only in some situations.

6.6 EXERCISES

1–8 ■ Find the determinant of the matrix, if it exists.

1. $[3]$

2. $[0]$

3. $\begin{bmatrix} 4 & 5 \\ 0 & -1 \end{bmatrix}$

4. $\begin{bmatrix} -2 & 1 \\ 3 & -2 \end{bmatrix}$

5. $[2 \quad 5]$

6. $\begin{bmatrix} 3 \\ 0 \end{bmatrix}$

7. $\begin{bmatrix} \frac{1}{2} & \frac{1}{8} \\ 1 & \frac{1}{2} \end{bmatrix}$

8. $\begin{bmatrix} 2.2 & -1.4 \\ 0.5 & 1.0 \end{bmatrix}$

9–14 ■ Evaluate the minor and cofactor using the matrix A.

$$A = \begin{bmatrix} 1 & 0 & \frac{1}{2} \\ -3 & 5 & 2 \\ 0 & 0 & 4 \end{bmatrix}$$

9. M_{11}, A_{11}

10. M_{33}, A_{33}

11. M_{12}, A_{12}

12. M_{13}, A_{13}

13. M_{23}, A_{23}

14. M_{32}, A_{32}

15–20 ■ Find the determinant of the matrix. Determine whether the matrix has an inverse, but do not calculate the inverse.

15. $\begin{bmatrix} 1 & 3 & 7 \\ 2 & 0 & -1 \\ 0 & 2 & 6 \end{bmatrix}$

16. $\begin{bmatrix} -2 & -\frac{3}{2} & \frac{1}{2} \\ 2 & 4 & 0 \\ \frac{1}{2} & 2 & 1 \end{bmatrix}$

17. $\begin{bmatrix} 30 & 0 & 20 \\ 0 & -10 & -20 \\ 40 & 0 & 10 \end{bmatrix}$

18. $\begin{bmatrix} 1 & 2 & 5 \\ -2 & -3 & 2 \\ 3 & 5 & 3 \end{bmatrix}$

19. $\begin{bmatrix} 1 & 3 & 3 & 0 \\ 0 & 2 & 0 & 1 \\ -1 & 0 & 0 & 2 \\ 1 & 6 & 4 & 1 \end{bmatrix}$

20. $\begin{bmatrix} 1 & 2 & 0 & 2 \\ 3 & -4 & 0 & 4 \\ 0 & 1 & 6 & 0 \\ 1 & 0 & 2 & 0 \end{bmatrix}$

21–24 ■ Evaluate the determinant, using row or column operations whenever possible to simplify your work.

21. $\begin{vmatrix} 0 & 0 & 4 & 6 \\ 2 & 1 & 1 & 3 \\ 2 & 1 & 2 & 3 \\ 3 & 0 & 1 & 7 \end{vmatrix}$

22. $\begin{vmatrix} -2 & 3 & -1 & 7 \\ 4 & 6 & -2 & 3 \\ 7 & 7 & 0 & 5 \\ 3 & -12 & 4 & 0 \end{vmatrix}$

23. $\begin{vmatrix} 1 & 2 & 3 & 4 & 5 \\ 0 & 2 & 4 & 6 & 8 \\ 0 & 0 & 3 & 6 & 9 \\ 0 & 0 & 0 & 4 & 8 \\ 0 & 0 & 0 & 0 & 5 \end{vmatrix}$

24. $\begin{vmatrix} 2 & -1 & 6 & 4 \\ 7 & 2 & -2 & 5 \\ 4 & -2 & 10 & 8 \\ 6 & 1 & 1 & 4 \end{vmatrix}$

25. Let

$$B = \begin{bmatrix} 4 & 1 & 0 \\ -2 & -1 & 1 \\ 4 & 0 & 3 \end{bmatrix}$$

(a) Evaluate $|B|$ by expanding by the second row.
(b) Evaluate $|B|$ by expanding by the third column.
(c) Do your results in parts (a) and (b) agree?

26. Consider the system

$$\begin{cases} x + 2y + 6z = 5 \\ -3x - 6y + 5z = 8 \\ 2x + 6y + 9z = 7 \end{cases}$$

(a) Verify that $x = -1$, $y = 0$, $z = 1$ is a solution of the system.
(b) Find the determinant of the coefficient matrix.
(c) Without solving the system, determine whether there are any other solutions.
(d) Can Cramer's Rule be used to solve this system? Why or why not?

27–44 ■ Use Cramer's Rule to solve the system.

27. $\begin{cases} 2x - y = -9 \\ x + 2y = 8 \end{cases}$

28. $\begin{cases} 6x + 12y = 33 \\ 4x + 7y = 20 \end{cases}$

29. $\begin{cases} x - 6y = 3 \\ 3x + 2y = 1 \end{cases}$

30. $\begin{cases} \frac{1}{2}x + \frac{1}{3}y = 1 \\ \frac{1}{4}x - \frac{1}{6}y = -\frac{3}{2} \end{cases}$

31. $\begin{cases} 0.4x + 1.2y = 0.4 \\ 1.2x + 1.6y = 3.2 \end{cases}$

32. $\begin{cases} 10x - 17y = 21 \\ 20x - 31y = 39 \end{cases}$

33. $\begin{cases} x - y + 2z = 0 \\ 3x + z = 11 \\ -x + 2y = 0 \end{cases}$

34. $\begin{cases} 5x - 3y + z = 6 \\ 4y - 6z = 22 \\ 7x + 10y = -13 \end{cases}$

35. $\begin{cases} 2x_1 + 3x_2 - 5x_3 = 1 \\ x_1 + x_2 - x_3 = 2 \\ 2x_2 + x_3 = 8 \end{cases}$

36. $\begin{cases} -2a + c = 2 \\ a + 2b - c = 9 \\ 3a + 5b + 2c = 22 \end{cases}$

37. $\begin{cases} \frac{1}{3}x - \frac{1}{5}y + \frac{1}{2}z = \frac{7}{10} \\ -\frac{2}{3}x + \frac{2}{5}y + \frac{3}{2}z = \frac{11}{10} \\ x - \frac{4}{5}y + z = \frac{9}{5} \end{cases}$

38. $\begin{cases} 2x - y = 5 \\ 5x + 3z = 19 \\ 4y + 7z = 17 \end{cases}$

39. $\begin{cases} 3y + 5z = 4 \\ 2x - z = 10 \\ 4x + 7y = 0 \end{cases}$

40. $\begin{cases} 2x - 5y = 4 \\ x + y - z = 8 \\ 3x + 5z = 0 \end{cases}$

41. $\begin{cases} 3r - s + 3t = 7 \\ 4r + 5s - 2t = 0 \\ 9r + s + t = 0 \end{cases}$

42. $\begin{cases} \theta + \phi + \psi = 2 \\ 2\theta - \phi + \psi = 4 \\ \theta - 3\phi + 2\psi = 0 \end{cases}$

43. $\begin{cases} x + y + z + w = 0 \\ 2x + w = 0 \\ y - z = 0 \\ x + 2z = 1 \end{cases}$

44. $\begin{cases} x + y = 1 \\ y + z = 2 \\ z + w = 3 \\ w - x = 4 \end{cases}$

45. (a) Show that the equation

$$\begin{vmatrix} x_1 & y_1 & 1 \\ x_2 & y_2 & 1 \\ x & y & 1 \end{vmatrix} = 0$$

is an equation for the line that passes through the points (x_1, y_1) and (x_2, y_2).
(b) Use the result of part (a) to find an equation for the line that passes through the points $(20, 50)$ and $(-10, 25)$.

46. Evaluate the determinant

$$\begin{vmatrix} a & a & a & a & a \\ 0 & a & a & a & a \\ 0 & 0 & a & a & a \\ 0 & 0 & 0 & a & a \\ 0 & 0 & 0 & 0 & a \end{vmatrix}$$

47–50 ■ Solve for x.

47. $\begin{vmatrix} x & 12 & 13 \\ 0 & x-1 & 23 \\ 0 & 0 & x-2 \end{vmatrix} = 0$

48. $\begin{vmatrix} x & 1 & 1 \\ 1 & 1 & x \\ x & 1 & x \end{vmatrix} = 0$

49. $\begin{vmatrix} 1 & 0 & x \\ x^2 & 1 & 0 \\ x & 0 & 1 \end{vmatrix} = 0$

50. $\begin{vmatrix} a & b & x-a \\ x & x+b & x \\ 0 & 1 & 1 \end{vmatrix} = 0$

6.7 NONLINEAR SYSTEMS

Up to this point we have been studying systems of *linear* equations. As we have seen, mathematicians have developed several techniques for handling such systems. In calculus and the sciences, however, one often encounters systems of nonlinear equations, so we study them in this section. There are no general techniques for solving nonlinear systems like the ones we have been applying to linear systems. We have to approach each problem individually and solve it using whatever improvised method or combination of methods happens to work in that particular situation.

The technique we use most often is simple substitution. If the system we are dealing with consists of a linear equation and a quadratic polynomial in two variables (as in Example 1), then this method always gives us the complete solution.

EXAMPLE 1 ■ The Substitution Method

Find all solutions of the following system.

$$\begin{cases} x^2 + y^2 = 100 \\ 3x - y = 10 \end{cases}$$

SOLUTION

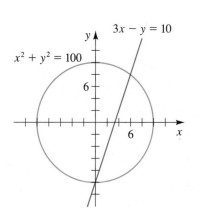

FIGURE 1

The graph of the first equation is a circle and the graph of the second is a line (see Figure 1). The figure shows that the graphs intersect in two points, so the system has two solutions. We solve the system by solving for y in the second equation and then substituting into the first.

$y = 3x - 10$	Solve for y in second equation
$x^2 + (3x - 10)^2 = 100$	Substitute $y = 3x - 10$ in first equation
$x^2 + (9x^2 - 60x + 100) = 100$	Expand
$10x^2 - 60x = 0$	Simplify
$10x(x - 6) = 0$	Factor

$$x = 0 \quad \text{or} \quad x = 6$$

If $x = 0$, then $y = 3(0) - 10 = -10$, and if $x = 6$, then $y = 3(6) - 10 = 8$. Thus, the solutions are $(0, -10)$ and $(6, 8)$.

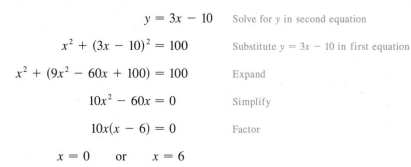

EXAMPLE 2 ■ The Elimination Method

Solve the following system.

$$\begin{cases} x^2 + 2y^2 = 11 \\ 3x^2 + 4y = 23 \end{cases}$$

SOLUTION

Here we could solve for y in the second equation and substitute into the first, just as we did in Example 1. But this would lead to a fourth-degree equation in x involving fractions, which might be difficult to solve. Instead, we multiply the first equation by 3 and subtract the second from it to eliminate the x term:

$$\begin{array}{r} 3x^2 + 6y^2 = 33 \\ \underline{3x^2 + 4y = 23} \\ 6y^2 - 4y = 10 \end{array}$$

We now solve this new equation by factoring:

$$6y^2 - 4y - 10 = 0$$

$$2(3y^2 - 2y - 5) = 0 \qquad \text{Factor out 2}$$

$$2(y + 1)(3y - 5) = 0 \qquad \text{Factor the quadratic}$$

$$y = -1 \qquad \text{or} \qquad y = \tfrac{5}{3}$$

We can now solve for the corresponding values of x by substituting into either of the two original equations. If $y = -1$, then using the second equation, we get

$$3x^2 + 4(-1) = 23$$

$$3x^2 = 27$$

$$x^2 = 9$$

$$x = 3 \qquad \text{or} \qquad x = -3$$

If $y = \tfrac{5}{3}$, then

$$3x^2 + 4\left(\tfrac{5}{3}\right) = 23$$

$$3x^2 = 23 - \tfrac{20}{3} = \tfrac{49}{3}$$

$$x^2 = \tfrac{49}{9}$$

$$x = \tfrac{7}{3} \qquad \text{or} \qquad x = -\tfrac{7}{3}$$

The solutions are the intersection points of an *ellipse* and a *parabola*, curves which are studied in Chapter 7.

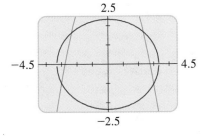

This means that the system has four solutions:

$$(3, -1) \qquad (-3, -1) \qquad \left(\tfrac{7}{3}, \tfrac{5}{3}\right) \qquad \left(-\tfrac{7}{3}, \tfrac{5}{3}\right)$$

You should check that each of these ordered pairs satisfies both original equations. ∎

Some nonlinear systems can be changed to linear ones using substitution, like the system in the next example. We can solve these using any of the available linear methods.

EXAMPLE 3 ∎ **Using Substitution to Change a Nonlinear System to a Linear System**

Solve the following system.

$$\begin{cases} 4x^3 + 6y^2 = 22 \\ 5x^3 + 8y^2 = 32 \end{cases}$$

SOLUTION

If we substitute $u = x^3$ and $v = y^2$, then the system becomes linear:

$$\begin{cases} 4u + 6v = 22 \\ 5u + 8v = 32 \end{cases}$$

We can solve this using Gaussian elimination, matrix inversion, Cramer's Rule, or any method that works on linear systems. Omitting the details of the calculations, we have

$$u = -8 \quad \text{and} \quad v = 9$$

Since $u = x^3$ and $v = y^2$, this gives us

$$x^3 = -8 \quad \text{and} \quad y^2 = 9$$

$$x = -2 \qquad\qquad y = \pm 3$$

So, the solutions of the original system are

$$(-2, 3) \quad \text{and} \quad (-2, -3)$$ ∎

EXAMPLE 4 ∎ **Using Factoring and Substitution to Solve a Nonlinear System**

Solve the following system.

$$\begin{cases} 3x^2 + 2y^2 + 15x = 0 \\ xy + y^2 = 0 \end{cases}$$

Julia Robinson (1919–1985) was born in St. Louis, Missouri, and grew up at Point Loma, California. Due to an illness, Robinson missed two years of school but later, with the aid of a tutor, she completed fifth, sixth, seventh, and eighth grades, all in one year. At San Diego State University she became especially interested in mathematics after reading biographies of mathematicians in the book *Men of Mathematics* by E. T. Bell. She said, "I cannot overemphasize the importance of such books... in the intellectual life of a student." Robinson is famous for her work on Hilbert's tenth problem (page 490), which asks for a general procedure for determining whether an equation has integer solutions. Her ideas led to a complete answer to the problem. Interestingly, the answer involved certain properties of the Fibonacci numbers (page 592) discovered by the then-22-year-old Russian mathematician Yuri Matijasevič. As a result of her brilliant work on Hilbert's tenth problem, Robinson was offered a professorship at the University of California, Berkley, and became the first *(continued)*

SOLUTION

We begin by tackling the second equation, since it looks more manageable. Factoring, we get

$$(x + y)y = 0$$

so $x = -y$ or $y = 0$

We now substitute each of these possibilities into the first equation to determine the consequences. First, if $x = -y$, then

$$3(-y)^2 + 2y^2 + 15(-y) = 0 \qquad \text{Substitute } x = -y$$

$$5y^2 - 15y = 0 \qquad \text{Simplify}$$

$$5y(y - 3) = 0 \qquad \text{Factor}$$

$$y = 0 \quad \text{or} \quad y = 3$$

Since these values for y were obtained from the assumption that $x = -y$, we get the following solutions of the original system:

$$(0, 0) \quad \text{and} \quad (-3, 3)$$

Second, we check what happens in the first equation if $y = 0$:

$$3x^2 + 2(0)^2 + 15x = 0 \qquad \text{Substitute } y = 0$$

$$3x^2 + 15x = 0 \qquad \text{Simplify}$$

$$3x(x + 5) = 0 \qquad \text{Factor}$$

$$x = 0 \quad \text{or} \quad x = -5$$

This leads to the solutions $(0, 0)$ and $(-5, 0)$. We have already found the first of these, so all the solutions are

$$(0, 0) \qquad (-3, 3) \qquad (-5, 0) \qquad \blacksquare$$

The next example involves three equations and three variables.

EXAMPLE 5 ■ A Nonlinear System with Three Variables

Solve the following system.

$$\begin{cases} x + yz = 0 \\ y + 4xz = 0 \\ x^2 + y^2 = 20 \end{cases}$$

woman mathematician elected to the National Academy of Sciences. She also served as president of the American Mathematical Society.

SOLUTION

If we solve for x in the first equation, we get $x = -yz$. Substituting into the second equation, we get

$$y + 4(-yz)z = 0 \qquad \text{Substitute } x = -yz$$

$$y(1 - 4z^2) = 0 \qquad \text{Factor}$$

$$y = 0 \quad \text{or} \quad z^2 = \tfrac{1}{4}$$

Substituting $y = 0$ into the first equation leads to $x = 0$, but this does not satisfy the third equation. So, it must be true that $z^2 = \tfrac{1}{4}$, or $z = \pm\tfrac{1}{2}$. Letting $z = \pm\tfrac{1}{2}$ in the second equation, we get $y = \mp 2x$. The third equation then becomes

$$x^2 + 4x^2 = 20 \qquad \text{Substitute } y = \mp 2x$$

$$5x^2 = 20 \qquad \text{Simplify}$$

$$x^2 = 4 \qquad \text{Divide by 5}$$

$$x = \pm 2$$

This leads to four solutions:

$$\left(2, 4, -\tfrac{1}{2}\right) \qquad \left(2, -4, \tfrac{1}{2}\right) \qquad \left(-2, 4, \tfrac{1}{2}\right) \qquad \left(-2, -4, -\tfrac{1}{2}\right) \qquad \blacksquare$$

We now consider an application that arises from a problem in geometry.

EXAMPLE 6 ■ An Application of Nonlinear Systems to Geometry

A right triangle has an area of 120 ft^2 and a perimeter of 60 ft. Find the lengths of its sides.

SOLUTION

Let x and y be the lengths of the sides adjacent to the right angle. Then, by the Pythagorean Theorem, the hypotenuse has length $\sqrt{x^2 + y^2}$ (see Figure 2). Since the area is 120 ft^2, we have

$$\tfrac{1}{2}xy = 120$$

or

$$xy = 240$$

Also, since the perimeter is 60 ft,

$$x + y + \sqrt{x^2 + y^2} = 60$$

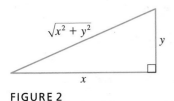

FIGURE 2

We simplify this equation as follows:

$$\sqrt{x^2 + y^2} = 60 - x - y$$

$$\left(\sqrt{x^2 + y^2}\right)^2 = (60 - x - y)^2 \qquad \text{Square each side}$$

$$x^2 + y^2 = 3600 + x^2 + y^2 - 120x - 120y + 2xy \qquad \text{Expand}$$

$$120x + 120y = 3600 + 2xy \qquad \text{Combine like terms}$$

$$60x + 60y = 1800 + xy \qquad \text{Divide by 2}$$

Thus, we must solve the system

$$\begin{cases} xy = 240 \\ 60x + 60y = 1800 + xy \end{cases}$$

Adding these equations (to eliminate the xy term) and simplifying, we get

$$60x + 60y = 2040$$

$$x + y = 34 \qquad \text{Divide by 60}$$

$$y = 34 - x \qquad \text{Solve for } y$$

Substituting this into the equation $xy = 240$ gives

$$x(34 - x) = 240$$

$$x^2 - 34x + 240 = 0$$

$$(x - 24)(x - 10) = 0$$

So, either $x = 24$ or $x = 10$. Since $y = 240/x$, we get $y = 10$ or $y = 24$, respectively. In either case, the hypotenuse is

$$\sqrt{(10)^2 + (24)^2} = \sqrt{676} = 26$$

The lengths of the sides of the triangle are 10 ft, 24 ft, and 26 ft. ■

CHECK YOUR ANSWER

Area of triangle $= \frac{1}{2}xy$

$$= \frac{1}{2} \cdot 10 \cdot 24$$

$$= 120 \text{ ft}^2$$

Perimeter $= 10 + 24 + 26$

$$= 60 \text{ ft} \qquad ✓$$

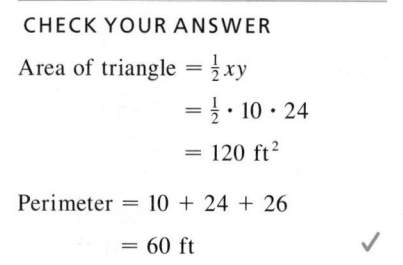

USING GRAPHING DEVICES TO SOLVE NONLINEAR SYSTEMS

Graphing devices are sometimes useful in solving systems of equations that involve just two variables. Note that with most graphing devices, any equation must first be expressed in terms of one or more functions of the form $y = f(x)$ before we can use the calculator to graph it. Not all equations can be readily expressed in this way, so not all systems can be solved by this technique.

EXAMPLE 7 ■ Solving a Nonlinear System with a Graphing Calculator

Find all solutions of the following system, correct to one decimal.

$$\begin{cases} \dfrac{x^2}{12} + \dfrac{y^2}{7} = 1 \\ y = 3x^2 - 6x + \frac{1}{2} \end{cases}$$

SOLUTION

The graph of the first equation has x-intercepts $x = \pm\sqrt{12} \approx \pm3.46$ and y-intercepts $y = \pm\sqrt{7} \approx \pm2.65$. We therefore select the viewing rectangle $[-4, 4]$ by $[-3, 3]$ to be sure that our graph contains the intercepts. Solving for y in terms of x, we get

$$\frac{y^2}{7} = 1 - \frac{x^2}{12}$$

$$y^2 = 7\left(1 - \frac{x^2}{12}\right)$$

$$y = \pm\sqrt{7\left(1 - \frac{x^2}{12}\right)}$$

To graph the entire curve, we must graph both of the functions

$$y = \sqrt{7[1 - (x^2/12)]} \quad \text{and} \quad y = -\sqrt{7[1 - (x^2/12)]}$$

We graph the second equation in the system in the same viewing rectangle as the first (see Figure 3). The two equations in the system appear to intersect at three points. First, we zoom in to the two points above the x-axis. Their approximate coordinates are $(-0.3, 2.6)$ and $(2.2, 2.0)$. There also appears to be an intersection point in quadrant IV. However, when we zoom in we see that the curves come close to each other here but do not intersect (see Figure 4). Thus the system has only two solutions; correct to the nearest tenth, they are

$$(-0.3, 2.6) \quad \text{and} \quad (2.2, 2.0)$$

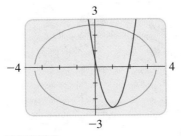

FIGURE 3
$\dfrac{x^2}{12} + \dfrac{y^2}{7} = 1, \ y = 3x^2 - 6x + \frac{1}{2}$

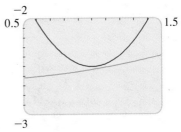

FIGURE 4

■

6.7 EXERCISES

1–26 ■ Find all solutions (x, y) of the system of equations.

1. $\begin{cases} y = x^2 \\ y = x + 6 \end{cases}$

2. $\begin{cases} x^2 + y^2 = 25 \\ y = \frac{3}{4}x \end{cases}$

3. $\begin{cases} x^2 + y^2 = 8 \\ x + y = 0 \end{cases}$

4. $\begin{cases} x^2 + y = 9 \\ x - y + 3 = 0 \end{cases}$

5. $\begin{cases} y + x^2 = 4x \\ y + 4x = 16 \end{cases}$

6. $\begin{cases} x - y^2 = 0 \\ y - x^2 = 0 \end{cases}$

7. $\begin{cases} x - 2y = 2 \\ y^2 - x^2 = 2x + 4 \end{cases}$

8. $\begin{cases} y = 4 - x^2 \\ y = x^2 - 4 \end{cases}$

9. $\begin{cases} x - y = 4 \\ xy = 12 \end{cases}$

10. $\begin{cases} x^3 - y = 0 \\ -2x + y = 4 \end{cases}$

11. $\begin{cases} 3x^2 - y^2 = 11 \\ x^2 + 4y^2 = 8 \end{cases}$

12. $\begin{cases} xy = 24 \\ 2x^2 - y^2 + 4 = 0 \end{cases}$

13. $\begin{cases} x^2y = 16 \\ x^2 + 4y + 16 = 0 \end{cases}$

14. $\begin{cases} 2x^2 + 4y = 13 \\ x^2 - y^2 = \frac{7}{2} \end{cases}$

15. $\begin{cases} x + \sqrt{y} = 0 \\ y^2 - 4x^2 = 12 \end{cases}$

16. $\begin{cases} \sqrt{x} - 2\sqrt{y} = 1 \\ 2x - 4(y + \sqrt{y}) = 2 \end{cases}$

17. $\begin{cases} x^2 + y^2 = 9 \\ x^2 - y^2 = 1 \end{cases}$

18. $\begin{cases} x^2 + 2y^2 = 2 \\ 2x^2 - 3y = 15 \end{cases}$

19. $\begin{cases} 2x^2 - 8y^3 = 19 \\ 4x^2 + 16y^3 = 34 \end{cases}$

20. $\begin{cases} x^4 - y^3 = 17 \\ 3x^4 + 5y^3 = 53 \end{cases}$

21. $\begin{cases} \dfrac{2}{x} - \dfrac{3}{y} = 1 \\ -\dfrac{4}{x} + \dfrac{7}{y} = 1 \end{cases}$

22. $\begin{cases} \dfrac{4}{x^2} + \dfrac{6}{y^4} = \dfrac{7}{2} \\ \dfrac{1}{x^2} - \dfrac{2}{y^4} = 0 \end{cases}$

23. $\begin{cases} 3\sqrt{x} + 5\sqrt{y} = 19 \\ 2\sqrt{x} + 7\sqrt{y} = 20 \end{cases}$

24. $\begin{cases} 2\sqrt{x} - \sqrt{y} = 3 \\ 4\sqrt{x} + 3\sqrt{y} = 1 \end{cases}$

25. $\begin{cases} x^2 - xy + 2y^2 = 8 \\ x^3 - xy^2 = 0 \end{cases}$

26. $\begin{cases} xy - 3x = 0 \\ x^3 - y + 11 = 0 \end{cases}$

27–32 ■ Find all solutions (x, y, z) of the system of equations.

27. $\begin{cases} x - y = 2 \\ y + z = 0 \\ x^2 + y^2 + z^2 = 4 \end{cases}$

28. $\begin{cases} xy + z = 0 \\ yz + x = 0 \\ x^2 + y^2 = 2 \end{cases}$

29. $\begin{cases} x^2 + yz = 0 \\ y + xz = 2 \\ xyz = 1 \end{cases}$

30. $\begin{cases} xy - xz = 0 \\ y + yz = 2 \\ x^2 + y^2 = 5 \end{cases}$

31. $\begin{cases} x^2 + y + z = 0 \\ 2x^2 - y + 3z = -4 \\ y^2 + yz = 0 \end{cases}$

32. $\begin{cases} x^2 + y^3 + z^4 = 4 \\ 3x^2 - y^3 - z^4 = 12 \\ 2x^2 - 3y^3 + 2z^4 = 13 \end{cases}$

33–40 ■ Use a graphing device to find all solutions (x, y) of the system of equations.

33. $\begin{cases} x^2 + y^2 = 25 \\ x + 3y = 2 \end{cases}$

34. $\begin{cases} x^2 + y^2 = 17 \\ x^2 - 2x + y^2 = 13 \end{cases}$

35. $\begin{cases} \dfrac{x^2}{9} + \dfrac{y^2}{18} = 1 \\ y = -x^2 + 6x - 2 \end{cases}$

36. $\begin{cases} x^2 - y^2 = 3 \\ y = x^2 - 2x - 8 \end{cases}$

37. $\begin{cases} x^4 + 16y^4 = 32 \\ x^2 + 2x + y = 0 \end{cases}$

38. $\begin{cases} x^2 - 2x - y^2 = 0 \\ x - y^5 = 0 \end{cases}$

39. $\begin{cases} y = e^x + e^{-x} \\ y = 5 - x^2 \end{cases}$

40. $\begin{cases} \ln x + \ln y = 3 \\ (\ln x)^2 - \ln y = 0 \end{cases}$

41. A right triangle has a perimeter of 40 cm and an area of 60 cm². What are the lengths of its sides?

42. A rectangle has an area of 180 cm² and a perimeter of 54 cm. What are its dimensions?

43. A right triangle has an area of 54 in². The product of the lengths of the three sides is 1620 in³. What are the lengths of its sides?

44. A right triangle has an area of 84 ft² and a hypotenuse 25 ft long. What are the lengths of its other two sides?

45. The perimeter of a rectangle is 70 and its diagonal is 25. Find its length and width.

46. A circular piece of sheet metal has a diameter of 20 in. The edges are to be cut off to form a rectangle of area 160 in² (see the figure). What are the dimensions of the rectangle?

47. A hill is inclined so that its "slope" is $\frac{1}{2}$, as shown in the figure. We introduce a coordinate system with the origin at the base of the hill and with the scales on the axes measured in meters. A rocket is fired from the base of the hill in such a way that its trajectory is the parabola $y = -x^2 + 401x$. At what point does the rocket strike the hillside? How far is this point from the base of the hill (to the nearest centimeter)?

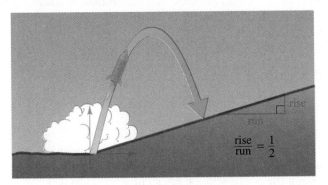

$$\frac{\text{rise}}{\text{run}} = \frac{1}{2}$$

48. A rectangular piece of sheet metal with an area of 1200 in² is to be bent into a cylindrical length of stovepipe having a volume of 600 in³. What are the dimensions of the sheet metal?

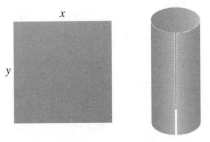

49. Find an equation for the line that passes through the points of intersection of the circles $x^2 + y^2 = 25$ and $x^2 - 3x + y^2 + y = 30$.

50. (a) For what value of k does the following system have exactly one solution?

$$\begin{cases} y = x^2 \\ y = x + k \end{cases}$$

(b) Graph both equations in the system on the same set of axes, using the value of k you chose in part (a).

(c) Based on your graph, how many solutions will the system have if k is smaller than your value from part (a)? How many will it have if k is larger?

51–54 ■ Find all solutions of the system.

51. $\begin{cases} x - y = 3 \\ x^3 - y^3 = 387 \end{cases}$ [*Hint:* Factor the left side of the second equation.]

52. $\begin{cases} x^2 + xy + xz = 1 \\ xy + y^2 + yz = 3 \\ xz + yz + z^2 = 5 \end{cases}$ [*Hint:* Add the equations and factor the result.]

53. $\begin{cases} 2^x + 2^y = 10 \\ 4^x + 4^y = 68 \end{cases}$

54. $\begin{cases} \log x + \log y = \frac{3}{2} \\ 2\log x - \log y = 0 \end{cases}$

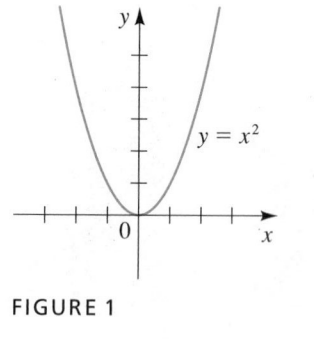

FIGURE 1

6.8   SYSTEMS OF INEQUALITIES

In this section we study systems of inequalities in two variables from a graphical point of view. First we consider the graph of a single inequality. We already know that the graph of $y = x^2$, for example, is the *parabola* in Figure 1. If we replace the equal sign by the symbol $\geq$, we obtain the **inequality**

$$y \geq x^2$$

Its graph consists of not just the parabola in Figure 1, but also every point whose y-coordinate is *larger* than x^2. We indicate the solution in Figure 2 by shading the points *above* the parabola.

Similarly, the graph of $y \leq x^2$ in Figure 3 consists of all points on and *below* the parabola, whereas the graphs of $y > x^2$ and $y < x^2$ do not include the points on the parabola itself, as indicated by the dashed curves in Figures 4 and 5.

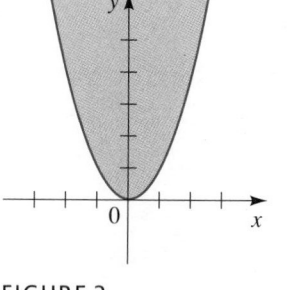

FIGURE 2
$y \geq x^2$

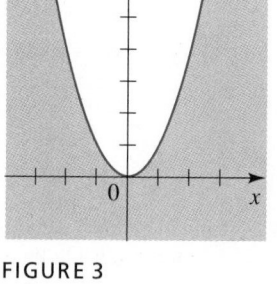

FIGURE 3
$y \leq x^2$

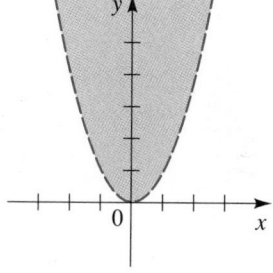

FIGURE 4
$y > x^2$

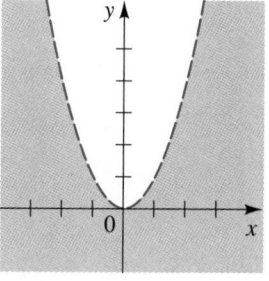

FIGURE 5
$y < x^2$

The graph of an inequality, in general, consists of a region in the plane whose boundary is the graph of the equation obtained by replacing the inequality sign ($\geq$, $\leq$, $>$, or $<$) with an equal sign. To determine which side of the graph gives the solution set of the inequality, we need to check only **test points,** as illustrated in the next example.

EXAMPLE 1 ■ Graphs of Nonlinear Inequalities

Graph each of the following inequalities.

(a) $x^2 + y^2 < 25$ (b) $x + 2y \geq 5$

SOLUTION

(a) The graph of $x^2 + y^2 = 25$ is a circle of radius 5 centered at the origin. The points on the circle itself do not satisfy the inequality because it is of the form $<$, so we graph the circle with a dashed curve in Figure 6.

To determine whether the inside or the outside of the circle satisfies the inequality, we use the test points $(0,0)$ on the inside and $(6,0)$ on the

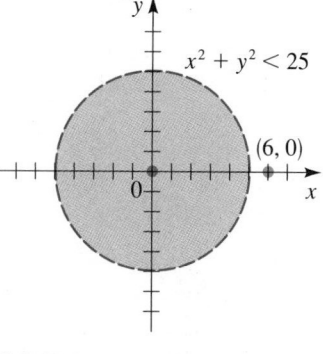

FIGURE 6

outside. (Note that *any* point inside or outside the circle can serve as a test point. We chose these for simplicity.)

Check: Does $(0, 0)$ satisfy $x^2 + y^2 < 25$?

$$0^2 + 0^2 \overset{?}{<} 25$$

$$0 \overset{?}{<} 25 \qquad \text{Yes}$$

Check: Does $(6, 0)$ satisfy $x^2 + y^2 < 25$

$$6^2 + 0^2 \overset{?}{<} 25$$

$$36 \overset{?}{<} 25 \qquad \text{No}$$

Thus, the graph of $x^2 + y^2 < 25$ is the set of all points inside the circle (see Figure 6).

(b) The graph of $x + 2y = 5$ is the line shown in Figure 7. We use the test points $(0, 0)$ and $(5, 5)$ on opposite sides of the line.

Check: Does $(0, 0)$ satisfy $x + 2y \geq 5$?

$$(0) + 2(0) \overset{?}{\geq} 5$$

$$0 \overset{?}{\geq} 5 \qquad \text{No}$$

Check: Does $(5, 5)$ satisfy $x + 2y \geq 5$?

$$(5) + 2(5) \overset{?}{\geq} 5$$

$$15 \overset{?}{\geq} 5 \qquad \text{Yes}$$

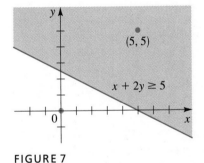

FIGURE 7

Our check shows that the points *above* the line satisfy the inequality.

Alternatively, we could put the inequality into slope-intercept form and graph it directly:

$$x + 2y \geq 5$$

$$2y \geq -x + 5$$

$$y \geq -\tfrac{1}{2}x + \tfrac{5}{2}$$

From this form, we see that the graph includes all points whose y-coordinates are *greater* than those on the line $y = -\tfrac{1}{2}x + \tfrac{5}{2}$; that is, the graph consists of the points *on or above* this line, as shown in Figure 7. ∎

EXAMPLE 2 ■ A System of Two Inequalities

Graph the solution set of the following pair of inequalities.

$$\begin{cases} x^2 + y^2 < 25 \\ x + 2y \geq 5 \end{cases}$$

SOLUTION

These are the two inequalities of Example 1. In this example we wish to graph only those points that simultaneously satisfy both inequalities. The solution thus consists of the intersection of the graphs in Example 1 (see Figure 8). The points $(-3, 4)$ and $(5, 0)$ in Figure 8 are called the **vertices** of the solution set. They are obtained by simultaneously solving the *equations*

$$\begin{cases} x^2 + y^2 = 25 \\ x + 2y = 5 \end{cases}$$

We solve this system of equations by substitution. Solving for x in the second equation gives $x = 5 - 2y$, and substituting this into the first equation gives

$$(5 - 2y)^2 + y^2 = 25 \quad \text{Substitute } x = 5 - 2y$$

$$(25 - 20y + 4y^2) + y^2 = 25 \quad \text{Expand}$$

$$-20y + 5y^2 = 0 \quad \text{Simplify}$$

$$-5y(4 - y) = 0 \quad \text{Factor}$$

Thus, $y = 0$ or $y = 4$. When $y = 0$, we have $x = 5 - 2(0) = 5$, and when $y = 4$, we have $x = 5 - 2(4) = -3$. So the points of intersection of these curves are $(5, 0)$ and $(-3, 4)$.

Note that in this case the vertices are not part of the solution set, since they do not satisfy the inequality $x^2 + y^2 < 25$ (and so they are graphed as open circles in the figure). They simply show where the "corners" of the solution set lie. ■

An inequality is **linear** if it can be put into one of the following forms:

$$ax + by \geq c \qquad ax + by \leq c \qquad ax + by > c \qquad ax + by < c$$

In the next example we graph the solution set of a system of linear inequalities.

EXAMPLE 3 ■ A System of Four Linear Inequalities

Graph the solution set of the following system.

$$\begin{cases} x + 3y \leq 12 \\ x + y \leq 8 \\ x \geq 3 \\ y \geq 0 \end{cases}$$

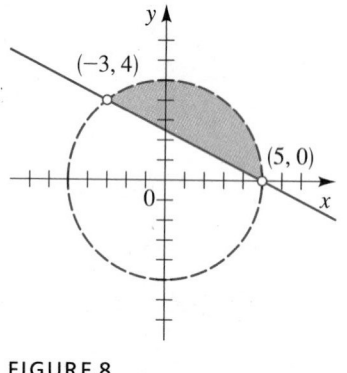

FIGURE 8

$$\begin{cases} x^2 + y^2 < 25 \\ x + 2y \geq 5 \end{cases}$$

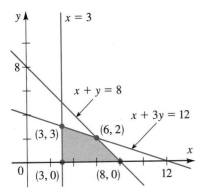

FIGURE 9

SOLUTION

In Figure 9 we first graph the lines given by the equations that correspond to each of the inequalities. The shaded region is the set of points that satisfy all four inequalities simultaneously. The coordinates of each vertex are obtained by simultaneously solving the equations of the lines that intersect at that vertex. For example, the vertex $(6, 2)$ lies on both the lines

$$x + 3y = 12$$

and

$$x + y = 8$$

In this case all the vertices *are* part of the solution set. ∎

EXAMPLE 4 ■ An Application of Systems of Linear Inequalities

A manufacturer of insulating materials produces two different brands: Foamboard and Plastiflex. Each cubic yard of Foamboard weighs 8 lb, and each cubic yard of Plastiflex weighs 24 lb. The products are moved from the plant to the loading dock on carts that have a maximum capacity of 25 yd^3 and 432 lb. Find a system of inequalities that describes all possible combinations of Foamboard and Plastiflex that can be carried on such a cart. Graph the solution set of this system.

SOLUTION

First we let

$$x = \text{number of cubic yards of Foamboard on a cart}$$

$$y = \text{number of cubic yards of Plastiflex on a cart}$$

Since a cart can carry no more than 25 yd^3, we have

$$x + y \leq 25$$

In addition, the total weight carried on a cart cannot exceed 432 lb. Since each cart carries $8x$ pounds of Foamboard and $24y$ pounds of Plastiflex, this means that

$$8x + 24y \leq 432$$

Dividing each side of this inequality by 8 simplifies it to

$$x + 3y \leq 54$$

Finally, negative amounts would be meaningless in this context, so

$$x \geq 0 \quad \text{and} \quad y \geq 0$$

Thus, the possible amounts of material that a cart can hold are given by the system

$$\begin{cases} x + y \le 25 \\ x + 3y \le 54 \\ x \ge 0 \\ y \ge 0 \end{cases}$$

The graph is shown in Figure 10.

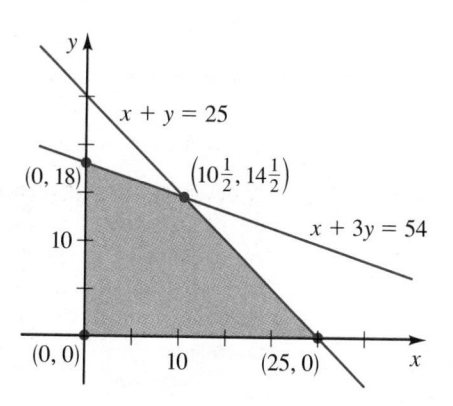

FIGURE 10

When a region in the plane can be covered by a (sufficiently large) circle, it is said to be **bounded.** A region that is not bounded is called **unbounded.** For example, the regions graphed in Figures 6, 8, 9, and 10 are bounded, whereas those in Figures 2, 3, 4, 5, and 7 are unbounded. An unbounded region cannot be "fenced in"—it extends infinitely far in some direction.

6.8 EXERCISES

1–14 ■ Graph the inequality.

1. $x \le 2$

2. $y > -3$

3. $y > x$

4. $y < x + 2$

5. $y \ge 2x + 2$

6. $y < -x + 5$

7. $2x - y \le 8$

8. $3x + 4y + 12 > 0$

9. $4x + 5y < 25$

10. $-x^2 + y \ge 10$

11. $y > x^3 + 1$

12. $x - y^2 \le 0$

13. $x^2 + y^2 \ge 5$

14. $xy < 4$

15–36 ■ Graph the solution of the system of inequalities. In each case, find the coordinates of all vertices, and determine whether the solution set is bounded.

15. $\begin{cases} x + y \le 4 \\ y \ge x \end{cases}$

16. $\begin{cases} 2x + 3y > 12 \\ 3x - y < 21 \end{cases}$

17. $\begin{cases} y < \frac{1}{4}x + 2 \\ y \ge 2x - 5 \end{cases}$

18. $\begin{cases} x - y > 0 \\ 4 + y \le 2x \end{cases}$

19. $\begin{cases} x \ge 0 \\ y \ge 0 \\ 3x + 5y \le 15 \\ 3x + 2y \le 9 \end{cases}$

20. $\begin{cases} x > 2 \\ y < 12 \\ 2x - 4y > 8 \end{cases}$

21. $\begin{cases} y < 9 - x^2 \\ y \geqslant x + 3 \end{cases}$

22. $\begin{cases} x \geqslant y^2 \\ x + y \geqslant 6 \end{cases}$

23. $\begin{cases} x^2 + y^2 \leqslant 4 \\ x - y > 0 \end{cases}$

24. $\begin{cases} x > 0 \\ y > 0 \\ x + y < 10 \\ x^2 + y^2 > 9 \end{cases}$

25. $\begin{cases} x^2 - y \leqslant 0 \\ 2x^2 + y \leqslant 12 \end{cases}$

26. $\begin{cases} x^2 + y^2 < 9 \\ 2x + y^2 \geqslant 1 \end{cases}$

27. $\begin{cases} x + 2y \leqslant 14 \\ 3x - y \geqslant 0 \\ x - y \geqslant 2 \end{cases}$

28. $\begin{cases} y < x + 6 \\ 3x + 2y \geqslant 12 \\ x - 2y \leqslant 2 \end{cases}$

29. $\begin{cases} x \geqslant 0 \\ y \geqslant 0 \\ x \leqslant 5 \\ x + y \leqslant 7 \\ x + 2y \geqslant 4 \end{cases}$

30. $\begin{cases} x \geqslant 0 \\ y \geqslant 0 \\ y \leqslant 4 \\ 2x + y \leqslant 8 \\ 20x + 3y \leqslant 66 \end{cases}$

31. $\begin{cases} y > x + 1 \\ x + 2y \leqslant 12 \\ x + 1 > 0 \end{cases}$

32. $\begin{cases} x + y > 12 \\ y < \frac{1}{2}x - 6 \\ 3x + y < 6 \end{cases}$

33. $\begin{cases} x^2 + y^2 \leqslant 8 \\ x \geqslant 2 \\ y \geqslant 0 \end{cases}$

34. $\begin{cases} x^2 - y \geqslant 0 \\ x + y < 6 \\ x - y < 6 \end{cases}$

35. $\begin{cases} x^2 + y^2 < 9 \\ x + y > 0 \\ x \leqslant 0 \end{cases}$

36. $\begin{cases} y \geqslant x^3 \\ y \leqslant 2x + 4 \\ x + y \geqslant 0 \end{cases}$

37. A publishing company publishes a total of no more than 100 books every year. At least 20 of these are nonfiction, but the company always publishes at least as much fiction as nonfiction. Find a system of inequalities that describes the possible numbers of fiction and nonfiction books that the company can produce each year consistent with these policies. Graph the solution set.

38. A man and his daughter manufacture unfinished tables and chairs. Each table requires 3 h of sawing and 1 h of assembly. Each chair requires 2 h of sawing and 2 h of assembly. The two of them can do a total of up to 12 h of sawing and 8 h of assembly work each day. Find a system of inequalities that describes all possible combinations of tables and chairs that they can make daily. Graph the solution set.

6.9 APPLICATION: LINEAR PROGRAMMING

Linear programming is a mathematical technique used to determine the optimal allocation of resources in business, the military, and other areas of human endeavor. For example, a manufacturer who makes several different products from the same raw materials can use linear programming to determine how much of each product should be produced to maximize the profit. This technique is probably the most important practical application of systems of linear inequalities. In 1975 Leonid Kantorovich and T. C. Koopmans won the Nobel Prize in economics for their work in the development of this subject.

Although linear programming can be applied to very complex problems with hundreds or even thousands of variables, we will consider only a few simple examples to which the graphical methods of the preceding section can be applied. We introduce the technique with a typical problem.

EXAMPLE 1 ■ Manufacturing for Maximum Profit

A small shoe manufacturer makes two styles of shoes: oxfords and loafers. Two machines are used in the process: a cutting machine and a sewing

Linear programming is used by the telephone industry to determine the most efficient way to route telephone calls. The routing decisions have to be made very rapidly by computer so as not to keep callers waiting on the line for the telephone connection. Since the data in such problems are huge, a very fast method for solving linear programming problems is essential. In 1984 the 28-year-old mathematician Narendra Karmarkar, working at Bell Labs in Murray Hill, New Jersey, discovered just such a method. His idea is so ingenious and his method so fast that the discovery caused a sensation in the mathematical world. His technique is now also used by airlines in scheduling passengers, flight personnel, fuel, baggage, and maintenance workers so as to minimize costs. Although mathematical discoveries rarely make the news, this one was reported in *Time*, on December 3, 1984.

machine. Each type of shoe requires 15 min per pair on the cutting machine. Oxfords require 10 min of sewing per pair, and loafers require 20 min of sewing per pair. Because the manufacturer can hire only one operator for each machine, each process is available for just 8 h per day. If the profit on each pair of oxfords is $15 and on each pair of loafers is $20, how many pairs of each type should be produced per day for maximum profit?

SOLUTION

First we organize the given information into a table. To be consistent, we convert all times to hours.

	Oxfords	Loafers	Time available
Time on cutting machine (h)	$\frac{1}{4}$	$\frac{1}{4}$	8
Time on sewing machine (h)	$\frac{1}{6}$	$\frac{1}{3}$	8
Profit	$15	$20	

Let

$$x = \text{number of pairs of oxfords made daily}$$

$$y = \text{number of pairs of loafers made daily}$$

The total number of cutting hours needed is then $\frac{1}{4}x + \frac{1}{4}y$. Since only 8 h are available on the cutting machine, we have

$$\tfrac{1}{4}x + \tfrac{1}{4}y \leq 8$$

Similarly, by considering the amount of time needed and available on the sewing machine, we get

$$\tfrac{1}{6}x + \tfrac{1}{3}y \leq 8$$

We cannot produce a negative number of shoes, so we also have

$$x \geq 0 \quad \text{and} \quad y \geq 0$$

Thus x and y must satisfy the system of inequalities

$$\begin{cases} \tfrac{1}{4}x + \tfrac{1}{4}y \leq 8 \\ \tfrac{1}{6}x + \tfrac{1}{3}y \leq 8 \\ \qquad\qquad x \geq 0 \\ \qquad\qquad y \geq 0 \end{cases}$$

If we multiply the first inequality by 4 and the second by 6, we obtain the simplified system

$$\begin{cases} x + y \leq 32 \\ x + 2y \leq 48 \\ \ x \geq 0 \\ \ y \geq 0 \end{cases}$$

The solution of this system (with vertices labeled) is graphed in Figure 1.

We wish to determine which values for x and y give maximum profit. The only values that satisfy the restrictions of the problem are the ones that correspond to points of the shaded region in Figure 1. This is called the *feasible region* for the problem. Since each pair of oxfords provides $15 profit and each pair of loafers $20, the total profit will be

$$P = 15x + 20y$$

As x or y increases, profit will increase as well. Thus, it seems reasonable that the maximum profit will occur at a point on one of the outside edges of the feasible region, where it is impossible to increase x or y without going outside the region. In fact, it can be shown that the maximum value will occur at a vertex. This means that we need to check the profit only at the vertices.

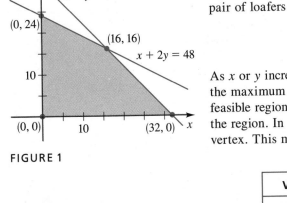

FIGURE 1

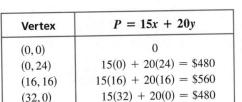

Vertex	$P = 15x + 20y$	
(0, 0)	0	
(0, 24)	$15(0) + 20(24) = \$480$	
(16, 16)	$15(16) + 20(16) = \$560$	$\leftarrow$ maximum profit
(32, 0)	$15(32) + 20(0) = \$480$	

The largest value of P occurs at the point $(16, 16)$, where $P = \$560$. Thus the manufacturer should make 16 pairs of oxfords and 16 pairs of loafers, for a maximum daily profit of $560. ∎

All the linear programming problems that we consider follow the pattern of this example. Each problem involves two variables, and certain restrictions described in the problem lead to a system of linear inequalities that involve these variables. The graph of this system is called the **feasible region.** We consider only bounded feasible regions. The function we are trying to maximize or minimize is called the **objective function.** This function will always attain its largest and smallest values at the **vertices** of the feasible region, so checking its value at all vertices gives the solution to the problem.

EXAMPLE 2 ■ A Shipping Problem

A car dealer has warehouses in Millville and Trenton and dealerships in Camden and Atlantic City. Every car sold at the dealerships must be delivered from one of the warehouses. On a certain day the Camden dealers sell 10 cars and the Atlantic City dealers sell 12. The Millville warehouse has 15 cars available and the Trenton warehouse has 10. The cost of shipping one car is $50 from Millville to Camden, $40 from Millville to Atlantic City, $60 from Trenton to Camden, and $55 from Trenton to Atlantic City. How many cars should be moved from each warehouse to each dealership to fill the orders at minimum cost?

SOLUTION

The first step is to organize the given information. Rather than construct a table, we draw a diagram to show the flow of cars from the warehouses to the dealerships (see Figure 2). The diagram shows the number of cars available at each warehouse or required at each dealership and the cost of shipping between these locations.

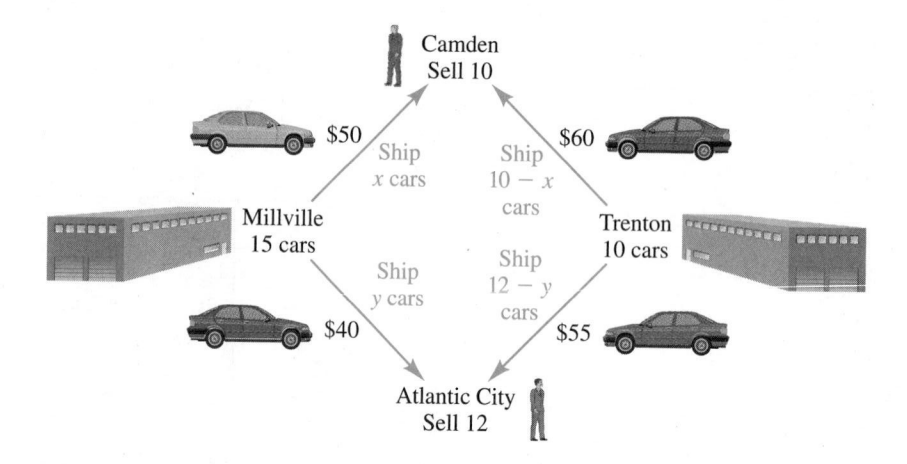

FIGURE 2

The arrows in Figure 2 indicate four possible routes, so the problem seems to involve four variables. But if we let x be the number of cars to be shipped from Millville to Camden, then $10 - x$ cars would be shipped from Trenton to Camden, because the Camden dealership needs 10 cars in all. Similarly, if y cars are shipped from Millville to Atlantic City, then $12 - y$ cars would be shipped from Trenton to Atlantic City.

We now derive the inequalities that define the feasible region. First, the number of cars shipped on each route cannot be negative, so we have

$$x \geq 0 \qquad\qquad y \geq 0$$

$$10 - x \geq 0 \qquad\qquad 12 - y \geq 0$$

Second, the total number of cars shipped from each warehouse cannot exceed the number of cars available there, so

$$x + y \le 15$$

$$(10 - x) + (12 - y) \le 10$$

Simplifying the latter inequality, we get

$$22 - x - y \le 10$$

$$-x - y \le -12$$

$$x + y \ge 12$$

The inequalities $10 - x \ge 0$ and $12 - y \ge 0$ can be rewritten as $x \le 10$ and $y \le 12$, respectively. Thus, the feasible region is described by the system of inequalities

$$\begin{cases} x \ge 0 \\ y \ge 0 \\ x \le 10 \\ y \le 12 \\ x + y \le 15 \\ x + y \ge 12 \end{cases}$$

FIGURE 3

The feasible region is graphed in Figure 3.

From Figure 2, we see that the total cost of shipping the cars is

$$C = 50x + 40y + 60(10 - x) + 55(12 - y)$$

$$= 50x + 40y + 600 - 60x + 660 - 55y$$

$$= 1260 - 10x - 15y$$

This is the objective function. We check its value at each vertex.

Vertex	$C = 1260 - 10x - 15y$	
$(0, 12)$	$1260 - 10(0) - 15(12) = \$1080$	
$(3, 12)$	$1260 - 10(3) - 15(12) = \$1050$	← minimum cost
$(10, 5)$	$1260 - 10(10) - 15(5) = \$1085$	
$(10, 2)$	$1260 - 10(10) - 15(2) = \$1130$	

The lowest cost is incurred at the point $(3, 12)$. Thus the dealer should ship

3 cars from Millville to Camden

12 cars from Millville to Atlantic City

7 cars from Trenton to Camden

0 cars from Trenton to Atlantic City   ■

In the 1940s mathematicians developed matrix methods for solving linear programming problems that involve more than two variables. These methods were first used by the Allies in World War II to solve supply problems similar to (but of course much more complicated than) Example 2. Improving these matrix methods is an active and exciting area of current mathematical research.

6.9 EXERCISES

1–4 ■ Find the maximum and minimum values of the given objective function on the feasible region sketched in the figure.

1. $M = 2x + 3y$

2. $N = 100x - 50y$

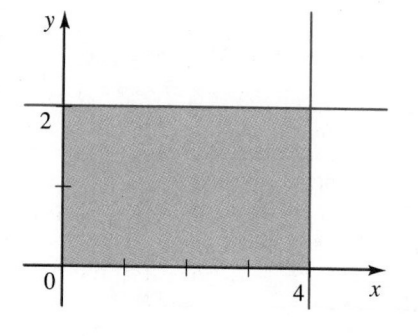

3. $P = 200 - x - y$

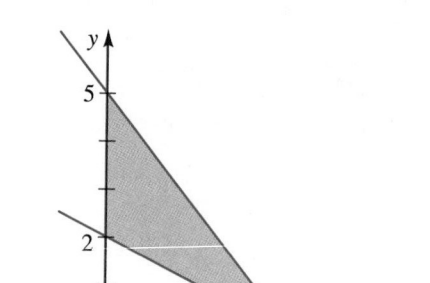

4. $R = \frac{1}{2}x + \frac{1}{4}y + 40$

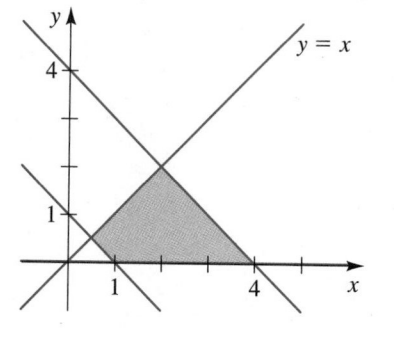

5–8 ■ The given set of inequalities describes a feasible region and an objective function. Graph the feasible region, determine the coordinates of its vertices, and then find the maximum and minimum values of the objective function on the feasible region by checking its values at the vertices.

5. $\begin{cases} x \geq 0, & y \geq 0 \\ 2x + y \leq 10 \\ 2x + 4y \leq 28 \end{cases}$

$P = 140 - x + 3y$

6. $\begin{cases} x \geq 0, & y \geq 0 \\ x \leq 10, & y \leq 20 \\ x + y \geq 5 \\ x + 2y \leq 18 \end{cases}$

$Q = 70x + 82y$

7. $\begin{cases} x \geq 3, & y \geq 4 \\ 2x + y \leq 24 \\ 2x + 3y \leq 36 \end{cases}$

$R = 12 + x + 4y$

8. $\begin{cases} x \leq 36, & y \leq 40 \\ x \leq 2y \\ 2x + y \geq 60 \end{cases}$

$S = 200 + 3x - y$

9. A furniture manufacturer makes wooden tables and chairs. The production process involves two basic types of labor: carpentry and finishing. Making a table requires 2 h of carpentry and 1 h of finishing, whereas making a chair requires 3 h of carpentry and $\frac{1}{2}$ h of finishing. The profit is $35 per table and $20 per chair. The manufacturer's employees can supply a maximum of 108 h of carpentry work and 20 h of finishing work per day. How many tables and chairs should be made each day to maximize profit?

10. A housing contractor has subdivided a farm into 100 building lots. He has designed two types of homes for these lots: colonial and ranch style. A colonial requires $30,000 of capital and produces a profit of $4000 when sold. A ranch-style house requires $40,000 of capital and provides an $8000 profit. If he has $3.6 million of capital on hand, how many houses of each type should he build for maximum profit? Will any of the lots be left vacant?

11. A trucker is planning to carry citrus fruit from Florida to Montreal. Each crate of oranges is 4 ft^3 in volume and weighs 80 lb. Each crate of grapefruit has a volume of 6 ft^3 and weighs 100 lb. Her truck has a maximum capacity of 300 ft^3 and can carry no more than 5600 lb. Moreover, she is not permitted to carry more crates of grapefruit than crates of oranges. If she makes a profit of $2.50 on each crate of oranges and $4 on each crate of grapefruit, how many crates of each type of fruit should she carry for maximum profit?

12. A manufacturer of calculators produces two models: standard and scientific. Long-term demand for the two models mandates that the company manufacture at least 100 standard and 80 scientific calculators each day. However, because of limitations on production capacity, no more than 200 standard and 170 scientific calculators can be made daily. To satisfy a shipping contract, a total of at least 200 calculators must be shipped every day.
(a) If it costs $5 to produce a standard calculator and $7 for a scientific one, how many of each model should be made each day to minimize production cost?
(b) If each standard calculator results in a $2 loss but each scientific one produces a $5 profit, how many of each model should be made each day to maximize profit?

13. An electronics discount chain has a sale on a certain brand of stereo. The chain has stores in Santa Monica and El Toro and warehouses in Long Beach and Pasadena. To satisfy rush orders, 15 sets must be shipped from the warehouses to the Santa Monica store, and 19 must be shipped to the El Toro store. The cost of shipping a set is $5 from Long Beach to Santa Monica, $6 from Long Beach to El Toro, $4 from Pasadena to Santa Monica, and $5.50 from Pasadena to El Toro. If the Long Beach warehouse has 24 sets and the Pasadena warehouse has 18 sets in stock, how many sets should be shipped from each warehouse to each store to fill the orders at a minimum shipping cost?

14. A man owns two building supply stores, one on the east side and one on the west side of a city. Two customers order some $\frac{1}{2}$-inch plywood. Customer A needs 50 sheets and customer B needs 70 sheets. The east-side store has 80 sheets and the west-side store has 45 sheets of this plywood in stock. Delivery cost per sheet is $0.50 from the east-side store to customer A, $0.60 from the east-side store to customer B, $0.40 from the west-side store to A, and $0.55 from the west-side store to B. How many sheets should be shipped from each store to each customer to minimize delivery cost?

15. Refer to Example 4 of Section 6.8. Suppose that a worker can load 12 yd^3 of Foamboard per minute and 10 yd^3 of Plastiflex per minute. If each cubic yard of Foamboard produces $0.50 profit and each cubic yard of Plastiflex produces $0.70, how should the carts be loaded to provide maximum profit per minute for each cartload?

16. A confectioner sells two types of nut mixtures. Each package of the standard mixture contains 100 g of cashews and 200 g of peanuts and sells for $1.95. Each package of the deluxe mixture contains 150 g of cashews and 50 g of peanuts and sells for $2.25. The confectioner has 15 kg of cashews and 20 kg of peanuts available. Based on past sales statistics, he wishes to have at least as many standard as deluxe packages available. How many bags of each type of mixture should he package to maximize his revenue?

17. A furniture manufacturer has factories in two locations. The Vancouver factory can produce 20 sofas, 20 chairs, and 35 ottomans each day, whereas the Seattle factory can produce 50 sofas, 25 chairs, and 25 ottomans each day. The daily operating cost is $3000 at the Vancouver factory and $4000 at the Seattle factory. An order for 400 sofas, 300 chairs, and 375 ottomans is received. The order is to be filled in no more than 30 days. How many days should each factory be operated to fill the order at minimum cost?

18. A biologist wishes to feed laboratory rabbits a mixture of two types of foods. Type I contains 8 g of fat, 12 g of carbohydrate, and 2 g of protein per ounce, whereas type II contains 12 g of fat, 12 g of carbohydrate, and 1 g of protein per ounce. Type I costs $0.20 per ounce and type II costs $0.30 per ounce. Each rabbit is to receive a daily minimum of 24 g of fat, 36 g of carbohydrate, and 4 g of protein, but should get no more than 5 oz of food per day. How many ounces of each type of food should be fed to each rabbit daily to satisfy the dietary requirements at minimum cost?

19. A woman wishes to invest $12,000 in three types of bonds: municipal bonds paying 7% interest per year,

bank investment certificates paying 8%, and high-risk bonds paying 12%. For tax reasons, she wants the amount invested in municipal bonds to be at least three times the amount invested in bank certificates. To keep her level of risk manageable, she will invest no more than $2000 in high-risk bonds. How much should she invest in each type of bond to maximize her annual interest yield? [*Hint:* Let $x =$ amount in municipal bonds and $y =$ amount in bank certificates. Then the amount in high-risk bonds will be $12,000 - x - y$.]

20. Refer to Exercise 19. Suppose the investor decides to increase to $3000 the maximum amount she will allow to be invested in high-risk bonds but leaves the other conditions unchanged. By how much will her maximum possible interest yield increase?

21. A small software company publishes computer games and educational and utility software. Their policy is to market a total of 36 new programs each year, with at least four of these being games. The number of utility programs published is never more than twice the number of educational programs. On average, the company can expect to make an annual profit of $5000 on each computer game, $8000 on each educational program, and $6000 on each utility program. How many of each type of program should they publish annually for maximum profit?

22. All parts of this problem refer to the following feasible region and objective function.

$$\begin{cases} x \ge 0 \\ x \ge y \\ x + 2y \le 12 \\ x + y \le 10 \end{cases}$$

$$P = x + 4y$$

(a) Graph the feasible region.

(b) On your graph from part (a), sketch the graphs of the linear equations obtained by setting P equal to 40, 36, 32, and 28.

(c) If we continue to decrease the value of P, at which vertex of the feasible region will these lines first touch the feasible region?

(d) Verify that the maximum value of P on the feasible region occurs at the vertex you chose in part (c).

6.10	**APPLICATION: PARTIAL FRACTIONS**

The process of adding and subtracting fractions by first writing them with a common denominator is familiar for both numerical fractions and rational functions. Thus, it is easy to add

$$\frac{1}{3} + \frac{1}{4} = \frac{4}{12} + \frac{3}{12} = \frac{4+3}{12} = \frac{7}{12}$$

or to subtract

$$\frac{1}{x-1} - \frac{1}{x+1} = \frac{x+1}{(x+1)(x-1)} - \frac{x-1}{(x+1)(x-1)} = \frac{2}{x^2-1}$$

The ancient Egyptians used a number notation that required all fractions be written as sums of reciprocals of whole numbers. For them it was therefore important to know how to reverse this process—for example, to be able to write 7/12 as the sum of the more elementary fractions 1/3 and 1/4. For us this skill is of little use in the context of numerical fractions. For rational functions, however, the process opposite to "bringing to a common denominator" turns out to be very important in the applications of algebra. In calculus, for example, one must know how to split up $2/(x^2 - 1)$ into the difference of the simpler functions $1/(x - 1)$ and $1/(x + 1)$. These simpler functions are called **partial fractions,** and since the process of finding them involves solving linear equations, we study it in this section.

Let r be the rational function

$$r(x) = \frac{P(x)}{Q(x)}$$

where the degree of P is less than the degree of Q. It can be shown, using advanced algebra techniques, that every polynomial with real coefficients can be factored completely into linear and irreducible quadratic factors; that is, factors of the form $ax + b$ and $ax^2 + bx + c$, where a, b, and c are real numbers. For instance,

$$x^4 - 1 = (x^2 - 1)(x^2 + 1) = (x - 1)(x + 1)(x^2 + 1)$$

After we have completely factored the denominator Q of r, we will be able to express $r(x)$ as a sum of **partial fractions** of the form

$$\frac{A}{(ax + b)^i} \qquad \text{and} \qquad \frac{Ax + B}{(ax^2 + bx + c)^j}$$

This sum is called the **partial fraction decomposition** of r. We now explain the details in the four cases that occur.

The Rhind papyrus is the oldest known mathematical document. It is an Egyptian scroll written in 1650 B.C. by the scribe Ahmes, who explains that it is an exact copy of a scroll written 200 years earlier. Ahmes claims that his papyrus contains "a thorough study of all things, insight into all that exists, knowledge of all obscure secrets." Actually, the document contains rules for doing arithmetic, including multiplication and division of fractions and several exercises with solutions. The exercise shown below reads: A heap and its seventh make nineteen; how large is the heap? In solving problems of this sort the Egyptians used partial fractions. In their notation legs "going" meant add; legs "coming" meant subtract.

The papyrus gives a correct formula for the volume of a truncated pyramid (page 67). It also gives the formula $A = \left(\frac{8}{9}d\right)^2$ for the area of a circle with diameter d. How close is this to the actual area?

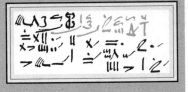

CASE 1: THE DENOMINATOR IS A PRODUCT OF DISTINCT LINEAR FACTORS

This means that we can write

$$Q(x) = (a_1x + b_1)(a_2x + b_2) \cdots (a_nx + b_n)$$

with no factor repeated. In this case the partial fraction decomposition of r takes the form

$$r(x) = \frac{P(x)}{Q(x)} = \frac{A_1}{a_1x + b_1} + \frac{A_2}{a_2x + b_2} + \cdots + \frac{A_n}{a_nx + b_n}$$

where the constants $A_1, A_2, \ldots, A_n$ are determined as in the following example.

EXAMPLE 1 ■ **Distinct Linear Factors**

Find the partial fraction decomposition of $\dfrac{5x + 7}{x^3 + 2x^2 - x - 2}$.

SOLUTION

The denominator factors as follows.

$$x^3 + 2x^2 - x - 2 = x^2(x + 2) - (x + 2) = (x^2 - 1)(x + 2)$$

$$= (x - 1)(x + 1)(x + 2)$$

This gives us the partial fraction decomposition

$$\frac{5x + 7}{x^3 + 2x^2 - x - 2} = \frac{A}{x - 1} + \frac{B}{x + 1} + \frac{C}{x + 2}$$

Multiplying each side by the common denominator, $(x - 1)(x + 1)(x + 2)$, we get

$$5x + 7 = A(x + 1)(x + 2) + B(x - 1)(x + 2) + C(x - 1)(x + 1)$$

$$= A(x^2 + 3x + 2) + B(x^2 + x - 2) + C(x^2 - 1) \qquad \text{Expand}$$

$$= (A + B + C)x^2 + (3A + B)x + (2A - 2B - C) \qquad \text{Combine like terms}$$

If two polynomials are equal, then their coefficients are equal. Thus, since $5x + 7$ has no x^2 term, we have that $A + B + C = 0$. Similarly, by comparing the coefficients of x, we see that $3A + B = 5$, and by comparing constant terms, we get that $2A - 2B - C = 7$. This leads to the following system of linear equations for A, B, and C.

$$\begin{cases} A + B + C = 0 \\ 3A + B = 5 \\ 2A - 2B - C = 7 \end{cases}$$

We solve this system using the methods developed in Section 6.2.

$$\begin{bmatrix} 1 & 1 & 1 & 0 \\ 3 & 1 & 0 & 5 \\ 2 & -2 & -1 & 7 \end{bmatrix} \xrightarrow[\substack{R_3 - 2R_1 \rightarrow R_3}]{R_2 - 3R_1 \rightarrow R_2} \begin{bmatrix} 1 & 1 & 1 & 0 \\ 0 & -2 & -3 & 5 \\ 0 & -4 & -3 & 7 \end{bmatrix} \xrightarrow[\substack{-R_2}]{R_3 - 2R_2 \rightarrow R_3}$$

$$\begin{bmatrix} 1 & 1 & 1 & 0 \\ 0 & 2 & 3 & -5 \\ 0 & 0 & 3 & -3 \end{bmatrix} \xrightarrow[\substack{\frac{1}{3}R_3}]{R_2 - R_3 \rightarrow R_2} \begin{bmatrix} 1 & 1 & 1 & 0 \\ 0 & 2 & 0 & -2 \\ 0 & 0 & 1 & -1 \end{bmatrix}$$

Thus, we see that $C = -1$, $B = -1$, and $A = 2$, so the required partial fraction decomposition is

$$\frac{5x + 7}{x^3 + 2x^2 - x - 2} = \frac{2}{x - 1} + \frac{-1}{x + 1} + \frac{-1}{x + 2} \qquad \blacksquare$$

The same method of attack works in each of the remaining cases. We set up the partial fraction decomposition with the unknown constants A, B, C, ... We then multiply each side of the resulting equation by the common denominator, simplify the right-hand side of the equation, and equate coefficients. This gives a set of linear equations that will always have a unique solution (provided the partial fraction decomposition has been set up correctly).

CASE 2: THE DENOMINATOR IS A PRODUCT OF LINEAR FACTORS, SOME OF WHICH ARE REPEATED

Suppose the complete factorization of $Q(x)$ contains the linear factor $ax + b$ repeated k times; that is, $(ax + b)^k$ is a factor of $Q(x)$. Then, corresponding to each such factor the partial fraction decomposition for $P(x)/Q(x)$ will contain

$$\frac{A_1}{ax + b} + \frac{A_2}{(ax + b)^2} + \cdots + \frac{A_k}{(ax + b)^k}$$

EXAMPLE 2 ■ **Repeated Linear Factors**

Find the partial fraction decomposition of $\dfrac{x^2 + 1}{x(x - 1)^3}$.

SOLUTION

Because the factor $x - 1$ is repeated three times in the denominator, the partial fraction decomposition has the form

$$\frac{x^2 + 1}{x(x - 1)^3} = \frac{A}{x} + \frac{B}{x - 1} + \frac{C}{(x - 1)^2} + \frac{D}{(x - 1)^3}$$

Multiplying each side by the common denominator, $x(x - 1)^3$, gives

$$x^2 + 1 = A(x - 1)^3 + Bx(x - 1)^2 + Cx(x - 1) + Dx$$

$$= A(x^3 - 3x^2 + 3x - 1) + B(x^3 - 2x^2 + x) + C(x^2 - x) + Dx \qquad \text{Expand}$$

$$= (A + B)x^3 + (-3A - 2B + C)x^2 + (3A + B - C + D)x - A \qquad \text{Combine like terms}$$

Equating coefficients, we get the equations

$$\begin{cases} A + B && = 0 \\ -3A - 2B + C && = 1 \\ 3A + B - C + D & = 0 \\ -A && = 1 \end{cases}$$

If we rearrange these equations by putting the last one in the first position, we can easily see (without having to use matrix techniques) that the solution to the system is $A = -1$, $B = 1$, $C = 0$, and $D = 2$, so the partial fraction decomposition is

$$\frac{x^2 + 1}{x(x - 1)^3} = \frac{-1}{x} + \frac{1}{x - 1} + \frac{2}{(x - 1)^3} \qquad \blacksquare$$

CASE 3: THE DENOMINATOR HAS IRREDUCIBLE QUADRATIC FACTORS, NONE OF WHICH IS REPEATED

If the complete factorization of the denominator $Q(x)$ contains the quadratic factor $ax^2 + bx + c$ (which cannot be factored further), then corresponding to this the partial fraction decomposition of $P(x)/Q(x)$ will have a term of the form

$$\frac{Ax + B}{ax^2 + bx + c}$$

EXAMPLE 3 ■ Distinct Quadratic Factors

Find the partial fraction decomposition of $\dfrac{2x^2 - x + 4}{x^3 + 4x}$.

SOLUTION

Since $x^3 + 4x = x(x^2 + 4)$, which cannot be factored further, we write

$$\frac{2x^2 - x + 4}{x^3 + 4x} = \frac{A}{x} + \frac{Bx + C}{x^2 + 4}$$

Multiplying by $x(x^2 + 4)$, we get

$$2x^2 - x + 4 = A(x^2 + 4) + (Bx + C)x$$

$$= (A + B)x^2 + Cx + 4A$$

Equating coefficients gives the equations

$$\begin{cases} A + B = 2 \\ C = -1 \\ 4A = 4 \end{cases}$$

and so $A = 1$, $B = 1$, and $C = -1$. The required partial fraction decomposition is

$$\frac{2x^2 - x + 4}{x^3 + 4x} = \frac{1}{x} + \frac{x - 1}{x^2 + 4}$$

∎

CASE 4: THE DENOMINATOR HAS A REPEATED IRREDUCIBLE QUADRATIC FACTOR

If the complete factorization of $Q(x)$ contains the factor $(ax^2 + bx + c)^k$, where $ax^2 + bx + c$ cannot be factored further, then corresponding to this the partial fraction decomposition of $P(x)/Q(x)$ will have the terms

$$\frac{A_1 x + B_1}{ax^2 + bx + c} + \frac{A_2 x + B_2}{(ax^2 + bx + c)^2} + \cdots + \frac{A_k x + B_k}{(ax^2 + bx + c)^k}$$

EXAMPLE 4 ■ **Repeated Quadratic Factors**

Write out the form of the partial fraction decomposition of

$$\frac{x^5 - 3x^2 + 12x - 1}{x^3(x^2 + x + 1)(x^2 + 2)^3}$$

SOLUTION

$$\frac{x^5 - 3x^2 + 12x - 1}{x^3(x^2 + x + 1)(x^2 + 2)^3}$$

$$= \frac{A}{x} + \frac{B}{x^2} + \frac{C}{x^3} + \frac{Dx + E}{x^2 + x + 1} + \frac{Fx + G}{x^2 + 2} + \frac{Hx + I}{(x^2 + 2)^2} + \frac{Jx + K}{(x^2 + 2)^3}$$

∎

   In order to find the values of A, B, C, D, E, F, G, H, I, J, and K in Example 4, we would have to solve a system of 11 linear equations. Although certainly possible, this would involve a great deal of work!

   The techniques we have described in this section apply only to rational functions $P(x)/Q(x)$ in which the degree of P is less than the degree of Q. If this is not the case, we must first use long division to divide Q into P.

EXAMPLE 5 ■ **Using Long Division to Prepare for Partial Fractions**

Find the partial fraction decomposition of

$$\frac{2x^4 + 4x^3 - 2x^2 + x + 7}{x^3 + 2x^2 - x - 2}$$

SOLUTION

Since the degree of the numerator is larger than the degree of the denominator, we use long division to obtain

$$\frac{2x^4 + 4x^3 - 2x^2 + x + 7}{x^3 + 2x^2 - x - 2} = 2x + \frac{5x + 7}{x^3 + 2x^2 - x - 2}$$

The remainder term now satisfies the requirement that the degree of the numerator is less than the degree of the denominator. At this point we would proceed as in Example 1 to obtain the decomposition

$$\frac{2x^4 + 4x^3 - 2x^2 + x + 7}{x^3 + 2x^2 - x - 2} = 2x + \frac{2}{x - 1} + \frac{-1}{x + 1} + \frac{-1}{x + 2} \quad ■$$

6.10 EXERCISES

1–10 ■ Write out the form of the partial fraction decomposition of the function (as in Example 4). Do not determine the numerical values of the coefficients.

1. $\dfrac{1}{(x - 1)(x + 2)}$

2. $\dfrac{x}{x^2 + 3x - 4}$

3. $\dfrac{x^2 - 3x + 5}{(x - 2)^2(x + 4)}$

4. $\dfrac{1}{x^4 - x^3}$

5. $\dfrac{x^2}{(x - 3)(x^2 + 4)}$

6. $\dfrac{1}{x^4 - 1}$

7. $\dfrac{x^3 - 4x^2 + 2}{(x^2 + 1)(x^2 + 2)}$

8. $\dfrac{x^4 + x^2 + 1}{x^2(x^2 + 4)^2}$

9. $\dfrac{x^3 + x + 1}{x(2x - 5)^3(x^2 + 2x + 5)^2}$

10. $\dfrac{1}{(x^6 - 1)(x^4 - 1)}$

11–40 ■ Find the partial fraction decomposition of the rational function.

11. $\dfrac{5}{(x - 1)(x + 4)}$

12. $\dfrac{x + 6}{x(x + 3)}$

13. $\dfrac{12}{x^2 - 9}$

14. $\dfrac{x - 12}{x^2 - 4x}$

15. $\dfrac{4}{x^2 - 4}$

16. $\dfrac{2x + 1}{x^2 + x - 2}$

17. $\dfrac{x + 14}{x^2 - 2x - 8}$

18. $\dfrac{8x - 3}{2x^2 - x}$

19. $\dfrac{x}{8x^2 - 10x + 3}$

20. $\dfrac{7x - 3}{x^3 + 2x^2 - 3x}$

21. $\dfrac{9x^2 - 9x + 6}{2x^3 - x^2 - 8x + 4}$

22. $\dfrac{-3x^2 - 3x + 27}{(x + 2)(2x^2 + 3x - 9)}$

23. $\dfrac{x^2 + 1}{x^3 + x^2}$

24. $\dfrac{3x^2 + 5x - 13}{(3x + 2)(x^2 - 4x + 4)}$

25. $\dfrac{2x}{4x^2 + 12x + 9}$

26. $\dfrac{x - 4}{(2x - 5)^2}$

27. $\dfrac{4x^2 - x - 2}{x^4 + 2x^3}$

28. $\dfrac{x^3 - 2x^2 - 4x + 3}{x^4}$

29. $\dfrac{-10x^2 + 27x - 14}{(x - 1)^3(x + 2)}$

30. $\dfrac{-2x^2 + 5x - 1}{x^4 - 2x^3 + 2x - 1}$

31. $\dfrac{3x^3 + 22x^2 + 53x + 41}{(x + 2)^2(x + 3)^2}$

32. $\dfrac{3x^2 + 12x - 20}{x^4 - 8x^2 + 16}$

33. $\dfrac{x - 3}{x^3 + 3x}$

34. $\dfrac{3x^2 - 2x + 8}{x^3 - x^2 + 2x - 2}$

35. $\dfrac{2x^3 + 7x + 5}{(x^2 + x + 2)(x^2 + 1)}$

36. $\dfrac{x^2 + x + 1}{2x^4 + 3x^2 + 1}$

37. $\dfrac{x^4 + x^3 + x^2 - x + 1}{x(x^2 + 1)^2}$

38. $\dfrac{2x^2 - x + 8}{(x^2 + 4)^2}$

39. $\dfrac{x^5 - 2x^4 + x^3 + x + 5}{x^3 - 2x^2 + x - 2}$

40. $\dfrac{x^5 - 3x^4 + 3x^3 - 4x^2 + 4x + 12}{(x - 2)^2(x^2 + 2)}$

41. Determine A and B in terms of a and b:

$$\frac{ax + b}{x^2 - 1} = \frac{A}{x - 1} + \frac{B}{x + 1}$$

42. Determine A, B, C, and D in terms of a and b:

$$\frac{ax^3 + bx^2}{(x^2 + 1)^2} = \frac{Ax + B}{x^2 + 1} + \frac{Cx + D}{(x^2 + 1)^2}$$

---

6 | REVIEW

KEY TOPICS ■ Define, state, or discuss each of the following.

1. System of equations
2. Linear equation
3. Gaussian elimination
4. Triangular form
5. Elementary row operations
6. Matrix
7. Echelon form
8. Reduced echelon form
9. Leading variables
10. Inconsistent system of equations
11. Dependent system of equations
12. Addition and subtraction of matrices
13. Product of matrices
14. Identity matrix
15. Inverse of a matrix
16. Minor
17. Cofactor
18. Determinant
19. Invertibility Criterion
20. Row and column transformations of a determinant
21. Cramer's Rule
22. Nonlinear system of equations
23. System of inequalities
24. Bounded and unbounded regions
25. Vertex
26. Linear programming
27. Feasible region
28. Objective function
29. Partial fraction decomposition

EXERCISES

1–6 ■ Solve the system of equations and graph the lines.

1. $\begin{cases} 3x - y = 5 \\ 2x + y = 5 \end{cases}$

2. $\begin{cases} y = 2x + 6 \\ y = -x + 3 \end{cases}$

3. $\begin{cases} 2x - 7y = 28 \\ y = \frac{2}{7}x - 4 \end{cases}$

4. $\begin{cases} 6x - 8y = 15 \\ -\frac{3}{2}x + 2y = -4 \end{cases}$

5. $\begin{cases} 2x - y = 1 \\ x + 3y = 10 \\ 3x + 4y = 15 \end{cases}$

6. $\begin{cases} 2x + 5y = 9 \\ -x + 3y = 1 \\ 7x - 2y = 14 \end{cases}$

7–14 ■ Find the complete solution of the system using Gaussian elimination, or show that the system has no solution.

7. $\begin{cases} x + y + 2z = 6 \\ 2x \quad\;\; + 5z = 12 \\ x + 2y + 3z = 9 \end{cases}$

8. $\begin{cases} x - 2y + 3z = 1 \\ x - 3y - z = 0 \\ 2x \quad\quad - 6z = 6 \end{cases}$

9. $\begin{cases} x - 2y + 3z = 1 \\ 2x - y + z = 3 \\ 2x - 7y + 11z = 2 \end{cases}$

10. $\begin{cases} x + y + z + w = 2 \\ 2x \quad\;\; - 3z = 5 \\ x - 2y \quad\;\; + 4w = 9 \\ x + y + 2z + 3w = 5 \end{cases}$

11. $\begin{cases} x - 3y + z = 4 \\ 4x - y + 15z = 5 \end{cases}$

12. $\begin{cases} 2x - 3y + 4z = 3 \\ 4x - 5y + 9z = 13 \\ 2x \quad\;\; + 7z = 0 \end{cases}$

13. $\begin{cases} -x + 4y + z = 8 \\ 2x - 6y + z = -9 \\ x - 6y - 4z = -15 \end{cases}$

14. $\begin{cases} x \quad\quad - z + w = 2 \\ 2x + y \quad\;\; - 2w = 12 \\ \quad\;\; 3y + z + w = 4 \\ x + y - z \quad\;\; = 10 \end{cases}$

15. A man invests his savings in two accounts, one paying 6% interest per year and the other paying 7%. He has twice as much invested in the 7% account as in the 6% account, and his annual interest income is $600. How much is invested in each account?

16. Find the values of a, b, and c if the parabola

$$y = ax^2 + bx + c$$

passes through the points $(1, 0)$, $(-1, -4)$, and $(2, 11)$.

17–28 ■ Let

$$A = [2 \quad 0 \quad -1] \qquad B = \begin{bmatrix} 1 & 2 & 4 \\ -2 & 1 & 0 \end{bmatrix}$$

$$C = \begin{bmatrix} \frac{1}{2} & 3 \\ 2 & \frac{3}{2} \\ -2 & 1 \end{bmatrix} \qquad D = \begin{bmatrix} 1 & 4 \\ 0 & -1 \\ 2 & 0 \end{bmatrix}$$

$$E = \begin{bmatrix} 2 & -1 \\ -\frac{1}{2} & 1 \end{bmatrix} \qquad F = \begin{bmatrix} 4 & 0 & 2 \\ -1 & 1 & 0 \\ 7 & 5 & 0 \end{bmatrix}$$

$$G = [5]$$

Carry out the indicated operation, or explain why it cannot be performed.

17. $A + B$ **18.** $C - D$ **19.** $2C + 3D$

20. $5B - 2C$ **21.** GA **22.** AG

23. BC **24.** CB **25.** BF

26. FC **27.** $(C + D)E$ **28.** $F(2C - D)$

29–34 ■ Find the determinant and, if possible, the inverse of the matrix.

29. $\begin{bmatrix} 1 & 4 \\ 2 & 9 \end{bmatrix}$ **30.** $\begin{bmatrix} 2 & 2 \\ 1 & -3 \end{bmatrix}$

31. $\begin{bmatrix} 4 & -12 \\ -2 & 6 \end{bmatrix}$ **32.** $\begin{bmatrix} 2 & 4 & 0 \\ -1 & 1 & 2 \\ 0 & 3 & 2 \end{bmatrix}$

33. $\begin{bmatrix} 3 & 0 & 1 \\ 2 & -3 & 0 \\ 4 & -2 & 1 \end{bmatrix}$ **34.** $\begin{bmatrix} 1 & 0 & 0 & 1 \\ 0 & 2 & 0 & 2 \\ 0 & 0 & 3 & 3 \\ 0 & 0 & 0 & 4 \end{bmatrix}$

35–36 ■ Express the system of linear equations as a matrix equation. Then solve the matrix equation by multiplying each side by the inverse of the coefficient matrix.

35. $\begin{cases} 12x - 5y = 10 \\ 5x - 2y = 17 \end{cases}$ **36.** $\begin{cases} 2x + y + 5z = \frac{1}{3} \\ x + 2y + 2z = \frac{1}{4} \\ x \quad\quad + 3z = \frac{1}{6} \end{cases}$

37–40 ■ Solve the system using Cramer's Rule.

37. $\begin{cases} 2x + 7y = 13 \\ 6x + 16y = 30 \end{cases}$ **38.** $\begin{cases} 12x - 11y = 140 \\ 7x + 9y = 20 \end{cases}$

39. $\begin{cases} 2x - y + 5z = 0 \\ -x + 7y = 9 \\ 5x + 4y + 3z = -9 \end{cases}$ **40.** $\begin{cases} 3x + 4y - z = 10 \\ x - 4z = 20 \\ 2x + y + 5z = 30 \end{cases}$

41–44 ■ Find all solutions of the system.

41. $\begin{cases} x^2 + y^2 + 6y = 0 \\ x - 2y = 3 \end{cases}$ **42.** $\begin{cases} x^2 + y^2 = 10 \\ x^2 + 2y^2 - 7y = 0 \end{cases}$

43. $\begin{cases} 3x^4 + \dfrac{4}{y} = 50 \\ x^4 - \dfrac{8}{y} = 12 \end{cases}$ **44.** $\begin{cases} x^2 + yz = 0 \\ y^2 + xz = 0 \\ x^2 + xy + y^2 = 3 \end{cases}$

45–48 ■ Graph the solution set of the system of inequalities. Find the coordinates of all vertices, and determine whether the solution set is bounded or unbounded.

45. $\begin{cases} x^2 + y^2 < 9 \\ x + y < 0 \end{cases}$ **46.** $\begin{cases} y - x^2 \geq 4 \\ y < 20 \end{cases}$

47. $\begin{cases} x \geq 0, \quad y \geq 1 \\ x + 2y \leq 12 \\ y \leq x + 4 \end{cases}$ **48.** $\begin{cases} x \geq 4 \\ x + y \geq 24 \\ x \leq 2y + 12 \end{cases}$

49. Find the maximum and minimum values of the function $P = 3x + 4y$ on the region described by the inequalities in Exercise 47.

50. (a) Find the minimum value of the function

$$Q = 60 + 3x + 5y$$

on the region described by the inequalities in Exercise 48.

(b) Explain why Q has no maximum value on this region.

51. A farmer wishes to plant oats and barley on 400 acres of land. Each acre yields 40 bushels of oats or 50 bushels of barley. After harvest, the farmer will have to store the grain for several months in order to get the best price for it, and he has facilities to store no more than 18,000 bushels of grain.

(a) If he can get $2.05 per bushel for oats and $1.80 per bushel for barley, how many acres of each grain should be planted for maximum revenue?

(b) If, instead, the price per bushel is $1.20 for oats and $1.60 for barley, how many acres of each grain should be planted to maximize revenue?

52. A woman wishes to invest $12,000, some in a high-risk stock with an expected annual dividend of 15%, some in long-term bonds yielding 10% interest per year, and the remainder in a money-market account paying an annual yield of 6% interest. She wishes to put at least $4000 into the money-market account. Also, the amount invested in the high-risk stock should be no more than half the total of her other two investments. How much should she put in each investment to maximize her annual interest and dividend yield?

53–54 ■ Solve for x, y, and z in terms of a, b, and c.

53. $\begin{cases} -x + y + z = a \\ x - y + z = b \\ x + y - z = c \end{cases}$

54. $\begin{cases} ax + by + cz = a - b + c \\ bx + by + cz = c \\ cx + cy + cz = c \end{cases}$
$(a \neq b, \ b \neq c, \ c \neq 0)$

55. For what values of k do the following three lines have a common point of intersection?

$$x + y = 12$$
$$kx - y = 0$$
$$y - x = 2k$$

56. For what value of k does the following system have infinitely many solutions?

$$\begin{cases} kx + y + z = 0 \\ x + 2y + kz = 0 \\ -x + 3z = 0 \end{cases}$$

57–60 ■ Use a graphing device to solve the system, correct to the nearest hundredth.

57. $\begin{cases} 0.32x + 0.43y = 0 \\ 7x - 12y = 341 \end{cases}$

58. $\begin{cases} \sqrt{12}\,x - 3\sqrt{2}\,y = 660 \\ 7137x + 3931y = 20{,}000 \end{cases}$

59. $\begin{cases} x - y^2 = 10 \\ x = \frac{1}{22}y + 12 \end{cases}$ **60.** $\begin{cases} y = 5^x + x \\ y = x^5 + 5 \end{cases}$

61–64 ■ Find the partial fraction decomposition of the rational function.

61. $\dfrac{3x + 1}{x^2 - 2x - 15}$ **62.** $\dfrac{8}{x^3 - 4x}$

63. $\dfrac{2x - 4}{x(x - 1)^2}$ **64.** $\dfrac{x + 6}{x^3 - 2x^2 + 4x - 8}$

1. In $2\frac{1}{2}$ hours an airplane travels 600 km against the wind. It takes 50 min to travel 300 km with the wind. Find the speed of the wind and the speed of the airplane in still air.

2–5 ■ Find all solutions of the system. Determine whether the system is linear or nonlinear. If it is linear, state whether it is inconsistent, dependent, or neither.

2. $\begin{cases} 3x - y = 10 \\ 2x + 5y = 1 \end{cases}$

3. $\begin{cases} x - y + 9z = -8 \\ x \quad - 4z = 7 \\ 3x - y + z = 5 \end{cases}$

4. $\begin{cases} 2x - y + z = 0 \\ 3x + 2y - 3z = 1 \\ x - 4y + 5z = -1 \end{cases}$

5. $\begin{cases} 2x^2 + y^2 = 6 \\ 3x^2 - 4y = 11 \end{cases}$

6–13 ■ Let

$$A = \begin{bmatrix} 2 & 3 \\ 2 & 4 \end{bmatrix} \qquad B = \begin{bmatrix} 2 & 4 \\ -1 & 1 \\ 3 & 0 \end{bmatrix} \qquad C = \begin{bmatrix} 1 & 0 & 4 \\ -1 & 1 & 2 \\ 0 & 1 & 3 \end{bmatrix}$$

Carry out the indicated operation, or explain why it cannot be performed.

6. $A + B$ **7.** AB **8.** $BA - 3B$ **9.** CBA

10. A^{-1} **11.** B^{-1} **12.** $|B|$ **13.** $|C|$

14. Write a matrix equation equivalent to the system

$$\begin{cases} 4x - 3y = 10 \\ 3x - 2y = 30 \end{cases}$$

Find the inverse of the coefficient matrix, and use it to solve the system.

15. Solve using Cramer's Rule:

$$\begin{cases} 2x \quad - z = 14 \\ 3x - y + 5z = 0 \\ 4x + 2y + 3z = -2 \end{cases}$$

16. Only one of the following matrices has an inverse. Find the determinant of each matrix, and use the determinants to identify the one that has an inverse. Then find the inverse.

$$A = \begin{bmatrix} 1 & 4 & 1 \\ 0 & 2 & 0 \\ 1 & 0 & 1 \end{bmatrix} \qquad B = \begin{bmatrix} 1 & 4 & 0 \\ 0 & 2 & 0 \\ -3 & 0 & 1 \end{bmatrix}$$

17. Graph the following system of inequalities, and state the coordinates of the vertices.

$$\begin{cases} x^2 - 2x - y + 5 \le 0 \\ y \le 5 + 2x \end{cases}$$

18. Find the partial fraction decomposition of the rational function

$$\frac{4x - 1}{(x - 1)^2(x + 2)}$$

19. A farmer grows wheat and barley on 200 acres of land. He can borrow no more than $10,000 at the beginning of the season for production costs, and he has no other financial resources. The growing cost per acre is $60 for wheat and $40 for barley, and the land yields either 50 bushels of wheat per acre or 40 bushels of barley per acre. The market prediction for the fall harvest indicates that the farmer can make a profit of $2.50 per bushel of wheat and $2.00 per bushel of barley. How many acres of each grain should he plant to maximize profit?

20. Use a graphing calculator to find all solutions of the following system, correct to two decimals.

$$\begin{cases} 2x^2 + y^2 = 16 \\ y = x^4 - 4x^3 + 6x^2 - 4x \end{cases}$$

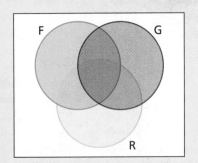

F = French   G = German
R = Russian

FIGURE 1

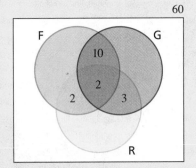

FIGURE 2

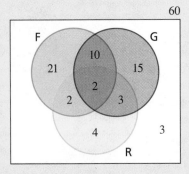

FIGURE 3

FOCUS ON PROBLEM SOLVING

For many problems the solution involves **drawing a diagram,** even when no diagram is immediately suggested by the problem itself. To illustrate this we present a situation where drawing a diagram is the key to solving the problem.

PROBLEM ■ Counting the Elements of Intersecting Sets

In a survey of 60 students it is found that 35 study French, 30 study German, 11 study Russian, 12 study French and German, 4 French and Russian, 5 German and Russian, and 2 study all three languages. How many of the 60 students study none of these languages?

SOLUTION

At first glance it may seem that all we need to do is add up the number of students studying French, German, and Russian and then subtract from 60. But, since $35 + 30 + 11 = 76$, which is more than 60, this will clearly not work. To solve this problem we must keep track of how many students study each possible combination of these languages. This is difficult to do without organizing the information in some pictorial form. One way is to draw a **Venn diagram,** which consists of circles drawn within a rectangle, with the circles representing the various sets that occur in the problem. (This idea is attributed to the English logician John Venn). In the Venn diagram in Figure 1, the circles represent the students who study the three given languages and the rectangle represents all the students who were interviewed.

Since 2 students study all three languages, we put the number 2 in the intersection of all three circles in Figure 2. Of the 12 students who study French and German, 2 study all three languages, so just 10 study *only* French and German. Thus, we put the number 10 in the portion of the diagram that belongs to F and G but not to R. Similarly, 2 students study only French and Russian and 3 study only German and Russian. This gives the distribution shown in Figure 3.

Now we use the Venn diagram in Figure 3 to determine how many students study just one language. Since 35 students study French, and since we have already accounted for $10 + 2 + 2 = 14$ of them in the diagram of Figure 3, the number of students who study only French must be $35 - 14 = 21$. We put this number in the appropriate section in circle F in Figure 3. By similar reasoning, we see that 15 study only German and 4 study only Russian.

To find the total number of students who study at least one of the three languages, we add all the numbers inside the circles in Figure 3:

$$21 + 15 + 4 + 10 + 3 + 2 + 2 = 57$$

Thus, the number of students who study none of these languages is $60 - 57 = 3$. ∎

PROBLEMS

1. Among 100 business executives, 68 have an American Express card, 52 have a Visa card, and 52 have a MasterCard. Some executives have more than one card: 33 have both an American Express card and a MasterCard, 35 have both an American Express and a Visa card, and 37 have both a MasterCard and a Visa card, and 30 have all three cards. How many of the business executives have none of these three cards?

2. A city has two newspapers, the *Morning Telegraph* and the *Evening Standard*. In a survey of 100 people, 54 say they read the *Morning Telegraph*, 42 read the *Evening Standard*, and 14 read neither. How many of the 100 people read both newspapers?

3. In a group of 95 pet owners, 65 own a dog, 43 own a cat, and 30 own a bird. We know that 15 own a dog and a bird, 8 own a cat and a bird, and 23 own a dog and a cat, and that 10 own a bird but do not own a cat or a dog.
   (a) Of these pet owners, how many own neither a dog nor a cat nor a bird?
   (b) How many own a cat and a dog but not a bird?

4. In a group of 400 college professors, it is found that 120 earn less than $25,000 and 260 earn $50,000 or less annually. In this group, 63 own a boat, and of these, 5 earn less than $25,000 a year. Of those who earn more than $50,000, it is known that 100 do not own a boat. How many of those professors who earn between $25,000 and $50,000 (inclusive) do not own a boat?

5. How many integers are there from one to one million (inclusive) that are either perfect squares or perfect cubes (or both)?

6. How many positive integers less than 1000 are divisible by neither 5 nor 7?

7. Justin, Sasha, and Vanessa each have a bag of marbles. First Justin gives Sasha and Vanessa each as many marbles as they already have. Then Sasha gives Justin and Vanessa as many marbles as they now have. Finally, Vanessa gives Justin and Sasha as many marbles as they now have. Everyone ends up with 16 marbles. How many did each person have to begin with?

8. Write 116 as a sum of four or fewer perfect squares in three different ways.

9. The ancient Egyptians considered rectangles whose perimeters and areas were numerically the same to be special. Find all rectangles with integer sides whose perimeter and area are the same integer.

10. This problem is adapted from one of the first textbooks on algebra, written by Leonhard Euler (1707–1783).

Suppose that a sheep is worth 5 dollars, a goose is worth 3, and three chickens are worth 1. A farmer bought 100 of these animals and spent 100 dollars. How many of each kind did he buy, assuming that he bought less than 10 geese?

11. Solve the following system of equations:

$$\begin{cases} 3x + 5y + z + 7t = -8 \\ 4x + 8y + 2z + 6t = -8 \\ 6x + 2y + 8z + 4t = 8 \\ 7x + y + 5z + 3t = 8 \end{cases}$$

[*Hint:* Instead of immediately using the methods of Chapter 6, look for a shortcut.]

12. Find the complete solution of the following system of equations.

$$\begin{cases} x + y + z = 2 \\ x^2 + y^2 + z^2 = 2 \\ xy = z^2 \end{cases}$$

13. (a) Suppose $x + y = 1$ and $x^2 + y^2 = 4$. Find $x^3 + y^3$.
(b) Suppose $x + y = 4$ and $xy = 1$. Find $x^3 + y^3$.

14. A pair of pulleys is connected by a belt, as shown in each figure. Find the length of the belt.

(a)   **(b)**

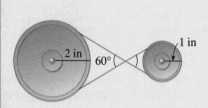

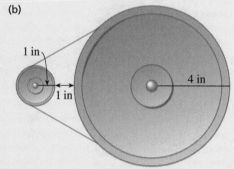

Hint:

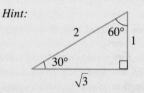

15. Find the equation of the circle that passes through the points $(-19, 62)$, $(45, 70)$, and $(20, -55)$.

16. Show that the area of the triangle shaded in the figure is

$$\pm \frac{1}{2} \begin{vmatrix} x_1 & y_1 & 1 \\ x_2 & y_2 & 1 \\ x_3 & y_3 & 1 \end{vmatrix}$$

[*Hint:* Use the additional triangles in the following figure.]

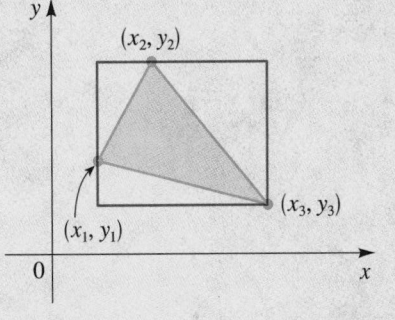

17. Show that $\begin{vmatrix} 1 & a & a^2 \\ 1 & b & b^2 \\ 1 & c & c^2 \end{vmatrix} = (a - b)(b - c)(c - a)$. This is called a *Vandermonde determinant*.

18. If A is a square matrix, then the square root symbol $\sqrt{A}$ refers to any matrix such that $\sqrt{A} \cdot \sqrt{A} = A$. Let $A = \begin{bmatrix} 1 & 0 \\ 0 & 1 \end{bmatrix}$. Find three different square roots of A.

19. The *characteristic polynomial* of a square matrix A is the polynomial $P(x)$ given by the determinant of $A - xI$, that is, $P(x) = |A - xI|$. Let $A = \begin{bmatrix} 3 & 5 \\ -2 & 1 \end{bmatrix}$. Show that $P(A) = 0$. (In general, a matrix is a zero of its own characteristic polynomial.)

CONIC SECTIONS

The path of a projectile, such as a basketball, a missile, or a comet, is a conic section.

In studying the procedures of geometric thought we may hope to reach what is most essential in the human mind.

HENRI POINCARÉ

In this chapter we study the geometry of **conic sections** (or simply **conics**). Conic sections are the curves formed by the intersection of a plane with a pair of circular cones. These curves have four basic shapes, called **circles, ellipses, parabolas,** and **hyperbolas,** as illustrated in the figure.

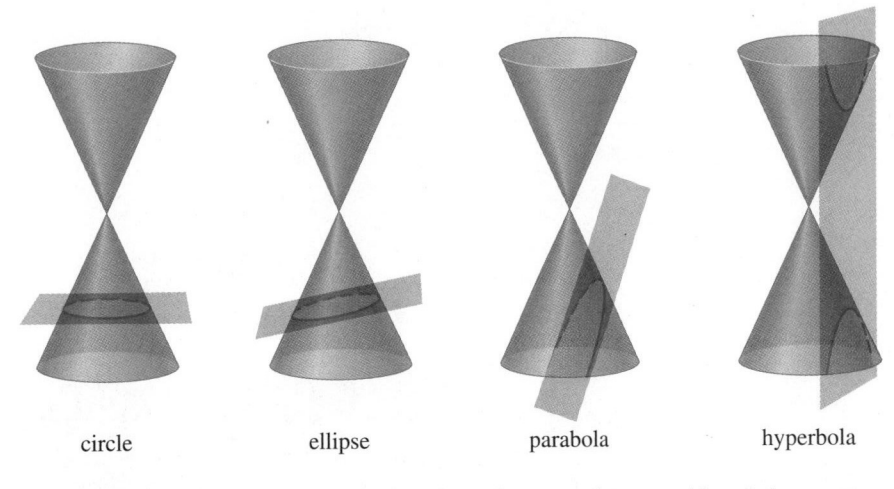

circle   ellipse   parabola   hyperbola

The ancient Greeks studied these curves because they considered the geometry of conic sections to be very beautiful. The mathematician Apollonius (262–190 B.C.) wrote a definitive eight-volume work on the subject. In more modern times, conics were found to be useful as well as beautiful. Galileo discovered in 1590 that the path of a missile shot upward at an angle is a parabola. In 1609, Kepler found that the planets move in elliptical orbits around the sun. In 1668, Newton was the first to build a reflecting telescope, whose principle is based on the properties of parabolas and hyperbolas. In this century, many further applications of conic sections have been developed. One important application is the LORAN radio navigation system, which uses the intersection points of hyperbolas to pinpoint the location of ships and aircraft. Another application is to the medical procedure of lithotripsy, a method of removing kidney stones without surgery; this method uses a property of ellipses. Other applications of conic sections will also be considered in this chapter.

7.1   PARABOLAS

We have seen in Section 3.9 that the graph of the equation $y = ax^2 + bx + c$ is a U-shaped curve called a *parabola* that opens either upward or downward, depending on whether the sign of a is positive or negative (see Figure 1). The lowest or highest point of the parabola is called the *vertex* and the parabola is symmetric about its *axis*.

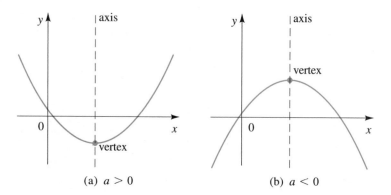

FIGURE 1
$y = ax^2 + bx + c$

(a) $a > 0$ (b) $a < 0$

In this section we study parabolas from a geometric rather than an algebraic point of view. We begin with the geometric definition of a parabola and show how this leads to the algebraic formula that we are already familiar with.

GEOMETRIC DEFINITION OF A PARABOLA

A **parabola** is the set of points in the plane equidistant from a fixed point F (called the **focus**) and a fixed line l (called the **directrix**).

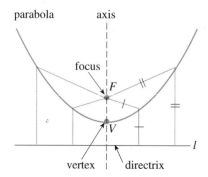

FIGURE 2

This definition is illustrated in Figure 2. Note that the vertex V of the parabola lies halfway between the focus and the directrix and that the axis of symmetry is the line that runs through the focus perpendicular to the directrix.

In this section we restrict our attention to parabolas that are situated with the vertex at the origin and that have a vertical or horizontal axis of symmetry. (Parabolas in more general positions will be considered in Section 7.4.) If the focus of such a parabola is the point $F(0, p)$, then the axis of symmetry must be vertical and the directrix has the equation $y = -p$. Figure 3 illustrates the case $p > 0$.

If $P(x, y)$ is any point on the parabola, then the distance from P to the focus F (using the Distance Formula) is

$$\sqrt{x^2 + (y - p)^2}$$

and the distance from P to the directrix is

$$|y - (-p)| = |y + p|$$

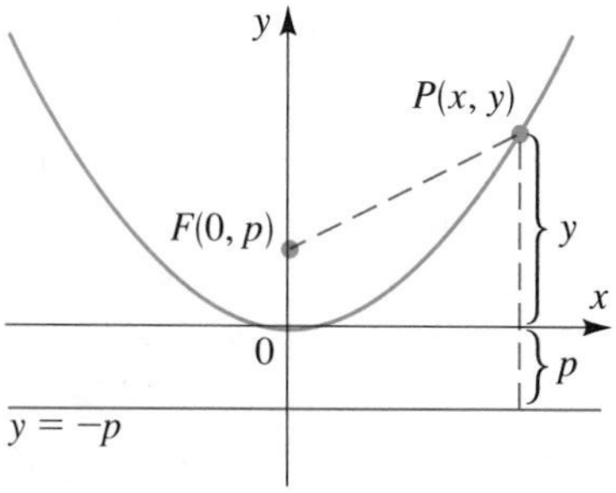

By the definition of a parabola, these two distances must be equal:

$$\sqrt{x^2 + (y - p)^2} = |y + p|$$

$$x^2 + (y - p)^2 = |y + p|^2 = (y + p)^2 \quad \text{Square each side}$$

$$x^2 + y^2 - 2py + p^2 = y^2 + 2py + p^2 \quad \text{Expand}$$

FIGURE 3

$y = -p$

$$x^2 - 2py = 2py \qquad \text{Simplify}$$
$$x^2 = 4py$$

If $p > 0$, then the parabola opens upward, but if $p < 0$, it opens downward. When x is replaced by $-x$, the equation remains unchanged, so the graph is symmetric about the y-axis. We summarize what we have proved in the following box.

PARABOLA WITH VERTICAL AXIS

The graph of the equation

$$x^2 = 4py$$

is a parabola with the following properties.

VERTEX	$V(0, 0)$
FOCUS	$F(0, p)$
DIRECTRIX	$y = -p$

The parabola opens upward if $p > 0$ or downward if $p < 0$.

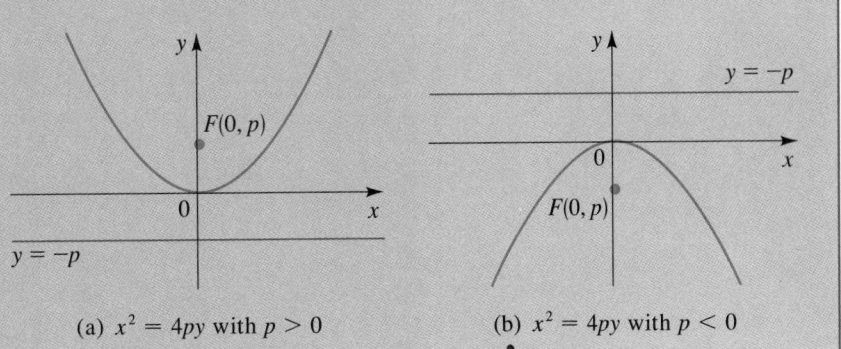

(a) $x^2 = 4py$ with $p > 0$ (b) $x^2 = 4py$ with $p < 0$

EXAMPLE 1 ■ Finding the Equation of a Parabola

Find the equation of the parabola with vertex $V(0, 0)$ and focus $F(0, 2)$, and sketch its graph.

SOLUTION

Since the focus is $F(0, 2)$, we conclude that $p = 2$ (and so the directrix is $y = -2$). Thus, the equation of the parabola is

$$x^2 = 4(2)y \qquad x^2 = 4py \text{ with } p = 2$$
$$x^2 = 8y$$

Since $p = 2 > 0$, the parabola opens upward. The graph is shown in Figure 4. ■

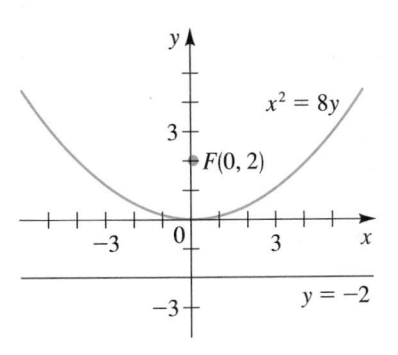

FIGURE 4

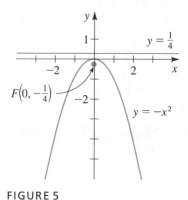

FIGURE 5

EXAMPLE 2 ■ Finding the Focus and Directrix of a Parabola from its Equation

Find the focus and directrix of the parabola $y = -x^2$, and sketch its graph.

SOLUTION

Comparing the equation $y = -x^2$ with the general equation $x^2 = 4py$, we see that $4p = -1$, so $p = -\frac{1}{4}$. Thus, the focus is $F(0, -\frac{1}{4})$ and the directrix is $y = \frac{1}{4}$.

The graph is shown in Figure 5.   ■

Reflecting the graph in Figure 3 about the diagonal line $y = x$ has the effect of interchanging the roles of x and y. This results in a parabola with horizontal axis. By the same method as before, we can prove the following.

PARABOLA WITH HORIZONTAL AXIS

The graph of the equation

$$y^2 = 4px$$

is a parabola with the following properties.

VERTEX	$V(0,0)$
FOCUS	$F(p,0)$
DIRECTRIX	$x = -p$

The parabola opens to the right if $p > 0$ or to the left if $p < 0$.

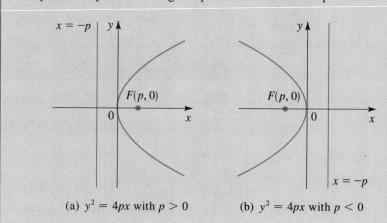

(a) $y^2 = 4px$ with $p > 0$ (b) $y^2 = 4px$ with $p < 0$

EXAMPLE 3 ■ A Parabola with Horizontal Axis

Find the focus and directrix of the parabola $6x + y^2 = 0$, and sketch the graph.

SOLUTION

We first write the equation as $y^2 = -6x$. Comparing this with the general equation $y^2 = 4px$, we see that $-6 = 4p$, so $p = -\frac{3}{2}$. Thus, the focus is $(-\frac{3}{2}, 0)$ and the directrix is $x = \frac{3}{2}$.

Since $p = -\frac{3}{2} < 0$, the parabola opens to the left. The graph is shown in Figure 6.

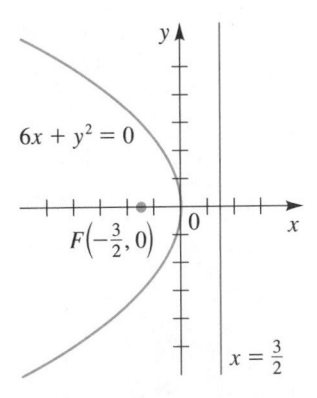

FIGURE 6

We can use the coordinates of the focus to estimate the "width" of a parabola when sketching its graph. The line segment that runs through the focus perpendicular to the axis, with endpoints on the parabola, is called the **latus rectum,** and its length is the **focal diameter** of the parabola. From Figure 7 we can see that the distance from an endpoint Q of the latus rectum to the directrix is $|2p|$. Thus, the distance from Q to the focus must be $|2p|$ as well (by the definition of a parabola), and so the focal diameter is $|4p|$. In the next example we use the focal diameter to determine the "width" of a parabola when graphing it.

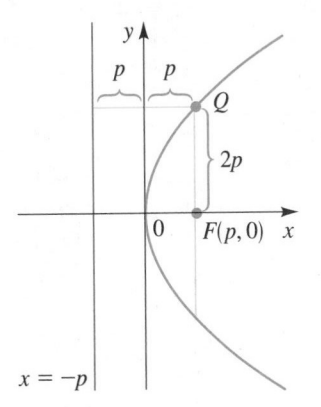

FIGURE 7

EXAMPLE 4 ■ Using the Focal Diameter to Sketch a Parabola

Find the focus, directrix, and focal diameter of the parabola $y = \frac{1}{2}x^2$, and sketch its graph.

SOLUTION

We first put the equation in the form $x^2 = 4py$.

$$y = \tfrac{1}{2}x^2$$

$$x^2 = 2y \qquad \text{Multiply each side by 2}$$

From this equation we see that $4p = 2$, so the focal diameter is 2. Solving for p gives $p = \tfrac{1}{2}$, so the focus is $(0, \tfrac{1}{2})$ and the directrix is $y = -\tfrac{1}{2}$. Since the focal diameter is 2, the latus rectum extends one unit to the left and one unit to the right of the focus. This enables us to sketch the graph in Figure 8.

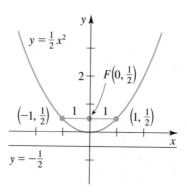

FIGURE 8

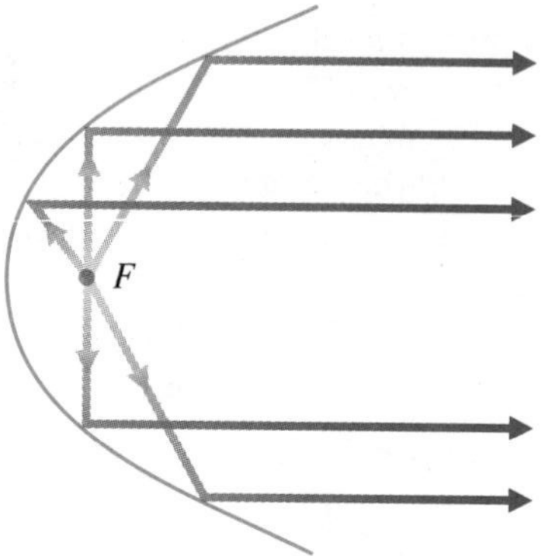

Parabolas have an important property that makes them useful as reflectors for lamps and telescopes. Light from a source placed at the focus of a surface with parabolic cross section will be reflected in such a way that it travels parallel to the axis of the parabola (see Figure 9). Thus, a parabolic mirror reflects the light into a beam of parallel rays. Conversely, light approaching the reflector in rays parallel to its axis of symmetry is concentrated to the focus. This principle, which can be proved using calculus, is used in the construction of reflecting telescopes.

EXAMPLE 5 ■ Finding the Focal Point of a Searchlight Reflector

A searchlight has a parabolic reflector that forms a "bowl," which is 12 in. wide from rim to rim and 8 in. deep, as shown in Figure 10. If the bulb is located at the focus, how far from the vertex of the reflector is it?

FIGURE 9

Parabolic reflector

FIGURE 10

A parabolic reflector

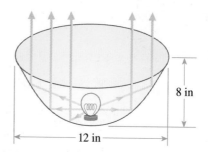

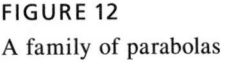

FIGURE 11

SOLUTION

We introduce a coordinate system and place a parabolic cross section of the reflector so that its vertex is at the origin and its axis is vertical (see Figure 11). Then the equation of this parabola has the form $x^2 = 4py$. From Figure 11 we see that the point $(6, 8)$ lies on the parabola. We use this to find p.

$$6^2 = 4p(8) \qquad \text{The point } (6, 8) \text{ satisfies the equation } x^2 = 4py$$

$$36 = 32p$$

$$p = \tfrac{9}{8}$$

Thus, the focus is $F(0, \tfrac{9}{8})$. So the distance between the vertex and the focus is $\tfrac{9}{8} = 1\tfrac{1}{8}$ in. Because the bulb is at the focus, it is located $1\tfrac{1}{8}$ in. from the vertex of the bowl. ■

In the next example we graph a family of parabolas, to show how changing the distance between the focus and the vertex affects the "width" of a parabola.

EXAMPLE 6 ■ A Family of Parabolas

(a) Find equations for the parabolas with vertex at the origin and foci $F_1(0, \tfrac{1}{8})$, $F_2(0, \tfrac{1}{2})$, $F_3(0, 1)$, and $F_4(0, 4)$.

(b) Draw the graphs of the parabolas in part (a). What do you conclude?

SOLUTION

(a) Since the foci are on the positive y-axis, the parabolas open upward and have equations of the form $x^2 = 4py$. This leads to the following equations.

Focus	p	Equation $x^2 = 4py$	Form of the equation for graphing calculator
$F_1(0, \tfrac{1}{8})$	$p = \tfrac{1}{8}$	$x^2 = \tfrac{1}{2}y$	$y = 2x^2$
$F_2(0, \tfrac{1}{2})$	$p = \tfrac{1}{2}$	$x^2 = 2y$	$y = 0.5x^2$
$F_3(0, 1)$	$p = 1$	$x^2 = 4y$	$y = 0.25x^2$
$F_4(0, 4)$	$p = 4$	$x^2 = 16y$	$y = 0.0625x^2$

FIGURE 12

A family of parabolas

(b) The graphs are drawn in Figure 12. We see that the closer the focus to the vertex, the narrower the parabola.

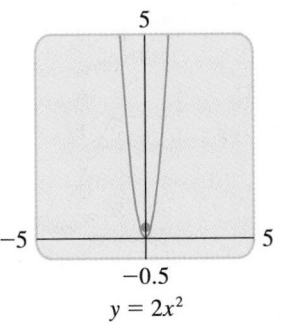

$y = 2x^2$

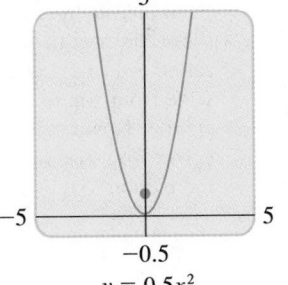

$y = 0.5x^2$

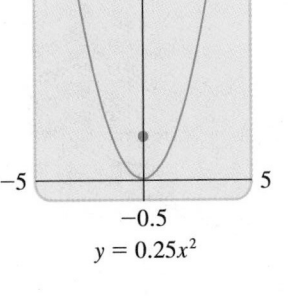

$y = 0.25x^2$

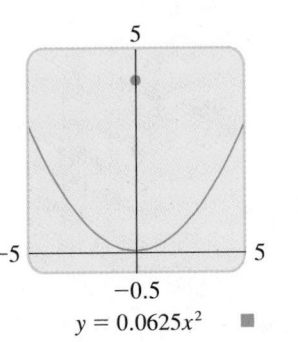

$y = 0.0625x^2$ ■

| 7.1 | **EXERCISES** |

1–12 ■ Find the focus, directrix, and focal diameter of the parabola, and sketch its graph.

1. $y^2 = 4x$ **2.** $x^2 = y$

3. $x^2 = 9y$ **4.** $y^2 = 3x$

5. $y = 5x^2$ **6.** $y = -2x^2$

7. $x = -8y^2$ **8.** $x = \frac{1}{2}y^2$

9. $x^2 + 6y = 0$ **10.** $x - 7y^2 = 0$

11. $5x + 3y^2 = 0$ **12.** $8x^2 + 12y = 0$

13–24 ■ Find an equation for the parabola described by the given information. Each parabola has its vertex at the origin.

13. Focus $F(0, 2)$ **14.** Focus $F\left(0, -\frac{1}{2}\right)$

15. Focus $F(-8, 0)$ **16.** Focus $F(5, 0)$

17. Directrix $x = 2$ **18.** Directrix $y = 6$

19. Directrix $y = -10$ **20.** Directrix $x = -\frac{1}{8}$

21. Focus on the positive x-axis, two units away from the directrix

22. Directrix has y-intercept 6

23. Opens upward with focus five units from the vertex

24. Focal diameter 8 and focus on the negative y-axis

25–32 ■ Use the sketch to find an equation of the parabola.

25.

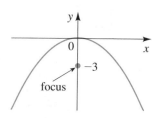

26.

directrix

$x = -2$

27.

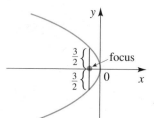

28.

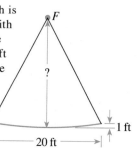

29.

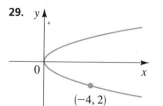

$(-4, 2)$

30.

square has area 16

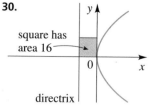

directrix

31.

focus   shaded region has area 8

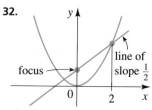

32.

focus   line of slope $\frac{1}{2}$

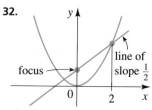

33. A lamp with a parabolic reflector is shown in the figure. The bulb is placed at the focus and the focal diameter is 12 cm.
(a) Find an equation of the parabola.
(b) Find the diameter $d(C, D)$ of the opening, 20 cm from the vertex.

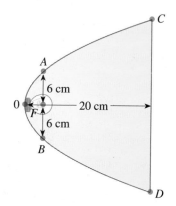

34. A reflector for a satellite dish is parabolic in cross section, with the receiver at the focus. The reflector is 1 ft deep and 20 ft wide from rim to rim (see the figure). How far is the receiver from the vertex of the parabolic reflector?

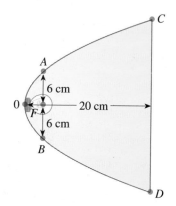

35. In a suspension bridge the shape of the suspension cables is parabolic. The bridge shown in the figure has towers that are 600 m apart and the lowest point of the suspension cables is 150 m below the top of the towers. Find the equation of the parabolic part of the cables, placing the origin of the coordinate system at the vertex.

NOTE: This equation is used to find the length of cable needed in the construction of the bridge.

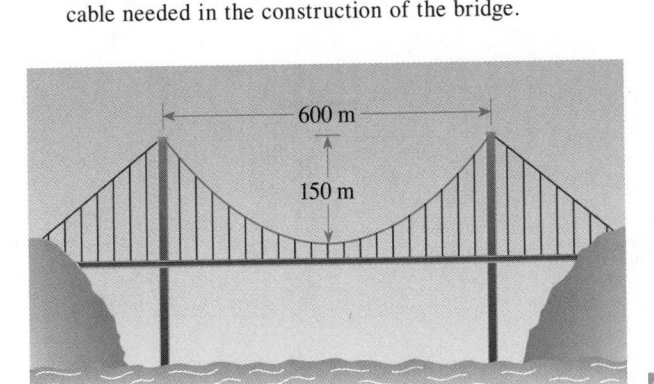

36. The Hale telescope at the Mount Palomar Observatory has a 200-inch mirror (see the figure). The mirror is constructed in a parabolic shape that collects light from the stars and focuses it at the *prime focus,* that is, the focus of the parabola. The mirror is 3.79 in deep at its center. Find the *focal length* of this parabolic mirror, that is, the distance from the vertex to the focus.

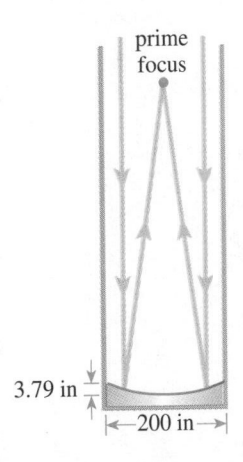

37. (a) Find equations for the family of parabolas with vertex at the origin and with directrices $y = \frac{1}{2}$, $y = 1$, $y = 4$, and $y = 8$.
(b) Draw the graphs. What do you conclude?

7.2 ELLIPSES

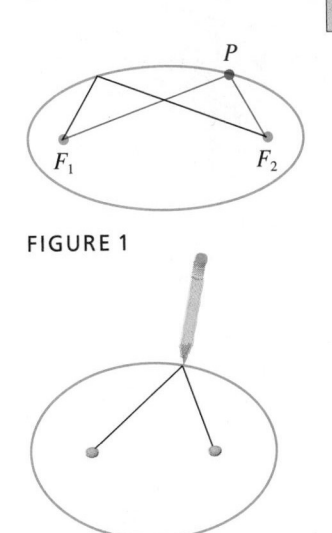

FIGURE 1

FIGURE 2

An ellipse is an oval curve that looks like an elongated circle. More precisely, we have the following definition.

> **GEOMETRIC DEFINITION OF AN ELLIPSE**
>
> An **ellipse** is the set of all points in the plane the sum of whose distances from two fixed points F_1 and F_2 is a constant. (See Figure 1.) These two fixed points are the **foci** (plural of **focus**) of the ellipse.

The geometric definition suggests a simple method for drawing an ellipse. Place a sheet of paper on a drawing board and insert thumbtacks at the two points that are to be the foci of the ellipse. Attach the ends of a string to the tacks, as shown in Figure 2. With the point of a pencil, hold the string taut. Then carefully move the pencil around the foci, keeping the string taut at all times. The pencil will trace out an ellipse, since the sum of the distances from the point of the pencil to the foci will always equal the length of the string, which is constant.

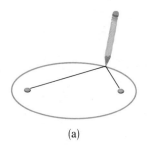

(a)

(b)

FIGURE 3

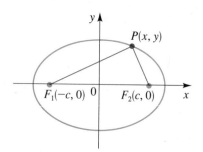

FIGURE 4

If the string is only slightly longer than the distance between the foci, then the ellipse traced out will be elongated in shape as in Figure 3(a), but if the foci are close together relative to the length of the string, the ellipse will be almost circular, as in Figure 3(b).

To obtain the simplest equation for an ellipse, we place the foci on the x-axis at $F_1(-c, 0)$ and $F_2(c, 0)$, so that the origin is halfway between them (see Figure 4). For later convenience, we let the sum of the distances from a point on the ellipse to the foci be $2a$. Then if $P(x, y)$ is any point on the ellipse, we have

$$d(P, F_1) + d(P, F_2) = 2a$$

so, from the Distance Formula,

$$\sqrt{(x + c)^2 + y^2} + \sqrt{(x - c)^2 + y^2} = 2a$$

or

$$\sqrt{(x - c)^2 + y^2} = 2a - \sqrt{(x + c)^2 + y^2}$$

Squaring each side and multiplying, we get

$$x^2 - 2cx + c^2 + y^2 = 4a^2 - 4a\sqrt{(x + c)^2 + y^2} + (x^2 + 2cx + c^2 + y^2)$$

which simplifies to

$$4a\sqrt{(x + c)^2 + y^2} = 4a^2 + 4cx$$

Dividing each side by 4 and squaring again, we get

$$a^2[(x + c)^2 + y^2] = (a^2 + cx)^2$$

$$a^2x^2 + 2a^2cx + a^2c^2 + a^2y^2 = a^4 + 2a^2cx + c^2x^2$$

$$(a^2 - c^2)x^2 + a^2y^2 = a^2(a^2 - c^2)$$

Since the sum of the distances from P to the foci must be larger than the distance between the foci, we have that $2a > 2c$, or $a > c$. Thus $a^2 - c^2 > 0$, and we can divide each side of the preceding equation by $a^2(a^2 - c^2)$ to get

$$\frac{x^2}{a^2} + \frac{y^2}{a^2 - c^2} = 1$$

For convenience, let $b^2 = a^2 - c^2$ (with $b > 0$). Since $b^2 < a^2$, it follows that $b < a$. The preceding equation then becomes

$$\frac{x^2}{a^2} + \frac{y^2}{b^2} = 1 \qquad \text{with } a > b$$

This is the equation of the ellipse. To graph it, we need to know the x- and y-intercepts. Setting $y = 0$, we get

$$\frac{x^2}{a^2} = 1$$

so $x^2 = a^2$, or $x = \pm a$. Thus, the ellipse crosses the x-axis at $(a, 0)$ and $(-a, 0)$.

These points are called the **vertices** of the ellipse, and the segment that joins them is called the **major axis.** Its length is $2a$.

Similarly, if we set $x = 0$, we get $y = \pm b$, so the ellipse crosses the y-axis at $(0, b)$ and $(0, -b)$. The segment that joins these points is called the **minor axis** and has length $2b$. Note that $2a > 2b$, so the major axis is longer than the minor axis.

In Section 3.2 we studied several tests that detect symmetry in a graph. If we replace x → $-x$ or y → $-y$ in the ellipse equation, it remains unchanged. Thus, the ellipse is symmetric about both the x- and y-axes, and hence about the origin as well. For this reason, the origin is called the **center** of the ellipse. The complete graph is shown in Figure 5.

If the foci of the ellipse are placed on the y-axis at $(0, \pm c)$ rather than on the x-axis, then the roles of x and y are reversed in the discussion leading up to the equation for a horizontal ellipse. Thus, we have the following description of ellipses.

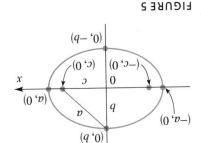

FIGURE 5

$$\frac{x^2}{a^2} + \frac{y^2}{b^2} = 1 \quad \text{with } a > b$$

ELLIPSE WITH CENTER AT THE ORIGIN

The graph of the equation

$$\frac{x^2}{a^2} + \frac{y^2}{b^2} = 1 \quad \text{or} \quad \frac{x^2}{b^2} + \frac{y^2}{a^2} = 1 \quad \text{with } a > b > 0$$

is an **ellipse** with center at the origin and having the following properties.

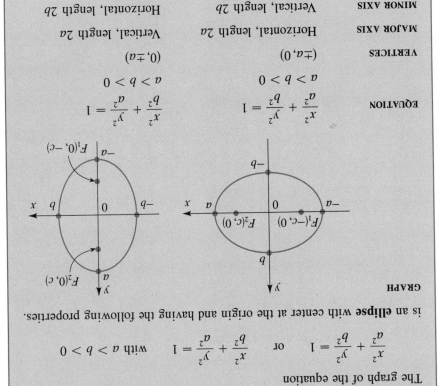

GRAPH		
EQUATION	$\dfrac{x^2}{a^2} + \dfrac{y^2}{b^2} = 1$ $a > b > 0$	$\dfrac{x^2}{b^2} + \dfrac{y^2}{a^2} = 1$ $a > b > 0$
VERTICES	$(\pm a, 0)$	$(0, \pm a)$
MAJOR AXIS	Horizontal, length $2a$	Vertical, length $2a$
MINOR AXIS	Vertical, length $2b$	Horizontal, length $2b$
FOCI	$(\pm c, 0),$ $c^2 = a^2 - b^2$	$(0, \pm c),$ $c^2 = a^2 - b^2$

In the standard equation for an ellipse, a^2 is the *larger denominator* and b^2 is the *smaller*. To find c^2 we subtract: larger denominator minus smaller denominator.

EXAMPLE 1 ■ Sketching an Ellipse

Find the foci, vertices, and the lengths of the major and minor axes for the following ellipse, and sketch its graph.

$$\frac{x^2}{9} + \frac{y^2}{4} = 1$$

SOLUTION

Since the denominator of x^2 is larger, the ellipse has horizontal major axis. This gives $a^2 = 9$ and $b^2 = 4$, so $c^2 = a^2 - b^2 = 9 - 4 = 5$. Thus, $a = 3$, $b = 2$, and $c = \sqrt{5}$.

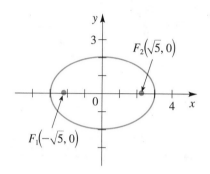

FOCI $(\pm\sqrt{5}, 0)$

VERTICES $(\pm 3, 0)$

LENGTH OF MAJOR AXIS   6

LENGTH OF MINOR AXIS   4

FIGURE 6
$$\frac{x^2}{9} + \frac{y^2}{4} = 1$$

The graph is shown in Figure 6.   ■

EXAMPLE 2 ■ Finding the Equation of an Ellipse

The vertices of an ellipse are $(\pm 4, 0)$ and the foci are at $(\pm 2, 0)$. Find its equation and sketch the graph.

SOLUTION

Since the vertices are $(\pm 4, 0)$, we have $a = 4$. The foci are $(\pm 2, 0)$, so $c = 2$. To write the equation we need to find b. Since $c^2 = a^2 - b^2$, we have

$$2^2 = 4^2 - b^2$$

$$b^2 = 16 - 4 = 12$$

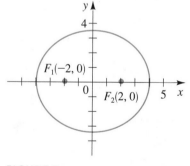

Thus the equation of the ellipse is

$$\frac{x^2}{16} + \frac{y^2}{12} = 1$$

FIGURE 7
$$\frac{x^2}{16} + \frac{y^2}{12} = 1$$

Its graph is shown in Figure 7.   ■

EXAMPLE 3 ■ Finding the Foci of an Ellipse

Find the foci of the ellipse $16x^2 + 9y^2 = 144$, and sketch the graph.

SOLUTION

Dividing through by 144, we get

$$\frac{x^2}{9} + \frac{y^2}{16} = 1$$

Since $16 > 9$, this is an ellipse with its foci on the y-axis, and with $a = 4$ and $b = 3$. We have

$$c^2 = a^2 - b^2 = 16 - 9 = 7$$

$$c = \sqrt{7}$$

Thus, the foci are $(0, \pm\sqrt{7}\,)$. The graph is shown in Figure 8.

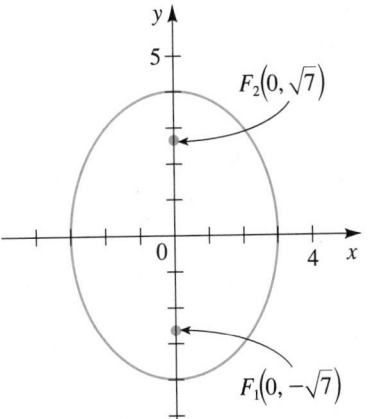

FIGURE 8
$16x^2 + 9y^2 = 144$

We saw earlier in this section (Figure 3) that if $2a$ is only slightly greater than $2c$, the ellipse is long and thin, whereas if $2a$ is much greater than $2c$, the ellipse is almost circular. We measure the deviation of an ellipse from being circular by the ratio of a and c.

DEFINITION OF ECCENTRICITY

For the ellipse $\dfrac{x^2}{a^2} + \dfrac{y^2}{b^2} = 1$ or $\dfrac{x^2}{b^2} + \dfrac{y^2}{a^2} = 1$ (with $a > b > 0$) the
eccentricity e is the number

$$e = \frac{c}{a}$$

where $c = \sqrt{a^2 - b^2}$. The eccentricity of every ellipse satisfies
$0 < e < 1$.

Johannes Kepler (1571–1630) was the first to give a correct description of the motion of the planets. The cosmology of his time postulated complicated systems of circles moving on circles to describe these motions. Kepler sought a simpler and more harmonious description. As official astronomer at the imperial court in Prague, he studied the astronomical observations of the Danish astronomer Tycho Brahe, whose data was at the time the most accurate available. After numerous attempts and failures at a theory, Kepler made the momentous discovery that the orbits of the planets are elliptical. His three great laws of planetary motion are:

1. The orbit of each planet is an ellipse with the sun at one focus.
2. The line segment that joins the sun to a planet sweeps out equal areas in equal time (see the figure).
3. The square of the period of revolution of a planet is proportional to the cube of the length of the major axis of its orbit.

His formulation of these laws is perhaps the most impressive deduction from empirical data in the history of science.

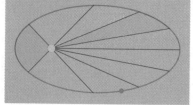

Thus, if e is close to 1, then c is almost equal to a, and the ellipse is elongated in shape, but if e is close to 0, then the ellipse is close to a circle in shape. The eccentricity is a measure of how "stretched" the ellipse is.

In Figure 9 we show a number of ellipses to demonstrate the effect of varying the eccentricity e.

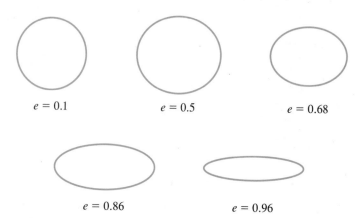

$e = 0.1$ $e = 0.5$ $e = 0.68$

$e = 0.86$ $e = 0.96$

FIGURE 9   Ellipses with various eccentricities

EXAMPLE 4 ■ Finding the Equation of an Ellipse from Its Eccentricity and Foci

Find the equation of the ellipse with foci $(0, \pm 8)$ and eccentricity $e = \frac{4}{5}$.

SOLUTION

We are given $e = \frac{4}{5}$ and $c = 8$. Thus,

$$\frac{4}{5} = \frac{8}{a} \qquad \text{Eccentricity } e = c/a$$

$$4a = 40 \qquad \text{Cross multiply}$$

$$a = 10$$

To find b we use the fact that $c^2 = a^2 - b^2$.

$$8^2 = 10^2 - b^2$$

$$b^2 = 10^2 - 8^2 = 36$$

$$b = 6$$

Thus, the equation of the ellipse is

$$\frac{x^2}{36} + \frac{y^2}{100} = 1$$

Because the foci are on the y-axis, the ellipse is oriented vertically. To sketch the ellipse we find the intercepts: The x-intercepts are ± 6 and the y-intercepts are ± 10. The graph is sketched in Figure 10.

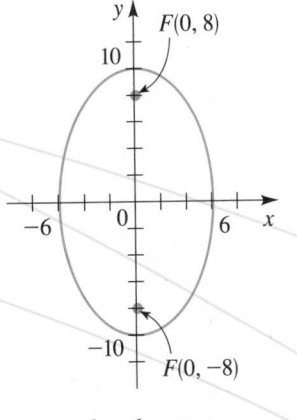

FIGURE 10

$$\frac{x^2}{36} + \frac{y^2}{100} = 1$$

Gravitational attraction causes the planets to move in elliptical orbits around the sun with the sun at one focus. This remarkable property was first observed by Johannes Kepler, and was later deduced by Isaac Newton from his inverse square law of gravity, using calculus. The orbits of the planets have different eccentricities, but most are nearly circular (see Exercises 33 and 34).

Ellipses, like parabolas, have an interesting reflection property that leads to a number of practical applications. If a light source is placed at one focus of a reflecting surface with elliptical cross sections, then all the light will be reflected off the surface to the other focus, as shown in Figure 11. This principle, which works for sound waves as well as for light, is used in *lithotripsy*, a treatment for kidney stones. The patient is placed in a tub of water with elliptical cross sections, in such a way that the kidney stone is accurately located at one focus. High-intensity sound waves generated at the other focus are reflected to the stone and destroy it with minimal damage to surrounding tissue. The patient is spared the trauma of surgery and recovers within days instead of weeks.

The reflection property of ellipses is also used in the construction of *whispering galleries*. Sound coming from one focus bounces off the walls and ceiling of an elliptical room and passes through the other focus. Thus, even quiet whispers spoken at one focus can be heard clearly at the other. Famous whispering galleries include the National Statuary Gallery of the U.S. Capitol in Washington, D.C., and the Mormon Tabernacle in Salt Lake City, Utah.

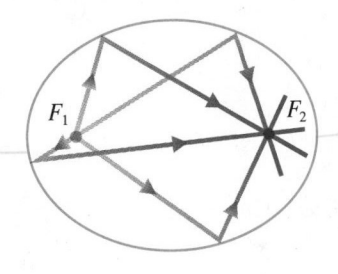

FIGURE 11

| 7.2 | **EXERCISES** |

1–14 ■ Find the vertices, foci, and eccentricity of the ellipse. Determine the lengths of the major and minor axes, and sketch the graph.

1. $\dfrac{x^2}{25} + \dfrac{y^2}{9} = 1$ **2.** $\dfrac{x^2}{16} + \dfrac{y^2}{25} = 1$

3. $9x^2 + 4y^2 = 36$ **4.** $4x^2 + 25y^2 = 100$

5. $x^2 + 4y^2 = 16$ **6.** $4x^2 + y^2 = 16$

7. $2x^2 + y^2 = 3$ **8.** $5x^2 + 6y^2 = 30$

9. $x^2 + 4y^2 = 1$ **10.** $9x^2 + 4y^2 = 1$

11. $\frac{1}{2}x^2 + \frac{1}{8}y^2 = \frac{1}{4}$ **12.** $x^2 = 4 - 2y^2$

13. $y^2 = 1 - 2x^2$ **14.** $20x^2 + 4y^2 = 5$

15–18 ■ Find an equation for the ellipse whose graph is shown.

15.
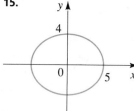

16.
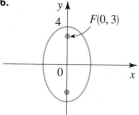

17.
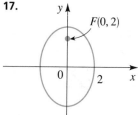

18.
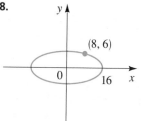

19–30 ■ Find an equation for the ellipse that satisfies the given conditions.

19. Foci $(\pm 4, 0)$, vertices $(\pm 5, 0)$

20. Foci $(0, \pm 3)$, vertices $(0, \pm 5)$

21. Length of major axis 4, length of minor axis 2, foci on y-axis

22. Length of major axis 6, length of minor axis 4, foci on x-axis

23. Foci $(0, \pm 2)$, length of minor axis 6

24. Foci $(\pm 5, 0)$, length of major axis 12

25. Endpoints of major axis $(\pm 10, 0)$, distance between foci 6

26. Endpoints of minor axis $(0, \pm 3)$, distance between foci 8

27. Length of major axis 10, foci on x-axis, ellipse passes through point $\left(\sqrt{5}, 2\right)$

28. Eccentricity $\frac{1}{9}$, foci $(0, \pm 2)$

29. Eccentricity 0.8, foci $(\pm 1.5, 0)$

30. Eccentricity $\sqrt{3}/2$, foci on y-axis, length of major axis 4

31–32 ■ Find the intersection points of the pair of ellipses. Sketch the graphs of each pair of equations on the same coordinate axes and label the points of intersection.

31. $\begin{cases} 4x^2 + y^2 = 4 \\ 4x^2 + 9y^2 = 36 \end{cases}$ **32.** $\begin{cases} \dfrac{x^2}{16} + \dfrac{y^2}{9} = 1 \\ \dfrac{x^2}{9} + \dfrac{y^2}{16} = 1 \end{cases}$

33. The planets move around the sun in elliptical orbits with the sun at one focus. The point in the orbit at which the planet is closest to the sun is called **peri-helion,** and the point at which it is farthest is called **aphelion.** These points are the vertices of the orbit. The earth's distance from the sun is 147,000,000 km at perihelion and 153,000,000 km at aphelion. Find an equation for the earth's orbit. (Place the origin at the center of the orbit with the sun on the x-axis.)

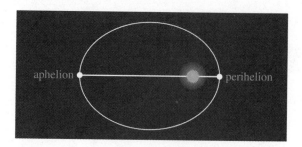

34. With an eccentricity of 0.25, Pluto's orbit is the most eccentric in the solar system. The length of the minor axis of its orbit is approximately 10,000,000,000 km.

Find the distance between Pluto and the sun at perihelion and at aphelion. (See Exercise 33.)

35. For an object in an elliptical orbit around the moon, the points in the orbit that are closest to and farthest from the center of the moon are called **perilune** and **apolune,** respectively. These are the vertices of the orbit. The center of the moon is at one of the foci of the orbit. The *Apollo 11* spacecraft was placed in a lunar orbit with perilune at 68 mi and apolune at 195 mi above the surface of the moon. Assuming the moon is a sphere of radius 1075 mi, find an equation for the orbit of the *Apollo 11*. (Place the coordinate axes so that the origin is at the center of the orbit and the foci are located on the *x*-axis.)

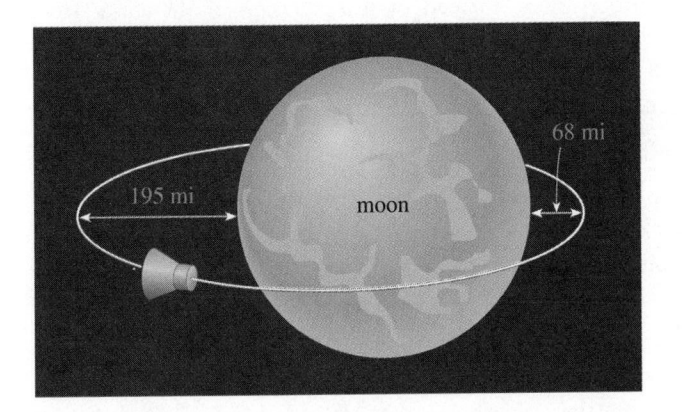

36. A carpenter wishes to construct an elliptical table top from a sheet of plywood, 4 ft by 8 ft. He will trace out the ellipse using the "thumbtack and string" method illustrated in Figures 2 and 3. What length of string should he use, and how far apart should the tacks be located, if the ellipse is to be the largest possible that can be cut out of the plywood sheet?

37. A "sunburst" window above a doorway is constructed in the shape of the top half of an ellipse, as shown in the figure. The window is 20 in. tall at its highest point and 80 in. wide at the bottom. Find the height of the window 25 in. from the center of the base.

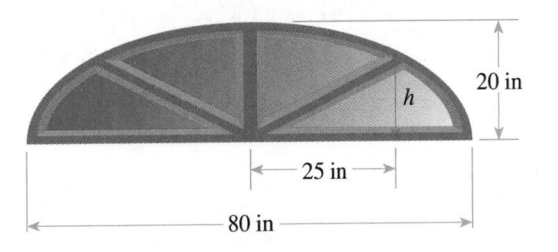

38. The **ancillary circle** of an ellipse is the circle with radius equal to half the length of the minor axis and center the same as the ellipse (see the figure). The ancillary circle is thus the largest circle that can fit within an ellipse.
   (a) Find an equation for the ancillary circle of the ellipse $x^2 + 4y^2 = 16$.
   (b) For the ellipse and ancillary circle of part (a), show that if (s, t) is a point on the ancillary circle, then $(2s, t)$ is a point on the ellipse.

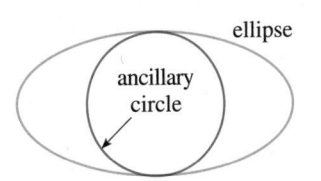

39. A **latus rectum** for an ellipse is a line segment perpendicular to the major axis at a focus, with endpoints on the ellipse, as shown in the figure. Show that the length of a latus rectum is $2b^2/a$ for the ellipse

$$\frac{x^2}{a^2} + \frac{y^2}{b^2} = 1 \qquad \text{with } a > b$$

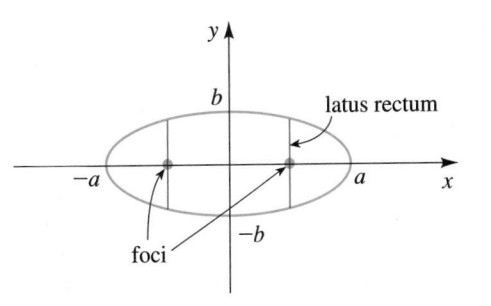

40. If $k > 0$, the following equation represents an ellipse:

$$\frac{x^2}{k} + \frac{y^2}{4 + k} = 1$$

Show that all the ellipses represented by this equation have the same foci, no matter what the value of k.

41. Use a graphing device to draw each of the following ellipses by solving for y and graphing both solutions.

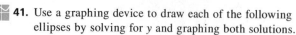

(a) $\dfrac{x^2}{25} + \dfrac{y^2}{20} = 1$ (b) $6x^2 + y^2 = 36$

 42. (a) Use a graphing device to sketch the top half (the portion in the first and second quadrants) of the family of ellipses $x^2 + ky^2 = 100$ for $k = 4, 10, 25,$ and 50.

(b) What do the members of this family of ellipses have in common? How do they differ?

7.3 HYPERBOLAS

Although ellipses and hyperbolas have completely different shapes, their definitions and equations are similar. Instead of using a *sum* of distances from two fixed foci, as in the case of an ellipse, we use the *difference* to define a hyperbola.

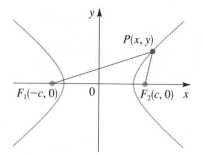

FIGURE 1

P is on the hyperbola if
$|d(P, F_1) - d(P, F_2)| = 2a$

GEOMETRIC DEFINITION OF A HYPERBOLA

A **hyperbola** is the set of all points in the plane, the difference of whose distances from two fixed points F_1 and F_2 is a constant. (See Figure 1.) These two fixed points are the **foci** of the hyperbola.

As in the case of the ellipse, we get the simplest equation for the hyperbola by placing the foci on the x-axis at $(\pm c, 0)$, as shown in Figure 1. From the definition, if $P(x, y)$ lies on the hyperbola, then either $d(P, F_1) - d(P, F_2)$ or $d(P, F_2) - d(P, F_1)$ must equal some positive constant, which we call $2a$. Thus we have

$$d(P, F_1) - d(P, F_2) = \pm 2a$$

or $$\sqrt{(x + c)^2 + y^2} - \sqrt{(x - c)^2 + y^2} = \pm 2a$$

Proceeding as we did in the case of the ellipse (Section 7.2), we simplify this to

$$(c^2 - a^2)x^2 - a^2 y^2 = a^2(c^2 - a^2)$$

From triangle PF_1F_2 in Figure 1, we see that $|d(P, F_1) - d(P, F_2)| < 2c$. It follows that $2a < 2c$, or $a < c$. Thus, $c^2 - a^2 > 0$, so we can set $b^2 = c^2 - a^2$. We then simplify the last displayed equation to get

$$\frac{x^2}{a^2} - \frac{y^2}{b^2} = 1$$

This is the *equation of the hyperbola*. If we replace x by $-x$ or y by $-y$ in this equation it remains unchanged, so the hyperbola is symmetric about both the x- and y-axes and about the origin. The x-intercepts are $\pm a$, and the points $(a, 0)$ and $(-a, 0)$ are the **vertices** of the hyperbola. There is no y-intercept, because setting $x = 0$ in the equation of the hyperbola leads to $-y^2 = b^2$, which is impossible. Furthermore, the equation of the hyperbola implies

$$\frac{x^2}{a^2} = 1 + \frac{y^2}{b^2} \geq 1$$

so $x^2/a^2 \geq 1$; thus $x^2 \geq a^2$, and hence $x \geq a$ or $x \leq -a$. This means that the hyperbola consists of two parts, called its **branches.** The segment joining the two vertices on the separate branches is the **transverse axis** of the hyperbola, and the origin is called its **center.**

If we place the foci of the hyperbola on the y-axis rather than on the x-axis, then this has the effect of reversing the roles of x and y in the derivation of the equation of the hyperbola. This leads to a hyperbola with a vertical transverse axis.

The main properties of hyperbolas are listed in the following box.

HYPERBOLA WITH CENTER AT THE ORIGIN

The graph of the equation

$$\frac{x^2}{a^2} - \frac{y^2}{b^2} = 1 \qquad \text{or} \qquad \frac{y^2}{a^2} - \frac{x^2}{b^2} = 1 \qquad \text{with } a > 0,\ b > 0$$

is a **hyperbola** with its center at the origin and with the following properties.

GRAPH	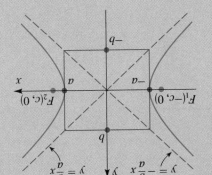	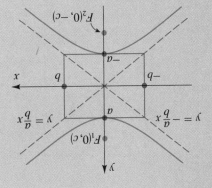
EQUATION	$\dfrac{x^2}{a^2} - \dfrac{y^2}{b^2} = 1$	$\dfrac{y^2}{a^2} - \dfrac{x^2}{b^2} = 1$
VERTICES	$(\pm a, 0)$	$(0, \pm a)$
TRANSVERSE AXIS	Horizontal, length $2a$	Vertical, length $2a$
ASYMPTOTES	$y = \pm\dfrac{b}{a}x$	$y = \pm\dfrac{a}{b}x$
FOCI	$(\pm c, 0), \quad c^2 = a^2 + b^2$	$(0, \pm c), \quad c^2 = a^2 + b^2$

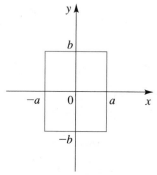

(a) Central box

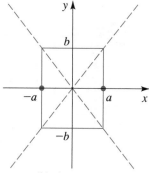

(b) Asymptotes

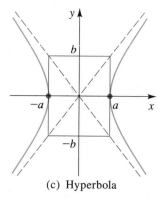

(c) Hyperbola

FIGURE 2

Steps in graphing the hyperbola
$$\frac{x^2}{a^2} - \frac{y^2}{b^2} = 1$$

The asymptotes mentioned in this box are lines that the hyperbola approaches for large values of x and y. (Asymptotes of rational functions are discussed in Section 4.8). We now show that the lines $y = \dfrac{b}{a}x$ and $y = -\dfrac{b}{a}x$ given in the first case in the box are asymptotes of the hyperbola

$$\frac{x^2}{a^2} - \frac{y^2}{b^2} = 1$$

To find the asymptotes we solve the equation for y, to get

$$y = \pm \frac{b}{a}\sqrt{x^2 - a^2}$$

$$= \pm \frac{b}{a}x \sqrt{1 - \frac{a^2}{x^2}}$$

As x gets large, a^2/x^2 gets closer to zero. In other words, as $x \to \infty$ we have $a^2/x^2 \to 0$. So, for large x the value of y can be approximated as $y = \pm(b/a)x$. This shows that these lines are asymptotes of the hyperbola.

Asymptotes are an essential aid for graphing a hyperbola; they help us determine its shape. A convenient way to find the asymptotes is to first plot the points $(a, 0)$, $(-a, 0)$, $(0, b)$, and $(0, -b)$. Then draw horizontal and vertical segments through these points to construct a rectangle, as shown in Figure 2(a). We call this rectangle the **central box** of the hyperbola. The slopes of the diagonals of the central box are $\pm b/a$, so by extending them we obtain the asymptotes $y = \pm(b/a)x$, as sketched in part (b) of the figure. Finally, we plot the vertices and use the asymptotes as a guide in sketching the hyperbola shown in part (c). (A similar procedure applies to graphing a hyperbola that has a vertical transverse axis.)

HOW TO DRAW A HYPERBOLA

1. DRAW THE CENTRAL BOX. This is the rectangle centered at the origin, with sides parallel to the axes, that crosses one axis at $\pm a$, the other at $\pm b$.

2. DRAW THE ASYMPTOTES. These are the lines obtained by extending the diagonals of the central box.

3. PLOT THE VERTICES. These are the two x-intercepts or the two y-intercepts.

4. DRAW THE HYPERBOLA. Start at a vertex and draw a branch of the hyperbola, approaching the asymptotes. Draw the other branch in the same way.

Galileo Galilei (1564–1642) was born in Pisa, Italy. He studied medicine, but later abandoned this in favor of science and mathematics. At the age of 25 he demonstrated that light objects fall at the same rate as heavier ones, by dropping cannonballs of various sizes from the Leaning Tower of Pisa. This contradicted the then-accepted view of Aristotle that heavier objects fall more quickly. He also showed that the distance an object falls is proportional to the square of the time it has been falling, and from this was able to prove that the path of a projectile is a parabola.

Galileo constructed the first telescope, and using it, discovered the moons of Jupiter. His advocacy of the Copernican view that the earth revolves around the sun (rather than being stationary) led to his being called before the Inquisition. By then an old man, he was forced to recant his views, but he is said to have muttered under his breath "the earth nevertheless does move." Galileo revolutionized science by expressing scientific principles in the language of mathematics. He said, "The great book of nature is written in mathematical symbols."

EXAMPLE 1 ■ A Hyperbola with Horizontal Transverse Axis

Find the vertices, foci, and asymptotes of the following hyperbola, and sketch its graph.

$$9x^2 - 16y^2 = 144$$

SOLUTION

First we divide both sides of the equation by 144 to put it into standard form:

$$\frac{x^2}{16} - \frac{y^2}{9} = 1$$

Because the x^2 term is positive, the hyperbola has a horizontal transverse axis; its vertices and foci are on the x-axis. Its vertices are $(\pm 4, 0)$. Since $a^2 = 16$ and $b^2 = 9$, we get $c = \sqrt{16 + 9} = 5$. Thus, the foci are $(\pm 5, 0)$. We have $a = 4$ and $b = 3$, so the asymptotes are $y = \pm \frac{3}{4}x$. After drawing the central box and asymptotes, we complete the sketch of the hyperbola as in Figure 3.

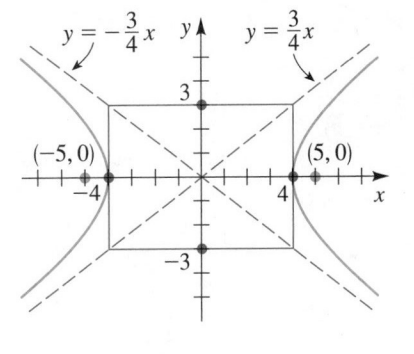

FIGURE 3
$9x^2 - 16y^2 = 144$

EXAMPLE 2 ■ A Hyperbola with Vertical Transverse Axis

Find the vertices, foci, and asymptotes of the following hyperbola, and sketch its graph.

$$x^2 - 9y^2 + 9 = 0$$

SOLUTION

We begin by writing the equation in the standard form for a hyperbola.

$$x^2 - 9y^2 = -9$$

$$y^2 - \frac{x^2}{9} = 1 \qquad \text{Divide by } -9$$

Because the y^2 term is positive, the hyperbola has a vertical transverse axis; its foci and vertices are on the y-axis. Its vertices are $(0, \pm 1)$. Since $a^2 = 1$ and

PATHS OF COMETS

The path of a comet is an ellipse, a parabola, or a hyperbola with the sun at a focus. This fact can be proved using calculus and Newton's laws of motion.* If the path is a parabola or a hyperbola, the comet will never return. If the path is an ellipse, it can be determined precisely when and where the comet can be seen again. Halley's comet has an elliptical path and returns every 75 years; it was last seen in 1987.

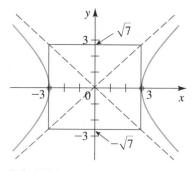

*J. Stewart, *Calculus, Third Edition*, pages 745–47.

$b^2 = 9$, we get $c = \sqrt{1 + 9} = \sqrt{10}$. Thus, the foci are $\left(0, \pm\sqrt{10}\,\right)$. We have $a = 1$ and $b = 3$, so the asymptotes are $y = \pm\frac{1}{3}x$. We draw the central box and asymptotes, then complete the graph as in Figure 4.

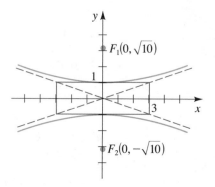

FIGURE 4
$x^2 - 9y^2 + 9 = 0$

EXAMPLE 3 ■ **Finding the Equation of a Hyperbola from Its Vertices and Foci**

Find the equation of the hyperbola with vertices $(\pm 3, 0)$ and foci $(\pm 4, 0)$. Sketch the graph.

SOLUTION

Since the vertices are on the x-axis, the hyperbola has a horizontal transverse axis. Its equation is of the form

$$\frac{x^2}{3^2} - \frac{y^2}{b^2} = 1$$

We have $a = 3$ and $c = 4$. To find b we use the relation $a^2 + b^2 = c^2$:

$$3^2 + b^2 = 4^2$$
$$b^2 = 4^2 - 3^2 = 7$$
$$b = \sqrt{7}$$

Thus, the equation of the hyperbola is

$$\frac{x^2}{9} - \frac{y^2}{7} = 1$$

FIGURE 5
$\dfrac{x^2}{9} - \dfrac{y^2}{7} = 1$

The graph is shown in Figure 5.

EXAMPLE 4 ■ **Finding the Equation of a Hyperbola from Its Vertices and Asymptotes**

Find the equation and the foci of the hyperbola with vertices $(0, \pm 2)$ and asymptotes $y = \pm 2x$. Sketch the graph.

SOLUTION

Since the vertices are on the *y*-axis, the hyperbola has a vertical transverse axis with $a = 2$. From the asymptote equation, we see that

$$\frac{a}{b} = 2$$

$$\frac{2}{b} = 2 \qquad \text{Since } a = 2$$

$$b = \frac{2}{2} = 1$$

Thus, the equation of the hyperbola is

$$\frac{y^2}{4} - x^2 = 1$$

To find the foci, we calculate $c^2 = a^2 + b^2 = 2^2 + 1^2 = 5$, so $c = \sqrt{5}$. Thus, the foci are $(0, \pm\sqrt{5})$. The graph is shown in Figure 6. ∎

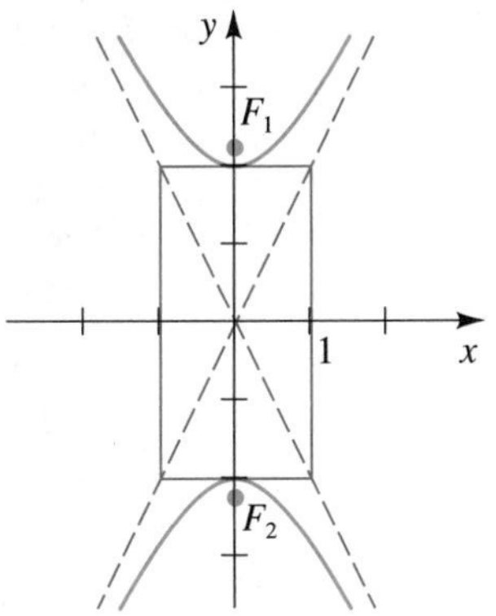

FIGURE 6

$$\frac{y^2}{4} - x^2 = 1$$

Like parabolas and ellipses, hyperbolas have an interesting reflection property. Light aimed at one focus of a hyperbolic mirror is reflected towards the other focus, as shown in Figure 7. This property is used in the construction of Cassegrain-type telescopes. A hyperbolic mirror is placed in the telescope tube so that light reflected from the primary parabolic reflector is aimed at one focus of the hyperbolic mirror. The light is then refocused at a more accessible point below the primary reflector (Figure 8).

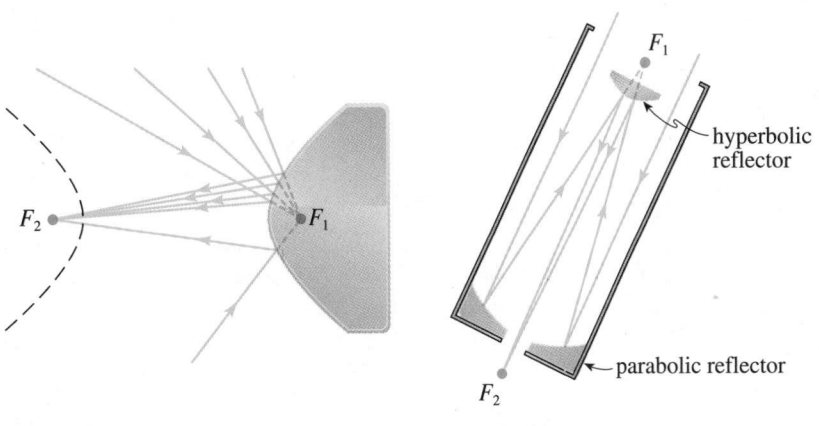

FIGURE 7

Reflection property of hyperbolas

FIGURE 8

Cassegrain-type telescope

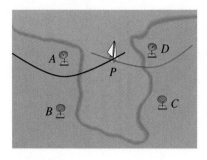

FIGURE 9

LORAN system for finding the location of a ship

In the LORAN (LOng RAnge Navigation) system, hyperbolas are used on-board a ship to determine its location. In Figure 9, radio stations at A and B transmit signals simultaneously for reception by the ship at P. The onboard computer converts the time difference in reception of these signals into a distance difference $d(P, A) - d(P, B)$. By the definition of a hyperbola, this locates the ship on one branch of a hyperbola with foci at A and B (sketched in black in the figure). The same procedure is carried out with two other radio stations at C and D, and this locates the ship on a second hyperbola (shown in red in the figure). (In practice, only three stations are needed because one station can be used as a focus for both hyperbolas.) The coordinates of the intersection point of these two hyperbolas, which can be calculated precisely by the computer, give the location of P.

7.3 EXERCISES

1–12 ■ Find the vertices, foci, and asymptotes of the hyperbola, and sketch its graph.

1. $\dfrac{x^2}{4} - \dfrac{y^2}{16} = 1$

2. $\dfrac{y^2}{9} - \dfrac{x^2}{16} = 1$

3. $y^2 - \dfrac{x^2}{25} = 1$

4. $\dfrac{x^2}{2} - y^2 = 1$

5. $x^2 - y^2 = 1$

6. $9x^2 - 4y^2 = 36$

7. $25y^2 - 9x^2 = 225$

8. $x^2 - y^2 + 4 = 0$

9. $x^2 - 4y^2 - 8 = 0$

10. $x^2 - 2y^2 = 3$

11. $4y^2 - x^2 = 1$

12. $9x^2 - 16y^2 = 1$

13–16 ■ Find the equation for the hyperbola whose graph is shown.

13.

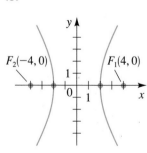

14.

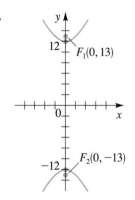

15.

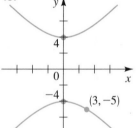

16.

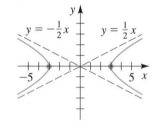

17–28 ■ Find an equation for the hyperbola that satisfies the given conditions.

17. Foci $(\pm 5, 0)$, vertices $(\pm 3, 0)$

18. Foci $(0, \pm 10)$, vertices $(0, \pm 8)$

19. Foci $(0, \pm 2)$, vertices $(0, \pm 1)$

20. Foci $(\pm 6, 0)$, vertices $(\pm 2, 0)$

21. Vertices $(\pm 1, 0)$, asymptotes $y = \pm 5x$

22. Vertices $(0, \pm 6)$, asymptotes $y = \pm \tfrac{1}{3}x$

23. Foci $(0, \pm 8)$, asymptotes $y = \pm \tfrac{1}{2}x$

24. Vertices $(0, \pm 6)$, hyperbola passes through $(-5, 9)$

25. Asymptotes $y = \pm x$, hyperbola passes through $(5, 3)$

26. Foci $(\pm 3, 0)$, hyperbola passes through $(4, 1)$

27. Foci $(\pm 5, 0)$, length of transverse axis 6

28. Foci $(0, \pm 1)$, length of transverse axis 1

29. (a) Show that the asymptotes of the hyperbola
$x^2 - y^2 = 5$ are perpendicular to each other.
   (b) Find an equation for the hyperbola with vertices
$(\pm c, 0)$ and with asymptotes perpendicular to each other.

30. The hyperbolas

$$\frac{x^2}{a^2} - \frac{y^2}{b^2} = 1 \quad \text{and} \quad \frac{x^2}{a^2} - \frac{y^2}{b^2} = -1$$

are said to be **conjugate** to each other.
   (a) Show that the hyperbolas

$$x^2 - 4y^2 + 16 = 0 \quad \text{and} \quad 4y^2 - x^2 + 16 = 0$$

   are conjugate to each other, and graph them on the same coordinate axes.
   (b) What do the hyperbolas of part (a) have in common?
   (c) Show that any pair of conjugate hyperbolas have the relationship you discovered in part (b).

31. In the derivation of the equation of the hyperbola at the beginning of this section, we said that the equation

$$\sqrt{(x + c)^2 + y^2} - \sqrt{(x - c)^2 + y^2} = \pm 2a$$

simplifies to

$$(c^2 - a^2)x^2 - a^2y^2 = a^2(c^2 - a^2)$$

Supply the steps needed to show this.

32. (a) For the hyperbola

$$\frac{x^2}{9} - \frac{y^2}{16} = 1$$

   determine the values of a, b, and c, and find the coordinates of the foci F_1 and F_2.
   (b) Show that the point $P\left(5, \frac{16}{3}\right)$ lies on this hyperbola.
   (c) Find $d(P, F_1)$ and $d(P, F_2)$.
   (d) Verify that the difference between $d(P, F_1)$ and $d(P, F_2)$ is $2a$.

33. Refer to Figure 9 in the text. Suppose that the radio stations at A and B are 500 miles apart, and that the

ship at P receives Station A's signal 2640 microseconds (μs) before it receives the signal from B.
   (a) Assuming that radio signals travel at 980 ft/μs, find $d(P, A) - d(P, B)$.
   (b) Find an equation for the branch of the hyperbola indicated in black in the figure. (Place A and B on the y-axis with the origin halfway between them. Use miles as the unit of distance.)
   (c) If A is due north of B, and if P is due east of A, how far is P from A?

34. Some comets, such as Halley's comet, are a permanent part of the solar system, traveling in elliptical orbits around the sun. Others pass through the solar system only once, following a hyperbolic path with the sun at a focus. The figure shows the path of such a comet. Find an equation for the path, assuming that the closest the comet comes to the sun is 2×10^9 mi and that the path the comet was taking before it neared the solar system is at a right angle to the path it continues on after leaving the solar system.

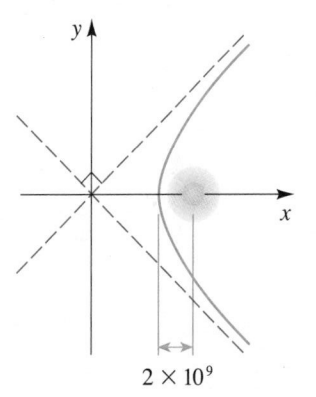

$$2 \times 10^9$$

35. Hyperbolas are called **confocal** if they have the same foci.
   (a) Show that the hyperbolas

$$\frac{y^2}{k} - \frac{x^2}{16 - k} = 1 \quad \text{with } 0 < k < 16$$

   are confocal.
   (b) Use a graphing device to draw the top branches of the family of hyperbolas in part (a) for $k = 1, 4, 8$, and 12. How does the shape change as k increases?

7.4 SHIFTED CONICS

In the preceding sections we have studied parabolas with vertices at the origin and ellipses and hyperbolas with centers at the origin. We restricted ourselves to these cases because the equations then have the simplest form. In this section we consider conics whose vertices and centers are not necessarily at the origin, and we determine how this affects their equations.

In Section 3.8 we studied transformations of functions that have the effect of shifting the graphs. In general, for any equation in x and y, if we replace x by $x - h$ or by $x + h$, the graph of the new equation is simply the old graph shifted horizontally; if y is replaced by $y - k$ or by $y + k$, the graph is shifted vertically. The following box gives the details.

SHIFTING GRAPHS OF EQUATIONS

If h and k are positive real numbers, then replacing x by $x - h$ or by $x + h$ and replacing y by $y - k$ or by $y + k$ has the following effect(s) on the graph of any equation in x and y.

Replacement	How the graph is shifted
1. x replaced by $x - h$	h units right
2. x replaced by $x + h$	h units left
3. y replaced by $y - k$	k units upward
4. y replaced by $y + k$	k units downward

For example, consider the ellipse with equation

$$\frac{x^2}{a^2} + \frac{y^2}{b^2} = 1$$

which is shown in Figure 1. If we shift it so that its center is at the point (h, k) instead of at the origin, then its equation becomes

$$\frac{(x - h)^2}{a^2} + \frac{(y - k)^2}{b^2} = 1$$

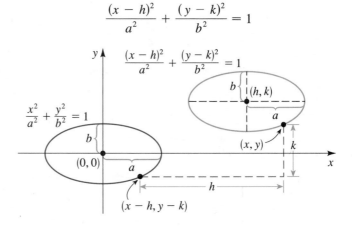

FIGURE 1

Shifted ellipse

EXAMPLE 1 ■ Sketching the Graph of a Shifted Ellipse

Sketch the graph of the ellipse

$$\frac{(x+1)^2}{4} + \frac{(y-2)^2}{9} = 1$$

and determine the coordinates of the foci.

SOLUTION

The ellipse

$$\frac{(x+1)^2}{4} + \frac{(y-2)^2}{9} = 1 \qquad \text{(Shifted ellipse)}$$

is shifted so that its center is at $(-1, 2)$. It is obtained from the ellipse

$$\frac{x^2}{4} + \frac{y^2}{9} = 1 \qquad \text{(Ellipse with center at origin)}$$

by shifting it left 1 unit and upward 2 units. The endpoints of the minor and major axes of the unshifted ellipse are $(2, 0)$, $(-2, 0)$, $(0, 3)$, $(0, -3)$. We apply the required shifts to these points to obtain the corresponding points on the shifted ellipse:

$$(2, 0) \;\rightarrow\; (2-1, 0+2) = (1, 2)$$

$$(-2, 0) \;\rightarrow\; (-2-1, 0+2) = (-3, 2)$$

$$(0, 3) \;\rightarrow\; (0-1, 3+2) = (-1, 5)$$

$$(0, -3) \;\rightarrow\; (0-1, -3+2) = (-1, -1)$$

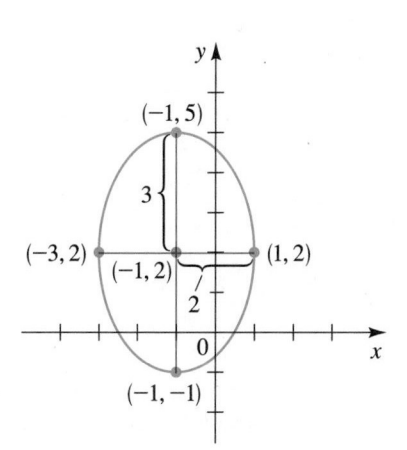

FIGURE 2
$$\frac{(x+1)^2}{4} + \frac{(y-2)^2}{9} = 1$$

This helps us sketch the graph in Figure 2.

To find the foci of the shifted ellipse, we first find the foci of the ellipse with center at the origin. Since $a^2 = 9$ and $b^2 = 4$, we have $c^2 = 9 - 4 = 5$, so $c = \sqrt{5}$. So the foci are $(0, \pm\sqrt{5})$. Shifting left 1 unit and upward 2 units, we get

$$(0, \sqrt{5}) \;\rightarrow\; (0-1, \sqrt{5}+2) = (-1, 2+\sqrt{5})$$

$$(0, -\sqrt{5}) \;\rightarrow\; (0-1, -\sqrt{5}+2) = (-1, 2-\sqrt{5})$$

Thus, the foci of the shifted ellipse are

$$(-1, 2+\sqrt{5}) \qquad \text{and} \qquad (-1, 2-\sqrt{5})$$

■

Applying shifts to parabolas and hyperbolas leads to the equations and their graphs shown in Figures 3 and 4.

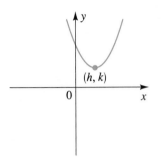

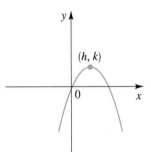

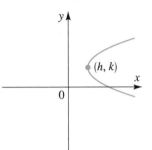

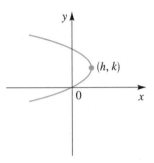

(a) $(x - h)^2 = 4p(y - k)$
$\quad\quad p > 0$

(b) $(x - h)^2 = 4p(y - k)$
$\quad\quad p < 0$

(c) $(y - k)^2 = 4p(x - h)$
$\quad\quad p > 0$

(d) $(y - k)^2 = 4p(x - h)$
$\quad\quad p < 0$

FIGURE 3
Shifted parabolas

FIGURE 4
Shifted hyperbolas

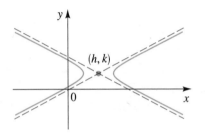

(a) $\dfrac{(x - h)^2}{a^2} - \dfrac{(y - k)^2}{b^2} = 1$

(b) $\dfrac{(y - k)^2}{a^2} - \dfrac{(x - h)^2}{b^2} = 1$

EXAMPLE 2 ■ Graphing a Shifted Parabola

Determine the vertex, focus, and directrix and sketch the graph of the following parabola.

$$x^2 - 4x = 8y - 28$$

SOLUTION

We complete the square in x to put this equation into one of the forms in Figure 3.

$$x^2 - 4x + 4 = 8y - 28 + 4 \quad\quad \text{Add 4 to complete the square}$$

$$(x - 2)^2 = 8y - 24$$

$$(x - 2)^2 = 8(y - 3) \quad\quad \text{(Shifted parabola)}$$

This is a parabola that opens upwards with vertex at $(2, 3)$. It is obtained from the parabola

$$x^2 = 8y \quad\quad \text{(Parabola with vertex at origin)}$$

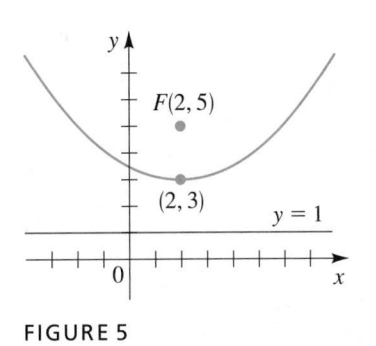

FIGURE 5

$x^2 - 4x = 8y - 28$

by shifting right 2 units and upward 3 units. Since $4p = 8$, we have $p = 2$, so the focus is 2 units above the vertex and the directrix is 2 units below the vertex. Thus, the focus is $(2, 5)$ and the directrix is $y = 1$. The graph is shown in Figure 5. ∎

EXAMPLE 3 ■ **Graphing a Shifted Hyperbola**

Show that the following equation represents a hyperbola:

$$9x^2 - 72x - 16y^2 - 32y = 16$$

Find its center, vertices, foci, and asymptotes, and sketch its graph.

SOLUTION

We first complete the square in both x and y:

$$9(x^2 - 8x \qquad) - 16(y^2 + 2y \qquad) = 16$$

$$9(x^2 - 8x + 16) - 16(y^2 + 2y + 1) = 16 + 9 \cdot 16 - 16 \cdot 1 \qquad \text{Complete the squares}$$

$$9(x - 4)^2 - 16(y + 1)^2 = 144$$

$$\frac{(x - 4)^2}{16} - \frac{(y + 1)^2}{9} = 1 \qquad \text{Divide by 144}$$

This is a hyperbola with center $(4, -1)$ and with a horizontal transverse axis. Its graph will have the same shape as the unshifted hyperbola

$$\frac{x^2}{16} - \frac{y^2}{9} = 1 \qquad \text{(Hyperbola with center at origin)}$$

Since $a^2 = 16$ and $b^2 = 9$, we have $a = 4$, $b = 3$, and $c = \sqrt{a^2 + b^2} = \sqrt{16 + 9} = 5$. Thus, the foci lie 5 units to the left and to the right of the center, and the vertices lie 4 units to either side of the center.

$$\text{FOCI} \quad (9, -1) \text{ and } (-1, -1)$$

$$\text{VERTICES} \quad (8, -1) \text{ and } (0, -1)$$

The asymptotes of the unshifted hyperbola are $y = \pm\frac{3}{4}x$, so the asymptotes of the shifted parabola are

$$(y + 1) = \pm\tfrac{3}{4}(x - 4)$$

$$y + 1 = \pm\tfrac{3}{4}x \mp 3$$

$$y = \tfrac{3}{4}x - 4 \qquad \text{and} \qquad y = -\tfrac{3}{4}x + 2$$

To help us sketch the hyperbola we draw the central box; it extends 4 units left and right of the center and 3 units upward and downward from the center. We then draw the asymptotes and complete the graph of the shifted hyperbola as shown in Figure 6.

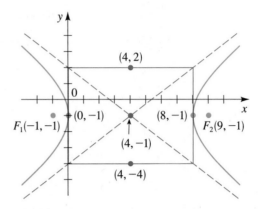

FIGURE 6

$9x^2 - 72x - 16y^2 - 32y = 16$

If we expand and simplify the equations of any of the shifted conics illustrated in Figures 1, 3, and 4, then we will always obtain an equation of the form

$$Ax^2 + By^2 + Cx + Dy + E = 0$$

where A and B are not both zero. Conversely, if we begin with an equation of this form, then we can complete the square in x and y to see which type of conic section the equation represents. In some cases the graph of the equation turns out to be just a pair of lines, a single point, or there may be no graph at all. These are called the **degenerate cases.** The next example illustrates such a case.

EXAMPLE 4 ■ An Equation that Leads to a Degenerate Conic

Sketch the graph of the equation

$$9x^2 - y^2 + 18x + 6y = 0$$

SOLUTION

Because the coefficients of x^2 and y^2 are of opposite sign, this equation looks as if it should represent a hyperbola (like the equation of Example 3). To see if this is in fact the case, we complete the square:

$$9(x^2 + 2x \quad\quad) - (y^2 - 6y \quad\quad) = 0$$

$$9(x^2 + 2x + 1) - (y^2 - 6y + 9) = 0 + 9 - 9$$

$$9(x + 1)^2 - (y - 3)^2 = 0$$

$$(x + 1)^2 - \frac{(y - 3)^2}{9} = 0 \qquad \text{Divide by 9}$$

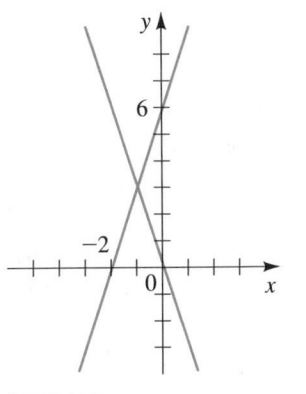

FIGURE 7

$9x^2 - y^2 + 18x + 6y = 0$

For this to fit the form of the equation of a hyperbola, we would need a non-zero constant to the right of the equal sign. In fact, further analysis shows that this is the equation of a pair of intersecting lines:

$$(y - 3)^2 = 9(x + 1)^2$$

$$y - 3 = \pm 3(x + 1) \qquad \text{Take square root}$$

$$y = 3(x + 1) + 3 \qquad \text{or} \qquad y = -3(x + 1) + 3$$

$$y = 3x + 6 \qquad\qquad\qquad y = -3x$$

These lines are graphed in Figure 7. ▪

Because the equation in Example 4 looked at first glance like the equation of a hyperbola but, in fact, turned out to represent simply a pair of lines, we refer to its graph as a **degenerate hyperbola.** Degenerate ellipses and parabolas can also arise when we complete the square in an equation that seems to represent a conic. For example, the equation

$$4x^2 + y^2 - 8x + 2y + 6 = 0$$

looks as if it should represent an ellipse, because the coefficients of x^2 and y^2 have the same sign. But completing the square leads to

$$(x - 1)^2 + \frac{(y + 1)^2}{4} = -\frac{1}{4}$$

which has no solution at all (since the sum of two squares cannot be negative). This equation is therefore degenerate.

To summarize, we have the following theorem.

GENERAL EQUATION OF A CONIC SECTION

The graph of the equation

$$Ax^2 + By^2 + Cx + Dy + E = 0$$

where A and B are not both zero, is a conic or a degenerate conic. In the nondegenerate cases, the graph is

1. a parabola if A or B is zero
2. an ellipse if A and B have the same sign (or a circle if $A = B$)
3. a hyperbola if A and B have opposite signs.

7.4 **EXERCISES**

1–4 ■ Find the center, foci, and vertices of the ellipse, and determine the lengths of the major and minor axes. Then sketch the graph.

1. $\dfrac{(x-2)^2}{9} + \dfrac{(y-1)^2}{4} = 1$

2. $\dfrac{(x-3)^2}{16} + (y+3)^2 = 1$

3. $\dfrac{x^2}{9} + \dfrac{(y+5)^2}{25} = 1$

4. $\dfrac{(x+2)^2}{4} + y^2 = 1$

5–8 ■ Find the vertex, focus, and directrix of the parabola, and sketch the graph.

5. $(x-3)^2 = 8(y+1)$

6. $(y+5)^2 = -6x + 12$

7. $-4\left(x + \tfrac{1}{2}\right)^2 = y$

8. $y^2 = 16x - 8$

9–12 ■ Find the center, foci, vertices, and asymptotes of the hyperbola. Then sketch the graph.

9. $\dfrac{(x+1)^2}{9} - \dfrac{(y-3)^2}{16} = 1$

10. $(x-8)^2 - (y+6)^2 = 1$

11. $y^2 - \dfrac{(x+1)^2}{4} = 1$

12. $\dfrac{(y-1)^2}{25} - (x+3)^2 = 1$

13–18 ■ Find an equation for the conic whose graph is shown.

13.

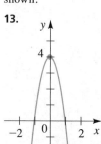

14.

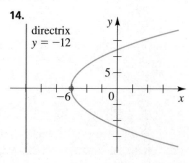

15.

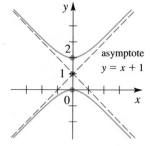

16.

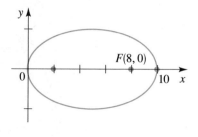

17.

18.

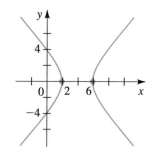

19–30 ■ Complete the square to determine whether the equation represents an ellipse, a parabola, a hyperbola, or a degenerate conic. Then sketch the graph of the equation. If the graph is an ellipse, find the center, foci, vertices, and lengths of the major and minor axes. If it is a parabola, find the vertex, focus, and directrix. If it is a hyperbola, find the center, foci, vertices, and asymptotes. If the equation has no graph, explain why.

19. $9x^2 - 36x + 4y^2 = 0$

20. $y^2 = 4(x + 2y)$

21. $x^2 - 4y^2 - 2x + 16y = 20$

22. $x^2 + 6x + 12y + 9 = 0$

23. $4x^2 + 25y^2 - 24x + 250y + 561 = 0$

24. $2x^2 + y^2 = 2y + 1$

25. $16x^2 - 9y^2 - 96x + 288 = 0$

26. $4x^2 - 4x - 8y + 9 = 0$

27. $x^2 + 16 = 4(y^2 + 2x)$

28. $x^2 - y^2 = 10(x - y) + 1$

29. $3x^2 + 4y^2 - 6x - 24y + 39 = 0$

30. $x^2 + 4y^2 + 20x - 40y + 300 = 0$

31. Determine what the value of F must be if the graph of the equation

$$4x^2 + y^2 + 4(x - 2y) + F = 0$$

is **(a)** an ellipse, **(b)** a single point, or **(c)** the empty set.

32. Find an equation for the ellipse that shares a vertex and a focus with the parabola $x^2 + y = 100$ and has its other focus at the origin.

33. This exercise deals with **confocal parabolas,** that is, families of parabolas that have the same focus.
 (a) Draw the graphs of the family of parabolas

$$x^2 = 4p(y + p)$$

 for $p = -2, -\frac{3}{2}, -1, -\frac{1}{2}, \frac{1}{2}, 1, \frac{3}{2}, 2$.
 (b) Show that each parabola in this family has its focus at the origin.
 (c) Describe the effect on the graph of moving the vertex closer to the origin.

34. Consider the family of conics that have a vertex at the origin and a focus at $(0, 1)$, as shown in the figure.
 (a) Show that there is precisely one parabola satisfying these properties, and find its equation.
 (b) Find equations of two different ellipses that satisfy these properties.
 (c) Find equations of two different hyperbolas that satisfy these properties.
 (d) Graph the conics you found in parts (a), (b), and (c) on the same coordinate axes. (For the hyperbolas, graph the top branches only.)
 (e) Describe how the ellipses and hyperbolas are related to the parabola.

---

7 | REVIEW

KEY TOPICS ■ Define, state, or discuss each of the following.

1. Conic sections

2. Parabola

3. Focus, directrix, and vertex of a parabola

4. Focal diameter of a parabola

5. Latus rectum of a parabola

6. Reflection property of parabolas

7. Ellipse

8. Foci of an ellipse

9. Major and minor axes of an ellipse

10. Eccentricity of an ellipse

11. Reflection property of an ellipse

12. Hyperbola

13. Foci of a hyperbola

14. Asymptotes of a hyperbola

15. Reflection property of a hyperbola

16. Shifted conics

17. The general equation of a conic:
$$Ax^2 + By^2 + Cx + Dy + E = 0$$

EXERCISES

1–4 ■ Find the vertex, focus, and directrix of the parabola, and sketch the graph.

1. $x^2 + 8y = 0$ **2.** $2x - y^2 = 0$

3. $x - y^2 + 4y - 2 = 0$

4. $2x^2 + 6x + 5y + 10 = 0$

5–8 ■ Find the center, vertices, foci, and the lengths of the major and minor axes of the ellipse, and sketch the graph.

5. $x^2 + 4y^2 = 16$

6. $9x^2 + 4y^2 = 1$

7. $4x^2 + 9y^2 = 36y$

8. $2x^2 + y^2 = 2 + 4(x - y)$

9–12 ■ Find the center, vertices, foci, and asymptotes of the hyperbola, and sketch the graph.

9. $x^2 - 2y^2 = 16$

10. $x^2 - 4y^2 + 16 = 0$

11. $9y^2 + 18y = x^2 + 6x + 18$

12. $y^2 = x^2 + 6y$

13–18 ■ Find an equation for the conic whose graph is shown.

13.

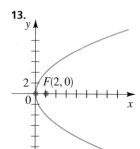

14.

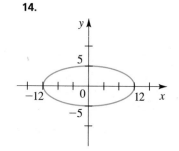

15.

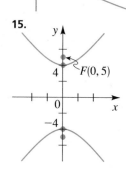

16.

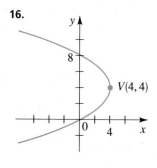

17.

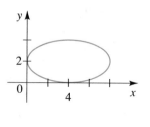

18.

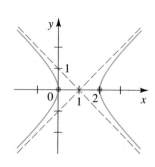

19–30 ■ Determine the type of curve represented by the equation. Find the foci and vertices (if any), and sketch the graph.

19. $\dfrac{x^2}{12} + y = 1$

20. $\dfrac{x^2}{12} + \dfrac{y^2}{144} = \dfrac{y}{12}$

21. $x^2 - y^2 + 144 = 0$

22. $x^2 + 6x = 9y^2$

23. $4x^2 + y^2 = 8(x + y)$

24. $3x^2 - 6(x + y) = 10$

25. $x = y^2 - 16y$

26. $2x^2 + 4 = 4x + y^2$

27. $2x^2 - 12x + y^2 + 6y + 26 = 0$

28. $36x^2 - 4y^2 - 36x - 8y = 31$

29. $9x^2 + 8y^2 - 15x + 8y + 27 = 0$

30. $x^2 + 4y^2 = 4x + 8$

31–38 ■ Find an equation for the conic section with the given properties.

31. The parabola with focus $F(0, 1)$ and directrix $y = -1$

32. The ellipse with center $C(0, 4)$, foci $F_1(0, 0)$ and $F_2(0, 8)$, and major axis of length 10

33. The hyperbola with vertices $V(0, \pm 2)$ and asymptotes $y = \pm \frac{1}{2}x$

34. The hyperbola with center $C(2, 4)$, foci $F_1(2, 7)$ and $F_2(2, 1)$, and vertices $V_1(2, 6)$ and $V_2(2, 2)$

35. The ellipse with foci $F_1(1, 1)$ and $F_2(1, 3)$, and with one vertex on the x-axis

36. The parabola with vertex $V(5, 5)$ and directrix the y-axis

37. The ellipse with vertices $V_1(7, 12)$ and $V_2(7, -8)$, and passing through the point $P(1, 8)$

38. The parabola with vertex $V(-1, 0)$ and horizontal axis of symmetry, and crossing the y-axis at $y = 2$

39. A cannon fires a cannonball as shown in the figure. The path of the cannonball is a parabola with vertex at the highest point of the path. If the cannonball lands 1600 ft from the cannon and the highest point it reaches is 3200 ft above the ground, find an equation for the path of the cannonball. Place the origin at the location of the cannon.

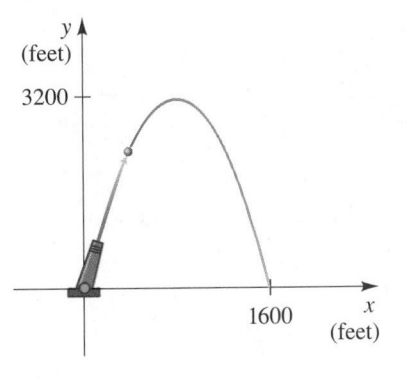

40. A satellite is in an elliptical orbit around the earth with the center of the earth at one focus. The height of the satellite above the earth varies between 140 mi and 440 mi. Assume the earth is a sphere with radius 3960 mi. Find an equation for the path of the satellite with the origin at the center of the earth.

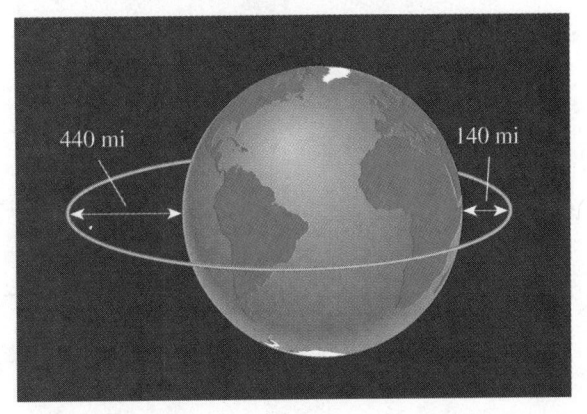

440 mi   140 mi

41. The path of the earth around the sun is an ellipse with the sun at one focus. The ellipse has major axis 186,000,000 mi and eccentricity 0.017. Find the distance between the earth and the sun when the earth is **(a)** closest to the sun and **(b)** furthest from the sun.

42. **(a)** Draw graphs of the following family of ellipses for $k = 1, 2, 4, 8$.

$$\frac{x^2}{16 + k^2} + \frac{y^2}{k^2} = 1$$

(b) Prove that all the ellipses in part (a) have the same foci.

43. **(a)** Draw graphs of the following family of parabolas for $k = \frac{1}{2}, 1, 2, 4$.

$$y = kx^2$$

(b) Find the foci of the parabolas in part (a).
(c) How does the location of the focus change as k increases?

1. Find the focus and directrix of the parabola $x^2 = -12y$, and sketch its graph.

2. Find the vertices, foci, and the lengths of the major and minor axes for the following ellipse. Then sketch its graph.

$$\frac{x^2}{16} + \frac{y^2}{4} = 1$$

3. Find the vertices, foci, and asymptotes of the following hyperbola. Then sketch its graph.

$$\frac{y^2}{9} - \frac{x^2}{16} = 1$$

4–6 ■ Find an equation for the conic whose graph is shown.

4.

$(-4, 2)$

5.

$(4, 3)$

6.

$F(4, 0)$

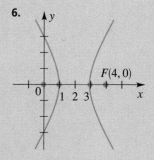

7–9 ■ Sketch the graph of the equation.

7. $16x^2 + 36y^2 - 96x + 36y + 9 = 0$

8. $9x^2 - 8y^2 + 36x + 64y = 92$

9. $2x + y^2 + 8y + 8 = 0$

10. Find an equation for the hyperbola with foci $(0, \pm 5)$ and with asymptotes $y = \pm\frac{3}{4}x$.

11. Find an equation for the parabola with focus $(2, 4)$ and directrix the x-axis.

12. A parabolic reflector for a car headlight forms a bowl shape that is 6 in. wide at its opening and 3 in. deep, as shown in the figure. How far from the vertex should the bulb be placed if it is to be located at the focus?

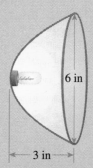

6 in

3 in

FOCUS ON PROBLEM SOLVING

In the Focus on Problem Solving after Chapter 2, we considered the strategy of **taking cases.** A classic use of this strategy is in the classification of the regular polyhedra. These are called *Platonic solids* because they were first studied by Plato.

A *regular polygon* is one in which all sides and all angles are equal. There are infinitely many regular polygons, as indicated in the following figure:

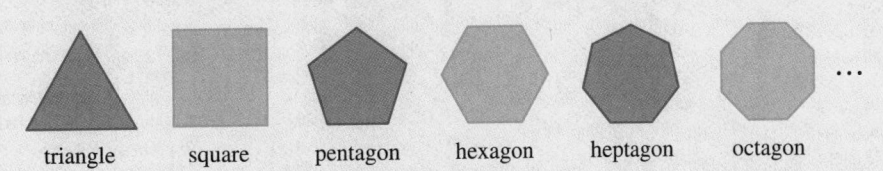

The regular polygons   triangle   square   pentagon   hexagon   heptagon   octagon

For three-dimensional shapes the analogous concept is that of regular polyhedra. A *regular polyhedron* is a solid in which all faces are congruent regular polygons, and the same number of polygons meet at each corner, or *vertex*. We would like to find all possible regular polyhedra. It might seem at first that there should be infinitely many, just as for regular polygons. But we will see that there are just finitely many polyhedra.

PROBLEM 1 ■ Classifying the Regular Polyhedra

In this problem we prove that there are exactly five regular polyhedra:

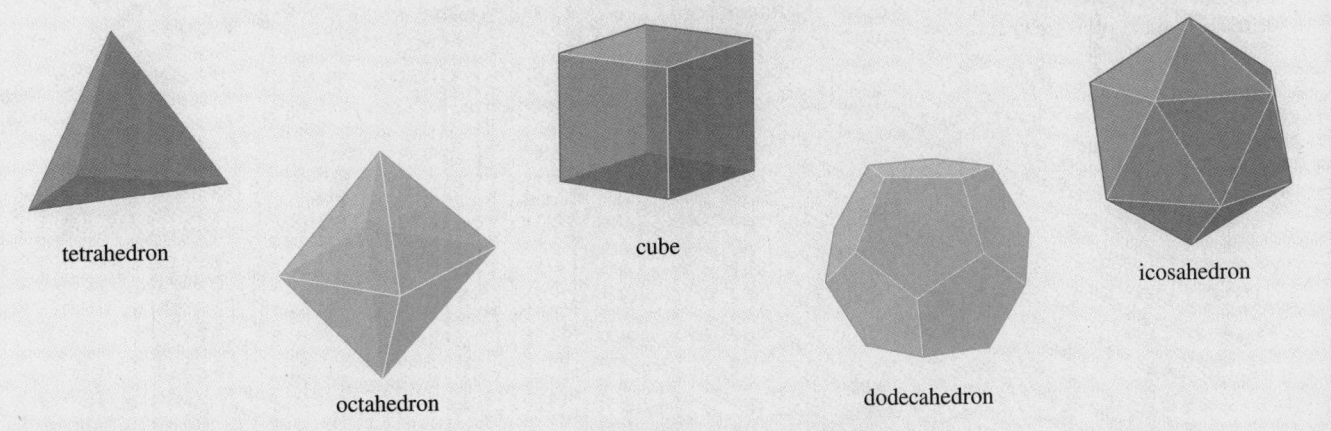

tetrahedron

cube

icosahedron

octahedron

dodecahedron

The regular polyhedra

To show this, we consider all possible cases.

■ **CASE 1** Suppose the faces of a regular polyhedron are equilateral triangles. How many such triangles can meet at a corner? To make a corner

Three, four, and five equilateral triangles can be folded up to make a corner, but six such triangles lie flat.

It is interesting to note that there are infinitely many 2-dimensional regular polygons but only five 3-dimensional regular polyhedra. Although it is impossible to draw 4-dimensional regular polyhedra, mathematicians have shown that there are six. Surprisingly, in all higher dimensions there are exactly three regular polyhedra—the *n*-dimensional cube, tetrahedron, and octahedron.

there must be at least three triangles. We can also have four or five. But six equilateral triangles cannot meet at a point to make a corner. (Why?) If three triangles meet at each vertex we can complete the polyhedron by adding one more triangle to make a *tetrahedron*. If four triangles meet at each vertex we have an *octahedron*. If five triangles meet at each vertex, the resulting regular polyhedron is an *icosahedron*. Thus, we have found all the regular polyhedra with triangular faces.

■ **CASE 2** Suppose the faces of a regular polygon are squares. If three squares meet at each point, then the polyhedron is a *cube*. It is impossible for four or more squares to meet at a point to make a corner. (Why?) Thus, the only regular polyhedron with square faces is the cube.

■ **CASE 3** Suppose the faces of a regular polygon are pentagons. If three pentagons meet at each vertex, the resulting polyhedron is a *dodecahedron*. Since the angles of a regular pentagon are 108°, it is impossible for more than three regular pentagons to meet at a vertex. Thus, the only regular polyhedron with pentagonal faces is the dodecahedron.

■ **CASE 4** Is it possible for the faces of a regular polygon to be regular hexagons? Since the angle of a regular hexagon is 120°, when three such hexagons meet at a point they do not form a corner. Thus, it is impossible for a regular polyhedron to have hexagonal faces. The same reasoning shows that no other regular polygon can be the face of a regular polyhedron.

Since these four cases account for all the possibilities, we have shown that there are exactly five regular polyhedra.

PROBLEM 2 ■ Euler's Formula

How many faces, edges, and vertices does a regular polyhedron have? In the 1700s Euler observed that

$$F - E + V = 2 \qquad \text{Euler's formula}$$

where F is the number of faces, E is the number of edges, and V is the number of vertices. We can use Euler's formula to answer the question. For example, the icosahedron is assembled from F equilateral triangles:

In these triangles the total number of sides is $3F$ and the total number of angles is also $3F$. In an icosahedron, five angles of these triangles meet to form a vertex, so the total number of vertices must be

$$V = \frac{3F}{5}$$

Since two sides of a triangle meet to form one edge of the polyhedron, the number of edges must be

$$E = \frac{3F}{2}$$

Substituting into Euler's formula gives

$$F - \frac{3F}{2} + \frac{3F}{5} = 2$$

Solving gives $F = 20$. Substituting this value of F into the formulas for edges and faces gives $E = 30$ and $V = 12$. Thus, for the icosahedron, $F = 20$, $E = 30$, and $V = 12$. Using similar reasoning we can find the number of faces, edges, and vertices for the other regular polyhedra.

PROBLEMS

1. Find the number of faces, edges, and vertices for each of the regular polyhedra, not by counting them, but by using Euler's formula.

2. Describe the polyhedron whose edges are the line segments joining the centers of the faces of an octahedron, as shown in the figure at the left. Do the same for the other Platonic solids.

3. As the following figure indicates, it is possible to *tile* the plane (that is, completely cover it) with equilateral triangles and with squares. Find all other regular polygons that tile the plane. Prove your answer.

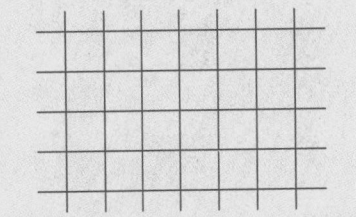

4. Consider a 6×6 grid as shown in the figure.

(a) Find the number of squares of all sizes in this grid. Generalize your result to an $n \times n$ grid.

(b) Find the number of rectangles of all sizes in this grid. Generalize your result to an $n \times n$ grid.

5. A bug is sitting at point A in one corner of the room and wants to crawl to point B in the corner diagonally opposite. Find the shortest path for the bug.

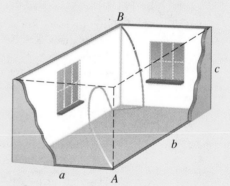

6. For what values of the real number k does the circle $(x - k)^2 + y^2 = 4$ intersect the ellipse

$$x^2 + \frac{y^2}{9} = 1$$

in exactly 0, 1, 2, 3, 4, or 5 points?

7. In a right triangle, the hypotenuse has length 5 cm and another side has length 3 cm. What is the length of the altitude that is perpendicular to the hypotenuse?

8. The perimeter of a right triangle is 60 cm and the altitude perpendicular to the hypotenuse is 12 cm. Find the lengths of the three sides.

9. A point P is located in the interior of a rectangle so that the distance from P to one corner is 5 cm, from P to the opposite corner is 14 cm, and from P to a third corner is 10 cm. What is the distance from P to the fourth corner?

Pierre de Fermat (1601–1665) was a French lawyer who became interested in mathematics at the age of 30. Because of his job as a magistrate Fermat had little time to write complete proofs of his discoveries and often wrote them in the margin of whatever book he was reading at the time. After his death his copy of Diophantus' *Arithmetica* (see page 5) was found to contain a particularly tantalizing comment. Where Diophantus discusses the solutions of $x^2 + y^2 = z^2$ (for example, $x = 3$, $y = 4$, $z = 5$; see Chapter 1 Review Exercises, Exercise 94), Fermat states in the margin that for $n \geq 3$ there are no natural number solutions to the equation $x^n + y^n = z^n$. In other words, it is impossible for a cube to equal the sum of two cubes, a fourth power to equal the sum of two fourth powers, and so on. Fermat writes "I have discovered a truly wonderful proof for this but the margin is too small to contain it." All the other margin comments in Fermat's copy of *Arith-*

(continued)

metica have proved to be true. This one, however, remained unproved, and came to be known as "Fermat's Last Theorem." Finally, in 1994, Andrew Wiles of Princeton University announced a proof of this theorem, an astounding 350 years after it was conjectured. His proof is one of the most widely reported mathematical results in the popular press.

10. (a) Write 13 as the sum of two squares. Then do the same for 41.
(b) Verify that $(a^2 + b^2)(c^2 + d^2) = (ac + bd)^2 + (ad - bc)^2$.
(c) Express 533 as the sum of two squares in two different ways.
 [*Hint:* Factor 533 and use parts (a) and (b).]

11. This problem was first stated by Fermat:

> Prove that every prime number is the leg
> of exactly one right triangle with integer sides.

p

12. Show that the equation $x^2 + y^2 = 4z + 3$ has no solution in integers. [*Hint:* Recall that an even number is of the form $2n$ and an odd number is of the form $2n + 1$. Consider all possible cases for x and y even or odd.]

13. (a) Find all prime numbers p such that $2p + 1$ is a perfect square.
 [*Hint:* Write the equation $2p + 1 = n^2$ as $2p = n^2 - 1$ and factor. Then consider cases.]
(b) Find all prime numbers p such that $2p + 1$ is a perfect cube.

14. Find every positive integer that gives a perfect square if 132 is added to it and another perfect square if 200 is added to it.

15. A group of n pulleys, all of radius 1, is fixed so that their centers form a convex n-gon of perimeter P. (The figure at the left shows the case $n = 4$.) Find the length of the belt that fits around the pulleys. [*Hint:* Try fitting together the sectors of the pulleys that touch the belt.]

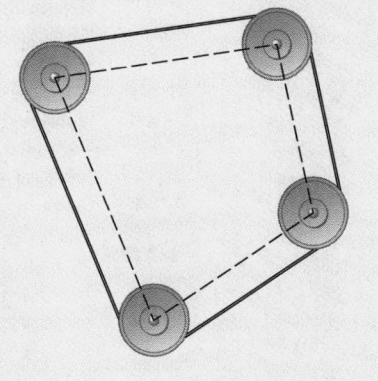

16. On the parabola $x^2 = 4py$, let P and Q be two points with the property that $\angle POQ$ is a right angle. Show that the y-intercept of the segment PQ is the same for any such pair of points P and Q.

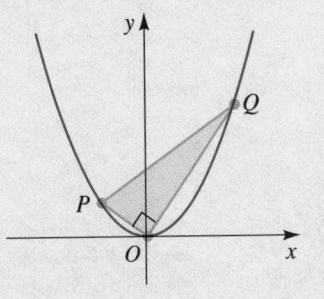

8

COUNTING AND PROBABILITY

Probability is the mathematical study of chance and random processes. The laws of probability are essential for understanding genetics, opinion polls, games of chance, and many other fields.

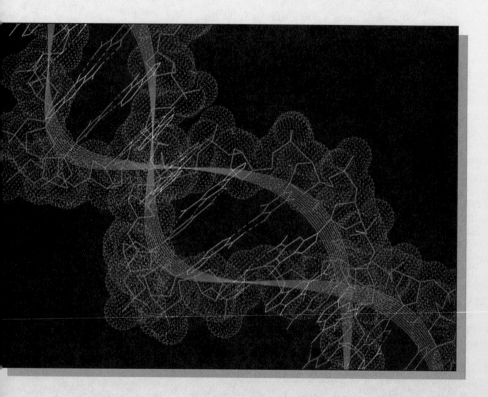

When it is not in our power to determine what is true, we ought to follow what is most probable.

RENÉ DESCARTES

Many questions in mathematics involve counting. For example, in how many ways can a committee of two men and three women be chosen from a group of 35 men and 40 women? How many different license plates have three letters followed by three numbers? How many different poker hands are possible?

Closely related to the problem of counting is that of probability. We consider questions such as these: If a committee of five people is chosen randomly from a group of 35 men and 40 women, what are the chances that no women will be chosen for the committee? What is the likelihood of getting a straight flush in a poker game? In studying probability we give precise mathematical meaning to phrases such as "what are the chances..." and "what is the likelihood...".

COUNTING PRINCIPLES

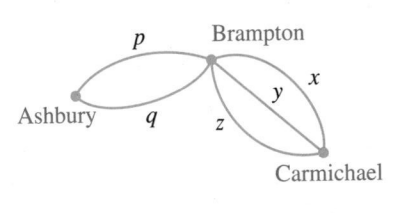

Suppose that the three towns Ashbury, Brampton, and Carmichael are located in such a way that two roads connect Ashbury to Brampton and three roads connect Brampton to Carmichael. How many different routes can one take to travel from Ashbury to Carmichael via Brampton? The key idea in answering this question is to consider the problem in stages. At the first stage—from Ashbury to Brampton—there are two choices to make. For each one of these choices, there are three choices to make at the second stage—from Brampton to Carmichael. Thus, the number of different routes is $2 \times 3 = 6$. These routes are conveniently enumerated by a *tree diagram* as in Figure 1.

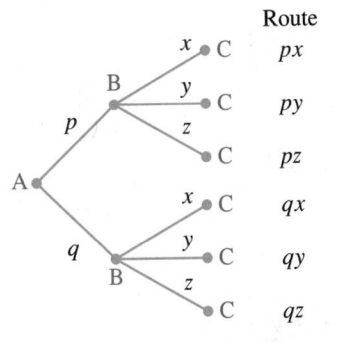

FIGURE 1
Tree diagram

The method used to solve this problem leads to the following principle.

FUNDAMENTAL COUNTING PRINCIPLE

Suppose that two events occur in order. If the first can occur in m ways and the second in n ways (after the first has occurred), then the two events can occur in order in $m \times n$ ways.

Ronald Graham was born in Taft, California, in 1935. He has been described as the world's leading mathematician in the field of combinatorics, the branch of mathematics that deals with counting. For many years Graham was head of the Mathematical Studies Center at Bell Laboratories in Murray Hill, New Jersey. There he faced and solved numerous problems that arose in the telephone industry. During the *Apollo* program, NASA needed to evaluate mission schedules so that the three astronauts aboard a spacecraft could find the time to perform all the necessary tasks. The number of ways to allot these tasks was astronomical— too vast for even a computer to sort out. Graham, using his knowledge of combinatorics, was able to reassure NASA that there were easy ways of solving their problem that were not too far from the theoretically best possible solutions. Besides being a prolific mathematician, Graham is an accomplished juggler and has been president of the International Jugglers Association.

There is an immediate consequence of this principle for any number of events: If $E_1, E_2, \ldots, E_k$ are events that occur in order and if E_1 can occur in n_1 ways, E_2 in n_2 ways, and so on, then the events can occur in order in $n_1 \times n_2 \times \cdots \times n_k$ ways.

EXAMPLE 1 ■ Using the Fundamental Counting Principle

An ice-cream store offers three types of cones and 31 flavors. How many different single-scoop ice-cream cones is it possible to buy at this store?

SOLUTION

There are two choices to make: type of cone and flavor of ice cream. At the first stage we choose a type of cone, and at the second stage we choose a flavor. We can think of the different stages as boxes:

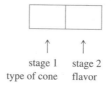

stage 1   stage 2
type of cone   flavor

The first box can be filled in three ways and the second in 31 ways:

stage 1   stage 2

Thus, by the Fundamental Counting Principle, there are $3 \times 31 = 93$ ways of choosing a single-scoop ice-cream cone at this store.   ■

EXAMPLE 2 ■ Using the Fundamental Counting Principle

In a certain state, automobile license plates display three letters followed by three digits. How many such plates are possible if repetition of the letters

(a) is allowed?   (b) is not allowed?

SOLUTION

(a) There are six choices to be made, one choice for each letter or digit on the license plate. As in the preceding example, we sketch a box for each stage:

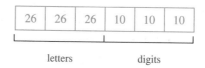

letters   digits

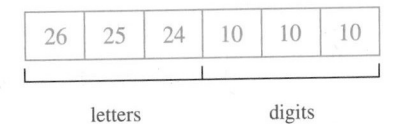

At the first stage we choose a letter (from 26 possible choices), at the second stage another letter (again from 26 choices), at the third stage another letter (26 choices), at the fourth stage a digit (from 10 possible choices), at the fifth stage a digit (again from 10 choices), and at the sixth stage another digit (10 choices). By the Fundamental Counting Principle, the number of possible license plates is

$$26 \times 26 \times 26 \times 10 \times 10 \times 10 = 17{,}576{,}000$$

(b) If no repetition of letters is allowed, then we can arrange the choices as follows:

26	25	24	10	10	10

      letters   digits

At the first stage we have 26 letters to choose from, but once the first letter is chosen there are only 25 letters to choose from at the second stage. Once the first two letters are chosen, 24 letters are left to choose from for the third stage. The digits are chosen as before. Thus, the number of possible license plates in this case is

$$26 \times 25 \times 24 \times 10 \times 10 \times 10 = 15{,}600{,}000$$ ∎

EXAMPLE 3 ■ Using the Fundamental Counting Principle

There are nine players on a baseball team. In how many different ways can the manager choose the batting order?

SOLUTION

The manager can choose the first batter in nine ways, the second in eight ways (since there are only eight players left to choose from), the third in seven ways, and so on. So, by the Fundamental Counting Principle, the number of different batting orders is

$$9 \times 8 \times 7 \times 6 \times 5 \times 4 \times 3 \times 2 \times 1 = 362{,}880$$ ∎

Long multiplications of the type in Example 3 occur often in counting problems. So we introduce notation that helps us write these more succinctly.

FACTORIAL NOTATION

The product of the first n natural numbers is denoted by $n!$ and is called n **factorial**:

$$n! = 1 \cdot 2 \cdot 3 \cdot \cdots \cdot (n - 1) \cdot n$$

We also define 0! as follows:

$$0! = 1$$

This definition of 0! makes many formulas involving factorials shorter and easier to write. An example of this is the binomial formula in Section 9.8.

EXAMPLE 4 ■ Factorial Notation

(a) $4! = 1 \cdot 2 \cdot 3 \cdot 4 = 24$

(b) $7! = 1 \cdot 2 \cdot 3 \cdot 4 \cdot 5 \cdot 6 \cdot 7 = 5040$

(c) $10! = 1 \cdot 2 \cdot 3 \cdot 4 \cdot 5 \cdot 6 \cdot 7 \cdot 8 \cdot 9 \cdot 10 = 3{,}628{,}800$

(d) $\dfrac{7!}{5!} = \dfrac{1 \cdot 2 \cdot 3 \cdot 4 \cdot 5 \cdot 6 \cdot 7}{1 \cdot 2 \cdot 3 \cdot 4 \cdot 5} = 6 \cdot 7 = 42$

(e) $\dfrac{6!\,100!}{2!\,102!} = \dfrac{(1 \cdot 2 \cdot 3 \cdot 4 \cdot 5 \cdot 6)(1 \cdot 2 \cdot 3 \cdot \cdots \cdot 100)}{(1 \cdot 2)(1 \cdot 2 \cdot 3 \cdot \cdots \cdot 100 \cdot 101 \cdot 102)} = \dfrac{3 \cdot 4 \cdot 5 \cdot 6}{101 \cdot 102} = \dfrac{60}{1717}$

(f) $\dfrac{n!}{(n-2)!} = \dfrac{1 \cdot 2 \cdot 3 \cdot \cdots \cdot (n-2)(n-1)n}{1 \cdot 2 \cdot 3 \cdot \cdots \cdot (n-2)} = (n-1)n = n^2 - n$ ■

EXAMPLE 5 ■ Using Factorial Notation

In how many different ways can a race with six runners be completed? Assume there is no tie.

SOLUTION

There are six possible choices for first place, five choices for second place (since only five runners are left after first place has been decided), four choices for third place, and so on. So, by the Fundamental Counting Principle, the number of different ways this race can be completed is

$$6 \times 5 \times 4 \times 3 \times 2 \times 1 = 6! = 720$$ ■

8.1 **EXERCISES**

1–9 ■ Evaluate the expression.

1. $8!$

2. $2!\,4!$

3. $\dfrac{12!}{10!}$

4. $\dfrac{5!}{2!\,4!}$

5. $\dfrac{100!}{98!}$

6. $\dfrac{11!\,12!}{10!\,11!}$

7. $\dfrac{8! + 9!}{8!}$

8. $\dfrac{1001!}{1000!}$

9. $\dfrac{100! - 99!}{98!}$

10–13 ■ Simplify the expression. In each case, n is a positive integer.

10. $\dfrac{n!}{(n-1)!}$

11. $\dfrac{(n-1)!}{n!}$

12. $\dfrac{(n+1)!}{n!}$

13. $\dfrac{(n+1)!}{(n-1)!\,n}$

14. How many three-letter "words" (strings of letters) can be formed using the 26 letters of the alphabet if repetition of letters
(a) is allowed?   (b) is not allowed?

15. How many three-letter "words" (strings of letters) can be formed using the letters *WXYZ* if repetition of letters
(a) is allowed?   (b) is not allowed?

16. Eight horses are entered in a race.
(a) How many different orders are possible for completing the race?
(b) In how many different ways can first, second, and third places be decided? (Assume there is no tie.)

17. A multiple-choice test has five questions with four choices for each question. In how many different ways can the test be completed?

18. How many different seven-digit phone numbers are possible if the first digit cannot be 0 or 1?

19. In how many different ways can a race with five runners be completed? (Assume there is no tie.)

20. In how many ways can five people be seated in a row of five seats?

21. A restaurant offers six different main courses, eight types of drinks, and three kinds of desserts. How many different meals consisting of a main course, a drink, and a dessert does the restaurant offer?

22. In how many ways can five different mathematics books be placed next to each other on a shelf?

23. Towns A, B, C, and D are located in such a way that there are four roads from A to B, five roads from B to C, and six roads from C to D. How many routes are there from town A to town D via towns B and C?

24. In a family of four children, how many different boy-girl birth-order combinations are possible? (The birth orders *BBBG* and *BBGB* are different.)

25. A coin is flipped five times, and the resulting sequence of heads and tails is recorded. How many such sequences are possible?

26. A red die and a white die are rolled, and the numbers showing are recorded. How many different outcomes are possible? (The singular form of the word *dice* is *die*.)

 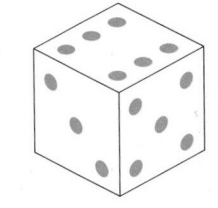

27. A red die, a blue die, and a white die are rolled, and the numbers showing are recorded. How many different outcomes are possible?

28. Two cards are chosen in order from a deck. In how many ways can this be done if
(a) the first card must be a spade and the second must be a heart?
(b) both cards must be spades?

29. A girl has five skirts, eight blouses, and 12 pairs of shoes. How many different skirt-blouse-shoe outfits can she wear? (Assume that each item matches all the others, so she is willing to wear any combination.)

30. A company's employee ID number system consists of one letter followed by three digits. How many different ID numbers are possible with this system?

31. A company has 2844 employees. Each employee is to be given an ID number that consists of one letter followed by two digits. Is it possible to give each employee a different ID number using this scheme? Explain.

32. An all-star baseball team has a roster of seven pitchers and three catchers. How many pitcher-catcher pairs can the manager select from this roster?

33. Standard automobile license plates in California display a nonzero digit, followed by three letters, followed by three digits. How many different standard plates are possible in this system?

34. A combination lock has 60 different positions. In order to open the lock, the dial is turned to a certain number in the clockwise direction, then to a number in the counterclockwise direction, and finally to a third number in the clockwise direction. If successive numbers in the combination cannot be the same, how many different combinations are possible?

35. A true-false test contains ten questions. In how many different ways can this test be completed?

36. An automobile dealer offers five models. Each model comes in a choice of four colors, three types of stereo equipment, with or without air conditioning, and with or without a sunroof. In how many different ways can a customer order an auto from this dealer?

37. The registrar at a certain university classifies students according to major, minor, year (1, 2, 3, 4), and sex (M, F). Each student must choose one major and either one or no minor from the 32 fields taught at this university. How many different student classifications are possible?

38. Explain why in any group of 677 people at least two must have the same first and last initials.

39. A state has registered 8 million automobiles. In order to simplify the license plate system, a state employee suggests that each plate display only two letters followed by three digits. Will this system create enough different license plates for all the vehicles registered?

40. A state license plate design has six places. Each plate begins with a fixed number of letters, and the remaining places are filled with digits. (For example, one letter followed by five digits or two letters followed by four digits.) The state has registered 17 milllion vehicles.
  (a) The state decides to change to a system consisting of one letter followed by five digits. Will this design allow for enough different plates to accommodate all the vehicles registered?

  (b) Find a system that will be sufficient if the smallest possible number of letters is to be used.

41. In how many ways can a president, vice president, and secretary be chosen from a class of 30 students?

42. In how many ways can a president, vice president, and secretary be chosen from a class of 20 females and 30 males if the president must be a female and the vice president a male?

43. A senate subcommittee consists of ten Democrats and seven Republicans. In how many ways can a chairman, vice chairman, and secretary be chosen if the chairman must be a Democrat and the vice chairman must be a Republican?

44. Before 1990, telephone area codes were chosen according to the following rule. The first digit must not be 0 or 1, the middle digit must be 0 or 1, and the last digit must not be 0. How many different area codes were possible in this system?

45. Five-letter "words" are formed using the letters A, B, C, D, E, F, G. How many such words are possible for each of the following conditions?
  (a) Repetition of letters is allowed.
  (b) No letter can be repeated in a word.
  (c) Each word must begin with the letter A.
  (d) The letter C must be in the middle.
  (e) The middle letter must be a vowel.

46. How many five-letter palindromes are possible? (A *palindrome* is a string of letters that reads the same backward and forward, such as the string *XCZCX*.)

47. A certain computer programming language allows names of variables to consist of two characters, the first being any letter and the second any letter or digit. How many names of variables are possible?

48. How many different three-character code words consisting of letters or digits are possible for each of the following code designs?
  (a) The first entry must be a letter.
  (b) The first entry cannot be zero.

49. In how many ways can four men and four women be seated in a row of eight seats for each of the following situations?
  (a) The women are to be seated together and the men are to be seated together.
  (b) They are to be seated alternately by gender.

50. In how many ways can five different mathematics books be placed on a shelf if the two algebra books are to be placed next to each other?

51. Eight mathematics books and three chemistry books are to be placed on a shelf. In how many ways can this be done if the mathematics books are to be next to each other and the chemistry books are next to each other?

52. Three-digit numbers are formed using the digits 2, 4, 5, and 7, with repetition of digits allowed. How many such numbers can be formed if
(a) The numbers are less than 700?
(b) The numbers are even?
(c) The numbers are divisible by 5?

53. How many three-digit odd numbers can be formed using the digits 1, 2, 4, and 6 if no digit may be used more than once?

8.2 **PERMUTATIONS**

Permutations of three colored squares

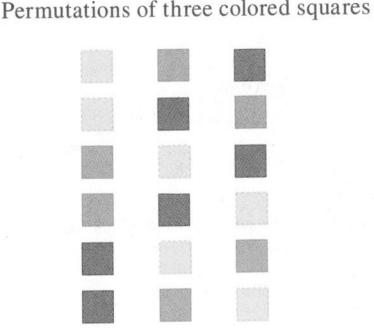

In this section we single out an important special case of the multiplication principle.

A **permutation** of a set of distinct objects is an ordering of these objects. For example, some of the permutations of the letters *ABCDWXYZ* are

$$XAYBZWCD \qquad ZAYBCDWX \qquad DBWAZXYC \qquad YDXAWCZB$$

How many such permutations are possible? Since there are eight choices for the first position, seven for the second (after the first has been chosen), six for the third (after the first two have been chosen), and so on, the Fundamental Counting Principle tells us that the number of possible permutations is

$$8 \times 7 \times 6 \times 5 \times 4 \times 3 \times 2 \times 1 = 40{,}320$$

This same reasoning with 8 replaced by *n* leads to the following observation.

> The number of permutations of *n* objects is *n*!.

How many permutations consisting of five letters can be made from these same eight letters? Some of these permutations are

$$XYZWC \qquad AZDWX \qquad AZXYB \qquad WDXZB$$

Again, there are eight ways to choose the letter in the first position, seven ways for the second, six for the third, five for the fourth, and four for the fifth. By the Fundamental Counting Principle, the number of such permutations is

$$8 \times 7 \times 6 \times 5 \times 4 = 6720$$

In general, if a set has n elements, then the number of ways of ordering r elements from the set is denoted by $P(n, r)$ and is called **the number of permutations of n objects taken r at a time.**

We have just shown that $P(8, 5) = 6720$. The same reasoning used to find $P(8, 5)$ will help us to find a general formula for $P(n, r)$. Indeed, there are n objects and r positions to place them in. Thus, there are n choices for the first position, $n - 1$ choices for the second, $n - 2$ choices for the third, and so on. The last position can be filled in $n - r + 1$ ways. By the Fundamental Counting Principle,

$$P(n, r) = n(n - 1)(n - 2) \cdots (n - r + 1)$$

This formula can be written more compactly using factorial notation:

$$P(n, r) = n(n - 1)(n - 2) \cdots (n - r + 1)$$

$$= \frac{n(n - 1)(n - 2) \cdots (n - r + 1)(n - r) \cdots 3 \cdot 2 \cdot 1}{(n - r) \cdots 3 \cdot 2 \cdot 1} = \frac{n!}{(n - r)!}$$

PERMUTATIONS OF n OBJECTS TAKEN r AT A TIME

The number of permutations of n objects taken r at a time is

$$P(n, r) = \frac{n!}{(n - r)!}$$

EXAMPLE 1 ■ Finding the Number of Permutations

A club has nine members. In how many ways can a president, vice president, and secretary be chosen from the members of this club?

SOLUTION

We need the number of ways of selecting three members *in order* for the positions of president, vice president, and secretary. So, we want the number of permutations of nine objects (the club members) taken three at a time. This number is

$$P(9, 3) = \frac{9!}{(9 - 3)!} = \frac{9!}{6!} = 9 \times 8 \times 7 = 504$$

■

EXAMPLE 2 ■ Finding the Number of Permutations

From 20 raffle tickets in a hat, four tickets are to be selected in order. The holder of the first ticket wins a car, the second a motorcycle, the third a bicycle, and the fourth a skateboard. In how many different ways can these prizes be awarded?

SOLUTION

The order in which the tickets are chosen determines who wins each prize. So, we need to find the number of ways of ordering 20 objects (the tickets) taken four at a time. This number is

$$P(20, 4) = \frac{20!}{(20 - 4)!} = \frac{20!}{16!} = 20 \times 19 \times 18 \times 17 = 116{,}280 \qquad \blacksquare$$

DISTINGUISHABLE PERMUTATIONS

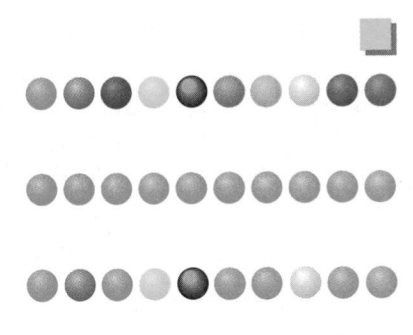

If we have a collection of ten balls, each a different color, then the number of permutations of these balls is $P(10, 10) = 10!$. If all ten balls are red, then we have just one distinguishable permutation because all the ways of ordering these balls look the same. In general, when considering a set of objects, some of which are of the same kind, then two permutations are **distinguishable** if one cannot be obtained from the other by interchanging the positions of elements of the same kind. For example, if we have ten balls, of which six are red and the other four are each a different color, then how many distinguishable permutations are possible? The key point here is that balls of the same color are not distinguishable. So each rearrangement of the red balls, keeping all the other balls fixed, gives essentially the same permutation. Since there are 6! rearrangements of the red balls for each fixed position of the other balls, the total number of distinguishable permutations is 10!/6!. The same type of reasoning gives the following general rule.

> ## DISTINGUISHABLE PERMUTATIONS
>
> If a set of n objects consists of k different kinds of objects with n_1 objects of the first kind, n_2 objects of the second kind, n_3 objects of the third kind, and so on, where $n_1 + n_2 + \cdots + n_k = n$, then the number of distinguishable permutations of these objects is
>
> $$\frac{n!}{n_1! \, n_2! \, n_3! \cdots n_k!}$$

EXAMPLE 3 ■ Finding the Number of Distinguishable Permutations

Find the number of different ways of placing 15 balls in a row given that 4 are red, 3 are yellow, 6 are black, and 2 are blue.

SOLUTION

We want to find the number of distinguishable permutations of these balls. By the formula, this number is

$$\frac{15!}{4! \, 3! \, 6! \, 2!} = 6{,}306{,}300 \qquad \blacksquare$$

Suppose we have 15 wooden balls in a row and four colors of paint: red, yellow, black, and blue. In how many different ways can the 15 balls be painted in such a way that we have 4 red, 3 yellow, 6 black, and 2 blue balls? A little thought will show that this number is exactly the same as that calculated in Example 3. This way of looking at the problem is somewhat different, however. Here we think of the number of **partitions** of the balls into four groups containing 4, 3, 6, and 2 balls to be painted red, yellow, black, and blue, respectively. The next example shows how this reasoning is used.

EXAMPLE 4 ■ Finding the Number of Partitions

Fourteen construction workers are to be assigned to three different tasks. Seven workers are needed for mixing cement, five for laying bricks, and two for carrying the bricks to the brick layers. In how many different ways can the workers be assigned to these tasks?

SOLUTION

We need to partition the workers into three groups containing 7, 5, and 2 workers, respectively. This number is

$$\frac{14!}{7!\,5!\,2!} = 72{,}072$$

■

8.2 EXERCISES

1–9 ■ Evaluate the expression.

1. $P(8, 3)$

2. $P(9, 2)$

3. $P(11, 4)$

4. $P(10, 5)$

5. $P(100, 1)$

6. $P(99, 3)$

7. $P(15, 5)$

8. $P(n, 1)$

9. $P(n, n)$

10. In how many different ways can a president, vice president, and a secretary be chosen from a class of 15 students?

11. In how many different ways can first, second, and third prizes be awarded in a game with eight contestants?

12. In how many different ways can three of eight people be seated in a row of three chairs?

13. In how many different ways can six people be seated in a row of six chairs?

14. In how many different ways can six of ten people be seated in a row of six chairs?

15. How many three-letter "words" can be made from the letters *FGHIJK*?

16. How many permutations are possible from the letters of the word *LOVE*?

17. How many different three-digit whole numbers can be formed using the digits 1, 3, 5, and 7 if no repetition of digits is allowed?

18. A pianist plans to play eight pieces at a recital. In how many ways can she arrange these pieces in the program?

19. In how many different ways can a race with nine runners be completed, assuming there is no tie?

20. A ship carries five signal flags of different colors. How many different signals can be sent by hoisting these flags on the ship's flagpole in different orders?

21. In how many ways can first, second, and third prizes be awarded in a contest with 1000 contestants?

22. In how many ways can a president, vice president, secretary, and treasurer be chosen from a class of 30 students?

23. In how many ways can five students be seated in a row of five chairs if Jack insists on sitting in the first chair?

Jack

24. In how many ways can the students in Exercise 23 be seated if Jack insists on sitting in the middle chair?

25. How many arrangements of the word *LOVE* begin with the letter *L*? How many end with the letter *L*?

26. Ten seats on an airplane are vacant. In how many ways can three passengers be assigned to these seats?

27. How many different whole numbers can be formed using the digits 2, 3, 5, and 8 if no repetition of digits is allowed within each number?

28. How many five letter "words" can be formed from the first ten letters of the alphabet?

29–32 ■ Find the number of distinguishable permutations of the given letters.

29. *AAABBC*

30. *AAABBBCCC*

31. *AABCD*

32. *ABCDDDEE*

33. In how many ways can two blue marbles and four red marbles be arranged in a row?

34. In how many different ways can five red balls, two white balls, and seven blue balls be arranged in a row?

35. In how many different ways can four pennies, three nickels, two dimes, and three quarters be arranged in a row?

36. In how many different ways can the letters of the word *VACATION* be arranged?

37. In how many different ways can the letters of the word *ELEEMOSYNARY* be arranged?

38. In how many ways can the algebraic expression a^3b^2 be written without using exponents? (For example, *abbaa* is one way.)

39. In how many ways can the algebraic expression $a^4b^5c^2d$ be written without using exponents?

40. A man bought three vanilla ice-cream cones, two chocolate cones, four strawberry cones, and five butterscotch cones for his 14 children. In how many ways can he distribute the cones among his children?

41. When seven students take a trip, they find a hotel with three rooms available—a room for one person, a room for two people, and a room for three people. In how many different ways can the students be assigned to these rooms? (One student has to sleep in the car.)

42. Eight workers are cleaning a large house. Five are needed to clean windows, two to clean the carpets, and one to clean the rest of the house. In how many different ways can these tasks be assigned to the eight workers?

43. A jogger jogs every morning to his health club, which is eight blocks east and five blocks north of his home. He always takes a shortest route, but he likes to vary the path he follows (see the figure). How many different paths can he take? [*Hint:* The path shown can be thought of as *ENNEEENENEENE*, where *E* is East and *N* is North.]

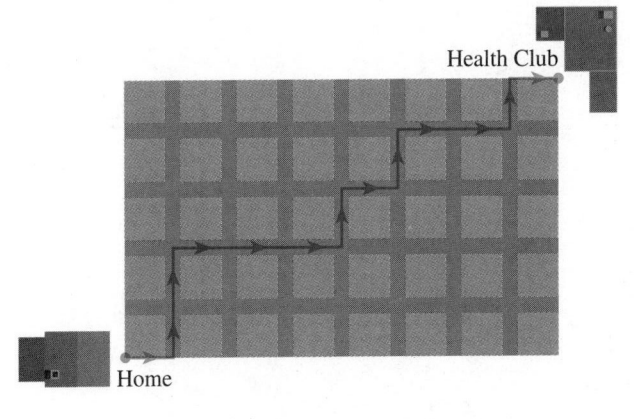

Health Club

Home

44. A computer company is awarding a million dollars to whomever guesses the order in which they have rearranged the letters of the name of their computer, *ULTRALOGICAL*. Each entry must be mailed in a separate envelope with a first-class 32-cent stamp. How much would it cost in postage to submit all possible entries?

8.3 COMBINATIONS

In the preceding section we were interested in the number of ways of ordering elements of a set. In many counting problems, however, order is *not* important. For example, a poker hand is the same hand, regardless of how it is ordered. A poker player interested in the number of possible hands wants to know the number of ways of drawing five cards from 52 cards, without regard to the order in which the cards of a given hand are dealt. In this section we develop a formula for counting in situations such as this, where order does not matter.

A **combination** of *r* elements of a set is any subset of *r* elements from the set (without regard to order). If the set has *n* elements, then the number of combinations of *r* elements is denoted by $C(n, r)$ and is called the **number of combinations of *n* elements taken *r* at a time**.

For example, consider a set with the four elements *A*, *B*, *C*, and *D*. The combinations of these four elements taken three at a time are

ABC	*ABD*	*ACD*	*BCD*

The permutations of these elements taken three at a time are

ABC	*ABD*	*ACD*	*BCD*
ACB	*ADB*	*ADC*	*BDC*
BAC	*BAD*	*CAD*	*CBD*
BCA	*BDA*	*CDA*	*CDB*
CAB	*DAB*	*DAC*	*DBC*
CBA	*DBA*	*DCA*	*DCB*

We notice that the number of combinations is a lot fewer than the number of permutations. In fact, each combination of three elements generates 3! permutations. So

$$C(4, 3) = \frac{P(4, 3)}{3!} = \frac{4!}{3!\,(4 - 3)!} = 4$$

In general, each combination of *r* objects gives rise to *r*! permutations of these objects. Thus

$$C(n, r) = \frac{P(n, r)}{r!} = \frac{n!}{r!\,(n - r)!}$$

> ## COMBINATIONS OF n OBJECTS TAKEN r AT A TIME
>
> The number of combinations of n objects taken r at a time is
>
> $$C(n, r) = \frac{n!}{r!\,(n - r)!}$$

The key difference between permutations and combinations is *order*. If we are interested in ordered arrangements, then we are counting permutations; but if we are concerned with subsets without regard to order, then we are counting combinations. Compare Examples 1 and 2 below (where order does not matter) with the corresponding examples in Section 8.2 (where order does matter).

EXAMPLE 1 ■ Finding the Number of Combinations

A club has nine members. In how many ways can a committee of three be chosen from the members of this club?

SOLUTION

We need the number of ways of choosing three of the nine members. Order is not important here, because the committee is the same no matter how its members are ordered. So, we want the number of combinations of nine objects (the club members) taken three at a time. This number is

$$C(9, 3) = \frac{9!}{3!\,(9 - 3)!} = \frac{9!}{3!\,6!} = \frac{9 \times 8 \times 7}{3 \times 2 \times 1} = 84$$

EXAMPLE 2 ■ Finding the Number of Combinations

From 20 raffle tickets in a hat, four tickets are to be chosen at random. The holders of the winning tickets are to be awarded free trips to the Bahamas. In how many ways can the four winners be chosen?

SOLUTION

We need to find the number of ways of choosing four winners from 20 entries. The order in which the tickets are chosen does not matter, because the same prize is awarded to all four winners. So, we want the number of combinations of 20 objects (the tickets) taken four at a time. This number is

$$C(20, 4) = \frac{20!}{4!\,(20 - 4)!} = \frac{20!}{4!\,16!} = \frac{20 \times 19 \times 18 \times 17}{4 \times 3 \times 2 \times 1} = 4845$$

DISTINGUISHABLE COMBINATIONS

Patti has four pennies, two nickels, three dimes, and one quarter. How many different combinations of these can she throw into a fountain? She could, of course, throw in all, none, or any number of them. For instance, she could decide on the combination of two pennies and a nickel. Since it doesn't matter which two pennies and which nickel she chooses, all such combinations are essentially the same. So, what we need to count is the number of distinguishable combinations—combinations that we can tell apart. To do this we count the number of ways each type of coin can be selected. She could choose to throw four, three, two, one, or none of the pennies—five choices in all. She could choose two, one, or none of the nickels—three choices in all. Similarly, she has four choices for selecting the dimes and two choices for selecting the quarter. By the Fundamental Counting Principle, the number of distinguishable combinations of the coins is

$$5 \times 3 \times 4 \times 2 = 120$$

If a collection of objects is made up of groups of like kind, then two combinations are **indistinguishable** if they contain the same number of objects of each kind. Otherwise, the combinations are **distinguishable.** Reasoning as in the preceding paragraph, we get the following general result.

DISTINGUISHABLE COMBINATIONS

If a set of n objects consists of k different kinds of objects, with n_1 objects of the first kind, n_2 objects of the second kind, n_3 objects of the third kind, and so on, where $n_1 + n_2 + \cdots + n_k = n$, then the number of distinguishable combinations of these objects is

$$(n_1 + 1)(n_2 + 1)(n_3 + 1)\cdots(n_k + 1)$$

EXAMPLE 3 ■ Finding the Number of Distinguishable Combinations

Jack has 15 marbles, of which four are red, three are blue, six are black, and two are yellow. In how many ways can Jack give some of his marbles to Jill?

SOLUTION

Jack may give none, some, or all of his marbles to Jill. Since marbles of the same color are not distinguishable, we want to find the number of distinguishable combinations of these marbles. By the formula, this number is

$$(4 + 1)(3 + 1)(6 + 1)(2 + 1) = 5 \times 4 \times 7 \times 3 = 420 \qquad ■$$

If each of the objects in a set is of a different kind, then the number of distinguishable combinations is the same as the number of all subsets of that set. If the set has n elements, then using the formula with each $n_i = 1$ for $1 \le i \le n$, we get that the number of subsets is

$$\underbrace{(1 + 1)(1 + 1)(1 + 1) \cdots (1 + 1)}_{n \text{ times}} = 2^n$$

> A set with n elements has 2^n subsets.

EXAMPLE 4 ■ Finding the Number of Subsets of a Set

A pizza parlor offers the basic cheese pizza and a choice of 16 toppings. How many different kinds of pizza can be ordered at this pizza parlor?

SOLUTION

We need the number of possible subsets of the 16 toppings (including the empty set, which corresponds to a plain cheese pizza). Thus, $2^{16} = 65,536$ different pizzas can be ordered. ■

8.3 EXERCISES

1–9 ■ Evaluate the expression.

1. $C(8, 3)$ **2.** $C(9, 2)$ **3.** $C(11, 4)$

4. $C(10, 5)$ **5.** $C(100, 1)$ **6.** $C(99, 3)$

7. $C(15, 5)$ **8.** $C(n, 1)$ **9.** $C(n, n)$

10. In how many ways can a group of three students be chosen from a class of twelve?

11. In how many ways can three books be chosen from a group of six?

12. In how many ways can three pizza toppings be chosen from 12 available toppings?

13. In how many ways can six people be chosen from a group of ten?

14. In how many ways can a committee of three members be chosen from a club of 25 members?

15. How many five-card hands can be dealt from a deck of 52 cards?

16. How many seven-card hands can be picked from a deck of 52 cards?

17. A student must answer seven of the ten questions on an exam. In how many ways can she choose the seven questions?

18. A pizza parlor offers a choice of 16 different toppings. How many three-topping pizzas are possible?

19. A violinist has practiced 12 pieces. In how many ways can he choose eight of these pieces for a recital?

20. If a woman has eight skirts, in how many ways can she choose five of these to take on a weekend trip?

21. In how many ways can seven students from a class of 30 be chosen for a field trip?

22. In how many ways can the seven students in Exercise 21 be chosen if Jack must go on the field trip?

23. In how many ways can the seven students in Exercise 21 be chosen if Jack may not go on the field trip?

24. In the 6/49 lottery game, a player picks six numbers from 1 to 49. How many different choices does the player have?

25. In the California Lotto game, a player chooses six numbers from 1 to 53. It costs one dollar to play this game. How much would it cost to buy every possible combination of six numbers to ensure picking the winning six numbers?

26. In the Pennsylvania Super-7 lottery, a player chooses seven different numbers from 1 to 74. The Lottery Commission chooses ten numbers randomly. A player is a winner if his seven numbers are among the ten chosen by the commission.
    (a) In how many ways can a player choose seven numbers for his lottery ticket?
    (b) How many winning combinations are possible on each draw?

27. A class has 20 students, of which 12 are females and 8 are males. In how many ways can a committee of five students be picked from this class under each of the following conditions?
    (a) No restriction is placed on the number of males or females on the committee.
    (b) No males are to be included on the committee.
    (c) The committee must have three females and two males.

28. A student is required to answer nine of 12 questions on a test. In how many ways can he do this under each of the following conditions?
    (a) No restriction is made on which questions he must answer.
    (b) He definitely cannot answer the first and last questions.
    (c) He must answer three of the first five questions.

29–32 ■ Find the number of distinguishable combinations of the given letters.

29. *AAABBC*   **30.** *AAABBBCCC*

31. *AABCD*   **32.** *ABCDDDEE*

33. A set has eight elements.
    (a) How many subsets containing five elements does this set have?
    (b) How many subsets does this set have?

34. Maria has four pennies, three nickels, two dimes, and three quarters. In how many different ways can she give some of these coins to her brother?

35. Manuel has four pennies, a nickel, a dime, and a quarter. How many different sums of money can he make with these coins? Why would this problem be more complicated if Manuel had 14 pennies?

36. Jane's bag of marbles has five red, two white, and seven blue marbles. If she reaches in to pick some without looking, how many different selections might she make?

37. The prime factorization of the number 540 is $2 \times 2 \times 3 \times 3 \times 3 \times 5$. How many different divisors does 540 have?

38. Find the number of divisors of 3240 that fit each of the following descriptions.
    (a) The divisors are odd numbers.
    (b) The divisors are even numbers.
    (c) The divisors are multiples of 5.
    (d) The divisors are multiples of 3.

39. A travel agency has limited numbers of eight different free brochures about Australia. The agent tells you to take any that you like, but no more than two of any kind. How many different ways can you choose the brochures?

40. A hamburger chain gives their customers a choice of ten different hamburger toppings. In how many different ways can a customer order a hamburger?

41. Each of 20 shoppers in a shopping mall chooses to enter or not to enter the Dressfastic clothing store. How many different outcomes of their decisions are possible?

42. (a) Calculate $C(100, 0)$ and $C(100, 100)$.
    (b) Show that, in general, $C(n, 0) = C(n, n) = 1$.
    (c) Give a counting argument that explains why the formula in part (b) is true.

43. (a) Calculate $C(100, 3)$ and $C(100, 97)$.
    (b) Show that, in general, $C(n, r) = C(n, n - r)$.
    (c) Give a counting argument that explains why the formula in part (b) is true.

8.4 PROBLEM SOLVING WITH PERMUTATIONS AND COMBINATIONS

In the preceding three sections we studied some important counting techniques. Here are some broad guidelines we need to remember.

> 1. The Fundamental Counting Principle is used when there are consecutive choices to be made.
>
> 2. When we want to find the number of ways of choosing r objects from n objects we need to ask ourselves: Does the order of these objects matter? If the order does matter, then we are counting permutations; if order does not matter, then we are counting combinations.

In many situations a problem reduces to deciding which formula to use. But often no formula applies directly. In other cases the solution may require more than one formula. So, each problem must be analyzed carefully to determine what is involved. The examples in this section show how these various principles work together to solve some interesting counting problems.

EXAMPLE 1 ■ A Problem involving Permutations

In how many ways can 12 students stand in a row for a picture if Jane and John insist on standing next to each other?

Jane John

SOLUTION

Since the order in which the students stand is important, we are interested in permutations. But we can't use the formula for permutations directly. Since Jane and John insist on standing next to each other, we can think of them as one object. Now, we want to arrange 11 objects in a row, and there are 11! ways of doing this. For each one of these arrangements, there are two ways of having Jane and John stand together—namely, Jane-John or John-Jane. Thus, by the Fundamental Counting Principle, the total number of arrangements is $2 \times 11! = 79{,}833{,}600$. ■

It is often difficult to count the number of elements in a set directly, but easy to count the **complement** of the set, that is, the number of elements *not* in the set. The next example illustrates this very useful idea.

EXAMPLE 2 ■ A Problem involving Permutations

In how many ways can 12 students stand in a row for a picture if Jane and John refuse to stand next to each other?

Jane   John

SOLUTION

First we note that there are 12! ways in which these 12 students can stand in a row. We want to count those ways that have Jane and John standing apart. It is difficult to enumerate all the ways in which this can happen. So let us look at the complement—all the ways in which Jane and John stand together. From Example 1 we know that this number is $2 \times 11!$. Since in all the other ways of arranging the students Jane and John stand apart, we conclude that this number is

$$12! - 2 \times 11! = 11!(12 - 2) = 11! \times 10 = 399{,}168{,}000$$ ■

EXAMPLE 3 ■ Permutations of Objects in a Circle

In a game of ring-around-the-rosie, six children hold hands and dance around in a circle. In how many different ways can the children be arranged in a circle?

SOLUTION

Because the children are in a circle, only their relative positions to each other are significant. To deal with this, we choose one child as a reference point and then consider all possible arrangements of the remaining five children to complete the circle. Thus the total number is $5! = 120$.

Another way of looking at this problem is to notice that there are 6! ways for the children to stand in a line. If the line is closed to form a circle, then many different lines would give the same circle. For example, if each child in the line moved one place to the right and the child at the end went to the beginning, then the circle formed from this line would be the same as the original one. This process of moving to the right can be done six times before returning to the original line. Since the six different lines formed in this way produce the same circle, the number of ways for the children to form a circle is

$$\frac{6!}{6} = 5! = 120$$ ■

EXAMPLE 4 ■ A Problem involving Combinations

A group of 25 campers includes 15 women and 10 men. In how many ways can a scouting party of five be chosen from this group if it must consist of three women and two men?

SOLUTION

Three women can be chosen from the 15 women in the group in $C(15, 3)$ ways, and two men can be chosen from the 10 men in the group in $C(10, 2)$ ways. Thus, by the Fundamental Counting Principle, the total number of ways of choosing the scouting party is

$$C(15, 3) \times C(10, 2) = 455 \times 45 = 20{,}475 \qquad ■$$

EXAMPLE 5 ■ A Problem involving Combinations

In how many ways can the scouting party of Example 4 be chosen if the party must include

(a) at least four women?   (b) at most four women?

SOLUTION

(a) The phrase "at least four women" means four or more women. In this case, four women or five women must be included in the scouting party. By the method of Example 4, the number of ways that four women and one man can be chosen is

$$C(15, 4) \times C(10, 1) = 1365 \times 10 = 13{,}650$$

The number of ways that five women and no man can be chosen is

$$C(15, 5) \times C(10, 0) = 3003 \times 1 = 3003$$

Thus, the total number of ways that the scouting party can be chosen is

$$C(15, 4) \times C(10, 1) + C(15, 5) \times C(10, 0) = 13{,}650 + 3003 = 16{,}653$$

(b) The phrase "at most four women" means four or fewer women. So, the scouting party may have four, three, two, or one woman, or none. As in part (a), we can calculate each of these possibilities and add the results. But since this is a long and tedious process, let us consider the complement of this set—namely, the number of ways in which the party can have more than four women. In this case, we must have five women and no men, and the number of ways this can be done is

$$C(15, 5) \times C(10, 0) = 3003 \times 1 = 3003$$

Now, the total number of ways in which the scouting party can be chosen from the 25 campers is

$$C(25, 5) = 53,130$$

Of these 53,130 possibilities, 3003 have five women, so it follows that the rest have four or fewer women. Thus, the number of scouting parties that include at most four women is

$$53,130 - 3003 = 50,127$$ ∎

EXAMPLE 6 ■ A Problem involving Permutations and Combinations

A class of 20 students is going to choose a committee of seven, consisting of a chairman, a vice chairman, a secretary, and four other members. In how many ways can this committee be chosen?

SOLUTION

In choosing the three officers, order is important. So, the number of ways of choosing them is

$$P(20, 3) = 6840$$

Next, we need to choose four other students from the 17 remaining. Since order is immaterial in this case, the number of ways of doing this is

$$C(17, 4) = 2380$$

Thus, by the Fundamental Counting Principle, the number of ways of choosing this committee is

$$P(20, 3) \times C(17, 4) = 6840 \times 2380 = 16,279,200$$

Note that we could have first chosen the four unordered members of the committee—in $C(20, 4)$ ways—and then the three officers from the remaining 16 members, in $P(16, 3)$ ways. Check to see that this gives the same answer. ∎

EXAMPLE 7 ■ A Problem involving a Sum of Permutations

A ship carries four signal flags of different colors. How many different signals can be sent using at least two flags for each signal?

SOLUTION

Since it is the order of the flags that determines the signal, order is important in this problem.

The phrase "at least two flags" means two or more flags—in this case, two, three, or four flags. The number of signals that can be made using two flags is

$P(4, 2)$, using three flags is $P(4, 3)$, and using four flags is $P(4, 4)$. Thus, the total number of possible signals using at least two flags is

$$P(4, 2) + P(4, 3) + P(4, 4) = 12 + 24 + 24 = 60$$ ■

EXAMPLE 8 ■ A Problem involving a Product of Permutations

A railcar compartment has ten seats, five facing the front of the train and five facing the back. Of ten passengers, three prefer to face frontward, one prefers to face backward, and the others have no preference. In how many ways can the passengers be seated?

SOLUTION

First we seat the picky passengers. There are $P(5, 3)$ ways to seat the three passengers wishing to face the front and $P(5, 1)$ ways to seat the passenger wishing to face the back. The remaining six passengers must be arranged in the remaining six seats; there are $P(6, 6)$ ways of doing this. Thus, by the Fundamental Counting Principle, the total number of ways of seating the ten passengers is

$$P(5, 3) \times P(5, 1) \times P(6, 6) = 60 \times 5 \times 720 = 216{,}000$$ ■

8.4 EXERCISES

1. In how many ways can ten people be arranged in a row if two insist on standing next to each other?

2. In how many ways can ten people be arranged in a row if three insist on standing together?

3. In how many ways can eight books be arranged on a shelf if the four mathematics books are to be placed next to each other?

4. In how many ways can ten books be arranged on a shelf if the three mathematics books are to be placed together and the four chemistry books are to be placed together?

5. In how many ways can six runners be arranged on starting blocks if two refuse to be placed next to each other?

6. In how many ways can seven different keys be arranged on a key ring?

7. An artisan has 12 different stones to be strung as a necklace. How many different ways can he design it if the necklace is to have the given design?
   (a) The necklace is a continuous loop with no clasp, as shown in the figure on the left.

(b) The necklace has a clasp, as shown in the figure on the right.

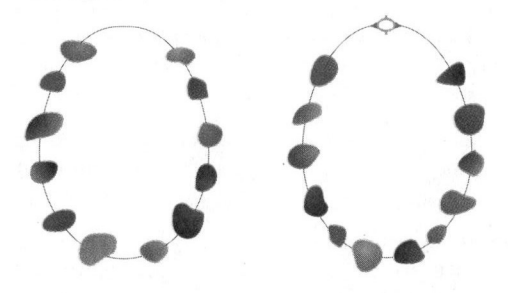

8. In how many ways can a committee consisting of four Republicans and five Democrats be chosen from a group of 20 Republicans and 16 Democrats?

9. A group of 13 workers includes seven welders and six machinists. In how many different ways can five of these workers be assigned to a task that requires the following skills?
   (a) Three welders and two machinists
   (b) At least one welder
   (c) At most two welders

10. From a group of ten male and ten female tennis players, two men and two women are to face each other in a men-versus-women doubles match. In how many different ways can this match be arranged?

11. A school dance committee is to consist of two freshmen, three sophomores, four juniors, and five seniors. If six freshmen, eight sophomores, twelve juniors, and ten seniors are eligible to be on the committee, in how many ways can the committee be chosen?

12. In how many ways can ten students be arranged in a row for a class picture if John and Jane want to stand next to each other, and Mike and Molly also insist on standing next to each other?

13. In how many ways can the ten students in Exercise 12 be arranged if Mike and Molly insist on standing together but John and Jane refuse to stand next to each other?

14. In how many ways can seven different keys be arranged on a key ring if the house key and the car key are to be next to each other?

15. In how many ways can seven children hold hands in a circle if two refuse to hold hands with each other?

16. A group of 22 aspiring thespians contains ten men and twelve women. For the next play the director wants to choose a leading man, a leading lady, a supporting male role, a supporting female role, and eight extras—three women and five men. In how many ways can the cast be chosen?

17. In how many ways can a group of six boys and six girls be divided into two groups of three boys and three girls?

18. A hockey team has 20 players of which twelve play forward, six play defense, and two are goalies. In how many ways can the coach pick a starting lineup consisting of three forwards, two defense players, and one goalie?

19. A four-record album is to be played on a turntable. In how many ways can the eight sides be played so that at least one side is played out of its correct order?

20. A pizza parlor offers four sizes of pizza (small, medium, large, and colossus), two types of crust (thick and thin), and 14 different toppings. How many different pizzas can be made with these choices?

21. An ice-cream parlor offers 12 flavors of ice cream and 8 toppings. Banana splits are made with three scoops of ice cream and three spoonfuls of toppings. How many different ways can a banana split be made if flavors and toppings may be repeated?

22. Suppose that the ice-cream parlor in Exercise 21 offers "super" banana splits with four or fewer spoonfuls of toppings and five or fewer scoops of ice cream (but not less than one scoop). Each scoop must be a different flavor and each spoonful must be a different topping. How many different ways can a super banana split be made with these choices?

23–24 ■ In how many ways can four men and four women be seated in a row of eight seats for each of the following arrangements?

23. (a) The first seat is to be occupied by a man.
   (b) The first and last seats are to be occupied by women.

24. (a) The women are to be seated together.
   (b) The men and women are to be seated alternately by gender.

25. In how many ways can four men and four women be seated around a circular table for each of the following arrangements?
   (a) The women are to be seated together.
   (b) The men and women are to be seated alternately by gender.

26. From a group of 30 contestants, six are to be chosen as semifinalists, then two of those are chosen as finalists, and then the top prize is awarded to one of the finalists. In how many ways can these choices be made in sequence?

27. In how many different ways can eight students be grouped in pairs?

28. The eight students in a chemistry lab are to be grouped in pairs, and each pair is to work on a different experiment. In how many ways can this be done? (Compare to Exercise 27.)

29. Three delegates are to be chosen from a group of four lawyers, a priest, and three professors. In how many ways can the delegation be chosen if it must include at least one professor?

30. In how many ways can a committee of four be chosen from a group of ten if two people refuse to serve together on the same committee?

31. Twelve dots are drawn on a page in such a way that no three are collinear. How many straight lines can be formed by joining the dots?

32. How many triangles can be formed by connecting the dots in Exercise 31?

33. How many diagonals does a regular dodecagon have? (A dodecagon is a 12-sided figure.)

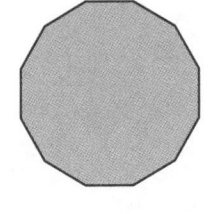

34. A five-person committee consisting of students and teachers is being formed to study the issue of student parking privileges. Of those who have expressed an interest in serving on the committee, 12 are teachers and 14 are students. In how many ways can the committee be formed if at least one student and one teacher must be included?

35. A ship has five different signal flags that may be hoisted in order up the ship's flagpole. How many different signals can the ship's captain make with these flags if the given number of flags is to be used?
(a) At most three flags
(b) At least four flags

36. A row of eight beach chairs are set out on a beach; four have umbrellas and four do not. In how many ways can eight people be seated in these seats if two insist on having umbrellas and three insist on not having umbrellas?

37. In a movie theater, only 13 seats are vacant—four in the first row, three in the second row, and six in the last row. In how many ways can a group of 13 friends be seated if two insist on sitting in the first row, one in the second row, three in the last row, and the others have no preference?

38. In the situation of Exercise 37, a group of eight friends enters the theater. One insists on sitting in the first row, one in the second row, and two in the last row. In how many ways can they be seated in the 13 vacant seats?

39. A bookshelf holds seven mystery novels and eight biographies. Jane chooses a book from the shelf and then John chooses five books—two mysteries and three biographies. Does John have more choices if the book Jane chooses is a mystery or a biography?

40–42 ■ A five-card poker hand is dealt from a standard deck of 52 cards. In how many ways can each of the described hands be dealt?

40. Full house (three of a kind, such as kings, and a pair)

41. Flush (five cards in the same suit, but not in sequence)

42. Straight (five cards in sequence, but not in the same suit)

8.5 PROBABILITY

If you roll a pair of dice, what are the chances of rolling a double six? What is the likelihood of winning a state lottery? The subject of probability was invented to give precise answers to questions like these. It is now an indispensable tool for making decisions in such diverse areas as business, manufacturing, psychology, genetics, and in many of the sciences. Probability is used to determine the effectiveness of new medicines, assess a fair price for an insurance policy, decide on the likelihood of a candidate winning an election, determine the opinion of many people on a certain topic (without interviewing everyone), and answer many other questions that involve a measure of uncertainty.

To discuss probability we begin by defining some terms. An **experiment** is a process, such as tossing a coin or rolling a die, that gives definite results, called the **outcomes** of the experiment. For tossing a coin the possible outcomes are

The mathematical theory of probability began in 1654 in a series of letters between Pascal (see page 630) and Fermat (see page 532). Their correspondence was prompted by a question raised by the experienced gambler the Chevalier de Méré. The Chevalier was interested in the equitable distribution of the stakes of an interrupted gambling game (see Problem 3, page 585).

"heads" and "tails"; for rolling a die the outcomes are 1, 2, 3, 4, 5, and 6. The **sample space** of an experiment is the set of all possible outcomes. If we let H stand for heads and T for tails, then the sample space of the coin-tossing experiment is

$$S = \{H, T\}$$

The sample space for rolling a die is

$$S = \{1, 2, 3, 4, 5, 6\}$$

We will be concerned only with experiments for which all the outcomes are "equally likely." We already have an intuitive feeling for what this means. When tossing a perfectly balanced coin, heads and tails are equally likely outcomes in the sense that if this experiment is repeated many times, we expect that about half the results will be heads and half will be tails.

In any given experiment we are often concerned with a particular set of outcomes. We might be interested in a die showing an even number or in picking an ace from a deck of cards. Any particular set of outcomes is a subset of the sample space. This leads to the following definition.

DEFINITION OF AN EVENT

If S is the sample space of an experiment, then an **event** is any subset of the sample space.

EXAMPLE 1 ■ Events in a Sample Space

If an experiment consists of tossing a coin three times and recording the results in order, the sample space is

$$S = \{HHH, HHT, HTH, THH, TTH, THT, HTT, TTT\}$$

The event E of showing "exactly two heads" is the subset of S that consists of all outcomes with two heads. Thus

$$E = \{HHT, HTH, THH\}$$

The event F of showing "at least two heads" is

$$F = \{HHH, HHT, HTH, THH\}$$

and the event of showing no heads is $G = \{TTT\}$. ■

We are now ready to define the notion of probability. Intuitively, we know that rolling a die may result in any of six equally likely outcomes, so the chance of any particular outcome occurring is $\frac{1}{6}$. What is the chance of showing an even

Persi Diaconis (b. 1945) is currently professor of statistics at Stanford University in California. He was born in New York City into a musical family and studied violin until the age of 14. At that time he left home to become a magician. He was a magician (apprentice and master) for ten years. Magic is still his major passion, and if there were a professorship for magic he would qualify for such a post! His interest in card tricks led him to a study of probability and statistics. He is now one of the leading statisticians in the world. With his background he approaches mathematics with an undeniable flair. He says "Statistics is the physics of numbers. Numbers seem to arise in the world in an orderly fashion. When we examine the world, the same regularities seem to appear again and again." Among his many original contributions to mathematics is a probabilistic study of the perfect card shuffle.

number? Of the six equally likely outcomes possible, three are even numbers. So, it is reasonable to say that the chance of showing an even number is $\frac{3}{6} = \frac{1}{2}$. This reasoning is the intuitive basis for the following definition of probability.

DEFINITION OF PROBABILITY

Let S be the sample space of an experiment and E an event. The probability of E, written $P(E)$, is

$$P(E) = \frac{n(E)}{n(S)} = \frac{\text{number of elements in } E}{\text{number of elements in } S}$$

Notice that $0 \leq n(E) \leq n(S)$, so the probability $P(E)$ of an event is a number between 0 and 1, that is,

$$0 \leq P(E) \leq 1$$

The closer the probability of an event is to 1, the more likely the event is to happen; the closer to 0, the less likely. If $P(E) = 1$, then E is called the **certain event** and if $P(E) = 0$, then E is called the **impossible event.**

EXAMPLE 2 ■ Finding the Probability of an Event

A coin is tossed three times and the results are recorded. What is the probability of getting exactly two heads? at least two heads? no heads?

SOLUTION

By the results of Example 1, the sample space S of this experiment contains eight outcomes and the event E of getting "exactly two heads" contains three outcomes, $\{HHT, HTH, THH\}$, so by the definition of probability,

$$P(E) = \frac{n(E)}{n(S)} = \frac{3}{8}$$

Similarly, the event F of getting "at least two heads" has four outcomes, $\{HHH, HHT, HTH, THH\}$, and so

$$P(F) = \frac{n(F)}{n(S)} = \frac{4}{8} = \frac{1}{2}$$

The event G of getting "no heads" has one element, so

$$P(G) = \frac{n(G)}{n(S)} = \frac{1}{8}$$

■

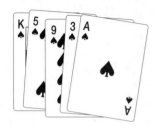

To find the probability of an event, we do not need to list all the elements in the sample space and the event. All we do need is the *number* of elements in these sets. The counting techniques we have learned in the preceding sections will be very useful here.

EXAMPLE 3 ■ Finding the Probability of an Event

A five-card poker hand is drawn from a standard deck of 52 cards. What is the probability that all five cards are spades?

SOLUTION

The experiment here consists of choosing five cards from the deck, and the sample space S consists of all possible choices. Thus, the number of elements in the sample space is

$$n(S) = C(52, 5) = \frac{52!}{5!\,(52 - 5)!} = 2{,}598{,}960$$

The event E we are interested in consists of choosing five spades. Since the deck contains only 13 spades, the number of ways of choosing five spades is

$$n(E) = C(13, 5) = \frac{13!}{5!\,(13 - 5)!} = 1287$$

Thus, the probability of drawing five spades is

$$P(E) = \frac{n(E)}{n(S)} = \frac{1287}{2{,}598{,}960} \approx 0.0005 \qquad \blacksquare$$

What does the answer to Example 3 tell us? Since $0.0005 = \frac{1}{2000}$, this means that if you play poker many, many times, on average you will get a hand consisting of only spades about once in every 2000 hands.

EXAMPLE 4 ■ Finding the Probability of an Event

A bag contains 20 tennis balls, of which four are defective. If two balls are selected at random from the bag, what is the probability that both are defective?

SOLUTION

The experiment consists of choosing two balls from 20. Thus, the number of elements in the sample space S is $C(20, 2)$. Since there are four defective balls, the number of ways of picking two defective balls is $C(4, 2)$. Thus, the probability of the event E of picking two defective balls is

$$P(E) = \frac{n(E)}{n(S)} = \frac{C(4, 2)}{C(20, 2)} = \frac{6}{190} \approx 0.032 \qquad \blacksquare$$

The **complement** of an event E is the set of outcomes in the sample space that are not in E. We denote the complement of an event E by E'. We can calculate the probability of E' using the definition and the fact that $n(E') = n(S) - n(E)$:

$$P(E') = \frac{n(E')}{n(S)} = \frac{n(S) - n(E)}{n(S)} = \frac{n(S)}{n(S)} - \frac{n(E)}{n(S)} = 1 - P(E)$$

PROBABILITY OF THE COMPLEMENT OF AN EVENT

Let S be the sample of an experiment and E an event. Then

$$P(E') = 1 - P(E)$$

This is an extremely useful result, since it is often difficult to calculate the probability of an event E but easy to find that of E', from which $P(E)$ can be calculated immediately using this formula.

EXAMPLE 5 ■ Finding the Probability of the Complement of an Event

An urn contains 10 red balls and 15 blue balls. Six balls are drawn at random from the urn. What is the probability that at least one ball is red?

SOLUTION

It is tedious to count all the possible ways in which one or more of the balls drawn are red. So, let us consider the complement of this event—namely, that none of the balls chosen is red. The number of ways of choosing 6 blue balls from the 15 blue balls is $C(15, 6)$; the number of ways of choosing 6 balls from the 25 balls is $C(25, 6)$. Thus,

$$P(E') = \frac{n(E')}{n(S)} = \frac{C(15, 6)}{C(25, 6)} = \frac{5005}{177,100} = \frac{13}{460}$$

By the formula for the complement of an event, we have

Since
$$P(E') = 1 - P(E)$$
we have
$$P(E) = 1 - P(E')$$

$$P(E) = 1 - P(E') = 1 - \frac{13}{460} = \frac{447}{460} \approx 0.97$$

■

8.5 **EXERCISES**

1. An experiment consists of tossing a coin twice.
  (a) Find the sample space.
  (b) Find the probability of getting heads exactly two times.
  (c) Find the probability of getting heads at least one time.
  (d) Find the probability of getting heads exactly one time.

2. An experiment consists of tossing a coin and rolling a die.
   (a) Find the sample space.
   (b) Find the probability of getting heads and an even number.
   (c) Find the probability of getting heads and a number greater than 4.
   (d) Find the probability of getting tails and an odd number.

3–4 ■ A die is rolled. Find the probability of the given event.

3. (a) The number showing is a six.
   (b) The number showing is an even number.
   (c) The number showing is greater than 5.

4. (a) The number showing is a two or a three.
   (b) The number showing is an odd number.
   (c) The number showing is a number divisible by 3.

5–6 ■ A card is drawn randomly from a standard 52-card deck. Find the probability of the given event.

5. (a) The card drawn is a king.
   (b) The card drawn is a face card.
   (c) The card drawn is not a face card.

6. (b) The card drawn is a heart.
   (b) The card drawn is either a heart or a spade.
   (c) The card drawn is a heart, a diamond, or a spade.

7–8 ■ A ball is drawn randomly from a jar that contains five red balls, two white balls, and one yellow ball. Find the probability of the given event.

7. (a) A red ball is drawn.
   (b) The ball drawn is not yellow.
   (c) A black ball is drawn.

8. (a) Neither a white nor yellow ball is drawn.
   (b) A red, white, or yellow ball is drawn.
   (c) The ball drawn is not white.

9. A game is played with a dodecahedral die, that is, a die with 12 faces. What is the probability of rolling a number greater than 8?

10. A drawer contains an unorganized collection of 18 socks—three pairs are red, two pairs are white, and four pairs are black.
   (a) If one sock is drawn at random from the drawer, what is the probability that it is red?

   (b) Once a sock is drawn and discovered to be red, what is the probability of drawing another red sock to make a matching pair?

11. Find the probability that a number chosen at random from 1 to 100 has the property described.
   (a) The number is divisible by 3.
   (b) The number is not divisible by 3.
   (c) The number is divisible by 3 and by 5.

12. A child's game has a spinner as shown in the figure. Find the probability of the given event.
   (a) The spinner stops on an even number.
   (b) The spinner stops on an odd number or a number greater than 3.

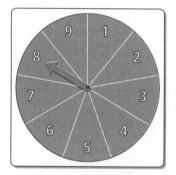

13. A letter is chosen at random from the word *EXTRATERRESTRIAL*. Find the probability of the given event.
   (a) The letter *T* is chosen.
   (b) The letter chosen is a vowel.
   (c) The letter chosen is a consonant.

14–21 ■ A poker hand, consisting of five cards, is dealt from a standard deck of 52 cards. Find the probability that the hand contains the cards described.

14. Five hearts

15. Five cards of the same suit

16. Five face cards

17. Four cards of a kind (such as kings)

18. An ace, king, queen, jack, and 10 of the same suit (royal flush)

19. Three hearts and two diamonds

20. At least one ace

21. At most four hearts

22. A pair of dice is rolled, and the numbers showing are observed.
 (a) List the sample space of this experiment.
 (b) Find the probability of getting a sum of 7.
 (c) Find the probability of getting a sum of 9.
 (d) Find the probability that the two dice show doubles (the same number).
 (e) Find the probability that the two dice show different numbers.
 (f) Find the probability of getting a sum of 9 or higher.

23. A couple intends to have four children. Assume that having a boy or a girl is an equally likely event.
 (a) List the sample space of this experiment.
 (b) Find the probability that the couple has only boys.
 (c) Find the probability that the couple has two boys and two girls.
 (d) Find the probability that the couple has four children of the same sex.
 (e) Find the probability that the couple has at least two girls.

24. What is the probability that a 13-card bridge hand consists of all cards from the same suit?

25. An American roulette wheel has 38 slots: two slots are numbered 0 and 00, and the remaining slots are numbered from 1 to 36. Find the probability that the ball lands in an odd-numbered slot.

26. A toddler has wooden blocks showing the letters C, E, F, H, N, and R. Find the probability that the child arranges the letters in the indicated order.
 (a) In the order *FRENCH*
 (b) In alphabetical order

27. In the 6/49 lottery game, a player selects six numbers from 1 to 49. What is the probability of picking the six winning numbers?

28. The California Lotto game used to be played as described in Exercise 27. In order to reduce the number of winners and so increase the size of the jackpots from draw to draw (and, hence, to also increase interest in the game), officials changed to a 6/53 game. What is the probability of winning this new game? Compare the chances of winning for the 6/49 and 6/53 games.

29. A student has locked her locker with a combination lock, showing numbers from 1 to 40, but she has forgotten the three-number combination that opens the lock. In order to open the lock, she decides to try all possible combinations. If she can try ten different combinations every minute, what is the probability that she will open the lock within one hour?

30. The president of a large company selects six employees to receive a special bonus. He claims that the six employees are chosen randomly from among the 30 employees, of which 19 are women and 11 are men. What is the probability that no woman is chosen?

31. An exam has ten true-false questions. A student who has not studied answers all ten questions by just guessing. Find the probability that the student correctly answers the given number of questions.
 (a) All ten questions
 (b) Exactly seven questions

32. To control the quality of their product, the Bright-Light Company inspects three light bulbs out of each batch of ten bulbs manufactured. If a defective bulb is found, the batch is discarded. Suppose that a batch contains two defective bulbs. What is the probability that the batch will be discarded?

33. Twenty students are arranged randomly in a row for a class picture. Paul wants to stand next to Phyllis. Find the probability that he gets his wish.

34. Eight boys and 12 girls are arranged in a row. What is the probability that all the boys will be standing at one end of the line and all the girls at the other end?

35. An often-quoted example of an event of extremely low probability is that a monkey types the Shakespearean play *Hamlet* by randomly striking keys on a typewriter. Assume that the typewriter has 48 keys (including the space bar) and that the monkey is equally likely to hit any key.
 (a) Find the probability that such a monkey will actually correctly type just the title of the play as his first word.
 (b) What is the probability that the monkey will type the phrase "To be or not to be" as his first words?

36. A monkey is trained to arrange wooden blocks in a straight line. He is then given six blocks showing the letters A, E, H, L, M, T. What is the probability that he will arrange them to spell the word *HAMLET*?

37. A monkey is trained to arrange wooden blocks in a straight line. He is then given 11 blocks showing the letters A, B, B, I, I, L, O, P, R, T, Y. What is the probability that the monkey will arrange the blocks to spell the word *PROBABILITY*?

38. Eight horses are entered in a race. You randomly predict a particular order for the horses to complete the race. What is the probability that your prediction is correct?

39. Many genetic traits are controlled by two genes, one dominant and one recessive. In Gregor Mendel's original experiments with peas, the genes controlling the height of the plant are denoted by T (tall) and t (short). The gene T is dominant, so a plant with the genotype (genetic makeup) TT or Tt is tall, whereas one with genotype tt is short. By a statistical analysis of the offspring in his experiments, Mendel concluded that offspring inherit one gene from each parent, and each possible combination of the two genes is equally likely. If each parent has the genotype Tt, then the following chart gives the possible genotypes of the offspring:

		Parent 2	
		T	**t**
Parent 1	**T**	TT	Tt
	t	Tt	tt

Find the probability that a given offspring of these parents will be **(a)** tall or **(b)** short.

40. Refer to Exercise 39. Make a chart of the possible genotypes of the offspring if one parent has genotype Tt and the other tt. Find the probability that a given offspring will be **(a)** tall or **(b)** short.

8.6 THE UNION OF EVENTS

Many of the problems of probability involve events that are described by more than one condition. For example, in drawing a card from a deck, what is the probability of drawing an ace *or* a spade? If two cards are drawn, what is the probability that the first is a king *and* the second is a queen? Each of these questions involves events joined by the words *or* or *and*. In this section we study the probability of events that are joined by the word *or*—in other words, the union of events.

MUTUALLY EXCLUSIVE EVENTS

Two events that have no outcome in common are said to be **mutually exclusive** (see Figure 1). For example, in drawing a card from a deck, the events

E: The card is an ace

F: The card is a queen

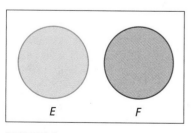

FIGURE 1

are mutually exclusive, because a card cannot be both an ace and a queen.

If E and F are mutually exclusive events, what is the probability that E or F occurs? The word *or* indicates that we want the probability of $E \cup F$. Since E

and F have no element in common,

$$n(E \cup F) = n(E) + n(F)$$

Thus

$$P(E \cup F) = \frac{n(E \cup F)}{n(S)} = \frac{n(E) + n(F)}{n(S)} = \frac{n(E)}{n(S)} + \frac{n(F)}{n(S)} = P(E) + P(F)$$

We have proved the following formula.

PROBABILITY OF THE UNION OF MUTUALLY EXCLUSIVE EVENTS

If E and F are mutually exclusive events in a sample space S, then the probability of E *or* F is

$$P(E \cup F) = P(E) + P(F)$$

There is a natural extension of this formula for any number of mutually exclusive events: If $E_1, E_2, \ldots, E_n$ are mutually exclusive events, then

$$P(E_1 \cup E_2 \cup \cdots \cup E_n) = P(E_1) + P(E_2) + \cdots + P(E_n)$$

EXAMPLE 1 ■ The Probability of Mutually Exclusive Events

A card is drawn at random from a standard deck of 52 cards. What is the probability that the card is either a seven or a face card?

SOLUTION

Let E and F denote the following events.

$$E: \quad \text{The card is a seven}$$

$$F: \quad \text{The card is a face card}$$

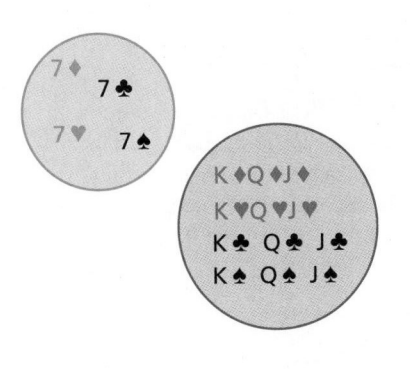

Since a card cannot be both a seven and a face card, the events are mutually exclusive. We want the probability of E *or* F; in other words, the probability of $E \cup F$. By the formula,

$$P(E \cup F) = P(E) + P(F) = \frac{4}{52} + \frac{12}{52} = \frac{4}{13}$$

■

EXAMPLE 2 ■ The Probability of Mutually Exclusive Events

A student who has not studied for a ten-question true-false quiz guesses the answer to each question. What is the probability that he earns a score of 80% or higher on the quiz?

SOLUTION

Since there are two ways of answering each of the ten questions, there are 2^{10} ways of completing the quiz. So the sample space has $2^{10} = 1024$ outcomes. Now let E_8, E_9, and E_{10} denote the following events:

E_8:   The student answers exactly eight questions correctly

E_9:   The student answers exactly nine questions correctly

E_{10}:   The student answers exactly ten questions correctly

There are $C(10, 8)$ ways for the student to answer exactly eight questions correctly. Thus, $P(E_8) = C(10, 8)/1024$. Similarly, $P(E_9) = C(10, 9)/1024$, and $P(E_{10}) = C(10, 10)/1024$. Since the events E_8, E_9, and E_{10} are mutually exclusive, we get $P(E_8 \cup E_9 \cup E_{10}) = P(E_8) + P(E_9) + P(E_{10})$. Thus, the probability that the student will answer eight or more questions correctly by guessing is

$$P(E_8 \cup E_9 \cup E_{10}) = P(E_8) + P(E_9) + P(E_{10})$$

$$= \frac{C(10, 8)}{1024} + \frac{C(10, 9)}{1024} + \frac{C(10, 10)}{1024}$$

$$= \frac{45 + 10 + 1}{1024} = \frac{7}{128} \approx 0.055$$

■

THE PROBABILITY OF THE UNION OF TWO EVENTS

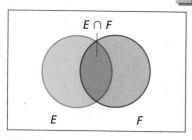

FIGURE 2

If two events E and F are not mutually exclusive, then they share outcomes in common. The situation is described graphically in Figure 2. The overlap of the two sets is $E \cap F$. Again, we are interested in the event E or F, so we must count the elements in $E \cup F$. If we simply added the number of elements in E to the number of elements in F, then we would be counting the elements in the overlap twice—once in E and once in F. So, to get the correct total we must subtract the number of elements in $E \cap F$. Thus

$$n(E \cup F) = n(E) + n(F) - n(E \cap F)$$

Using the formula for probability, we get

$$P(E \cup F) = \frac{n(E \cup F)}{n(S)} = \frac{n(E) + n(F) - n(E \cap F)}{n(S)}$$

$$= \frac{n(E)}{n(S)} + \frac{n(F)}{n(S)} - \frac{n(E \cap F)}{n(S)}$$

$$= P(E) + P(F) - P(E \cap F)$$

We have proved the following.

PROBABILITY OF THE UNION OF TWO EVENTS

If E and F are events in a sample space S, then the probability of *E or F* is

$$P(E \cup F) = P(E) + P(F) - P(E \cap F)$$

EXAMPLE 3 ■ The Probability of the Union of Events

What is the probability that a card drawn at random from a standard 52-card deck is either a face card or a spade?

SOLUTION

We let E and F denote the following events:

E: The card is a face card

F: The card is a spade

There are 12 face cards and 13 spades in a 52-card deck, so

$$P(E) = \frac{12}{52} \quad \text{and} \quad P(F) = \frac{13}{52}$$

Since there are 3 cards that are both face cards and spades,

$$P(E \cap F) = \frac{3}{52}$$

Thus, by the formula for the probability of the union of two events we have

$$P(E \cup F) = P(E) + P(F) - P(E \cap F)$$

$$= \frac{12}{52} + \frac{13}{52} - \frac{3}{52} = \frac{11}{26}$$ ■

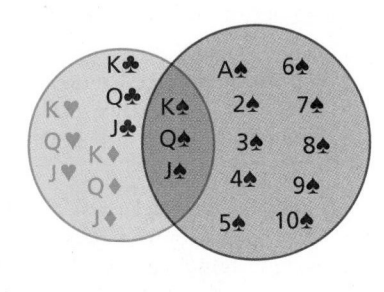

8.6 EXERCISES

1–2 ■ Determine whether the events E and F in the given experiment are mutually exclusive.

1. The experiment consists of selecting a person at random.
   (a) E: The person is male
       F: The person is female
   (b) E: The person is tall
       F: The person is blond

2. The experiment consists of choosing at random a student from your class.
   (a) E: The student is female
       F: The student wears glasses
   (b) E: The student has long hair
       F: The student is male

3–4 ■ A die is rolled and the number showing is observed. Determine whether the events E and F are mutually exclusive. Then find the probability of the event $E \cup F$.

3. (a) E: The number is even
       F: The number is odd
   (b) E: The number is even
       F: The number is greater than 4

4. (a) E: The number is greater than 3
       F: The number is less than 5
   (b) E: The number is divisible by 3
       F: The number is less than 3

5–6 ■ A card is drawn at random from a standard 52-card deck. Determine whether the events E and F are mutually exclusive. Then find the probability of the event $E \cup F$.

5. (a) E: The card is a face card
       F: The card is a spade
   (b) E: The card is a heart
       F: The card is a spade

6. (a) E: The card is a club
       F: The card is a king
   (b) E: The card is an ace
       F: The card is a spade

7–8 ■ Refer to the spinner shown in the figure at the top of the next column. Find the probability of the given event.

7. (a) The spinner stops on red.
   (b) The spinner stops on an even number.
   (c) The spinner stops on red or an even number.

8. (a) The spinner stops on blue.
   (b) The spinner stops on an odd number.
   (c) The spinner stops on blue or an odd number.

9. An American roulette wheel has 38 slots: two of the slots are numbered 0 and 00, and the rest are numbered from 1 to 36. Find the probability that the ball lands in an odd-numbered slot or in a slot with a number higher than 31.

10. The probability that a teenager enjoys rock music is 0.8 and the probability that a teenager enjoys country and western music is 0.2. If only one in ten teenagers enjoys both types of music, what is the probability that a teenager chosen at random enjoys one or the other of these types of music?

11. A toddler has eight wooden blocks showing the letters A, E, I, G, L, N, T, and R. What is the probability that the child will arrange the letters to spell one of the words *TRIANGLE* or *INTEGRAL*?

12. A monkey is trained to arrange wooden blocks in a straight line. He is then given blocks with the letters A, E, H, L, M, T. What is the probability that he will arrange them to spell one of the words *HAMLET* or *THELMA*?

13. A committee of five is chosen randomly from a group of six males and eight females. What is the probability that the committee includes either all males or all females?

14. An exam has ten true-false questions. Find the probability that a student can pass the test by just guessing. (A passing score on this test is 70%.)

15. A couple intends to have four children. Find the probability that the couple will have three or more girls. (Assume that the event of having a child of one gender is equally as likely as having a child of the other gender.)

16. In the 6/49 lottery game a player selects six numbers from 1 to 49. What is the probability of selecting at least five of the six winning numbers?

17. A jar contains six red marbles numbered 1 to 6 and ten blue marbles numbered 1 to 10. A marble is drawn at random from the jar. Find the probability that the given event occurs.

(a) The marble is red.

(b) The marble is odd-numbered.

(c) The marble is red or odd-numbered.

(d) The marble is blue or even-numbered.

8.7 THE INTERSECTION OF EVENTS

In the preceding section we considered the probability of events joined by the word *or*. Here we study the probability of events joined by the word *and*—in other words, the intersection of events.

CONDITIONAL PROBABILITY

Often when we consider probabilities we have additional information available that affects the calculation of the probability of an event. For example, suppose we are interested in the probability that a person chosen at random wears a skirt. If we are then given the additional information that the person already chosen is a woman, we might want to revise our estimate of the probability to make it higher. As another example, let us consider the experiment of drawing a card from a standard 52-card deck, and let E and F denote the following events:

$$E: \quad \text{The card is a face card}$$

$$F: \quad \text{The card is a king}$$

We know that the probability of drawing a king is $\frac{4}{52} = \frac{1}{13}$. Suppose Paul has already drawn a card from a deck and tells us (additional information) that the card is a face card. Now, what is the probability that the card Paul holds is a king? Since there are only 12 face cards in the deck and four of these are kings, we can see that the probability in question is $\frac{4}{12} = \frac{1}{3}$.

What we are calculating here is the probability of an event F given that another event E occurs. We denote this probability by $P(F \mid E)$. Figure 1 helps us see how to calculate this probability. Since we are given that E has already occurred, E serves as our sample space. Thus, $P(F \mid E) = \dfrac{n(F \cap E)}{n(E)}$.

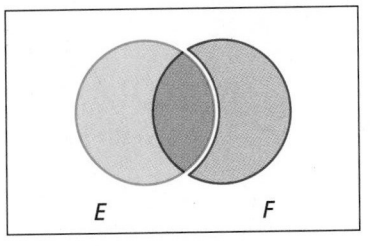

FIGURE 1

CONDITIONAL PROBABILITY

If E and F are events in a sample space S, then $P(F \mid E)$ is the **conditional probability of F given that E occurs.** We have

$$P(F \mid E) = \frac{n(F \cap E)}{n(E)}$$

EXAMPLE 1 ■ Finding Conditional Probability

A mathematics class consists of 30 students, 12 of whom study French, eight study German, four study both of these languages, and the rest do not study any foreign language. If a student is chosen at random from this class, find the probability of each of the following events.

(a) The student studies French.
(b) The student studies French, given that he studies German.
(c) The student studies French, given that he studies a foreign language.

SOLUTION

We let F, G, and L denote the following events.

$$F: \quad \text{The student studies French}$$

$$G: \quad \text{The student studies German}$$

$$L: \quad \text{The student studies a foreign language}$$

(a) There are 30 students in the class, 12 of whom study French, so

$$P(F) = \frac{12}{30} = \frac{2}{5}$$

(b) We are asked for $P(F \mid G)$, the probability that a student who studies German also studies French. Since eight students study German, and four of these study French, it is clear that this probability is $\frac{4}{8} = \frac{1}{2}$. Our formula for conditional probability confirms this:

$$P(F \mid G) = \frac{n(F \cap G)}{n(G)} = \frac{4}{8} = \frac{1}{2}$$

(c) The number of students who study a foreign language is $12 + 8 - 4 = 16$. (Why?) Since 12 of these study French, we have

$$P(F \mid L) = \frac{n(F \cap L)}{n(L)} = \frac{12}{16} = \frac{3}{4}$$

■

PROBABILITY OF THE INTERSECTION OF TWO EVENTS

We use the expression for conditional probability to get a formula for calculating the probability of *E and F*, that is, the probability of $E \cap F$.

$$P(F \mid E) = \frac{n(E \cap F)}{n(E)} = \frac{\dfrac{n(E \cap F)}{n(S)}}{\dfrac{n(E)}{n(S)}} = \frac{P(E \cap F)}{P(E)}$$

Rearranging this last expression gives the following important formula.

PROBABILITY OF THE INTERSECTION OF TWO EVENTS

If *E* and *F* are events in a sample space *S*, then the probability of *E and F* is

$$P(E \cap F) = P(E) P(F \mid E)$$

EXAMPLE 2 ■ The Probability of the Intersection of Events

Two cards are drawn at random, without replacement, from a deck of 52 cards. What is the probability that the first is an ace and the second is a king?

SOLUTION

Let *E* be the event "the card is an ace" and *F* the event "the card is a king." Clearly, $P(E) = \frac{4}{52}$. Now, $P(F \mid E)$ is the probability that the card drawn is an ace given that a king has already been drawn. Since 51 cards now remain in the deck, and four of these are aces, we have $P(F \mid E) = \frac{4}{51}$. By the formula for the probability of the intersection of two events, we have

$$P(E \cap F) = P(E) P(F \mid E) = \frac{4}{52} \times \frac{4}{51} \approx 0.006$$ ■

EXAMPLE 3 ■ The Probability of the Intersection of Events

A jar contains five red balls and four white balls. Two balls are drawn, one at a time, without replacement. What is the probability that the first is red and the second is also red?

SOLUTION

The probability that the first ball is red is $\frac{5}{9}$. The probability that the second is red, given that the first is red, is $\frac{4}{8}$. Thus, the probability that both the first and second balls are red is

$$\frac{5}{9} \times \frac{4}{8} = \frac{5}{18}$$ ■

In Example 3 it is also possible to consider the two balls drawn at once. Looking at the problem in this way, the probability of picking two red balls is

$$\frac{C(5,2)}{C(9,2)} = \frac{10}{36} = \frac{5}{18}$$

INDEPENDENT EVENTS

When the occurrence of one event does not affect the probability of another event, we say that the events are **independent.** Thus, E and F are independent if $P(E \mid F) = P(E)$ and $P(F \mid E) = P(F)$. For instance, if a balanced coin is tossed, the probability of showing heads on the second toss is $\frac{1}{2}$, regardless of what was obtained on the first toss. So, any two tosses of a coin are independent.

From the formula for the probability of the intersection of two events, we get the following formula.

PROBABILITY OF THE INTERSECTION OF INDEPENDENT EVENTS

If E and F are independent events in a sample space S, then

$$P(E \cap F) = P(E) P(F)$$

EXAMPLE 4 ■ The Probability of Independent Events

A jar contains five red balls and four black balls. A ball is drawn at random from the jar and then replaced; then another ball is picked. What is the probability that both balls are red?

SOLUTION

The events are independent. The probability that the first ball is red is $\frac{5}{9}$. The probability that the second is red is also $\frac{5}{9}$. Thus, the probability that both balls are red is

$$\frac{5}{9} \times \frac{5}{9} = \frac{25}{81} \approx 0.31$$

■

EXAMPLE 5 ■ The Birthday Problem

What is the probability that in a class of 35 students at least two have the same birthday?

SOLUTION

It is reasonable to assume that the 35 birthdays are independent and that each day of the 365 days in a year is equally likely as a date of birth. (We ignore February 29.)

Number of people in a group	Probability that at least two have the same birthday
5	.02714
10	.11695
15	.25290
20	.41144
22	.47569
23	.50730
24	.53834
25	.56870
30	.70631
35	.81438
40	.89123
50	.97037

Let E be the event that two of the students have the same birthday. It is tedious to list all the possible ways in which at least two of the students have matching birthdays. So, we consider the complementary event E', that is, that *no* two students have the same birthday. To find this probability we consider the students one at a time. The probability that the first student has a birthday is 1, the probability that the second has a birthday different from the first is $\frac{364}{365}$, the probability that the third has a birthday different from the first two is $\frac{363}{365}$, the probability that the fourth has a birthday different from the first three is $\frac{362}{365}$, and so on. Thus

$$P(E') = 1 \cdot \frac{364}{365} \cdot \frac{363}{365} \cdot \frac{362}{365} \cdot \ldots \cdot \frac{331}{365} \approx 0.186$$

So
$$P(E) = 1 - P(E') \approx 1 - 0.186 = 0.814$$

Most people find it very surprising that this probability is so high. For this reason this problem is sometimes called the "birthday paradox." The table in the margin gives the probability that two people in a group will share the same birthday for groups of various sizes. ■

8.7 EXERCISES

1. A coin is tossed twice. Let E be the event "the first toss shows heads" and F the event "the second toss shows heads."
(a) Are the events E and F independent?
(b) Find the probability of showing heads on both tosses.

2. A die is rolled twice. Let E be the event "the first roll shows a six" and F the event "the second roll shows a six."
(a) Are the events E and F independent?
(b) Find the probability of showing a six on both rolls.

3. A die is rolled. What is the probability that a 5 shows given that the number showing is greater than 3?

4. A die is rolled. What is the probability that a 3 shows given that the number showing is odd?

5. A card is drawn from a deck. What is the probability that it is a queen given that it is a face card?

6. A card is drawn from a deck. What is the probability that it is a king given that it is a spade?

7. A card is drawn from a deck. What is the probability that it is a spade given that it is a king?

8. A jar contains five red balls, numbered 1 to 5, and seven green balls, numbered 1 to 7. A ball is chosen at

random from the jar.
(a) Find the probability that the ball is red given that it shows the number 3.
(b) Find the probability that the ball is green given that it shows the number 7.
(c) Find the probability that the ball is red given that it shows an even number.
(d) Find the probability that the ball shows an even number given that it is red.

9–10 ■ Refer to the spinner shown in the figure.

9. Find the probability that the spinner has stopped on an even number given that it has stopped on red.

10. Find the probability that the spinner has stopped on a number divisible by 3 given that it has stopped on blue.

11–12 ■ Spinners A and B shown in the figure are spun at the same time.

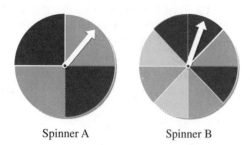

Spinner A   Spinner B

11. (a) Are the events "spinner A stops on red" and "spinner B stops on yellow" independent?
(b) Find the probability that spinner A stops on red and spinner B stops on yellow.

12. (a) Find the probability that both spinners stop on purple.
(b) Find the probability that both spinners stop on blue.

13. Two balls are drawn at random and without replacement from the jar in Exercise 8.
(a) Find the probability that the second ball is red given that the first is red.
(b) Find the probability that the second ball is red given that the first is green.
(c) Find the probability that the second ball shows an even number given that the first shows an odd number.
(d) Find the probability that the second ball shows an even number given that the first shows an even number.

14. In a class of 20 students, six are girls. Two students are chosen at random from the class, without replacement. Find the probability of each event.
(a) The first student chosen is a boy and the second is a girl.
(b) Two girls are chosen.

15. A die is rolled twice. What is the probability that the sum of the rolls is less than 6 given that the first roll shows a 1?

16. A die is rolled twice. What is the probability of showing a 1 on both rolls?

17. A die is rolled twice. What is the probability of showing a 1 on the first roll and an even number on the second roll?

18. A card is drawn from a deck and replaced, and then a second card is drawn.
(a) What is the probability that both cards are aces?
(b) What is the probability that the first is an ace and the second a spade?

19. A card is drawn from a deck and then a second card is drawn without replacing the first card.
(a) What is the probability that both cards are aces?
(b) What is the probability that the first is the ace of spades and the second is a spade?

20. Two balls are drawn, one at a time, from a jar containing five red balls and seven white balls. Find the probability of getting two red balls if the balls are chosen as follows.
(a) The first ball is replaced before the second is drawn.
(b) The first ball is not replaced before the second is drawn.

21. A roulette wheel has 38 slots: Two slots are numbered 0 and 00, and the rest are numbered 1 to 36. A player places a bet on a number between 1 and 36 and wins if a ball thrown into the spinning roulette wheel lands in the slot with the same number. Find the probability of winning on two consecutive spins of the roulette wheel.

22. A researcher claims that she has taught a monkey to spell the word *MONKEY* using the five wooden letters *E, O, K, M, N, Y*. If the monkey has not actually learned anything and is merely arranging the blocks randomly, what is the probability that he will spell the word correctly three consecutive times?

23. What is the probability of rolling "snake eyes" (double ones) three times in a row with a pair of dice?

24. In the 6/49 lottery game, a player selects six numbers from 1 to 49 and wins if he selects the winning six numbers. What is the probability of winning the lottery two times in a row?

25. A family is chosen at random from the set of all families with exactly two children. Find the probability of each event.
(a) The family has two boys if it is known that the first child is a boy.

(b) The family has two boys if it is known that one child is a boy.

26. Jar A contains three red balls and four white balls. Jar B contains five red balls and two white balls. Which one of the following ways of randomly selecting balls gives the greatest probability of drawing two red balls?
    (i) Draw two balls from jar B.
    (ii) Draw one ball from each jar.
    (iii) Put all the balls in one jar, and then draw two balls.

27. Balls are drawn, one at a time, without replacement from a bag containing four black and three white balls. What is the probability that the first is black, the second white, the third black, and so on, alternating colors?

28. Balls are drawn, one at a time, without replacement from a bag containing four black and three white balls. What is the probability that the second ball drawn is black?

29. A slot machine has three wheels: Each wheel has 11 positions—a bar and the digits 0, 1, 2, ..., 9. When the handle is pulled the three wheels spin independently before coming to rest. Find the probability that the wheels stop on each of the following positions.
    (a) Three bars

(b) The same number on each wheel
(c) At least one bar

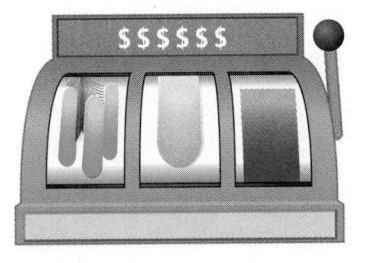

30. Find the probability that in a group of eight students at least two people have the same birthday.

31. What is the probability that in a group of six students at least two have birthdays in the same month?

32. The World Series is a best-four-out-of-seven series. This means that the teams play at most seven games until one team wins four games. Teams A and B are playing and it is estimated that the probability of team A winning any given game against team B is 0.6. Find the probability of each of the following events.
    (a) Team A wins the series in four games.
    (b) Team B wins the series in four games.
    (c) Team A wins the series in five games.
    (d) The series ends in five games.

8.8 EXPECTED VALUE

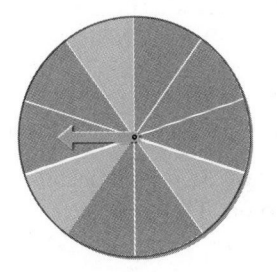

FIGURE 1

In the game shown in Figure 1 you pay \$1 to spin the arrow. If the arrow stops in a red region, you get \$3 (the dollar you paid plus \$2); otherwise, you lose the dollar you paid. If you play this game many times, how much would you expect to win? Or lose? To answer these questions, let us consider the probabilities of winning and losing. Since three of the regions are red, the probability of winning is $\frac{3}{10} = 0.3$ and that of losing is $\frac{7}{10} = 0.7$. Remember, this means that if you play this game many times, you expect to win "on average" three out of ten times. So, suppose you play the game 1000 times. Then you would expect to win 300 times and lose 700 times. Since we win \$2 or lose \$1 in each game, our expected payoff in 1000 games is

$$2(300) + (-1)(700) = -100$$

So the average expected return per game is $\frac{-100}{1000} = -0.1$. In other words, we expect to lose, on average, 10 cents per game. Another way to view this average is to divide each side of the preceding equation by 1000. Writing E for the result, we get

$$E = \frac{2(300) + (-1)(700)}{1000}$$

$$= 2\left(\frac{300}{1000}\right) + (-1)\frac{700}{1000}$$

$$= 2(0.3) + (-1)(0.7)$$

Thus the expected return, or *expected value,* per game is

$$E = a_1 p_1 + a_2 p_2$$

where a_1 is the payoff that occurs with probability p_1 and a_2 is the payoff that occurs with probability p_2. This example leads us to the following definition of expected value.

DEFINITION OF EXPECTED VALUE

A game gives payoffs of $a_1, a_2, \ldots, a_n$ with probabilities $p_1, p_2, \ldots, p_n$. The **expected value** (or **expectation**) E of this game is

$$E = a_1 p_1 + a_2 p_2 + \cdots + a_n p_n$$

The expected value is an average expectation per game if the game is played many times. In general, E is not one of the possible payoffs. In the preceding example the expected value is -10 cents, but it is impossible to lose exactly 10 cents in any given trial of the game.

EXAMPLE 1 ■ Finding an Expected Value

A die is rolled and you receive $1 for each point that shows. What is your expectation?

SOLUTION

Each face of the die has probability $\frac{1}{6}$ of showing. So you get $1 with probability $\frac{1}{6}$, $2 with probability $\frac{1}{6}$, $3 with probability $\frac{1}{6}$, and so on. Thus, the expected value is

$$E = 1\left(\frac{1}{6}\right) + 2\left(\frac{1}{6}\right) + 3\left(\frac{1}{6}\right) + 4\left(\frac{1}{6}\right) + 5\left(\frac{1}{6}\right) + 6\left(\frac{1}{6}\right) = \frac{21}{6} = 3.5$$

This means that if you play this game many times, you will make, on average, $3.50 per game. ■

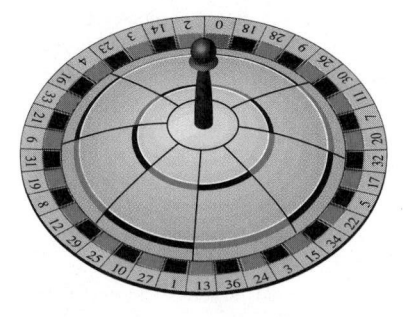

You gain \$35 with probability $\frac{1}{37}$.
You lose \$1 with probability $\frac{36}{37}$.

EXAMPLE 2 ■ Finding an Expected Value

In Monte Carlo the game of roulette is played on a wheel with slots numbered 0, 1, 2, ..., 36. The wheel is spun, and a ball dropped in the wheel is equally likely to end up in any one of the slots. To play the game you bet \$1 on any number other than zero. (For example, you may bet \$1 on number 23.) If the ball stops in your slot, you get \$36 (the \$1 you bet plus \$35). The expected value of this game is

$$E = (35)\frac{1}{37} + (-1)\frac{36}{37} \approx -0.027$$

In other words, if you play this game many times you would expect to lose 2.7 cents on every dollar you bet (on average). Consequently, the house expects to gain 2.7 cents on every dollar that is bet. This expected value is what makes gambling very profitable for the gaming house and very unprofitable for the gambler. ■

8.8 EXERCISES

1–10 ■ Find the expected value (or expectation) of the game described.

1. Mike wins \$2 if a coin toss shows heads and \$1 if it shows tails.

2. Jane wins \$10 if a die roll shows a 6, and she loses \$1 otherwise.

3. The game consists of drawing a card from a deck. You win \$100 if you draw the ace of spades or lose \$1 if you draw any other card.

4. Tim wins \$3 if a coin toss shows heads or \$2 if it shows tails.

5. Carol wins \$3 if a die roll shows a six and she wins \$0.50 otherwise.

6. A coin is tossed twice. Albert wins \$2 for each heads and must pay \$1 for each tails.

7. A die is rolled. Tom wins \$2 if the die shows an even number and he pays \$2 otherwise.

8. A card is drawn from a deck. You win \$104 if the card is an ace, \$26 if it is a face card, and \$13 if it is the 8 of clubs.

9. A bag contains two silver dollars and eight slugs. You pay 50 cents to reach into the bag and take a coin, which you get to keep.

10. A bag contains eight white balls and two black balls. John picks two balls at random from the bag, and he wins \$5 if he does not pick a black ball.

11. In the game of roulette as played in Las Vegas, the wheel has 38 slots: Two slots are numbered 0 and 00, and the rest are numbered 1 to 36. A \$1 bet on any number other than 0 and 00 wins \$36 (\$35 plus the \$1 bet). Find the expected value of this game.

12. A sweepstakes offers a first prize of one million dollars, second prize of \$100,000, and third prize of \$10,000. Suppose that two million people enter the contest and three names are drawn randomly for the three prizes.
(a) Find the expected winnings for a person participating in this contest.
(b) Is it worth paying a dollar to enter this sweepstakes?

13. A box contains 100 envelopes. Ten envelopes contain \$10 each, ten contain \$5 each, two are "unlucky," and the rest are empty. A player draws an envelope from the box and keeps whatever is in it. If a person draws an unlucky envelope, however, he must pay $100. What is the expectation of a person playing this game?

14. A safe containing one million dollars is locked with a combination lock. You pay $1 for one guess at the 6-digit combination. If you open the lock, you get to keep the million dollars. What is your expectation?

15. An investor buys 1000 shares of a risky stock for $5 a share. She estimates that the probability the stock will rise in value to $20 a share is 0.1 and the probability that it will fall to $1 a share is 0.9. If the only criterion for her decision to buy this stock was the expected value of her profit, did she make a wise investment?

16. A slot machine has three wheels, and each wheel has 11 positions—the digits 0, 1, 2, . . . , 9 and the picture of a watermelon. When a quarter is placed in the machine and the handle is pulled, the three wheels spin independently and come to rest. When three watermelons show, the payout is $5; otherwise, nothing is paid. What is the expected value of this game?

17. Find the expected value of playing the slot machine in Exercise 16 if the payouts are as follows: Three watermelons pays $5, two watermelons pays $1, and one watermelon pays $0.50.

18. In a 6/49 lottery game, a player pays $1 and selects six numbers from 1 to 49. Any player who has chosen the six winning numbers wins one million dollars. Assuming this is the only way to win, what is the expected value of this game?

19. A bag contains two silver dollars and six slugs. A game consists of reaching into the bag and drawing a coin, which you get to keep. Determine the "fair price" of playing this game, that is, a price at which the player can be expected to break even if he plays the game many times (in other words, the price at which his expectation is 0).

20. A game consists of drawing a card from a deck. You win $13 if you draw an ace. What is a "fair price" to pay to play this game? (See Exercise 19.)

21. A bag contains five white balls and two black balls. John selects balls from the bag, one at a time, without replacement. The game stops when he picks a black ball. If he is to receive $1 for each white ball drawn, what is his expectation?

22. A bag contains eight white balls and four black balls. If we draw three balls at random from the bag, how many white balls should we expect to get?

---

8 | REVIEW

KEY TOPICS ■ Define, state, or discuss each of the following.

1. The Fundamental Counting Principle

2. Factorial notation

3. Permutations

4. Distinguishable permutations

5. Combinations

6. Distinguishable combinations

7. The number of subsets of a set

8. Outcomes of an experiment

9. Sample space of an experiment

10. Event

11. Probability

12. Probability of the complement of an event

13. Mutually exclusive events

14. Probability of the union of mutually exclusive events

15. Probability of the union of two events

16. The conditional probability of the event E given the event F

17. The probability of the intersection of two events

18. Independent events

19. The probability of the intersection of independent events

20. Expected value

EXERCISES

1. A coin is tossed, a die is rolled, and a card is drawn from a deck. How many possible outcomes does this experiment have?

2. How many three-digit numbers can be formed using the digits 1, 2, 3, 4, 5, 6 if each digit can be used the given number of times?
   (a) Only once
   (b) Any number of times

3. A group of friends have one tennis court. They find that there are ten different ways in which two of them can play a singles game on this court. How many friends are in this group?

4. A pizza parlor advertises that they prepare 2048 different types of pizza. How many toppings does this parlor offer?

5. (a) How many different two-element subsets does the set $\{A, E, I, O, U\}$ have?
   (b) How many different two-letter "words" can be made using the letters from the set in part (a)?

6. A group of students determines that they can stand in a row for their class picture in 120 different ways. How many students are in this class?

7. A coin is tossed ten times. In how many different ways can the result be three heads and seven tails?

8. An airline company overbooks a particular flight and seven passengers are "bumped" from the flight. If 120 passengers are booked on this flight, in how many ways can the airline choose the seven passengers to be bumped?

9. A quiz has ten true-false questions. How many different ways is it possible to earn a score of exactly 70% on this quiz?

10. A test has ten true-false questions and five multiple-choice questions with four choices for each. In how many ways can this test be completed?

11. If you must answer only eight of ten questions on a test, how many ways do you have of choosing the questions you will omit?

12. An ice-cream store offers 15 flavors of ice cream. The specialty is a banana split with four scoops of ice cream. If each scoop must be a different flavor, how many different banana splits may be ordered?

13. A company uses a different three-letter security code for each of its employees. What is the maximum number of codes this security system can generate?

14. The Yukon Territory in Canada uses a license-plate system for automobiles that consists of two letters followed by three numbers. Explain how we can know that fewer than 700,000 autos are licensed in the Yukon.

15. Given 16 subjects from which to choose, in how many ways can a student select fields of study as follows?
   (a) A major and a minor
   (b) A major, a first minor, and a second minor
   (c) A major and two minors

16. (a) How many three-digit numbers can be formed using the digits 0, 1, ..., 9? (Remember, a three-digit number cannot have 0 as the leftmost digit.)
   (b) If a number is chosen randomly from the set $\{0, 1, 2, \ldots, 1000\}$, what is the probability that the number chosen is a three-digit number?

17–20 ■ An **anagram** of a word is a permutation of the letters of that word. For example, anagrams of the word *triangle* include *griantle, integral,* and *tenalgir.*

17. How many anagrams of the word *TRIANGLE* are possible?

18. How many anagrams of the word *TRIANGLE* satisfy each of the following conditions?
   (a) The vowels are in alphabetical order.
   (b) All the vowels are together.
   (c) All the letters are in alphabetical order.

19. How many anagrams are possible from the word *MISSISSIPPI*?

20. How many anagrams are possible from the word *MISSISSIPPI* if all the vowels must be together?

21. In Morse code, each letter is represented by a sequence of dots and dashes, with repetition allowed. How many letters can be represented using Morse code if three or fewer symbols are used?

22. The genetic code is based on the four nucleotides adenine (A), cytosine (C), guanine (G), and thymine (T). These are connected together in long strings to form DNA molecules. For example, a sequence in the DNA may look like CAGTGGTACC.... The code uses "words," all the same length, that are composed of the nucleotides A, C, G, and T. It is known that at least 20 different words exist. What is the minimum word length necessary to generate 20 words?

23. A committee of seven is to be chosen from a group of ten men and eight women. In how many ways can the committee be chosen using each of the following selection requirements?
   (a) No restriction is placed on the number of men and women on the committee.
   (b) The committee must have exactly four men and three women.
   (c) Susie refuses to serve on the committee.
   (d) At least five women must serve on the committee.
   (e) At most two men can serve on the committee.
   (f) The committee is to have a chairman, a vice chairman, a secretary, and four other members.

24. A jar contains ten red balls labeled 0, 1, 2, ..., 9 and five white balls labeled 0, 1, 2, 3, 4. If a ball is drawn from the jar, find the probability of the given event.
   (a) The ball is red.
   (b) The ball is even-numbered.
   (c) The ball is white and odd-numbered.
   (d) The ball is red or odd-numbered.

25. If two balls are drawn from the jar in Exercise 24, find the probability of the given event.
   (a) Both balls are red.
   (b) One ball is white and the other is red.
   (c) At least one ball is red.
   (d) Both balls are red and even-numbered.
   (e) Both balls are white and odd-numbered.

26. If three balls are drawn, one at a time, without replacement from a jar that contains five white balls and three black balls, find the probability of the given event.
   (a) The first ball drawn is white, the second black, and the third black.
   (b) The first ball drawn is black, the second white, and the third black.
   (c) The first ball drawn is black, the second black, and the third white.
   (d) Exactly one white ball is drawn.

27. Find the probability that the indicated card is drawn at random from a 52-card deck.
   (a) An ace
   (b) An ace or a jack
   (c) An ace or a spade
   (d) An ace and a red card

28. A card is drawn from a 52-card deck, a die is rolled, and a coin is tossed. Find the probability of each of the indicated outcomes.
   (a) The ace of spades, a 6, and heads
   (b) A spade, a 6, and heads
   (c) A face card, a number greater than 3, and heads

29. Two dice are rolled. Find the probability of each outcome.
   (a) The dice show the same number.
   (b) The dice show different numbers.

30. Two dice are rolled. John gets $5 if they show the same number or he pays $1 if they show different numbers. What is the expected value of this game?

31. Three dice are rolled. John gets $5 if they all show the same number or he pays $1 if they show different numbers. What is the expected value of this game?

32. Three dice are rolled. Find the probability of each event.
   (a) The dice show a sum of 4.
   (b) The dice show a sum of less than 5.

33. Mary will win one million dollars if she can name the 13 original states in the order in which they ratified the U.S. constitution. Mary has no knowledge of this order, so she makes a guess. What is her expectation?

34. A pizza parlor offers 12 different toppings, one of which is anchovies. If a pizza is ordered at random, what is the probability that anchovies is used as one of the toppings?

35. A drawer contains an unorganized collection of 50 socks—20 are red and 30 are blue. Suppose the lights go out so Kathy cannot distinguish the color of the socks.
  (a) What is the minimum number of socks Kathy must take out of the drawer to be sure of getting a matching pair?
  (b) If two socks are taken at random from the drawer, what is the probability that they make a matching pair?

36. (a) A volleyball team has nine players. In how many ways can a starting lineup be chosen if it consists of two forward players and three defense players?
  (b) Use a counting argument to explain why
$$C(n, 2)\, C(n - 2, 3) = C(n, 3)\, C(n - 3, 2)$$

37. Zip codes consist of five digits.
  (a) How many different zip codes are possible?
  (b) How many different zip codes can be read when the envelope is turned upside down? (An upside down 9 is a 6; and 0, 1, and 8 are the same when read upside down.)
  (c) What is the probability that a randomly chosen zip code can be read upside down?
  (d) How many zip codes read the same upside down as right side up?

38. In the Zip+4 postal code system, zip codes consist of nine digits.
  (a) How many different Zip+4 codes are possible?
  (b) How many Zip+4 codes are palindromes? (A palindrome is a number that reads the same from left to right as right to left.)
  (c) What is the probability that a randomly chosen Zip+4 code is a palindrome?

39. Let $N = 3,600,000$. (Note that $N = 2^7 3^2 5^5$.)
  (a) How many divisors does N have?
  (b) How many even divisors does N have?
  (c) How many divisors of N are multiples of 6?
  (d) What is the probability that a randomly chosen divisor of N is even?

40. (a) How many factors does the expression $a^4 b^3 c^7$ have?
  (b) How many different ways can the expression $a^4 b^3 c^7$ be written without using exponents?

41. In how many ways can a group of ten men and eight women be divided into two groups consisting of five men and four women?

42. The U.S. Senate has two senators from each of the 50 states. In how many ways can a committee of five senators be chosen if no state is to have two members on the committee?

1. How many five-letter "words" can be made from the letters *A, B, C, D, E, F, G, H, I, J* if repetition (a) is allowed? (b) is not allowed?

2. A restaurant offers five main courses, three types of desserts, and four kinds of drinks. In how many ways can a customer order a meal consisting of one choice from each category?

3. A board of directors consisting of eight members is to be chosen from a pool of 30 candidates. The board is to have a chairman, a treasurer, a secretary, and five other members. In how many ways can the board of directors be chosen?

4. A commuter must travel from Ajax to Barrie and back every day. Four roads join the two cities. The commuter likes to vary the trip as much as possible, so he always leaves and returns by different roads. In how many different ways can he make the round trip?

5. A pizza parlor offers four sizes of pizza and 14 different toppings. A customer may choose any number of toppings (or no topping at all). How many different pizzas does this parlor offer?

6. Ten students are to stand in a row for a class picture. In how many ways can this be done for each of the following conditions?
   (a) No restriction is placed on position.
   (b) The two tallest students are to stand next to each other.
   (c) The two tallest students refuse to stand next to each other.

7. An *anagram* of a word is a rearrangement of the letters of the word.
   (a) How many anagrams of the word *LOVE* are possible?
   (b) How many different anagrams of the word *KISSES* are possible?

8. What is the probability that a number chosen at random between 1 and 100 is (a) divisible by 3? (b) not divisible by 3?

9. Three people are chosen at random from a group of five men and ten women. What is the probability that all three are men?

10. Two dice are rolled. What is the probability of getting doubles?

11. Two cards are drawn from a deck. Find the probability of the given event.
    (a) Both cards are face cards.
    (b) The second card drawn is an ace, given that the first card drawn was an ace.

12. A jar contains five red balls, numbered 1 to 5, and eight white balls, numbered 1 to 8. A ball is chosen at random from the jar. Find the probability of the given event.
    (a) The ball is red.
    (b) The ball is even-numbered.
    (c) The ball is red or even-numbered.

(d) The ball is even-numbered given that it is red.

(e) The ball is red given that it is even-numbered.

(f) The ball is red and even-numbered.

13. You are to draw one card from a deck. If it is an ace you win $10; if it is a face card you win $1; otherwise, you lose $0.50. What is the expected value of this game?

14. In a group of four students, what is the probability that at least two have the same astrological sign?

15. A mathematics professor decides on the number of problems he will assign for homework each day by drawing a card from a deck. If the card is a face card, the students do 12 problems; otherwise they do as many problems as the number on the card drawn (with the ace counting as 1). How many problems should the students expect to be assigned each day?

FOCUS ON PROBLEM SOLVING

We arrive at knowledge in two ways—by experience and by reasoning. In mathematics we use reasoning. But sometimes we prove a statement that is so startling or apparently paradoxical that we may be tempted to doubt the validity of our reasoning. The paradoxical nature of a statement, however, often reflects our lack of experience with the concepts involved.

A good way to familiarize ourselves with a fact is to **experiment** with it. For instance, to convince ourselves that the earth is a sphere (which was considered a major paradox at one time), we could go up in a space shuttle a few times and *feel* that it is so; to see whether a given equation is an identity, we might try some special cases to make sure there are no obvious counterexamples. In problems involving probability we can sometimes actually perform an experiment many times and see whether the results match the probabilities we calculated. Here is an example.

PROBLEM ■ The Contestant's Dilemma

In a TV game show a contestant chooses one of three doors. Behind one of them is a valuable prize—the other two doors have nothing behind them. After the contestant has made her choice, the host opens one of the other two doors, one that he knows does not conceal a prize, and then gives her the opportunity to change her choice.

Should the contestant switch, stay, or does it matter? In other words, by switching doors, does she increase, decrease, or leave unchanged her probability of winning? At first it may seem that switching doors doesn't make any difference. After all, two doors are left—one with the prize and one without—so it seems reasonable that the contestant has an equal chance of winning or losing. But this reasoning is wrong. Here is the correct way of looking at this problem:

1. When the contestant first made her choice, she had a $\frac{1}{3}$ chance of winning. If she does not switch, no matter what the host does, her probability of winning remains $\frac{1}{3}$.

2. If the contestant does decide to switch, she will switch to the winning door if she had initially chosen a losing one, or to a losing door if she had initially chosen the winning one. Since the probability of having initially selected a losing door is $\frac{2}{3}$, by switching the probability of winning then becomes $\frac{2}{3}$.

We conclude that the contestant should switch, because her probability of winning is $\frac{2}{3}$ if she switches and $\frac{1}{3}$ if she doesn't. Put simply, there is a much greater chance that she had initially chosen an empty door (since there are more of these), so she should switch.

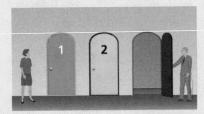

Contestant: "I choose door number 2."

Contestant: "Oh no, what should I do?"

Here is an excellent way to see that this is all true: Play the game many times and keep track of the results. One of the authors used a computer to simulate this game. In the first 100 games played on the computer, the simulated contestant (who always switches) won 67 games! (Almost *too* close to the calculated probability.)

PROBLEMS

1. In a game show like the one described in the preceding problem, a prize is concealed behind one of ten doors. After the contestant chooses a door, the host opens eight losing doors, and then gives the contestant the opportunity to switch to the other unopened door. What is the probability of winning if the contestant chooses (a) not to switch? (b) to switch?

2. A couple intend to have two children. What is the probability that they will have one child of each sex? The French mathematician D'Alembert analyzed this problem (incorrectly) by reasoning that three outcomes are possible: two boys, or two girls, or one child of each sex. He concluded that the probability of having one of each sex is $\frac{1}{3}$, mistakingly assuming that the three outcomes are "equally likely."
   (a) What is the correct probability of having one child of each sex?
   (b) Model this problem by using coins, and perform the experiment 40 times. Compare the results to the probability you calculated in part (a).

3. A game between two players consists of tossing a coin. Player A gets a point if the coin shows heads and player B gets a point if it shows tails. The player to first get six points wins an $8000 jackpot. As it happens, the police raid the place when player A has five points and B has three points. After everyone has calmed down, how should the jackpot be divided between the two players? In other words, what is the probability of A winning (and that of B winning) if the game were to continue?

   The French mathematicians Pascal and Fermat corresponded about this problem and both came to the same correct conclusion (though by very different reasonings). Their friend Roberval disagreed with both of them. He argued that player A has probability $\frac{3}{4}$ of winning since, he claimed, the game can end in the four ways H, TH, TTH, TTT, and in three of these A wins. Roberval was wrong.
   (a) Calculate the correct probability of player A winning. (Roberval's list does not consist of equally likely outcomes. Calculate correctly the probability of each of the outcomes in his list.)
   (b) Continue the game from the point that it was interrupted, using a balanced coin. Do this 80 or more times, and compare the results to the probability you found in part (a).

4. In the World Series the top teams in the National League and the American League play a best-of-seven series—that is, they play until one team has won four games. (No tie is allowed, so this results in a maximum of seven games.) Suppose that the teams are evenly matched, so that the probability that either team wins a given game is $\frac{1}{2}$. Find the probability that the series ends after the given number of games.

(a) Four   (b) Five   (c) Six   (d) Seven

(e) Find the expected value for the number of games until the series ends. [*Hint:* This will be P(four games) $\times$ 4 + P(five) $\times$ 5 + P(six) $\times$ 6 + P(seven) $\times$ 7.]

5. Four customers check their hats in a Vienna opera house. The hat-check clerk fails to properly identify the hats and returns them to the four customers at random. What is the probability that no one gets his own hat back?

6. A *diagonal* of a regular polyhedron is a line segment that joins two of its vertices and that lies in the interior of the polyhedron, on neither a face nor an edge. For example, a cube has four diagonals, one from each top corner to the bottom corner opposite it, as shown in the figure. How many diagonals does a regular icosahedron have?   [*Hint:* First count the number of diagonals that emanate from any given vertex.]

7. The Canadian constitution allows each province to vote YES, vote NO, or ABSTAIN on any proposed constitutional amendment. In order for the amendment to pass, seven or more of the ten provinces must vote YES, and at least one of Ontario and Quebec must be among the YES votes. In how many different ways is it possible for the provinces to cast their votes in order for an amendment to pass?

8. In a singles "knock-out" tennis tournament, a player is eliminated as soon as he or she loses a match. A prearranged schedule determines who plays whom initially, and the winner of each match advances to the next round, as shown in the following sample.

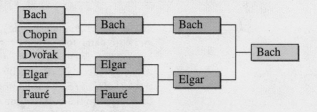

If an odd number of players participates in any round, then the round will have at least one *bye*, that is, a player who does not play but advances automatically to the next round. Suppose n players participate in the tournament.

(a) For what values of n is it possible to avoid a bye?

(b) Show that exactly $n - 1$ matches must be played, no matter how the schedule is arranged. [*Hint:* There is an easy way to see this. Consider the losers of matches.]

9. The number 4 can be written as the sum of natural numbers in several ways:

| 4 | 1 + 3 | 3 + 1 | 2 + 2 |

| 1 + 1 + 2 | 1 + 2 + 1 | 2 + 1 + 1 | 1 + 1 + 1 + 1 |

Find a formula for the number of ways a natural number n can be written in this way. Give an argument using counting techniques to prove that your formula is true.

10. (a) How many zeros are at the end of the number 100!?
(b) If $S = 1! + 2! + 3! + \cdots + 99!$, what is the last digit in the value of S?

11. Suppose 10 lines are drawn in the coordinate plane.
(a) What is the maximum number of points at which these lines can intersect?
(b) What is the maximum number of regions into which the lines divide the plane? [*Hint:* Experiment by drawing diagrams with fewer points. The diagram at the left shows that three lines intersect at three points and divide the plane into seven regions. Try to find a pattern.]

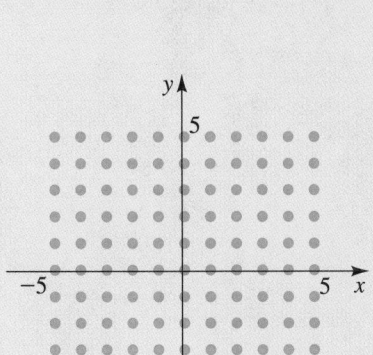

12. Many whole numbers can be written as the difference of two squares. For example,

$$4 = 2^2 - 0^2 \qquad 5 = 3^2 - 2^2$$
$$8 = 3^2 - 1^2 \qquad 13 = 7^2 - 6^2$$

(a) Which of the whole numbers $0, 1, 2, 3, \ldots, 20$ can be expressed as the difference of two squares?
(b) Generalize your result.

13. A point $P(x, y)$ is selected at random from the points in the plane shown in the figure. Find the probability of the given event.
(a) P has x-coordinate 2.
(b) P lies on the line $y = x$.
(c) P lies on the circle $x^2 + y^2 = 25$.
(d) P lies inside (but not on) the circle $x^2 + y^2 = 25$.
(e) P lies above (but not on) the line $y = 2x - 1$.

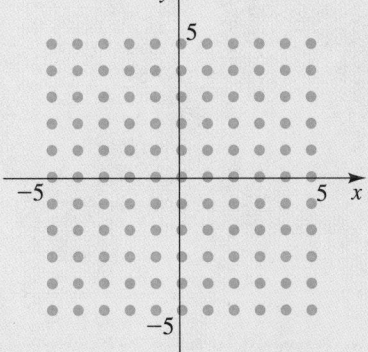

14. Choose two numbers at random from the interval $[0, 1]$. What is the probability that the sum of the numbers is less than 1? [*Hint:* Call the numbers x and y. Choosing these two numbers is the same as choosing an ordered pair (x, y) in the unit square $\{(x, y) \mid 0 \leqslant x \leqslant 1, 0 \leqslant y \leqslant 1\}$. What proportion of the points in this square corresponds to $x + y$ being less than 1?]

9

SEQUENCES AND SERIES

Sequences and series abound in nature; the Fibonacci sequence, for example, is hidden in the intricate structure of the nautilus.

A mathematician, like a painter or a poet, is a maker of patterns.

G. H. HARDY

In this chapter we study sequences and series of numbers. Roughly speaking, a sequence is a list of numbers written in a specific order and a series is what one gets by adding the numbers in a sequence. Sequences and series have many theoretical and practical uses. Among other applications, we consider how series are used to calculate the value of an annuity.

Section 9.7 introduces a special kind of proof called *mathematical induction*. In Section 9.8 we use mathematical induction to prove a formula for expanding $(a + b)^n$ for any natural number n.

9.1 SEQUENCES

A *sequence* is a set of numbers written in a specific order:

$$a_1, a_2, a_3, a_4, \ldots, a_n, \ldots$$

The number a_1 is called the *first term*, a_2 is the *second term*, and in general a_n is the *nth term*. Since for every natural number n there is a corresponding number a_n, we can define a sequence as a function.

DEFINITION OF A SEQUENCE

A **sequence** is a function f whose domain is the set of natural numbers. The values $f(1), f(2), f(3), \ldots$ are called the **terms** of the sequence.

We usually write a_n instead of the function notation $f(n)$ for the value of the function at the number n.

Here is a simple example of a sequence:

$$2, 4, 6, 8, 10, \ldots$$

The dots indicate that the sequence continues indefinitely. We can write a sequence in this way when it is clear what the subsequent terms of the sequence are. This sequence consists of even numbers. To be more accurate, however, we need to specify a procedure for finding *all* the terms of the sequence. This can be done by giving a formula for the nth term a_n of the sequence. In this case,

$$a_n = 2n$$

Another way to write this sequence is to use function notation:

$$a(n) = 2n$$

so $a(1) = 2$, $a(2) = 4$, $a(3) = 6, \ldots$

and the sequence can be written as

$$
\begin{array}{ccccccc}
2, & 4, & 6, & 8, & \ldots, & 2n, & \ldots \\
\uparrow & \uparrow & \uparrow & \uparrow & & \uparrow & \\
\text{1st} & \text{2nd} & \text{3rd} & \text{4th} & & n\text{th} & \\
\text{term} & \text{term} & \text{term} & \text{term} & & \text{term} &
\end{array}
$$

Notice how the formula $a_n = 2n$ gives all the terms of the sequence. For instance, substituting 1, 2, 3, and 4 for n gives the first four terms:

$$a_1 = 2 \cdot 1 = 2 \qquad a_2 = 2 \cdot 2 = 4$$

$$a_3 = 2 \cdot 3 = 6 \qquad a_4 = 2 \cdot 4 = 8$$

To find the 103rd term of this sequence, we use $n = 103$ to get

$$a_{103} = 2 \cdot 103 = 206$$

EXAMPLE 1 ■ Finding the Terms of a Sequence

Find the first five terms and the 100th term of the sequence defined by each formula.

(a) $a_n = 2n - 1$ (b) $c_n = n^2 - 1$ (c) $t_n = \dfrac{n}{n+1}$ (d) $r_n = \dfrac{(-1)^n}{2^n}$

SOLUTION

(a) Using the formula for this sequence, we have

$$a_1 = 2(1) - 1 = 1 \qquad a_2 = 2(2) - 1 = 3 \qquad a_3 = 2(3) - 1 = 5$$

$$a_4 = 2(4) - 1 = 7 \qquad a_5 = 2(5) - 1 = 9 \qquad a_{100} = 2(100) - 1 = 199$$

The sequence can be written as

$$1, 3, 5, 7, 9, \ldots, 2n - 1, \ldots$$

This is the sequence of odd numbers.

(b) We have

$$c_1 = 1^2 - 1 = 0 \qquad c_2 = 2^2 - 1 = 3 \qquad c_3 = 3^2 - 1 = 8$$

$$c_4 = 4^2 - 1 = 15 \qquad c_5 = 5^2 - 1 = 24 \qquad c_{100} = 100^2 - 1 = 9999$$

This sequence can be written as

$$0, 3, 8, 15, 24, \ldots, n^2 - 1, \ldots$$

(c) From the formula for t_n, we get $t_1 = \frac{1}{2}$, $t_2 = \frac{2}{3}$, $t_3 = \frac{3}{4}$, $t_4 = \frac{4}{5}$, $t_5 = \frac{5}{6}$, and $t_{100} = \frac{100}{101}$. This sequence can be written as

$$\frac{1}{2}, \frac{2}{3}, \frac{3}{4}, \frac{4}{5}, \frac{5}{6}, \ldots, \frac{n}{n+1}, \ldots$$

(d) We have $r_1 = -\frac{1}{2}$, $r_2 = \frac{1}{4}$, $r_3 = -\frac{1}{8}$, $r_4 = \frac{1}{16}$, $r_5 = -\frac{1}{32}$, and $r_{100} = \dfrac{1}{2^{100}}$.

(The number 2^{100} has 31 digits, so we will not write it here.) The sequence can be written as

$$-\frac{1}{2}, \frac{1}{4}, -\frac{1}{8}, \frac{1}{16}, -\frac{1}{32}, \ldots, \frac{(-1)^n}{2^n}, \ldots$$ ∎

In Example 1(d) the presence of $(-1)^n$ in the sequence has the effect of making successive terms alternately positive and negative.

It is often useful to picture a sequence by sketching its graph. Since a sequence is a function whose domain is the natural numbers, we can draw its graph in the Cartesian plane. For instance, the graph of the sequence

$$1, \frac{1}{2}, \frac{1}{3}, \frac{1}{4}, \frac{1}{5}, \frac{1}{6}, \ldots, \frac{1}{n}, \ldots$$

is shown in Figure 1. Compare this to the graph of

$$1, -\frac{1}{2}, \frac{1}{3}, -\frac{1}{4}, \frac{1}{5}, -\frac{1}{6}, \ldots, \frac{(-1)^{n+1}}{n}, \ldots$$

shown in Figure 2. The graph of every sequence consists of isolated points that are *not* connected.

Some sequences do not have simple defining formulas like those of Example 1. The *n*th term of a sequence may depend on some or all of the terms preceding it. A sequence defined in this way is called **recursive.** Here are two examples.

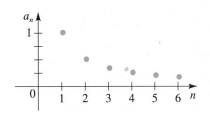

FIGURE 1

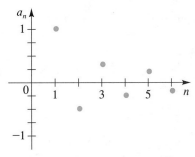

FIGURE 2

EXAMPLE 2 ■ Finding the Terms of a Recursive Sequence

Find the first five terms of the sequence defined recursively by

$$a_n = 3(a_{n-1} + 2)$$

and $a_1 = 1$.

SOLUTION

The defining formula for this sequence is recursive. It allows us to find the *n*th term a_n if we know the preceding term a_{n-1}. Thus, we can find the second term from the first term, the third term from the second term, the fourth term from the third term, and so on. Since we are given the first term $a_1 = 1$, we can proceed as follows.

$$a_2 = 3(a_1 + 2) = 3(1 + 2) = 9$$
$$a_3 = 3(a_2 + 2) = 3(9 + 2) = 33$$
$$a_4 = 3(a_3 + 2) = 3(33 + 2) = 105$$
$$a_5 = 3(a_4 + 2) = 3(105 + 2) = 321$$

Thus, the first five terms of this sequence are

$$1, 9, 33, 105, 321, \ldots$$ ■

Note that in order to find the 100th term of the sequence in Example 2 we must first find all 99 preceding terms.

EXAMPLE 3 ■ The Fibonacci Sequence

Find the first 11 terms of the sequence defined recursively by

$$F_n = F_{n-1} + F_{n-2}$$

with $F_1 = 1$ and $F_2 = 1$.

SOLUTION

To find F_n we need to find the two preceding terms F_{n-1} and F_{n-2}. Since we are given F_1 and F_2, we proceed as follows.

$$F_3 = F_2 + F_1 = 1 + 1 = 2$$

$$F_4 = F_3 + F_2 = 2 + 1 = 3$$

$$F_5 = F_4 + F_3 = 3 + 2 = 5$$

It is clear what is happening here. Each term is simply the sum of the two terms that precede it, so we can easily write down as many terms as we please. Here are the first 11 terms:

$$1, 1, 2, 3, 5, 8, 13, 21, 34, 55, 89, \ldots$$ ■

FIGURE 3

The Fibonacci sequence in the branching of a tree

A prime number is a natural number with no factor other than 1 and itself. By convention, 1 is not considered prime.

The sequence in Example 3 is called the **Fibonacci sequence,** named after the 13th-century Italian mathematician who used it to solve a problem about the breeding of rabbits (see Exercise 39). The sequence also occurs in numerous other applications in nature. In fact, so many phenomena behave like the Fibonacci sequence that one mathematical journal, the *Fibonacci Quarterly,* is devoted entirely to its properties. Figures 3 and 4 show two applications.

The sequences we have considered so far are defined by either a formula or a recursive procedure. But not all sequences can be defined in this way. For example, there is no known formula that produces the sequence of prime numbers:

$$2, 3, 5, 7, 11, 13, 17, 19, 23, \ldots$$

If we let a_n be the digit in the nth decimal place of the number π, we get the sequence

$$1, 4, 1, 5, 9, 2, 6, 5, 4, \ldots$$

Again, no simple formula can be given for finding the terms of this sequence.

Fibonacci (1175–1250) was born in Pisa, Italy, and educated in North Africa. He traveled widely in the Mediterranean area and learned the various methods then in use for writing numbers. On returning to Pisa in 1202, Fibonacci advocated the use of the Hindu-Arabic decimal system, the one we use today, over the Roman numeral system used in Europe in his time. His most famous book *Liber Abaci* is devoted to expounding the advantages of the Hindu-Arabic numerals. In fact, multiplication and division are so complicated using Roman numerals that it required a college degree to master these skills. Interestingly, in 1299 the city of Florence outlawed the use of the decimal system for merchants and other businesses, allowing numbers to be written using only Roman numerals or words. One can only speculate about the reasons for this.

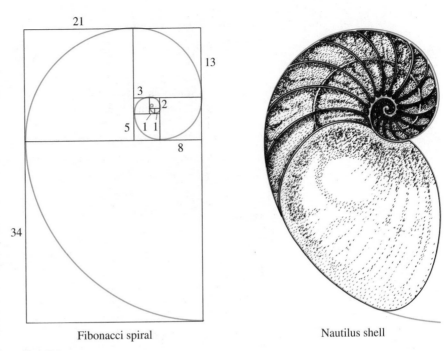

Fibonacci spiral   Nautilus shell

FIGURE 4

Finding patterns is a very important part of mathematics. Consider a sequence

$$1, 4, 9, 16, \ldots$$

Can you detect a pattern in these numbers? In other words, can you define a sequence whose first four terms are these numbers? The answer to this question seems easy; these numbers are the squares of the numbers 1, 2, 3, 4. Thus, the sequence we are looking for is defined by $a_n = n^2$. We point out, however, that this is not the *only* sequence whose first four terms are 1, 4, 9, 16. In other words, the answer to our problem is not unique (see Exercises 33 and 34). In the next example we are interested in finding an *obvious* sequence whose first few terms agree with the given ones.

EXAMPLE 4 ■ Finding the *n*th Term of a Sequence

Find the *n*th term of a sequence whose first several terms are given.

(a) $\frac{1}{2}, \frac{3}{4}, \frac{5}{6}, \frac{7}{8}, \ldots$ (b) $-2, 4, -8, 16, -32, \ldots$

SOLUTION

(a) We notice that the numerators of these fractions are the odd numbers and the denominators are the even numbers. Even numbers are of the form $2n$ and odd numbers are of the form $2n - 1$ (an odd number differs from an even number by 1). So, a sequence that has these numbers for its first four terms is given by

$$a_n = \frac{2n - 1}{2n}$$

(b) These numbers are powers of 2 and they alternate in sign, so a sequence that agrees with these terms is given by

$$a_n = (-1)^n 2^n$$

You should check that these formulas do indeed generate the given terms. ■

9.1   EXERCISES

1–14 ■ Find the first four terms and the 1000th term of the sequence.

1. $a_n = n + 1$

2. $a_n = 2n + 3$

3. $a_n = \dfrac{1}{n + 1}$

4. $a_n = n^2 + 1$

5. $a_n = \dfrac{(-1)^n}{n^2}$

6. $a_n = \dfrac{1}{n^2}$

7. $a_n = 1 + (-1)^n$

8. $a_n = 1 - \dfrac{1}{2^n}$

9. $a_n = (-2)^n$

10. $a_n = \dfrac{n}{n^2 + 1}$

11. $a_n = (-1)^{n+1} \dfrac{n}{n + 1}$

12. $a_n = 3n - 2$

13. $a_n = n^n$

14. $a_n = 3$

15–22 ■ Find the first five terms of the given sequence, which is defined recursively.

15. $a_n = 2(a_{n-1} - 2)$ and $a_1 = 3$

16. $a_n = \dfrac{a_{n-1}}{2}$ and $a_1 = -8$

17. $a_n = 2a_{n-1} + 1$ and $a_1 = 1$

18. $a_n = (a_{n-1})^2$ and $a_1 = 2$

19. $a_n = a_{n-1} - a_{n-2}$ and $a_1 = 0, a_2 = 1$

20. $a_n = \dfrac{1}{1 + a_{n-1}}$ and $a_1 = 1$

21. $a_n = a_{n-1} + a_{n-2}$ and $a_1 = 1, a_2 = 2$

22. $a_n = a_{n-1} + a_{n-2} + a_{n-3}$ and $a_1 = a_2 = a_3 = 1$

23–32 ■ Find the nth term of a sequence whose first several terms are given.

23. $2, 4, 8, 16, \ldots$

24. $-\frac{1}{3}, \frac{1}{9}, -\frac{1}{27}, \frac{1}{81}, \ldots$

25. $1, 4, 7, 10, \ldots$

26. $5, -25, 125, -625, \ldots$

27. $1, \frac{3}{4}, \frac{5}{9}, \frac{7}{16}, \frac{9}{25}, \ldots$

28. $\frac{3}{4}, \frac{4}{5}, \frac{5}{6}, \frac{6}{7}, \ldots$

29. $r, r^2, r^3, r^4, \ldots$

30. $a, a + d, a + 2d, a + 3d, \ldots$

31. $0, 2, 0, 2, 0, 2, \ldots$

32. $1, \frac{1}{2}, 3, \frac{1}{4}, 5, \frac{1}{6}, \ldots$

33–34 ■ These exercises explain why a finite number of terms does not uniquely determine a sequence.

33. (a) Show that the first four terms of the sequence $a_n = n^2$ are

$$1, 4, 9, 16, \ldots$$

   (b) Show that the first four terms of the sequence $a_n = n^2 + (n - 1)(n - 2)(n - 3)(n - 4)$ are also

$$1, 4, 9, 16, \ldots$$

   (c) Find a sequence whose first six terms are the same as those of $a_n = n^2$ but whose succeeding terms differ from this sequence.

34. Find two different sequences that begin

$$2, 4, 8, 16, \ldots$$

35. Find a formula for the nth term of the sequence

$$\sqrt{2}, \ \sqrt{2\sqrt{2}}, \ \sqrt{2\sqrt{2\sqrt{2}}}, \ \sqrt{2\sqrt{2\sqrt{2\sqrt{2}}}}, \ \ldots$$

[*Hint*: Write each term as a power of 2.]

36. Find the first 100 terms of the sequence defined by

$$a_{n+1} = \begin{cases} \dfrac{a_n}{2} & \text{if } a_n \text{ is an even number} \\ 3a_n + 1 & \text{if } a_n \text{ is an odd number} \end{cases}$$

and $a_1 = 11$.

37. Repeat Exercise 36 with $a_1 = 25$.

38. Find the first ten terms of the sequence defined recursively by

$$a_n = a_{n-a_{n-1}} + a_{n-a_{n-2}}$$

and having the following first two terms.

(a) $a_1 = 1, \quad a_2 = 1$

(b) $a_1 = 1, \quad a_2 = 2$

39. Fibonacci posed the following problem: Suppose that rabbits live forever and that every month each pair produces a new pair that becomes productive at age two months. If we start with one newborn pair, how many pairs of rabbits will we have in the nth month? Show that the answer is F_n, where F_n is the nth term of the Fibonacci sequence.

9.2 ARITHMETIC AND GEOMETRIC SEQUENCES

In this section we study two special kinds of sequences: arithmetic sequences, whose terms are generated by successively adding a fixed constant, and geometric sequences, whose terms are generated by successively multiplying by a fixed constant.

ARITHMETIC SEQUENCES

Perhaps the simplest way to generate a sequence is to start with a number a and add to it a fixed constant d, over and over again.

DEFINITION OF ARITHMETIC SEQUENCE

An **arithmetic sequence** is a sequence of the form

$$a, \ a + d, \ a + 2d, \ a + 3d, \ a + 4d, \ \ldots$$

The number a is the **first term,** and d is the **common difference** of the sequence.

The number d is called the common difference because any two consecutive terms of an arithmetic sequence differ by d.

EXAMPLE 1 ■ Arithmetic Sequences

(a) If $a = 2$ and $d = 3$, then we have the arithmetic sequence

$$2, \ 2 + 3, \ 2 + 6, \ 2 + 9, \ \ldots$$

or

$$2, \ 5, \ 8, \ 11, \ \ldots$$

Any two consecutive terms of this sequence differ by $d = 3$.

(b) Consider the sequence

$$9, 4, -1, -6, -11, \ldots$$

Here the common difference is $d = -5$. The terms of an arithmetic sequence decrease if the common difference is negative. ∎

It is easy to find a formula for the nth term of an arithmetic sequence:

the 1st term is $a + 0d$

the 2nd term is $a + 1d$

the 3rd term is $a + 2d$

the 4th term is $a + 3d$

$$\vdots \qquad\qquad \vdots$$

Continuing in this manner, we get the following formula.

THE nth TERM OF AN ARITHMETIC SEQUENCE

The nth term of the arithmetic sequence $a, a + d, a + 2d, a + 3d, \ldots$ is

$$a_n = a + (n - 1)d$$

For example, the nth terms of the sequences in parts (a) and (b) of Example 1 are

$$a_n = 2 + 3(n - 1)$$

and

$$a_n = 9 - 5(n - 1)$$

An arithmetic sequence is determined completely by the first term a and the common difference d. Thus, if we know the first two terms of an arithmetic sequence, then we can find a formula for the nth term, as the following example shows.

EXAMPLE 2 ∎ Finding Terms of an Arithmetic Sequence

Find the first six terms and the 300th term of the arithmetic sequence

$$13, 7, \ldots$$

SOLUTION

Since the first term is 13, we have $a = 13$. The common difference is $d = 7 - 13 = -6$. Thus, the nth term of this sequence is

$$a_n = 13 - 6(n - 1)$$

From this we find the first six terms:

$$13, 7, 1, -5, -11, -17, \ldots$$

The 300th term is $a_{300} = 13 - 6(299) = -1781$.

The next example shows that an arithmetic sequence is determined completely by *any* two of its terms.

EXAMPLE 3 ■ Finding Terms of an Arithmetic Sequence

The 11th term of an arithmetic sequence is 52 and the 19th term is 92. Find the 1000th term.

SOLUTION

To find the nth term of this sequence, we need to find a and d in the formula

$$a_n = a + (n - 1)d$$

From this formula, we get

$$a_{11} = a + (11 - 1)d = a + 10d$$

$$a_{19} = a + (19 - 1)d = a + 18d$$

Since $a_{11} = 52$ and $a_{19} = 92$, we get the two equations:

$$\begin{cases} 52 = a + 10d \\ 92 = a + 18d \end{cases}$$

Solving this system for a and d, we get $a = 2$ and $d = 5$. (Verify this.) Thus, the nth term of this sequence is

$$a_n = 2 + 5(n - 1)$$

The 1000th term is $a_{1000} = 2 + 5(999) = 4997$.

GEOMETRIC SEQUENCES

Another simple way of generating a sequence is to start with a number a and repeatedly multiply by a fixed nonzero constant r.

DEFINITION OF A GEOMETRIC SEQUENCE

A **geometric sequence** is a sequence of the form

$$a, ar, ar^2, ar^3, ar^4, \ldots$$

The number a is the **first term,** and r is the **common ratio** of the sequence.

The number r is called the common ratio because the ratio of any two consecutive terms of the sequence is r.

EXAMPLE 4 ■ Geometric Sequences

(a) If $a = 3$ and $r = 2$, then we have the geometric sequence

$$3, 3 \cdot 2, 3 \cdot 2^2, 3 \cdot 2^3, 3 \cdot 2^4, \ldots$$

or $\qquad\qquad$ 3, 6, 12, 24, 48, ...

Notice that the ratio of any two consecutive terms is $r = 2$.

(b) The sequence

$$2, -10, 50, -250, 1250, \ldots$$

is a geometric sequence with $a = 2$ and $r = -5$. When r is negative, the terms of the sequence alternate in sign.

(c) The sequence

$$1, \tfrac{1}{3}, \tfrac{1}{9}, \tfrac{1}{27}, \tfrac{1}{81}, \ldots$$

is a geometric sequence with $a = 1$ and $r = \tfrac{1}{3}$.

If $0 < r < 1$, then the terms of the sequence decrease, but if $r > 1$, then the terms increase. (What happens if $r = 1$?) ■

To find a formula for the nth term of a geometric sequence, we consider the pattern generated by the first few terms:

$$\begin{aligned} &\text{the 1st term is} \quad ar^0 \\ &\text{the 2nd term is} \quad ar^1 \\ &\text{the 3rd term is} \quad ar^2 \\ &\text{the 4th term is} \quad ar^3 \\ &\qquad\qquad \vdots \qquad\quad \vdots \end{aligned}$$

Since this pattern continues, we get the following formula.

THE nth TERM OF A GEOMETRIC SEQUENCE

The nth term of the geometric sequence $a, ar, ar^2, ar^3, ar^4, \ldots$ is

$$a_n = ar^{n-1}$$

Thus, the nth terms of the sequences in parts (a), (b), and (c) of Example 4 are, respectively,

$$a_n = 3(2)^{n-1} \qquad a_n = 2(-5)^{n-1} \qquad a_n = 1\left(\tfrac{1}{3}\right)^{n-1}$$

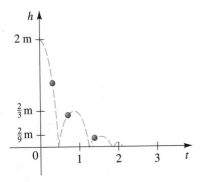

FIGURE 1

Geometric sequences occur naturally. Here is a simple example: Suppose that a ball has elasticity such that when it is dropped it bounces up one-third of the distance it has fallen. If this ball is dropped from a height of 2 m, then it bounces up to a height of $2(\frac{1}{3}) = \frac{2}{3}$ m. On its second bounce it returns to a height of $(\frac{2}{3})(\frac{1}{3}) = \frac{2}{9}$ m, and so on (see Figure 1). Thus, the height h_n that the ball reaches on its nth bounce is given by the geometric sequence

$$h_n = \tfrac{2}{3}\left(\tfrac{1}{3}\right)^{n-1} = 2\left(\tfrac{1}{3}\right)^n$$

We can find the nth term of a geometric sequence if we know any two terms, as the following examples show.

EXAMPLE 5 ■ **Finding Terms of a Geometric Sequence**

Find the eighth term of the geometric sequence 5, 15, 45, ...

SOLUTION

To find a formula for the nth term of this sequence, we need to find a and r. Clearly, $a = 5$. To find r we find the ratio of any two consecutive terms. For instance,

$$r = \tfrac{45}{15} = 3$$

Thus
$$a_n = 5(3)^{n-1}$$

The eighth term is

$$a_8 = 5(3)^{8-1} = 5(3)^7 = 10{,}935$$
■

EXAMPLE 6 ■ **Finding Terms of a Geometric Sequence**

The third term of a geometric series is $\frac{63}{4}$ and the sixth term is $\frac{1701}{32}$. Find the fifth term.

SOLUTION

Since this series is geometric, its nth term is given by the formula

$$a_n = ar^{n-1}$$

Thus
$$a_3 = ar^{3-1} = ar^2$$

and
$$a_6 = ar^{6-1} = ar^5$$

From the values we are given for these two terms, we get the system of equations:

$$\begin{cases} \frac{63}{4} = ar^2 \\ \frac{1701}{32} = ar^5 \end{cases}$$

One way to solve this system is to divide:

$$\frac{ar^5}{ar^2} = \frac{\frac{1701}{32}}{\frac{63}{4}}$$

So

$$r^3 = \frac{27}{8}$$

$$r = \frac{3}{2}$$

Substituting for r in the first equation, $\frac{63}{4} = ar^2$, gives

$$\frac{63}{4} = a\left(\frac{3}{2}\right)^2$$

$$a = 7$$

It follows that the nth term of this sequence is

$$a_n = 7\left(\frac{3}{2}\right)^{n-1}$$

and so the fifth term is

$$a_5 = 7\left(\frac{3}{2}\right)^{5-1} = 7\left(\frac{3}{2}\right)^4 = \frac{567}{16} \qquad \blacksquare$$

9.2   EXERCISES

1–10 ■ Determine the common difference, the fifth term, the nth term and the 100th term of the arithmetic sequence.

1. 2, 5, 8, 11, ...

2. 1, 5, 9, 13, ...

3. 4, 9, 14, 19, ...

4. 11, 8, 5, 2, ...

5. $-12, -8, -4, 0, \dots$

6. $\frac{7}{6}, \frac{5}{3}, \frac{13}{6}, \frac{8}{3}, \dots$

7. 25, 26.5, 28, 29.5, ...

8. 15, 12.3, 9.6, 6.9, ...

9. $2, 2 + s, 2 + 2s, 2 + 3s, \dots$

10. $-t, -t + 3, -t + 6, -t + 9, \dots$

11–20 ■ Determine the common ratio, the fifth term, and the nth term of the geometric sequence.

11. 2, 6, 18, 54, ...

12. $7, \frac{14}{3}, \frac{28}{9}, \frac{56}{27}, \dots$

13. $0.3, -0.09, 0.027, -0.0081, \dots$

14. $1, \sqrt{2}, 2, 2\sqrt{2}, \dots$

15. $144, -12, 1, -\frac{1}{12}, \dots$

16. $-8, -2, -\frac{1}{2}, -\frac{1}{8}, \dots$

17. $3, 3^{5/3}, 3^{7/3}, 27, \dots$

18. $t, \dfrac{t^2}{2}, \dfrac{t^3}{4}, \dfrac{t^4}{8}, \dots$

19. $1, s^{2/7}, s^{4/7}, s^{6/7}, \dots$

20. $5, 5^{c+1}, 5^{2c+1}, 5^{3c+1}, \dots$

21–32 ■ The first four terms of a sequence are given. Determine whether these terms can be the terms of an arithmetic sequence, a geometric sequence, or neither. Find the next term if the sequence is arithmetic or geometric.

21. $5, -3, 5, -3, \dots$

22. $\frac{1}{3}, 1, \frac{5}{3}, \frac{7}{3}, \dots$

23. $\sqrt{3}, 3, 3\sqrt{3}, 9, \dots$

24. $1, -1, 1, -1, \dots$

25. $2, -1, \frac{1}{2}, 2, \dots$

26. $-3, 1, 5, 8, \dots$

27. $x - 1, x, x + 1, x + 2, \dots$

28. $\dfrac{\sqrt{2}}{\sqrt{2} + 1}, \dfrac{2}{\sqrt{2} + 1}, \dfrac{4}{\sqrt{2} + 2}, \dfrac{4}{\sqrt{2} + 1}, \dots$

29. $16, 8, 4, 1, \dots$

30. $-3, -\frac{3}{2}, 0, \frac{3}{2}, \dots$

31. $1, \frac{3}{2}, 2, \frac{5}{2}, \dots$

32. $\sqrt{5}, \sqrt[3]{5}, \sqrt[6]{5}, 1, \dots$

33. The tenth term of an arithmetic sequence is $\frac{55}{2}$ and the second term is $\frac{7}{2}$. Find the first term.

34. The 12th term of an arithmetic sequence is 32 and the fifth term is 18. Find the 20th term.

35. The 100th term of an arithmetic sequence is 98 and the common difference is 2. Find the first three terms.

36. The first term of a geometric sequence is 8, and the second term is 4. Find the fifth term.

37. The first term of a geometric sequence is 3 and the third term is $\frac{4}{3}$. Find the fifth term.

38. The first term of a geometric sequence is 1 and the fifth term is 7^4. Find the common ratio, assuming it is positive.

39. The common ratio in a geometric sequence is $\frac{2}{5}$ and the fourth term is $\frac{5}{2}$. Find the third term.

40. The common ratio in a geometric sequence is $\frac{3}{2}$ and the fifth term is 1. Find the first three terms.

41. Which term of the arithmetic sequence 1, 4, 7, . . . is 88?

42. Which term of the geometric sequence 2, 6, 18, . . . is 118,098?

43. The first term of an arithmetic sequence is 1 and the common difference is 4. Is 11,937 a term of this sequence? If so, which term is it?

44. The second and the fifth terms of a geometric sequence are 10 and 1250, respectively. Is 31,250 a term of this sequence? If so, which term is it?

45. A ball is dropped from a height of 80 ft. The elasticity of this ball is such that it rebounds three-fourths of the distance it has fallen. How high does the ball go on the fifth bounce? Find a formula for how high the ball goes on the nth bounce.

46. A culture initially has 5000 bacteria and its size increases by 8% every hour. How many bacteria are present at the end of 5 h? Find a formula for the number of bacteria present after n hours.

47. Suppose that the value of a certain machine depreciates 15% each year. What is the value of the machine after six years if its original cost is $12,500?

48. A certain radioactive substance decays so that at the end of each year only 85% as much is present as at the beginning of the year. If 1 kg of the substance was present originally, find the amount that remains after five years. Find a formula for the amount that remains after n years.

49. A truck radiator holds five gallons and is filled with water. A gallon of water is removed from the radiator and replaced with a gallon of antifreeze; then, a gallon of the mixture is removed from the radiator and again

replaced by a gallon of antifreeze. This process is repeated indefinitely. How much water remains in the tank after this process is repeated three times? five times? n times?

50. Show that the sequence defined recursively by

$$a_{n+1} = \frac{3a_n + 1}{3} \quad \text{and} \quad a_1 = c$$

is an arithmetic sequence and find the common difference.

51. Show that the sequence defined recursively by $a_{n+1} = 3a_n$ and $a_1 = c$ is a geometric sequence and find the common ratio.

52. If $a_1, a_2, a_3, \ldots$ is a geometric sequence with common ratio r, show that the sequence

$$\frac{1}{a_1}, \frac{1}{a_2}, \frac{1}{a_3}, \ldots$$

is also a geometric sequence and find the common ratio.

53. If $a_1, a_2, a_3, \ldots$ is a geometric sequence with common ratio r, show that the sequence

$$a_1^2, a_2^2, a_3^2, \ldots$$

is also a geometric sequence and find the common ratio.

54. If $a_1, a_2, a_3, \ldots$ is a geometric sequence with common ratio $r > 0$ and $a_1 > 0$, show that the sequence

$$\log a_1, \log a_2, \log a_3, \ldots$$

is an arithmetic sequence and find the common difference.

55. If $a_1, a_2, a_3, \ldots$ is an arithmetic sequence with common difference d, show that the sequence

$$10^{a_1}, 10^{a_2}, 10^{a_3}, \ldots$$

is a geometric sequence and find the common ratio.

56. Show that a right triangle whose sides are in arithmetic progression is similar to a 3–4–5 triangle.

57. If the sum of three consecutive terms in an arithmetic sequence is 15 and their product is 80, find the three terms. [*Hint:* Let x denote the middle term.]

58. The sum of five consecutive terms of an arithmetic sequence is 260. Find the middle term.

59. The first three terms of an arithmetic sequence are $x - 1$, $x + 1$, and $3x + 3$. Find x.

60. If the product of three consecutive terms in a geometric sequence is 216 and their sum is 21, find the three terms.   [*Hint:* Let x denote the middle term.]

61. Let $a_1, a_2, a_3, \ldots$ be a geometric sequence with positive terms that satisfies

$$a_n = a_{n+1} + a_{n+2}$$

Find the common ratio.

62. If the numbers $a_1, a_2, \ldots, a_n$ form an arithmetic sequence, then $a_2, a_3, \ldots, a_{n-1}$ are *arithmetic means* between a_1 and a_n. Insert three arithmetic means between 2 and 14. (If only one arithmetic mean is inserted between two numbers, then it is their average.)

63. If the numbers $a_1, a_2, \ldots, a_n$ form a geometric sequence, then $a_2, a_3, \ldots, a_{n-1}$ are *geometric means* between a_1 and a_n. Insert three geometric means between 5 and 80.

64. A sequence is *harmonic* if the reciprocals of the terms of the sequence form an arithmetic sequence. Determine whether the following sequence is harmonic:

$$1, \tfrac{3}{5}, \tfrac{3}{7}, 3, \ldots$$

65. The *harmonic mean* of two numbers is the reciprocal of the arithmetic mean (see Exercise 62) of the reciprocals of the two numbers. Find the harmonic mean of 3 and 5.

9.3 SERIES

In this section we are interested in adding the terms of a sequence. For example, we might want to add the first 100 terms of the sequence

$$1, 2, 3, 4, \ldots$$

to find

$$1 + 2 + 3 + 4 + \cdots + 100$$

In order to tackle these problems we need a more compact way of writing such long sums.

SIGMA NOTATION

Given a sequence

$$a_1, a_2, a_3, a_4, \ldots$$

we can write the sum of the first n terms using **sigma notation.** This notation derives its name from the Greek letter Σ (capital sigma, corresponding to our S for sum). Sigma notation is used as follows:

$$\sum_{k=1}^{n} a_k = a_1 + a_2 + a_3 + a_4 + \cdots + a_n$$

The left side of this expression is read "The sum of a_k from $k = 1$ to $k = n$." The letter k is called the **index of summation,** or the **summation variable,** and the idea is to replace k in the expression after the sigma by the integers $1, 2, 3, \ldots, n$, and add the resulting expressions, arriving at the right side of the equation. For example, the sum of the squares of the first five integers can be written as

$$\sum_{k=1}^{5} k^2 = 1^2 + 2^2 + 3^2 + 4^2 + 5^2 = 55$$

Although we often use the letter k for the index of summation, any other letter can be used without affecting the sum. So this last sum can also be written using, say, j, for the index:

$$\sum_{j=1}^{5} j^2 = 1^2 + 2^2 + 3^2 + 4^2 + 5^2$$

The index of summation need not start at 1. For example,

$$\sum_{j=3}^{7} j^2 = 3^2 + 4^2 + 5^2 + 6^2 + 7^2$$

The following examples illustrate these concepts.

EXAMPLE 1 ■ Sigma Notation

Find each sum.

(a) $\displaystyle\sum_{k=1}^{3} k^3(k-1)$ (b) $\displaystyle\sum_{j=3}^{5} \frac{1}{j}$ (c) $\displaystyle\sum_{i=5}^{10} i$ (d) $\displaystyle\sum_{i=1}^{6} 2$

SOLUTION

(a) $\displaystyle\sum_{k=1}^{3} k^3(k-1) = 1^3(1-1) + 2^3(2-1) + 3^3(3-1) = 0 + 8 + 54 = 62$

(b) $\displaystyle\sum_{j=3}^{5} \frac{1}{j} = \frac{1}{3} + \frac{1}{4} + \frac{1}{5} = \frac{47}{60}$

(c) $\displaystyle\sum_{i=5}^{10} i = 5 + 6 + 7 + 8 + 9 + 10 = 45$

(d) $\displaystyle\sum_{i=1}^{6} 2 = 2 + 2 + 2 + 2 + 2 + 2 = 12$

EXAMPLE 2 ■ Writing Series in Sigma Notation

Write each sum using sigma notation.

(a) $1^3 + 2^3 + 3^3 + 4^3 + 5^3 + 6^3 + 7^3$

(b) $\sqrt{3} + \sqrt{4} + \sqrt{5} + \cdots + \sqrt{77}$

SOLUTION

(a) We can write

$$1^3 + 2^3 + 3^3 + 4^3 + 5^3 + 6^3 + 7^3 = \sum_{k=1}^{7} k^3$$

The ancient Greeks considered a line segment to be divided into **the golden ratio** if the ratio of the shorter part to the longer part is the same as the ratio of the longer part to the whole segment.

Thus, the segment shown is divided into the golden ratio if

$$\frac{1}{x} = \frac{x}{1+x}$$

This leads to a quadratic equation whose positive solution is

$$x = \frac{1 + \sqrt{5}}{2} \approx 1.618$$

This ratio occurs naturally in many places. For instance, psychological experiments show that the most pleasing shape of rectangle is one whose sides are in golden ratio. The ancient Greeks agreed with this, and built their temples in this ratio.

The golden ratio is related to Fibonacci numbers (see page 592). In fact, it can be shown using calculus (see Stewart, *Calculus, Third Edition*, page 607) that the ratio of two successive Fibonacci numbers

$$\frac{F_{n+1}}{F_n}$$

gets closer to the golden ratio the larger the value of n. Try finding this ratio for $n = 10$.

(b) A natural way to write this sum is

$$\sqrt{3} + \sqrt{4} + \sqrt{5} + \cdots + \sqrt{77} = \sum_{k=3}^{77} \sqrt{k}$$

However, there is no unique way of writing a sum in sigma notation. We also could write this last sum as

$$\sqrt{3} + \sqrt{4} + \sqrt{5} + \cdots + \sqrt{77} = \sum_{k=0}^{74} \sqrt{k+3}$$

or $$\sqrt{3} + \sqrt{4} + \sqrt{5} + \cdots + \sqrt{77} = \sum_{k=1}^{75} \sqrt{k+2}$$ ∎

The following properties of sums are natural consequences of properties of the real numbers.

PROPERTIES OF SUMS

Let $a_1, a_2, a_3, a_4, \ldots$ and $b_1, b_2, b_3, b_4, \ldots$ be sequences. Then for every positive integer n and any real number c,

1. $\displaystyle\sum_{k=1}^{n} (a_k + b_k) = \sum_{k=1}^{n} a_k + \sum_{k=1}^{n} b_k$

2. $\displaystyle\sum_{k=1}^{n} (a_k - b_k) = \sum_{k=1}^{n} a_k - \sum_{k=1}^{n} b_k$

3. $\displaystyle\sum_{k=1}^{n} ca_k = c\left(\sum_{k=1}^{n} a_k\right)$

To prove Property 1 we write out the left side of the equation to get

$$\sum_{k=1}^{n} (a_k + b_k) = (a_1 + b_1) + (a_2 + b_2) + (a_3 + b_3) + \cdots + (a_n + b_n)$$

Because addition is commutative and associative, we can rearrange the terms on the right side to read

$$\sum_{k=1}^{n} (a_k + b_k) = (a_1 + a_2 + a_3 + \cdots + a_n) + (b_1 + b_2 + b_3 + \cdots + b_n)$$

Rewriting the right side using sigma notation gives Property 1. Property 2 is proved in a similar manner. To prove Property 3 we use the Distributive Property:

$$\sum_{k=1}^{n} ca_k = ca_1 + ca_2 + ca_3 + \cdots + ca_n$$

$$= c(a_1 + a_2 + a_3 + \cdots + a_n)$$

$$= c\left(\sum_{k=1}^{n} a_k\right)$$

SERIES

When we add some of the terms of a sequence we get a *series* (or a *finite series*). For example,

$$\sum_{k=1}^{1,000,000} a_k$$

is the series that consists of the first one million terms of the sequence a_1, a_2, a_3, a_4, ... added together.

> ### DEFINITION OF A SERIES
>
> Let $a_1, a_2, a_3, a_4, \ldots$ be a sequence. A sum of the form
>
> $$a_1 + a_2 + a_3 + a_4 + \cdots + a_N$$
>
> is called a **series.** The number a_1 is called the **first term** of the series, a_2 the **second term,** and so on. The number that the series adds to is called the **sum** of the series.

In this section and the next we find the sums of series that consist of many terms. For example, suppose we want to find the sum of the series

$$\sum_{k=1}^{1000} \left(\frac{1}{k} - \frac{1}{k+1}\right)$$

Since this series has a thousand terms, it would take a long time to write down all the terms and add them together. We need a better way of doing this. So, we start by adding a few terms of the series and trying to detect a pattern as we add more and more terms.

To describe the sums we need some notation. We write S_1 for the first term of a series, S_2 for the sum of the first two terms, S_3 for the sum of the first three terms, and so on. These are called *partial sums* because we are only "partially" adding the terms of the series.

THE PARTIAL SUMS OF A SERIES

For the series

$$a_1 + a_2 + a_3 + a_4 + \cdots + a_N$$

the **partial sums** are

$$S_1 = a_1$$
$$S_2 = a_1 + a_2$$
$$S_3 = a_1 + a_2 + a_3$$
$$S_4 = a_1 + a_2 + a_3 + a_4$$
$$\cdot$$
$$\cdot$$
$$\cdot$$
$$S_n = a_1 + a_2 + a_3 + \cdots + a_n$$
$$\cdot$$
$$\cdot$$
$$\cdot$$

S_1 is called the **first partial sum**, S_2 is the **second partial sum**, and so on. S_n is called the *n*th partial sum. The sequence $S_1, S_2, S_3, \ldots, S_n, \ldots$ is called the **sequence of partial sums.**

EXAMPLE 3 ■ Finding the Sum of a Series

Find the sum of the series $\displaystyle\sum_{k=1}^{1000} \left(\frac{1}{k} - \frac{1}{k+1} \right)$.

SOLUTION

We begin by finding the first few partial sums of this series.

$$S_1 = \left(1 - \frac{1}{2} \right) \hspace{5cm} = 1 - \frac{1}{2}$$

$$S_2 = \left(1 - \frac{1}{2} \right) + \left(\frac{1}{2} - \frac{1}{3} \right) \hspace{3.5cm} = 1 - \frac{1}{3}$$

$$S_3 = \left(1 - \frac{1}{2} \right) + \left(\frac{1}{2} - \frac{1}{3} \right) + \left(\frac{1}{3} - \frac{1}{4} \right) \hspace{2cm} = 1 - \frac{1}{4}$$

$$S_4 = \left(1 - \frac{1}{2} \right) + \left(\frac{1}{2} - \frac{1}{3} \right) + \left(\frac{1}{3} - \frac{1}{4} \right) + \left(\frac{1}{4} - \frac{1}{5} \right) = 1 - \frac{1}{5}$$

Do we detect a pattern here? Of course, we have

$$S_n = 1 - \frac{1}{n+1}$$

It is now easy to find the sum of as many terms of this series as we please. For instance, the sum of the first 100 terms is

$$S_{100} = 1 - \frac{1}{101} = \frac{100}{101}$$

The sum of the series is the sum of all 1000 terms. Thus, the sum of this series is

$$S_{1000} = 1 - \frac{1}{1001} = \frac{1000}{1001}$$

∎

EXAMPLE 4 ■ Finding the Sum of a Series

Find the sum of the series $\displaystyle\sum_{k=1}^{100} \frac{1}{2^k}$.

SOLUTION

We first find a formula for the nth partial sum of this series. We do this by writing down the first few partial sums and trying to see a pattern:

$$S_1 = \frac{1} {2} \qquad\qquad\qquad\quad = \frac{1}{2}$$

$$S_2 = \frac{1}{2} + \frac{1}{4} \qquad\qquad\quad = \frac{3}{4}$$

$$S_3 = \frac{1}{2} + \frac{1}{4} + \frac{1}{8} \qquad\quad = \frac{7}{8}$$

$$S_4 = \frac{1}{2} + \frac{1}{4} + \frac{1}{8} + \frac{1}{16} = \frac{15}{16}$$

Notice that in the value of each partial sum the denominator is a power of 2 and the numerator is one less than the denominator. In general,

$$S_n = \frac{2^n - 1}{2^n} = 1 - \frac{1}{2^n}$$

Now we see that the sum of the given series is

$$S_{100} = 1 - \frac{1}{2^{100}}$$

∎

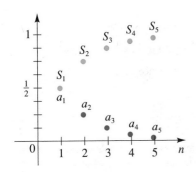

Graph of the sequence a_n and the sequence of partial sums S_n

9.3 **EXERCISES**

1–12 ■ Find the sum.

1. $\displaystyle\sum_{k=1}^{4} k$

2. $\displaystyle\sum_{k=1}^{4} k^2$

3. $\displaystyle\sum_{k=1}^{3} \frac{1}{k}$

4. $\displaystyle\sum_{j=1}^{100} (-1)^j$

5. $\displaystyle\sum_{i=1}^{8} [1 + (-1)^i]$

6. $\displaystyle\sum_{i=4}^{12} 10$

7. $\displaystyle\sum_{k=1}^{5} 2^{k-1}$

8. $\displaystyle\sum_{j=1}^{10} \frac{3}{j+2}$

9. $\displaystyle\sum_{m=0}^{4} (-3)^{m+2}$

10. $\displaystyle\sum_{i=1}^{3} i 2^i$

11. $\displaystyle\sum_{m=3}^{5} (2^m + m^2)$

12. $\displaystyle\sum_{k=1}^{1} k^{100}$

13–20 ■ Write the sum without using sigma notation.

13. $\displaystyle\sum_{k=1}^{5} \sqrt{k}$

14. $\displaystyle\sum_{i=0}^{4} \frac{2i-1}{2i+1}$

15. $\displaystyle\sum_{k=0}^{6} \sqrt{k+4}$

16. $\displaystyle\sum_{k=3}^{100} x^k$

17. $\displaystyle\sum_{i=1}^{8} i x^{i+1}$

18. $\displaystyle\sum_{k=6}^{9} k(k+3)$

19. $\displaystyle\sum_{j=1}^{n} (-1)^{j+1} x^j$

20. $\displaystyle\sum_{j=1}^{8} \frac{x^j}{j^2}$

21–30 ■ Write the sum using sigma notation.

21. $1 + 2 + 3 + 4 + \cdots + 100$

22. $2 + 4 + 6 + \cdots + 2n$

23. $\dfrac{1}{2\ln 2} - \dfrac{1}{3\ln 3} + \dfrac{1}{4\ln 4} - \dfrac{1}{5\ln 5} + \cdots + \dfrac{1}{100\ln 100}$

24. $1 + x + x^2 + x^3 + \cdots + x^{100}$

25. $1 - \dfrac{x}{3} + \dfrac{x^2}{9} - \dfrac{x^3}{27} + \dfrac{x^4}{81} - \dfrac{x^5}{243}$

26. $\dfrac{10}{15} + \dfrac{11}{16} + \dfrac{12}{17} + \cdots + \dfrac{100}{105}$

27. $1 - 2x + 3x^2 - 4x^3 + 5x^4 + \cdots - 100x^{99}$

28. $\dfrac{1}{1 \cdot 2} + \dfrac{1}{2 \cdot 3} + \dfrac{1}{3 \cdot 4} + \cdots + \dfrac{1}{999 \cdot 1000}$

29. $1 \cdot 2 \cdot 3 + 2 \cdot 3 \cdot 4 + 3 \cdot 4 \cdot 5 + \cdots + 97 \cdot 98 \cdot 99$

30. $\dfrac{\sqrt{1}}{1^2} + \dfrac{\sqrt{2}}{2^2} + \dfrac{\sqrt{3}}{3^2} + \cdots + \dfrac{\sqrt{n}}{n^2}$

31–34 ■ Find the first six partial sums $S_1, S_2, S_3, S_4, S_5, S_6$ of the series.

31. $1 + 3 + 5 + 7 + \cdots + 1001$

32. $1^2 + 2^2 + 3^3 + \cdots + 600^2$

33. $\displaystyle\sum_{k=1}^{100} \frac{1}{3^k}$

34. $\displaystyle\sum_{j=1}^{20} (-1)^j$

35–41 ■ Find a formula for the nth partial sum S_n of the series, and then find the sum of the series. (See Examples 3 and 4.)

35. $\displaystyle\sum_{k=1}^{1000} \left(\frac{1}{k+1} - \frac{1}{k+2} \right)$

36. $\displaystyle\sum_{k=1}^{100} \left(\frac{1}{2k-1} - \frac{1}{2k+1} \right)$

37. $\displaystyle\sum_{k=1}^{20} \frac{2}{3^k}$

38. $\displaystyle\sum_{j=1}^{20} \frac{4}{5^j}$

39. $\displaystyle\sum_{i=1}^{99} \left(\sqrt{i} - \sqrt{i+1} \right)$

40. $\displaystyle\sum_{k=1}^{20} (2^{k-1} - 2^k)$

41. $\displaystyle\sum_{k=1}^{999999} \log\left(\frac{k}{k+1} \right)$ ⠀⠀ [*Hint:* Use a property of logarithms to write the kth term as a difference.]

42. Let $a_1, a_2, a_3, \ldots$ be a sequence.
(a) Show that

$$\sum_{k=1}^{n} [a_k - a_{k+1}] = a_1 - a_{n+1}$$

⠀⠀⠀A series of this form is called a **telescoping series**.
(b) Which of the series in Exercises 35–41 are telescoping?

9.4 ARITHMETIC AND GEOMETRIC SERIES

In this section we find formulas for the sums of series whose terms form an arithmetic or geometric sequence.

ARITHMETIC SERIES

An **arithmetic series** is a series whose terms form an arithmetic sequence. We will find a formula for the nth partial sum of an arithmetic series.

We begin with a simple example: Suppose we want to find the sum of the numbers 1, 2, 3, 4, . . . , 100, that is,

$$\sum_{k=1}^{100} k$$

When the famous mathematician C. F. Gauss was a schoolboy his teacher asked the class this question and expected that it would keep the students busy for a long time. But Gauss answered the question almost immediately. His idea was that, since we are adding numbers that are produced according to a fixed pattern, there must also be a pattern (or formula) for finding the sum. He started by writing the numbers from 1 to 100 and below them the same numbers in reverse order. Writing S for the sum and adding corresponding terms gives

$$
\begin{aligned}
S &= 1 + 2 + 3 + \cdots + 98 + 99 + 100 \\
S &= 100 + 99 + 98 + \cdots + 3 + 2 + 1 \\
\hline
2S &= 101 + 101 + 101 + \cdots + 101 + 101 + 101
\end{aligned}
$$

It follows that $2S = 100(101) = 10{,}100$ and so $S = 5050$.

Of course, the sequence of natural numbers 1, 2, 3, . . . is an arithmetic sequence (with $a = 1$ and $d = 1$), and the method used for summing the first 100 terms of this series can be used to find a formula for the nth partial sum of any arithmetic series. We want to find the sum of the first n terms of the arithmetic sequence whose terms are $a_k = a + (k - 1)d$, that is, we want to find

$$S_n = \sum_{k=1}^{n} [a + (k - 1)d]$$

$$= a + (a + d) + (a + 2d) + (a + 3d) + \cdots + [a + (n - 1)d]$$

Using Gauss's method, we write

$$
\begin{aligned}
S_n &= a + (a + d) + \cdots + [a + (n - 2)d] + [a + (n - 1)d] \\
S_n &= [a + (n - 1)d] + [a + (n - 2)d] + \cdots + (a + d) + a \\
\hline
2S_n &= [2a + (n - 1)d] + [2a + (n - 1)d] + \cdots + [2a + (n - 1)d] + [2a + (n - 1)d]
\end{aligned}
$$

There are n identical terms on the right side of this equation, so

$$2S_n = n[2a + (n - 1)d]$$

$$S_n = \frac{n}{2}[2a + (n - 1)d]$$

Notice that $a_n = a + (n - 1)d$ is the last term of this series. So, we can write

$$S_n = \frac{n}{2}[a + a + (n - 1)d] = n\left(\frac{a + a_n}{2}\right)$$

This last formula says that the sum of the first n terms of an arithmetic series is the average of the first and last terms multiplied by n, the number of terms in the series. We now summarize this result.

SUM OF AN ARITHMETIC SERIES

The sum S_n of the first n terms of an arithmetic series

$$S_n = a + (a + d) + (a + 2d) + (a + 3d) + \cdots + [a + (n - 1)d]$$

is given by either of the following formulas.

1. $S_n = \dfrac{n}{2}[2a + (n - 1)d]$

2. $S_n = n\left(\dfrac{a + a_n}{2}\right)$

EXAMPLE 1 ■ Finding the Sum of an Arithmetic Series

Find the sum of the first 50 odd numbers.

SOLUTION

The odd numbers form an arithmetic sequence whose first term is $a = 1$ and whose nth term is $a_n = 2n - 1$, so the 50th odd number is $a_{50} = 2(50) - 1 = 99$. Substituting in Formula 2 for the sum of an arithmetic series, we get

$$S_{50} = 50\left(\frac{a + a_{50}}{2}\right) = 50\left(\frac{1 + 99}{2}\right) = 50 \cdot 50 = 2500$$

■

EXAMPLE 2 ■ Finding the Sum of an Arithmetic Series

Find the sum of the first 40 terms of the arithmetic sequence

$$3, 7, 11, 15, \ldots$$

SOLUTION

For this arithmetic sequence, $a = 3$ and $d = 4$. Using Formula 1 for the sum of an arithmetic series, we get

$$S_{40} = \frac{40}{2}[2(3) + (40 - 1)4] = 20(6 + 156) = 3240$$

∎

EXAMPLE 3 ■ A Problem involving Arithmetic Series

An arithmetic series has first term 5 and 50th term 103. How many terms of this series must be added to get 572?

SOLUTION

We first find the common difference. Since, for an arithmetic sequence, $a_n = a + (n - 1)d$, we get

$$a_{50} = a + (50 - 1)d$$

so $$103 = 5 + 49d$$

Solving for d gives $d = 2$.

We are asked to find n when $S_n = 572$. Using Formula 1 for the sum of an arithmetic series and substituting for S_n, a, and d gives

$$572 = \frac{n}{2}[2 \cdot 5 + (n - 1)2]$$

Solving for n, we have

$$572 = 5n + n(n - 1)$$

$$n^2 + 4n - 572 = 0$$

$$(n - 22)(n + 26) = 0$$

This gives $n = 22$ or $n = -26$. But since n is a *number* of terms in a sequence, we must have $n = 22$.

∎

GEOMETRIC SERIES

A **geometric series** is a series whose terms form a geometric sequence. So, adding the first n terms of the geometric sequence

$$a, ar, ar^2, ar^3, ar^4, \ldots, ar^{n-1}, \ldots$$

gives us the geometric series

$$S_n = \sum_{k=1}^{n} ar^{k-1} = a + ar + ar^2 + ar^3 + ar^4 + \cdots + ar^{n-1}$$

To find a formula for S_n, we multiply S_n by r and subtract from S_n, to get

$$S_n = a + ar + ar^2 + ar^3 + ar^4 + \cdots + ar^{n-1}$$

$$rS_n = \quad\ ar + ar^2 + ar^3 + ar^4 + \cdots + ar^{n-1} + ar^n$$

$$\overline{S_n - rS_n = a - ar^n}$$

So,

$$S_n(1 - r) = a(1 - r^n)$$

$$S_n = \frac{a(1 - r^n)}{1 - r} \qquad (r \neq 1)$$

We summarize this result.

SUM OF A GEOMETRIC SERIES

The sum S_n of the first n terms of a geometric series

$$S_n = a + ar + ar^2 + ar^3 + ar^4 + \cdots + ar^{n-1} \qquad (r \neq 1)$$

is given by

$$S_n = a\frac{1 - r^n}{1 - r}$$

EXAMPLE 4 ■ Finding the Sum of a Geometric Series

Find the sum of the first five terms of the geometric sequence

$$1, 0.7, 0.49, 0.343, \ldots$$

SOLUTION

The required sum is the sum of the first five terms of a geometric series with $a = 1$ and $r = 0.7$. Using the formula for the sum of a geometric series with $n = 5$, we get

$$S_5 = 1\frac{1 - (0.7)^5}{1 - 0.7} = 2.7731$$

Thus, the sum of the first five terms of this sequence is 2.7731. ■

EXAMPLE 5 ■ Finding the Sum of a Geometric Series

Find the sum of the series $\displaystyle\sum_{k=1}^{5} 7\left(-\frac{2}{3}\right)^k$.

SOLUTION

The given series is a geometric series with first term $a = 7(-\frac{2}{3}) = -\frac{14}{3}$ and common ratio $r = -\frac{2}{3}$, and the series has five terms. Thus, by the formula for the sum of a geometric series, we have

$$S_5 = -\frac{14}{3} \frac{\left[1 - \left(-\frac{2}{3}\right)^5\right]}{1 - \left(-\frac{2}{3}\right)} = -\frac{14}{3} \frac{1 + \frac{32}{243}}{\frac{5}{3}} = -\frac{770}{243}$$

9.4 EXERCISES

1–8 ■ Find the sum S_n of the arithmetic series that satisfies the given conditions.

1. $a = 1$, $d = 2$, $n = 10$

2. $a = 3$, $d = 2$, $n = 12$

3. $a = 4$, $d = 2$, $n = 20$

4. $a = 100$, $d = -5$, $n = 8$

5. $a_1 = 55$, $d = 12$, $n = 10$

6. $a_2 = 8$, $a_5 = 9.5$, $n = 15$

7. $a_3 = 980$, $a_{10} = 910$, $n = 5$

8. $a_4 = 21$, $d = 3$, $n = 10$

9–14 ■ Find the sum S_n of the geometric series that satisfies the given conditions.

9. $a = 5$, $r = 2$, $n = 6$

10. $a = \frac{2}{3}$, $r = \frac{1}{3}$, $n = 4$

11. $a_3 = 28$, $a_6 = 224$, $n = 6$

12. $a_2 = \frac{10}{3}$, $a_4 = \frac{40}{27}$, $r < 0$, $n = 4$

13. $a_3 = 0.18$, $r = 0.3$, $n = 5$

14. $a_2 = 0.12$, $a_5 = 0.00096$, $n = 4$

15–20 ■ Find the sum of the arithmetic series.

15. $1 + 5 + 9 + \cdots + 401$

16. $-3 + \left(-\frac{3}{2}\right) + 0 + \frac{3}{2} + 3 + \cdots + 30$

17. $0.7 + 2.7 + 4.7 + \cdots + 56.7$

18. $-10 - 9.9 - 9.8 - \cdots - 0.1$

19. $\displaystyle\sum_{k=0}^{10} (3 + 0.25k)$ **20.** $\displaystyle\sum_{n=0}^{20} (1 - 2n)$

21–26 ■ Find the sum of the geometric series.

21. $1 + 3 + 9 + \cdots + 2187$

22. $1 - \frac{1}{2} + \frac{1}{4} - \frac{1}{8} + \cdots - \frac{1}{512}$

23. $0.7 + 0.49 + 0.343 + \cdots + 0.16807$

24. $1 - \sqrt{2} + 2 - 2\sqrt{2} + \cdots + 32$

25. $\displaystyle\sum_{k=0}^{10} 3\left(\frac{1}{2}\right)^k$ **26.** $\displaystyle\sum_{j=0}^{5} 7\left(\frac{3}{2}\right)^j$

27–38 ■ Determine whether the series is arithmetic or geometric, and find its sum.

27. $4 + 2.4 + 1.44 + \cdots + 0.5184$

28. $2 + 5 + 8 + \cdots + 32$

29. $1 - x + x^2 - x^3 + \cdots + x^{20}$

30. $1 - \sqrt{3} + 3 - 3\sqrt{3} + \cdots + 243$

31. $2 + 4 + 6 + \cdots + 1000$

32. $\sqrt{5} + 2\sqrt{5} + 3\sqrt{5} + \cdots + 100\sqrt{5}$

33. $\frac{1}{2} + 1 + \frac{3}{2} + \cdots + 64$

34. $\frac{1}{2} + 1 + 2 + \cdots + 64$

35. $\displaystyle\sum_{i=0}^{8} (1 + \sqrt{2}\,i)$ **36.** $\displaystyle\sum_{k=0}^{8} 2\left(\sqrt{3}\right)^k$

37. $\displaystyle\sum_{n=0}^{8} 5^{n/3}$ **38.** $\displaystyle\sum_{i=0}^{8} \sqrt{5} \cdot 2^i$

39. An arithmetic sequence has first term $a = 5$ and common difference $d = 2$. How many terms of this sequence must be added to get 2700?

40. A geometric sequence has first term $a = 1$ and common ratio $r = -\frac{1}{2}$. How many terms of this sequence must be added to get $\frac{341}{512}$?

41. The sum of the first four terms of a geometric series is 50 and the common ratio is $r = \frac{1}{2}$. Find the first term.

42. The sum of the first ten terms of an arithmetic series is 100 and the first term is 1. Find the tenth term.

43. The sum of the first 20 terms of an arithmetic series is 155 and the first term is 3. Find the common difference.

44. The sum of the first and 20th terms of an arithmetic sequence is 182. Find the sum of the first 20 terms.

45. The second term in a geometric series is $\frac{14}{3}$ and the fifth term is $\frac{112}{81}$. Find the sum of the first four terms.

46. The common ratio in a certain geometric sequence is $r = 0.2$ and the sum of the first four terms is 1248. Find the first term.

47. Find the product of the numbers

$$10^{1/10},\ 10^{2/10},\ 10^{3/10},\ 10^{4/10},\ \ldots,\ 10^{19/10}$$

48. Find the sum of the first ten terms of the sequence

$$a + b,\ a^2 + 2b,\ a^3 + 3b,\ a^4 + 4b,\ \ldots$$

49. A very patient women wishes to become a billionaire. She decides to follow a simple scheme: She puts aside 1 cent the first day, 2 cents the second day, 4 cents the third day, and so on, doubling the number of cents each day. How much money will she have at the end of 30 days? How many days will it take this woman to realize her wish?

50. A ball is dropped from a height of 9 ft. The elasticity of the ball is such that it always bounces up one-third of the distance it has fallen.
(a) Find the total distance the ball has traveled at the instant it hits the ground the fifth time.

(b) Find a formula for the total distance the ball has traveled at the instant it hits the ground the nth time.

51. When an object is allowed to fall freely near the surface of the earth, the gravitational pull is such that the object falls 16 ft in the first second, 48 ft in the next second, 80 ft in the next second, and so on.
(a) Find the total distance the ball falls in 6 seconds.
(b) Find a formula for the total distance the ball falls in n seconds.

52. In the well-known song "The Twelve Days of Christmas," a person gives his sweetheart k gifts on the kth day for each of the 12 days of Christmas. The person also repeats each gift identically on each subsequent day. Thus, on the twelfth day the sweetheart receives a gift for the first day, 2 gifts for the second, 3 gifts for the third, and so on. Show that the number of gifts received on the 12th day is an arithmetic series, and find its sum.

53. The following is a well-known children's rhyme:

> As I was going to St. Ives
> I met a man with seven wives;
> Every wife had seven sacks;
> Every sack had seven cats;
> Every cat had seven kits;
> Kits, cats, sacks, and wives,
> How many were going to St. Ives?

Assuming that the entire group is actually going to St. Ives, show that the answer to the question in the children's rhyme is the sum of a geometric series, and find its sum.

9.5   ANNUITIES AND INSTALLMENT BUYING

Many financial transactions involve payments that are made at regular intervals. For example, if you deposit $100 each month in an interest-bearing account, what will the value of the account be at the end of five years? If you borrow $100,000 to buy a house, how much will the monthly payments be in order to pay off the loan in 30 years? These questions involve the sum of a series of numbers, and we use the results of the preceding section to answer them here.

THE AMOUNT OF AN ANNUITY

An **annuity** is a sum of money that is paid in regular equal payments. Although the word *annuity* suggests annual (or yearly) payments, they can be made semi-annually, quarterly, monthly, or at some other regular interval. Payments are usually made at the end of the payment interval. The **amount of an annuity** is the sum of all the individual payments from the time of the first payment until the last payment is made, together with all the interest. We denote this sum by A_f (the subscript f here is used to denote *final* amount).

EXAMPLE 1 ■ Calculating the Amount of an Annuity

An investor deposits $400 every December 15 and June 15 for ten years into an account that earns interest at the rate of 8% per year, compounded semi-annually. How much will be in the account immediately after the last payment?

SOLUTION

We need to find the amount of an annuity consisting of 20 semiannual payments of $400 each. Since the interest rate is 8% per year, compounded semi-annually, the interest rate per time period is $i = 0.04$. The first payment is in the account for 19 time periods, the second for 18 time periods, and so on. The last payment receives no interest. The situation can be illustrated by the time line in Figure 1.

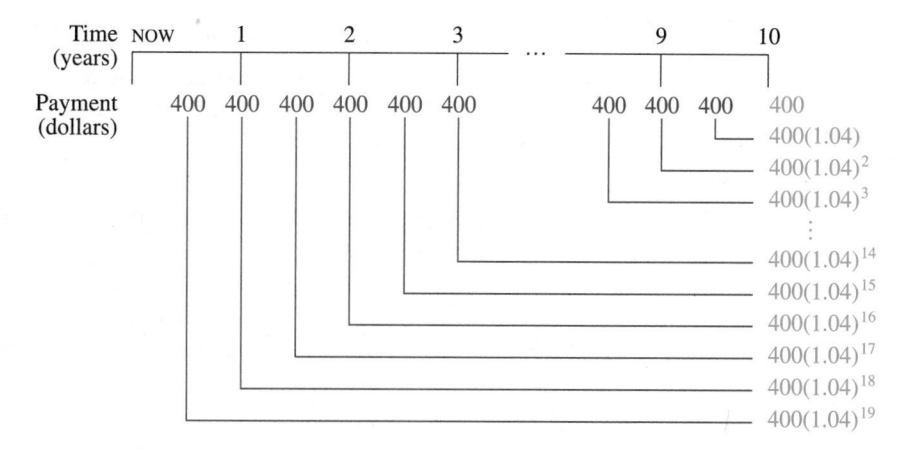

FIGURE 1

The amount A_f of the annuity is the sum of these 20 amounts. Thus

$$A_f = 400 + 400(1.04) + 400(1.04)^2 + \cdots + 400(1.04)^{19}$$

But this is a geometric series with $a = 400$, $r = 1.04$, and $n = 20$, so

$$A_f = 400 \, \frac{1 - (1.04)^{20}}{1 - 1.04} \approx 11{,}911.23$$

Thus, the amount of the annuity after the last payment is $11,911.23. ■

In general, the regular annuity payment is called the **periodic rent** and is denoted by R. We also let i denote the interest rate per time period and n the number of payments. *We always assume that the time period in which interest is compounded is equal to the time between payments.* By the same reasoning as in Example 1, we see that the amount A_f of an annuity is

$$A_f = R + R(1 + i) + R(1 + i)^2 + \cdots + R(1 + i)^{n-1}$$

Since this is a geometric series with n terms, $a = R$, and $r = 1 + i$, the formula for the sum of a geometric series gives

$$A_f = R \frac{1 - (1 + i)^n}{1 - (1 + i)} = R \frac{1 - (1 + i)^n}{-i} = R \frac{(1 + i)^n - 1}{i}$$

AMOUNT OF AN ANNUITY

The amount A_f of an annuity consisting of n regular equal payments of size R with interest rate i per time period is given by

$$A_f = R \frac{(1 + i)^n - 1}{i}$$

EXAMPLE 2 ■ Calculating the Amount of an Annuity

How much money should be invested every month at 12% per year, compounded monthly, in order to have $4000 in 18 months?

SOLUTION

In this problem $i = 0.12/12 = 0.01$, $A_f = 4000$, and $n = 18$. We need to find the amount R of each payment. By the formula for the amount of an annuity,

$$4000 = R \frac{(1 + 0.01)^{18} - 1}{0.01}$$

Solving for R, we get

$$R = \frac{4000(0.01)}{(1 + 0.01)^{18} - 1} \approx \frac{40}{1.196147 - 1} \approx 203.928$$

Thus, the monthly investment should be $203.93. ■

THE PRESENT VALUE OF AN ANNUITY

If you were to receive $10,000 five years from now, it would be worth much less than getting $10,000 right now. This is because of the interest you could accumulate during that time. What smaller amount would you be willing to accept *now* instead of receiving $10,000 in five years? This is the amount of money that, together with interest, would be worth $10,000 in five years. The amount we are looking for here is called the *discounted value,* or *present value.* If the interest rate is 8% per year, compounded quarterly, then the interest per time period is $i = 0.08/4 = 0.02$, and there are $4 \times 5 = 20$ time periods. If we let PV denote the present value, then by the formula for compound interest (Section 5.2) we have

$$10,000 = PV(1 + i)^n = PV(1 + 0.02)^{20}$$

so $$PV = 10,000(1 + 0.02)^{-20} \approx 6729.713$$

Thus, in this situation, the present value of $10,000 is $6729.71. This reasoning leads to a general formula for present value:

$$PV = A(1 + i)^{-n}$$

Similarly, the **present value of an annuity** is the amount A_p that must be invested now at the interest rate i per time period in order to provide n payments, each of amount R. Clearly, A_p is the sum of the present values of each of the individual payments (see Exercise 22). Another way of finding A_p is to note that A_p is the present value of A_f:

$$A_p = A_f(1 + i)^{-n} = R\,\frac{(1 + i)^n - 1}{i}(1 + i)^{-n} = R\,\frac{1 - (1 + i)^{-n}}{i}$$

THE PRESENT VALUE OF AN ANNUITY

The **present value** A_p of an annuity consisting of n regular equal payments of size R and interest rate i per time period is given by

$$A_p = R\,\frac{1 - (1 + i)^{-n}}{i}$$

EXAMPLE 3 ■ Calculating the Present Value of an Annuity

A person wins $10,000,000 in the California lottery, and the amount is paid in yearly installments of half a million dollars each for 20 years. What is the present value of his winnings? We assume that he can earn 10% interest, compounded annually.

SOLUTION

Since the amount won is paid as an annuity, we need to find its present value. Here $i = 0.1$, $R = \$500,000$, and $n = 20$. Thus,

$$A_p = 500,000\,\frac{1 - (1 + 0.1)^{-20}}{0.1} \approx 4{,}256{,}781.859$$

This means that the winner really won only $4,256,781.86 if it were paid immediately. ■

INSTALLMENT BUYING

When you buy a house or a car by installment, the payments you make are an annuity whose present value is the amount of the loan.

EXAMPLE 4 ■ The Amount of a Loan

A student wishes to buy a car. He can afford to pay $200 per month but has no money for a down payment. If he can make these payments for four years and if the interest rate is 12%, what price car can he buy?

SOLUTION

The payments the student makes constitute an annuity whose present value is the price of the car (which is also the amount of the loan, in this case). Here we have $i = 0.12/12 = 0.01$, $R = 200$, $n = 12 \times 4 = 48$. Thus,

$$A_p = R\,\frac{1 - (1 + i)^{-n}}{i} = 200\,\frac{1 - (1 + 0.01)^{-48}}{0.01} \approx 7594.792$$

Thus, the student can buy a car worth $7594.79. ■

When a bank makes a loan that is to be repaid with regular equal payments R, then the payments form an annuity whose present value A_p is the amount of the loan. So, to find the size of the payments we solve for R in the formula for the amount of an annuity. This gives the following formula for R.

INSTALLMENT BUYING

If a loan A_p is to be repaid in n regular equal payments with interest rate i per time period, then the size R of each payment is given by

$$R = \frac{iA_p}{1 - (1 + i)^{-n}}$$

EXAMPLE 5 ■ Calculating Monthly Mortgage Payments

A couple borrows $100,000 at 9% interest as a mortgage loan on a house. They expect to make monthly payments for 30 years to repay the loan. What is the size of each payment?

SOLUTION

The payments form an annuity with present value $A_p = \$100{,}000$. Also, $i = 0.09/12 = 0.0075$, and $n = 12 \times 30 = 360$. We are looking for the amount R of each payment. From the formula for installment buying, we get

$$R = \frac{iA_p}{1 - (1 + i)^{-n}}$$

$$= \frac{(0.0075)(100,000)}{1 - (1 + 0.0075)^{-360}} \approx 804.623$$

Thus, the monthly payments are \$804.62. ∎

We now give an example that illustrates the use of graphing devices in solving problems related to installment buying.

EXAMPLE 6 ■ Calculating the Interest Rate from the Size of Monthly Payments

A car dealer sells a new car for \$18,000. He also offers to sell the same car for payments of \$405 per month for five years. What interest rate is this car dealer charging?

SOLUTION

The payments form an annuity with present value $A_p = \$18,000$, $R = 405$, and $n = 12 \times 5 = 60$. To find the interest rate, we must solve for i in the equation

$$R = \frac{iA_p}{1 - (1 + i)^{-n}}$$

A little experimentation will convince you that it is not possible to algebraically solve this equation for i. So, to find i we use a graphing device to graph R as a function of the interest rate x, and we then use the graph to find the interest rate corresponding to the value of R we are interested in (\$405 in this case). Since $i = x/12$, we graph the function

$$R(x) = \frac{\dfrac{x}{12}(18,000)}{1 - \left(1 + \dfrac{x}{12}\right)^{-60}}$$

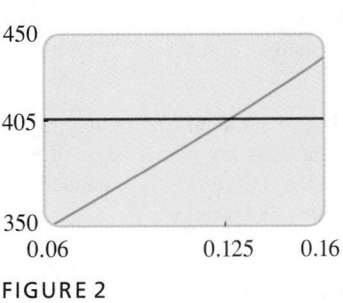

FIGURE 2

in the viewing rectangle $[0.06, 0.16] \times [350, 450]$. We also graph the horizontal line $R(x) = 405$ in the same viewing rectangle. Then, by moving the cursor to the point of intersection of the two graphs, we find that the corresponding x-value is approximately 0.125. Thus, the interest rate is about $12\frac{1}{2}\%$ per year. ∎

9.5 EXERCISES

1. Find the amount of an annuity that consists of 10 annual payments of $1000 each into an account that pays 6% interest per year.

2. Find the amount of an annuity that consists of 24 monthly payments of $500 each into an account that pays 8% interest per year.

3. Find the amount of an annuity that consists of 20 annual payments of $5000 each into an account that pays interest of 12% per year.

4. Find the amount of an annuity that consists of 20 semi-annual payments of $500 each into an account that pays 6% interest per year, compounded semiannually.

5. Find the amount of an annuity that consists of 16 quarterly payments of $300 each into an account that pays 8% interest per year, compounded quarterly.

6. How much money should be invested every quarter at 10% per year, compounded quarterly, in order to have $5000 in 2 years?

7. How much money should be invested monthly at 6% per year, compounded monthly, in order to have $2000 in eight months?

8. What is the present value of an annuity that consists of 20 semiannual payments of $1000 at the interest rate of 9% per year, compounded semiannually?

9. How much money must be invested now at 9% per year, compounded semiannually, to fund an annuity of 20 payments of $200 each, paid every 6 months, the first payment being 6 months from now?

10. A 55-year-old man deposits $50,000 to fund an annuity with an insurance company. The money will be invested at 8% per year, compounded semiannually. He is to draw semiannual payments until he reaches age 65. What is the amount of each payment?

11. A woman wants to borrow $12,000 in order to buy a car. She wants to repay the loan by monthly installments for four years. If the interest rate on this loan is $10\frac{1}{2}$% per year, compounded monthly, what is the amount of each payment?

12. What is the monthly payment on a 30-year mortgage of $80,000 at 9% interest? What is the monthly payment

on this same mortgage if it is to be repaid over a 15-year period?

13. What is the monthly payment on a 30-year mortgage of $100,000 at 8% interest per year, compounded monthly? What is the total amount paid on this loan over the 30-year period?

14. A couple can afford to make a monthly mortgage payment of $650. If the mortgage rate is 9% and the couple intends to secure a 30-year mortgage, how much can they borrow?

15. A couple secures a 30-year loan of $100,000 at $9\frac{3}{4}$% per year, compounded monthly, to buy a house.
   (a) What is the amount of the monthly payment?
   (b) What is the total amount that will be paid by the couple over the 30-year period?
   (c) If, instead of taking the loan, the couple deposits the monthly payments in an account that pays $9\frac{3}{4}$% interest per year, compounded monthly, how much will be in the account at the end of the 30-year period?

16. Jane agrees to buy a car by making a down payment of $2000 and payments of $220 per month for three years. It the interest rate is 8% per year, compounded monthly, what is the actual purchase price of the car?

17. Mike buys a ring for his fiancee by paying $30 per month for one year. If the interest rate is 10% per year, compounded monthly, what is the price of the ring?

18. Janet decides to buy a $12,500 car by paying $420 a month for three years. Assuming that interest is compounded monthly, what interest rate is she paying on the car loan?

19. John buys a stereo system that is advertised for $640. He agrees to pay $32 per month for two years. Assuming that interest is compounded monthly, what interest rate is he paying?

20. A woman purchases a $2000 diamond ring by paying a down payment of $200 and monthly installments of $88 for two years. Assuming that interest is compounded monthly, what interest rate is she paying?

21. An item at a department store is priced at $189.99 and can be bought by making 20 payments of $10.50. Find the interest rate, assuming that interest is compounded monthly.

$$A_p = \frac{R}{1+i} + \frac{R}{(1+i)^2} + \frac{R}{(1+i)^3} + \cdots + \frac{R}{(1+i)^n}$$

22. (a) Draw a time line as in Example 1 to show that the present value of an annuity is the sum of the present values of each of the payments, that is,

(b) Use part (a) to derive the formula for A_p given in the text.

9.6 INFINITE GEOMETRIC SERIES

So far we have been discussing series with a finite number of terms. Let us write down a series with an infinite number of terms:

$$a_1 + a_2 + a_3 + a_4 + \cdots$$

The dots mean that we are to continue the addition indefinitely. A series of this kind is called an **infinite series.**

WHAT IS AN INFINITE SERIES?

What meaning can we attach to the sum of infinitely many numbers? It seems at first that it is not possible to add infinitely many numbers and arrive at a finite number. But consider the following problem. You have a cake and you want to eat it by first eating half the cake, then eating half of what remains, then again eating half of what remains. This process can continue indefinitely because at each stage some of the cake remains. (See Figure 1.)

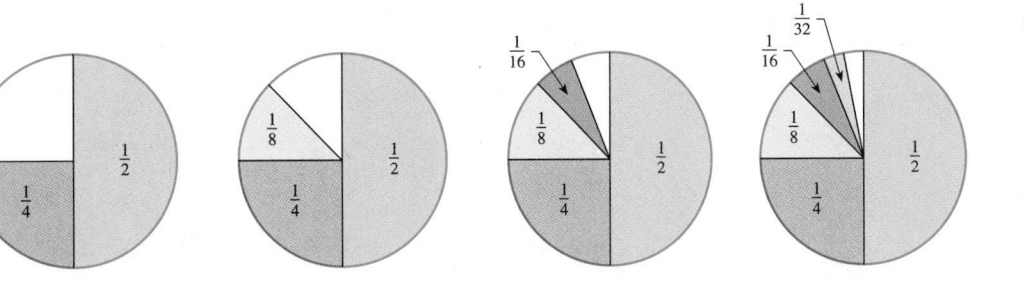

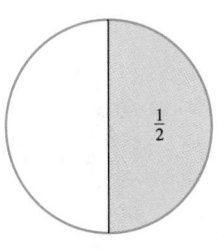
FIGURE 1

Does this mean that it is impossible to eat all of the cake? Of course not. Let us write down what you have eaten from this cake:

$$\frac{1}{2} + \frac{1}{4} + \frac{1}{8} + \frac{1}{16} + \cdots + \frac{1}{2^n} + \cdots$$

This is an infinite series, and we note two things about it: First, from Figure 1 it is clear that no matter how many terms of this series we add, the total will never exceed 1. Second, the more terms of this series we add, the closer the sum is to 1 (see Figure 1). This suggests that the number 1 can be written as the sum of infinitely many smaller numbers:

$$1 = \frac{1}{2} + \frac{1}{4} + \frac{1}{8} + \frac{1}{16} + \cdots + \frac{1}{2^n} + \cdots$$

To make this more precise, let us look at the partial sums of this series:

$$S_1 = \frac{1}{2} \qquad\qquad\qquad = \frac{1}{2}$$

$$S_2 = \frac{1}{2} + \frac{1}{4} \qquad\qquad = \frac{3}{4}$$

$$S_3 = \frac{1}{2} + \frac{1}{4} + \frac{1}{8} \qquad = \frac{7}{8}$$

$$S_4 = \frac{1}{2} + \frac{1}{4} + \frac{1}{8} + \frac{1}{16} = \frac{15}{16}$$

$$\vdots$$

and, in general (see Example 4 of Section 9.3),

$$S_n = 1 - \frac{1}{2^n}$$

As n gets larger and larger, we are adding more and more of the terms of this series. Intuitively, as n gets larger, S_n gets closer to the sum of the series. Now notice that as n gets large, $1/2^n$ gets closer and closer to 0. Thus, S_n gets close to $1 - 0 = 1$. Using the notation of Section 4.8, we can write

$$S_n \to 1 \qquad \text{as} \qquad n \to \infty$$

In general, if S_n gets close to a finite number S as n gets large, we say that S is the **sum of the infinite series.**

INFINITE GEOMETRIC SERIES

We call an infinite series of the form

$$a + ar + ar^2 + ar^3 + ar^4 + \cdots + ar^{n-1} + \cdots$$

an **infinite geometric series.** We can apply the reasoning used earlier to find the sum of an infinite geometric series. The nth partial sum of such a series is given by the formula for the sum of a geometric series in Section 9.4:

$$S_n = a\,\frac{1 - r^n}{1 - r} \qquad (r \neq 1)$$

It can be shown that if $|r| < 1$, then r^n gets close to 0 as n gets large (you can easily convince yourself of this using a calculator). It follows that S_n gets close to $a/(1 - r)$ as n gets large, or

$$S_n \to \frac{a}{1 - r} \quad \text{as} \quad n \to \infty$$

Thus, the sum of this infinite geometric series is $a/(1 - r)$. We summarize this result.

SUM OF AN INFINITE GEOMETRIC SERIES

If $|r| < 1$, then the infinite geometric series

$$a + ar + ar^2 + ar^3 + ar^4 + \cdots + ar^{n-1} + \cdots$$

has the sum

$$S = \frac{a}{1 - r}$$

EXAMPLE 1 ■ **Finding the Sum of an Infinite Geometric Series**

Find the sum of the infinite geometric series

$$2 + \frac{2}{5} + \frac{2}{25} + \frac{2}{125} + \cdots + \frac{2}{5^n} + \cdots$$

SOLUTION

We use the formula for the sum of an infinite geometric series. In this case, $a = 2$ and $r = \frac{1}{5}$. Thus, the sum of this infinite series is

$$S = \frac{2}{1 - \frac{1}{5}} = \frac{5}{2}$$

■

EXAMPLE 2 ■ **Writing a Repeated Decimal as a Fraction**

Find the fraction that represents the rational number $2.3\overline{51}$.

SOLUTION

This repeating decimal can be written as a series:

$$\frac{23}{10} + \frac{51}{1000} + \frac{51}{100,000} + \frac{51}{10,000,000} + \frac{51}{1,000,000,000} + \cdots$$

The terms of this series after the first term form an infinite geometric series with

$$a = \frac{51}{1000} \qquad \text{and} \qquad r = \frac{1}{100}$$

Thus, the sum of this part of the series is

$$S = \frac{\frac{51}{1000}}{1 - \frac{1}{100}} = \frac{\frac{51}{1000}}{\frac{99}{100}} = \frac{51}{1000} \cdot \frac{100}{99} = \frac{51}{990}$$

So,

$$2.3\overline{51} = \frac{23}{10} + \frac{51}{990} = \frac{2328}{990} = \frac{388}{165} \qquad \blacksquare$$

9.6 EXERCISES

1–10 ■ Find the sum of the infinite geometric series.

1. $1 + \dfrac{1}{3} + \dfrac{1}{9} + \dfrac{1}{27} + \cdots$

2. $1 - \dfrac{1}{2} + \dfrac{1}{4} - \dfrac{1}{8} + \cdots$

3. $1 - \dfrac{1}{3} + \dfrac{1}{9} - \dfrac{1}{27} + \cdots$

4. $\dfrac{2}{5} + \dfrac{4}{25} + \dfrac{8}{125} + \cdots$

5. $\dfrac{1}{3^6} + \dfrac{1}{3^8} + \dfrac{1}{3^{10}} + \dfrac{1}{3^{12}} + \cdots$

6. $3 - \dfrac{3}{2} + \dfrac{3}{4} - \dfrac{3}{8} + \cdots$

7. $-\dfrac{100}{9} + \dfrac{10}{3} - 1 + \dfrac{3}{10} - \cdots$

8. $\dfrac{1}{\sqrt{2}} + \dfrac{1}{2} + \dfrac{1}{2\sqrt{2}} + \dfrac{1}{4} + \cdots$

9. $5^{4/3} - 5^{5/3} + 5^{6/3} - 5^{7/3} + \cdots$

10. $\dfrac{1}{1 + \sqrt{2}} - 1 - \dfrac{1}{1 - \sqrt{2}} - \cdots$

11–16 ■ Express the repeating decimal as a fraction.

11. $0.777\ldots$ **12.** $0.2\overline{53}$

13. $0.030303\ldots$

14. $2.11\overline{25}$

15. $0.\overline{112}$

16. $0.123123123\ldots$

17. The elasticity of a ball is such that it rebounds two-thirds the distance that it falls. Use an infinite geometric series to approximate the total distance the ball travels, after being dropped from 12 ft above the ground, until it stops bouncing.

18. A certain ball rebounds to half the height from which it is dropped. Use an infinite geometric series to approximate the total distance the ball travels, after being dropped from 1 m above the ground, until it comes to rest.

19. If the ball in Exercise 18 is dropped from a height of 8 ft, then 1 s is required for its first complete bounce—from the instant it first touches the ground until it next touches the ground. Each subsequent complete bounce requires $1/\sqrt{2}$ as long as the preceding complete bounce. Use an infinite geometric series to estimate the time from the instant the ball first touches the ground until it stops bouncing.

20. The midpoints of the sides of a square of side 1 are joined to form a new square. This procedure is repeated for each new square. (See the figure.)
    (a) Find the sum of the areas of all the squares.
    (b) Find the sum of the perimeters of all the squares.

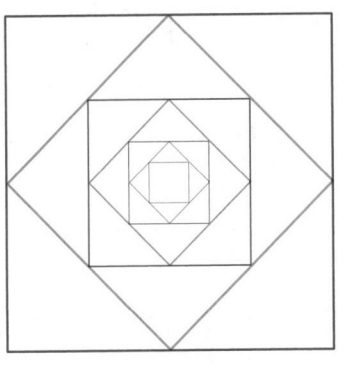

21. A circular disk of radius R is cut out of paper, as shown in figure (a). Two disks of radius $\frac{1}{2}R$ are cut out of paper and placed on top of the first disk, as in figure (b), and then four disks of radius $\frac{1}{4}R$ are placed on these two disks [figure (c)]. Assuming that this process can be repeated indefinitely, find the total area of all the disks.

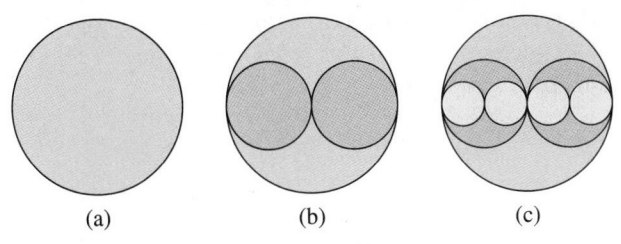

(a)   (b)   (c)

22. An **annuity in perpetuity** is one that continues forever. Such annuities are useful in setting up scholarship funds to ensure that the award continues.
    (a) Draw a time line (as in Example 1 of Section 9.5) to show that the amount of money to be invested now (A_p) at interest rate i per time period in order to set up an annuity in perpetuity of amount R per time period is

    $$A_p = \frac{R}{1 + i} + \frac{R}{(1 + i)^2} + \frac{R}{(1 + i)^3} + \cdots + \frac{R}{(1 + i)^n} + \cdots$$

    (b) Find the sum of the infinite series in part (a) to show that

    $$A_p = \frac{R}{i}$$

23. How much money must be invested now at 10% per year, compounded annually, to provide an annuity of $5000 every year in perpetuity? The first payment is due in one year. (Refer to Exercise 22.)

24. How much money must be invested now at 8% per year, compounded quarterly, in order to provide an annuity of $3000 per year in perpetuity? The first payment is due in one year. (Refer to Exercise 22.)

25. A yellow square of side 1 is divided into nine smaller squares and the middle square is colored blue as shown in the figure. Each of the smaller yellow squares is in turn divided into nine squares and each middle square is colored blue. If this process is continued indefinitely, what is the total area colored blue?

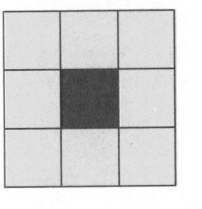  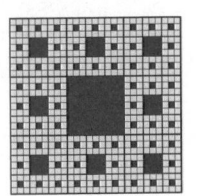

9.7 **MATHEMATICAL INDUCTION**

There are two aspects to mathematics—discovery and proof—and both are of equal importance. It is necessary to discover something before attempting to prove it, and we can only be certain of its truth once it has been proved. In this section we look more carefully at the relationship between these two parts of mathematics.

CONJECTURE AND PROOF

Let us try a simple experiment. We add up more and more of the odd numbers as follows:

$$1 = 1$$

$$1 + 3 = 4$$

$$1 + 3 + 5 = 9$$

$$1 + 3 + 5 + 7 = 16$$

$$1 + 3 + 5 + 7 + 9 = 25$$

What do you notice about the numbers on the right side of these equations? They are in fact all perfect squares. These equations say that

The sum of the first 1 odd number is 1^2.

The sum of the first 2 odd numbers is 2^2.

The sum of the first 3 odd numbers is 3^2.

The sum of the first 4 odd numbers is 4^2.

The sum of the first 5 odd numbers is 5^2.

This leads naturally to the following question: Is it true that for every natural number n, the sum of the first n odd numbers is n^2? Could this remarkable property be true? We could try a few more numbers and find that the pattern persists for the first 6, 7, 8, 9, and 10 odd numbers. At this point we feel quite sure that this is always true, so we make a conjecture:

The sum of the first n odd numbers is n^2.

Since we know that the nth odd number is $2n - 1$, we can write this statement more precisely as

$$1 + 3 + 5 + \cdots + (2n - 1) = n^2$$

It is important to realize that this is still a conjecture. We cannot conclude by checking a finite number of cases that a property is true for all numbers (there are infinitely many). To see this more clearly, suppose someone tells us that he has added up the first trillion odd numbers and found out that they do *not* add up to one trillion squared. What would you tell this person? It would be silly to tell him that you are sure it is true because you have already checked the first five cases. You could, however, take out paper and pencil and start checking it yourself, but this task would probably take the rest of your life. The tragedy would be that after completing this task you would still not be sure of the truth of the conjecture. Do you see why?

Herein lies the power of mathematical proof. A **proof** is a clear argument that demonstrates the truth of a statement beyond doubt. We consider here a special kind of proof called *mathematical induction* that will help us prove statements like the one we have just considered. But before we do this, let us try another experiment.

Consider the polynomial

$$p(n) = n^2 - n + 41$$

Let us find some of the values of $p(n)$ for natural numbers n:

$$p(1) = 41 \qquad p(2) = 43 \qquad p(3) = 47$$

$$p(4) = 53 \qquad p(5) = 61 \qquad p(6) = 71$$

$$p(7) = 83 \qquad p(8) = 97 \qquad p(9) = 113$$

We notice this time that all the values that we have calculated are prime numbers. We might want to conclude at this point that all the values of this polynomial are prime. But let's try a few more values:

$$p(10) = 131 \quad \text{(prime)} \qquad p(11) = 151 \quad \text{(prime)} \qquad p(12) = 173 \quad \text{(prime)}$$

$$p(13) = 197 \quad \text{(prime)} \qquad p(14) = 223 \quad \text{(prime)} \qquad p(15) = 251 \quad \text{(prime)}$$

At this point we are getting tired of calculating values. We make the following conjecture:

> For every natural number n, $p(n)$ is a prime number.

If we try values for n from 16 to 40, we will find that again $p(n)$ is prime. But our conjecture is too hasty. It is easily seen that $p(41) = 41^2 - 41 + 41 = 41^2$ is *not* prime. This is the first n for which $p(n)$ is not prime! So our conjecture is false.

This illustrates clearly that we cannot be certain of the truth of a statement by checking special cases. We need a convincing argument to determine the truth of a statement—a proof.

MATHEMATICAL INDUCTION

We consider a special kind of proof called **mathematical induction.** Here is how it works: Suppose we have a statement that says something about all natural numbers n. Let's call this statement P. For example, we could consider the statement

P: For every natural number n, the sum of the first n odd numbers is n^2.

Since this statement is about *all* natural numbers, it contains infinitely many statements; we will call them $P(1)$, $P(2)$, ...

$P(1)$: The sum of the first 1 odd number is 1^2.

$P(2)$: The sum of the first 2 odd numbers is 2^2.

$P(3)$: The sum of the first 3 odd numbers is 3^2.

       .   .
       .   .
       .   .

How can we prove all of these statements at once? Mathematical induction is a clever way of doing just that.

The crux of the idea is this: Suppose we can prove that whenever one of these statements is true, then the one following it in the list is also true. In other words,

For every k, if $P(k)$ is true, then $P(k + 1)$ is true.

This is called the **induction step** because it leads us from the truth of one statement to the next. Now, suppose that we can also prove that

$P(1)$ is true.

The induction step now leads us through the following chain of statements:

$P(1)$ is true, so $P(2)$ is true.

$P(2)$ is true, so $P(3)$ is true.

$P(3)$ is true, so $P(4)$ is true.

$$\vdots \qquad \vdots$$

So we see that if both of these statements are proved, then statement P is proved for all n. We summarize this important method of proof.

PRINCIPLE OF MATHEMATICAL INDUCTION

For each natural number n, let $P(n)$ be a statement depending on n. Suppose that the following two conditions are satisfied.

1. $P(1)$ is true.

2. For every natural number k, if $P(k)$ is true, then $P(k + 1)$ is true.

Then $P(n)$ is true for all natural numbers n.

To apply this principle, there are two steps:

Step 1 Prove that $P(1)$ is true.

Step 2 Assume that $P(k)$ is true and use this assumption to prove that $P(k + 1)$ is true.

Notice that in Step 2 we do not prove that $P(k)$ is true. We only show that *if* $P(k)$ is true, *then* $P(k + 1)$ is also true. The assumption that $P(k)$ is true is called the **induction hypothesis.**

FOOTIES AT THE INDUCTION STEP

IF I WERE ON RUNG NUMBER K THEN IT WOULD BE EASY TO GET TO THE NEXT RUNG.

SURE, AND THEN YOU COULD CLIMB THE WHOLE LADDER.

BUT ONE HAS TO MANAGE TO REACH THE FIRST RUNG TO BEGIN WITH.

© 1979 National Council of Teachers of Mathematics. Used by permission. Courtesy of Andrejs Dunkels, Sweden.

We now use mathematical induction to prove that the conjecture we made at the beginning of this section is true.

EXAMPLE 1 ■ A Proof by Mathematical Induction

Prove that for all natural numbers n,

$$1 + 3 + 5 + \cdots + (2n - 1) = n^2$$

SOLUTION

Let $P(n)$ denote the statement $1 + 3 + 5 + \cdots + (2n - 1) = n^2$.

Step 1 We need to show that $P(1)$ is true. But $P(1)$ is simply the statement that $1 = 1^2$, which is of course true.

Step 2 We assume that $P(k)$ is true. Thus, our induction hypothesis is

$$1 + 3 + 5 + \cdots + (2k - 1) = k^2$$

We want to use this to show that $P(k + 1)$ is true, that is,

$$1 + 3 + 5 + \cdots + (2k - 1) + [2(k + 1) - 1] = (k + 1)^2$$

Blaise Pascal (1623–1662) is considered one of the most versatile minds in modern history. He was a writer and philosopher as well as a gifted mathematician and physicist. Among his contributions that appear in this book are the theory of probability, Pascal's triangle, and the Principle of Mathematical Induction. Pascal's father, himself a mathematician, believed that his son should not study mathematics until he was 15 or 16. But at age 12 Blaise insisted on learning geometry, and proved most of its elementary theorems himself. At 19 he invented the first mechanical adding machine. In 1647, after writing a major treatise on the conic sections, he abruptly abandoned mathematics because he felt that his intense studies were contributing to his ill health. He devoted himself instead to frivolous recreations such as gambling, but this only served to pique his interest in probability. In 1654 he miraculously survived a carriage accident in which his horses ran off a bridge. Taking this to be a sign from God, he entered a monastery, where he pursued theology and philosophy, writing his

(continued)

[Note that we get $P(k + 1)$ by substituting $k + 1$ for each n in the statement $P(n)$.] We start with the left side and use the induction hypothesis to obtain the right side of the equation:

$$1 + 3 + 5 + \cdots + (2k - 1) + [2(k + 1) - 1]$$

$$= [1 + 3 + 5 + \cdots + (2k - 1)] + [2(k + 1) - 1] \qquad \text{Group the first } k \text{ terms}$$

$$= k^2 + [2(k + 1) - 1] \qquad \text{Induction hypothesis}$$

$$= k^2 + [2k + 2 - 1] \qquad \text{Distributive Property}$$

$$= k^2 + 2k + 1 \qquad \text{Simplify}$$

$$= (k + 1)^2 \qquad \text{Factor}$$

Thus, $P(k + 1)$ follows from $P(k)$ and this completes the induction step.

Having proved Steps 1 and 2, we conclude by the Principle of Mathematical Induction that $P(n)$ is true for all natural numbers n. ∎

EXAMPLE 2 ■ A Proof by Mathematical Induction

Prove that for every natural number n,

$$1 + 2 + 3 + \cdots + n = \frac{n(n + 1)}{2}$$

SOLUTION

We let $P(n)$ denote the statement $1 + 2 + 3 + \cdots + n = n(n + 1)/2$. We want to show that $P(n)$ is true for all natural numbers n.

Step 1 We need to show that $P(1)$ is true. But $P(1)$ says that

$$1 = \frac{1(1 + 1)}{2}$$

and this statement is clearly true.

Step 2 Assume that $P(k)$ is true. Thus, our induction hypothesis is

$$1 + 2 + 3 + \cdots + k = \frac{k(k + 1)}{2}$$

We want to use this to show that $P(k + 1)$ is true, that is

$$1 + 2 + 3 + \cdots + k + (k + 1) = \frac{(k + 1)[(k + 1) + 1]}{2}$$

To show this we start with the left side and use the induction

famous *Pensées*. He also continued his mathematical research. He valued faith and intuition more than reason as the source of truth, declaring that "the heart has its own reasons, which reason cannot know."

hypothesis to obtain the right side:

$$1 + 2 + 3 + \cdots + k + (k + 1)$$

$$= [1 + 2 + 3 + \cdots + k] + (k + 1) \qquad \text{Group the first } k \text{ terms}$$

$$= \frac{k(k + 1)}{2} + (k + 1) \qquad \text{Induction hypothesis}$$

$$= (k + 1)\left(\frac{k}{2} + 1\right) \qquad \text{Factor } k + 1$$

$$= (k + 1)\left(\frac{k + 2}{2}\right) \qquad \text{Common denominator}$$

$$= \frac{(k + 1)[(k + 1) + 1]}{2} \qquad \text{Write } k + 2 \text{ as } k + 1 + 1$$

Thus, $P(k + 1)$ follows from $P(k)$ and this completes the induction step.

Having proved Steps 1 and 2, we conclude by the Principle of Mathematical Induction that $P(n)$ is true for all natural numbers n. ■

It might happen that a statement $P(n)$ is false for the first few natural numbers, but true from some number on. For example, we may want to prove that $P(n)$ is true for $n \geq 5$. Notice that if we prove that $P(5)$ is true, then this fact, together with the induction step, would imply the truth of $P(5), P(6), P(7), \ldots$. The next example illustrates this point.

EXAMPLE 3 ■ Proving an Inequality by Mathematical Induction

Prove that $4n < 2^n$ for all $n \geq 5$.

SOLUTION

Let $P(n)$ denote the statement $4n < 2^n$.

Step 1 $P(5)$ is the statement that $4 \cdot 5 < 2^5$, or $20 < 32$, which is true.

Step 2 Assume that $P(k)$ is true. Thus, our induction hypothesis is

$$4k < 2^k$$

We want to use this to show that $P(k + 1)$ is true, that is,

$$4(k + 1) < 2^{k+1}$$

To show this we start with the left side of the inequality and use the induction hypothesis to show that it is less than the right side. For

$k \geq 5$, we have

$$4(k + 1) = 4k + 4$$

$$< 2^k + 4 \qquad \text{Induction hypothesis}$$

$$< 2^k + 4k \qquad 4 < 4k$$

$$< 2^k + 2^k \qquad \text{Induction hypothesis}$$

$$= 2 \cdot 2^k$$

$$= 2^{k+1} \qquad \text{Property of exponents}$$

Thus, $P(k + 1)$ follows from $P(k)$ and this completes the induction step.

Having proved Steps 1 and 2, we conclude by the Principle of Mathematical Induction that $P(n)$ is true for all natural numbers n. ∎

9.7   EXERCISES

1–12 ■ Use mathematical induction to prove that the formula is true for all natural numbers n.

1. $2 + 4 + 6 + \cdots + 2n = n(n + 1)$

2. $1 + 4 + 7 + \cdots + 3(n - 2) = \dfrac{n(3n - 1)}{2}$

3. $5 + 8 + 11 + \cdots + (3n + 2) = \dfrac{n(3n + 7)}{2}$

4. $1^2 + 2^2 + 3^2 + \cdots + n^2 = \dfrac{n(n + 1)(2n + 1)}{6}$

5. $1 \cdot 2 + 2 \cdot 3 + 3 \cdot 4 + \cdots + n(n + 1)$
$= \dfrac{n(n + 1)(n + 2)}{3}$

6. $1 \cdot 3 + 2 \cdot 4 + 3 \cdot 5 + \cdots + n(n + 2)$
$= \dfrac{n(n + 1)(2n + 7)}{6}$

7. $1^3 + 2^3 + 3^3 + \cdots + n^3 = \dfrac{n^2(n + 1)^2}{4}$

8. $1^3 + 3^3 + 5^3 + \cdots + (2n - 1)^3 = n^2(2n^2 - 1)$

9. $2^3 + 4^3 + 6^3 + \cdots + (2n)^3 = 2n^2(n + 1)^2$

10. $\dfrac{1}{1 \cdot 2} + \dfrac{1}{2 \cdot 3} + \dfrac{1}{3 \cdot 4} + \cdots + \dfrac{1}{n(n + 1)} = \dfrac{n}{(n + 1)}$

11. $1 \cdot 2 + 2 \cdot 2^2 + 3 \cdot 2^3 + 4 \cdot 2^4 + \cdots + n \cdot 2^n$
$= 2[1 + (n - 1)2^n]$

12. $1 + 2 + 2^2 + \cdots + 2^{n-1} = 2^n - 1$

13. Show that $n^2 + n$ is divisible by 2 for all natural numbers n.

14. Show that $5^n - 1$ is divisible by 4 for all natural numbers n.

15. Show that $n^2 - n + 41$ is odd for all natural numbers n.

16. Show that $n^3 - n + 3$ is divisible by 3 for all natural numbers n.

17. Show that $8^n - 3^n$ is divisible by 5 for all natural numbers n.

18. Show that $3^{2n} - 1$ is divisible by 8 for every natural number n.

19. Prove that $n < 2^n$ for all natural numbers n.

20. Prove that $(n + 1)^2 < 2n^2$ for all natural numbers $n \geq 3$.

21. Prove that if $x > -1$, then $(1 + x)^n \geq 1 + nx$ for all natural numbers n.

22. Show that $100n \leq n^2$ for all $n \geq 100$.

23. Let $a_{n+1} = 3a_n$ and $a_1 = 5$. Show that $a_n = 5 \cdot 3^{n-1}$ for all natural numbers n.

24. A sequence is defined recursively by $a_{n+1} = 3a_n - 8$ and $a_1 = 4$. Find an explicit formula for a_n and then use mathematical induction to prove that the formula you found is true.

25. Show that $x - y$ is a factor of $x^n - y^n$ for all natural numbers n.
[*Hint:* $x^{k+1} - y^{k+1} = x^k(x - y) + (x^k - y^k)y$]

26. Show that $x + y$ is a factor of $x^{2n-1} + y^{2n-1}$ for all natural numbers n.

27. Determine whether each of the following statements is true or false. If the statement is true, prove it. If it is false, give an example where it fails.
(a) $p(n) = n^2 - n + 11$ is prime for all n.
(b) $n^2 > n$ for all $n \geq 2$.
(c) $2^{2n+1} + 1$ is divisible by 3 for all $n \geq 1$.
(d) $n^3 \geq (n + 1)^2$ for all $n \geq 2$.
(e) $n^3 - n$ is divisible by 3 for all $n \geq 2$.
(f) $n^3 - 6n^2 + 11n$ is divisible by 6 for all $n \geq 1$.

28–32 ■ F_n denotes the nth term of the Fibonacci sequence discussed in Section 9.1. Use mathematical induction to prove the statement.

28. F_{3n} is even for all natural numbers n.

29. $F_1 + F_2 + F_3 + \cdots + F_n = F_{n+2} - 1$

30. $F_1^2 + F_2^2 + F_3^2 + \cdots + F_n^2 = F_n F_{n+1}$

31. If $a_{n+2} = a_{n+1} \cdot a_n$ and $a_1 = a_2 = 2$, then $a_n = 2^{F_n}$ for all natural numbers n.

32. For all $n \geq 2$,
$$\begin{bmatrix} 1 & 1 \\ 1 & 0 \end{bmatrix}^n = \begin{bmatrix} F_{n+1} & F_n \\ F_n & F_{n-1} \end{bmatrix}$$

33. Let a_n be the nth term of the sequence defined recursively by
$$a_{n+1} = \frac{1}{1 + a_n}$$
and $a_1 = 1$. Find a formula for a_n in terms of the Fibonacci numbers F_n. Prove that the formula you found is valid for all natural numbers n.

34. Let F_n be the nth term of the Fibonacci sequence. Find and prove an inequality relating n and F_n for natural numbers n.

35. Find and prove an inequality relating $100n$ and n^3.

36. What is wrong with the following "proof" by mathematical induction that all girls have blond hair? Let $P(n)$ denote the statement: In any group of n girls, if one of them has blond hair, then they all do.

Step 1 The statement is clearly true for $n = 1$.

Step 2 Suppose that $P(k)$ is true. We show that $P(k + 1)$ is true. Consider a group of $k + 1$ girls, one of whom has blond hair; we call her *Ex*. Remove a girl, call her *Es*, from the group. Now, we have a group of k girls, one of whom has blond hair, and by our induction hypothesis all k girls have blond hair. Now put *Es* back in the group and remove another girl. Again, by our induction hypothesis, all k girls in this group have blond hair, and so *Es* also has blond hair. It follows that all $k + 1$ girls in the group have blond hair and this completes the induction step.

Since everyone knows at least one girl with blond hair, it follows that all girls have blond hair.

9.8 ## THE BINOMIAL THEOREM

An expression of the form $a + b$ is called a **binomial.** Although in principle it is easy to raise $a + b$ to any power, raising it to a very high power would be a tedious task. In this section we find a formula that gives the expansion of $(a + b)^n$ for any natural number n.

Pascal's triangle appears in this Chinese document by Chu Shi-kie, dated 1303. The title reads "The Old Method Chart of the Seven Multiplying Squares." The triangle was rediscovered by Pascal (see page 630).

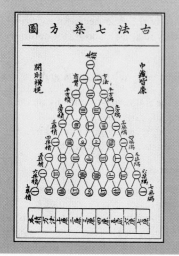

We discover this formula by finding a pattern for the successive powers of $a + b$. So, first we look at some special cases:

$$(a + b)^1 = a + b$$

$$(a + b)^2 = a^2 + 2ab + b^2$$

$$(a + b)^3 = a^3 + 3a^2b + 3ab^2 + b^3$$

$$(a + b)^4 = a^4 + 4a^3b + 6a^2b^2 + 4ab^3 + b^4$$

$$(a + b)^5 = a^5 + 5a^4b + 10a^3b^2 + 10a^2b^3 + 5ab^4 + b^5$$

$$\vdots$$

The following simple patterns emerge for the expansion of $(a + b)^n$:

1. There are $n + 1$ terms, the first being a^n and the last b^n.
2. The exponents of a decrease by 1 from term to term while the exponents of b increase by 1.
3. The sum of the exponents of a and b in each term is n.

For instance, notice how the exponents of a and b behave in the expansion of $(a + b)^5$.

The exponents of a decrease:

$$(a + b)^5 = a^{⑤} + 5a^{④}b^1 + 10a^{③}b^2 + 10a^{②}b^3 + 5a^{①}b^4 + b^5$$

The exponents of b increase:

$$(a + b)^5 = a^5 + 5a^4b^{①} + 10a^3b^{②} + 10a^2b^{③} + 5a^1b^{④} + b^{⑤}$$

With these observations we can write the form of the expansion of $(a + b)^n$ for any natural number n. For example, writing a question mark for the missing coefficients, we have

$$(a + b)^8 = a^8 + ?a^7b + ?a^6b^2 + ?a^5b^3 + ?a^4b^4 + ?a^3b^5 + ?a^2b^6 + ?ab^7 + b^8$$

To complete the expansion we need to determine these coefficients. To find a pattern, let us write the coefficients in the expansion of $(a + b)^n$ for the first few values of n in a triangular array as shown in the following array, which is called **Pascal's triangle.**

$(a + b)^0$				1			
$(a + b)^1$			1		1		
$(a + b)^2$		1		2		1	
$(a + b)^3$		①	③		3		1
$(a + b)^4$	1	④		⑥		④	1
$(a + b)^5$	1	5	10		⑩	5	1

The row corresponding to $(a + b)^0$ is called the zeroth row and is included to show the symmetry of the array. The key observation about Pascal's triangle is the following property.

KEY PROPERTY OF PASCAL'S TRIANGLE

Every entry (other than a 1) is the sum of the two entries diagonally above it.

From this property it is easy to find any row of Pascal's triangle from the row above it. For instance, we find the sixth and seventh rows, starting with the fifth row:

$$
\begin{array}{lcccccccc}
(a + b)^5 & & 1 & 5 & 10 & 10 & 5 & 1 & \\
(a + b)^6 & 1 & 6 & 15 & 20 & 15 & 6 & 1 & \\
(a + b)^7 & 1 & 7 & 21 & 35 & 35 & 21 & 7 & 1
\end{array}
$$

To see why this property holds, let us consider the following expansions:

$$(a + b)^5 = a^5 + 5a^4b + 10a^3b^2 + 10a^2b^3 + 5ab^4 + b^5$$

$$(a + b)^6 = a^6 + 6a^5b + 15a^4b^2 + 20a^3b^3 + 15a^2b^4 + 6ab^5 + b^6$$

We arrive at the expansion of $(a + b)^6$ by multiplying $(a + b)^5$ by $(a + b)$. Now notice, for instance, that the circled term in the expansion of $(a + b)^6$ is obtained via this multiplication from the two circled terms above it. (In fact, we get this term when the two terms above it are multiplied by b and a, respectively.) Thus, its coefficient is the sum of the coefficients of these two terms. This is the observation we will use at the end of this section in proving the Binomial Theorem.

Having found these patterns, we can now easily obtain the expansion of any binomial, at least to relatively small powers.

EXAMPLE 1 ■ Expanding a Binomial Using Pascal's Triangle

Find the expansion of $(a + b)^7$ using Pascal's triangle.

SOLUTION

The first term in the expansion is a^7 and the last term is b^7. Using the fact that the exponent of a decreases by 1 from term to term and that of b increases by 1 from term to term, we have

$$(a + b)^7 = a^7 + ?a^6b + ?a^5b^2 + ?a^4b^3 + ?a^3b^4 + ?a^2b^5 + ?ab^6 + b^7$$

The appropriate coefficients appear in the seventh row of Pascal's triangle. Thus,

$$(a + b)^7 = a^7 + 7a^6b + 21a^5b^2 + 35a^4b^3 + 35a^3b^4 + 21a^2b^5 + 7ab^6 + b^7$$

∎

EXAMPLE 2 ∎ Expanding a Binomial using Pascal's Triangle

Use Pascal's triangle to expand $(2 - 3x)^5$.

SOLUTION

We find the expansion of $(a + b)^5$ and then substitute 2 for a and $-3x$ for b. Using Pascal's triangle for the coefficients, we get

$$(a + b)^5 = a^5 + 5a^4b + 10a^3b^2 + 10a^2b^3 + 5ab^4 + b^5$$

Substituting $a = 2$ and $b = -3x$ gives

$$(2 - 3x)^5 = (2)^5 + 5(2)^4(-3x) + 10(2)^3(-3x)^2 + 10(2)^2(-3x)^3 + 5(2)(-3x)^4 + (-3x)^5$$

$$= 32 - 240x + 720x^2 - 1080x^3 + 810x^4 - 243x^5$$

∎

THE BINOMIAL COEFFICIENTS AND PASCAL'S TRIANGLE

Although Pascal's triangle is useful in finding the binomial expansion for reasonably small values of n, it is not practical for finding $(a + b)^n$ for large values of n. The reason is that the method we use for finding the successive rows of Pascal's triangle is recursive. Thus, to find the 100th row of this triangle we must first find all the preceding rows.

We need to examine the pattern in the coefficients more carefully to develop a formula that will allow us to calculate directly any coefficient in the binomial expansion. Such a formula exists, and the rest of this section is devoted to finding and proving it. However, to state this formula we need some notation, which we discuss next.

THE BINOMIAL COEFFICIENT

Let n and r be nonnegative integers with $r \le n$. The **binomial coefficient** is denoted by $\binom{n}{r}$ and is defined by

$$\binom{n}{r} = \frac{n!}{r!(n-r)!}$$

In Section 8.3 we denoted this quantity by $C(n, r)$, but it is customary to use the symbol $\binom{n}{r}$ in the context of the binomial expansion, and we follow this custom. It is interesting that the formula for combinations is related to the binomial expansion—an explanation is given in Exercise 61.

EXAMPLE 3 ■ Calculating Binomial Coefficients

(a) $\displaystyle \binom{9}{4} = \frac{9!}{4!\,(9-4)!} = \frac{9!}{4!\,5!} = \frac{1 \cdot 2 \cdot 3 \cdot 4 \cdot 5 \cdot 6 \cdot 7 \cdot 8 \cdot 9}{(1 \cdot 2 \cdot 3 \cdot 4)\,(1 \cdot 2 \cdot 3 \cdot 4 \cdot 5)}$

$\displaystyle \qquad\qquad = \frac{6 \cdot 7 \cdot 8 \cdot 9}{1 \cdot 2 \cdot 3 \cdot 4} = 126$

(b) $\displaystyle \binom{100}{3} = \frac{100!}{3!\,(100-3)!} = \frac{1 \cdot 2 \cdot 3 \cdot \cdots \cdot 97 \cdot 98 \cdot 99 \cdot 100}{(1 \cdot 2 \cdot 3)\,(1 \cdot 2 \cdot 3 \cdot \cdots \cdot 97)}$

$\displaystyle \qquad\qquad = \frac{98 \cdot 99 \cdot 100}{1 \cdot 2 \cdot 3} = 161{,}700$

(c) $\displaystyle \binom{100}{97} = \frac{100!}{97!\,(100-97)!} = \frac{1 \cdot 2 \cdot 3 \cdot \cdots \cdot 97 \cdot 98 \cdot 99 \cdot 100}{(1 \cdot 2 \cdot 3 \cdot \cdots \cdot 97)\,(1 \cdot 2 \cdot 3)}$

$\displaystyle \qquad\qquad = \frac{98 \cdot 99 \cdot 100}{1 \cdot 2 \cdot 3} = 161{,}700$ ■

Although the binomial coefficient $\binom{n}{r}$ is defined in terms of a fraction, all the results of Example 3 are natural numbers. In fact, $\binom{n}{r}$ is always a natural number (see Exercise 60). Notice that the binomial coefficients in parts (b) and (c) of Example 3 are equal. This is a special case of the relation

$$\binom{n}{r} = \binom{n}{n-r}$$

which you are asked to prove in Exercise 56.

To see the connection between the binomial coefficients and the binomial expansion of $(a + b)^n$, let us calculate the following binomial coefficients:

$$\binom{5}{0} = 1 \qquad \binom{5}{1} = 5 \qquad \binom{5}{2} = 10$$

$$\binom{5}{3} = 10 \qquad \binom{5}{4} = 5 \qquad \binom{5}{5} = 1$$

$$\binom{5}{2} = \frac{5!}{2!\,(5-2)!} = 10$$

These are precisely the entries in the fifth row of Pascal's triangle. In fact it is true that, for all natural numbers n, the nth row of Pascal's triangle has the entries

$$\binom{n}{0} \qquad \binom{n}{1} \qquad \binom{n}{2} \cdots \binom{n}{n-1} \qquad \binom{n}{n}$$

So, we can write Pascal's triangle as follows.

$$\binom{0}{0}$$

$$\binom{1}{0} \qquad \binom{1}{1}$$

$$\binom{2}{0} \qquad \binom{2}{1} \qquad \binom{2}{2}$$

$$\binom{3}{0} \qquad \binom{3}{1} \qquad \binom{3}{2} \qquad \binom{3}{3}$$

$$\binom{4}{0} \qquad \binom{4}{1} \qquad \binom{4}{2} \qquad \binom{4}{3} \qquad \binom{4}{4}$$

$$\binom{5}{0} \qquad \binom{5}{1} \qquad \binom{5}{2} \qquad \binom{5}{3} \qquad \binom{5}{4} \qquad \binom{5}{5}$$

$$\binom{n}{0} \quad \binom{n}{1} \quad \binom{n}{2} \qquad \cdots \qquad \binom{n}{n-1} \quad \binom{n}{n}$$

To show that this pattern actually holds, we need to show that any entry in this version of Pascal's triangle is the sum of the two entries diagonally above it. In other words, we need to show that each entry satisfies the key property of Pascal's triangle. We now state this property in terms of the binomial coefficients.

KEY PROPERTY OF THE BINOMIAL COEFFICIENTS

For any nonnegative integers r and k with $r \leq k$,

$$\binom{k}{r-1} + \binom{k}{r} = \binom{k+1}{r}$$

Notice that the two terms on the left side of this equation are adjacent entries in the kth row of Pascal's triangle and the term on the right side is the entry diagonally below them, in the $(k+1)$st row. Thus, this equation is a restatement of the key property of Pascal's triangle in terms of the binomial coefficients. A proof of this formula is outlined in Exercise 59.

THE BINOMIAL THEOREM

We are now ready to state the Binomial Theorem.

THE BINOMIAL THEOREM

$$(a + b)^n = \binom{n}{0}a^n + \binom{n}{1}a^{n-1}b + \binom{n}{2}a^{n-2}b^2 + \cdots + \binom{n}{n-1}ab^{n-1} + \binom{n}{n}b^n$$

We prove this theorem at the end of this section. First, we show some of its applications.

EXAMPLE 4 ■ Expanding a Binomial using the Binomial Theorem

Use the Binomial Theorem to expand $(x + y)^4$.

SOLUTION

By the Binomial Theorem,

$$(x + y)^4 = \binom{4}{0}x^4 + \binom{4}{1}x^3y + \binom{4}{2}x^2y^2 + \binom{4}{3}xy^3 + \binom{4}{4}y^4$$

Verify that

$$\binom{4}{0} = 1 \qquad \binom{4}{1} = 4 \qquad \binom{4}{2} = 6 \qquad \binom{4}{3} = 4 \qquad \binom{4}{4} = 1$$

It follows that

$$(x + y)^4 = x^4 + 4x^3y + 6x^2y^2 + 4xy^3 + y^4$$

■

EXAMPLE 5 ■ Expanding a Binomial using the Binomial Theorem

Use the Binomial Theorem to expand $\left(\sqrt{x} - 1\right)^8$.

SOLUTION

We first find the expansion of $(a + b)^8$ and then substitute $\sqrt{x}$ for a and -1 for b. Using the Binomial Theorem, we have

$$(a + b)^8 = \binom{8}{0}a^8 + \binom{8}{1}a^7b + \binom{8}{2}a^6b^2 + \binom{8}{3}a^5b^3 + \binom{8}{4}a^4b^4$$

$$+ \binom{8}{5}a^3b^5 + \binom{8}{6}a^2b^6 + \binom{8}{7}ab^7 + \binom{8}{8}b^8$$

Verify that

$$\binom{8}{0} = 1 \qquad \binom{8}{1} = 8 \qquad \binom{8}{2} = 28 \qquad \binom{8}{3} = 56 \qquad \binom{8}{4} = 70$$

$$\binom{8}{5} = 56 \qquad \binom{8}{6} = 28 \qquad \binom{8}{7} = 8 \qquad \binom{8}{8} = 1$$

So

$$(a + b)^8 = a^8 + 8a^7b + 28a^6b^2 + 56a^5b^3 + 70a^4b^4 + 56a^3b^5$$
$$+ 28a^2b^6 + 8ab^7 + b^8$$

Performing the substitutions $a = x^{1/2}$ and $b = -1$ gives

$$(\sqrt{x} - 1)^8 = (x^{1/2})^8 + 8(x^{1/2})^7(-1) + 28(x^{1/2})^6(-1)^2 + 56(x^{1/2})^5(-1)^3$$
$$+ 70(x^{1/2})^4(-1)^4 + 56(x^{1/2})^3(-1)^5 + 28(x^{1/2})^2(-1)^6$$
$$+ 8(x^{1/2})(-1)^7 + (-1)^8$$

This simplifies to

$$(\sqrt{x} - 1)^8 = x^4 - 8x^{7/2} + 28x^3 - 56x^{5/2} + 70x^2 - 56x^{3/2} + 28x - 8x^{1/2} + 1$$

$\blacksquare$

The Binomial Theorem can be used to find a particular term of a binomial expansion without having to find the entire expansion.

GENERAL TERM OF THE BINOMIAL EXPANSION

The term that contains a^r in the expansion of $(a + b)^n$ is

$$\binom{n}{n - r}a^rb^{n-r}$$

Recall that $\binom{n}{n - r} = \binom{n}{r}$.

EXAMPLE 6 ■ **Finding a Particular Term in a Binomial Expansion**

Find the term that contains x^5 in the expansion of $(2x + y)^{20}$.

SOLUTION

The term that contains x^5 is given by the formula for the general term with $a = 2x$, $b = y$, $n = 20$ and $r = 5$. So, this term is

$$\binom{20}{15}a^5b^{15} = \frac{20!}{15!\,(20 - 5)!}(2x)^5y^{15} = \frac{20!}{15!\,5!}32x^5y^{15} = 496{,}128x^5y^{15}$$

$\blacksquare$

EXAMPLE 7 ■ Finding a Particular Term in a Binomial Expansion

Find the coefficient of x^8 in the expansion of $\left(x^2 + \dfrac{1}{x}\right)^{10}$.

SOLUTION

Both x^2 and $1/x$ are powers of x, so the power of x in each term of the expansion is determined by both terms of the binomial. To find the required coefficient we first find the general term in the expansion. By the formula, we have $a = x^2$, $b = 1/x$, and $n = 10$, so the general term is

$$\binom{10}{10-r}(x^2)^r\left(\frac{1}{x}\right)^{10-r} = \binom{10}{10-r}(x^2)^r(x^{-1})^{10-r} = \binom{10}{10-r}x^{3r-10}$$

Thus, the term that contains x^8 is the term in which

$$3r - 10 = 8$$

$$r = 6$$

So the required coefficient is

$$\binom{10}{10-6} = \binom{10}{4} = 210$$

    ■

PROOF OF THE BINOMIAL THEOREM

We now give a proof of the Binomial Theorem using mathematical induction.

■ **Proof**   Let $P(n)$ denote the statement

$$(a + b)^n = \binom{n}{0}a^n + \binom{n}{1}a^{n-1}b + \binom{n}{2}a^{n-2}b^2 + \cdots + \binom{n}{n-1}ab^{n-1} + \binom{n}{n}b^n$$

Step 1   We show that $P(1)$ is true. But $P(1)$ is just the statement

$$(a + b)^1 = \binom{1}{0}a^1 + \binom{1}{1}b^1 = 1a + 1b = a + b$$

which is certainly true.

Step 2   We assume that $P(k)$ is true and show that $P(k + 1)$ is true. The statement $P(k)$ reads

$$(a + b)^k = \binom{k}{0}a^k + \binom{k}{1}a^{k-1}b + \binom{k}{2}a^{k-2}b^2 + \cdots + \binom{k}{k-1}ab^{k-1} + \binom{k}{k}b^k$$

Multiplying each side of this equation by $(a + b)$ and collecting like terms gives the following.

$$(a + b)^{k+1} = (a + b)\left[\binom{k}{0}a^k + \binom{k}{1}a^{k-1}b + \binom{k}{2}a^{k-2}b^2 + \cdots + \binom{k}{k-1}ab^{k-1} + \binom{k}{k}b^k\right]$$

$$= a\left[\binom{k}{0}a^k + \binom{k}{1}a^{k-1}b + \binom{k}{2}a^{k-2}b^2 + \cdots + \binom{k}{k-1}ab^{k-1} + \binom{k}{k}b^k\right]$$

$$+ b\left[\binom{k}{0}a^k + \binom{k}{1}a^{k-1}b + \binom{k}{2}a^{k-2}b^2 + \cdots + \binom{k}{k-1}ab^{k-1} + \binom{k}{k}b^k\right]$$

$$= \binom{k}{0}a^{k+1} + \binom{k}{1}a^k b + \binom{k}{2}a^{k-1}b^2 + \cdots + \binom{k}{k-1}a^2 b^{k-1} + \binom{k}{k}ab^k$$

$$+ \binom{k}{0}a^k b + \binom{k}{1}a^{k-1}b^2 + \binom{k}{2}a^{k-2}b^3 + \cdots + \binom{k}{k-1}ab^k + \binom{k}{k}b^{k+1}$$

$$= \binom{k}{0}a^{k+1} + \left[\binom{k}{0} + \binom{k}{1}\right]a^k b + \left[\binom{k}{1} + \binom{k}{2}\right]a^{k-1}b^2 + \cdots + \left[\binom{k}{k-1} + \binom{k}{k}\right]ab^k + \binom{k}{k}b^{k+1}$$

Using the key property of the binomial coefficients, we can write each of the expressions in square brackets as a single binomial coefficient. Also, writing the first and last coefficients as $\binom{k+1}{0}$ and $\binom{k+1}{k+1}$ (these are equal to 1 by Exercise 54) gives

$$(a + b)^{k+1} = \binom{k+1}{0}a^{k+1} + \binom{k+1}{1}a^k b + \binom{k+1}{2}a^{k-1}b^2 + \cdots + \binom{k+1}{k}ab^k + \binom{k+1}{k+1}b^{k+1}$$

But this last equation is precisely $P(k + 1)$, and this completes the induction step.

Having proved Steps 1 and 2, we conclude by the Principle of Mathematical Induction that the theorem is true for all natural numbers n.   □

9.8   EXERCISES

1–12 ■ Use Pascal's triangle to expand the expression.

1. $(x + y)^6$

2. $(2x + 1)^4$

3. $\left(x + \dfrac{1}{x}\right)^4$

4. $(x - y)^5$

5. $(x - 1)^5$

6. $(\sqrt{a} + \sqrt{b})^6$

7. $(x^2 y - 1)^5$

8. $(1 + \sqrt{2})^6$

9. $(2x - 3y)^3$

10. $(1 + x^3)^3$

11. $\left(\dfrac{1}{x} - \sqrt{x}\right)^5$

12. $\left(2 + \dfrac{x}{2}\right)^5$

13–20 ■ Evaluate the expression.

13. $\binom{6}{4}$

14. $\binom{8}{3}$

15. $\binom{100}{98}$

16. $\binom{10}{5}$

17. $\binom{3}{1}\binom{4}{2}$

18. $\binom{5}{2}\binom{5}{3}$

19. $\binom{5}{0} + \binom{5}{1} + \binom{5}{2} + \binom{5}{3} + \binom{5}{4} + \binom{5}{5}$

20. $\dbinom{5}{0} - \dbinom{5}{1} + \dbinom{5}{2} - \dbinom{5}{3} + \dbinom{5}{4} - \dbinom{5}{5}$

21–24 ■ Use the Binomial Theorem to expand the expression.

21. $(x + 2y)^4$

22. $(1 - x)^5$

23. $\left(1 + \dfrac{1}{x}\right)^6$

24. $(2A + B^2)^4$

25. Find the first three terms in the expansion of $(x + 2y)^{20}$.

26. Find the first four terms in the expansion of $(x^{1/2} + 1)^{30}$.

27. Find the last two terms in the expansion of $(a^{2/3} + a^{1/3})^{25}$.

28. Find the first three terms in the expansion of $\left(x + \dfrac{1}{x}\right)^{40}$.

29. Find the middle term in the expansion of $(x^2 + 1)^{18}$.

30. Find the fifth term in the expansion of $(ab - 1)^{20}$.

31. Find the 24th term in the expansion of $(a + b)^{25}$.

32. Find the 28th term in the expansion of $(A - B)^{30}$.

33. Find the 100th term in the expansion of $(1 + y)^{100}$.

34. Find the second term in the expansion of
$$\left(x^2 - \dfrac{1}{x}\right)^{25}$$

35. Find the term containing x^4 in the expansion of $(x + 2y)^{10}$.

36. Find the term containing y^3 in the expansion of $(\sqrt{2} + y)^{12}$.

37. Find the term containing b^8 in the expansion of $(a + b^2)^{12}$.

38. Find the term containing a in the expansion of
$$\left(\sqrt{a} + \dfrac{1}{\sqrt{a}}\right)^{10}$$

39. Find the term that does not contain x in the expansion of
$$\left(8x + \dfrac{1}{2x}\right)^8$$

40. Find the term that does not contain m in the expansion of $(mn + m^{-4})^5$.

41. Find the term containing c^7 in the expansion of $(2c + \sqrt{c})^8$.

42. Find the coefficient of r^{-5} in the expansion of
$$\left(\dfrac{r^2}{4} - \dfrac{4}{r^3}\right)^5$$

43–46 ■ Find the number of distinct terms in the expansion of the expression.

43. $\left(x + \dfrac{1}{x}\right)^{20}$

44. $\left(x^2 + \dfrac{1}{x}\right)^6$

45. $(a^2 - 2ab + b^2)^5$

46. $[(x + y)^2(x - y)^2]^3$

47–50 ■ Simplify the expression.

47. $x^4 + 4x^3y + 6x^2y^2 + 4xy^3 + y^4$

48. $(x - 1)^5 + 5(x - 1)^4 + 10(x - 1)^3 + 10(x - 1)^2 + 5(x - 1) + 1$

49. $8a^3 + 12a^2b + 6ab^2 + b^3$

50. $x^8 + 4x^6y + 6x^4y^2 + 4x^2y^3 + y^4$

51. Expand $(a^2 + a + 1)^4$.
[*Hint:* $a^2 + a + 1 = a^2 + (a + 1)$]

52. Show that $(1.01)^{100} > 2$.
[*Hint:* Note that $(1.01)^{100} = (1 + 0.01)^{100}$ and use the Binomial Theorem to show that the sum of the first two terms of the expansion is greater than 2.]

53. Which is larger: $(100!)^{101}$ or $(101!)^{100}$?

54. Show that $\dbinom{n}{0} = 1$ and $\dbinom{n}{n} = 1$.

55. Show that $\dbinom{n}{1} = \dbinom{n}{n-1} = n$.

56. Show that $\dbinom{n}{r} = \dbinom{n}{n-r}$ for $0 \leqslant r \leqslant n$.

57. (a) Show that $\dbinom{n}{0} + \dbinom{n}{1} + \dbinom{n}{2} + \cdots + \dbinom{n}{n} = 2^n$.
[*Hint:* $2^n = (1 + 1)^n$]
(b) Give a counting argument that explains the equality in part (a).

58. Show that
$$\dbinom{n}{0} - \dbinom{n}{1} + \dbinom{n}{2} - \cdots + (-1)^k\dbinom{n}{k} +$$
$$\cdots + (-1)^n\dbinom{n}{n} = 0$$
[*Hint:* $0 = 1 - 1$]

59. In this exercise we prove the identity

$$\binom{n}{r-1} + \binom{n}{r} = \binom{n+1}{r}$$

(a) Write the left side of this equation as the sum of two fractions.

(b) Show that a common denominator of the expression you found in part (a) is $r!(n-r+1)!$.

(c) Add the two fractions using the common denominator in part (b), simplify the numerator, and note that the resulting expression is equal to the right side of the equation.

60. Prove that $\binom{n}{r}$ is an integer for all n and for $0 \le r \le n$. [*Suggestion:* Use induction to show that the statement is true for all n, and use Exercise 59 for the induction step.]

61. This exercise explains why the binomial coefficients $\binom{n}{r}$ that appear in the expansion of $(x+y)^n$ are the same

as $C(n, r)$, the number of ways of choosing r objects from n objects. First, note that expanding a binomial using only the Distributive Property gives

$$(x+y)^2 = (x+y)(x+y)$$
$$= (x+y)x + (x+y)y$$
$$= xx + xy + yx + yy$$

$$(x+y)^3 = (x+y)(xx + xy + yx + yy)$$
$$= xxx + xxy + xyx + xyy + yxx$$
$$\quad + yxy + yyx + yyy$$

(a) Expand $(x+y)^5$ using only the Distributive Property.

(b) Write all the terms that represent x^2y^3 together. These are all the terms that contain two x's and three y's.

(c) Note that the two x's appear in all possible positions. Conclude that the number of terms that represent x^2y^3 is $C(5, 2)$.

---

9   REVIEW

KEY TOPICS ■ Define, state, or discuss each of the following.

1. Sequence
2. Recursive sequence
3. Fibonacci sequence
4. Arithmetic sequence
5. Common difference
6. Geometric sequence
7. Common ratio
8. Sigma notation
9. Series
10. Arithmetic series
11. Sum of an arithmetic series

12. Geometric series
13. Sum of a geometric series
14. Annuity
15. Infinite series
16. Sum of an infinite geometric series
17. Mathematical induction
18. Pascal's triangle
19. Key property of Pascal's triangle
20. Binomial coefficients
21. The Binomial Theorem
22. General term of the binomial expansion

EXERCISES

1-6 ■ Find the first four terms as well as the tenth term of the sequence with the given nth term.

1. $a_n = \dfrac{n^2}{n + 1}$

2. $a_n = (-1)^n \dfrac{2^n}{n}$

3. $a_n = \dfrac{(-1)^n + 1}{n^3}$

4. $a_n = \dfrac{n(n + 1)}{2}$

5. $a_n = \dfrac{(2n)!}{2^n n!}$

6. $a_n = \dbinom{n + 1}{2}$

7-12 ■ A sequence is defined recursively. Find the first seven terms of the sequence.

7. $a_n = a_{n-1} + 2n - 1, \quad a_1 = 1$

8. $a_n = \dfrac{a_{n-1}}{n}, \quad a_1 = 1$

9. $a_n = a_{n-1} + 2a_{n-2}, \quad a_1 = 1, a_2 = 3$

10. $a_n = \sqrt{3a_{n-1}}, \quad a_1 = \sqrt{3}$

11. $a_n = (a_{n-1} - 1)!, \quad a_1 = 3$

12. $a_n = \dbinom{n + 1}{a_{n-1}}, \quad a_1 = 1$

13-24 ■ The first four terms of a sequence are given. Determine whether the given terms can be the terms of an arithmetic sequence, a geometric sequence, or neither. If the sequence is arithmetic or geometric, find the fifth term.

13. $5, 5.5, 6, 6.5, \ldots$

14. $1, -\frac{3}{2}, 2, -\frac{5}{2}, \ldots$

15. $\sqrt{2}, 2\sqrt{2}, 3\sqrt{2}, 4\sqrt{2}, \ldots$

16. $\sqrt{2}, 2, 2\sqrt{2}, 4, \ldots$

17. $t - 3, t - 2, t - 1, t, \ldots$ **18.** $t^3, t^2, t, 1, \ldots$

19. $\dfrac{3}{4}, \dfrac{1}{2}, \dfrac{1}{3}, \dfrac{2}{9}, \ldots$

20. $\dfrac{a}{c}, 1, \dfrac{c}{a}, \left(\dfrac{c}{a}\right)^2, \ldots$

21. $\ln a, \ln 2a, \ln 3a, \ln 4a, \ldots$

22. $a, 1, \dfrac{1}{a}, \dfrac{1}{a^2}, \ldots$

23. $a, abc^3, ab^2c^6, ab^3c^9, \ldots$

24. $a, a + b^2, a + 2b^2, a + 3b^2, \ldots$

25. Show that $3, 6i, -12, -24i, \ldots$ is a geometric sequence and find the common ratio.

26. Find the nth term of the geometric sequence $2, 2 + 2i, 4i, -4 + 4i, -8, \ldots$

27. The sixth term of an arithmetic sequence is 17 and the fourth term is 11. Find the second term.

28. The 20th term of an arithmetic sequence is 96 and the common difference is 5. Find the nth term.

29. The third term of a geometric sequence is 9 and the common ratio is $\frac{3}{2}$. Find the fifth term.

30. The second term of a geometric sequence is 10 and the fifth term is $\frac{1250}{27}$. Find the nth term.

31. The frequencies of musical notes (measured in cycles per second) form a geometric sequence. Middle C has a frequency of 256 and C an octave higher has a frequency of 512. Find the frequency of C two octaves below middle C.

32. A person has two parents, four grandparents, eight great-grandparents, and so on. How many ancestors does a person have 15 generations back?

33. A certain type of bacteria divides every five seconds. If three of these bacteria are put into a petri dish, how many bacteria are in the dish at the end of one minute?

34. If $a_1, a_2, a_3, \ldots$ and $b_1, b_2, b_3, \ldots$ are arithmetic sequences, show that $a_1 + b_1, a_2 + b_2, a_3 + b_3, \ldots$ is also an arithmetic sequence.

35. If $a_1, a_2, a_3, \ldots$ and $b_1, b_2, b_3, \ldots$ are geometric sequences, show that $a_1b_1, a_2b_2, a_3b_3, \ldots$ is also a geometric sequence.

36. (a) If $a_1, a_2, a_3, \ldots$ is an arithmetic sequence, is the sequence $a_1 + 2, a_2 + 2, a_3 + 2, \ldots$ arithmetic?
(b) If $a_1, a_2, a_3, \ldots$ is a geometric sequence, is the sequence $5a_1, 5a_2, 5a_3, \ldots$ geometric?

37. Find the values of x for which the sequence $6, x, 12, \ldots$ is
(a) arithmetic   (b) geometric

38. Find the values of x and y for which the sequence $2, x, y, 17, \ldots$ is
(a) arithmetic   (b) geometric

39–42 ■ Find the sum.

39. $\displaystyle\sum_{k=3}^{6} (k+1)^2$

40. $\displaystyle\sum_{i=1}^{4} \frac{2i}{2i-1}$

41. $\displaystyle\sum_{k=1}^{6} (k+1)2^{k-1}$

42. $\displaystyle\sum_{m=1}^{5} 3^{m-2}$

43–46 ■ Write the sum without using sigma notation. Do not evaluate.

43. $\displaystyle\sum_{k=1}^{10} (k-1)^2$

44. $\displaystyle\sum_{j=2}^{100} \frac{1}{j-1}$

45. $\displaystyle\sum_{k=1}^{50} \frac{3^k}{2^{k+1}}$

46. $\displaystyle\sum_{n=1}^{10} n^2 2^n$

47–50 ■ Write the sum using sigma notation. Do not evaluate.

47. $3 + 6 + 9 + 12 + \cdots + 99$

48. $1^2 + 2^2 + 3^2 + \cdots + 100^2$

49. $1 \cdot 2^3 + 2 \cdot 2^4 + 3 \cdot 2^5 + 4 \cdot 2^6 + \cdots + 100 \cdot 2^{102}$

50. $\dfrac{1}{1 \cdot 2} + \dfrac{1}{2 \cdot 3} + \dfrac{1}{3 \cdot 4} + \cdots + \dfrac{1}{999 \cdot 1000}$

51–58 ■ Determine whether the series is arithmetic or geometric, and find its sum.

51. $1 + 0.9 + (0.9)^2 + \cdots + (0.9)^5$

52. $3 + 3.7 + 4.4 + \cdots + 10$

53. $1 - \sqrt{5} + 5 - 5\sqrt{5} + \cdots + 625$

54. $\sqrt{5} + 2\sqrt{5} + 3\sqrt{5} + \cdots + 100\sqrt{5}$

55. $\dfrac{1}{3} + \dfrac{2}{3} + 1 + \dfrac{4}{3} + \cdots + 33$

56. $a + abc^3 + ab^2c^6 + ab^3c^9 + \cdots + ab^8c^{24}$

57. $\displaystyle\sum_{n=0}^{6} 3(-4)^n$

58. $\displaystyle\sum_{k=0}^{8} 7(5)^{k/2}$

59. The first term of an arithmetic sequence is $a = 7$ and the common difference is $d = 3$. How many terms of this sequence must be added to obtain 325?

60. A geometric sequence has first term $a = 81$ and common ratio $r = -\frac{2}{3}$. How many terms of this sequence must be added to get 55?

61. The sum of the first eight terms of an arithmetic series is 100 and the first term is 2. Find the tenth term.

62. The sum of the first three terms of a geometric series is 52 and the common ratio is $r = 3$. Find the first term.

63. A city has a population of 100,000. If the population is increasing at the rate of 10% per year, what will be the population of this city in 10 years? Find a formula for the population of the city after n years.

64. Refer to Exercise 32. What is the total number of ancestors of a person in 15 generations?

65. Find the amount of an annuity consisting of 16 annual payments of $1000 each into an account that pays 8% interest per year, compounded annually.

66. How much money should be invested every quarter at 12% per year, compounded quarterly, in order to have $10,000 in one year?

67. What are the monthly payments on a mortgage of $60,000 at 9% interest if the loan is to be repaid in (a) 30 years? (b) 15 years?

68–71 ■ Find the sum of the infinite geometric series.

68. $1 - \dfrac{2}{5} + \dfrac{4}{25} - \dfrac{8}{125} + \cdots$

69. $0.1 + 0.01 + 0.001 + 0.0001 + \cdots$

70. $1 + \dfrac{1}{3^{1/2}} + \dfrac{1}{3} + \dfrac{1}{3^{3/2}} + \cdots$

71. $a + ab^2 + ab^4 + ab^6 + \cdots$

72–75 ■ Use mathematical induction to prove that the formula is true for all natural numbers n.

72. $1 + 4 + 7 + \cdots + (3n - 2) = \dfrac{n(3n-1)}{2}$

73. $1^4 + 2^4 + 3^4 + \cdots + n^4$
$$= \frac{n(n+1)(2n+1)(3n^2 + 3n - 1)}{30}$$

74. $\dfrac{1}{1 \cdot 3} + \dfrac{1}{3 \cdot 5} + \dfrac{1}{5 \cdot 7} + \cdots + \dfrac{1}{(2n-1)(2n+1)}$
$$= \frac{n}{2n+1}$$

75. $\left(1 + \dfrac{1}{1}\right)\left(1 + \dfrac{1}{2}\right)\left(1 + \dfrac{1}{3}\right)\cdots\left(1 + \dfrac{1}{n}\right) = n + 1$

76. Show that $7^n - 1$ is divisible by 6 for all natural numbers n.

77. Show that $11^{n+2} + 12^{2n+1}$ is divisible by 133 for every natural number n.

78. Let $a_{n+1} = 3a_n + 4$ and $a_1 = 4$. Show that $a_n = 2 \cdot 3^n - 2$ for all natural numbers n.

79. Prove that the Fibonacci number F_{4n} is divisible by 3 for all natural numbers n.

80. Find and prove an inequality that relates 2^n and $n!$.

81–84 ■ Evaluate the expression.

81. $\binom{5}{2}\binom{5}{3}$

82. $\binom{10}{2} + \binom{10}{6}$

83. $\sum_{k=0}^{5} \binom{5}{k}$

84. $\sum_{k=0}^{8} \binom{8}{k}\binom{8}{8-k}$

85–86 ■ Expand the expression.

85. $(1 - x^2)^6$

86. $(2x + y)^4$

87. Find the 20th term in the expansion of $(a + b)^{22}$.

88. Find the first three terms in the expansion of $(b^{-2/3} + b^{1/3})^{20}$.

89. Find the coefficient of s^5 in the expansion of

$$\left(\frac{s^3}{2} - \frac{2}{s^2} \right)^5$$

90. Find the term containing A^6 in the expansion of $(A + 3B)^{10}$.

91–92 ■ Simplify the expression.

91. $(a + 1)^3 - 3(a + 1)^2 + 3(a + 1) - 1$

92. $x^5 + 5x^4y + 10x^3y^2 + 10x^2y^3 + 5xy^4 + y^5$

1. Find the tenth term of the sequence whose nth term is

$$a_n = \frac{n}{1 - n^2}$$

2. A sequence is defined recursively by

$$a_{n+2} = (a_n)^2 - a_{n+1}$$

If $a_1 = 1$ and $a_2 = 1$, find a_5.

3. Find the 30th term in the arithmetic sequence: 80, 76, 72, ...

4. The second term of a geometric sequence is 125 and the fifth term is 1. Is $\frac{1}{5}$ a term of this sequence? If so, which term is it?

5. Determine whether each of the following statements is true or false. If it is true, prove it. If it is false, give an example where it fails.
   (a) If $a_1, a_2, a_3, \ldots$ is an arithmetic sequence, then the sequence $a_1^2, a_2^2, a_3^2, \ldots$ is also arithmetic.
   (b) If $a_1, a_2, a_3, \ldots$ is a geometric sequence, then the sequence $a_1^2, a_2^2, a_3^2, \ldots$ is also geometric.

6. (a) Write the formula for the sum of a (finite) arithmetic series.
   (b) The first term of an arithmetic series is 10 and the tenth term is 2. Find the sum of the first ten terms.
   (c) Find the common difference and the 100th term of the series in part (b).

7. (a) Write the formula for the sum of a (finite) geometric series.
   (b) Find the sum of the geometric series

$$\frac{1}{3} + \frac{2}{3^2} + \frac{2^2}{3^3} + \frac{2^3}{3^4} + \cdots + \frac{2^9}{3^{10}}$$

8. Find the sum of the infinite geometric series: $1 + \dfrac{1}{2^{1/2}} + \dfrac{1}{2} + \dfrac{1}{2^{3/2}} + \cdots$

9. Use mathematical induction to prove that, for all natural numbers n,

$$1^2 + 2^2 + 3^2 + \cdots + n^2 = \frac{n(n + 1)(2n + 1)}{6}$$

10. Write the expression without using sigma notation, and then find the sum.
   (a) $\displaystyle\sum_{n=1}^{5} (1 - n^2)$ (b) $\displaystyle\sum_{n=3}^{6} (-1)^n 2^{n-2}$

11. Write the expansion of $(a + b)^n$ using sigma notation.

12. Expand $(2x + y^2)^5$.

13. Find the term that contains a^3 in the expansion of $(2a + b)^{100}$.

14. Find the term that does not contain x in the expansion of

$$\left(2x + \frac{1}{x}\right)^{10}$$

FOCUS ON PROBLEM SOLVING

The solutions to many of the problems of mathematics involve **finding patterns.** The algebraic formulas we have found in this book are compact ways of describing patterns. For example, the familiar equation $(a + b)^2 = a^2 + 2ab + b^2$ gives the pattern for squaring the sum of two numbers. Another pattern we have encountered is the pattern for the sum of the first n natural numbers:

$$1 + 2 + 3 + \cdots + n = \frac{n(n + 1)}{2}$$

How do we discover patterns? In many cases, a good way to start is to experiment with the problem. To prove that a pattern always holds, we can often use mathematical induction, but other proofs are possible. A geometrical method is used in the problem we give here. It is attributed to the 11th-century mathematician Abu Bekr Mohammed ibn Al Husain Al Karchi.

Problem ■ The Gnomons of Al Karchi

We prove the beautiful formula

$$1^3 + 2^3 + 3^3 + \cdots + n^3 = (1 + 2 + 3 + \cdots + n)^2$$

But first, let's see how this formula was discovered. The sum of the first n natural numbers is called a *triangular number*. The name "triangular" comes from the following figures:

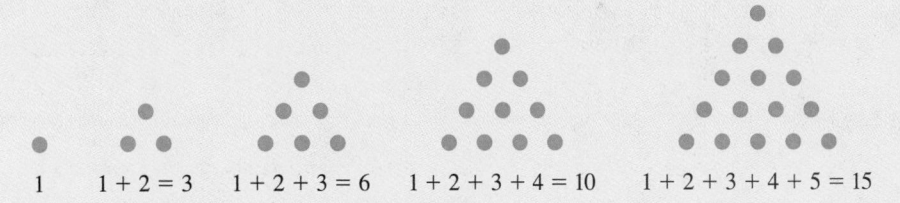

| 1 | $1 + 2 = 3$ | $1 + 2 + 3 = 6$ | $1 + 2 + 3 + 4 = 10$ | $1 + 2 + 3 + 4 + 5 = 15$ |

The first few triangular numbers are

$$1, 3, 6, 10, 15, \ldots$$

Now, let's look at the sums of the cubes:

$$1^3 = 1$$

$$1^3 + 2^3 = 9$$

$$1^3 + 2^3 + 3^3 = 36$$

$$1^3 + 2^3 + 3^3 + 4^3 = 100$$

$$1^3 + 2^3 + 3^3 + 4^3 + 5^3 = 225$$

We get the sequence

$$1, 9, 36, 100, 225, \ldots$$

It doesn't take long to notice that these are the squares of the triangular numbers. It appears that the sum of the first n cubes equals the square of the sum of the first n numbers.

To show that this pattern always holds, Al Karchi sketches the following diagram:

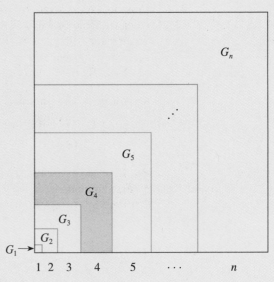

Each region G_n in the shape of an inverted L is called a *gnomon*. The area of gnomon G_4 is

$$\text{Area} = 4^2 + 2[4 \times (1 + 2 + 3)]$$
$$= 64$$

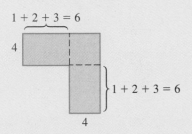

Similarly, the areas of gnomons G_1, G_2, G_3, G_4, G_5, ... are 1, 8, 27, 64, 125, ... These are the cubes of the natural numbers. The pattern persists here also, because the area of the nth gnomon G_n is n^3, as the following calculation shows:

$$n^2 + 2[n \times (1 + 2 + \cdots + (n - 1))] = n^2 + 2n\,\frac{(n - 1)n}{2}$$
$$= n^2 + n^3 - n^2 = n^3$$

Now comes Al Karchi's punchline: The first n gnomons form a square of side $1 + 2 + 3 + \cdots + n$ and so the sum of the areas of these gnomons equals the area of the square, that is, $1^3 + 2^3 + \cdots + n^3 = (1 + 2 + \cdots + n)^2$. ∎

PROBLEMS

1. Use the diagram to find and prove a formula for the sum of the first n odd numbers.

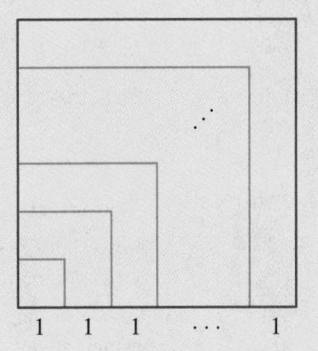

2. Use the figure to find a simple formula for $F_1^2 + F_2^2 + \cdots + F_n^2$, where F_k is the kth Fibonacci number.

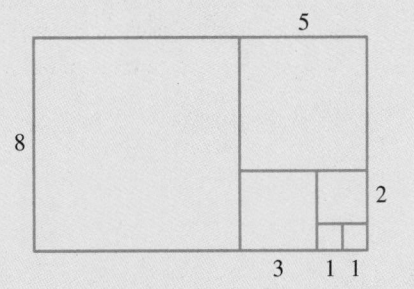

3. Show that every perfect cube is a difference of two squares.   [*Hint:* Use the formula for the sum of cubes that we have just proved.]

4. (a) Find the product $\left(1 - \dfrac{1}{2}\right)\left(1 - \dfrac{1}{3}\right)\left(1 - \dfrac{1}{4}\right) \cdots \left(1 - \dfrac{1}{200}\right)$.

   (b) Find the product $\left(1 - \dfrac{1}{4}\right)\left(1 - \dfrac{1}{9}\right)\left(1 - \dfrac{1}{16}\right) \cdots \left(1 - \dfrac{1}{n^2}\right)$.

The earliest known magic square is a 3 × 3 magic square from ancient China. According to Chinese legend it appeared on the back of a turtle emerging from the river Lo.

4	9	2
3	5	7
8	1	6

Here is an example of a 4 × 4 magic square:

16	3	2	13
5	10	11	8
9	6	7	12
4	15	14	1

5. Find a formula for the sums

$$S_n = 1 \cdot 1! + 2 \cdot 2! + 3 \cdot 3! + \cdots + n \cdot n!$$

and prove that your formula holds for all n.
[*Hint:* Show that $k \cdot k! = (k + 1)! - k!$.]

6. Find a formula for the sums

$$S_n = \frac{1}{2!} + \frac{2}{3!} + \frac{3}{4!} + \cdots + \frac{n}{(n + 1)!}$$

and prove that your formula holds for all n.

$$\left[\textit{Hint: } \text{Show that } \frac{k}{(k + 1)!} = \frac{1}{k!} - \frac{1}{(k + 1)!} . \right]$$

7. Prove that

$$\frac{n^5}{5} + \frac{n^4}{2} + \frac{n^3}{3} - \frac{n}{30}$$

is an integer for all natural numbers n.

8. Prove that the number of people who have shaken hands an odd number of times is an even number.

9. An $n \times n$ magic square consists of the numbers from 1 to n^2 arranged in a square in such a way that the sum of the numbers in any row, column, or diagonal is the same. Let us call this constant sum S. Find a formula for S for an $n \times n$ magic square.

10. The ancient Greeks studied the triangular numbers that we described in the Problem. In a similar way, we can define other *polygonal numbers,* such as the square and pentagonal numbers:

Square numbers:

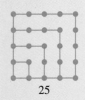 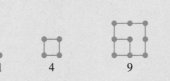

1   4   9   16   25

Pentagonal numbers:

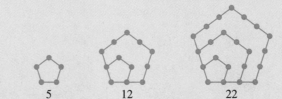

1   5   12   22   35

To find a pattern for such numbers, we construct the *difference table* by taking differences of successive terms in the sequence, then taking differences of the differences, and so on. For example, the triangular numbers give the following difference table:

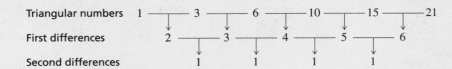

Triangular numbers   1 —— 3 —— 6 —— 10 —— 15 —— 21

First differences   2 —— 3 —— 4 —— 5 —— 6

Second differences   1   1   1   1

We stop at this point because we have obtained a constant sequence. Working backward from the bottom row, we can easily find more of the first differences, and from these, more of the triangular numbers.

(a) Construct the difference tables for the square numbers and the pentagonal numbers. Use the pattern to find the tenth pentagonal number.

(b) Using the patterns you have observed so far, what do you think the second differences would be for the *hexagonal numbers*? Use this, together with the fact that the first two hexagonal numbers are 1 and 6, to construct the difference table that gives the first eight hexagonal numbers.

(c) Sketch a dot pattern like those shown on page 652 to illustrate the first four hexagonal numbers.

11. Choose n different points on the circumference of a circle, and then connect them with line segments. We are interested in the number of regions into which these segments divide the circle. Let's investigate what happens for $n = 1, 2, 3,$ and 4.

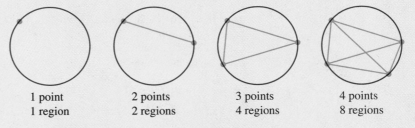

| 1 point | 2 points | 3 points | 4 points |
| 1 region | 2 regions | 4 regions | 8 regions |

(a) How many regions result from five points on the circle?

(b) Based on the pattern you have observed, how many regions do you think would arise from n different points on the circle?

(c) Draw a diagram of the regions that result from six points on the circle. (Do this very carefully.) Does this result fit with the pattern you conjectured in part (b)?

(d) Take the sequence of numbers of regions for $n = 1, 2, 3, 4, 5, 6$, and construct the difference table (see Problem 10). Use the difference table to determine the number of regions for $n = 7, 8, 9, 10$.

12. Starting with an equilateral triangle of side 1, we successively construct new figures with more and more sides as shown. The **snowflake curve** is the result of repeating this process indefinitely.

(a) Find the area enclosed by the snowflake curve.   [*Hint:* This area is the sum of an infinite series.]

(b) Observe that the length of the snowflake curve is infinite.

Parts (a) and (b) show that if your lawn has the shape of a snowflake curve, you could easily mow it, because it has finite area, but you could never put a fence around it.

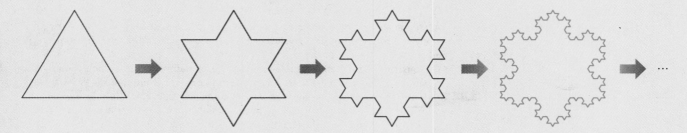

13. (a) Show that $1 + 2r + 3r^2 + 4r^3 + \cdots = \dfrac{1}{(1 - r)^2}$.   [*Hint:* Let S denote this sum, find $S - rS$, and use the formula for the sum of an infinite geometric series.]

(b) Find $\dfrac{1}{6} + \dfrac{2}{6^2} + \dfrac{3}{6^3} + \dfrac{4}{6^4} + \cdots$.

14. A statistician repeatedly rolls a die, observing the numbers that show when it comes to rest. She stops rolling when a 6 appears. What is the expected value for the number of times she has to roll the die?   [*Hint:* To find the expected number of rolls until a 6 appears, we calculate the probability p_n that this takes n rolls. Then the expected number of rolls is $1 \cdot p_1 + 2 \cdot p_2 + 3 \cdot p_3 + \cdots$. Use Problem 13.]

15. For fixed n and for $0 \le k \le n$, show that the sum of the binomial coefficients $\binom{n}{k}$ for k odd is equal to the sum of the binomial coefficients $\binom{n}{k}$ for k even. [*Hint:* Expand $(1 - 1)^n$.]

16. Let p be a prime number.

(a) Show that $\binom{p}{k}$ is divisible by p for $1 \le k \le p$.

(b) Use mathematical induction to prove that $n^p - n$ is divisible by p for all n. [*Hint:* For the induction step, use the Binomial Theorem to expand $(n + 1)^p$, and then use part (a).]

ANSWERS TO ODD-NUMBERED EXERCISES AND CHAPTER TESTS

CHAPTER 1

Section 1.1 ■ page 7

1. $ab = ba$ **3.** $(a - b)(a + b) = a^2 - b^2$
5. $(a + b) + c = a + (b + c)$ **7.** $(ab)^2 = a^2b^2$
9. $A = (a + b)/2$ **11.** $S = x + 2x^2$ **13.** $d = 7w$
15. $P = n(n + 1)$ **17.** $S = x^2 + y^2$ **19.** $t = d/r$
21. $A = x^2$ **23.** $V = 2x^3$ **25.** $S = 2lw + 2lh + 2wh$
27. $A = x^2$ **29.** $L = 2x + 2\pi r$
31. $V = \frac{4}{3}\pi(R^3 - r^3)$ **33.** (a) 84.5
(b) $A = (a + b + 2f)/4$ (c) 89   **35.** (a) 4500 ft^2
(b) $A = 150T$ ft^2 (c) 64 min   (d) 160 ft^2/min
37. (a) GPA $= (4a + 3b + 2c + d)/(a + b + c + d + f)$
(b) 2.84

Section 1.2 ■ page 18

1. Commutative Property for addition
3. Associative Property for addition
5. Distributive Property   **7.** Distributive Property
9. $3x + 3y$ **11.** $8m$ **13.** $-8y$ **15.** $-5x + 10y$
17. (a) $\frac{7}{13}$ (b) $\frac{23}{30}$ **19.** (a) $\frac{4}{9}$ (b) $\frac{15}{2}$
21. (a) False   (b) True   **23.** (a) True   (b) True
25. (a) True   (b) True
27. (a) $x > 0$ (b) $t < 4$ (c) $a \geq \pi$ (d) $-5 < x < \frac{1}{3}$
(e) $|p - 3| \leq 5$
29. (a) $\{1, 2, 3, 4, 5, 6, 8\}$ (b) $\{2, 4, 6\}$
31. (a) $\{1, 2, 3, 4, 5, 6, 7, 8, 9, 10\}$ (b) $\varnothing$
33. (a) $\{x \mid x \leq 5\}$ (b) $\{x \mid -1 < x < 4\}$
35. $-3 < x < 0$ **37.** $2 \leq x < 8$

39. $-1 \leq x \leq 1$ **41.** $x \geq 2$

43. $x \leq -2$ **45.** $(-\infty, 1]$

47. $[1, 2]$ **49.** $(-2, 1]$

51. $(-1, \infty)$ **53.**

55.   **57.**

59.

61. (a) 100   (b) 73   (c) 4
63. (a) $3 - \sqrt{3}$ (b) 2   (c) -1
65. (a) 15   (b) 24   (c) $\frac{67}{40}$
67. (a) $\frac{7}{9}$ (b) $\frac{13}{45}$ (c) $\frac{19}{33}$
69. (a)   (b)

(c)   (d)

71. (a)

x	$1/x$
1	1
2	0.5
10	0.1
100	0.01
1000	0.001

x	$1/x$
1	1
0.5	2
0.1	10
0.01	100
0.001	1000

(b) As x becomes larger, $\dfrac{1}{x}$ becomes smaller.

(c) As x becomes smaller, $\dfrac{1}{x}$ becomes larger.

73. (a) No   **(b)** No

Section 1.3 ■ page 32

1. (a) 16   **(b)** -16 **(c)** 1

3. (a) $\frac{625}{8}$ **(b)** 100,000   **(c)** 4096

5. (a) $\frac{2}{3}$ **(b)** 4   **(c)** $\frac{1}{2}$ **7. (a)** 6   **(b)** 4   **(c)** $\frac{3}{5}$

9. (a) $\frac{3}{2}$ **(b)** $\frac{9}{4}$ **(c)** $\frac{125}{512}$ **11.** $\sqrt[3]{4}$ **13.** $2\sqrt{5}$

15. (a) 2^7 **(b)** 2^{14} **(c)** 2^{36}

17. (a) 2^{-6} **(b)** $2^{5/2}$ **(c)** $2^{-1/2}$ **19.** t^5 **21.** $6x^7y^5$

23. $16x^{10}$ **25.** $4/b^2$ **27.** $64r^7s$ **29.** $648y^7$

31. $\dfrac{x^3}{y}$ **33.** $\dfrac{y^2z^9}{x^5}$ **35.** $\dfrac{s^3}{q^7r^6}$ **37.** $x^{13/15}$

39. $16b^{9/10}$ **41.** $\dfrac{1}{c^{2/3}d}$ **43.** $y^{1/2}$ **45.** $\dfrac{32x^{12}}{y^{16/15}}$

47. $\dfrac{x^{15}}{y^{15/2}}$ **49.** $\dfrac{4a^2}{3b^{1/3}}$ **51.** $\dfrac{3t^{25/6}}{s^{1/2}}$ **53.** $|x|$

55. $x\sqrt[3]{y}$ **57.** $ab\sqrt[5]{ab^2}$ **59.** $|xy^3|$ **61.** $2|x|$

63. (a) $\dfrac{\sqrt{6}}{6}$ **(b)** $\dfrac{\sqrt{3xy}}{3y}$ **(c)** $\dfrac{\sqrt{15}}{10}$

65. (a) $\dfrac{\sqrt[3]{x^2}}{x}$ **(b)** $\dfrac{\sqrt[5]{x^3}}{x}$ **(c)** $\dfrac{\sqrt[7]{x^4}}{x}$

67. (a) 6.93×10^7 **(b)** 2.8536×10^{-5} **(c)** 1.2954×10^8

69. (a) 319,000   **(b)** 0.0000000267

(c) 710,000,000,000,000

71. (a) 5.9×10^{12} mi   **(b)** 4×10^{-13} cm

(c) 3.3×10^{19} molecules

73. 1.3×10^{-20} **75.** 1.429×10^{19} **77.** 7.4×10^{-14}

79. $8\frac{1}{3}$ min   **81.** 10^{10} and 10^{53}

83. (a) 1.15   **(b)** The mass becomes larger

85. (a) 28 mi/h   **(b)** 167 ft   **87.** 1.5×10^{11} m

Section 1.4 ■ page 44

1. $6x + 6$ **3.** $3x^2 - 2x + 6$ **5.** $x^3 + 3x^2 - 6x + 11$

7. $-t^4 + t^3 - t^2 - 10t + 5$ **9.** $x^{3/2} - x$

11. $y^{7/3} - y^{1/3}$ **13.** $21t^2 - 29t + 10$

15. $3x^2 + 5xy - 2y^2$ **17.** $1 - 4y + 4y^2$

19. $2x^3 - 7x^2 + 7x - 5$ **21.** $x^3 + x^2 - 2x$

23. $30y^4 + y^5 - y^6$ **25.** $4x^4 + 12x^2y^2 + 9y^4$

27. $x^4 - a^4$ **29.** $1 + 3a^3 + 3a^6 + a^9$ **31.** $a - 1/b^2$

33. $x^5 + x^4 - 3x^3 + 3x - 2$ **35.** $1 - x^{2/3} + x^{4/3} - x^2$

37. $1 - 2b^2 + b^4$ **39.** $3x^4y^4 + 7x^3y^5 - 6x^2y^3 - 14xy^4$

41. $2x(1 + 6x^2)$ **43.** $3y^3(2y - 5)$ **45.** $(x + 6)(x + 1)$

47. $(x - 4)(x + 2)$ **49.** $(y - 3)(y - 5)$

51. $(2x + 3)(x + 1)$ **53.** $9(x - 2)(x + 2)$

55. $(3x + 2)(2x - 3)$ **57.** $3(x - 1)(x + 2)$

59. $y^4(y + 2)^3(y + 1)^2$ **61.** $(a + 1)(a - 1)(b + 2)(b - 2)$

63. $(t + 1)(t^2 - t + 1)$ **65.** $(2t - 3)^2$ **67.** $x(x + 1)^2$

69. $(2x + y)^2$ **71.** $x^2(x + 3)(x - 1)$

73. $(2x - 5)(4x^2 + 10x + 25)$ **75.** $(x^2 + 2)(x + 1)(x - 1)$

77. $(y + 2)(y - 2)(y - 3)$ **79.** $(2x^2 + 1)(x + 2)$

81. $(x + y)(x - y)(x^2 + xy + y^2)(x^2 - xy + y^2)$

83. $x^{1/2}(x + 1)(x - 1)$ **85.** $x^{-3/2}(x + 1)^2$

87. $(x^2 + 3)(x^2 + 1)^{-1/2}$

89. $(a + 2)(a - 2)(a + 1)(a - 1)$

91. $(x^2 - x + 2)(x^2 + x + 2)$

95. $(a + b + c)(a - b + c)(a + b - c)(-a + b + c)$

Section 1.5 ■ page 54

1. $\dfrac{x + 1}{x + 3}$ **3.** $\dfrac{-y}{y + 1}$ **5.** $\dfrac{x(2x + 3)}{2x - 3}$ **7.** $\dfrac{1}{t^2 + 9}$

9. $\dfrac{x + 4}{x + 1}$ **11.** $\dfrac{(2x + 1)(2x - 1)}{(x + 5)^2}$ **13.** $x^2(x + 1)$

15. $\dfrac{x}{yz}$ **17.** $\dfrac{3x + 7}{(x - 3)(x + 5)}$ **19.** $\dfrac{1}{(x + 1)(x + 2)}$

21. $\dfrac{3x + 2}{(x + 1)^2}$ **23.** $\dfrac{u^2 + 3u + 1}{u + 1}$ **25.** $\dfrac{2x + 1}{x^2(x + 1)}$

27. $\dfrac{2x + 7}{(x + 3)(x + 4)}$ **29.** $\dfrac{x - 2}{(x + 3)(x - 3)}$ **31.** $\dfrac{5x - 6}{x(x - 1)}$

33. $\dfrac{-5}{(x + 1)(x + 2)(x - 3)}$ **35.** $-xy$ **37.** $\dfrac{c}{c - 2}$

39. $\dfrac{3x + 7}{x^2 + 2x - 1}$ **41.** $\dfrac{y - x}{xy}$ **43.** $\dfrac{-1}{a(a + h)}$

45. $\dfrac{-3}{(2 + x)(2 + x + h)}$ **47.** $\dfrac{1}{\sqrt{1 - x^2}}$

49. $\dfrac{x + 2}{(x + 1)^{3/2}}$ **51.** $\dfrac{2x + 3}{(x + 1)^{4/3}}$ **53.** $\dfrac{3 - \sqrt{5}}{2}$

55. $\dfrac{2(\sqrt{7} - \sqrt{2})}{5}$ **57.** $\dfrac{-4}{3(1 + \sqrt{5})}$ **59.** $\dfrac{r - 2}{5(\sqrt{r} - \sqrt{2})}$

61. $\dfrac{-1}{\sqrt{x(x + h)}(\sqrt{x} + \sqrt{x + h})}$ **63.** $\dfrac{x + 1}{\sqrt{x^2 + x + 1} - x}$

65. True   **67.** False   **69.** False   **71.** True

73. False   **75.** (a) $\dfrac{R_1 R_2}{R_1 + R_2}$ (b) $\frac{20}{3} \approx 6.7$ ohms

Chapter 1 Review ■ page 57

1. Commutative Property for addition

3. Distributive Property

5. $-1 < x \le 3$

7. $(2, \infty)$

9. 6   **11.** $\frac{1}{72}$ **13.** $\frac{1}{6}$ **15.** 11   **17.** 4

19. x^{-2} **21.** x^{4m+2} **23.** x^{a+b+c} **25.** x^{5c-1}

27. $12x^5 y^4$ **29.** $9x^3$ **31.** $x^2 y^2$ **33.** $\dfrac{x(2 - \sqrt{x})}{4 - x}$

35. $\dfrac{4r^{5/2}}{s^7}$ **37.** 7.825×10^{10} **39.** 1.65×10^{-32}

41. $3xy^2(4xy^2 - y^3 + 3x^2)$ **43.** $(x - 2)(x + 5)$

45. $(4t + 3)(t - 4)$ **47.** $(5 - 4t)(5 + 4t)$

49. $(x - 1)(x^2 + x + 1)(x + 1)(x^2 - x + 1)$

51. $x^{-1/2}(x - 1)^2$ **53.** $(x - 2)(4x^2 + 3)$

55. $\sqrt{x^2 + 2}\,(x^2 + x + 2)^2$

57. $y(a + b)(a - b)$ **59.** x^2 **61.** $6x^2 - 21x + 3$

63. $4a^4 - 4a^2 b + b^2$ **65.** $x^3 - 6x^2 + 11x - 6$

67. $2x^{3/2} + x - x^{1/2}$ **69.** $2x^3 - 6x^2 + 4x$

71. $\dfrac{x - 3}{2x + 3}$ **73.** $\dfrac{3(x + 3)}{x + 4}$ **75.** $\dfrac{x + 1}{x - 4}$

77. $\dfrac{x + 1}{(x - 1)(x^2 + 1)}$ **79.** $\dfrac{1}{x + 1}$ **81.** $-\dfrac{1}{2x}$

83. $6x + 3h - 5$ **85.** (a) Yes   (b) 170   **87.** No

89. Yes   **91.** No   **93.** No

Chapter 1 Test ■ page 60

1. (a)

(b) $(-\infty, 5)$; $[-2, 1]$ (c) 53

2. (a) 81   (b) $\frac{1}{16}$ (c) $5^6 = 15,625$

3. (a) $\frac{3}{2}$ (b) 2   (c) $\frac{1}{8}$ **4.** $x^{b/2}$

5. (a) $8\sqrt{2}$ (b) $54a^6 b^{14}$ (c) $\dfrac{y^{32}}{x^8}$ (d) $\dfrac{8x^{1/4}}{y}$

6. (a) $\dfrac{x + 2}{x - 2}$ (b) $\dfrac{1}{x - 2}$ (c) $-x - y$

7. (a) 3.25×10^{11} (b) 8.931×10^{-6}

8. (a) $-3 - 7x$ (b) $2x^2 - 7x - 15$ (c) $x - y$

(d) $9t^2 + 24t + 16$ (e) $8 - 12x^2 + 6x^4 - x^6$

9. (a) $(3x - 5)(3x + 5)$ (b) $(3x + 5)(2x - 1)$

(c) $(x^2 - 3)(x - 4)$ (d) $x(x + 3)(x^2 - 3x + 9)$

(e) $3x^{-1/2}(x - 1)(x - 2)$

10. $\dfrac{x(\sqrt{x} + 2)}{x - 4}$

PRINCIPLES OF PROBLEM SOLVING ■ page 65

1. 37.5 mi/h   **3.** 150 mi   **5.** 2   **7.** 57 min

9. No   **11.** The same amount

13. (a) Uncle   (b) Father   (d) No   **15.** 8.49

CHAPTER 2

Section 2.1 ■ page 74

1. (a) Yes   (b) No   **3.** (a) No   (b) Yes

5. (a) No   (b) Yes   **7.** Not an identity   **9.** Identity

11. Identity   **13.** Yes, $x = 5$ **15.** No

17. Yes, $x = -20$ **19.** $x = 4$ **21.** $x = -9$

23. $w = -3$ **25.** $y = 12$ **27.** $x = -\frac{3}{4}$ **29.** $y = \frac{32}{9}$

31. $x = -\frac{1}{3}$ **33.** $t = -20$ **35.** $r = 3$ **37.** $x = -\frac{1}{2}$

39. $x = -\frac{4}{9}$ **41.** $x = \frac{13}{3}$ **43.** $x = \frac{29}{2}$ **45.** $x = 30$

47. $z = \frac{3}{97}$ **49.** $u = 2$ **51.** No solution

53. No solution   **55.** No solution   **57.** $x \approx 5.06$

59. $x \approx 43.66$ **61.** $R = \dfrac{PV}{nT}$ **63.** $R_1 = \dfrac{RR_2}{R_2 - R}$

65. $x = \dfrac{2d - b}{a - 2c}$ **67.** $x = \dfrac{1 - a}{a^2 - a - 1}$ **69.** $m = \dfrac{3a}{8}$

71. $k =$ any real number

73. Can't divide by an expression that contains the variable; correct solution $x = -1$

Section 2.2 ■ page 82

1. $0.14x$ **3.** $3n + 6$ **5.** $50w$ **7.** $3s + 15$

9. $\dfrac{3a - 8}{3}$ **11.** 9000 at $4\frac{1}{2}\%$ and 3000 at 4%

13. 45 ft   **15.** Plumber, 70 h; assistant, 35 h

17. 9 coins of each type   **19.** 7 years

21. 48 lb expensive, 32 lb cheaper   **23.** 65, 67, 69, 71

25. 4 in.   **27.** $3000   **29.** 8 min 20 s   **31.** $1\frac{1}{2}$ h

33. 450 mi   **35.** 4 h   **37.** 500 mi/h   **39.** 200 mL

41. 18 g   **43.** 0.6 L   **45.** 37 min 20 s   **47.** 3 h

49. 95 ft   **51.** 120 ft   **53.** 18 ft   **55.** 4.55 ft

Section 2.3 ■ page 95

1. ± 7 **3.** $\pm 2\sqrt{6}$ **5.** $\pm 2\sqrt{2}$ **7.** $-2, 3$ **9.** 2

11. $-\frac{1}{2}, -3$ **13.** $-\frac{4}{3}, \frac{1}{2}$ **15.** $-1 \pm \sqrt{3}$

17. $3 \pm 3\sqrt{2}$ **19.** $\dfrac{-4 \pm \sqrt{14}}{2}$ **21.** $0, \frac{1}{4}$ **23.** $-2, 4$

25. $-6 \pm 3\sqrt{7}$ **27.** $\dfrac{-3 \pm 2\sqrt{6}}{3}$ **29.** $\dfrac{1 \pm \sqrt{5}}{4}$

31. $-\frac{9}{2}, \frac{1}{2}$ **33.** $\dfrac{-5 \pm \sqrt{13}}{2}$ **35.** $\dfrac{\sqrt{5} \pm 1}{2}$

37. $-50, 100$ **39.** -4 **41.** No real solution

43. $1, 1 + \dfrac{1}{a}$ **45.** $-0.248, 0.259$ **47.** $1.200, 1.250$

49. $r = \pm\sqrt{\dfrac{3V}{\pi h}}$ **51.** $b = \pm\sqrt{c^2 - a^2}$

53. $i = 100\left(-1 \pm \sqrt{A/p}\right)$ **55.** $k = 15$ **57.** 2

59. 1 **61.** 2 **63.** $k = \pm 20$ **65.** 19 and 36

67. 25 ft by 35 ft   **69.** 13 in. by 13 in.

71. 120 ft by 126 ft

73. (a) After 1 s and $1\frac{1}{2}$ s   **(b)** Never   **(c)** 25 ft

(d) After $1\frac{1}{4}$ s   **(e)** After $2\frac{1}{2}$ s

75. (a) After 17 yr, on Jan. 1, 2009

(b) After 18.612 yr, on Aug. 12, 2010

77. 50 mi/h (or 240 mi/h)   **79.** 6 km/h

81. Irene 3 h, Henry $4\frac{1}{2}$ h   **83.** 215,000 mi

Section 2.4 ■ page 104

1. Real part 3, imaginary part -5

3. Real part 0, imaginary part 6

5. Real part $\sqrt{2}$, imaginary part $\sqrt{3}$

7. $9 + i$ **9.** $12 + i$ **11.** $-19 + 4i$ **13.** $-4 + 8i$

15. $30 + 10i$ **17.** $-33 - 56i$ **19.** $-i$ **21.** $\frac{8}{5} + \frac{1}{5}i$

23. $-5 + 12i$ **25.** $-4 + 2i$ **27.** $-i$ **29.** 1

31. $5i$ **33.** -6 **35.** $(3 + \sqrt{5}) + (3 - \sqrt{5})i$ **37.** 2

39. $-i\sqrt{2}$ **41.** $\pm 3i$ **43.** $2 \pm i$ **45.** $-\dfrac{1}{2} \pm \dfrac{\sqrt{3}}{2}i$

47. $\frac{1}{2} \pm \frac{1}{2}i$ **49.** $-\dfrac{3}{2} \pm \dfrac{\sqrt{3}}{2}i$ **51.** $-1 + 5i$

53. $-\frac{5}{2} + 2i$ **55.** $(-2 \pm \sqrt{5})i$ **57.** $-\dfrac{\sqrt{6}}{4} \pm \dfrac{\sqrt{26}}{4}i$

Section 2.5 ■ page 111

1. 3   **3.** ± 2 **5.** $-1, 0, 2$ **7.** $-\sqrt{2}, \sqrt{2}, 5$ **9.** 2

11. 4   **13.** 4   **15.** 21   **17.** No real solution

19. $-7, 0$ **21.** $-1, 0, 3$ **23.** $\pm 2\sqrt{2}, \pm 3\sqrt{3}$

25. 27, 729   **27.** $-\frac{1}{2}$ **29.** $-3, \dfrac{1 \pm \sqrt{13}}{2}$ **31.** 2

33. $-2, 1 \pm i\sqrt{3}$ **35.** $1, \dfrac{-1 \pm i\sqrt{3}}{2}$

37. $0, \dfrac{-1 \pm i\sqrt{3}}{2}$ **39.** $\pm\sqrt{2}, \pm 2$

41. $1, 2, \dfrac{-1 \pm i\sqrt{3}}{2}, -1 \pm i\sqrt{3}$

43. $\pm 3i$ **45.** 2 ft by 6 ft by 15 ft   **47.** 16 mi; no

49. 7.52 ft   **51.** 49 ft, 168 ft, and 175 ft

53. $\pm i\sqrt{a}, \pm 2i\sqrt{a}$ **55.** $\sqrt{a^2 + 36}$

Section 2.6 ■ page 116

1. $\left\{-1, 0, \frac{1}{2}, \sqrt{2}, 2\right\}$ **3.** $\left\{0, \frac{1}{2}, \sqrt{2}, 2\right\}$ **5.** $\{-1, 2\}$

7. $\left\{-1, 0, \frac{1}{2}\right\}$

9. $(-\infty, 4]$ **11.** $(-\infty, -5)$

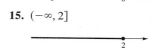

13. $(4, \infty)$ **15.** $(-\infty, 2]$

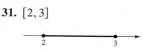

17. $\left(-\infty, -\frac{1}{2}\right)$ **19.** $[1, \infty)$

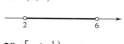

21. $[-1, \infty)$ **23.** $(-\infty, -1]$

25. $(2, 6)$ **27.** $(0, 1]$

29. $\left[-1, \frac{1}{2}\right)$ **31.** $[2, 3]$

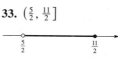

33. $\left(\frac{5}{2}, \frac{11}{2}\right]$ **35.** $\left[\frac{7}{3}, \infty\right)$

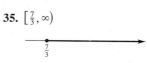

37. $(-\infty, 0) \cup \left(\frac{1}{4}, \infty\right)$ **39.** $\left(3, \frac{7}{2}\right]$

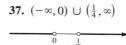

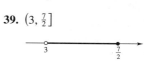

41. $68 \leqslant F \leqslant 86$

43. (a) $-\frac{1}{3}P + \frac{560}{3}$ **(b)** From \$215 to \$290

45. Between 12,000 mi and 14,000 mi

51. $x \geqslant \dfrac{c(a + b)}{ab}$ **53.** $x > \dfrac{c - b}{a}$

Section 2.7 ■ page 124

1. $(-\infty, -2) \cup (5, \infty)$

3. $[-3, 6]$

5. $(-\infty, -1] \cup \left[\frac{1}{2}, \infty\right)$

7. $(-1, 4)$

9. $(-\infty, -3) \cup (6, \infty)$

11. $(-2, 2)$

13. $(-\infty, \infty)$

15. $[-2, 0] \cup [2, \infty)$

17. $(-\infty, -1) \cup [3, \infty)$

19. $\left(-\infty, -\frac{3}{2}\right)$

21. $(-\infty, 5) \cup [16, \infty)$

23. $(-2, 0) \cup (2, \infty)$

25. $(-\infty, -2] \cup [2, \infty)$

27. $[-2, -1) \cup (0, 1]$

29. $\left(0, \frac{3}{4}\right] \cup (1, \infty)$

31. $(-\infty, -3) \cup (6, \infty)$

33. $\left[-8, -\frac{5}{2}\right)$

35. $[-2, 0) \cup (1, 3]$

37. $\left(-3, -\frac{1}{2}\right) \cup (2, \infty)$

39. $(-\infty, -1) \cup (1, \infty)$

41. From 0 s to 3 s   **43.** Distances greater than 30 m

45. $-\frac{4}{3} \leqslant x \leqslant \frac{4}{3}$ **47.** $(-\infty, -2) \cup (7, \infty)$

49. $(-\infty, -c) \cup [-a, b]$

Section 2.8 ■ page 129

1. 4   **3.** 0.1   **5.** $4 - a$

7. $|x - 3| = \begin{cases} x - 3 & \text{if } x \geqslant 3 \\ 3 - x & \text{if } x < 3 \end{cases}$ **9.** $3|x + 3|$

11. $\frac{1}{2}|x - 5|$ **13.** $x^2 + 9$ **15.** ± 3 **17.** 1.95, 2.05

19. $-4, -\frac{2}{5}$ **21.** $-\frac{3}{2}, -\frac{1}{4}$ **23.** $(-2, 2)$ **25.** $[2, 8]$

27. $(-\infty, -2] \cup [0, \infty)$ **29.** $(-\infty, -7] \cup [-3, \infty)$

31. $[1.3, 1.7]$ **33.** $(-4, 8)$ **35.** $(-6.001, -5.999)$

37. $[-4, -1] \cup [1, 4]$ **39.** $\left(-\frac{15}{2}, -7\right) \cup \left(-7, -\frac{13}{2}\right)$

41. $\left(\frac{1}{2}, \infty\right)$ **43.** $(-1, \infty)$

Chapter 2 Review ■ page 130

1. 4   **3.** 5   **5.** $\frac{15}{2}$ **7.** -6 **9.** 0   **11.** 2, 7

13. $-1, \frac{1}{2}$ **15.** $0, \pm\frac{5}{2}$ **17.** $\dfrac{-2 \pm \sqrt{7}}{3}$ **19.** $\dfrac{3 \pm \sqrt{6}}{3}$

21. ± 3 **23.** 1   **25.** 3, 11   **27.** 20 lb raisins, 30 lb nuts

29. $\frac{1}{4}(\sqrt{329} - 3) \approx 3.78$ mi/h   **31.** 12 cm, 16 cm

33. 23 ft by 46 ft by 8 ft   **35.** $-3 - 9i$ **37.** $19 + 40i$

39. $\dfrac{-5 - 12i}{13}$ **41.** i **43.** $(2 + 2\sqrt{3}) + (2 - 2\sqrt{3})i$

45. $\pm 4i$ **47.** $-3 \pm i$ **49.** $-1 - 2i$ **51.** $\pm 4, \pm 4i$

53. $\pm\dfrac{\sqrt{3}}{3}i$

55. $(-3, \infty)$

57. $(-3, -1]$

59. $(-\infty, -6) \cup (2, \infty)$

61. $[-4, -1)$

63. $(-\infty, -2) \cup (2, 4]$

65. $[2, 8]$

67. $(-\infty, -1] \cup [0, \infty)$

69. (a) $\left[-3, \frac{8}{3}\right]$ **(b)** $(0, 1)$

Chapter 2 Test ■ page 132

1. (a) -10 **(b)** 1 **2.** 120 mi

3. (a) $-1 - \frac{3}{2}i$ **(b)** $5 + i$ **(c)** $-1 + 2i$ **(d)** -1
(e) $6\sqrt{2}$

4. (a) $-3, 4$ **(b)** $\dfrac{-2 \pm i\sqrt{2}}{2}$ **(c)** No solution

(d) $0, \frac{1}{2}, 1$ **(e)** $0, -3, \dfrac{3 \pm 3i\sqrt{3}}{2}$

5. 50 ft by 120 ft

6. (a) $\left(-\frac{5}{2}, 3\right]$ **(b)** $(0,1) \cup (2, \infty)$ **(d)** $[-4, -1]$

7. 41 °F to 50 °F **8.** $0 \leqslant x \leqslant 4$

FOCUS ON PROBLEM SOLVING ■ page 134

3. $-7, -1, 3, 9$ **5.** $9, -\frac{7}{3}$ **7.** $-5 \leqslant x \leqslant 0$

9. $A = 2, B = 1, C = 9, D = 7, E = 8$

11. *Hint:* First divide the eggs into groups of three.

13. $2\pi \approx 6.28$ ft

15. *Hint:* Use the quadratic formula to find r_1 and r_2 in terms of p, and consider the discriminant.

CHAPTER 3

Section 3.1 ■ page 141

1.

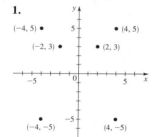

3. $A(5, 1), B(1, 2),$
$C(-2, 6), D(-6, 2),$
$E(-4, -1), F(-2, 0),$
$G(-1, -3), H(2, -2)$

5. 24 **7. (a)**

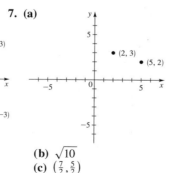

(b) $\sqrt{10}$
(c) $\left(\frac{7}{2}, \frac{5}{2}\right)$

9. (a)

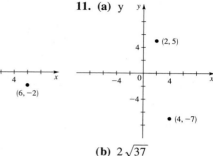

(b) $\sqrt{74}$
(c) $\left(\frac{5}{2}, \frac{1}{2}\right)$

11. (a) y

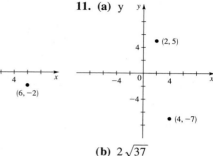

(b) $2\sqrt{37}$
(c) $(3, -1)$

13. (a)

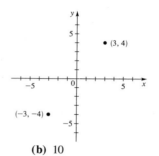

(b) 10
(c) $(0, 0)$

15. Trapezoid, area $= 9$

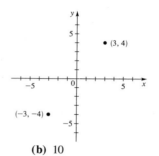

17.

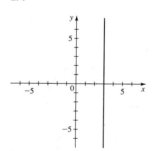

19.

21.

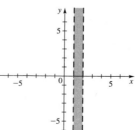

23.

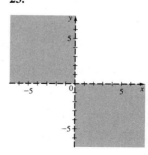

25.

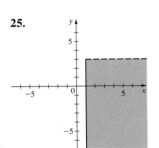

27.

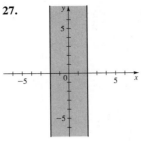

17. x-intercept 1,
y-intercept -1,
no symmetry

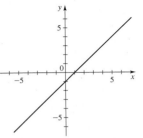

29.

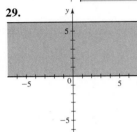

31. $A(6,7)$
33. $Q(-1,3)$
37. **(b)** 10
41. $(0,-4)$
43. $\left(1,\frac{7}{2}\right)$

19. x-intercept $\frac{5}{3}$,
y-intercept -5,
no symmetry

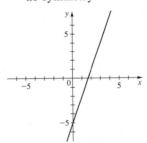

21. x-intercepts ±1,
y-intercept 1,
symmetry about y-axis

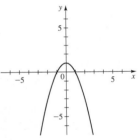

45. $(2,-3)$

47. **(a)**

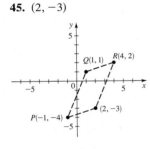

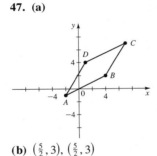

(b) $\left(\frac{5}{2},3\right)$, $\left(\frac{5}{2},3\right)$

49. **(a)** $(8,5)$ **(b)** $(a+3,b+2)$
   (c) $A'(-2,1)$, $B'(0,4)$, $C'(5,3)$

23. x-intercept 0,
y-intercept 0,
symmetry about y-axis

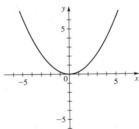

25. x-intercepts ±3,
y-intercept -9,
symmetry about y-axis

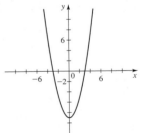

Section 3.2 ■ page 151

1. No, yes, no   **3.** No, yes, yes
5. Yes, no, yes
7. x-intercept 3, y-intercept -3
9. x-intercepts ±3, y-intercept -9
11. x-intercepts ±2, y-intercepts ±2
13. None
15. x-intercept 0, y-intercept 0,
   symmetry about origin

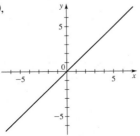

27. No intercepts,
   symmetry about origin

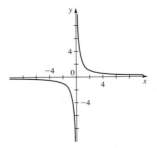

29. x-intercept 0,
   y-intercept 0,
   no symmetry

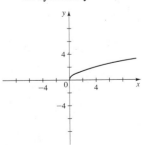

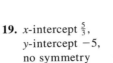
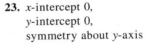

31. *x*-intercepts ±2,
   y-intercept 2,
   symmetry about *y*-axis

33. *x*-intercept 0,
   y-intercept 0,
   symmetry about *y*-axis

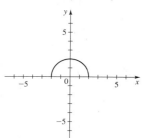

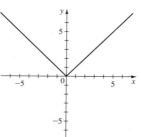

35. *x*-intercepts ±4,
   y-intercept 4,
   symmetry about *y*-axis

37. *x*-intercept 0,
   y-intercept 0,
   symmetry about origin

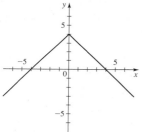

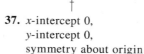

39. *x*-intercepts 0, *y*-intercept 0,
   symmetry about *y*-axis

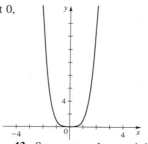

41. Symmetry about *y*-axis   **43.** Symmetry about origin
45. Symmetry about origin
47.

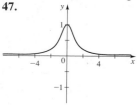

49.

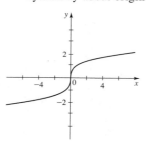

51. $(x - 2)^2 + (y + 1)^2 = 9$
53. $x^2 + y^2 = 65$
55. $(x - 2)^2 + (y - 3)^2 = 13$
57. $(x - 7)^2 + (y + 3)^2 = 9$
59. $(x + 2)^2 + (y - 2)^2 = 4$ **61.** $(1, -2), 2$

63. $(2, -5), 4$ **65.** $\left(-\frac{1}{2}, 0\right), \frac{1}{2}$ **67.** $\left(\frac{1}{4}, -\frac{1}{4}\right), \sqrt{5/8}$
69.   **71.**

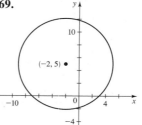

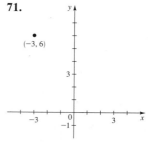

73.   **75.**

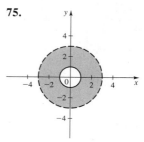

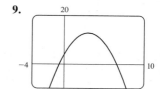

77. 12π
79. $a^2 + b^2 > 4c, \left(-\frac{1}{2}a, -\frac{1}{2}b\right), \frac{1}{2}\sqrt{a^2 + b^2 - 4ac}$

Section 3.3 ■ page 159

1. (c)   **3.** (c)   **5.** (c)
7.   **9.**

11.   **13.**

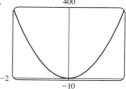

15.   **17.**

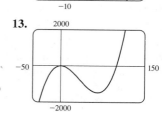

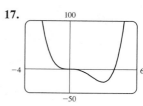

19.

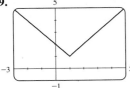

21.

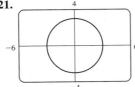

23.

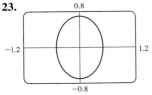

25. 3.0, 4.0   **27.** 1.0, 2.0, 3.0   **29.** 1.62
31. $-1.0, 0, 1.0$ **33.** $[-2.0, 5.0]$
35. $(-\infty, 1.0] \cup [2.0, 3.0]$ **37.** $(-1.0, 0) \cup (1.0, \infty)$
39. $(-\infty, 0)$

Section 3.4 ■ page 170

1. 4   **3.** $-\frac{9}{2}$ **5.** 1   **7.** $-2, \frac{1}{2}, 3, -\frac{1}{4}$
9. $x + y - 4 = 0$ **11.** $3x - 2y - 6 = 0$
13. $x - y + 1 = 0$ **15.** $2x - 3y + 19 = 0$
17. $5x + y - 11 = 0$ **19.** $3x - y - 2 = 0$
21. $3x - y - 3 = 0$ **23.** $y = 5$
25. $x + 2y + 11 = 0$ **27.** $x = -1$
29. $5x - 2y + 1 = 0$ **31.** $x - y + 6 = 0$
33. (a)   **(b)** $3x - 2y + 8 = 0$

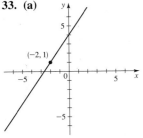

35. All lines pass through $(3, 2)$
37. $-1, 3$ **39.** $-\frac{1}{3}, 0$

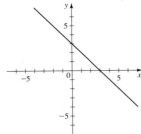
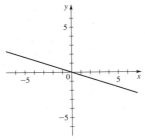

41. $\frac{3}{2}, 3$

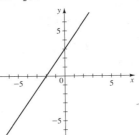

43. $0, 4$

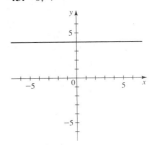

45. $\frac{3}{4}, -3$

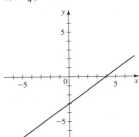

47. $-\frac{3}{4}, \frac{1}{4}$

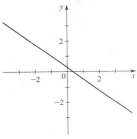

53. $x - y - 3 = 0$ **55. (b)** $4x - 3y - 24 = 0$
57. 16,667 ft
59. (a)

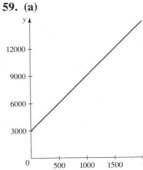

(b) The slope represents production cost per toaster; the y-intercept represents monthly fixed cost.

61. (a) $t = \frac{5}{24}n + 45$ **(b)** $76\,°F$
63. (a) $P = 0.434d + 15$, where P is pressure in lb/in^2 and d is depth in feet   **(b)** 196 ft
65. (a) $C = \frac{1}{4}d + 260$ **(b)** \$635
(c) The slope represents cost per mile. *See graph at right.*

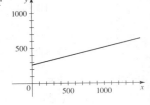

(d) The y-intercept represents annual fixed cost.

67. (a)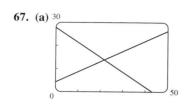

(b) $(21.82, 13.82)$
(c) Price $21.82, amount 13.82

Section 3.5 ■ page 179

1. $f(x) = 3x + 1$ **3.** $f(x) = (x + 2)^2$
5. Divide by 3, then subtract 5
7. Square, multiply by 2, then subtract 3
9.

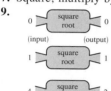 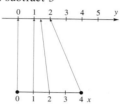

11. $3, \; -3, \; 2, \; 2\sqrt{5} + 1, \; 2a + 1, \; -2a + 1, \; 2(a + b) + 1$
13. $-\frac{1}{3}, \; -3, \; \dfrac{1 - \pi}{1 + \pi}, \; \dfrac{1 - a}{1 + a}, \; \dfrac{2 - a}{a}, \; \dfrac{1 + a}{1 - a}$
15. $-4, \; 10, \; 3\sqrt{2}, \; 5 + 7\sqrt{2}, \; 2x^2 - 3x - 4, \; 2x^2 + 7x + 1,$
$4x^2 + 6x - 8, \; 8x^2 + 6x - 4$
17. $3a + 2, \; 3a + 3h + 4, \; 3(a + h) + 2, \; 3$
19. $5, \; 10, \; 5, \; 0$
21. $3 - 5a + 4a^2, \; 6 - 5a - 5h + 4a^2 + 4h^2,$
$3 - 5a - 5h + 4a^2 + 8ah + 4h^2, \; -5 + 8a + 4h$
23. $15 + 8h, \; 8x + 8h - 1, \; 8$
25. $\dfrac{1}{2 + h}, \; \dfrac{1}{x + h}, \; \dfrac{-1}{x(x + h)}$
27. $(-\infty, \infty), \; (-\infty, \infty)$ **29.** $[-1, 5], \; [-2, 10]$
31. $[-2, 3], \; [-6, 14]$ **33.** $(-\infty, \infty), \; (-\infty, 2]$
35. $[\frac{5}{2}, \infty), \; [0, \infty)$ **37.** $[-1, 1], \; [3, 4]$ **39.** $\{x \mid x \neq 3\}$
41. $\{x \mid x \neq \pm 1\}$ **43.** $(-\infty, \infty)$ **45.** $[5, \infty)$
47. $(-\infty, \infty)$ **49.** $[-2, 3) \cup (3, \infty)$ **51.** $(10, \infty)$
53. $[0, 1]$ **55.** $(-\infty, -\frac{1}{2}] \cup [\frac{1}{2}, \infty)$
57. $(-\infty, 0] \cup [6, \infty)$ **59.** $[0, \pi)$

Section 3.6 ■ page 189

1. (a) $2, 0, 2, 3$ **(b)** $[-3, 3], \; [0, 3]$
(c) Increasing on $[0, 3]$; decreasing on $[-3, 0]$
3. (a) $3, 2, -2, 1, 0$ **(b)** $[-4, 4], \; [-2, 3]$
(c) Increasing on $[0, 2]$; decreasing on $[-4, 0], (2, 4]$
5. (a) $f(0)$ **(b)** $g(-3)$ **(c)** $-2, 2$
7. (a) Yes   **(b)** No   **(c)** Yes   **(d)** No
9. Function, domain $[-3, 2]$, range $[-2, 2]$
11. Not a function

13. (a)

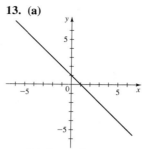

(b) $(-\infty, \infty)$
(c) Decreasing on $(-\infty, \infty)$

15. (a)

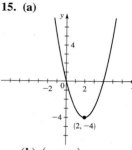

(b) $(-\infty, \infty)$
(c) Increasing on $[2, \infty)$, decreasing on $(-\infty, 2]$

17. (a)

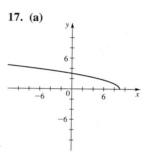

(b) $(-\infty, 9]$
(c) Decreasing on $(-\infty, 9]$

19. (a)

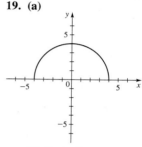

(b) $[-4, 4]$
(c) Increasing on $[-4, 0]$, decreasing on $[0, 4]$

21.

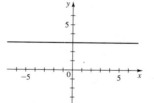

23.

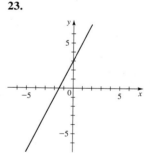

25.

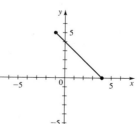

27.

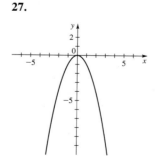

29.

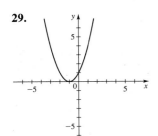

31.

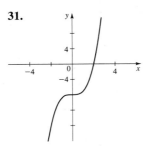

33.

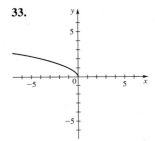

35.

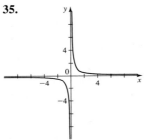

37.

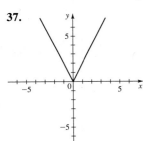

39.

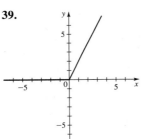

41.

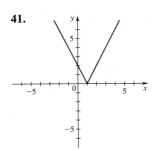

43.

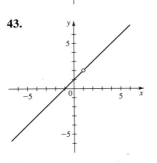

45. (a)

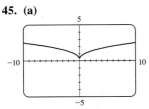

(b) Increasing on $[0, \infty)$, decreasing on $(-\infty, 0]$

47. (a)

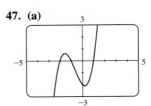

(b) Increasing on $(-\infty, -1.55]$, $[0.22, \infty)$; decreasing on $[-1.55, 0.22]$

49. (a)

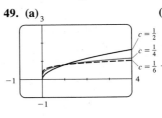

(b)

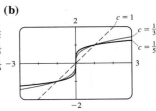

(c)

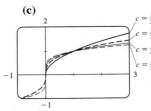

(d) Graphs of even roots are similar to $\sqrt{x}$, graphs of odd roots are similar to $\sqrt[3]{x}$. As c increases, the graph of $y = \sqrt[c]{x}$ becomes steeper near 0 and flatter when $x > 1$.

51. (a)

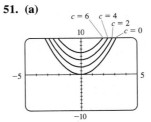

(b)

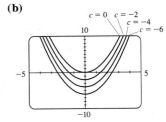

(c) If $c > 0$, then the graph of $f(x) = x^2 + c$ is the same as the graph of $y = x^2$ shifted upward c units. If $c < 0$, then the graph of $f(x) = x^2 + c$ is the same as the graph of $y = x^2$ shifted downward c units.

53. (a)

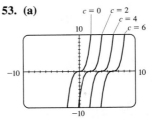

(b)

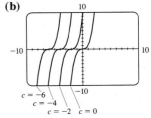

(c) If $c > 0$, then the graph of $f(x) = (x - c)^2$ is the same as the graph of $y = x^2$ shifted right c units. If $c < 0$, then the graph of $f(x) = (x - c)^2$ is the same as the graph of $y = x^2$ shifted left c units.

55. $g(x) = x^3/10$

57. (a)

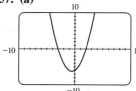 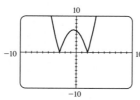

(b) To obtain the graph of g, reflect in the x-axis the part of the graph of f that is below the x-axis.

59. 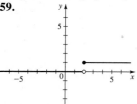   **61.**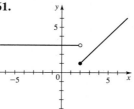

63. 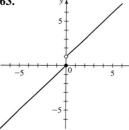   **65.**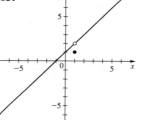

67. 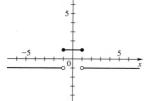   **69.**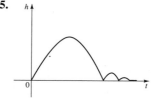

71. 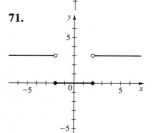   **73.**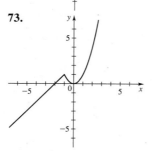

75.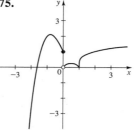

77.
$$C(x) = \begin{cases} 2 & 0 < x \le 1 \\ 2.2 & 1 < x \le 1.1 \\ 2.4 & 1.1 < x \le 1.2 \\ \vdots & \\ 4.0 & 1.9 < x < 2.0 \end{cases}$$

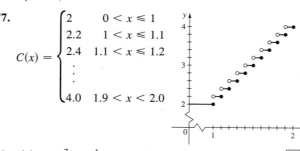

79. $f(x) = -\frac{7}{6}x - \frac{4}{3}, -2 \le x \le 4$ **81.** $f(x) = 1 - \sqrt{-x}$

Section 3.7 ■ page 200

1. This person's weight increases as he grows, then continues to increase; the person then goes on a crash diet (possibly) at age 31, then gains weight again, the weight gain eventually leveling off.

3. 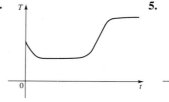   **5.**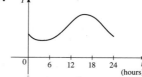

7.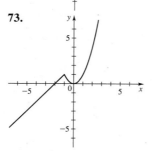

9. $A = 10x - x^2, 0 < x < 10$

11. $A = (\sqrt{3}x^2)/4, x > 0$

13. $r = \sqrt{A/\pi}, A > 0$

15. $A = x^2 + (48/x), x > 0$

17. $A = 15x - (\pi + 4)x^2/8, x > 0$

19. $A = 2x(1200 - x), 0 < x < 1200$

21. $d = 25t, t \ge 0$

23.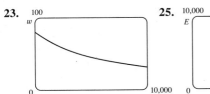

25.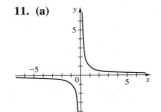

11. (a)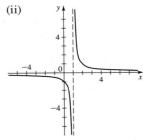

27. $R = kt$ **29.** $v = k/z$ **31.** $y = ks/t$
33. $z = k\sqrt{y}$ **35.** $y = 18x$ **37.** $M = 15x/y$
39. $W = 360/r^2$ **41. (a)** $F = kx$ **(b)** 8   **(c)** 32 N
43. (a) $C = \frac{1}{8}pn$ **(b)** \$57,500
45. (a) $R = kL/d^2$, $k = \frac{7}{2400} = 0.002916\overline{6}$ **(b)** $\approx137 \, \Omega$

Section 3.8 ■ page 214

1. (a) Shift downward 4 units   **(b)** Shift right 4 units
3. (a) Stretch vertically by a factor of 3
(b) Shrink vertically by a factor of 3
5. (a) Reflect in the x-axis and shift upward 5 units
(b) Reflect in the y-axis and shift upward 5 units
7. (a) Shift right 2 units and downward 3 units
(b) Shift right 3 units and stretch vertically by a factor of 2

(b) (i) 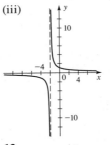   **(ii)**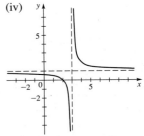

9. (a) (b) (c) (d) (e) (f)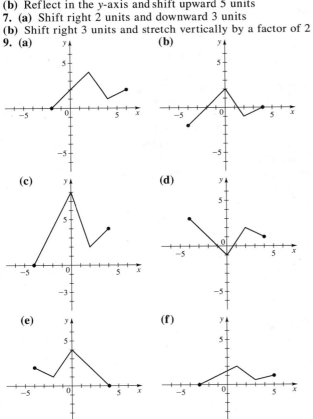

(iii) 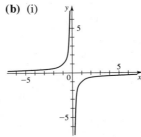   **(iv)**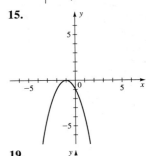

13. 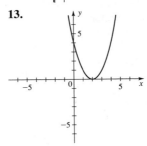   **15.**

17. 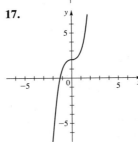   **19.**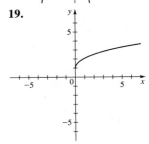

21.

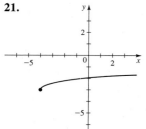

23.

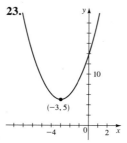

25.

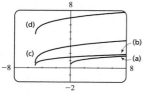

27.

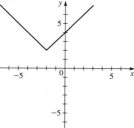

29.

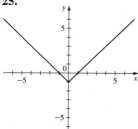

For part (b) shift the graph in (a) left 5 units; for part (c) shift the graph in (a) left 5 units and stretch vertically by a factor of 2; for part (d) shift the graph in (a) left 5 units, stretch vertically by a factor of 2, and then shift upward 4 units.

31.

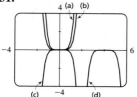

For part (b) shrink the graph in (a) vertically by a factor of 3; for part (c) shrink the graph in (a) vertically by a factor of 3 and reflect in the *x*-axis; for part (d) shift the graph in (a) left 4 units, shrink vertically by a factor of 3, and then reflect in the *x*-axis.

33. To obtain the graph of g, reflect in the *x*-axis the part of the graph of f that is below the *x*-axis.

35. (a)

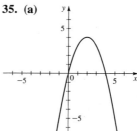

(b)

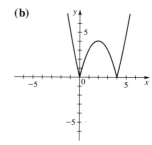

37. (a)

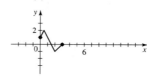

(b) Shrink the graph in (a) horizontally by a factor of a
(c) Stretch the graph in (a) horizontally by a factor of a

39.

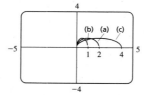

For part (b) shrink the graph in (a) horizontally by a factor of 2; for part (c) stretch the graph in (a) horizontally by a factor of 2.

41. Even   **43.** Neither

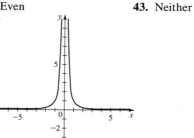

45. Odd   **47.** Neither

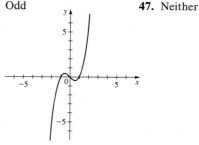

Section 3.9 ■ page 226

1. Vertex $(0, -8)$
x-intercept $\pm 2\sqrt{2}$
y-intercept -8

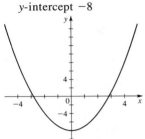

3. Vertex $(0, -2)$
no x-intercept
y-intercept -2

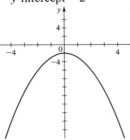

5. Vertex $\left(\frac{3}{2}, -\frac{9}{2}\right)$
x-intercepts $0, 3$
y-intercept 0

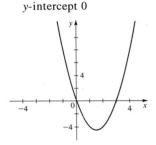

7. Vertex $(-3, -1)$
x-intercepts $-4, -2$
y-intercept 8

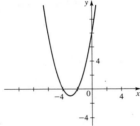

9. Vertex $(-1, 1)$
no x-intercept
y-intercept 3

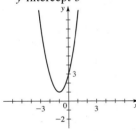

11. Vertex $(5, 7)$
no x-intercept
y-intercept 57

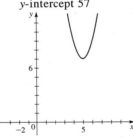

13.

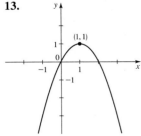

Maximum $f(1) = 1$

15.

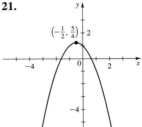

Minimum $f(-1) = -2$

17.

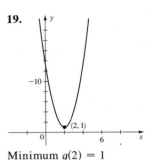

Maximum $f\left(-\frac{3}{2}\right) = \frac{21}{4}$

19.

Minimum $g(2) = 1$

21.

Maximum $h\left(-\frac{1}{2}\right) = \frac{5}{4}$

23. Minimum $f\left(-\frac{1}{2}\right) = \frac{3}{4}$

25. Maximum $f(-3.5) = 185.75$

27. Minimum $f(0.6) = 15.64$

29. Minimum $h(-2) = -8$

31. $f(x) = 2x^2 - 4x$ **33.** $(-\infty, \infty), (-\infty, 1]$ **35.** 25 ft

37. $50, -50$ **39.** $-12, -12$ **41.** 600 ft by 1200 ft

43. $14{,}062.5$ ft^2 **45. (a)** -4.01 **(b)** -4.011025

47. Local maximum ≈ 0.38 when $x \approx -0.58$;
local minimum ≈ -0.38 when $x \approx 0.58$

49. Local maximum 0 when $x = 0$;
local minimum ≈ -13.61 when $x \approx -1.71$;
local minimum ≈ -73.32 when $x \approx 3.21$

51. Local maximum ≈ 5.66 when $x \approx 4.00$

53. Local maximum ≈ 0.38 when $x \approx -1.73$;
local minimum ≈ -0.38 when $x \approx 1.73$

55. 7.5 mi/h

57. Height ≈ 1.44 ft, length and width of base ≈ 2.88 ft

59. Width ≈ 8.40 ft, straight height ≈ 4.20 ft

61. Width ≈ 3.27 ft, height ≈ 5.33 ft

63. (b) $9.23, 13.00$ **65.** $f(x) = x^2 + x - 3$

Section 3.10 ■ page 236

1. $(f + g)(x) = x^2 + 5, (-\infty, \infty)$;
$(f - g)(x) = x^2 - 2x - 5, (-\infty, \infty)$;
$(fg)(x) = x^3 + 4x^2 - 5x, (-\infty, \infty)$;
$(f/g)(x) = (x^2 - x)/(x + 5), (-\infty, -5) \cup (-5, \infty)$

3. $(f + g)(x) = \sqrt{1 + x} + \sqrt{1 - x}$, $[-1, 1]$;
$(f - g)(x) = \sqrt{1 + x} - \sqrt{1 - x}$, $[-1, 1]$;
$(fg)(x) = \sqrt{1 - x^2}$, $[-1, 1]$;
$(f/g)(x) = \sqrt{(1 + x)/(1 - x)}$, $[-1, 1)$
5. $(f + g)(x) = 8/(x(x + 4))$, $x \ne -4$, $x \ne 0$;
$(f - g)(x) = 4(x + 2)/(x(x + 4))$, $x \ne -4$, $x \ne 0$;
$(fg)(x) = -4/(x(x + 4))$, $x \ne -4$, $x \ne 0$;
$(f/g)(x) = -(x + 4)/x$, $x \ne -4$, $x \ne 0$
7. $[-3, -1) \cup (-1, 1) \cup (1, 4]$
9.

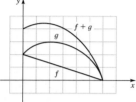

11.

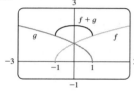

13.

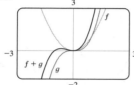

15. (a) 1　**(b)** -23
17. (a) -11　**(b)** -119
19. (a) $-3x^2 + 1$　**(b)** $-9x^2 + 30x - 23$
21. 4　**23.** 5　**25.** 4
27. $(f \circ g)(x) = 8x + 1$, $(-\infty, \infty)$;
$(g \circ f)(x) = 8x + 11$, $(-\infty, \infty)$;
$(f \circ f)(x) = 4x + 9$, $(-\infty, \infty)$; $(g \circ g)(x) = 16x - 5$,
$(-\infty, \infty)$
29. $(f \circ g)(x) = 3(6x^2 + 7x + 2)$, $(-\infty, \infty)$;
$(g \circ f)(x) = 6x^2 - 3x + 2$, $(-\infty, \infty)$;
$(f \circ f)(x) = 8x^4 - 8x^3 + x$, $(-\infty, \infty)$;
$(g \circ g)(x) = 9x + 8$, $(-\infty, \infty)$
31. $(f \circ g)(x) = \sqrt{x^2 - 1}$, $(-\infty, -1] \cup [1, \infty)$;
$(g \circ f)(x) = x - 1$, $[1, \infty)$;
$(f \circ f)(x) = \sqrt{\sqrt{x - 1} - 1}$, $[2, \infty)$;
$(g \circ g)(x) = x^4$, $(-\infty, \infty)$
33. $(f \circ g)(x) = -\frac{1}{2}(x + 1)$, $x \ne -1$;
$(g \circ f)(x) = (2 - x)/x$, $x \ne 0$, $x \ne 1$;
$(f \circ f)(x) = (x - 1)/(2 - x)$, $x \ne 1$, $x \ne 2$;
$(g \circ g)(x) = -1/x$, $x \ne -1$, $x \ne 0$
35. $(f \circ g)(x) = \sqrt[3]{1 - \sqrt{x}}$, $[0, \infty)$;
$(g \circ f)(x) = 1 - \sqrt[6]{x}$, $[0, \infty)$; $(f \circ f)(x) = \sqrt[9]{x}$, $(-\infty, \infty)$;
$(g \circ g)(x) = 1 - \sqrt{1 - \sqrt{x}}$, $[0, 1]$

37. $(f \circ g)(x) = (3x - 4)/(3x - 2)$, $x \ne 2$, $x \ne \frac{2}{3}$;
$(g \circ f)(x) = -(x + 2)/(3x)$, $x \ne -\frac{1}{2}$, $x \ne 0$;
$(f \circ f)(x) = (5x + 4)/(4x + 5)$, $x \ne -\frac{1}{2}$, $x \ne -\frac{5}{4}$;
$(g \circ g)(x) = x/(4 - x)$, $x \ne 2$, $x \ne 4$
39. $(f \circ g \circ h)(x) = \sqrt{x - 1} - 1$
41. $(f \circ g \circ h)(x) = (\sqrt{x} - 5)^4 + 1$
43. $g(x) = x - 9$, $f(x) = x^5$
45. $g(x) = x^2$, $f(x) = x/(x + 4)$
47. $g(x) = 1 - x^3$, $f(x) = |x|$
49. $h(x) = x^2$, $g(x) = x + 1$, $f(x) = 1/x$
51. $h(x) = \sqrt[3]{x}$, $g(x) = 4 + x$, $f(x) = x^9$
53. $A(t) = 3600\pi t^2$
55. (a) $d(t) = 350t$　**(b)** $s(d) = \sqrt{1 + d^2}$
(c) $s(t) = \sqrt{1 + 122,500t^2}$
57. $g(x) = x^2 + x - 1$　**59.** Yes

Section 3.11 ■ page 244

1. No　**3.** Yes　**5.** No　**7.** Yes　**9.** Yes
11. No　**13. (a)** 2　**(b)** 3　**15.** $f^{-1}(3) = 1$
23. $f^{-1}(x) = \frac{1}{2}(x - 1)$　**25.** $f^{-1}(x) = \frac{1}{4}(x - 7)$
27. $f^{-1}(x) = (1/x) - 2$, $x > 0$
29. $f^{-1}(x) = (5x - 1)/(2x + 3)$
31. $f^{-1}(x) = \frac{1}{5}(x^2 - 2)$, $x \ge 0$　**33.** $f^{-1}(x) = \sqrt{4 - x}$
35. $f^{-1}(x) = (x - 4)^3$　**37.** $f^{-1}(x) = x^2 - 2x$, $x \ge 1$
39. $f^{-1}(x) = \sqrt[4]{x}$

41. (a)　　　　　**(b)**

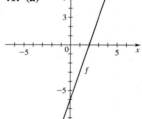

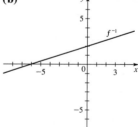

(c) $f^{-1}(x) = \frac{1}{3}(x + 6)$

43. (a)　　　　　**(b)**

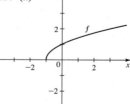
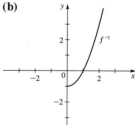

(c) $f^{-1}(x) = x^2 - 1$, $x \ge 0$

45. Not one-to-one   **47.** One-to-one

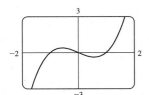

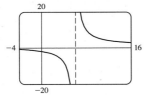

(d) $y = 2x - 15$

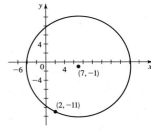

(e) $(x - 7)^2 + (y + 1)^2 = 125$

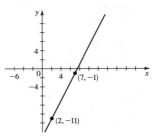

51. $m \neq 0, f^{-1}(x) = \dfrac{x - b}{m}$

Chapter 3 Review ■ page 246

1. (a)

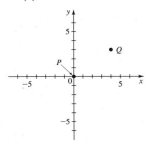

(b) 5   **(c)** $\left(2, \frac{3}{2}\right)$

(d) $y = \frac{3}{4}x$ **(e)** $x^2 + y^2 = 25$

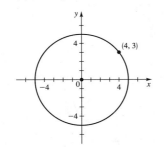

3. (a)

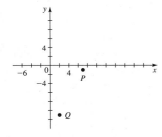

(b) $5\sqrt{5}$ **(c)** $\left(\frac{9}{2}, -6\right)$

5.

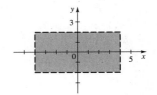

7. B

9. $(x - 2)^2 + (y + 5)^2 = 2$

11. $\left(x - \frac{1}{2}\right)^2 + \left(y - \frac{11}{2}\right)^2 = \frac{17}{2}$

13. Circle, center $(-1, 3)$, radius 1

15. No graph   **17.** $2x - 3y - 16 = 0$

19. $3x + y - 12 = 0$ **21.** $x + 5y = 0$

23. $x^2 + y^2 = 169, 5x - 12y + 169 = 0$

25. No symmetry   **27.** No symmetry

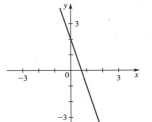

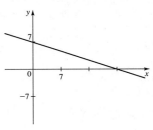

29. No symmetry   **31.** Symmetry about
                                   y-axis

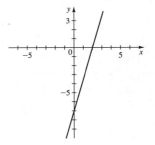

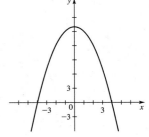

33. No symmetry

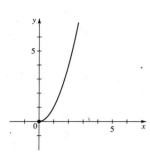

35. Symmetry about both axes and origin

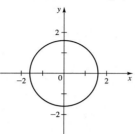

37. Symmetry about origin

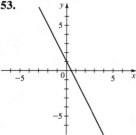

39. 1, 3, 7, $a^2 - a + 1$, $a^2 + a + 1$, $x^2 + x + 1$, $4x^2 - 2x + 1$, $2x^2 - 2x$

41. (a) $-1, 2$ (b) $[-4, 5]$ (c) $[-4, 4]$
(d) Increasing on $[-4, -2]$ and $[-1, 4]$; decreasing on $[-2, -1]$ and $[4, 5]$ (e) No
43. Domain $[-3, \infty)$, range $[0, \infty)$ **45.** $(-\infty, \infty)$
47. $[-4, \infty)$ **49.** $x \neq -2, x \neq -1, x \neq 0$
51. $(-\infty, -1] \cup [1, 4]$

53.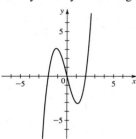

55.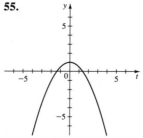

57.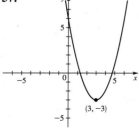

59.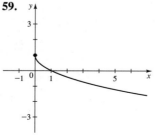

61.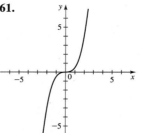

63.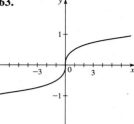

65.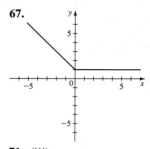

67.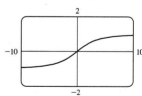

69.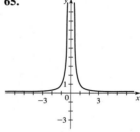

71. (iii)

73.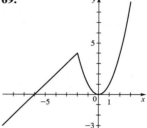

75.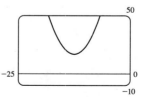

77. $x = -1.53, -0.35, 1.88$ **79.** $x \leq 1.03$
81. $[-2.1, 0.2] \cup [1.9, \infty)$
83. (a) Shift upward 8 units   (b) Shift left 8 units
(c) Stretch vertically by a factor 2, then shift upward 1 unit
(d) Shift right 2 units and downward 2 units
(e) Reflect in x-axis   (f) Reflect in line $y = x$
85. (a) Neither   (b) Odd   (c) Even   (d) Neither
87. $f(x) = (x + 2)^2 - 3$ **89.** $g(-1) = -7$
91. Local maximum ≈ 3.79 when $x \approx 0.46$;
local minimum ≈ 2.81 when $x \approx -0.46$

93. (a) $(f + g)(x) = x^2 - 6x + 6$
(b) $(f - g)(x) = x^2 - 2$
(c) $(fg)(x) = -3x^3 + 13x^2 - 18x + 8$
(d) $(f/g)(x) = (x^2 - 3x + 2)/(4 - 3x)$
(e) $(f \circ g)(x) = 9x^2 - 15x + 6$
(f) $(g \circ f)(x) = -3x^2 + 9x - 2$
95. $(f \circ g)(x) = -3x^2 + 6x - 1, (-\infty, \infty)$;
$(g \circ f)(x) = -9x^2 + 12x - 3, (-\infty, \infty)$;
$(f \circ f)(x) = 9x - 4, (-\infty, \infty)$;
$(g \circ g)(x) = -x^4 + 4x^3 - 6x^2 + 4x, (-\infty, \infty)$
97. $(f \circ g \circ h)(x) = 1 + \sqrt{x}$ **99.** Yes   **101.** No
103. No   **105.** $f^{-1}(x) = \frac{1}{3}(x + 2)$
107. (a), (b)

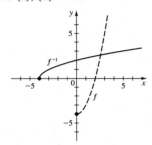

(c) $f^{-1}(x) = \sqrt{x + 4}$
109. $M = 8z$
111. (a) $I = k/d^2$ **(b)** 64,000   **(c)** 160 candles
113. $A = b\sqrt{4 - b}$
115. (a) $A(x) = (x^2/16) + (\sqrt{3}/36)(10 - x)^2$,
$0 \le x \le 10$ **(b)** $40\sqrt{3}/(9 + 4\sqrt{3}) \approx 4.35$ m
117. Square with side 10 m

Chapter 3 Test ■ page 252

1. (a) 20   **(b)** $(-1, -4)$ **(c)** $4x + 3y = -16$
(d) $3x - 4y - 13 = 0$ **(e)** $(x + 1)^2 + (y + 4)^2 = 100$
2. (a)   **(b)**

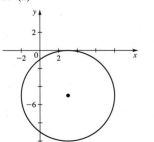

$(x - 3)^2 + (y + 5)^2 = 25$ $\left(x + \frac{3}{2}\right)^2 + \left(y + \frac{5}{2}\right)^2 = 0$

3. (a) $x + 3y - 7 = 0$ **(b)** $4x - y + 12 = 0$
4. (a)   **(b)**

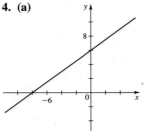
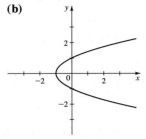

5. (a), (b) are graphs of functions, (a) is one-to-one
6. $[0, 1) \cup (1, \infty)$
7. (a)   **(b)**

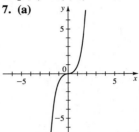
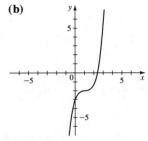

8. (a) Shift left 3 units, then reflect in y-axis
(b) Reflect in origin
9. (a) $f(x) = 2(x - 2)^2 + 5$ **(b)** $f(2) = 5$

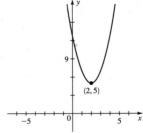

10. (a) $-3, 3$ **(b)**

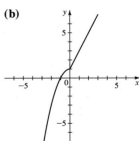

11. (a) $(f \circ g)(x) = 4x^2 - 8x + 2$
(b) $(g \circ f)(x) = 2x^2 + 4x - 5$ **(c)** 2   **(d)** 11
(e) $(g \circ g \circ g)(x) = 8x - 21$

12. (a) $f^{-1}(x) = 3 - x^2$, $x \geqslant 0$ **(b)**

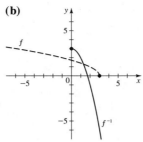

13. (a) $A(x) = 400x - 2x^2$ **(b)** 100 ft
14. (a)   **(b)** No

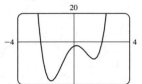

(c) Local minimum ≈ -27.18 when $x \approx -1.61$; local maximum ≈ -2.55 when $x \approx 0.18$; local minimum ≈ -11.93 when $x \approx 1.43$

(d) $[-27.18, \infty)$

(e) Increasing on $[-1.61, 0.18] \cup [1.43, \infty)$, decreasing on $(-\infty, -1.61] \cup [0.18, 1.43]$

FOCUS ON PROBLEM SOLVING ■ page 256

1. 9   **3.** $f_n(x) = x^{2^{n+1}}$
5. 15,999,999,999,992,000,000,000,001
7.   **9.**

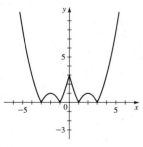

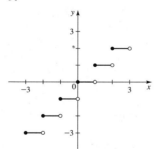

11.

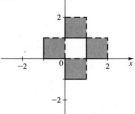

13. $1729 = 1^3 + 12^3 = 9^3 + 10^3$
15. Infinitely far   **17.** 427

19.   **21.**

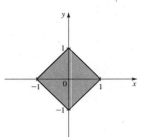

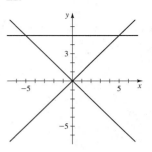

CHAPTER 4

Section 4.1 ■ page 265

1.   **3.**

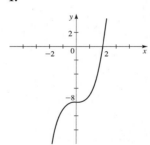

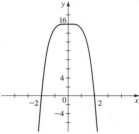

5.   **7.**

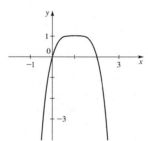

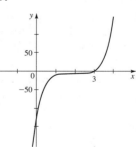

9.   **11.**

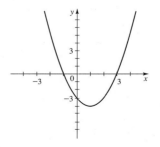

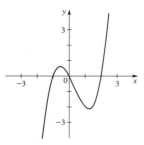

13.

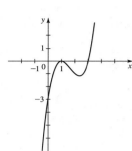

15.

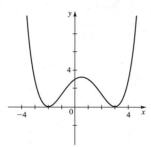

31. (a)

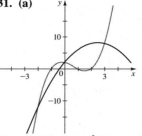

(b) Three
(c) $(0, 2), (3, 8), (-2, -12)$

17.

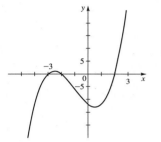

19.

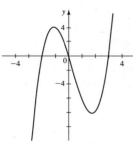

35. (a) $V(x) = 2x^2(18 - x)$ **(b)**

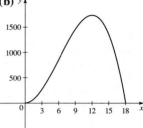

(c) $0 < x < 18$

21.

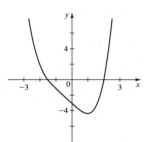

23.

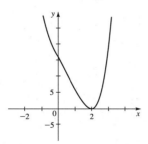

Section 4.2 ■ page 272

1.

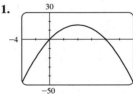

x-intercepts 0, 8
y-intercept 0
local maximum $(4, 16)$

3.

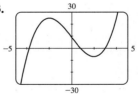

x-intercepts $-3.79, 0.79, 3$
y-intercept 9
local maximum $(-2, 25)$
local minimum $(2, -7)$

5.

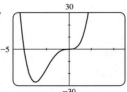

x-intercepts $-4, 0$
y-intercept 0
local minimum $(-3, -27)$

25.

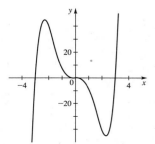

7.

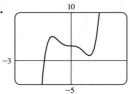

x-intercept -1.42
y-intercept 3
local maximum $(-1, 5)$
local minimum $(1, 1)$

27. $(-2, 0) \cup (2, \infty)$ **29.** $(-\infty, -3] \cup \{0\}$

9. Local maximum $(0.75, 6.13)$
11. Local maximum $(-0.33, 0.19)$,
local minimum $(1.00, -1.00)$
13. Local maximum $(0, 4)$;
local minima $(-1.58, -2.25)$, $(1.58, -2.25)$
15. No extrema
17. Local maximum $(-0.50, 0)$, local minimum $(2.50, -54)$
19. Local maximum: $(0.44, 0.33)$;
local minima $(1.09, -1.15)$, $(-1.12, -3.36)$
21. **23.**

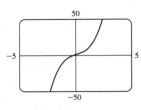

$y \to \infty$ as $x \to \infty$
$y \to -\infty$ as $x \to -\infty$

25. **27.**

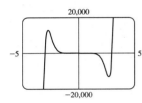

$y \to \infty$ as $x \to \infty$ $y \to -\infty$ as $x \to \infty$
$y \to -\infty$ as $x \to -\infty$ $y \to \infty$ as $x \to -\infty$

29. $(-3.38, 1.38)$ **31.** $[-2, 1] \cup [3, \infty)$
33. $(-\infty, -1.37) \cup (0.37, 1)$ **35.** $(0, 1.6)$
37.

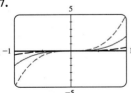

Increasing the value of c stretches the graph vertically.
39.

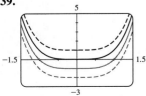

Increasing the value of c moves the graph down.

41.

Increasing the value of c causes a deeper dip in the graph, in the fourth quadrant, and moves the positive x-intercept to the right.

43.

(a) 26 blenders
(b) Maximum profit is $3,276.22 when producing 166 blenders

45. (a) Local maximum $(1.8, 2.1)$,
local minimum $(3.6, -0.6)$
(b) Local maximum $(1.8, 7.1)$, local minimum $(3.6, 4.4)$
47. (a) 3 x-intercepts, 2 local extrema
(b) 1 x-intercept, no local extrema
(c) 3, 1

Section 4.3 ■ page 279

In answers 1–25, the first polynomial given is the quotient and the second is the remainder.
1. $x - 2, -2$ **3.** $x^2 + 2, -3$ **5.** $x^2 - 3x + 1, -1$
7. $x^4 + x^3 + 4x^2 + 4x + 4, -2$ **9.** $x + 2, 8x - 1$
11. $3x + 1, 7x - 5$ **13.** $x^3 + 1, 0$ **15.** $x^2 - 4, 7$
17. $x^3 + x, x - 1$ **19.** $2x^2 + 4x, 1$
21. $x^{100} + x^{99} + x^{98} + \cdots + x^3 + x^2 + x + 1, 0$
23. $\frac{1}{2}x + \frac{1}{2}, -1$ **25.** $\frac{1}{2}x^3 - \frac{3}{4}x^2 - \frac{3}{8}x + \frac{5}{16}, \frac{37}{16}$ **27.** -3
29. 1.92 **31.** 20 **33.** -273 **35.** 100 **37.** $\frac{49}{64}$
39. 5 **45.** $-1 \pm \sqrt{6}$ **47.** $-\frac{3}{2}x^3 + 3x^2 + \frac{15}{2}x - 9$
49. $x^4 - 4x^3 - 7x^2 + 22x + 24$ **51.** No
53. $\dfrac{-1 \pm \sqrt{13}}{3}$

Section 4.4 ■ page 287

1. $\pm 1, \pm 3$ **3.** $\pm 1, \pm 2, \pm 3, \pm 6, \pm \frac{1}{2}, \pm \frac{3}{2}$ **5.** $-2, \frac{3}{2}$
7. 1 **9.** $-2, 2, 3$ **11.** $-1, 2, 3$ **13.** $1, 2, 4$
15. $-6, 1$ **17.** -1 **19.** ± 1 **21.** $-2, 1, 2$
23. $1, -1, -2, -4$ **25.** $\pm 2, \pm \frac{3}{2}$ **27.** $-\frac{3}{2}, \frac{1}{2}, 1$
29. $-1, \pm \frac{1}{2}$ **31.** $1, -1$
33. 1 positive, 2 or 0 negative; 3 or 1 real
35. 1 positive, 1 negative; 2 real
37. 2 or 0 positive, 0 negative; 3 or 1 real (because 0 is a root)
39. 5, 3, or 1 positive, 2 or 0 negative; 7, 5, 3, or 1 real

45. $3, -2$ **47.** $3, -1$ **49.** $-2, \frac{1}{2}, \pm 1$

51. $\pm \frac{1}{2}, \pm \sqrt{5}$ **53.** $-2, 1, 3, 4$ **55.** $-1, \frac{3}{4}, 4$

61. (a)   **(b)** No, there are too many turning points.

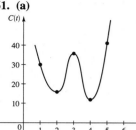

(c) Degree 4

67. $2, -1$

Section 4.5 ■ page 292

1. 0.7 **3.** 0.8 **5.** 0.79 **7.** -0.54 **9.** 1.63

11. 0.35 **13.** 0.62 **15.** $2, -2 \pm \sqrt{2}$

17. $\frac{1}{2}, -1, -1.47$ **19.** $-0.88, 1.35, 2.53$

21. $1, -2, \dfrac{-1 \pm \sqrt{5}}{2}$ **23.** $\frac{1}{2}, -0.71$

25. 2.626 ft by 3.808 ft   **27.** 2.76 m

Section 4.6 ■ page 298

1. $-2, 2, 3$ **3.** $-\frac{1}{2}$ **5.** $-\frac{3}{2}, -1, 1, 4$ **7.** $-1, \frac{5}{2}$

9. -2 **11.** 5.15 **13.** 0.67 **15.** $-1.00, 1.50$

17. $-0.93, 1.11, 5.82$ **19.** $-2.00, -1.24, 2.00, 3.24$

21. -1.50 **23.** $-1.41, 1.41, 3.00$ **25.** 11.3 ft

27. 10 in. by 10 in. by 4 in., 13.8 in. by 13.8 in. by 2.1 in.

29. (a) It began to snow again.   **(b)** No

(c) Just before midnight on Saturday night

Section 4.7 ■ page 307

1. $\pm 2i$ **3.** $\dfrac{1 \pm i \sqrt{3}}{2}$ **5.** $-2 \pm 2i$ **7.** $\dfrac{5 \pm i \sqrt{23}}{6}$

9. $4 \pm i$ **11.** $\dfrac{-3 \pm i \sqrt{3}}{2}$ **13.** $0, i$

15. $x^3 - 2x^2 + x - 2$ **17.** $x^4 - 8x^3 + 39x^2 - 62x + 50$

19. $2x^4 - 8x^3 + 16x^2 - 16x + 8$ **21.** $2, \pm 2i$

23. $\pm 2i, \pm i \sqrt{2}/2$ **25.** $\pm 1, \pm i$ **27.** $-2, 1 \pm i \sqrt{3}$

29. $\pm 2, \pm 2i$ **31.** $\pm 3, \dfrac{\pm 3 \pm 3 \sqrt{3} \, i}{2}$ **33.** $\pm i \sqrt{5}$

35. $-2, \pm 2i$ **37.** $1, \dfrac{1 \pm i \sqrt{3}}{2}$ **39.** $2, \dfrac{1 \pm i \sqrt{3}}{2}$

41. $-2, 1, \pm 3i$

43. $(x + 3)\left(x - \dfrac{3 + 3\sqrt{3}\, i}{2}\right)\left(x - \dfrac{3 - 3\sqrt{3}\, i}{2}\right)$

45. $(x - 2)(x + 2)\left[x + (1 + i\sqrt{3})\right]\left[x + (1 - i\sqrt{3})\right] \cdot$
$\left[x - (1 + i\sqrt{3})\right]\left[x - (1 - i\sqrt{3})\right]$

47. $2\left(x + \frac{3}{2}\right)\left(x + 1 - i\sqrt{2}\right)\left(x + 1 + i\sqrt{2}\right)$ **51.** $-22i$

53. (a) $x^4 - 2x^3 + 3x^2 - 2x + 2$

(b) $x^2 - (1 + 2i)x - 1 + i$

Section 4.8 ■ page 317

1. x-intercept 6, y-intercept -6

3. x-intercept 0, y-intercept 0

5. No x-intercept, no y-intercept

7. Vertical $x = -3$; horizontal $y = 0$

9. Vertical $x = 3$, $x = -2$; horizontal $y = 1$

11. Horizontal $y = 0$ **13.** Vertical $x = 1$; slant $y = x + 1$

15. Vertical $x = -3$

17.

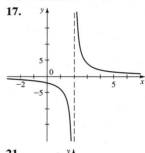

19.

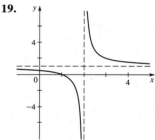

21.

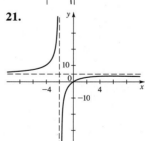

23.

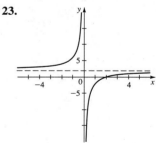

25.

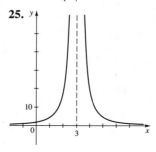

27.

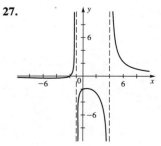

29.

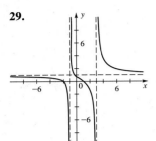

31.

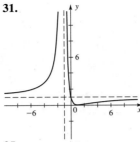

33.

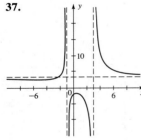

35.

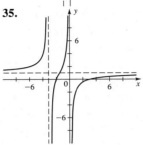

37.

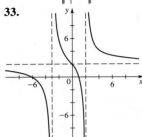

39.

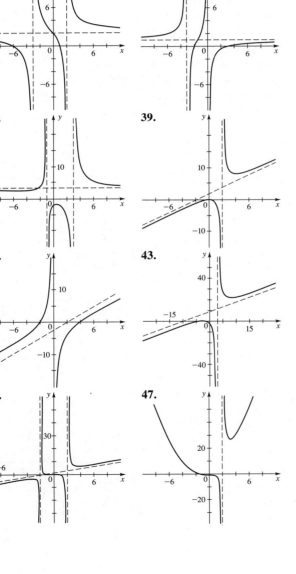

41.

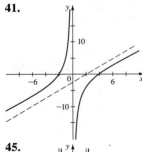

43.

45.

47.

51.

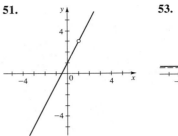

53.

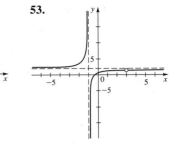

55. $y = \dfrac{x^2 - 5x + 6}{x^2 + 3x - 4}$ **57.** $y = \dfrac{6x^2 - 15x}{2x - 1}$

59. (a)

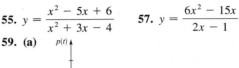

(b) It levels off at 3000.

Section 4.9 ■ page 322

1. Vertical $x = 3$, horizontal $y = 2$
3. Vertical $x = 0$, horizontal $y = -2$
5. Vertical $x = -3$, $x = 3$; horizontal $y = 3$
7.

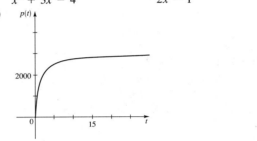

vertical $x = 1$
horizontal $y = 1$
x-intercept -1
y-intercept -1
no extrema

9.

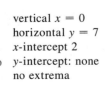

vertical $x = 0$
horizontal $y = 7$
x-intercept 2
y-intercept: none
no extrema

11.

vertical $x = -2$, $x = 2$
horizontal $y = 0$
x-intercept 0
y-intercept 0
no extrema

13.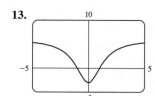

vertical: none
horizontal $y = 6$
x-intercepts ± 1
y-intercept -3
local minimum $(0, -3)$

15.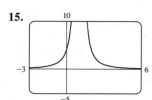

vertical $x = 1$
horizontal $y = 0$
x-intercept: none
y-intercept 4
no extrema

17.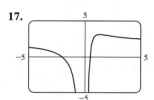

vertical $x = 0$
horizontal $y = 2$
x-intercepts $-2, \frac{1}{2}$
y-intercept: none
local maximum $(1.33, 3.13)$

19.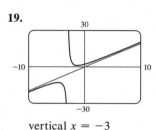

vertical $x = -3$

21.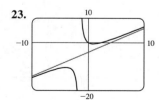

vertical $x = 2$

23.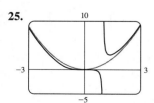

vertical $x = -1.5$
x-intercepts 0, 2.5
y-intercept 0
local maximum $(-3.9, -10.4)$
local minimum $(0.9, -0.6)$
end behavior: $y = x - 4$

25.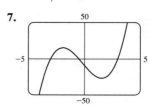

vertical $x = 1$
x-intercept 0
y-intercept 0
local minimum $(1.4, 3.1)$
end behavior: $y = x^2$

27. **(a)** 2.50 mg/L   **(b)** It decreases to 0.   **(c)** 16.61 h

29.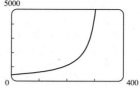

If the speed of the train approaches the speed of sound, then the pitch increases indefinitely (a sonic boom).

31. $y = x^2 + \dfrac{x}{x^2 - 1}$ (other answers are possible)

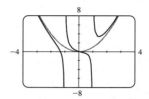

Chapter 4 Review ■ page 324

1.

3.

5.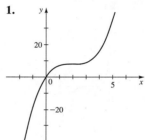

7.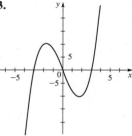

x-intercepts $-3, -0.5, 3$
y-intercept -9
local maximum $(-1.9, 15.1)$
local minimum $(1.6, -27.1)$
$y \to \infty$ as $x \to \infty$
$y \to -\infty$ as $x \to -\infty$

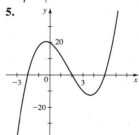

9.

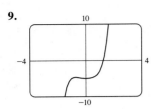

x-intercept 1.3
y-intercept -5
local maximum $(-0.7, -4.7)$
local minimum $(0, -5)$
$y \to \infty$ as $x \to \infty$
$y \to -\infty$ as $x \to -\infty$

In answers 11–17, the first polynomial given is the quotient and the second is the remainder.

11. $x^2 + 2x + 7, 10$ **13.** $x - 3, -9$
15. $x^3 - 5x^2 + 4, -5$
17. $x^3 + (\sqrt{3} + 1)x^2 + (\sqrt{3} + 1)x + \sqrt{3}, 2$
19. 3 **23.** 8 **25.** $\pm1, \pm2, \pm3, \pm6, \pm9, \pm18$
27. 1.62 **29.** 1.33 **31.** $4x^3 - 18x^2 + 14x + 12$
33. No; since the complex conjugates of imaginary zeros will also be zeros, the polynomial would have 8 zeros, contradicting the requirement that it have degree 4.
35. $-3, 1, 5$ **37.** $-1 \pm 2i, -2$ (multiplicity 2)
39. $\pm2, 1$ (multiplicity 3) **41.** $\pm2, \pm1 \pm i\sqrt{3}$
43. $1, 3, \dfrac{-1 \pm i\sqrt{7}}{2}$ **45.** $-1, 2, 2 \pm i\sqrt{2}$ **47.** 1.57
49. $x = -0.5, 3$ **51.** $x \approx -0.24, 4.24$
53.

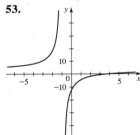

55.

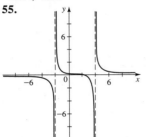

57.

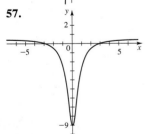

59.

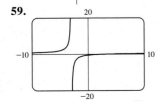

x-intercept 3
y-intercept -0.5
vertical $x = -3$
horizontal $y = 0.5$
no local extrema

61.

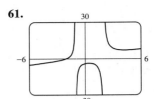

x-intercept -2
y-intercept -4
vertical $x = -1, x = 2$
slant $y = x + 1$
local maximum
$(0.425, -3.599)$
local minimum $(4.216, 7.175)$

Chapter 4 Test ■ page 326

1.

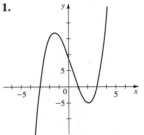

2. **(a)** $\pm1, \pm2, \pm3, \pm4, \pm6, \pm9, \pm12, \pm18, \pm36,$
$\pm\frac{1}{2}, \pm\frac{3}{2}, \pm\frac{9}{2}$
(b) 4, 2, or 0 positive real root(s); 0 negative real root
(d) $\frac{1}{2}, 1, \frac{3}{2}, 2, 3, 4, \frac{9}{2}, 6$ **(e)** $\frac{3}{2}, 2$ (double), 3
(f) $2(x - \frac{3}{2})(x - 2)^2(x - 3)$
3. $x^5 + x^4 + 2x^3 + 10x^2 + 13x + 5$
4. **(a)** P and Q: by Rational Roots Theorem; R: by Descartes' Rule **(b)** No; by Descartes' Rule
(c) Two (one positive, one negative); by Descartes' Rule
(d) Only possible rational roots are 1 and -1, neither of which is a root
5. 1.2 **6.** $4, \frac{1}{2}, 2 \pm \sqrt{3}$ **7.** $\pm2i, \dfrac{-1 \pm i\sqrt{3}}{2}$
8. **(a)** r, u **(b)** s **(c)** s **(d)**

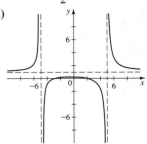

9.

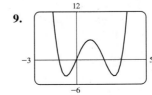

x-intercepts $-1.24, 0, 2,$
$\qquad 3.24$
local maximum $(1, 5)$
local minima $(-0.73, -4),$
$\qquad (2.73, -4)$

FOCUS ON PROBLEM SOLVING ■ **page 329**

1. Traversable **3.** Not traversable **5.** Traversable
7. Yes; tour is possible for odd n. **9.** Inside **11.** -36
13. No

CHAPTER 5

Section 5.1 ■ **page 338**

1.

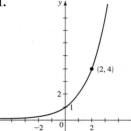

3.

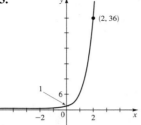

5.

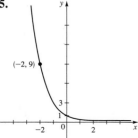

7.

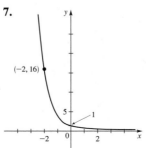

9.

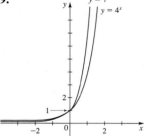

11. $f(x) = 3^x$ **13.** $f(x) = \left(\frac{1}{4}\right)^x$
15. III **17.** I **19.** II

21. $\mathbb{R}, (-\infty, 0), y = 0$

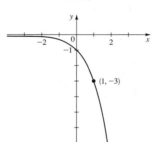

23. $\mathbb{R}, (-3, \infty), y = -3$

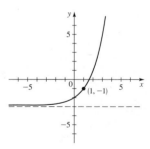

25. $\mathbb{R}, (4, \infty), y = 4$

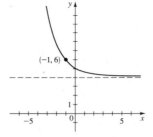

27. $\mathbb{R}, (0, \infty), y = 0$

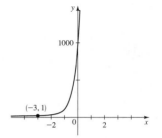

29. $\mathbb{R}, (-\infty, 0), y = 0$

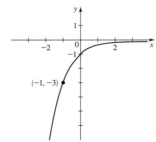

31. $\mathbb{R}, (0, \infty), y = 0$

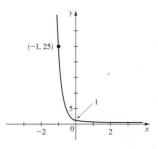

33. $\mathbb{R}, (-\infty, 5), y = 5$

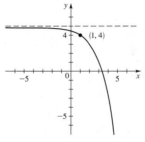

35. $\mathbb{R}, [1, \infty)$, no asymptote

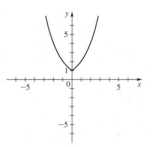

37. $y = 3(2^x)$

39. (a)

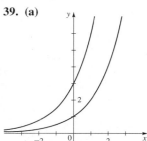

(b) The graph of g is steeper than that of f.

43. (ii)

45. (a)
(i)

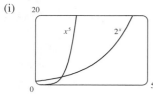

(ii)

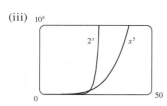

(iii)

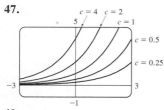

The graph of f ultimately grows much more quickly than g.

(b) 1.2, 22.4

47.

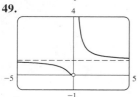

The larger the value of c, the more rapidly the graph increases.

49.

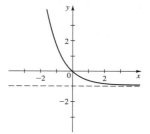

Vertical asymptote $x = 0$, horizontal asymptote $y = 1$

51. Local minimum $\approx (0.37, 0.69)$

53. (a) Increasing on $(-\infty, 0.50)$, decreasing on $[0.50, \infty)$

(b) $(0, 1.78]$

Section 5.2 ■ page 348

1.

x	$f(x) = 3e^x$
-2	0.4
-1.5	0.7
-1	1.1
-0.5	1.8
0	3
0.5	4.9
1	8.2
1.5	13.4
2	22.2

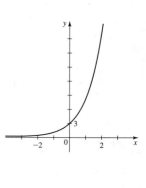

3. $\mathbb{R}, (-\infty, 0), y = 0$

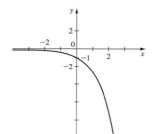

5. $\mathbb{R}, (-1, \infty), y = -1$

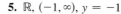

7. $\mathbb{R}, (0, \infty), y = 0$

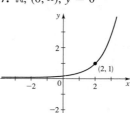

9. (a) \$16,288.95
(b) \$26,532.98
(c) \$43,219.42

11. (a) \$4,615.87 **(b)** \$4,658.91 **(c)** \$4,697.04
(d) \$4,703.11 **(e)** \$4,704.68 **(f)** \$4,704.93
(g) \$4,704.94
13. (i) **15.** \$7,678.96
17. (a) 45% **(b)** 500 **(c)** 4743
19. (a) $n(t) = 18,000e^{0.08t}$ **(b)** 34,137
(c)

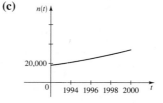

21. **(a)** 233 million   **(b)** 181 million
23. 5.87 billion   **25.** **(a)** About 500   **(b)** 361,000
27. **(a)** 6 g   **(b)** 1 g
29. **(a)** 2.7 lb   **(b)** 4.9 lb
(c)   **(d)** 15 lb

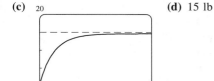

31. **(a)** 5164   **(b)**

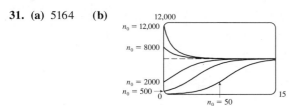

(c) 6000
33.   **35.**

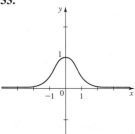 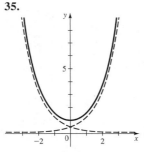

41.

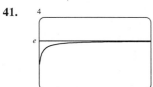

43. **(a)**

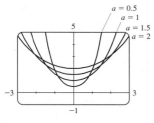

(b) The larger the value of a, the less steep the graph
and the higher its minimum value.
45. Local maximum $\approx (1.00, 0.37)$

Section 5.3 ■ page 359

1. **(a)** $2^5 = 32$ **(b)** $5^0 = 1$
3. **(a)** $4^{1/2} = 2$ **(b)** $2^{-4} = \frac{1}{16}$
5. **(a)** $e^x = 5$ **(b)** $e^5 = y$
7. **(a)** $\log_2 8 = 3$ **(b)** $\log_{10} 0.001 = -3$
9. **(a)** $\log_4 0.125 = -\frac{3}{2}$ **(b)** $\log_7 343 = 3$
11. **(a)** $\ln 2 = x$ **(b)** $\ln y = 3$
13. **(a)** 4   **(b)** 3   **(c)** 1
15. **(a)** 2   **(b)** 2   **(c)** 10
17. **(a)** -3 **(b)** $\frac{1}{2}$ **(c)** -1
19. **(a)** 37   **(b)** 8   **(c)** $\sqrt{5}$
21. **(a)** $-\frac{2}{3}$ **(b)** 4   **(c)** -1
23. **(a)** 32   **(b)** 4
25. **(a)** 100   **(b)** 25
27. **(a)** 2   **(b)** 4
29. **(a)** 0.3010   **(b)** 1.5465   **(c)** -0.1761
31. **(a)** 1.6094   **(b)** 3.2308   **(c)** 1.0051
33. $y = \log_5 x$ **35.** $y = \log_9 x$ **37.** II   **39.** III
41. VI   **43.**

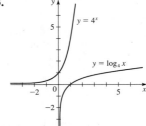

45. $(4, \infty)$, $\mathbb{R}$, $x = 4$ **47.** $(-\infty, 0)$, $\mathbb{R}$, $x = 0$

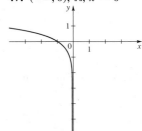 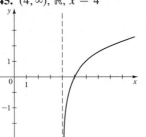

49. $(0, \infty)$, $\mathbb{R}$, $x = 0$ **51.** $(0, \infty)$, $\mathbb{R}$, $x = 0$

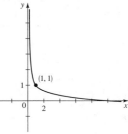

53. $(0, \infty)$, $[0, \infty)$, $x = 0$

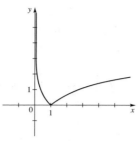

55. $\left(-\frac{2}{5}, \infty\right)$ **57.** $(-\infty, -1) \cup (1, \infty)$ **59.** $(0, 2)$

61.

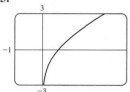

domain $= (-1, 1)$
vertical asymptotes
$x = 1$, $x = -1$
local maximum $(0, 0)$

63.

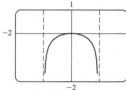

domain $= (0, \infty)$
vertical asymptote $x = 0$
no maximum or minimum

65.

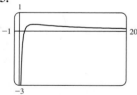

domain $= (0, \infty)$
vertical asymptote $x = 0$
horizontal asymptote $y = 0$
local maximum
$\approx (2.72, 0.37)$

67. The graph of f grows more slowly than g.

69. (a)

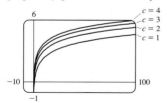

(b) The graph of $f(x) = \log(cx)$ is the graph of $f(x) = \log x$ shifted upward $\log c$ units.

71. (b) About 265 mi

73. (a) $(1, \infty)$ **(b)** $f^{-1}(x) = 10^{2^x}$

75. (a) $f^{-1}(x) = \log_2\left(\dfrac{x}{1 - x}\right)$ **(b)** $(0, 1)$

Section 5.4 ■ page 366

1. $\log_2 x + \log_2(x - 1)$ **3.** $23 \log 7$

5. $\log_2 A + 2 \log_2 B$ **7.** $\log_3 x + \frac{1}{2} \log_3 y$

9. $\frac{1}{3} \log_5(x^2 + 1)$ **11.** $\frac{1}{2}(\ln a + \ln b)$

13. $3 \log x + 4 \log y - 6 \log z$

15. $\log_2 x + \log_2(x^2 + 1) - \frac{1}{2} \log_2(x^2 - 1)$

17. $\ln x + \frac{1}{2}(\ln y - \ln z)$ **19.** $\frac{1}{4} \log(x^2 + y^2)$

21. $\frac{1}{2}[\log(x^2 + 4) + \log(x^2 + 1) - 2 \log(x^3 - 7)]$

23. $\frac{1}{2} \ln x + 4 \ln z - \frac{1}{3} \ln(y^2 + 6y + 17)$

25. $\frac{3}{2}$ **27.** 1 **29.** 3 **31.** $\ln 8$ **33.** 16

35. $\log_3 160$ **37.** $\log_2(AB/C^2)$ **39.** $\log\left[\dfrac{x^4(x - 1)^2}{\sqrt[3]{x^2 + 1}}\right]$

41. $\ln[5x^2(x^2 + 5)^3]$

43. $\log\left[\sqrt[3]{2x + 1}\sqrt{(x - 4)/(x^4 - x^2 - 1)}\right]$

45. No   **47.** Yes   **49.** No   **51.** Yes   **53.** Yes

55. 2.807355 **57.** 2.182658 **59.** 0.655407

61. 4.165458

63.

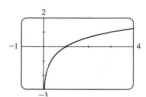

Section 5.5 ■ page 374

1. 1.7227 **3.** -0.5850 **5.** 1.2040 **7.** 0.0767

9. 0.2524 **11.** 1.9349 **13.** -43.06766

15. 2.1492 **17.** 6.2126 **19.** -2.9469

21. -2.4423 **23.** 14.0055 **25.** ± 1 **27.** $0, \frac{4}{3}$

29. $\ln 2 \approx 0.6931$ **31.** $\frac{1}{2} \ln 3 \approx 0.5493$

33. $e^{10} \approx 22026$ **35.** 0.01 **37.** $\frac{95}{3}$

39. $3 - e^2 \approx -4.3891$ **41.** 5 **43.** 5 **45.** $\frac{13}{12}$

47. 6 **49.** $\frac{3}{2}$ **51.** $1/\sqrt{5} \approx 0.4472$ **53.** 13 days

55. (a) $t = -\frac{5}{13} \ln\left(1 - \frac{13}{60} I\right)$ **(b)** 0.218 s

57. 2.21 **59.** $0.00, 1.14$ **61.** -0.57 **63.** 0.36

65. $2 < x < 4$ or $7 < x < 9$ **67.** $\log_{10} 2 < x < \log_{10} 5$

69. $101, 1.1$ **71.** $\log_2 3 \approx 1.58$

Section 5.6 ■ page 385

1. (a) \$12,870.19   **(b)** 8.24 yr   **3.** 5 yr   **5.** 8.15 yr

7. (a) 500   **(b)** 45%   **(c)** 1929   **(d)** 6.66 h

9. (a) $n(t) = 112,000e^{0.04t}$ **(b)** About 142,000   **(c)** 2008

11. (a) 20,000   **(b)** $n(t) = 20,000e^{0.1096t}$

(c) About 48,000   **(d)** 2004

13. (a) $n(t) = 1500e^{0.0231t}$ **(b)** About 24,000   **(c)** 42.5

15. (a) $n(t) = 8600e^{0.1508t}$ **(b)** About 11,600   **(c)** 4.6
17. (a) 2029   **(b)** 2049   **19.** 22.85 h
21. (a) $n(t) = 10e^{-0.0231t}$ **(b)** 1.6 g   **(c)** 70 yr
23. 16 yr   **25.** 149 h   **27.** 3560 yr
29. (a) 210°F   **(b)** 153°F   **(c)** 28 min
31. (a) 137°F   **(b)** 116 min
33. (a) 2.3   **(b)** 3.5   **(c)** 8.3
35. (a) 10^{-3} M   **(b)** 3.2×10^{-7} M   **37.** $4.8 \leqslant \text{pH} \leqslant 6.4$
39. $\log 20 \approx 1.3$ **41.** Twice as intense   **43.** 8.2
45. 6.3×10^{-3} watts/m^2 **47. (b)** 106 dB

Chapter 5 Review ■ page 389

1. $\mathbb{R}$, $(0, \infty)$, $y = 0$ **3.** $\mathbb{R}$, $(-\infty, 5)$, $y = 5$

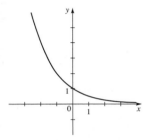

5. $(1, \infty)$, $\mathbb{R}$, $x = 1$ **7.** $(0, \infty)$, $\mathbb{R}$, $x = 0$

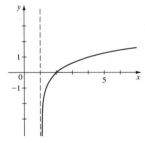

9. $\mathbb{R}$, $(-1, \infty)$, $y = -1$ **11.** $(0, \infty)$, $\mathbb{R}$, $x = 0$

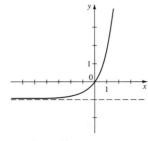

13. $\left(-\infty, \frac{1}{2}\right)$ **15.** $2^{10} = 1024$ **17.** $10^y = x$
19. $\log_2 64 = 6$ **21.** $\log 74 = x$ **23.** 7   **25.** 45

27. 6   **29.** -3 **31.** $\frac{1}{2}$ **33.** 2   **35.** 92   **37.** $\frac{2}{3}$
39. $\log A + 2 \log B + 3 \log C$
41. $\frac{1}{2}[\ln(x^2 - 1) - \ln(x^2 + 1)]$
43. $2 \log_5 x + \frac{3}{2} \log_5(1 - 5x) - \frac{1}{2} \log_5(x^3 - x)$
45. $\log 96$ **47.** $\log_2\left[\dfrac{(x - y)^{3/2}}{(x^2 + y^2)^2}\right]$ **49.** $\log\left(\dfrac{x^2 - 4}{\sqrt{x^2 + 4}}\right)$
51. -15 **53.** $\frac{1}{3}(5 - \log_5 26) \approx 0.99$
55. $\frac{4}{3} \ln 10 \approx 3.07$ **57.** 3   **59.** $-4, 2$ **61.** 0.430618
63. 2.303600
65.

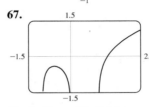

vertical asymptote
$x = -2$
horizontal asymptote
$y = 2.72$
no maximum of minimum

67.

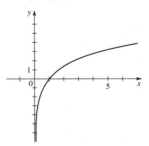

vertical asymptotes
$x = -1$, $x = 0$, $x = 1$
local maximum
$\approx (-0.58, -0.41)$

69. 2.42   **71.** $0.16 < x < 3.15$
73. Increasing on $(-\infty, 0]$ and $[1.10, \infty)$, decreasing on $[0, 1.10]$
75. 1.953445   **77.** $\log_4 258$
79. (a) \$16,081.15   **(b)** \$16,178.18   **(c)** \$16,197.64
(d) \$16,198.31
81. (a) $n(t) = 30e^{0.15t}$ **(b)** 55   **(c)** 19 yr
83. (a) 9.97 mg   **(b)** 1.39×10^5 yr
85. (a) $n(t) = 150e^{-0.0004359t}$ **(b)** 97.0 mg   **(c)** 2520 yr
87. (a) $n(t) = 1500e^{0.1515t}$ **(b)** 7940   **89.** 7.9, basic
91. 8.0

Chapter 5 Test ■ page 392

1.   **2.**

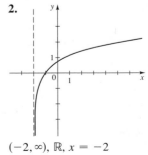

$(-2, \infty)$, $\mathbb{R}$, $x = -2$

3. (a) $\frac{3}{2}$ **(b)** 3 **(c)** $\frac{2}{3}$ **(d)** 2

4. $\frac{1}{2}[\log(x^2 - 1) - 3\log x - 5\log(y^2 + 1)]$

5. $\ln\left[\dfrac{x\sqrt{3 - x^4}}{(x^2 + 1)^2}\right]$

6. (a) 4.32 **(b)** 0.77 **(c)** 5.39 **(d)** 2

7. (a) $n(t) = 1000e^{2.07944t}$ **(b)** 22,627 **(c)** 1.3 h

(d)

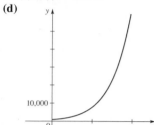

8. 8.33 yr **9.** $\left(-4, \frac{8}{5}\right)$

10. (a) **(b)** $x = 0,\ y = 0$

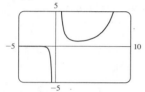

(c) Local minimum $\approx (3.00, 0.74)$

(d) $(-\infty, 0) \cup [0.74, \infty)$

(e) $-0.85,\ 0.96,\ 9.92$

FOCUS ON PROBLEM SOLVING ■ page 394

3. (a) 18 **(b)** 225 **7.** $\left[-1, 1 - \sqrt{3}\,\right) \cup \left(1 + \sqrt{3}, 3\right]$

9. (b) 151 **11.** 16 cm

CHAPTER 6

Section 6.1 ■ page 404

1. $(1, 2)$ **3.** $(3, 2)$

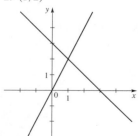 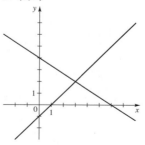

5. $\left(-\frac{5}{2}, 4\right)$

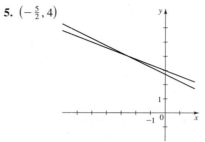

7. $(3, 5)$ **9.** $(1, 3)$ **11.** $(10, -9)$ **13.** $(2, 1)$

15. $\left(x, \frac{1}{3}x - \frac{5}{3}\right)$ **17.** $(-1, 4)$ **19.** $(-3, -7)$

21. $(12, -5)$ **23.** $\left(\frac{1}{5}, \frac{2}{5}\right)$ **25.** No solution **27.** $\left(\frac{7}{3}, 7\right)$

29. $(-1, 2)$ **31.** $\left(\frac{5}{2}, 4\right)$ **33.** $\left(\frac{2}{3}, -\frac{1}{6}\right)$

35. $\left(-\dfrac{1}{a - 1}, \dfrac{1}{a - 1}\right)$ **37.** $\left(\dfrac{1}{a + b}, \dfrac{1}{a + b}\right)$

39. 22, 12 **41.** 5 dimes, 9 quarters

43. Plane's speed 120 mi/h, wind speed 30 mi/h

45. Run 5 mi/h, cycle 20 mi/h

47. 200 g of A, 40 g of B **49.** 25%, 10%

51. 25 mi/h, 5:24 P.M. **53.** 72 **55.** $y = -5x^2 + 17x$

57. $(3.87, 2.74)$ **59.** $(61.00, 20.00)$ **61.** $(1.89, -0.28)$

Section 6.2 ■ page 414

1. Linear **3.** Not linear **5.** Not linear

7. $\begin{cases} 2x + 3y = 1 \\ 4x + 2y = 3 \end{cases}$ **9.** $\begin{cases} \quad\quad y \quad\ = 0 \\ x \quad + z = 0 \\ \quad -2y + 2z = 7 \end{cases}$ **11.** $(1, 1, 2)$

13. $(1, 0, 1)$ **15.** $(-1, 0, 1)$ **17.** $(-1, 5, 0)$

19. $(10, 3, -2)$ **21.** $(2, 0, -2)$ **23.** $\left(\frac{1}{3}, 1, -\frac{1}{2}\right)$

25. $(1, -3, 7)$ **27.** $(1, -1, -3)$ **29.** $(0, -3, 0, -3)$

31. $(1, -1, 2, -2, 0)$ **33.** 2 VitaMax, 1 Vitron, 2 VitaPlus

35. 11 pennies, 4 nickels, 5 dimes, 10 quarters

37. Standard \$40, deluxe \$55, first-class \$90

39. $y = 2x^2 - 5x + 6$ **41.** $(1, -2, 3)$

Section 6.3 ■ page 423

1. Reduced echelon form **3.** Not in echelon form

5. Not in echelon form **7.** No solution **9.** No solution

11. $x = 7,\ y = 1$ **13.** No solution

15. $x = 12 - 11z,\ y = 12 - 12z,\ z = $ any number

17. $x = -2z + 5,\ y = z - 2,\ z = $ any number

19. $x = -\frac{1}{2}y + z + 6,\ y = $ any number, $z = $ any number

21. $x = 2,\ y = w,\ z = -2w + 6,\ w = $ any number

23. $x = -12w + 20,\ y = -19w + 31,\ z = 2w - 2,$ $w = $ any number

25. $x = \frac{1}{3}z - \frac{2}{3}w,\ y = \frac{1}{3}z + \frac{1}{3}w,\ z = $ any number, $w = $ any number

27. No solution

29. x, y, and z are ounces of soya, millet, and milk powder, respectively: $x = -\frac{1}{3}z + \frac{13}{6}$, $y = -\frac{2}{3}z + \frac{4}{3}$, $0 \le z \le 2$

31. Impossible   **33.** 7 pennies, 5 nickels, 4 dimes

Section 6.4 ■ page 432

1. $\begin{bmatrix} 5 & -2 & 5 \\ 1 & 1 & 0 \end{bmatrix}$ **3.** $\begin{bmatrix} -1 & -3 & -5 \\ -1 & 3 & -6 \end{bmatrix}$ **5.** $\begin{bmatrix} 13 & -\frac{7}{2} & 15 \\ 3 & 1 & 3 \end{bmatrix}$

7. $\begin{bmatrix} -14 & -8 & -30 \\ -6 & 10 & -24 \end{bmatrix}$ **9.** Impossible   **11.** $\begin{bmatrix} 3 & \frac{1}{2} & 5 \\ 1 & -1 & 3 \end{bmatrix}$

13. $[28 \quad 21 \quad 28]$ **15.** $\begin{bmatrix} -1 \\ 8 \\ -1 \end{bmatrix}$ **17.** $\begin{bmatrix} 8 & -335 \\ 0 & 343 \end{bmatrix}$

19. Impossible   **21.** Impossible

23. $\begin{bmatrix} 2 & -5 \\ 3 & 2 \end{bmatrix}\begin{bmatrix} x \\ y \end{bmatrix} = \begin{bmatrix} 7 \\ 4 \end{bmatrix}$

25. $\begin{bmatrix} 3 & 2 & -1 & 1 \\ 1 & 0 & -1 & 0 \\ 0 & 3 & 1 & -1 \end{bmatrix}\begin{bmatrix} x_1 \\ x_2 \\ x_3 \\ x_4 \end{bmatrix} = \begin{bmatrix} 0 \\ 5 \\ 4 \end{bmatrix}$ **27.** $x = 2$, $y = 1$

29. $\begin{bmatrix} 3 & \frac{11}{2} \\ 2 & 5 \end{bmatrix}$ **31.** Impossible   **33.** No

37. (a) $A^2 = \begin{bmatrix} 1 & 2 \\ 0 & 1 \end{bmatrix}$, $A^3 = \begin{bmatrix} 1 & 3 \\ 0 & 1 \end{bmatrix}$, $A^4 = \begin{bmatrix} 1 & 4 \\ 0 & 1 \end{bmatrix}$

(b) $A^n = \begin{bmatrix} 1 & n \\ 0 & 1 \end{bmatrix}$

39. (a) $[4{,}690 \quad 1{,}690 \quad 13{,}310]$
(b) Total revenue in Santa Monica, Long Beach, and Anaheim, respectively.

41. Only ACB is defined. $\begin{bmatrix} -3 & -21 & 27 & -6 \\ -2 & -14 & 18 & -4 \end{bmatrix}$

Section 6.5 ■ page 443

1. $\begin{bmatrix} 1 & -2 \\ -\frac{3}{2} & \frac{7}{2} \end{bmatrix}$ **3.** $\begin{bmatrix} 2 & -3 \\ -3 & 5 \end{bmatrix}$ **5.** $\begin{bmatrix} 13 & 5 \\ -5 & -2 \end{bmatrix}$

7. No inverse   **9.** $\begin{bmatrix} 1 & 2 \\ -\frac{1}{2} & \frac{2}{3} \end{bmatrix}$ **11.** $\begin{bmatrix} -4 & -4 & 5 \\ 1 & 1 & -1 \\ 5 & 4 & -6 \end{bmatrix}$

13. No inverse   **15.** $\begin{bmatrix} -\frac{9}{2} & -1 & 4 \\ 3 & 1 & -3 \\ \frac{7}{2} & 1 & -3 \end{bmatrix}$

17. $\begin{bmatrix} 0 & 0 & -2 & 1 \\ -1 & 0 & 1 & 1 \\ 0 & 1 & -1 & 0 \\ 1 & 0 & 0 & -1 \end{bmatrix}$ **19.** $x = 8$, $y = -12$

21. $x = 126$, $y = -50$ **23.** $x = -38$, $y = 9$, $z = 47$

25. $x = -12$, $y = -21$, $z = 93$ **27.** $\begin{bmatrix} 7 & 2 & 3 \\ 10 & 3 & 5 \end{bmatrix}$

29. (a) $\begin{bmatrix} 0 & 1 & -1 \\ -2 & \frac{3}{2} & 0 \\ 1 & -\frac{3}{2} & 1 \end{bmatrix}$ **(b)** 1 oz A, 1 oz B, 2 oz C

(c) 2 oz A, 0 oz B, 1 oz C   **(d)** No

31. (a) $\begin{cases} x + y + 2z = 675 \\ 2x + y + z = 600 \\ x + 2y + z = 625 \end{cases}$

(b) $\begin{bmatrix} 1 & 1 & 2 \\ 2 & 1 & 1 \\ 1 & 2 & 1 \end{bmatrix}\begin{bmatrix} x \\ y \\ z \end{bmatrix} = \begin{bmatrix} 675 \\ 600 \\ 625 \end{bmatrix}$

(c) $A^{-1} = \begin{bmatrix} -\frac{1}{4} & \frac{3}{4} & -\frac{1}{4} \\ -\frac{1}{4} & -\frac{1}{4} & \frac{3}{4} \\ \frac{3}{4} & -\frac{1}{4} & -\frac{1}{4} \end{bmatrix}$

He earns \$125 on a standard set, \$150 on a deluxe set, and \$200 on a leather-bound set.

33. $\begin{bmatrix} 1/(2x) & -\frac{1}{2} \\ \frac{1}{2} & x/2 \end{bmatrix}$, $x \ne 0$ **35.** $\begin{bmatrix} \frac{1}{2} & \frac{1}{2}e^{-x} & 0 \\ \frac{1}{2}e^{-x} & -\frac{1}{2}e^{-2x} & 0 \\ 0 & 0 & \frac{1}{2} \end{bmatrix}$

37. $\begin{bmatrix} 1/a & 0 & 0 & 0 \\ 0 & 1/b & 0 & 0 \\ 0 & 0 & 1/c & 0 \\ 0 & 0 & 0 & 1/d \end{bmatrix}$

Section 6.6 ■ page 453

1. 3   **3.** -4 **5.** Does not exist   **7.** $\frac{1}{8}$ **9.** 20, 20
11. $-12, 12$ **13.** 0, 0   **15.** -6, has an inverse
17. 5000, has an inverse   **19.** -4, has an inverse
21. -18 **23.** 120   **25.** -2 **27.** $(-2, 5)$
29. $(0.6, -0.4)$ **31.** $(4, -1)$ **33.** $(4, 2, -1)$
35. $(1, 3, 2)$ **37.** $(0, -1, 1)$ **39.** $\left(\frac{189}{29}, -\frac{108}{29}, \frac{88}{29}\right)$
41. $\left(-\frac{49}{80}, \frac{77}{40}, \frac{287}{80}\right)$ **43.** $\left(\frac{1}{2}, \frac{1}{4}, \frac{1}{4}, -1\right)$
45. (b) $5x - 6y = -200$ **47.** 0, 1, 2   **49.** 1, -1

Section 6.7 ■ page 462

1. $(-2, 4), (3, 9)$ **3.** $(2, -2), (-2, 2)$ **5.** $(4, 0)$
7. $(-2, -2)$ **9.** $(6, 2), (-2, -6)$

11. $(2, 1), (2, -1), (-2, 1), (-2, -1)$
13. No solution   **15.** $\left(-\sqrt{6}, 6\right)$
17. $\left(\sqrt{5}, 2\right), \left(\sqrt{5}, -2\right), \left(-\sqrt{5}, 2\right), \left(-\sqrt{5}, -2\right)$
19. $\left(3, -\frac{1}{2}\right), \left(-3, -\frac{1}{2}\right)$ **21.** $\left(\frac{1}{5}, \frac{1}{3}\right)$ **23.** $(9, 4)$
25. $(0, 2), (0, -2), (2, 2), (-2, -2), \left(\sqrt{2}, -\sqrt{2}\right), \left(-\sqrt{2}, \sqrt{2}\right)$
27. $(2, 0, 0), \left(\frac{2}{3}, -\frac{4}{3}, \frac{4}{3}\right)$ **29.** $(-1, 1, -1)$
31. $(2, 0, -4), (-2, 0, -4), (0, 1, -1)$
33. $(-4.51, 2.17), (4.91, -0.97)$
35. $(1.23, 3.87), (-0.35, -4.21)$
37. $(-2.30, -0.70), (0.48, -1.19)$
39. $(1.19, 3.59), (-1.19, 3.59)$ **41.** 8 cm, 15 cm, 17 cm
43. 9 in., 12 in., 15 in.   **45.** 15, 20
47. $(400.50, 200.25), 447.77$ m   **49.** $y = 3x + 5$
51. $(8, 5), (-5, -8)$ **53.** $(1, 3), (3, 1)$

Section 6.8 ■ page 468

1.
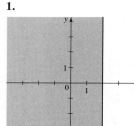

3.
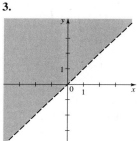

5.
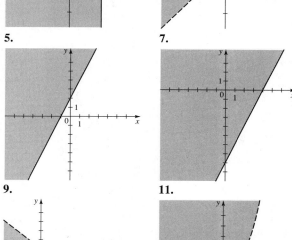

7.

9.

11.

13.
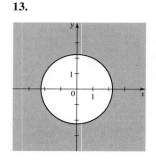

15.
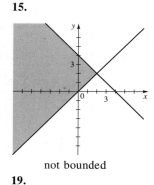
not bounded

17.
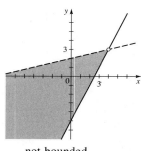
not bounded

19.
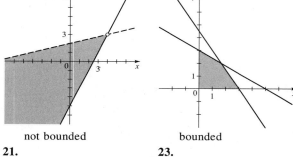
bounded

21.
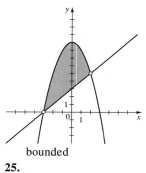
bounded

23.
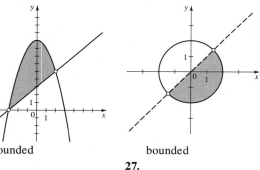
bounded

25.
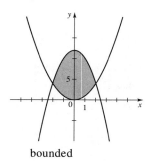
bounded

27.
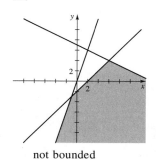
not bounded

29.

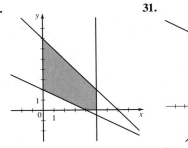

bounded

31.

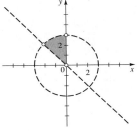

bounded

33.

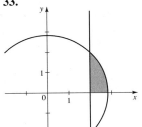

bounded

35.

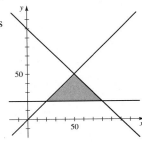

bounded

37. x = number of fiction books
y = number of nonfiction books

$$\begin{cases} x + y \le 100 \\ 20 \le y, \ x \ge y \\ x \ge 0, \ y \ge 0 \end{cases}$$

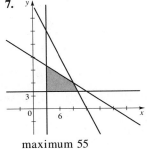

Section 6.9 ■ page 474

1. 6, 0   **3.** 198, 195

5.

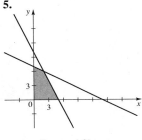

maximum 161
minimum 135

7.

maximum 55
minimum 31

9. 3 tables, 34 chairs
11. 30 grapefruit crates, 30 orange crates
13. 15 Pasadena to Santa Monica, 3 Pasadena to El Toro, 0 Long Beach to Santa Monica, 16 Long Beach to El Toro
15. $10\frac{1}{2}$ yd^3 Foamboard, $14\frac{1}{2}$ yd^3 Plastiflex
17. 10 days Vancouver, 4 days Seattle
19. \$7500 in municipal bonds, \$2500 in bank certificates, \$2000 in high-risk bonds
21. 4 games, 32 educational, 0 utility

Section 6.10 ■ page 482

1. $\dfrac{A}{x-1} + \dfrac{B}{x+2}$ **3.** $\dfrac{A}{x-2} + \dfrac{B}{(x-2)^2} + \dfrac{C}{x+4}$

5. $\dfrac{A}{x-3} + \dfrac{Bx+C}{x^2+4}$ **7.** $\dfrac{Ax+B}{x^2+1} + \dfrac{Cx+D}{x^2+2}$

9. $\dfrac{A}{x} + \dfrac{B}{2x-5} + \dfrac{C}{(2x-5)^2} + \dfrac{D}{(2x-5)^3}$
$+ \dfrac{Ex+F}{x^2+2x+5} + \dfrac{Gx+H}{(x^2+2x+5)^2}$

11. $\dfrac{1}{x-1} - \dfrac{1}{x+4}$ **13.** $\dfrac{2}{x-3} - \dfrac{2}{x+3}$

15. $\dfrac{1}{x-2} - \dfrac{1}{x+2}$ **17.** $\dfrac{3}{x-4} - \dfrac{2}{x+2}$

19. $\dfrac{-\frac{1}{2}}{2x-1} + \dfrac{\frac{3}{2}}{4x-3}$

21. $\dfrac{2}{x-2} + \dfrac{3}{x+2} - \dfrac{1}{2x-1}$

23. $\dfrac{2}{x+1} - \dfrac{1}{x} + \dfrac{1}{x^2}$ **25.** $\dfrac{1}{2x+3} - \dfrac{3}{(2x+3)^2}$

27. $\dfrac{2}{x} - \dfrac{1}{x^3} - \dfrac{2}{x+2}$

29. $\dfrac{4}{x+2} - \dfrac{4}{x-1} + \dfrac{2}{(x-1)^2} + \dfrac{1}{(x-1)^3}$

31. $\dfrac{3}{x+2} - \dfrac{1}{(x+2)^2} - \dfrac{1}{(x+3)^2}$

33. $\dfrac{x+1}{x^2+3} - \dfrac{1}{x}$

35. $\dfrac{2x-5}{x^2+x+2} + \dfrac{5}{x^2+1}$

37. $\dfrac{1}{x^2+1} - \dfrac{x+2}{(x^2+1)^2} + \dfrac{1}{x}$

39. $x^2 + \dfrac{3}{x-2} - \dfrac{x+1}{x^2+1}$

41. $A = \dfrac{a+b}{2}, \ B = \dfrac{a-b}{2}$

Chapter 6 Review ■ page 483

1. $(2, 1)$

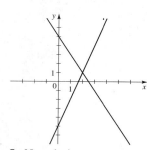

3. $x =$ any number
$y = \frac{2}{7}x - 4$

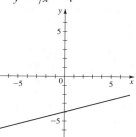

5. No solution

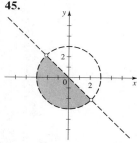

7. $(1, 1, 2)$

9. No solution

11. $x = -4z + 1,$
$y = -z - 1,$
$z =$ any number

13. $x = 6 - 5z,$
$y = \frac{1}{2}(7 - 3z),$
$z =$ any number

15. $3000 at 6%, $6000 at 7%   **17.** Impossible

19. $\begin{bmatrix} 4 & 18 \\ 4 & 0 \\ 2 & 2 \end{bmatrix}$ **21.** $\begin{bmatrix} 10 & 0 & -5 \end{bmatrix}$ **23.** $\begin{bmatrix} -\frac{7}{2} & 10 \\ 1 & -\frac{9}{2} \end{bmatrix}$

25. $\begin{bmatrix} 30 & 22 & 2 \\ -9 & 1 & -4 \end{bmatrix}$ **27.** $\begin{bmatrix} -\frac{1}{2} & \frac{11}{2} \\ \frac{15}{4} & -\frac{3}{2} \\ -\frac{1}{2} & 1 \end{bmatrix}$ **29.** $1, \begin{bmatrix} 9 & -4 \\ -2 & 1 \end{bmatrix}$

31. 0, no inverse   **33.** $-1, \begin{bmatrix} 3 & 2 & -3 \\ 2 & 1 & -2 \\ -8 & -6 & 9 \end{bmatrix}$ **35.** $(65, 154)$

37. $\left(\frac{1}{5}, \frac{9}{5}\right)$ **39.** $\left(-\frac{87}{26}, \frac{21}{26}, \frac{3}{2}\right)$ **41.** $(-3, -3), \left(\frac{9}{5}, -\frac{3}{5}\right)$

43. $(2, 2), (-2, 2)$

45.

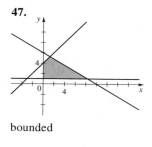

bounded

47.

bounded

49. Maximum 34, minimum 4

51. **(a)** 200 acres oats, 200 acres barley
(b) No oats, 360 acres barley

53. $x = \dfrac{b + c}{2}, y = \dfrac{a + c}{2}, z = \dfrac{a + b}{2}$ **55.** 2, 3

57. $(21.41, -15.93)$ **59.** $(11.94, -1.39), (12.07, 1.44)$

61. $\dfrac{2}{x - 5} + \dfrac{1}{x + 3}$ **63.** $\dfrac{-4}{x} + \dfrac{4}{x - 1} + \dfrac{-2}{(x - 1)^2}$

Chapter 6 Test ■ page 486

1. Wind speed 60 km/h, airplane 300 km/h
2. $(3, -1)$; linear, neither inconsistent nor dependent
3. No solution; linear, inconsistent
4. $x = \frac{1}{7}(z + 1), y = \frac{1}{7}(9z + 2), z =$ any number;
linear, dependent
5. $(\pm 1, -2), \left(\pm\frac{5}{3}, -\frac{2}{3}\right)$; nonlinear
6. Incompatible dimensions
7. Incompatible dimensions
8. $\begin{bmatrix} 6 & 10 \\ 3 & -2 \\ -3 & 9 \end{bmatrix}$ **9.** $\begin{bmatrix} 36 & 58 \\ 0 & -3 \\ 18 & 28 \end{bmatrix}$ **10.** $\begin{bmatrix} 2 & -\frac{3}{2} \\ -1 & 1 \end{bmatrix}$

11. B is not square   **12.** B is not square   **13.** -3
14. $(70, 90)$ **15.** $(5, -5, -4)$

16. $|A| = 0, |B| = 0, B^{-1} = \begin{bmatrix} 1 & -2 & 0 \\ 0 & \frac{1}{2} & 0 \\ 3 & -6 & 1 \end{bmatrix}$

17.

18. $\dfrac{1}{x - 1} + \dfrac{1}{(x - 1)^2} - \dfrac{1}{x + 2}$

19. He should grow $166\frac{2}{3}$ acres of wheat and no barley.
20. $(-0.49, 3.93), (2.34, 2.24)$

FOCUS ON PROBLEM SOLVING ■ page 489

1. 3   **3.** **(a)** None   **(b)** 20   **5.** 1090
7. Justin 26, Sasha 14, Vanessa 8
9. 3 ft by 6 ft, 4 ft by 4 ft   **11.** $x = z = 1, y = t = -1$
13. **(a)** $\frac{11}{2}$ **(b)** 52   **15.** $(x - 20)^2 + (y - 10)^2 = 65^2$

CHAPTER 7

Section 7.1 ■ page 500

Order of answers: focus; directrix; focal diameter

1. $F(1,0)$; $x = -1$; 4

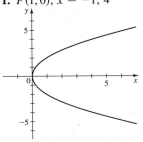

3. $F(0, \frac{9}{4})$; $y = -\frac{9}{4}$; 9

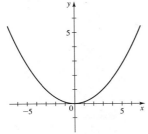

5. $F(0, \frac{1}{20})$; $y = -\frac{1}{20}$; $\frac{1}{5}$

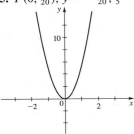

7. $F(-\frac{1}{32}, 0)$; $x = \frac{1}{32}$; $\frac{1}{8}$

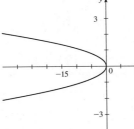

9. $F(0, -\frac{3}{2})$; $y = \frac{3}{2}$; 6

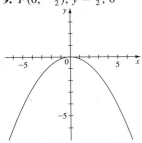

11. $F(-\frac{5}{12}, 0)$; $x = \frac{5}{12}$; $\frac{5}{3}$

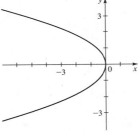

13. $x^2 = 8y$　**15.** $y^2 = -32x$　**17.** $y^2 = -8x$
19. $x^2 = 40y$　**21.** $y^2 = 4x$　**23.** $x^2 = 20y$
25. $x^2 = -12y$　**27.** $y^2 = -3x$　**29.** $x = y^2$
31. $x^2 = -4\sqrt{2}\, y$
33. (a) $y^2 = 12x$　(b) $8\sqrt{15} \approx 31$ cm　**35.** $x^2 = 600y$
37. (a) $x^2 = -4py$,
$p = \frac{1}{2}$, 1, 4, and 8
(b) The closer the directrix
to the vertex, the steeper the
parabola.

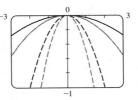

Section 7.2 ■ page 508

Order of answers: vertices; foci; eccentricity; major axis and minor axis

1. $V(\pm 5, 0)$; $F(\pm 4, 0)$;
$\frac{4}{5}$; 10, 6

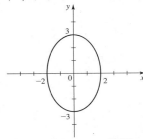

3. $V(0, \pm 3)$; $F(0, \pm\sqrt{5}\,)$;
$\sqrt{5}/3$; 6, 4

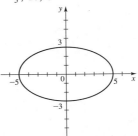

5. $V(\pm 4, 0)$; $F(\pm 2\sqrt{3}, 0)$;
$\sqrt{3}/2$; 8, 4

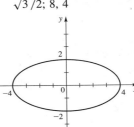

7. $V(0, \pm\sqrt{3}\,)$; $F(0, \pm\sqrt{3/2}\,)$;
$1/\sqrt{2}$; $2\sqrt{3}$, $\sqrt{6}$

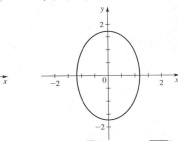

9. $V(\pm 1, 0)$; $F(\pm\sqrt{3}/2, 0)$;
$\sqrt{3}/2$; 2, 1

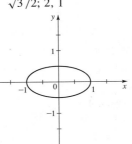

11. $V(0, \pm\sqrt{2}\,)$; $F(0, \pm\sqrt{3/2}\,)$;
$\sqrt{3}/2$; $2\sqrt{2}$, $\sqrt{2}$

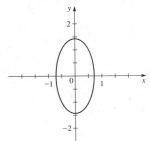

13. $V(0, \pm 1)$; $F(0, \pm 1/\sqrt{2}\,)$;
$1/\sqrt{2}$; 2, $\sqrt{2}$

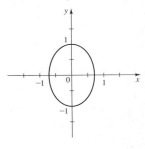

15. $\dfrac{x^2}{25} + \dfrac{y^2}{16} = 1$ **17.** $\dfrac{x^2}{4} + \dfrac{y^2}{8} = 1$

19. $\dfrac{x^2}{25} + \dfrac{y^2}{9} = 1$ **21.** $x^2 + \dfrac{y^2}{4} = 1$

23. $\dfrac{x^2}{9} + \dfrac{y^2}{13} = 1$ **25.** $\dfrac{x^2}{100} + \dfrac{y^2}{91} = 1$

27. $\dfrac{x^2}{25} + \dfrac{y^2}{5} = 1$ **29.** $\dfrac{64x^2}{225} + \dfrac{64y^2}{81} = 1$

31. $(0, \pm 2)$

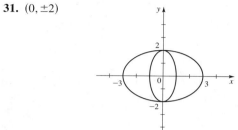

33. $\dfrac{x^2}{2.250 \times 10^{16}} + \dfrac{y^2}{2.249 \times 10^{16}} = 1$

35. $\dfrac{x^2}{1,455,642} + \dfrac{y^2}{1,451,610} = 1$

37. $5\sqrt{39}/2 \approx 15.6$ in.

41. (a)   **(b)**

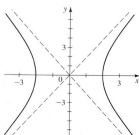   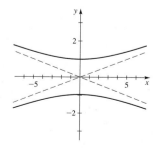

Section 7.3 ■ page 516

Order of answers: vertices; foci; asymptotes

1. $V(\pm 2, 0)$; $F(\pm 2\sqrt{5}, 0)$;
$y = \pm 2x$

3. $V(0, \pm 1)$; $F(0, \pm\sqrt{26})$;
$y = \pm\frac{1}{5}x$

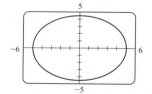   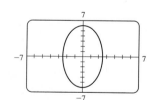

5. $V(\pm 1, 0)$; $F(\pm\sqrt{2}, 0)$;
$y = \pm x$

7. $V(0, \pm 3)$; $F(0, \pm\sqrt{34})$;
$y = \pm\frac{3}{5}x$

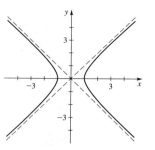   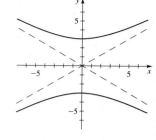

9. $V(\pm 2\sqrt{2}, 0)$;
$F(\pm\sqrt{10}, 0)$;
$y = \pm\frac{1}{2}x$

11. $V\left(0, \pm\frac{1}{2}\right)$;
$F(0, \pm\sqrt{5}/2)$;
$y = \pm\frac{1}{2}x$

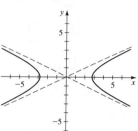   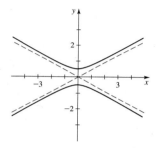

13. $\dfrac{x^2}{4} - \dfrac{y^2}{12} = 1$ **15.** $\dfrac{y^2}{16} - \dfrac{x^2}{16} = 1$

17. $\dfrac{x^2}{9} - \dfrac{y^2}{16} = 1$ **19.** $y^2 - \dfrac{x^2}{3} = 1$

21. $x^2 - \dfrac{y^2}{25} = 1$ **23.** $\dfrac{5y^2}{64} - \dfrac{5x^2}{256} = 1$

25. $\dfrac{x^2}{16} - \dfrac{y^2}{16} = 1$ **27.** $\dfrac{x^2}{9} - \dfrac{y^2}{16} = 1$

29. (b) $x^2 - y^2 = c^2/2$

33. (a) 490 mi   **(b)** $\dfrac{y^2}{60,025} - \dfrac{x^2}{2475} = 1$ **(c)** 10.1 mi

35. (b)

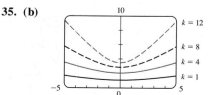

Section 7.4 ■ page 524

1. Center $C(2,1)$;
foci $F(2 \pm \sqrt{5}, 1)$;
vertices $V_1(-1,1)$,
$V_2(5,1)$; major axis 6,
minor axis 4

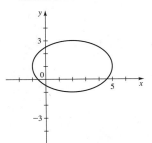

3. Center $C(0,-5)$;
foci $F_1(0,-1)$, $F_2(0,-9)$;
vertices $V_1(0,0)$, $V_2(0,-10)$;
major axis 10, minor axis 6

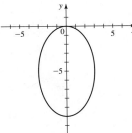

5. Vertex $V(3,-1)$;
focus $F(3,1)$;
directrix $y = -3$

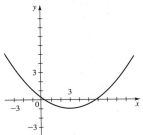

7. Vertex $V(-\frac{1}{2}, 0)$;
focus $F(-\frac{1}{2}, -\frac{1}{16})$;
directrix $y = \frac{1}{16}$

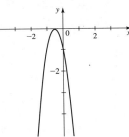

9. Center $C(-1,3)$;
foci $F_1(-6,3)$, $F_2(4,3)$;
vertices $V_1(-4,3)$,
$V_2(2,3)$;
asymptotes
$y = \pm \frac{4}{3}(x+1) + 3$

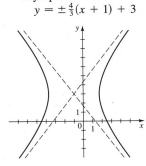

11. Center $C(-1,0)$;
foci $F(-1, \pm\sqrt{5})$;
vertices $V(-1, \pm 1)$;
asymptotes
$y = \pm \frac{1}{2}(x+1)$

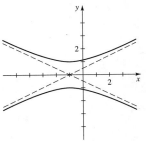

13. $x^2 = -\frac{1}{4}(y-4)$

15. $\dfrac{(x-5)^2}{25} + \dfrac{y^2}{16} = 1$

17. $(y-1)^2 - x^2 = 1$

19. Ellipse; $C(2,0)$;
$F(2, \pm\sqrt{5})$; $V(2, \pm 3)$;
major axis 6,
minor axis 4

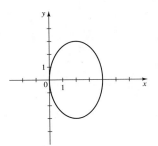

21. Hyperbola $C(1,2)$;
$F_1(-\frac{3}{2}, 2)$, $F_2(\frac{7}{2}, 2)$;
$V(1 \pm \sqrt{5}, 2)$;
asymptotes
$y = \pm \frac{1}{2}(x-1) + 2$

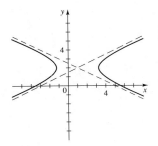

23. Ellipse; $C(3,-5)$;
$F(3 \pm \sqrt{21}, -5)$;
$V_1(-2,-5)$, $V_1(8,-5)$;
major axis 10,
minor axis 4

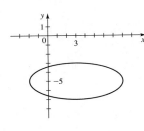

25. Hyperbola; $C(3,0)$;
$F(3, \pm 5)$; $V(3, \pm 4)$;
asymptotes
$y = \pm \frac{4}{3}(x-3)$

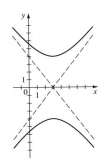

27. Degenerate conic
(pair of lines),
$y = \pm \frac{1}{2}(x-4)$

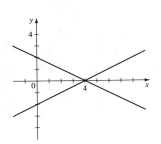

29. Point $(1,3)$

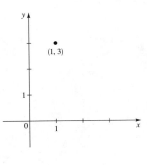

31. (a) $F < 17$ (b) $F = 17$ (c) $F > 17$

33. (a)

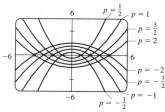

(c) The parabolas become narrower.

Chapter 7 Review ■ page 526

1. $V(0,0)$; $F(0,-2)$;
$y = 2$

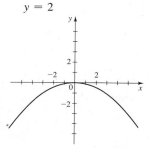

3. $V(-2,2)$; $F(-\frac{7}{4},2)$;
$x = -\frac{9}{4}$

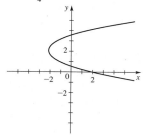

5. $C(0,0)$; $V(\pm 4,0)$;
$F(\pm 2\sqrt{3},0)$; axes 8, 4

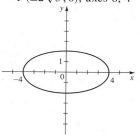

7. $C(0,2)$; $V(\pm 3,2)$;
$F(\pm\sqrt{5},2)$; axes 6, 4

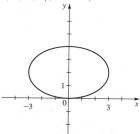

9. $C(0,0)$; $V(\pm 4,0)$;
$F(\pm 2\sqrt{6},0)$;
asymptotes $y = \pm\dfrac{1}{\sqrt{2}}x$

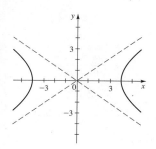

11. $C(-3,-1)$;
$V(-3,-1\pm\sqrt{2})$;
$F(-3,-1\pm 2\sqrt{5})$;
asymptotes $y = \frac{1}{3}x$,
$y = -\frac{1}{3}x - 2$

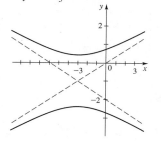

13. $y^2 = 8x$ **15.** $\dfrac{y^2}{16} - \dfrac{x^2}{9} = 1$

17. $\dfrac{(x-4)^2}{16} + \dfrac{(y-2)^2}{4} = 1$

19. Parabola;
$F(0,-2)$;
$V(0,1)$

21. Hyperbola;
$F(0,\pm 12\sqrt{2})$;
$V(0,\pm 12)$

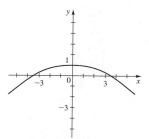

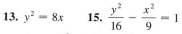

23. Ellipse;
$F(1, 4\pm\sqrt{15})$;
$V(1, 4\pm 2\sqrt{5})$

25. Parabola;
$F(-\frac{255}{4},8)$;
$V(-64,8)$

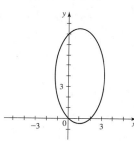

27. Ellipse;
$F(3, -3\pm 1/\sqrt{2})$;
$V_1(3,-4)$, $V_2(3,-2)$

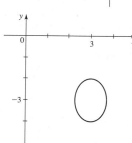

29. Has no graph   **31.** $x^2 = 4y$ **33.** $\dfrac{y^2}{4} - \dfrac{x^2}{16} = 1$

35. $\dfrac{(x-1)^2}{3} + \dfrac{(y-2)^2}{4} = 1$

37. $\dfrac{4(x-7)^2}{225} + \dfrac{(y-2)^2}{100} = 1$

39. $(x-800)^2 = -200(y-3200)$

41. (a) 91,419,000 mi   **(b)** 94,581,000 mi

43. (a)

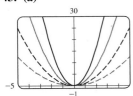

(b) $\left(0, \frac{1}{2}\right)$, $\left(0, \frac{1}{4}\right)$, $\left(0, \frac{1}{8}\right)$, $\left(0, \frac{1}{16}\right)$
(c) The focus approaches the point $(0, 0)$ as k increases.

8. $9(x + 2)^2 - 8(y - 4)^2 = 0$

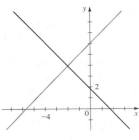

Chapter 7 Test ■ page 528

1. $F(0, -3)$, $y = 3$

2. $V(\pm 4, 0)$; $F(\pm 2\sqrt{3}, 0)$; 8, 4

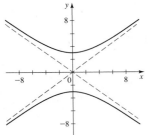

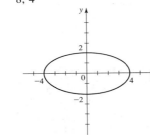

9. $(y + 4)^2 = -2(x - 4)$

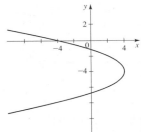

3. $V(0, \pm 3)$; $F(0, \pm 5)$; $y = \pm \frac{3}{4}x$

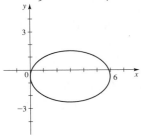

10. $\dfrac{y^2}{9} - \dfrac{x^2}{16} = 1$ **11.** $x^2 - 4x - 8y + 20 = 0$

12. $\frac{3}{4}$ in.

FOCUS ON PROBLEM SOLVING ■ Page 531

1. Tetrahedron 4, 6, 4; octahedron 8, 12, 6; cube 6, 12, 8; dodecahedron 12, 30, 20; icosahedron 20, 30, 12
3. Hexagons   **5.** $\sqrt{a^2 + (b + c)^2}$ **7.** 2.4 cm
9. 11 cm   **13. (a)** None   **(b)** 13   **15.** $P + 2\pi$

4. $y^2 = -x$ **5.** $\dfrac{x^2}{16} + \dfrac{(y - 3)^2}{9} = 1$

6. $(x - 2)^2 - \dfrac{y^2}{3} = 1$

7. $\dfrac{(x - 3)^2}{9} + \dfrac{\left(y + \frac{1}{2}\right)^2}{4} = 1$

CHAPTER 8

Section 8.1 ■ page 538

1. 40,320   **3.** 132   **5.** 9900   **7.** 10
9. 9801   **11.** $1/n$ **13.** $n + 1$
15. (a) 64   **(b)** 24   **17.** 1024   **19.** 120   **21.** 144
23. 120   **25.** 32   **27.** 216   **29.** 480   **31.** No
33. 158,184,000   **35.** 1024   **37.** 8192   **39.** No
41. 24,360   **43.** 1050
45. (a) 16,807   **(b)** 2520   **(c)** 2401   **(d)** 2401
(e) 4802
47. 936   **49. (a)** 1152   **(b)** 1152   **51.** 483,840
53. 6

Section 8.2 ■ page 544

1. 336   **3.** 7920   **5.** 100   **7.** 360,360   **9.** $n!$
11. 336   **13.** 720   **15.** 120   **17.** 24   **19.** 362,880
21. 997,002,000   **23.** 24   **25.** 6   **27.** 64   **29.** 60

31. 60   **33.** 15   **35.** 277,200   **37.** 39,916,800
39. 83,160   **41.** 420   **43.** 1287

Section 8.3 ■ page 549

1. 56   **3.** 330   **5.** 100   **7.** 3003   **9.** 1   **11.** 20
13. 210   **15.** 2,598,960   **17.** 120   **19.** 495
21. 2,035,800   **23.** 1,560,780   **25.** \$22,957,480
27. (a) 15,504   (b) 792   (c) 6160
29. 24   **31.** 24   **33.** (a) 56   (b) 256
35. 40   **37.** 24   **39.** 6561   **41.** 1,048,576
43. (a) 161,700 and 161,700

Section 8.4 ■ page 555

1. 725,760   **3.** 2880   **5.** 480
7. (a) 39,916,800   (b) 479,001,600
9. (a) 525   (b) 1281   (c) 531   **11.** 104,781,600
13. 564,480   **15.** 480   **17.** 200   **19.** 40,319
21. 884,736   **23.** (a) 20,160   (b) 8640
25. (a) 576   (b) 144   **27.** 84   **29.** 46   **31.** 66
33. 54   **35.** (a) 85   (b) 240   **37.** 21,772,800
39. A mystery   **41.** 5108

Section 8.5 ■ page 561

1. (a) $S = \{HH, HT, TH, TT\}$ (b) $\frac{1}{4}$ (c) $\frac{3}{4}$ (d) $\frac{1}{2}$
3. (a) $\frac{1}{6}$ (b) $\frac{1}{2}$ (c) $\frac{1}{6}$ **5.** (a) $\frac{1}{13}$ (b) $\frac{3}{13}$ (c) $\frac{10}{13}$
7. (a) $\frac{5}{8}$ (b) $\frac{7}{8}$ (c) 0   **9.** $\frac{1}{3}$
11. (a) 0.33   (b) 0.67   (c) 0.06
13. (a) $\frac{3}{16}$ (b) $\frac{3}{8}$ (c) $\frac{5}{8}$
15. $4C(13, 5)/C(52, 5) \approx 0.00198$
17. $13 \cdot 48/C(52, 5) = 1/4165$
19. $C(13, 3)C(13, 2)/C(52, 5) \approx 0.008583$
21. $1 - [C(13, 5)/C(52, 5)] \approx 0.999505$
23. (b) $\frac{1}{16}$ (c) $\frac{3}{8}$ (d) $\frac{1}{8}$ (e) $\frac{11}{16}$ **25.** $\frac{9}{19}$
27. $1/C(49, 6) \approx 7.15 \times 10^{-8}$ **29.** $600/P(40, 3) = 5/494$
31. (a) $\frac{1}{1024}$ (b) $\frac{15}{128}$ **33.** 0.1
35. (a) $1/48^6 \approx 8.18 \times 10^{-11}$ (b) $1/48^{18} \approx 5.47 \times 10^{-31}$
37. $4/11! \approx 1.00 \times 10^{-7}$ **39.** (a) $\frac{3}{4}$ (b) $\frac{1}{4}$

Section 8.6 ■ page 568

1. (a) Yes   (b) No
3. (a) Mutually exclusive; 1   (b) Not mutually exclusive; $\frac{2}{3}$
5. (a) Not mutually exclusive; $\frac{11}{26}$ (b) Mutually exclusive; $\frac{1}{2}$
7. (a) $\frac{3}{16}$ (b) $\frac{1}{2}$ (c) 1   **9.** $\frac{21}{38}$ **11.** $\frac{1}{20160}$
13. $\frac{31}{1001}$ **15.** $\frac{5}{16}$
17. (a) $\frac{3}{8}$ (b) $\frac{1}{2}$ (c) $\frac{11}{16}$ (d) $\frac{13}{16}$

Section 8.7 ■ page 573

1. (a) Yes   (b) $\frac{1}{4}$ **3.** $\frac{1}{3}$ **5.** $\frac{1}{3}$ **7.** $\frac{1}{4}$ **9.** $\frac{1}{3}$
11. (a) Yes   (b) $\frac{1}{8}$
13. (a) $\frac{4}{11}$ (b) $\frac{5}{11}$ (c) $\frac{5}{11}$ (d) $\frac{4}{11}$ **15.** $\frac{2}{3}$ **17.** $\frac{1}{12}$
19. (a) $\frac{4}{221}$ (b) $\frac{1}{221}$ **21.** $\frac{1}{1444}$
23. $\dfrac{1}{36^3} \approx 2.14 \times 10^{-5}$ **25.** (a) $\frac{1}{2}$ (b) $\frac{1}{3}$ **27.** $\frac{1}{35}$
29. (a) $\frac{1}{1331}$ (b) $\frac{10}{1331}$ (c) $\frac{331}{1331}$
31. $\frac{1343}{1728}$

Section 8.8 ■ page 577

1. \$1.50   **3.** \$0.94   **5.** \$0.92   **7.** 0   **9.** $-\$0.30$
11. \$0.0526   **13.** $-\$0.50$
15. No, she should expect to lose \$2.10 per stock.
17. $-\$0.11$ **19.** \$0.25   **21.** \$1.67

Chapter 8 Review ■ page 579

1. 624   **3.** 5   **5.** (a) 10   (b) 20   **7.** 120
9. 120   **11.** 45   **13.** 17,576
15. (a) 240   (b) 3360   (c) 1680
17. 40,320   **19.** 34,650   **21.** 14
23. (a) 31,824   (b) 11,760   (c) 19,448   (d) 2808
(e) 2808   (f) 6,683,040
25. (a) $\frac{3}{7}$ (b) $\frac{10}{21}$ (c) $\frac{19}{21}$ (d) $\frac{2}{21}$ (e) $\frac{1}{105}$
27. (a) $\frac{1}{13}$ (b) $\frac{2}{13}$ (c) $\frac{4}{13}$ (d) $\frac{1}{26}$
29. (a) $\frac{1}{6}$ (b) $\frac{5}{6}$ **31.** $-\$0.83$ **33.** \$0.00016
35. (a) 3   (b) 0.51
37. (a) 10^5 (b) 5^5 (c) $\frac{1}{32}$ (d) 75
39. (a) 144   (b) 126   (c) 84   (d) $\frac{7}{8}$ **41.** 8820

Chapter 8 Test ■ Page 582

1. (a) 10^5 (b) 30,240   **2.** 60
3. $30 \cdot 29 \cdot 28 \cdot C(27, 5) = 1,966,528,800$
4. $2C(4, 2) = 12$ **5.** $4 \cdot 2^{14} = 65,536$
6. (a) $10! = 3,628,800$ (b) $2 \cdot 9! = 725,760$
(c) $10! - 2 \cdot 9! = 2,903,040$
7. (a) $4! = 24$ (b) $6!/3! = 120$
8. (a) 0.33   (b) 0.67
9. $C(5, 3)/C(15, 3) \approx 0.022$ **10.** $\frac{1}{6}$
11. (a) $C(12, 2)/C(52, 2) \approx 0.0498$ (b) $\frac{3}{51} \approx 0.0588$
12. (a) $\frac{5}{13}$ (b) $\frac{6}{13}$ (c) $\frac{9}{13}$ (d) $\frac{2}{5}$ (e) $\frac{1}{3}$ (f) $\frac{2}{13}$
13. \$0.65   **14.** $1 - 1 \cdot \frac{11}{12} \cdot \frac{10}{12} \cdot \frac{9}{12} \approx 0.427$ **15.** 7

FOCUS ON PROBLEM SOLVING ■ Page 585

1. (a) $\frac{1}{10}$ (b) $\frac{9}{10}$ **3.** (a) $\frac{7}{8}$ **5.** $\frac{3}{8}$ **7.** 1093
9. 2^{n-1} **11.** (a) 45   (b) 56
13. (a) $\frac{1}{11}$ (b) $\frac{1}{11}$ (c) $\frac{12}{121}$ (d) $\frac{69}{121}$ (e) $\frac{63}{121}$

CHAPTER 9

Section 9.1 ■ page 594

1. $2, 3, 4, 5; 1001$ **3.** $\frac{1}{2}, \frac{1}{3}, \frac{1}{4}, \frac{1}{5}; \frac{1}{1001}$

5. $-1, \frac{1}{4}, -\frac{1}{9}, \frac{1}{16}; \frac{1}{1,000,000}$ **7.** $0, 2, 0, 2; 2$

9. $-2, 4, -8, 16; 2^{1000}$ **11.** $\frac{1}{2}, -\frac{2}{3}, \frac{3}{4}, -\frac{4}{5}; -\frac{1000}{1001}$

13. $1, 4, 27, 256; 1000^{1000}$ **15.** $3, 2, 0, -4, -12$

17. $1, 3, 7, 15, 31$ **19.** $0, 1, 1, 0, -1$ **21.** $1, 2, 3, 5, 8$

23. 2^n **25.** $3n - 2$ **27.** $(2n - 1)/n^2$ **29.** r^n

31. $1 + (-1)^n$

33. (c) $n^2 + (n - 1)(n - 2)(n - 3)(n - 4)(n - 5)(n - 6)$

35. $2^{(2^n-1)/2^n}$

37. $25, 76, 38, 19, 58, 29, 88, 44, 22, 11, 34, 17, 52, 26, 13,$
$40, 20, 10, 5, 16, 8, 4, 2, 1, 4, 2, 1, \ldots$

Section 9.2 ■ page 600

1. $3, 14, 2 + 3(n - 1), 299$ **3.** $5, 24, 4 + 5(n - 1), 499$

5. $4, 4, -12 + 4(n - 1), 384$

7. $1.5, 31, 25 + 1.5(n - 1), 173.5$

9. $s, 2 + 4s, 2 + (n - 1)s, 2 + 99s$

11. $3, 162, 2 \cdot 3^{n-1}$ **13.** $-0.3, 0.00243, (0.3)(-0.3)^{n-1}$

15. $-\frac{1}{12}, \frac{1}{144}, 144(-\frac{1}{12})^{n-1}$ **17.** $3^{2/3}, 3^{11/3}, 3^{(2n+1)/3}$

19. $s^{2/7}, s^{8/7}, s^{2(n-1)/7}$ **21.** Neither

23. Geometric, $9\sqrt{3}$ **25.** Neither

27. Arithmetic, $x + 3$ **29.** Neither

31. Arithemtic, 3 **33.** $\frac{1}{2}$ **35.** $-100, -98, -96$

37. $\frac{16}{27}$ **39.** $\frac{25}{4}$ **41.** 30th **43.** Yes, 2985th

45. 19 ft **47.** \$4714.37 **49.** $\frac{64}{25}, \frac{1024}{625}, 5(\frac{4}{5})^n$ **51.** 3

53. r^2 **55.** $r = 10^d$ **57.** $2, 5, 8$ **59.** 0

61. $(\sqrt{5} - 1)/2$ **63.** $10, 20, 40$ **65.** $\frac{15}{4}$

Section 9.3 ■ page 608

1. 10 **3.** $\frac{11}{6}$ **5.** 8 **7.** 31 **9.** 549 **11.** 106

13. $\sqrt{1} + \sqrt{2} + \sqrt{3} + \sqrt{4} + \sqrt{5}$

15. $\sqrt{4} + \sqrt{5} + \sqrt{6} + \sqrt{7} + \sqrt{8} + \sqrt{9} + \sqrt{10}$

17. $1 \cdot x^2 + 2 \cdot x^3 + \cdots + 8 \cdot x^9$

19. $x - x^2 + x^3 - \cdots + (-1)^{n+1}x^n$

21. $\sum_{k=1}^{100} k$ **23.** $\sum_{k=2}^{100} \frac{(-1)^k}{k \ln k}$ **25.** $\sum_{k=0}^{5} (-1)^k \frac{x^k}{3^k}$

27. $\sum_{k=1}^{100} (-1)^{k+1} k x^{k-1}$ **29.** $\sum_{k=1}^{97} k(k + 1)(k + 2)$

31. $1, 4, 9, 16, 25, 36$ **33.** $\frac{1}{3}, \frac{4}{9}, \frac{13}{27}, \frac{40}{81}, \frac{121}{243}, \frac{364}{729}$

35. $S_n = \frac{1}{2} - \frac{1}{n + 2}, \frac{250}{501}$ **37.** $S_n = 1 - \frac{1}{3^n}, 1 - \frac{1}{3^{20}}$

39. $S_n = 1 - \sqrt{n + 1}, -9$ **41.** $S_n = -\log(n + 1), -6$

Section 9.4 ■ page 613

1. 100 **3.** 460 **5.** 1090 **7.** 4900 **9.** 315

11. 441 **13.** 2.8502 **15.** 20,301 **17.** 832.3

19. 46.75 **21.** 3280 **23.** 1.94117

25. 5.997070313 **27.** Geometric, 9.2224

29. Geometric, $(1 + x^{21})/(1 + x)$

31. Arithmetic, 250,500 **33.** Arithmetic, 4128

35. Arithmetic, $9(1 + 4\sqrt{2})$

37. Geometric, $124/(\sqrt[3]{5} - 1)$ **39.** 50 **41.** $\frac{80}{3}$

43. $\frac{1}{2}$ **45.** $\frac{455}{27}$ **47.** 10^{19}

49. \$10,737,418.23; 37 days **51.** (a) 576 ft **(b)** $16n^2$ ft

53. 2801

Section 9.5 ■ page 620

1. \$13,180.79 **3.** \$360,262.21 **5.** \$5,591.79

7. \$245.66 **9.** \$2,601.59 **11.** \$307.24

13. \$733.76, \$264,153.60

15. (a) \$859.15 (b) \$309,294.00 (c) \$1,841,519.29

17. \$341.24 **19.** 18% **21.** 12%

Section 9.6 ■ page 624

1. $\frac{3}{2}$ **3.** $\frac{3}{4}$ **5.** $\frac{1}{648}$ **7.** $-\frac{1000}{117}$

9. $5^{4/3}/(1 + 5^{1/3})$ **11.** $\frac{7}{9}$ **13.** $\frac{1}{33}$ **15.** $\frac{112}{999}$

17. 60 ft **19.** $2 + \sqrt{2}s$ **21.** $2\pi R^2$ **23.** \$50,000

25. 1

Section 9.8 ■ page 642

1. $x^6 + 6x^5y + 15x^4y^2 + 20x^3y^3 + 15x^2y^4 + 6xy^5 + y^6$

3. $x^4 + 4x^2 + 6 + \frac{4}{x^2} + \frac{1}{x^4}$

5. $x^5 - 5x^4 + 10x^3 - 10x^2 + 5x - 1$

7. $x^{10}y^5 - 5x^8y^4 + 10x^6y^3 - 10x^4y^2 + 5x^2y - 1$

9. $8x^3 - 36x^2y + 54xy^2 - 27y^3$

11. $\frac{1}{x^5} - \frac{5}{x^{7/2}} + \frac{10}{x^2} - \frac{10}{x^{1/2}} + 5x - x^{5/2}$

13. 15 **15.** 4950 **17.** 18 **19.** 32

21. $x^4 + 8x^3y + 24x^2y^2 + 32xy^3 + 16y^4$

23. $1 + \frac{6}{x} + \frac{15}{x^2} + \frac{20}{x^3} + \frac{15}{x^4} + \frac{6}{x^5} + \frac{1}{x^6}$

25. $x^{20} + 40x^{19}y + 760x^{18}y^2$ **27.** $25a^{26/3} + a^{25/3}$

29. $48,620x^{18}$ **31.** $300a^2b^{23}$ **33.** $100y^{99}$

35. $13,440x^4y^6$ **37.** $495a^8b^8$ **39.** 17,920

41. $1792c^7$ **43.** 21 **45.** 11 **47.** $(x + y)^4$

49. $(2a + b)^3$

51. $a^8 + 4a^7 + 10a^6 + 16a^5 + 19a^4 + 16a^3 + 10a^2$
$+ 4a + 1$

53. $(101!)^{100}$

Chapter 9 Review ■ page 645

1. $\frac{1}{2}, \frac{4}{3}, \frac{9}{4}, \frac{16}{5}; \frac{100}{11}$ **3.** $0, \frac{1}{4}, 0, \frac{1}{32}; \frac{1}{500}$

5. 1, 3, 15, 105; 654,729,075   **7.** 1, 4, 9, 16, 25, 36, 49

9. 1, 3, 5, 11, 21, 43, 85   **11.** 3, 2, 1, 1, 1, 1, 1

13. Arithmetic, 7   **15.** Arithmetic, $5\sqrt{2}$

17. Arithmetic, $t + 1$ **19.** Geometric, $\frac{4}{27}$ **21.** Neither

23. Geometric, ab^4c^{12} **25.** $2i$ **27.** 5   **29.** $\frac{81}{4}$

31. 64   **33.** 12,288   **37. (a)** 9   **(b)** $\pm 6\sqrt{2}$

39. 126   **41.** 384   **43.** $0^2 + 1^2 + 2^2 + \cdots + 9^2$

45. $\dfrac{3}{2^2} + \dfrac{3^2}{2^3} + \dfrac{3^3}{2^4} + \cdots + \dfrac{3^{50}}{2^{51}}$ **47.** $\displaystyle\sum_{k=1}^{33} 3k$

49. $\displaystyle\sum_{k=1}^{100} k\,2^{k+2}$ **51.** Geometric; 4.68559

53. Geometric; $(1 + 5^{9/2})/(1 + 5^{1/2}) \approx 432.17$

55. Arithmetic, 1650   **57.** Geometric, 9831   **59.** 13

61. 29   **63. (a)** 259,374   **(b)** $100,000(1.1)^n$

65. \$30,324.28   **67. (a)** \$482.77   **(b)** \$608.56

69. $\frac{1}{9}$ **71.** $a/(1 - b^2)$ **81.** 100   **83.** 32

85. $1 - 6x^2 + 15x^4 - 20x^6 + 15x^8 - 6x^{10} + x^{12}$

87. $1540a^3b^{19}$ **89.** 5   **91.** a^3

Chapter 9 Test ■ Page 648

1. $-\frac{10}{99}$ **2.** -1 **3.** -36 **4.** Yes, 6th term

5. (a) False   **(b)** True

6. (a) $S_n = \dfrac{n}{2}[2a + (n - 1)d]$ or $S_n = n\left(\dfrac{a + a_n}{2}\right)$

(b) 60   **(c)** $-\frac{8}{9}, -78$

7. (a) $S_n = \dfrac{a(1 - r^n)}{1 - r}$ **(b)** 58,025/59,049

8. $2 + \sqrt{2}$ **10. (a)** -50 **(b)** 10

11. $(a + b)^n = \displaystyle\sum_{k=0}^{n} \binom{n}{k} a^{n-k}b^k$

12. $32x^5 + 80x^4y^2 + 80x^3y^4 + 40x^2y^6 + 10xy^8 + y^{10}$

13. $1,293,600a^3b^{97}$ **14.** 8064

FOCUS ON PROBLEM SOLVING ■ Page 651

1. $1 + 3 + 5 + \cdots + (2n - 1) = n^2$ **5.** $(n + 1)! - 1$

9. $\dfrac{n(n^2 + 1)}{2}$

11. (a) 16   **(c)** 31 regions   **(d)** 57, 99, 163, 256

13. (b) $\frac{6}{25}$

INDEX

PHOTO CREDITS